DNA REPAIR AND REPLICATION

MECHANISMS AND CLINICAL SIGNIFICANCE

DNA REPAIR AND REPLICATION

MECHANISMS AND CLINICAL SIGNIFICANCE

Edited by

Roger J. A. Grand

John J. Reynolds

A GARLAND SCIENCE BOOK

CRC Press
Taylor & Francis Group
6000 Broken Sound Parkway NW, Suite 300
Boca Raton, FL 33487-2742

Printed in Canada on acid-free paper

International Standard Book Number-13: 978-1-138-61291-4 (Hardback)
978-0-8153-4599-2 (Paperback)

Library of Congress Cataloging-in-Publication Data

Names: Grand, R. J. A. (Roger J. A.), editor. | Reynolds, John J. (John Joseph), 1983- editor.
Title: DNA repair and replication : mechanisms and clinical significance / editors, Roger J.A. Grand and John J. Reynolds.
Other titles: DNA repair and replication (Grand)
Description: Boca Raton : CRC Press/Taylor & Francis, 2018. | Includes bibliographical references.
Identifiers: LCCN 2018010405| ISBN 9780815345992 (pbk. : alk. paper) | ISBN 9781138612914 (hardback : alk. paper)
Subjects: | MESH: DNA Repair--genetics | DNA Replication--genetics | Genomic Instability--genetics |
Molecular Targeted Therapy
Classification: LCC QH467 | NLM QU 58.5 | DDC 572.8/6459--dc23
LC record available at https://lccn.loc.gov/2018010405

Visit the Taylor & Francis Web site at
http://www.taylorandfrancis.com

and the CRC Press Web site at
http://www.crcpress.com

CONTENTS

DETAILED CONTENTS

PREFACE

This book owes its origins to a decision made by the University of Birmingham Medical School to place an emphasis on recruiting young scientists with an interest in various aspects of DNA damage repair pathways and DNA replication. Most of the contributors to this book are part of that cohort. Their range of interests within these broad research areas has allowed us to cover most aspects of DNA repair and replication whilst maintaining a reasonably coherent viewpoint.

Research into the DNA damage response at Birmingham began more than half a century ago in the Department of Cancer Studies, where there was a very strong interest in cytogenetics and the chromosomal abnormalities seen in a number of inherited diseases. In particular, for many years, strong emphasis had been placed on the study of ataxia telangiectasia. When the ATM gene was cloned in 1995, and it became clear that the protein played a key role in the response to DNA damage, interest in the DNA damage response pathways increased and this was reflected in the expanding department, which has now become the Institute of Cancer and Genomic Sciences.

The articles in the first part of this book cover most aspects of the DNA damage response with an emphasis placed on their relationship to DNA replication stress. The second part concentrates on the relevance of this to human disease, with emphasis on both the causes and treatments which make use of DDR pathways.

EDITORS

Dr Roger J. A. Grand earned a degree in biochemistry at the University of Sheffield, followed by a PhD at the University of Leeds. After a short post-doctoral fellowship at Royal Holloway College, University of London, he moved to Birmingham to join the group of Professor Samuel V. Perry, FRS, in the Department of Biochemistry, studying proteins involved in the regulation of striated muscle contraction. This research, which formed the basis of a number of publications, helped to define how signals, in the form of calcium ions, were transmitted through the troponin complex to initiate muscle contraction. After a few years, Roger changed track and joined a group in the Department of Cancer Sciences, University of Birmingham, which he co-led with a virologist Professor Phil Gallimore. This work initially focussed on a study of the biochemistry of adenovirus oncoproteins, but later was more concerned with an investigation of the functions of the cellular proteins targeted by the virus. During his time in the department, which has since evolved into the Institute for Cancer and Genomic Sciences, Dr Grand has undertaken research into a wide range of biological processes, including apoptosis, the regulation of neuronal cell growth and differentiation, the DNA damage response, as well as viral infection. Dr Grand, a reader in Experimental Cancer Sciences, has now led a research group for many years and published widely. Over the past decade, he has specialised in the study of various aspects of the DNA damage response, both in normal cells and in those undergoing viral infection. Most recently, he has been interested in rare inherited diseases linked to mutations in DNA repair proteins.

Dr John J. Reynolds earned a degree in genetics and microbiology at the University of Sheffield (2006). He subsequently undertook his PhD studies with Professor Keith Caldecott at the Genome Damage and Stability Centre (University of Sussex), where he worked on characterising the molecular defects underlying rare human diseases caused by mutations in DNA single-strand break repair factors. Following the completion of his PhD in 2011, Dr Reynolds joined the laboratory of Professor Grant Stewart as a postdoctoral research fellow at the Institute of Cancer and Genomic Sciences (University of Birmingham). During his time at Birmingham, he has been working on identifying and characterising novel DNA damage response genes and investigating how defects in DNA repair and DNA replication factors give rise to human diseases such as microcephalic dwarfism. Dr Reynolds' main research interest is in understanding the mechanisms underlying DNA damage response and repair pathways, and investigating the roles they play in maintaining genome stability and preventing human disease. He has published 13 papers, including four first author research publications in high impact journals and three review articles, and has presented at 18 conferences, including eight international meetings. Dr Reynolds also lectures on DNA repair and human disease and reviews research articles and grants for journals and funding bodies.

CONTRIBUTORS

Rachel Bayley
Birmingham Centre for Genome Biology
Institute of Cancer and Genomic Sciences
University of Birmingham
Birmingham, United Kingdom
Contact: R.Bayley.1@bham.ac.uk

Clare Davies
Birmingham Centre for Genome Biology
Institute of Cancer and Genomic Sciences
University of Birmingham
Birmingham, United Kingdom
Contact: C.C.Davies@bham.ac.uk

Ruth M. Densham
Birmingham Centre for Genome Biology
Institute of Cancer and Genomic Sciences
University of Birmingham
Birmingham, United Kingdom
Contact: R.M.Densham@bham.ac.uk

Agnieszka Gambus
Birmingham Centre for Genome Biology
Institute of Cancer and Genomic Sciences
University of Birmingham
Birmingham, United Kingdom
Contact: A.Gambus@bham.ac.uk

Paloma Garcia
Birmingham Centre for Genome Biology
Institute of Cancer and Genomic Sciences
University of Birmingham
Birmingham, United Kingdom
Contact: P.Garcia@bham.ac.uk

Alexander J. Garvin
Birmingham Centre for Genome Biology
Institute of Cancer and Genomic Sciences
University of Birmingham
Birmingham, United Kingdom
Contact: A.J.Garvin@bham.ac.uk

Roger J. A. Grand
Birmingham Centre for Genome Biology
Institute of Cancer and Genomic Sciences
University of Birmingham
Birmingham, United Kingdom
Contact: R.J.A.GRAND@bham.ac.uk

Martin R. Higgs
Birmingham Centre for Genome Biology
Institute of Cancer and Genomic Sciences
University of Birmingham
Birmingham, United Kingdom
Contact: M.R.Higgs@bham.ac.uk

Rebecca M. Jones
Birmingham Centre for Genome Biology
Institute of Cancer and Genomic Sciences
University of Birmingham
Birmingham, United Kingdom
Contact: R.Jones.2@bham.ac.uk

Marwan Kwok
Birmingham Centre for Genome Biology
Institute of Cancer and Genomic Sciences
University of Birmingham
Birmingham, United Kingdom
Contact: M.Kwok@bham.ac.uk

A. Malcolm R. Taylor
Birmingham Centre for Genome Biology
Institute of Cancer and Genomic Sciences
University of Birmingham
Birmingham, United Kingdom
Contact: A.M.R.TAYLOR@bham.ac.uk

Michal Malewicz
MRC Toxicology Unit
Leicester, United Kingdom
Contact: mzm23@mrc-tox.cam.ac.uk

Sara Priego Moreno
Salk Institute for Biological Studies
La Jolla, San Diego, California
Contact: sarapriegomoreno@gmail.com

Joanna R. Morris
Birmingham Centre for Genome Biology
Institute of Cancer and Genomic Sciences
University of Birmingham
Birmingham, United Kingdom
Contact: J.Morris.3@bham.ac.uk

Eva Petermann
Birmingham Centre for Genome Biology
Institute of Cancer and Genomic Sciences
University of Birmingham
Birmingham, United Kingdom
Contact: E.Petermann@bham.ac.uk

John J. Reynolds
Birmingham Centre for Genome Biology
Institute of Cancer and Genomic Sciences
University of Birmingham
Birmingham, United Kingdom
Contact: j.j.reynolds@bham.ac.uk

Marco Saponaro
Birmingham Centre for Genome Biology
Institute of Cancer and Genomic Sciences
University of Birmingham
Birmingham, United Kingdom
Contact: M.Saponaro@bham.ac.uk

Tatjana Stankovic
Birmingham Centre for Genome Biology
Institute of Cancer and Genomic Sciences
University of Birmingham
Birmingham, United Kingdom
Contact: T.STANKOVIC@bham.ac.uk

Grant S. Stewart
Birmingham Centre for Genome Biology
Institute of Cancer and Genomic Sciences
University of Birmingham
Birmingham, United Kingdom
Contact: G.S.Stewart@bham.ac.uk

Cyrus Vaziri
Department of Pathology and Molecular Medicine
University of North Carolina
Chapel Hill, North Carolina
Contact: cyrus_vaziri@med.unc.edu

Alicja Winczura
Division of Biomedical Sciences
Warwick Medical School
University of Warwick
Coventry, United Kingdom
Contact: A.Winczura@warwick.ac.uk

Anastasia Zlatanou
Department of Pathology and Molecular Medicine
University of North Carolina
Chapel Hill, North Carolina
Contact: anastasia_zlatanou@med.unc.edu

Introduction

John J. Reynolds, Roger J. A. Grand, and Martin R. Higgs

1

It is now something of a cliché, much used by researchers working in the area of DNA damage and repair, to note that the genome in each human cell is constantly subject to vast quantities of DNA damage, with estimates upwards of 10^5 lesions per day. However, it is an indication of the effectiveness of cellular DNA repair pathways that, in spite of this continuous assault, almost all cells survive and divide with no long-lasting DNA damage or genome instability. How these repair pathways function and how they are coordinated has been the focus of research for over half a century. The subsequent chapters in this volume will describe the processes surrounding DNA repair in detail, focusing on two closely interlinked repair processes: the cellular response to replication stress and double-strand break repair. The first half of this book will focus more on the mechanisms underlying the DNA replication and repair pathways, whilst the second part will discuss the impact of loss/deficiency of these repair processes on human health and will explore the therapeutic potential of DNA damage and repair research. Indeed, many rare inherited human diseases are attributed to mutations in components of replication and repair pathways, the study of which has been invaluable in understanding the intricacies of the cellular response to DNA damage. To provide a background to these chapters, this short introduction will briefly outline the major forms of DNA damage, the DNA repair pathways which are used to deal with them, and the importance of efficient DNA replication and repair for human health. For the sake of simplicity, references have been omitted here but are included in chapters that form the body of the book.

DNA damage can take many forms, arising from both endogenous and exogenous sources (**Figure 1.1**), and a multitude of highly conserved, overlapping DNA damage signalling and repair pathways have evolved to deal with any type of DNA lesion that can arise. This coordinated cellular response is collectively called the DNA damage response (DDR). The impact of the DDR on normal cellular functions is reinforced by the fact that a single exposure to ionising radiation triggers in excess of 900 phosphorylation events involving more than 700 proteins, as well as a plethora of other post-translational modifications (PTMs), including SUMOylation, ubiquitylation and methylation. These PTMs play a critical role in regulating the function of many repair proteins, as well as modifying chromatin status, either as part of intracellular signalling cascades or to permit access to repair factors. Chapters 8 and 9, respectively, describe the roles of protein methylation and ubiquitylation/SUMOylation in the DNA damage response. Protein phosphorylation, the most widespread PTM after DNA damage, is discussed, where appropriate, throughout the book.

DNA double-strand breaks (DSBs) form when both strands of the DNA duplex are broken, and although they are a relatively rare event compared to other forms of damage, they are highly genotoxic if left unrepaired; indeed, it has been shown that a single unrepaired DSB can result in genomic instability, chromosomal rearrangements and/or cell death. DNA DSBs can be caused by exposure to exogenous agents (e.g. ionising radiation), but it is believed that most DSBs arise from endogenous sources, such as the collapse of replication forks following collision with unrepaired DNA lesions during DNA replication, or from the aberrant activity of topoisomerase II. DSBs can be repaired by four interlinked pathways: nonhomologous end joining (NHEJ), alternative NHEJ (alt-NHEJ), homologous recombination (HR) and single-strand annealing (SSA). The choice of pathway used to repair DSBs is dictated by several factors: the phase of cell cycle in which repair takes place, the type of DSB (one-ended versus

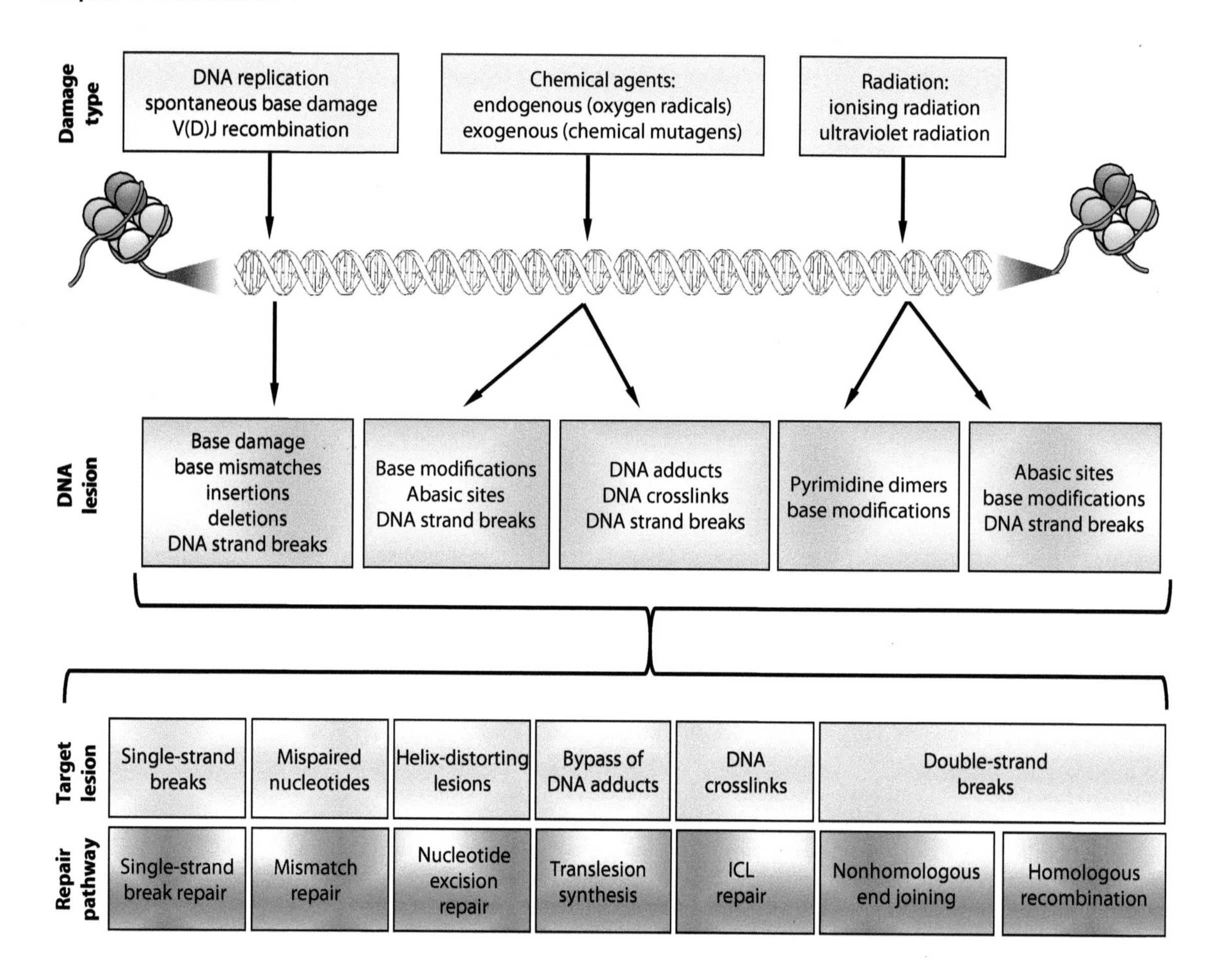

Figure 1.1 Diagram showing examples of the major DNA damaging agents (top row), the lesions they cause (second row), the resulting damage to the DNA (third row) and the repair pathways that deal with the damage (bottom row).

two-ended), the chromatin status where the DSB is located and the type/structure of DSB lesion. The factors governing pathway choice have received much attention over the last decade.

NHEJ is used to repair the majority of DSBs throughout the cell cycle, and although it can potentially introduce errors and is sometimes considered error-prone, on the whole it tends to be an accurate process. NHEJ is described in detail in Chapter 7 of this book. In comparison, HR is an error-free mechanism for the repair of DSBs and is restricted to the S and G2 phases of the cell cycle to allow the undamaged sister chromatid to be used as a template for strand invasion and DNA repair synthesis. HR, and its particular relevance to DNA replication, is discussed in Chapter 6. Both Alt-NHEJ and SSA provide alternative repair mechanisms to the primary DSB repair pathways (NHEJ and HR) and are also considered in Chapters 6 and 7, respectively. Alt-NHEJ (also known as microhomology-mediated end joining [MMEJ]) makes use of short regions of homology to repair DSBs but, as DNA sequences can be deleted at the repair junction, it is considered error-prone. In contrast, SSA is a subpathway of HR and involves long-range DNA resection of DSBs to uncover homologous DNA sequences. However, unlike canonical HR, there is no strand invasion event as the resected ends are annealed together; therefore SSA is error-prone as it typically results in loss of the genomic regions between homologous DNA sequences.

In contrast to DNA DSBs, DNA single-strand breaks (SSB) are one of the most commonly occurring types of DNA lesion in the cell and account for a large portion of the DNA damage a cell experiences each day. Although an individual SSB is not considered particularly deleterious, unrepaired SSBs can easily be converted into the more genotoxic DSBs upon collision with an ongoing replication fork, and therefore their rapid repair is essential to prevent genome instability. SSBs can arise in several different ways from endogenous sources. They can arise directly from the attack on DNA by reactive oxygen species (ROS) generated by cellular metabolism, indirectly following excision of base damage during the base excision repair (BER) pathway or as the result of abortive enzymatic activity. In all cases, the resulting SSB is repaired by the SSB repair pathway, which is described in Chapter 5.

A significant source of exogenous DNA damage is exposure to ultraviolet (UV) light within sunlight. The most common UV irradiation-induced DNA lesions are cyclic pyrimidine dimers and (6–4) photoproducts which are repaired by the nucleotide excision repair (NER) pathway, described in Chapter 5. Two pathways of NER have been identified, which differ in their mechanism of substrate lesion recognition but use converging repair pathways. Global genome NER (GG-NER) is used to rapidly recognise and repair lesions over the entire genome, whilst transcription-coupled NER (TC-NER) is only activated upon collision of RNA polymerase with a DNA lesion within an actively transcribed gene.

DNA cross-links are DNA lesions in which two DNA bases become covalently linked together, either within the same DNA strand (intrastrand) or between different strands (interstrand), and can arise from both endogenous and exogenous cross-linking agents. Interstrand cross-links (ICLs) are particularly cytotoxic as they pose a severe barrier to DNA replication and transcription, and require the coordinated action of a combination of different protein complexes to repair; of central importance are components of the Fanconi anaemia (FA), HR and translesion synthesis (TLS) pathways. In addition to ICL repair, the FA pathway also performs other vital roles during DNA replication, and is explored in Chapter 12. TLS also functions during DNA replication to promote the bypass of damaged bases by nonreplicative DNA polymerases, and is described in Chapter 4. DNA–protein cross-links (DPCs) are another type of replication blocking lesion that can arise from cross-linking agents, and are recognised and repaired in a recently discovered pathway mediated by the DNA-binding metalloprotease SPRTN, and are discussed alongside the FA pathway in Chapter 12.

Although the different forms of DNA damage discussed above are deleterious to genome integrity in their own right, they are especially important when viewed in the context of the two essential cellular processes that use DNA as a template: DNA replication and transcription. The process of cellular DNA replication is fundamental for the continuity of life, and thus the cell employs numerous tightly regulated mechanisms to ensure it proceeds in a timely and efficient manner. Cell cycle factors operate to ensure that initiation, progression and completion of replication occurs before the event of cell division and that the genome is only replicated once during each cell cycle (described in Chapter 2). Additionally, recent evidence has illustrated that a failure to terminate DNA replication successfully can also give rise to genome instability (detailed in Chapter 3). Furthermore, the cell faces the challenge that all enzymatic processes have a certain error rate, and therefore during the course of DNA replication, DNA polymerases will sometimes incorporate the wrong DNA base or produce insertion–deletion mispairs by polymerase slippage at repetitive DNA sequences. To correct these polymerase errors, and prevent them from being turned into permanent mutations in later rounds of the cell cycle, the DNA mismatch repair (MMR) pathway operates during the S phase to remove and repair any base mismatches or insertion–deletion loops (outlined in Chapter 12).

DNA replication also occurs in the face of numerous obstacles, including repetitive DNA sequences, transcription–replication conflicts, DNA lesions (small and bulky DNA adducts, SSBs, cross-links etc.), DNA-protein cross-links and DNA secondary structure. These all have the potential to impact negatively on the process of genome duplication, giving rise to a cellular state known as 'replication stress', which is broadly defined as a state in which DNA replication is impeded, leading to the slowing or stalling of DNA replication forks. It has been suggested that stalled replication forks represent one of the most dangerous lesions that cells encounter, as they will impact on chromosomal duplication and can result in chromosomal rearrangements, cell death or cellular transformation. In this context, it is clear that repair of DNA lesions by the combined actions of NER, SSBR, ICL repair and DSB repair are therefore crucial to maintain successful DNA replication, since forks that encounter these lesions will either stall or collapse to form a one-ended DSB, which requires HR for repair (detailed in Chapter 6). Moreover, prolonged fork stalling can also give rise to replication-associated DSBs. A cellular response to replication stress has therefore evolved to coordinate the processes of DNA replication, cell cycle progression/arrest, DNA repair and replication fork stability/restart, to ensure the successful completion of DNA replication despite the continued threat of DNA damage (described in Chapter 2).

Unrepaired DNA damage also poses a challenge to transcription. Additionally, as DNA and RNA polymerases share the same DNA template, collisions between the replication and transcription machinery also have an impact on transcription. Therefore, the cell employs repair pathways that specifically repair DNA lesions within actively transcribed genes (TC-NER), and uses several strategies/processes to resolve replication–transcription conflicts. The interplay between transcription, replication and DNA repair is addressed in Chapters 10 and 11.

Given all this, it is abundantly clear that the repair of genetic damage is of fundamental importance for the maintenance of genome stability and cellular integrity, and that these pathways have enormous clinical significance. It is therefore unsurprising that a large number of human disorders are caused by mutation or loss of proteins involved in DNA replication and DNA repair pathways. Although these human disorders exhibit a large amount of phenotypic variability, there are common clinical features to many of the diseases, including neurological dysfunction, immunodeficiency, growth retardation, developmental defects and cancer predisposition.

The link between DNA damage and human disease was established for the first time in 1969, when patients with xeroderma pigmentosum (XP), a disease characterised by sensitivity to sunlight and a 1000-fold increased risk of developing skin cancer, were found to be defective for the repair of DNA damage induced by UV light. Mutations in components of the NER pathway were subsequently found to be the cause of XP (Chapter 5). The next clear advance in the understanding of how DNA repair pathways maintain human health was in 1975 when ataxia telangiectasia (A-T) was the first inherited human disease characterised with cancer predisposition (particularly lymphoid tumours) that was also found to be associated with hypersensitivity to agents that induce DSBs. A-T was subsequently found to be caused by mutations in ATM, a critical protein kinase that functions to initiate the DDR signalling cascade upon induction of DSBs (Chapter 13). In the early 1990s, another important milestone in the DNA repair field was passed with the discovery of causative Fanconi anaemia complementation group C (FANCC) mutations in patients with Fanconi anaemia, a rare multigenic disorder characterised by bone marrow failure, cancer predisposition, growth retardation and microcephaly (Chapter 12).

Since these few early discoveries, a growing number of inherited diseases caused by mutations in DNA repair, DNA replication and replication stress factors have been identified (see Chapter 12), in part driven by recent improvements in whole genome and whole exome sequencing technologies. Importantly, the discovery and study of these diseases have led to great advances in our understanding of the process of tumourigenesis and have allowed the development and improvement of cancer therapies. In particular, deficiencies in DNA repair pathways are commonly associated with sporadic tumours, and it is now apparent that the exploitation of these DDR pathway aberrations in tumour cells represents an important mechanism for enhancing the efficacy of treatment. For example, mutations and epigenetic changes in the ATM gene have been identified in sporadic tumours in patients not suffering from A-T. Moreover, targeting the response to replication stress offers an exciting avenue of treating tumours and will be addressed in detail in Chapter 15. Furthermore, the repair of DNA damage is vitally important in stem cells, as the transmission of genetic lesions to daughter cells drives tumourigenesis, as well as ageing. In Chapter 14, we discuss the pathophysiological and therapeutic implications of the molecular pathways through which stem cells cope with DNA damage. Finally, the relationship of DNA damage response pathways to disease and treatments are also considered throughout the book where relevant.

Taken together, we hope that the following chapters offer an exciting insight into the mechanisms of DNA repair, the importance of these pathways in cellular and organism viability, and the therapeutic implications of DNA repair in human health and disease.

DNA Replication and Cell Cycle Control

2

Sara Priego Moreno, Rebecca M. Jones, and Agnieszka Gambus

INTRODUCTION

DNA replication is a universal process and its accurate execution is essential for the propagation of all existing forms of life, including bacteria, archaea and eukarya. As such, the basis of this process is strongly conserved throughout evolution, although the complexity of DNA replication machineries and their regulation increases in higher-order organisms with more complex genomes.

Large and complex eukaryotic genomes are confined into small nuclei and are formed by a set of chromosomes that have to be replicated during the S phase of the cell cycle. The difficulty of accurately completing this process does not just arise from the fact that duplication of large DNA molecules needs to occur extremely quickly in a very restricted space, but also from the fact that DNA replication is coordinated with many other related processes (i.e. DNA transcription, DNA repair, chromatin re-establishment, establishment of sister chromatid cohesion). Importantly, failures during DNA replication generate mutations and genome instabilities that can ultimately lead to cancer development and other pathological diseases (Branzei and Foiani 2010). DNA replication can be divided into three main stages: initiation, elongation and termination. In this chapter, we will give a general overview of the process of DNA replication and highlight its links with cell cycle.

INITIATION

DNA replication origins and features determining origin selection in mammalian cells

Replication origins are the genomic locations where DNA replication starts. In eukaryotes, the number of origins ranges from hundreds in simple organisms such as yeast to tens of thousands in mammalian cells. Such a high number of origins is essential to ensure the timely duplication of the large eukaryotic chromosomes prior to cell division (Leonard and Mechali 2013).

In contrast to the well-characterised sequence elements of replication origins in yeast, no consensus sequences have been found to determine origin selection in higher-order organisms. A number of procedures have been used to study mammalian origins, including the enrichment and isolation of molecular initiation intermediates such as short nascent fragments of leading strands, Okazaki fragments and replication bubbles. Alternatively, other methods have relied on the power of biochemistry to isolate origins by immunoprecipitating DNA cross-linked initiator factors. All these techniques coupled to microarray-based approaches and next-generation sequencing have extensively contributed to the current understanding of origin determinants as well as the genomic distribution of origins in mammalian cells. Interestingly, the consensus reached from these studies is that mammalian origins tend to be enriched in regions of open chromatin, such as transcription start sites and enhancers, and are not defined by a consensus sequence, although they normally exhibit high GC content since they tend to localize within CpG islands (Prioleau and MacAlpine 2016).

The first step of origin selection is the recruitment of the eukaryotic initiator factor, Origin Recognition Complex (ORC), to origin DNA, which then orchestrates the formation of pre-replicative complexes (pre-RCs) with the concerted action of other initiator proteins (explained in more detail in the following section) (**Figure 2.1**). What seems clear from all the knowledge generated about origins of replication in all eukaryotic organisms is that origin selection is favoured in areas of open chromatin, which would provide access for the ORC complex to recognise the origin DNA and for the assembly of the pre-RC. However, the molecular mechanisms by which ORC is recruited to origins remain unknown. While sequence features seem to play an essential role in origin selection by ORC in yeast (Bell and Labib 2016), other elements, including chromatin structure and epigenetic marks, seem to be the main regulators of ORC binding in mammalian cells (Prioleau and MacAlpine 2016). For instance, it is thought that methylated histone H4 influences origin selection by specifically interacting with Orc1 and facilitating ORC recruitment (Kuo et al. 2012). Furthermore, tethering of the H4 methyltransferase Set8 to specific genomic loci was shown to facilitate pre-RC assembly (Tardat et al. 2010). Another example is the ORC associated protein ORCA, which binds to chromatin regions containing repressive epigenetic marks such as methylation of H3 and CpG islands, and regulates pre-RC assembly in these regions by recruiting ORC (Wang et al. 2017).

Initiation of DNA replication: A concerted two-step mechanism

Initiation of eukaryotic DNA replication is a process comprising two steps, which operate at different stages during the cell cycle (Deegan and Diffley 2016).

Origin licensing

The first step takes part during late mitosis and G1; it is known as origin licensing and consists of the loading of double hexamers of Mcm2-7 complexes onto origins (**Figures 2.1** and **2.2**). Biochemical reconstitution with purified *S. cerevisiae* proteins has shown that three proteins are required for this reaction: ORC and Cdc6, which both belong to the AAA^+ ATPase protein family, and Cdt1 (Evrin et al. 2009, Remus et al. 2009). Mcm2-7 is the core of the replicative helicase and is also a AAA^+ ATPase, whose enzymatic activity depends on the specific arrangement of its subunits in a ring-shaped complex (Davey et al. 2003). The process starts with the binding of ORC to origin DNA (Bell and Stillman 1992), which in turn recruits Cdc6 (Wang et al. 1999). This generates a platform for the recruitment of the first Mcm2-7 hexamer. Importantly, the final product of the licensing reaction is an enzymatically inactive double hexamer of Mcm2-7 associated through the N-termini of each hexamer, with the C-terminal helicase domains facing the outside. This complex is also known as pre-RC and is topologically encircling dsDNA (Evrin et al. 2009, Remus et al. 2009, Gambus et al. 2011). Interestingly, the mechanism by which the second Mcm2-7 is recruited remains unknown. While some models favour the recruitment of the second Mcm2-7 to an already assembled one, others support the assembly of two Mcm2-7 by two independent ORC-Cdc6 (Yardimci and Walter 2014). Critically, Mcm2-7 complexes need to undergo several conformational changes during the licensing reaction, including opening and re-closing of the ring to engage dsDNA and transition from soluble single to DNA-bound double hexamers. ATP binding and hydrolysis by Mcm2-7 is required for double hexamer formation (Coster et al. 2014, Kang et al. 2014), which means that the conformational changes experienced during licensing are most likely driven, at least in part, by its own ATPase activity.

In contrast to mitotic events, the licensing reaction requires low levels of CDK activity. This environmental switch is achieved by the ubiquitin ligase action of the anaphase promoting complex or cyclosome (APC/C) associated to either Cdc20 or Cdh1 substrate receptors. The concerted action of APC/C-Cdc20 and APC/C-Cdh1 targets mitotic cyclins and the CDK activator tyrosine phosphatase Cdc25 for proteasomal degradation, and accumulates CDK inhibitors (CKIs) through proteasomal degradation of their negative regulator Skp2 (Moreno and Gambus 2015).

Apart from regulating CDK levels, in metazoans, APC/C-Cdh1 stabilises the presence of licensing factors, such as Orc1 (Narbonne-Reveau et al. 2008), Cdc6 (Mailand and Diffley 2005) and Cdt1 by targeting Cdt1 inhibitor geminin for proteasomal degradation (McGarry and Kirschner 1998). All these ubiquitin-driven proteasomal degradation events promoted by APC/C generate favourable conditions for the execution of origin licensing.

Helicase activation

The second step of DNA replication initiation occurs only in S phase; it is known as origin firing and consists of the conversion of the inactive double hexamer of Mcm2-7 into two active

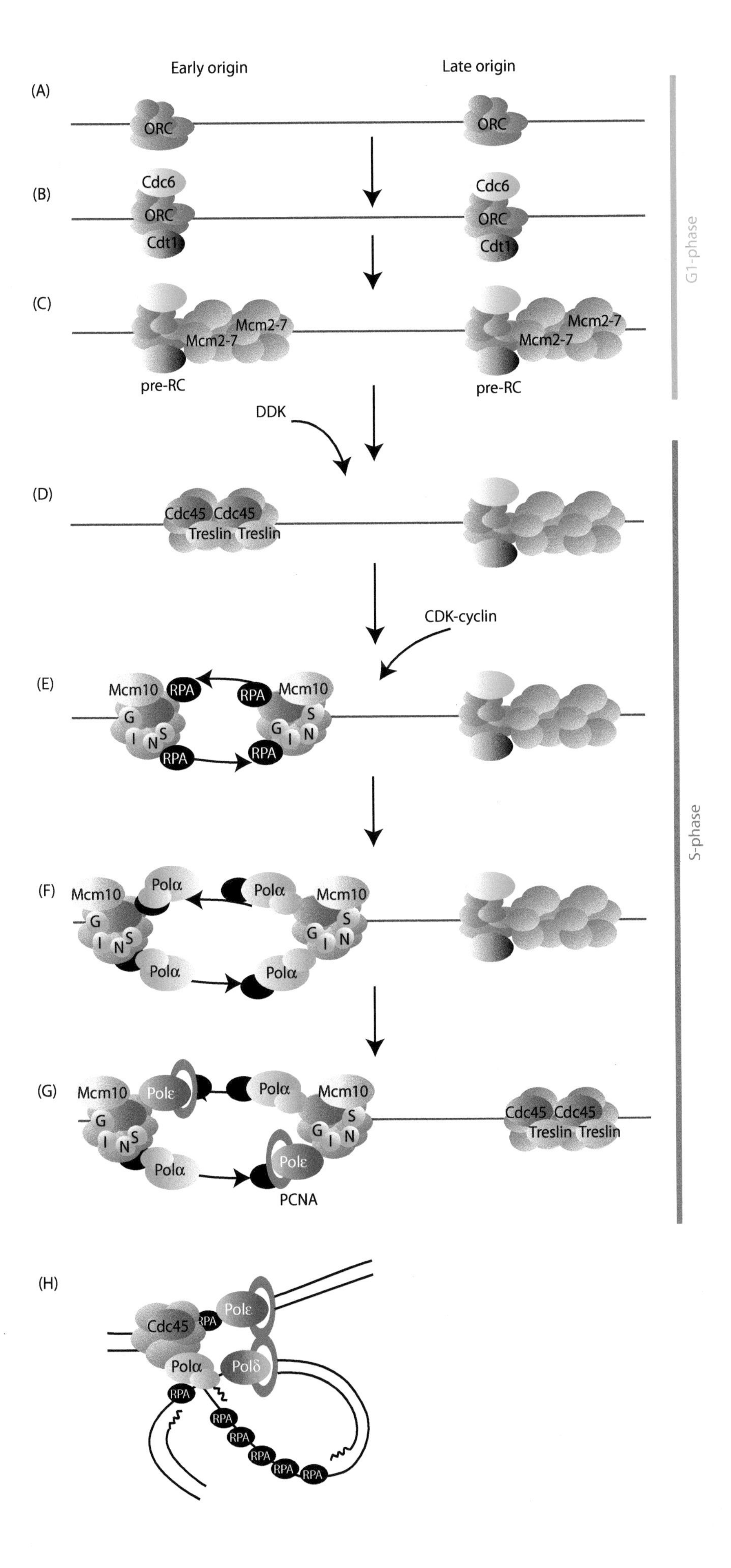

Figure 2.1 Origin licensing and firing. (A) The initiator factor Origin Recognition Complex (ORC) is recruited to the origin DNA. (B–C) With the help of proteins Cdc6 and Cdt1, origins become licensed in late mitosis and G1 as double hexamers of Mcm2-7 are loaded. This complex of proteins, which encircles the DNA, then becomes known as the pre-replicative complex (pre-RC). (D–E) In S phase, pre-RC complexes are activated by kinases DDK and CDK, leading to recruitment of Treslin, Cdc45, GINS and Mcm10. The origin DNA melts so that the two CMG complexes encircle the leading strand templates and unwind the DNA to create ssDNA, which is coated by RPA. (F–H) Initial DNA synthesis is performed by Polα at the bi-directional forks, with the leading strand then copied in a continuous manner by Polε and the lagging strand copied discontinuously by both Polα and Polδ.

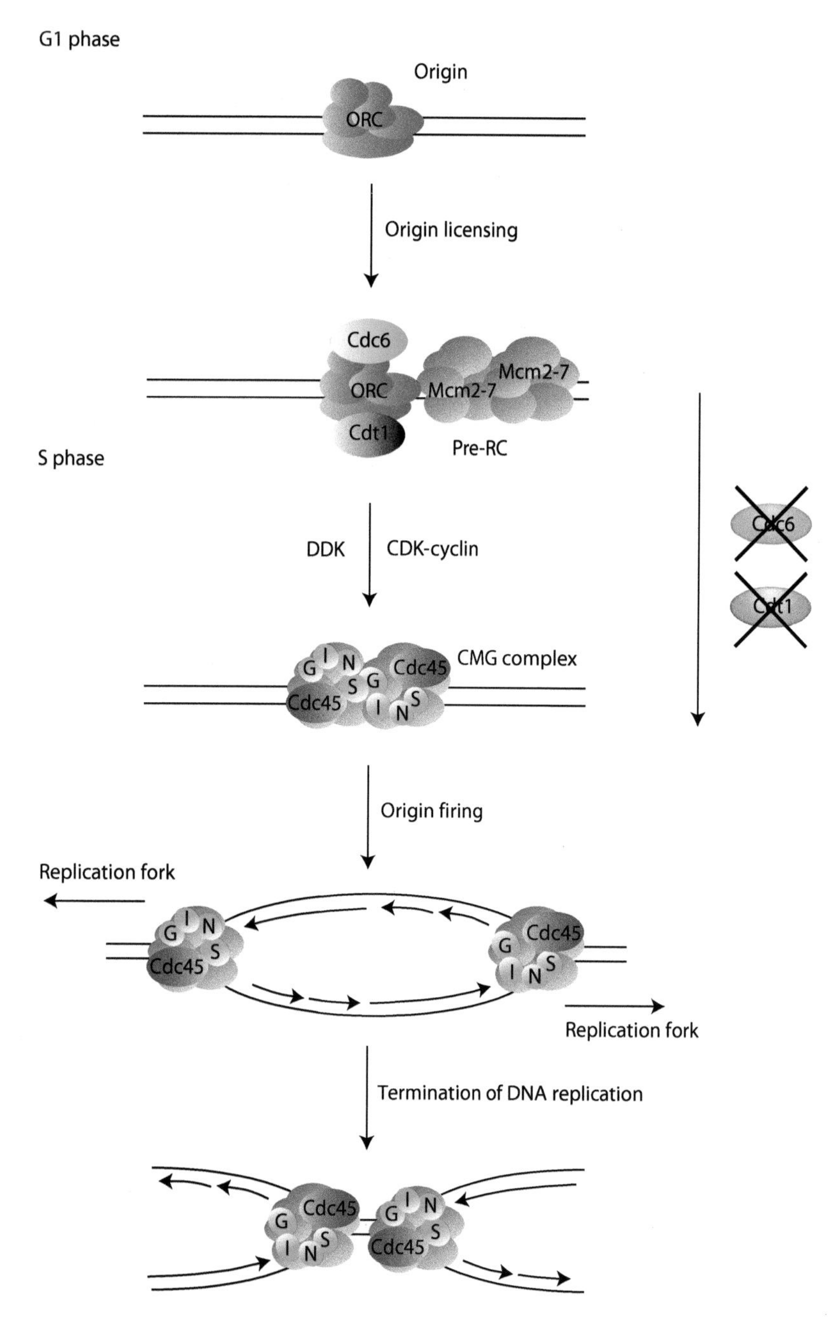

Figure 2.2 Origin firing and replication forks termination. As in Figure 2.1, origin DNA is licensed during G1 and activated in S phase. The final image depicts the convergence of two CMG complexes at replication termination.

replicative helicases or CMG complexes formed by the stable association of Mcm2-7 with Cdc45 and GINS (complex of Psf1, Psf2, Psf3 and Sld5 or GINS1, GINS2, GINS3 and GINS4) (Figures 2.1 and 2.2) (Gambus et al. 2006). Essential requirements for the initiation of DNA replication are the activation of cyclin-dependent kinase 2 (Cdk2)-Cyclin E and Dbf4/Drf1-dependent cell division cycle 7 (CDC7) kinase (DDK) and the expression of S phase genes. This is in part achieved by expression of the G1-specific cyclin D, which is not targeted by APC/C. In human cells, Cdk4/6-cyclin D expressed in G1 stage of the cell cycle activates transcription by partially phosphorylating the retinoblastoma family proteins (Rb, p130, p107). This allows the transcription factor E2F1-3 to switch from actively repressing transcription when associated with un-phosphorylated Rb to partially inducing genes required for DNA replication and cell cycle progression (reviewed in Bartek and Lukas 2001). p27 and p21 proteins, whose levels are high in G1 phase, bind to Cdk2-cyclin E and inhibit its activity, until they cannot stoichiometrically block the CDK activity anymore. This is accelerated by the activity of the transcription factor Myc (reviewed in Bretones et al. 2014). After reaching a critical level, Cdk2-cyclin E then phosphorylates p27, stimulating its interaction with ubiquitin ligase SCF^{Skp2} and its degradation (Sheaff et al. 1997, Sutterluty et al. 1999). Following full activation of Cdk2-cyclin E, the Rb proteins become hyper-phosphorylated and fully inactivated. This promotes the synthesis of genes required for S-phase progression such as the S-phase cyclin A, leading in turn to the inhibition of APC/C activity.

The successful in vitro reconstitution of the firing reaction using *S. cerevisiae* recombinant proteins revealed the requirement of nine factors: CDK, DDK, Sld3-Sld7 (homologue of human Treslin/TICCR-MTBP), Dpb11 (human TopBP1), Sld2 (human RecQL4), Cdc45, DNA polymerase ε, GINS and Mcm10 (Yeeles et al. 2015). Although the molecular mechanisms that underlay the different phases of this reaction are still poorly understood, it is clear that the activities of the S phase kinases DDK and CDK play different essential roles. While DDK phosphorylates the MCM complex, which in turn promotes the recruitment of Treslin/TICCR and Cdc45, CDK phosphorylates Treslin/TICCR-MTBP, resulting in their association with TopBP1 and subsequent recruitment of GINS, giving rise to the CMG complex (**Figure 2.1**) (Boos et al. 2011, 2013, Kumagai et al. 2011, Deegan and Diffley 2016). Importantly, Mcm10 was not found to be required for CMG assembly (Yeeles et al. 2015); instead, it seems to promote the activation of the Mcm2-7 helicase activity, since it was found to be necessary for origin DNA unwinding after loading of the CMG complex (Kanke et al. 2012). The completion of the firing reaction results in origin DNA melting and two CMGs encircling each of the leading strand templates, which translocate away from the origin unwinding DNA (**Figures 2.1** and **2.2**). Importantly, these CMGs comprise the platforms for the assembly of two replisomes that will assist with bi-directional replication of the origin flanking DNA (Deegan and Diffley 2016).

Replication timing program

Importantly, only a small proportion of licensed origins undergo activation during S phase in each cell, and not all of these origins are activated at the same time. Instead, they fire according to a specific replication-timing program. The main determinants of the timing program are chromatin structure and organization within the nucleus, the regulation of transcription programmes, and the availability of rate-limiting replication factors (Fragkos et al. 2015).

The genome can be divided into different chromosomal domains that are replicated at defined times during S phase in each cell type. Early replicating domains tend to coincide with actively transcribed regions of open chromatin containing a high density of licensed origins, and are generally localised in the interior of the nucleus. On the contrary, late replicating domains are observed in regions of heterochromatin with repressive epigenetic marks and low levels of licensed origins. They are localised in the nuclear periphery and are associated with the nuclear lamina. Critically, several replication factors required for origin firing, including Cdc45, Cdc7, Sld2/RecQL4, Sld3/Treslin and Dpb11/TopBP1, have been found to be rate limiting, which means that full duplication of the genome depends on their recycling during the replication cycle, becoming more available at the end of S phase and therefore allowing the activation of origin-poor, late-replicating domains (Fragkos et al. 2015).

Controlling DNA replication to only one round per cell cycle

In order to maintain genome integrity, it is important that only one copy of each chromosome is generated prior to cell division. Both the re-replication and the incomplete replication of genomic regions can lead to very toxic chromosomal aberrations. Hence, eukaryotic cells have evolved numerous mechanisms to control these situations (Blow et al. 2011, Siddiqui et al. 2013).

Avoiding re-replicated DNA

With the aim of avoiding re-replication, numerous mechanisms exist to strictly split the processes of origin licensing and firing into different cell cycle stages.

In summary, high APC/C and low CDK activity restricts origin licensing to late M phase and G1, in part by allowing the stability of licensing factors. During this stage, diverse mechanisms contribute to the low levels of CDK, including CDK inhibitors such as p27 (Chellappan et al. 1998) and degradation of specific cyclins. Importantly, this lack of CDK activity also inhibits origin firing, avoiding premature origin activation during G1 which could result in re-licensing and re-replication (Siddiqui et al. 2013).

Conversely, low APC/C and high CDK activity allows progression into S phase and origin firing. Apart from the already-discussed role of CDK in promoting origin firing, high levels of this kinase activity also result in the inhibition of licensing factors, again avoiding origins

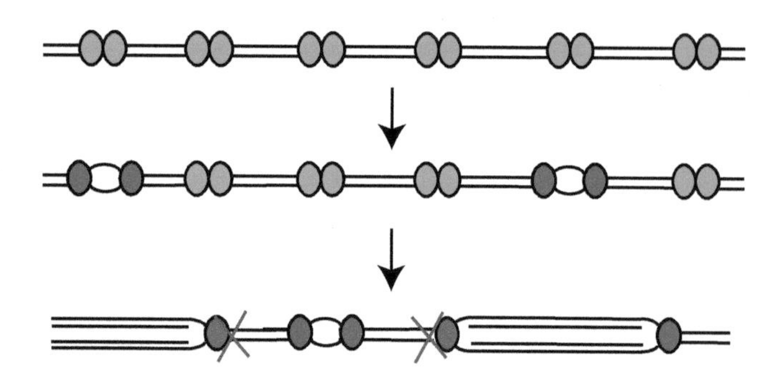

Figure 2.3 **Dormant origin firing.** In G1 phase, many origins are licensed by loading of the double hexamer Mcm2-7 (blue). In S phase, a small percentage of these pre-RCs are activated at which bi-directional replication will begin (red). Should two converging replication forks stall with a dormant origin lying between them, then this dormant origin can be activated in order to rescue and complete DNA synthesis.

re-licensing (Siddiqui et al. 2013). Critically, CDK activity promotes ubiquitin-driven proteasomal degradation of factors involved in origin selection and licensing including Set8, Cdt1, Cdc6 and Orc1 (Moreno and Gambus 2015). Particularly important in higher eukaryotes is the regulation of Cdt1, which is inhibited also in a CDK-independent manner by geminin (Wohlschlegel et al. 2000).

Avoiding under-replicated DNA: The importance of dormant origins

Theoretically, only two CMG helicases are enough to establish bi-directional replication forks at each origin. However, the levels of chromatin-bound Mcm2-7 in eukaryotic cells can be up to 40 times more than those of active CMGs, which hinders the visualisation of Mcm2-7 complexes at replication factories (Takahashi et al. 2005). Moreover, in *Xenopus laevis* egg extracts 10 to 20 molecules of Mcm2-7 are loaded per molecule of ORC, suggesting that this excess of Mcm2-7 loading could not only be a consequence of licensing more origins than required, but also that more than one double hexamer of Mcm2-7 is loaded per origin. Decreasing the levels of Mcm2-7 in *Xenopus laevis* egg extracts to the levels of minimal licensing (around two molecules per origin) did not impair DNA replication during unperturbed S phase (Oehlmann et al. 2004). When detecting MCM2-7 complexes on chromatin in replicating cells, they do not overlap with sites of DNA synthesis visualised by BrdU incorporation. This observation stems from the fact that there is a high level of Mcm2-7 on not-yet-replicated chromatin while they are depleted from already-replicated chromatin. This phenomenon was called an MCM paradox and raised the question of what the function was of such an excess of Mcm2-7 complexes loaded during licensing (Takahashi et al. 2005).

The explanation for the MCM paradox comes from the fact that origins cannot be licensed during S phase; therefore, the main way to ensure full duplication of the genome is by over-loading pre-RCs during G1. In a normal S phase, only a small percentage of the origins licensed in G1 will be activated, which means that the majority of them will remain as inactive or dormant origins. Importantly, these dormant origins have a backup role, and they come into action in situations of replication stress to rescue the defects of stalled replication forks (Figure 2.3) (Blow et al. 2011). Since double hexamers of Mcm2-7 have the ability to slide along dsDNA (Evrin et al. 2009, Remus et al. 2009), they may be displaced by progressing replication forks during elongation. Excessive loading of Mcm2-7 is critically important during the first embryonic cell cycles, as exemplified in organisms such as *Xenopus*, to ensure the duplication of the entire genome during the first cleavage divisions, which are required to proceed very quickly without apparent gap phases and with a weak G2/M checkpoint that delays entry into mitosis if genomic instabilities arise (Hyrien et al. 2003).

ELONGATION

During elongation, two replication forks emerge from each fired origin as a result of replisome assembly around individual CMGs. Replication fork progression through the chromatin is in part facilitated by the CMG complex, which moves at the head of the fork and unwinds dsDNA by translocating along the leading strand template with a 3′-5′ polarity (Georgescu et al. 2017). This dsDNA unwinding generates ssDNA, which is rapidly coated by the ssDNA binding protein RPA and provides the templates required for DNA synthesis by the replicative polymerases (Figure 2.1H). Because of the antiparallel structure of DNA and the 3′ end extension specificity of the replicative polymerases, DNA replication proceeds in a semi-discontinuous way: the leading strand is copied continuously in the direction of fork progression, while the lagging strand is copied discontinuously in the opposite direction via short Okazaki fragments.

Replicative polymerases

Eukaryotic replisomes depend on three multi-subunit replicative polymerases: Polα, Polδ and Polε (Kunkel and Burgers 2008). Polα has both RNA primase and DNA polymerase activity, and is the polymerase responsible for synthesising de novo 5′RNA-3′DNA primers that can be extended by Polδ or Polε (Pellegrini and Costa 2016). Due to the continuous requirement for Polα priming events for lagging strand synthesis, Polα is an important component of the replisome and is linked to it by the replication factor Ctf4 (human And-1) (Simon et al. 2014).

Polδ and Polε are responsible for the bulk synthesis of DNA. Genetic studies in yeast using mutants of Polδ or Polε that inserted specific mutations suggested that synthesis of the leading and lagging strands were carried out by Polε and Polδ, respectively (Pursell et al. 2007, Nick McElhinny et al. 2008, Larrea et al. 2010). The same specific strand assignment was obtained from biochemical studies in which replication of the leading and lagging strands were reconstituted in vitro with a minimal replisome and a forked template (Georgescu et al. 2014, 2015). In fact, a role for Polε in the continuous leading strand extension makes biochemical sense, since it has a greater processivity (Hogg et al. 2014) and it is physically connected to the CMG complex through interaction between the Dpb2 subunit of Polε and GINS (Sengupta et al. 2013).

Replication clamp and clamp loader

Replicative polymerases on their own can only synthesise small stretches of DNA in one go. In order to increase their processivity, they associate with ring-shaped sliding clamps that encircle the DNA template. In eukaryotic cells, the replication clamp is the homo-trimeric ring PCNA (Figure 2.1G and H) (Hedglin et al. 2013). The closed structure of these clamps requires an active mechanism to load them onto DNA, and clamp loaders assist this task. RFC is the eukaryotic replication clamp loader, and it works by recruiting PCNA to sites where DNA synthesis is initiated, also referred to as PT junctions (3′ end of primer–template junctions) (Hedglin et al. 2013). Once loaded, PCNA associates with the replicative polymerases as well as a multitude of other replication factors in part through PCNA-interacting protein (PIP box) motifs contained in these proteins (Mailand et al. 2013).

Okazaki fragments maturation

Okazaki fragments generated as a result of lagging strand synthesis need to be processed in order to form an RNA-free continuous dsDNA. Importantly, at least two pathways lead this task (Balakrishnan and Bambara 2013).

Short flap pathway

During the short flap pathway, elongating Polδ will encounter and displace the RNA–DNA primer of the downstream Okazaki fragment until it encounters the first nucleosome (Devbhandari et al. 2017), generating a short 5′ flap. Subsequently, the 5′-3′ endonuclease FEN1 will recognise the flap and cleave it at the base, creating a nick between the two Okazaki fragments that will then be ligated by DNA ligase I (Figure 2.4) (Bambara et al. 1997, Lieber 1997).

Long flap pathway

There are situations in which the 5′ flap displaced by Polδ is too long and becomes coated by RPA, which inhibits the ability of FEN1 to process the flap. These situations require the multi-functional protein Dna2, which has 5′-3′ endonuclease activity. Dna2 removes RPA and cleaves the flap, reducing it to a shorter size that can be processed by FEN1 (Figure 2.4) (Bae et al. 2001).

DNA topological stress and the role of topoisomerases

Unwinding of dsDNA during elongation generates topological stress ahead of the replication fork known as positive supercoiling, which needs to be relaxed in order to enable smooth replication fork progression. This torsional stress is resolved by topoisomerases I and II: Topo I cleaves one strand allowing free rotation of the other strand, while Topo II nicks both strands followed by passing another section of dsDNA through the break before assisting re-ligation (Figure 2.5) (Keszthelyi et al. 2016, Pommier et al. 2016). In particular, Topo I is thought to be the major player in the relaxation of positive supercoiling within unreplicated regions (Postow et al. 2001), which makes biochemical sense since Topo I is part of the replisome progression complex (RPC) built around CMG (Gambus et al. 2006); thus, it is perfectly

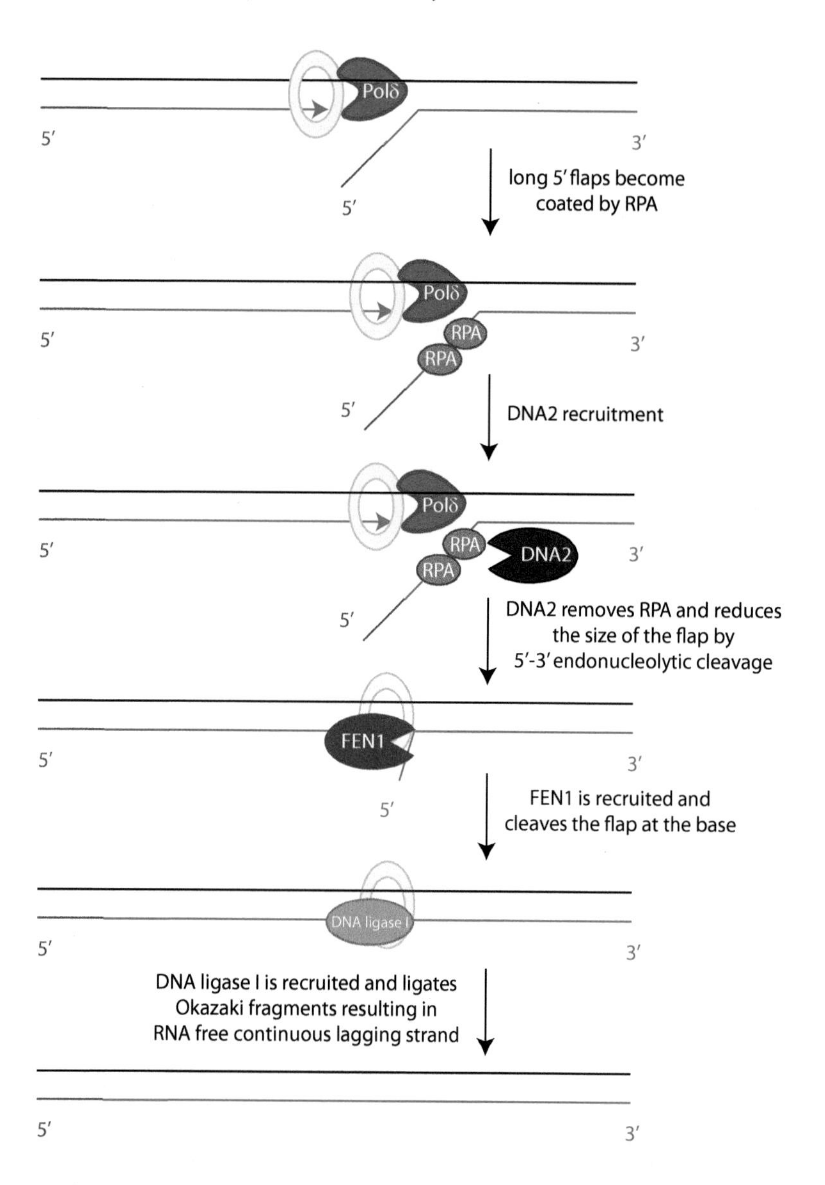

Figure 2.4 Okazaki fragment processing (short flap and long flap). During Okazaki fragment processing, the elongating Polδ will displace the RNA-DNA primer to create a 5′ flap. A short flap will be recognised and cleaved at the base by the 5′-3′ endonuclease FEN1. A long flap, on the other hand, will initially be coated with RPA before being processed by DNA2 so as to remove RPA and shorten the flap. At this point, FEN1 can then complete cleavage of the flap. Following both circumstances, DNA ligase I performs DNA end ligation.

positioned for such a role ahead of the fork. However, there are situations in which the accessibility of topoisomerases to these regions of positive supercoiling is sterically inhibited, requiring another mechanism to relieve the topological stress. Importantly, fork rotation can release the topological tension ahead of the fork by transmitting it to the replicated DNA, but this mechanism generates intertwined sister chromatids behind the fork, also known

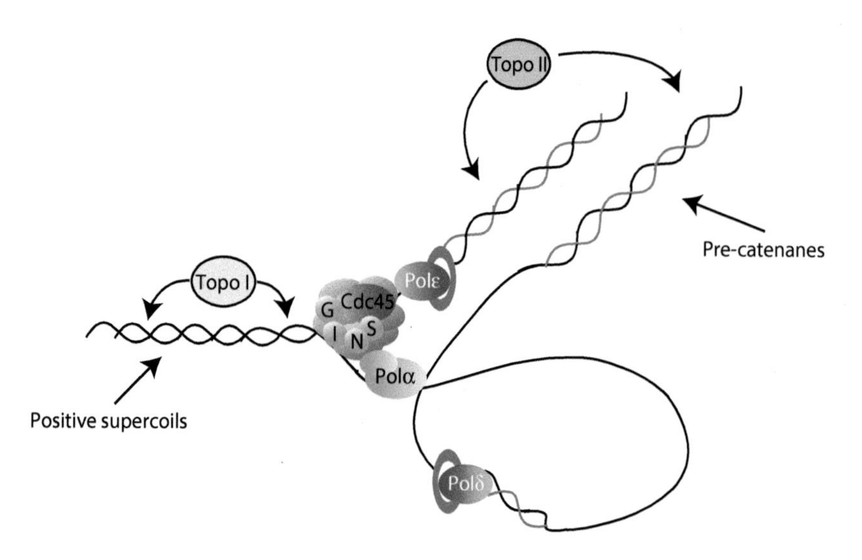

Figure 2.5 DNA supercoils. During DNA replication, topological stress that accumulates ahead of the replication fork can be resolved by Topoisomerase I (Topo I), facilitating fork progression. On the other hand, topological stress at terminating forks can be resolved by fork rotation. As a result, sister chromatids behind the replication forks become intertwined and end up as pre-catenanes. In this circumstance, Topoisomerase II (Topo II) is responsible for resolving the intertwined sister chromatids, to enable correct chromosome segregation during mitosis.

as pre-catenanes (Figure 2.5) (Keszthelyi et al. 2016). The fork rotation mechanism is not common during elongation, and in budding yeast is inhibited by the replisome components Tof1-Csm3 (Timeless-Tipin) (Schalbetter et al. 2015), ensuring its usage only under extreme conditions where topoisomerases cannot access the parental DNA ahead of the replication fork. A very good example of these situations is during convergence of replication forks during termination, where converging replisomes sterically block topoisomerases accessibility, and fork rotation is the main mechanism for releasing positive supercoiling between converging forks (Sundin and Varshavsky 1980, 1981). Critically, Topo II plays a crucial role in the resolution of intertwined sister chromatids upon completion of DNA synthesis by resolving the generated pre-catenanes, allowing faithful chromosome segregation during mitosis (Baxter and Diffley 2008).

TERMINATION

Replication fork termination occurs when two forks arising from neighbouring origins converge between two replicons (Figure 2.2). At this stage, DNA unwinding and synthesis between the two converging forks need to be completed, the replisome needs to be disassembled, and sister chromatids need to be efficiently decatenated to ensure faithful chromosome segregation. In contrast to the initiation and elongation stages, termination has been poorly understood. However, the last few years have brought insights that have allowed us to better understand this stage.

In eukaryotic cells, replication fork termination occurs throughout the entire S phase, whenever and wherever two forks coming from neighbouring origins converge. Genome-wide studies have shown that termination events are generally located between two adjacent activated origins (Petryk et al. 2016). Innovative plasmid setup allowing site-specific and synchronised termination of replication forks designed by Walter's laboratory provided evidence that when simple plasmid templates are replicated, terminating replication forks do not collide with each other but rather pass each other as each helicase travels on opposite strands of DNA (Yardimci et al. 2012, Dewar et al. 2015). The progression of the polymerases and nascent DNA synthesis was not slowed or paused when replisomes were passing each other, and DNA synthesis progressed until the end of the last Okazaki fragment on the lagging strand. Finally, on such templates, the removal of the replisome was observed to be the last step of termination – executed once DNA synthesis was completed (Dewar et al. 2015).

The unloading of the post-termination replisome is at present the most understood stage of replication fork termination. Upon termination, the Mcm7 subunit of the active helicase is polyubiquitylated with lysine-48–linked ubiquitin chains by the Cullin2-based ubiquitin ligase and Lrr1 as a substrate receptor ($CRL2^{Lrr1}$) (Dewar et al. 2017, Sonneville et al. 2017). The ubiquitin ligase acting in lower eukaryotes (budding yeast) is different – it is SCF^{Dia2} (Maric et al. 2014). This ubiquitylation is then followed by removal of the helicase from chromatin in a manner dependent on p97/VCP segregase. As the helicase forms the organising centre of the replisome, its removal leads to the disassembly of the remaining replication machinery (Moreno et al. 2014). Further details of this process are discussed in Chapter 3.

REPLICATION-COUPLED PROCESSES

Chromatin re-establishment

In order to enable its storage into the nucleus, eukaryotic DNA is packed into chromatin by association with multiple proteins. The nucleosome is the basic unit of chromatin and consists of a 147-bp-long fragment of dsDNA wrapped around an octamer formed by two copies of each histone: H2A, H2B, H3 and H4 (MacAlpine and Almouzni 2013). This chromatin organization needs to be flexible enough to allow access to regulatory elements and fulfil the correct transcription programs. The epigenetic code allows chromatin flexibility and is determined by the multiple post-translational modifications of histone tails (Kouzarides 2007) and DNA methylation patterns (Bird 2002). During replication fork progression, nucleosomes ahead of the fork are evicted and recycled to be positioned behind the fork on the newly synthesised sister chromatids (Figure 2.6) (MacAlpine and Almouzni 2013). In addition to this, new histones need to be produced and loaded onto chromatin during S phase in order to accomplish the complete chromatin re-establishment of two copies of genetic material (Marzluff et al. 2008).

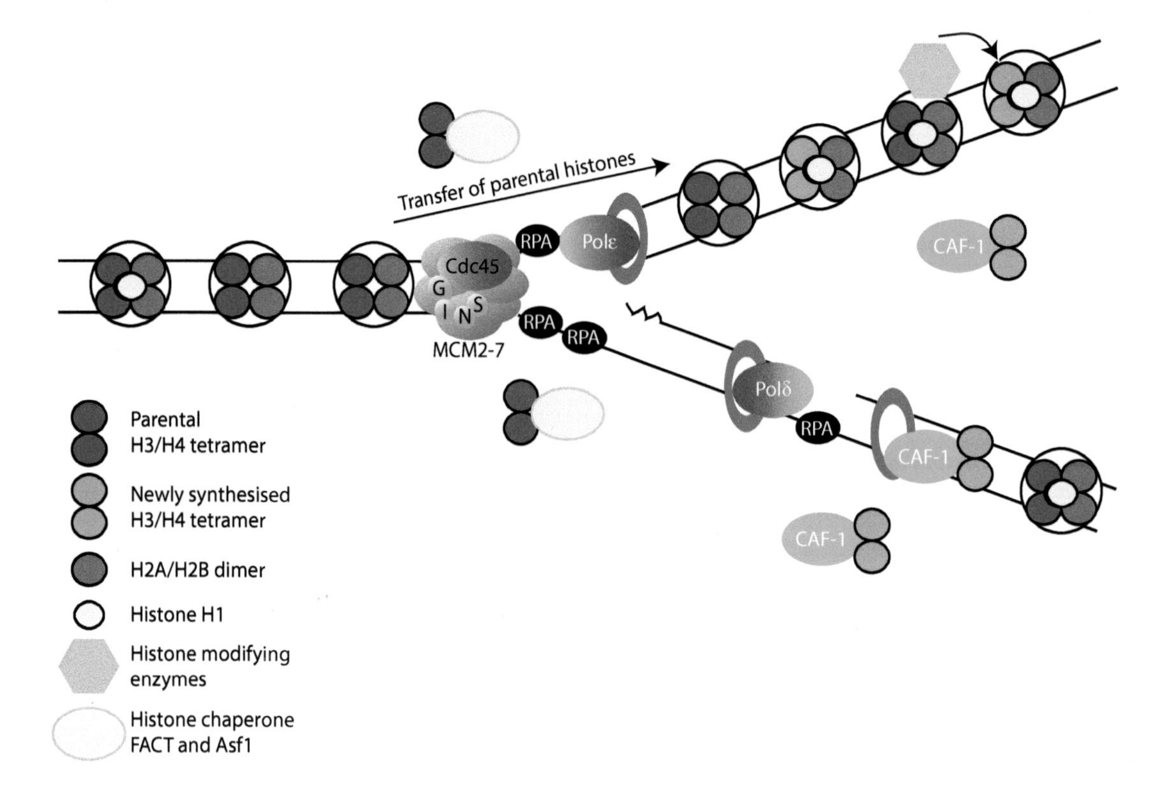

Figure 2.6 Histone re-positioning. As replication forks proceed, nucleosomes ahead must be displaced and then reconstituted onto the daughter DNA strands behind. Shuttling of parental histones and positioning of newly synthesised histones is performed by replication fork-associated histone chaperones such as the FACT complex, Asf1 and CAF-1.

The mechanism of nucleosome disruption ahead of the replication fork is not well understood, but it is likely to be promoted by forces that compromise their stability, such as direct collision with the replicative helicase or the positive supercoiling generated ahead of the fork during elongation. The current model is that parental $(H3\text{-}H4)_2$ tetramers are randomly distributed between the two newly synthesised sister chromatids, followed by the deposition of two new or parental (H2A-H2B) dimers (Figure 2.6). The recycling route of parental histones during replication is not well defined, but it is thought to be promoted by replication fork–associated histone chaperones such as the Mcm2 subunit of the CMG helicase itself, the FACT complex and Asf1. These histone chaperones could function as molecular platforms for receiving parental histones from evicted nucleosomes ahead of the fork, and facilitate their deposition into newly synthesised DNA (Hammond et al. 2017). On the other hand, the deposition of newly synthesised histones involves additional factors. For instance, shortly after being synthesised, H3 and H4 form dimers that are then shuttled to the nucleus through a complex transport pathway involving multiple factors such as the histone chaperones NASP1 and Asf1. Once in the nucleus, Asf1 transfers the H3-H4 dimers to CAF1, which is recruited to newly synthesised DNA through interaction with PCNA and facilitates deposition of new $(H3\text{-}H4)_2$ tetramers. Finally, H2A-H2B dimers are added, probably by NASP1 or FACT chaperones, leading to completion of nucleosome assembly (Alabert and Groth 2012).

Critically, chromatin epigenetic states need to be inherited during cell division in order to maintain cell type–specific transcription programs. For instance, histone marks are maintained during replication-coupled recycling (Alabert et al. 2015), and they can recruit their cognate enzymes to promote modification of neighbouring histones that have been de novo deposited (Hansen et al. 2008, Margueron et al. 2009). Finally, the propagation of DNA methylation patterns also seems to be, at least in part, replication coupled, and it is mediated by the DNA methyltransferase DNMT1 (Nishiyama et al. 2013).

Replication-coupled establishment of sister chromatid cohesion

Sister chromatids generated during replication need to be held together until mitosis in order to ensure faithful chromosome segregation. This is achieved by the establishment of cohesin-mediated sister chromatid cohesion during S phase. Cohesin is a tetrameric complex formed by the proteins Smc1, Smc3, Scc1/Rad21 and Scc3/SA, which adopt a ring-shaped structure that has the ability to topologically embrace DNA (Figure 2.7A). Cohesin loading onto chromatin occurs during G1 and is facilitated by the heterodimer Scc2-Scc4

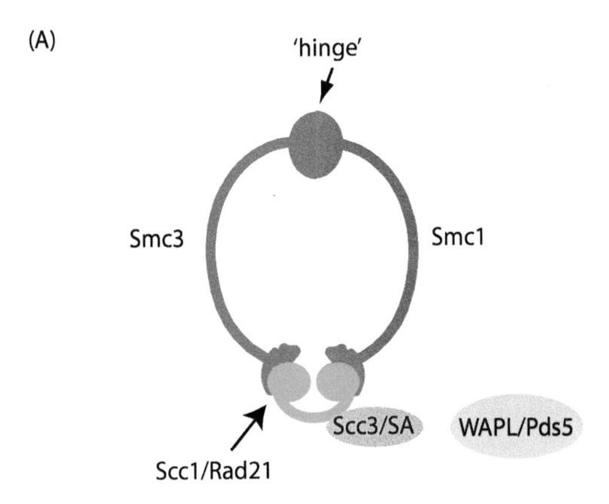

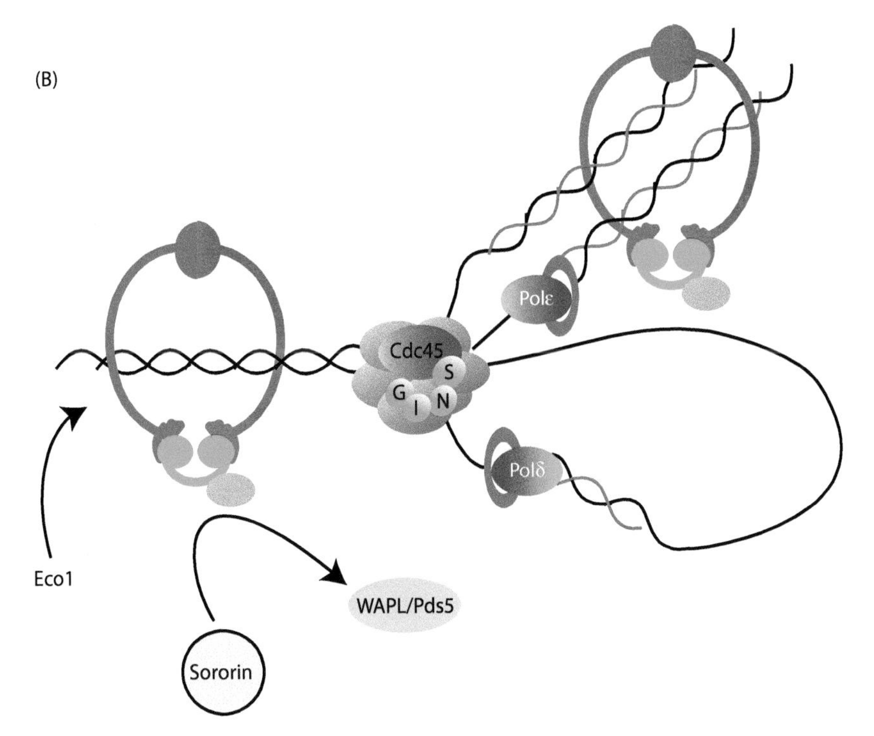

Figure 2.7 Cohesin. (A) Cohesin is a ring-shaped tetrameric complex composed of Smc1, Smc3, Scc1/Rad21 and Scc3/SA. Binding of cohesin to the DNA in G1 phase is unstable due to the action of WAPL/Pds5, which causes its unloading. (B) In S phase, binding of cohesin to DNA is stabilised as Eco1 acetylates the cohesin complex, causing recruitment of sororin, which in turn acts to displace WAPL/Pds5. The complex encircles sister chromatids throughout DNA replication until mitosis so as to ensure correct chromosome segregation.

(human NIPBL-MAU2). At this stage, cohesin association with chromatin is very dynamic due to the counteracting action of WAPL-Pds5, which promotes cohesin unloading. During S phase, cohesin rings become cohesive, entrapping newly synthesised sister chromatids together in a very stable way (**Figure 2.7B**). This cohesive state requires acetylation of the cohesin ring by the replication fork-associated Eco1 acetyltransferase, which in turn recruits sororin, a protein that displaces WAPL, hence avoiding cohesin unloading (Losada 2014, Uhlmann 2016). Two non-mutually exclusive models have been proposed for the co-entrapment of sister chromatids by cohesin rings. One possibility is that the replication fork passes through the cohesin ring, promoting its acetylation by Eco1, resulting in the stable capture of replication products. Alternatively, if the replication fork is unable to pass through cohesin rings, then they would need to be reloaded in the wake of the replication fork and be acetylated at this point to ensure efficient entrapment of replication products (Uhlmann 2016).

During prophase, cohesin rings start being unloaded from chromosome arms. This is directed by mitotic kinases which promote phosphorylation of sororin, destabilising its association with cohesin and allowing WAPL-mediated cohesin unloading. Importantly, at this stage, sister chromatid cohesion is protected at the centromeres by the phosphatase action of shugoshin-PPA2. However, in anaphase, the activity of the APC/C complex triggers activation of separase, which cleaves the Scc1/Rad21 subunit of the remaining chromosome-associated cohesin complexes. This mechanism allows accurate segregation of sister chromatids (Losada 2014, Uhlmann 2016).

DNA damage response pathways and the key role of ATR in controlling replication stress

During the cell cycle, our cells are constantly exposed to both extrinsic and intrinsic genotoxic agents that can potentially generate DNA lesions, including ultraviolet light-induced thymine dimers, interstrand cross-links generated by chemicals, base mismatches introduced during DNA replication, and single- and double-strand breaks (SSBs and DSBs) produced by ionizing radiation and chemical reactions. In order to ensure the transmission of intact genomes, our cells have evolved very complex DNA damage response signalling pathways to sense this broad spectrum of DNA lesions and promote their repair. Very often, induction of DDR pathways triggers checkpoint activation, whose main role is to delay cell cycle progression with the aim of providing the cell with enough time to solve the problem and complete faithful genome duplication prior to entering into mitosis (Blackford and Jackson 2017).

In the same manner as for many other cellular signalling pathways, the DDR is mediated by protein phosphorylation. Three related kinases control the DDR in vertebrate cells: DNA-PK, ATM and ATR. While DNA-PK and ATM become activated by and mediate the response to DSBs, ATR responds to basically any kind of insult that generates long stretches of RPA-coated ssDNA, which is the main signature of replication stress. Such long tracks of RPA bound to ssDNA can be generated as a result of nucleolytic processing in the repair of certain DNA lesions, such as DNA end resection at DSBs during homologous recombination-mediated repair, or when helicase and polymerase activities become uncoupled at stalled replication forks (Figure 2.8) (Blackford and Jackson 2017). This uncoupling can be induced by compromising DNA synthesis, for example, through inhibition of replicative polymerases with aphidicolin or upon depletion of available dNTPs by blocking the activity of the ribonucleotide reductase with hydroxyurea (Byun et al. 2005).

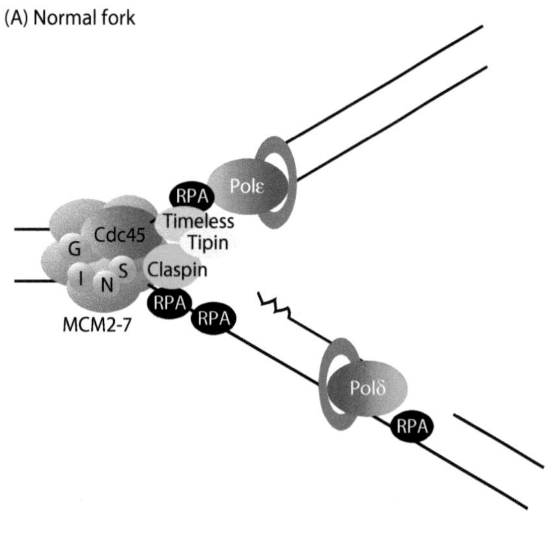

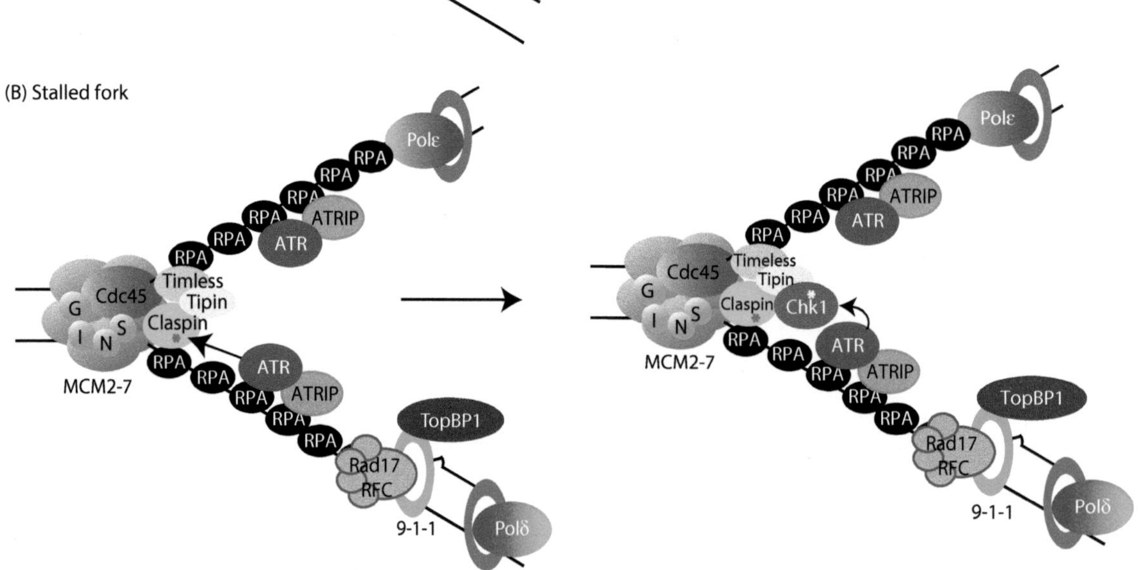

Figure 2.8 Checkpoint activation. Should the helicase and polymerase become uncoupled during DNA replication, long strands of ssDNA are generated, which become coated in RPA. These long RPA nucleofilaments and the subsequent interaction of TopBP1 with the 9-1-1 complex, which is loaded by the Rad17-RFC complex at ssDNA–dsDNA junctions, cause activation of the kinase ATR and its partner ATRIP. The replication fork–associated protein Claspin is then phosphorylated, causing recruitment and phosphorylation of the kinase Chk1. These events result in inactivation of CDK so as to reduce origin firing and to prevent collapse of the stalled fork into a DSB.

ATR then senses this replication stress with the help of its interacting partner ATRIP, and both are recruited to the long tracks of RPA-coated ssDNA. Importantly, recruitment of ATR-ATRIP to regions of replication stress is not enough for optimal activation of ATR signalling; other elements are also needed, such as the protein activator TopBP1. An important event for ATR activation by TopBP1 is the interaction between TopBP1 and the 9-1-1 clamp complex, which is loaded onto RPA ssDNA/dsDNA junctions by the Rad17-RFC complex (**Figure 2.8**) (Blackford and Jackson 2017). Once activated, ATR then orchestrates a complex cascade of phosphorylation events, which ultimately activate a plethora of effector proteins that promote cell cycle arrest and prevent stalled forks from collapsing (Cortez 2015). The protein kinase Chk1 is one of the most well-known effectors of the ATR cascade. ATR-dependent phosphorylation of the replication fork–associated protein Claspin transiently recruits Chk1, bringing it in close proximity to ATR. This recruitment results in ATR-mediated phosphorylation of Chk1 and its consequent activation (**Figure 2.8**) (Cimprich and Cortez 2008).

An important role of the effector protein Chk1 is to promote proteasomal degradation of the oncogene and protein phosphatase Cdc25A, which removes inhibitory phosphates on CDKs. The major consequence of Cdc25A degradation is therefore a decrease in CDK activity which results in the global inactivation of origin firing and cell cycle arrest (Blackford and Jackson 2017). Effector proteins also work to prevent collapse of stalled replication forks into DSBs, as these lesions are very toxic to cells. Mechanisms to avoid this collapse include the aforementioned global inactivation of origin firing, which prevents exhaustion of RPA and thus stabilises RPA-coating of the generated ssDNA (Toledo et al. 2013). Also, DNA-processing enzymes, such as those promoting fork reversal, are activated: the RecQ helicases BLM and WRN, the nucleases Dna2 and Exo1 and the DNA translocase SMARCAL1 (Cortez 2015). Moreover, ATR signalling regulates the availability of dNTPs by upregulating the ribonucleotide reductase subunit RRM2 in response to DNA damage (Buisson et al. 2015). Finally, dormant origins play an important role during replication stress by firing within the local environment of a stalled fork and rescuing the replication of that particular area. The mechanism by which dormant origins are activated and escape from the inhibitory action of the replication checkpoint remains unknown, although different models have been proposed (Yekezare et al. 2013).

G2 phase checkpoints and control of mitotic entry

Once the DNA replication stress has been resolved, DNA replication forks can restart and replication is completed so that the cell reaches G2 phase of the cell cycle. At this stage, further checkpoints exist which prevent the cell from entering mitosis should there be any remaining DNA damage (G2 checkpoint) or indeed defects in decatenation present (decatenation checkpoint). These two checkpoints are distinct from one another in their mode of activation, as exemplified by the fact that the G2 DNA damage checkpoint can be activated by ionising radiation (IR), while the decatenation checkpoint can be activated by the topoisomerase II inhibitor ICRF-193 (Nakagawa et al. 2004).

Despite their differences, both G2 checkpoints inhibit mitotic progression by blocking the activity of the cyclin B1-Cdk1 complex, also known as the maturation-promoting factor (MPF), and both are essential for correct chromosome segregation and maintenance of genome stability. One key downstream effector of the G2 DNA damage checkpoint is Chk1, as introduced earlier. In this capacity it competes with Polo-like kinase 1 (Plk1), causes activation of the protein kinase Wee1 and also downregulates the protein phosphatase Cdc25C. These events in turn maintain the inhibitory phosphates on Cdk1 (Thr-14 and Tyr-15) and as a consequence prevent mitotic progression (de Gooijer et al. 2017).

The decatenation checkpoint, on the other hand, specifically recognises DNA catenanes between sister chromatids, which form as a result of DNA replication but which are normally resolved by DNA topoisomerase II. Unlike the G2 DNA damage checkpoint, the decatenation checkpoint does not entail Chk1 activity but does require ATR, BRCA1 and MDC1 (Deming et al. 2001, Luo et al. 2009). Interestingly, the topoisomerase II protein itself is also required for this checkpoint to work. Treating topoisomerase II–depleted cells with ICRF-193 does not lead to G2 arrest but instead cells proceed through mitosis and accumulate genomic damage as a result of chromosomal instabilities (Bower et al. 2010).

Finally, once the perceived damages are resolved, accumulating evidence suggests that the protein kinase Plk1 is important for recovery from both forms of G2 checkpoint, allowing cells to proceed into mitosis (van Vugt et al. 2004, Ando et al. 2013).

CONCLUSION

The aim of this chapter was to give a general overview of the eukaryotic chromosomal replication process and its coordination with other processes throughout the cell cycle. Importantly, the accurate execution of DNA replication is essential for the maintenance of genome integrity and the prevention of cancer and other genetic disorders.

The process of DNA replication is rigorously regulated at every stage, starting with the two-step mechanism of origin activation. This mechanism starts with origin licensing at the end of mitosis and G1. Interestingly, origin selection during licensing has become more flexible throughout evolution in order to satisfy the requirements of more complex genomes. This is exemplified by the fact that very well-defined sequence elements determine origin selection in budding yeast, while other elements, such as chromatin structure and epigenetic marks, determine origin selection in mammalian cells. Despite all the work focused on understanding origin selection, several questions remain open.

The second step of origin activation is activation of the helicase during S phase, which is also tightly regulated. The in vitro reconstitution of origin firing using purified budding yeast proteins has identified the minimal set of proteins required for helicase activation.

The two-step mechanism of origin activation is strictly regulated largely by proteasomal degradation, ensuring that origin licensing and firing are restricted to the end of mitosis/G1 and S phase, respectively. The appropriate regulation of this process is essential for the maintenance of genome integrity, as it prevents re-replication.

During elongation, the progression of the replication fork needs to be properly coupled to other processes happening in parallel to replication, such as the re-establishment of chromatin and sister chromatid cohesion. As a result, a large number of proteins with roles in these processes are connected to the replisome machinery.

The termination stage of DNA replication has received special attention within the last 3 years and is discussed in detail in Chapter 3 of this book.

Finally, the replication fork needs to be prepared to deal with obstacles found in the DNA during elongation. These obstacles are not only caused by exogenous insults, but can also arise from the nature of DNA and other processes that co-occur with replication and use DNA as a template. Such obstacles include areas of condensed chromatin, R-loops generated during DNA transcription and DNA secondary structures such as G4 motifs. Upon encountering a DNA block, the replication fork will stall, and the replication checkpoint plays a crucial role in preventing fork collapse and promoting fork re-start. Moreover, dormant origins also play important roles in rescuing replication within the vicinity of stalled forks.

Due to their ability to dynamically regulate protein behaviour, post-translational modifications play essential roles in the regulation of many cellular processes, including DNA replication. Well-established examples are the role of phosphorylation in helicase activation during initiation and ubiquitylation for CMG disassembly during termination or in promoting protein degradation to allow cell cycle progression. An important area of research is the study of novel post-translational modifications that modulate DNA replication.

REFERENCES

Alabert, C., Barth, T. K., Reveron-Gomez, N., Sidoli, S., Schmidt, A., Jensen, O. N., Imhof, A., Groth, A. 2015. Two distinct modes for propagation of histone PTMs across the cell cycle. *Genes Dev* 29(6): 585–590.

Alabert, C., Groth, A. 2012. Chromatin replication and epigenome maintenance. *Nat Rev Mol Cell Biol* 13(3): 153–167.

Ando, K., Ozaki, T., Hirota, T., Nakagawara, A. 2013. NFBD1/MDC1 is phosphorylated by PLK1 and controls G2/M transition through the regulation of a TOPOIIalpha-mediated decatenation checkpoint. *PLOS ONE* 8(12): e82744.

Bae, S. H., Bae, K. H., Kim, J. A., Seo, Y. S. 2001. RPA governs endonuclease switching during processing of Okazaki fragments in eukaryotes. *Nature* 412(6845): 456–461.

Balakrishnan, L., Bambara, R. A. 2013. Okazaki fragment metabolism. *Cold Spring Harb Perspect Biol* 5(2): a010173.

Bambara, R. A., Murante, R. S., Henricksen, L. A. 1997. Enzymes and reactions at the eukaryotic DNA replication fork. *J Biol Chem* 272(8): 4647–4650.

Bartek, J., Lukas, J. 2001. Pathways governing G1/S transition and their response to DNA damage. *FEBS Lett* 490(3): 117–122.

Baxter, J., Diffley, J. F. 2008. Topoisomerase II inactivation prevents the completion of DNA replication in budding yeast. *Mol Cell* 30(6): 790–802.

Bell, S. P., Labib, K. 2016. Chromosome duplication in *Saccharomyces cerevisiae*. *Genetics* 203(3): 1027–1067.

Bell, S. P., Stillman, B. 1992. ATP-dependent recognition of eukaryotic origins of DNA replication by a multiprotein complex. *Nature* 357(6374): 128–134.

Bird, A. 2002. DNA methylation patterns and epigenetic memory. *Genes Dev* 16(1): 6–21.

Blackford, A. N., Jackson, S. P. 2017. ATM, ATR, and DNA-PK: The trinity at the heart of the DNA damage response. *Mol Cell* 66(6): 801–817.
Blow, J. J., Ge, X. Q., Jackson, D. A. 2011. How dormant origins promote complete genome replication. *Trends Biochem Sci* 36(8): 405–414.
Boos, D., Sanchez-Pulido, L., Rappas, M., Pearl, L. H., Oliver, A. W., Ponting, C. P., Diffley, J. F. 2011. Regulation of DNA replication through Sld3-Dpb11 interaction is conserved from yeast to humans. *Curr Biol* 21(13): 1152–1157.
Boos, D., Yekezare, M., Diffley, J. F. 2013. Identification of a heteromeric complex that promotes DNA replication origin firing in human cells. *Science* 340(6135): 981–984.
Bower, J. J., Karaca, G. F., Zhou, Y., Simpson, D. A., Cordeiro-Stone, M., Kaufmann, W. K. 2010. Topoisomerase IIalpha maintains genomic stability through decatenation G(2) checkpoint signaling. *Oncogene* 29(34): 4787–4799.
Branzei, D., Foiani, M. 2010. Maintaining genome stability at the replication fork. *Nat Rev Mol Cell Biol* 11(3): 208–219.
Bretones, G., Delgado, M. D., Le, J. 2014. Myc and cell cycle control. *Biochim Biophys Acta* 1849(5): 506–516.
Buisson, R., Boisvert, J. L., Benes, C. H., Zou, L. 2015. Distinct but concerted roles of ATR, DNA-PK, and Chk1 in countering replication stress during S phase. *Mol Cell* 59(6): 1011–1024.
Byun, T. S., Pacek, M., Yee, M. C., Walter, J. C., Cimprich, K. A. 2005. Functional uncoupling of MCM helicase and DNA polymerase activities activates the ATR-dependent checkpoint. *Genes Dev* 19(9): 1040–1052.
Chellappan, S. P., Giordano, A., Fisher, P. B. 1998. Role of cyclin-dependent kinases and their inhibitors in cellular differentiation and development. *Curr Top Microbiol Immunol* 227: 57–103.
Cimprich, K. A., Cortez, D. 2008. ATR: An essential regulator of genome integrity. *Nat Rev Mol Cell Biol* 9(8): 616–627.
Cortez, D. 2015. Preventing replication fork collapse to maintain genome integrity. *DNA Repair (Amst)* 32: 149–157.
Coster, G., Frigola, J., Beuron, F., Morris, E. P., Diffley, J. F. 2014. Origin licensing requires ATP binding and hydrolysis by the MCM replicative helicase. *Mol Cell* 55(5): 666–677.
Davey, M. J., Indiani, C., O'Donnell, M. 2003. Reconstitution of the Mcm2-7p heterohexamer, subunit arrangement, and ATP site architecture. *J Biol Chem* 278(7): 4491–4499.
Deegan, T. D., Diffley, J. F. 2016. MCM: One ring to rule them all. *Curr Opin Struct Biol* 37: 145–151.
de Gooijer, M. C., van den Top, A., Bockaj, I., Beijnen, J. H., Wurdinger, T., van Tellingen, O. 2017. The G2 checkpoint—A node-based molecular switch. *FEBS Open Bio* 7(4): 439–455.
Deming, P. B., Cistulli, C. A., Zhao, H., Graves, P. R., Piwnica-Worms, H., Paules, R. S., Downes, C. S., Kaufmann, W. K. 2001. The human decatenation checkpoint. *Proc Natl Acad Sci USA* 98(21): 12044–12049.
Devbhandari, S., Jiang, J., Kumar, C., Whitehouse, I., Remus, D. 2017. Chromatin constrains the initiation and elongation of DNA replication. *Mol Cell* 65(1): 131–141.
Dewar, J. M., Budzowska, M., Walter, J. C. 2015. The mechanism of DNA replication termination in vertebrates. *Nature* 525(7569): 345–350.
Dewar, J. M., Low, E., Mann, M., Raschle, M., Walter, J. C. 2017. CRL2Lrr1 promotes unloading of the vertebrate replisome from chromatin during replication termination. *Genes & Development* 31(3): 275–290.
Evrin, C., Clarke, P., Zech, J., Lurz, R., Sun, J., Uhle, S., Li, H., Stillman, B., Speck, C. 2009. A double-hexameric MCM2-7 complex is loaded onto origin DNA during licensing of eukaryotic DNA replication. *Proc Natl Acad Sci USA* 106(48): 20240–20245.
Fragkos, M., Ganier, O., Coulombe, P., Mechali, M. 2015. DNA replication origin activation in space and time. *Nat Rev Mol Cell Biol* 16(6): 360–374.
Gambus, A., Jones, R. C., Sanchez-Diaz, A., Kanemaki, M., van Deursen, F., Edmondson, R. D., Labib, K. 2006. GINS maintains association of Cdc45 with MCM in replisome progression complexes at eukaryotic DNA replication forks. *Nat Cell Biol* 8(4): 358–366.
Gambus, A., Khoudoli, G. A., Jones, R. C., Blow, J. J. 2011. MCM2-7 form double hexamers at licensed origins in *Xenopus* egg extract. *J Biol Chem* 286(13): 11855–11864.
Georgescu, R., Z. Yuan, L. Bai, R. de Luna Almeida Santos, J. Sun, D. Zhang, O. Yurieva, H. Li and M. E. O'Donnell. 2017. Structure of eukaryotic CMG helicase at a replication fork and implications to replisome architecture and origin initiation. *Proc Natl Acad Sci USA* 114(5): E697–E706.
Georgescu, R. E., Langston, L., Yao, N. Y., Yurieva, O., Zhang, D., Finkelstein, J., Agarwal, T., O'Donnell, M. E. 2014. Mechanism of asymmetric polymerase assembly at the eukaryotic replication fork. *Nat Struct Mol Biol* 21(8): 664–670.
Georgescu, R. E., Schauer, G. D., Yao, N. Y., Langston, L. D., Yurieva, O., Zhang, D., Finkelstein, J., O'Donnell, M. E. 2015. Reconstitution of a eukaryotic replisome reveals suppression mechanisms that define leading/lagging strand operation. *eLife* 4: e04988.
Hammond, C. M., Stromme, C. B., Huang, H., Patel, D. J., Groth, A. 2017. Histone chaperone networks shaping chromatin function. *Nat Rev Mol Cell Biol* 18(3): 141–158.
Hansen, K. H., Bracken, A. P., Pasini, D., Dietrich, N., Gehani, S. S., Monrad, A., Rappsilber, J., Lerdrup, M., Helin, K. 2008. A model for transmission of the H3K27me3 epigenetic mark. *Nat Cell Biol* 10(11): 1291–1300.
Hedglin, M., Kumar, R., Benkovic, S. J. 2013. Replication clamps and clamp loaders. *Cold Spring Harb Perspect Biol* 5(4): a010165.
Hogg, M., Osterman, P., Bylund, G. O., Ganai, R. A., Lundstrom, E. B., Sauer-Eriksson, A. E., Johansson, E. 2014. Structural basis for processive DNA synthesis by yeast DNA polymerase varepsilon. *Nat Struct Mol Biol* 21(1): 49–55.
Hyrien, O., Marheineke, K., Goldar, A. 2003. Paradoxes of eukaryotic DNA replication: MCM proteins and the random completion problem. *Bioessays* 25(2): 116–125.
Kang, S., Warner, M. D., Bell, S. P. 2014. Multiple functions for Mcm2-7 ATPase motifs during replication initiation. *Mol Cell* 55(5): 655–665.
Kanke, M., Kodama, Y., Takahashi, T. S., Nakagawa, T., Masukata, H. 2012. Mcm10 plays an essential role in origin DNA unwinding after loading of the CMG components. *EMBO J* 31(9): 2182–2194.
Keszthelyi, A., Minchell, N. E., Baxt, J. 2016. The causes and consequences of topological stress during DNA replication. *Gen. Basel* 7(12): 134.
Kouzarides, T. 2007. Chromatin modifications and their function. *Cell* 128(4): 693–705.
Kumagai, A., Shevchenko, A., Shevchenko, A., Dunphy, W. G. 2011. Direct regulation of Treslin by cyclin-dependent kinase is essential for the onset of DNA replication. *J Cell Biol* 193(6): 995–1007.
Kunkel, T. A., Burgers, P. M. 2008. Dividing the workload at a eukaryotic replication fork. *Trends Cell Biol* 18(11): 521–527.
Kuo, A. J., Song, J., Cheung, P., Ishibe-Murakami, S., Yamazoe, S., Chen, J. K., Patel, D. J., Gozani, O. 2012. The BAH domain of ORC1 links H4K20me2 to DNA replication licensing and Meier-Gorlin syndrome. *Nature* 484(7392): 115–119.
Larrea, A. A., Lujan, S. A., Nick McElhinny, S. A., Mieczkowski, P. A., Resnick, M. A., Gordenin, D. A., Kunkel, T. A. 2010. Genome-wide model for the normal eukaryotic DNA replication fork. *Proc Natl Acad Sci USA* 107(41): 17674–17679.
Leonard, A. C., Mechali, M. 2013. DNA replication origins. *Cold Spring Harb Perspect Biol* 5(10): a010116.
Lieber, M. R. 1997. The FEN-1 family of structure-specific nucleases in eukaryotic DNA replication, recombination and repair. *Bioessays* 19(3): 233–240.
Losada, A. 2014. Cohesin in cancer: Chromosome segregation and beyond. *Nat Rev Cancer* 14(6): 389–393.
Luo, K., Yuan, J., Chen, J., Lou, Z. 2009. Topoisomerase IIalpha controls the decatenation checkpoint. *Nat Cell Biol* 11(2): 204–210.
MacAlpine, D. M., Almouzni, G. 2013. Chromatin and DNA replication. *Cold Spring Harb Perspect Biol* 5(8): a010207.
Mailand, N., Diffley, J. F. 2005. CDKs promote DNA replication origin licensing in human cells by protecting Cdc6 from APC/C-dependent proteolysis. *Cell* 122(6): 915–926.
Mailand, N., Gibbs-Seymour, I., Bekker-Jensen, S. 2013. Regulation of PCNA-protein interactions for genome stability. *Nat Rev Mol Cell Biol* 14(5): 269–282.
Margueron, R., Justin, N., Ohno, K., Sharpe, M. L., Son, J., Drury, 3rd, W. J., Voigt, P. et al. 2009. Role of the polycomb protein EED in the propagation of repressive histone marks. *Nature* 461(7265): 762–767.

Maric, M., Maculins, T., De Piccoli, G., and Labib, K. 2014. Cdc48 and a ubiquitin ligase drive disassembly of the CMG helicase at the end of DNA replication. *Science* 346(6208): 1253596.

Marzluff, W. F., Wagner, E. J., Duronio, R. J. 2008. Metabolism and regulation of canonical histone mRNAs: Life without a poly(A) tail. *Nat Rev Genet* 9(11): 843–854.

McGarry, T. J., Kirschner, M. W. 1998. Geminin, an inhibitor of DNA replication, is degraded during mitosis. *Cell* 93(6): 1043–1053.

Moreno, S. P., Bailey, R., Campion, N., Herron, S., Gambus, A. 2014. Polyubiquitylation drives replisome disassembly at the termination of DNA replication. *Science* 346(6208): 477–481.

Moreno, S. P., Gambus, A. 2015. Regulation of unperturbed DNA replication by ubiquitylation. *Genes (Basel)* 6(3): 451–468.

Nakagawa, T., Y. Hayashita, K. Maeno, A. Masuda, N. Sugito, H. Osada, K. Yanagisawa, H. Ebi, K. Shimokata and T. Takahashi. 2004. Identification of decatenation G2 checkpoint impairment independently of DNA damage G2 checkpoint in human lung cancer cell lines. *Cancer Res* 64(14): 4826–4832.

Narbonne-Reveau, K., Senger, S., Pal, M., Herr, A., Richardson, A., Asano, M., Deak, P., Lilly, M. A. 2008. APC/CFzr/Cdh1 promotes cell cycle progression during the *Drosophila* endocycle. *Development* 135(8): 1451–1461.

Nick McElhinny, S. A., Gordenin, D. A., Stith, C. M., Burgers, P. M., Kunkel, T. A. 2008. Division of labor at the eukaryotic replication fork. *Mol Cell* 30(2): 137–144.

Nishiyama, A., Yamaguchi, L., Sharif, J., Johmura, Y., Kawamura, T., Nakanishi, K., Shimamura, S. et al. 2013. Uhrf1-dependent H3K23 ubiquitylation couples maintenance DNA methylation and replication. *Nature* 502(7470): 249–253.

Oehlmann, M., Score, A. J., Blow, J. J. 2004. The role of Cdc6 in ensuring complete genome licensing and S phase checkpoint activation. *J Cell Biol* 165(2): 181–190.

Pellegrini, L., Costa, A. 2016. New insights into the mechanism of DNA duplication by the eukaryotic replisome. *Trends Biochem Sci* 41(10): 859–871.

Petryk, N., Kahli, ,M. d'Aubenton-Carafa, Y., Jaszczyszyn, Y., Shen, Y., Silvain, M., Thermes, C., Chen, C. L., Hyrien, O. 2016. Replication landscape of the human genome. *Nat Commun* 7: 10208.

Pommier, Y., Sun, Y., Huang, S. N., Nitiss, J. L. 2016. Roles of eukaryotic topoisomerases in transcription, replication and genomic stability. *Nat Rev Mol Cell Biol* 17(11): 703–721.

Postow, L., Crisona, N. J., Peter, B. J., Hardy, C. D., Cozzarelli, N. R. 2001. Topological challenges to DNA replication: Conformations at the fork. *Proc Natl Acad Sci USA* 98(15): 8219–8226.

Prioleau, M. N., MacAlpine, D. M. 2016. DNA replication origins—where do we begin? *Genes Dev* 30(15): 1683–1697.

Pursell, Z. F., Isoz, I., Lundstrom, E. B., Johansson, E., Kunkel, T. A. 2007. Yeast DNA polymerase epsilon participates in leading-strand DNA replication. *Science* 317(5834): 127–130.

Remus, D., Beuron, F., Tolun, G., Griffith, J. D., Morris, E. P., Diffley, J. F. 2009. Concerted loading of Mcm2-7 double hexamers around DNA during DNA replication origin licensing. *Cell* 139(4): 719–730.

Schalbetter, S. A., Mansoubi, S., Chambers, A. L., Downs, J. A., Baxter, J. 2015. Fork rotation and DNA precatenation are restricted during DNA replication to prevent chromosomal instability. *Proc Natl Acad Sci USA* 112(33): E4565–4570.

Sengupta, S., van Deursen, F., de Piccoli, G., Labib, K. 2013. Dpb2 integrates the leading-strand DNA polymerase into the eukaryotic replisome. *Curr Biol* 23(7): 543–552.

Sheaff, R. J., Groudine, M., Gordon, M., Roberts, J. M., Clurman, B. E. 1997. Cyclin E-CDK2 is a regulator of p27Kip1. *Genes Dev* 11(11): 1464–1478.

Siddiqui, K., On, K. F., Diffl, J. F. 2013. Regulating DNA replication in eukarya. *Cold Spring Harb Perspect Biol* 5(9): a012930.

Simon, A. C., Zhou, J. C., Perera, R. L., van Deursen, F., Evrin, C., Ivanova, M. E., Kilkenny, M. L. et al. 2014. A Ctf4 trimer couples the CMG helicase to DNA polymerase alpha in the eukaryotic replisome. *Nature* 510(7504): 293–297.

Sonneville, R., Moreno, S. P., Knebel, A., Johnson, C., Hastie, C. J., Gartner, A., Gambus, A., Lab, K. 2017. CUL-2LRR-1 and UBXN-3 drive replisome disassembly during DNA replication termination and mitosis. *Nat Cell Biol* 19(5)468–479.

Sundin, O., Varshavsky, A. 1980. Terminal stages of SV40 DNA replication proceed via multiply intertwined catenated dimers. *Cell* 21(1): 103–114.

Sundin, O., Varshavsky, A. 1981. Arrest of segregation leads to accumulation of highly intertwined catenated dimers: Dissection of the final stages of SV40 DNA replication. *Cell* 25(3): 659–669.

Sutterluty, H., Chatelain, E., Marti, A., Wirbelauer, C., Senften, M., Muller, U., Krek, W. 1999. p45SKP2 promotes p27Kip1 degradation and induces S phase in quiescent cells. *Nat Cell Biol* 1(4): 207–214.

Takahashi, T. S., Wigley, D. B., Walter, J. C. 2005. Pumps, paradoxes and ploughshares: Mechanism of the MCM2-7 DNA helicase. *Trends Biochem Sci* 30(8): 437–444.

Tardat, M., Brustel, J., Kirsh, O., Lefevbre, C., Callanan, M., Sardet, C., Julien, E. 2010. The histone H4 Lys 20 methyltransferase PR-Set7 regulates replication origins in mammalian cells. *Nat Cell Biol* 12(11): 1086–1093.

Toledo, L. I., Altmeyer, M., Rask, M. B., Lukas, C., Larsen, D. H., Povlsen, L. K., Bekker-Jensen, S., Mailand, N., Bartek, J., Lukas, J. 2013. ATR prohibits replication catastrophe by preventing global exhaustion of RPA. *Cell* 155(5): 1088–1103.

Uhlmann, F. 2016. SMC complexes: From DNA to chromosomes. *Nat Rev Mol Cell Biol* 17(7): 399–412.

van Vugt, M. A., Bras, A., Medema, R. H. 2004. Polo-like kinase-1 controls recovery from a G2 DNA damage-induced arrest in mammalian cells. *Mol Cell* 15(5): 799–811.

Wang, B., Feng, L., Hu, Y., Huang, S. H., Reynolds, C. P., Wu, L., Jong, A. Y. 1999. The essential role of *Saccharomyces cerevisiae* CDC6 nucleotide-binding site in cell growth, DNA synthesis, and Orc1 association. *J Biol Chem* 274(12): 8291–8298.

Wang, Y., Khan, A., Marks, A. B., Smith, O. K., Giri, S., Lin, Y. C., Creager, R., MacAlpine, D. M. et al. 2017. Temporal association of ORCA/LRWD1 to late-firing origins during G1 dictates heterochromatin replication and organization. *Nucleic Acids Res* 45(5): 2490–2502.

Wohlschlegel, J. A., Dwyer, B. T., Dhar, S. K., Cvetic, C., Walter, J. C., Dutta, A. 2000. Inhibition of eukaryotic DNA replication by geminin binding to Cdt1. *Science* 290(5500): 2309–2312.

Yardimci, H., Walter, J. C. 2014). Prereplication-complex formation: A molecular double take? *Nat Struct Mol Biol* 21(1): 20–25.

Yardimci, H., Wang, X., Loveland, A. B., Tappin, I., Rudner, D. Z., Hurwitz, J., van Oijen, A. M., Walter, J. C. 2012. Bypass of a protein barrier by a replicative DNA helicase. *Nature* 492(7428): 205–209.

Yeeles, J. T., Deegan, T. D., Janska, A., Early, A., Diffley, J. F. 2015. Regulated eukaryotic DNA replication origin firing with purified proteins. *Nature* 519(7544): 431–435.

Yekezare, M., Gomez-Gonzalez, B., Diffley, J. F. 2013. Controlling DNA replication origins in response to DNA damage–Inhibit globally, activate locally. *J Cell Sci* 126(Pt 6): 1297–1306.

DNA Replication Termination and Genomic Instability

3

Rebecca M. Jones, Sara Priego Moreno, and Agnieszka Gambus

INTRODUCTION

DNA replication can be divided into three stages: initiation, elongation and termination (as described in Chapter 2). During initiation, replication origins fire and two replication forks move away from each origin, unwinding and replicating DNA. The elongation stage encompasses progression of forks through chromatin, while termination happens when forks from neighbouring origins meet each other. Research over the last 20 to 30 years has brought about a good understanding of the first two stages of replication (initiation and elongation), but the termination stage has remained a mystery until very recently. This chapter will focus specifically on this final stage and discuss the connections between aberrant DNA replication termination and disease onset. To maintain genomic stability, it is essential that every step of DNA replication be perfectly executed, including termination. We will present current knowledge of termination events and discuss the ways that their dysregulation can lead to chromosomal mistakes and disease.

TERMINATION OF REPLICATION FORKS: WHEN AND WHERE?

Although one could regard replication termination as a process linked with the end of S phase, in reality replication forks termination occurs whenever two replication forks coming from neighbouring origins meet each other. Replication origins fire throughout S phase, regulated by a strict replication timing program (Ryba et al. 2010, Rivera-Mulia and Gilbert 2016). Whenever the forks they produce meet, termination takes place. Early firing replication origins will result in forks terminating early in S phase and so we can think of termination as occurring throughout S phase.

In bacteria, the position of the termination zone is site specific. These so-called *Ter* sites ensure that termination occurs opposite the origin of replication. When considering the sites of eukaryotic replication termination, they depend on the location of initiation sites of replication and thus termination sites generally reside midway between two adjacent replication origins. Two independent genome-wide assays performed in yeast found termination events to be spread throughout the genome but mostly in the zones midway between two neighbouring origins of replication (Hawkins et al. 2013, McGuffee et al. 2013). Importantly, the precise position of termination depended on the relative activation time of each origin and their efficiency but not on any other chromatin features. A recent study in human cells also confirmed that termination sites occur midpoint between origins (Petryk et al. 2016).

Although stochastic, it seems that in many cases early replication initiation sites localise to open chromatin regions – often regions flanking active genes. Indeed, research using the *D. melanogaster* system found that the Mcm2-7 complexes are excluded from actively transcribed genes (Powell et al. 2015, Petryk et al. 2016). Subsequently, in early S phase, termination sites are often located *within* transcribed genes. In late S phase, when heterochromatin is mostly replicated, termination zones are then often found in large non-expressed DNA regions (Petryk et al. 2016).

TERMINATION OF REPLICATION FORKS: AN OVERVIEW

DNA replication termination occurs when two replication forks from neighbouring origins converge and the final fragment of DNA that lies between them is fully replicated. Indeed, as we learn more about this process, we can describe several steps leading to this event (**Figure 3.1**). First, replication forks converge, which requires removal of all nucleosomes, relaxation of torsional stress and two replisomes passing each other (**Figure 3.1A** to **C**). Second, the final sequences of DNA between the two converging replisomes must be synthesised and the last Okazaki fragment matured (**Figure 3.1D**). Third, the DNA ends are ligated (**Figure 3.1E**). Fourth, a specific protein within the replicative helicase (CMG complex) is modified to signal for disassembly of the replication machinery from chromatin (**Figure 3.1E**). Finally, the catenated sister chromatids are detangled by topoisomerase activity, allowing for correct chromosome segregation during mitosis (**Figure 3.1F,G**).

The precise order of these events is still being investigated. This chapter will discuss in turn the five key stages of replication termination and highlight how defects within each could lead to genomic instability.

REPLICATION FORK CONVERGENCE

When analysing fork convergence, seminal work by Dewar et al. found that replication forks could converge without any indication of slowing down and that converging replisomes can pass one another. To synchronise termination events and facilitate their analysis, Dewar et al. constructed plasmids with an array of *lac* repressors (LacRs) bound to *lac* operators (LacOs), which can be disrupted by IPTG. Such plasmids replicated in cell-free *Xenopus laevis* egg extracts accumulated blocked forks at the edges of the array. The blocked forks were then released by addition of IPTG, and proceeded to terminate within the DNA fragment comprising the array. Using this system, Dewar et al. could monitor unwinding of DNA as forks approach each other, synthesis of DNA, ligation of the replicated DNA and decatenation of daughter molecules. Strikingly, the rate of DNA synthesis within the array was almost perfectly linear after IPTG addition and resembled the fork progression speed reported in the same extracts. It suggests, therefore, that converging forks do not slow significantly before they meet; they do not collide with each other or stall, but rather pass each other (Dewar et al. 2015) (**Figure 3.1C** and **D**). Such passage can be possible as CMGs encircle the leading strand of the replication fork and, therefore, approach each other on opposite strands when converging at termination (Costa et al. 2011, Du et al. 2016). Interestingly, recent reports suggest that large protein barriers on lagging strands can slow down the CMG helicase (Duxin et al. 2014, Langston and O'Donnell 2017). An approaching replisome from a neighbouring origin would provide such a barrier. It is possible that stalling of forks at the LacR barrier in order to synchronise termination events means that the replisomes studied here are enriched with factors that promote barrier passage. Further studies are required to explain this issue.

As two neighbouring DNA replication forks move towards one another, intertwining of the parental strands between them increases and positive supercoiling tension is created. This is a large hindrance to the progression of the helicases. In order to maintain fork progression, this stress is relieved mostly by the activity of topoisomerase enzymes that temporarily cleave the supercoiled DNA. When considering DNA replication, two topoisomerases have been found to be important: topoisomerase I (Top1) and topoisomerase II (Top2). Each of these enzymes cuts the DNA temporarily in order to relax supercoiling stress, with Top1 making single-strand DNA cuts and Top2 making double-strand cuts. As explained in Chapter 2, Top1 is responsible for relaxing most of the tension during replication and is a part of the replisome progression complex, positioning it in front of the fork (Gambus et al. 2006). However, in the absence of Top1, Top2 can provide the relaxing function, most likely due to the ability of the fork to rotate. Rotating the fork releases the tension at the front, but as a consequence the daughter chromatids become intertwined behind the replication forks, creating precatenanes (Keszthelyi et al. 2016). Interestingly, fork rotation during replication is downregulated by components of the replisome progression complex, i.e. Tof1/Timeless and Csm3/Tipin and the helicase Rrm3 (Ivessa et al. 2003).

Although both Top1 and Top2 enzymes are important for fork progression, evidence from studies with SV40 viral DNA suggests that their accessibility to the DNA during fork convergence at termination becomes limited (Sundin and Varshavsky 1980). As such, the current theory is that fork rotation functions to relieve torsional stress at termination,

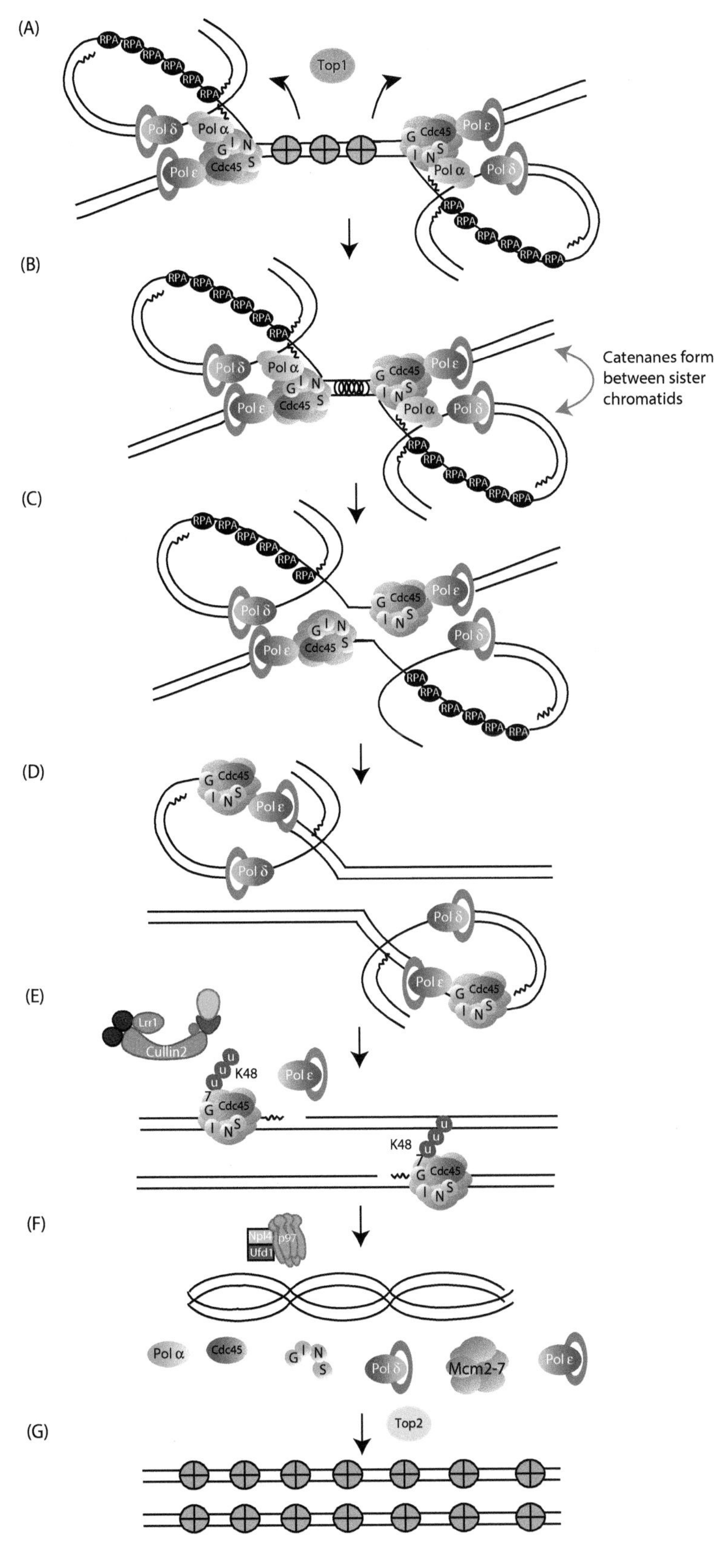

Figure 3.1 Mechanism of replication termination. (A) As replication forks converge, nucleosomes between them are displaced. Topoisomerase I (Top1) acts ahead of the replication forks to aid this convergence by relaxing the tension in the DNA during replication. (B) Upon convergence, space between the replisomes becomes limited for topoisomerase access and so torsional stress can be relieved by fork rotation instead. As a consequence, however, sister chromatids become intertwined and catenanes are formed. (C) Approaching forks consist of CMGs that encircle the leading strands and so at termination they are able to pass one another. (D) The final fragments of DNA on the leading strand are then synthesised up to a few bases away from the RNA–DNA primers of the last Okazaki fragment on the lagging strand. As yet, it is unclear how maturation of these RNA–DNA primers occurs. (E) Once the replisomes have passed one another, Mcm7 becomes polyubiquitylated by CUL2-Lrr1 in metazoa. The current model is that the replisomes then slide onto the dsDNA. (F) Ubiquitylated Mcm7 is subsequently recognised by the p97/VCP/Cdc48 protein remodeler with major cofactors Npl4 and Ufd1, which work to remove the entire complexes from the chromatin. (G) Topoisomerase II (Top2) is responsible at this point for resolving the catenanes so that sister chromatids can be separated effectively during mitosis.

enabling duplication of the final fragments of DNA (Postow et al. 2001). Although fork rotation provides the short-term solution for completing DNA synthesis, once replication is complete, the created precatenanes are converted into catenanes and must be unlinked by Top2 prior to mitosis; otherwise, chromosomal instabilities will arise (Lucas et al. 2001). Fork rotation also occurs when replication forks encounter protein–DNA barriers; presumably because topoisomerase access here is also spatially restricted. Although limiting fork rotation causes slowing of DNA replication at these obstacles (and perhaps also as terminating forks converge), this is important to protect against DNA damage and chromosome instability (Schalbetter et al. 2015). Interestingly, the results from the plasmid termination system described earlier suggest that Top2 is not required for fork convergence but is essential for decatenation of replicated plasmids (Dewar et al. 2015).

Besides the problem of superhelical tension building between the converging forks, another obstacle to overcome is the removal of all nucleosomes between the converging forks. Nucleosomes are essential for protecting DNA and packaging it into the small nucleus, and also play an essential role in regulating transcription and replication. Importantly, to fulfil their regulatory role, histone proteins within the nucleosomes are modified by a variety of post-translational modifications. The pattern of histone modifications throughout the DNA must be passed on to daughter cells for the correct regulation of gene transcription to be inherited. During replication, nucleosomes are removed from the chromatin ahead of the replication fork, and are distributed between the two daughter strands when replaced back behind the fork, while newly synthesised histones fill in the remaining space. When considering fork convergence in the context of nucleosomes, one can imagine that the displacement of the final nucleosome(s) would also be hindered due to lack of space for histone chaperones. Indeed, when Devbhandari et al. reconstituted DNA replication in vitro with plasmid DNA and purified budding yeast proteins, they found that although the bulk of DNA replication progressed unhindered, termination intermediates accumulated when using chromatinised plasmid DNA. These intermediates indicated an inability to complete the termination stage of replication and provide evidence that nucleosomal packaging hinders smooth replication termination (Devbhandari et al. 2017). This aspect of termination could not be assessed using the plasmid system explained earlier due to lack of nucleosomes in the termination area after release from the LacR barrier (Dewar et al. 2015).

DEFECTS IN FORK CONVERGENCE PRODUCE GENOMIC INSTABILITY

One can easily imagine how problems with fork convergence could occur. There may be a defect in removal of the last nucleosomes, a defect in fork rotation and release of positive supercoiling between converging forks, or two replisomes could find it difficult to pass each other smoothly. These situations would appear as two forks having stopped on either side of a barrier. Indeed, replication forks halted at barriers are often helped by additional helicases such as Rrm3. In yeast lacking Rrm3, Ivessa et al. observed a tenfold accumulation of termination structures (*X*-shaped DNA structures in 2D DNA gels), while they only saw a twofold accumulation of forks pausing at barriers (e.g. tRNA genes) (Ivessa et al. 2000, 2003). This suggests that although Rrm3 is not required for bulk replisome unloading during normal termination (Maric et al. 2014), and so is not essential for all fork convergence events, it is required for fork convergence and termination in rare situations when one fork is paused.

Another phenomenon that can occur when fork convergence is inhibited is accumulation of torsional stress and fork reversal. Inhibition of Top1 activity in human cells, mouse embryonic fibroblasts and *Xenopus laevis* egg extracts frequently induces replication fork reversal. Fork reversal can have physiological roles during replication but can also have pathological consequences and contribute to genome instability in neurodegenerative syndromes and cancer. Of note, a small but reproducible number of reversed forks has been detected in various unchallenged human cell lines. This number of reversed forks can be markedly increased in the absence of genotoxic replication stress by deregulation of poly(ADP-ribose) metabolism, which regulates fork reversal and restart (reviewed in Neelsen and Lopes 2015). Fork reversal is also very frequent in mouse embryonic stem cells (Ahuja et al. 2016). The question arises as to where these reversed forks come from, and one possibility is that they can derive from sites of problematic termination.

This observation of fork reversal at converging forks has also been examined by Wendel et al. Analysis of plasmid DNA replication in *E. coli* revealed that proteins normally implicated in DNA damage repair are important for processing termination DNA intermediates. It seems that in order to complete DNA synthesis and to avoid under-replicated DNA at termination, plasmid DNA is transiently over-replicated at the *Ter* site. This is followed by the formation of a reversed fork structure and exonuclease processing before resolution to complete replication (Wendel et al. 2014). These intermediate structures can only be visualised when the processing enzymes are depleted. It would be interesting to determine whether fork reversal occurs at replication termination sites in eukaryotes and, if so, what the importance of this process is for the completion of DNA synthesis.

SYNTHESIS OF DNA

As the replication forks converge, the final fragment of DNA between them is synthesised and the last Okazaki fragment needs to be matured. Research by Dewar et al. (2015) revealed that DNA of the leading strand is replicated up to a few bases away from the end of the last Okazaki fragment on the lagging strand. Interestingly, no gaps in the DNA were detected, so it is clear that a polymerase works to complete synthesis of these last few bases. It is also likely that the RNA–DNA primers of the last Okazaki fragments are removed, but the mechanism of this process is, as yet, unclear. In the final model proposed by Dewar et al. they theorised that after replisomes have passed one another, the helicases change conformation so that they can encircle the double-stranded DNA (dsDNA) at the RNA–DNA primers of the last Okazaki fragments (**Figure 3.1**). Of note, Kang et al. (2012) found that the CMG complex can slide onto dsDNA.

When considering maturation of *standard* Okazaki fragments, we know that polymerase delta (Pol δ) performs strand-displacement synthesis through the 5′ end of the preceding Okazaki fragment in order to remove the RNA–DNA primer sequence produced by the error-prone polymerase alpha (Pol α) (see Chapter 2). In fact, Pol δ will continue to cause strand displacement only until a DNA-bound protein barrier becomes an obstacle. For example, the nucleosome repositioned behind the ongoing replication fork will create resistance to Pol δ and cause its dissociation. This explains why Okazaki fragment termini, that is, the junctions between Okazaki fragments, correlate more with nucleosome midpoints rather than being inter-nucleosomal (Smith and Whitehouse 2012, Devbhandari et al. 2017). Strand displacement subsequently leads to the production of a single-stranded DNA (ssDNA) flap, which is then cleaved by the action of endonucleases Fen1 and Dna2, depending on the length of the flap (reviewed in Zheng and Shen 2011, and discussed in Chapter 2). Finally, the enzyme DNA ligase I catalyses the ligation of Okazaki fragments (Howes and Tomkinson 2012). It is unknown at present whether the last Okazaki fragment is matured in an analogous way.

Sonneville et al. identified Pol ε interacting with the post-termination replisome in *C. elegans* and *X. laevis* through mass spectrometry analysis (Sonneville et al. 2017), suggesting that Pol ε may mature the last Okazaki fragment. However, Pol ε on its own does not have the ability to mature Okazaki fragments on the lagging strand, as it does not have the strand displacement activity unless its 3′-5′ exonuclease domain is removed (Ganai et al. 2016, Devbhandari et al. 2017). However, it has not yet been tested whether Pol ε attached to CMG helicase (CMGE) can displace strands. Indeed, in support of a role for Pol ε in termination, Daigaku et al. (2015) found that Pol ε in yeast is more prone to incorporating NTPs at the mid-zone between the origins. In order to see this, they used a mutant of Pol ε, which incorporated excess ribonucleotides, and these were then directly detected in the synthesised DNA.

Alternative theories implicate Pol δ for completion, however. Dewar and Walter (2017) propose that after the CMG helicases jump onto the dsDNA after passing one another, Pol ε remains associated with the CMG and so is removed from the termination site due to CMG translocase activity. That would create space for Pol δ to complete the job (see the elegant review by Dewar and Walter 2017). The question remains as to how far the translocating post-termination CMGE can move away from the termination site in the context of chromatinised DNA. Further work is clearly required to resolve this mechanism, especially as several publications have highlighted the fact that polymerase switching during DNA replication is more common than previously anticipated (Georgescu et al. 2014, Yeeles et al. 2017). Whichever polymerase matures the last Okazaki fragment, it will have to deal with nucleosomes assembled on this fragment. We know that nucleosomes pose a barrier for Pol δ (see earlier), but maybe they are cleared away by translocating, post-termination CMGE.

SYNTHESIS OF DNA AND GENOMIC INSTABILITY

Enrichment of under-replicated DNA during DNA synthesis, which can be caused by a large number of defects, for example, defective polymerase function and defective topoisomerase activity, is the most common and well-known trigger for the development of cancer. Termination problems can easily create situations of defective DNA synthesis completion. Should these gaps in the nascent DNA persist through mitosis, chromosome segregation will be unsuccessful and cells will undergo mitotic catastrophe and genomic instability.

Areas within the genome which are known to accumulate this under-replicated DNA under conditions of mild replication stress (e.g. a low concentration of the DNA polymerase inhibitor aphidicolin), are called common fragile sites (CFSs). These regions are especially difficult to replicate, as they tend to be large genes, contain AT-rich repetitive sequences and consist of fewer than average replication origin sites (Letessier et al. 2011). Finally, they tend to be replicated late in S phase (Le Beau et al. 1998) and so termination at these sites is less likely to be completed. The lack of dormant origins and the late timing of replication also mean that should replication forks stall irreversibly within these areas, dormant origin firing is less able to compensate and DNA synthesis cannot be completed in a timely fashion. Of note, active transcription also contributes to CFS expression, with the theory that RNA polymerase II can, in fact, displace inactive pre-RC complexes from the DNA, thus reducing the number of dormant origins available for rescue from replication stress (reviewed in Glover et al. 2017).

Interestingly, it seems that DNA replication can occur in early mitosis as a last attempt to fill in these gaps before chromosome segregation (Minocherhomji et al. 2015). This DNA synthesis has been seen to occur at CFS sites under conditions of mild replication stress; it requires the TLS polymerase eta (Pol η) (Bergoglio et al. 2013) and is also dependent on the endonucleases MUS81 and SLX4 (Minocherhomji et al. 2015). Indeed, a similar form of DNA synthesis in early mitosis has also been detected in BRCA2-deficient cells and is also dependent on MUS81 (Lai et al. 2017). The need for these endonucleases arises because stalled replication forks are regularly processed by homologous recombination pathways. These pathways can lead to the formation of problematic double Holliday junctions between chromosomes (Germann et al. 2014). As a consequence, MUS81 and SLX4 endonucleases become important to resolve these connections, enabling correct chromosome segregation. Minocherhomji et al. (2015) proposed that these cleavages at the stalled replication forks then also promote further rescuing of DNA synthesis. The role of HR in replication stress is discussed in detail in Chapter 6.

Should the attempt to complete DNA synthesis in early mitosis be unsuccessful or ineffective, connections between chromosomes at anaphase can be detected. These connections are known as ultra-fine bridges (UFBs) and are visibly stretched under tension in anaphase before they are quickly resolved (Baumann et al. 2007). These bridges are believed to consist of ssDNA structures that result from under-replication and become coated in the Plk1-interacting checkpoint helicase (PICH), which works to protect the DNA from denaturation during such stretching (Biebricher et al. 2013). Once the bridges are resolved and chromosome segregation has taken place, regions of under-replicated DNA can still be passed to each new daughter cell and, in fact, these areas then become coated in G1 phase by the DNA repair protein 53BP1 to form nuclear bodies (Lukas et al. 2011, Moreno et al. 2016). Importantly, an increase in 53BP1 nuclear bodies is strongly associated with replication defects and is now commonly used as a marker of such stress.

DNA LIGATION

Ligation of DNA is carried out by DNA ligase: Cdc9 in *S. cerevisiae*, Cdc17 in *S. pombe* and DNA Ligase 1 (LIG1) in human cells. Both Cdc9 and Cdc17 are essential for cell growth, and different deletions of LIG1 in mice lead to embryonic death (Nasmyth 1977, Johnston and Nasmyth 1978, Ellenberger and Tomkinson 2008). DNA ligase is crucial for ligation of all of the Okazaki fragments together during replication of the lagging strand. Therefore, inactivation/deletion of LIG1 in all organisms leads to a defect in lagging strand synthesis (Ellenberger and Tomkinson 2008) and inducible inactivation of Cdc9 leads to an accumulation of short nascent DNA fragments of Okazaki fragments size (Smith and Whitehouse 2012). DNA ligase is essential not only for DNA replication but also for repair and recombination. It is most likely that LIG1 is also responsible for ligation of terminated DNA upon convergence of replication forks. Interestingly, Dewar et al. (2015) observed that in the plasmid termination system, ligation occurred very swiftly upon fork convergence and was not affected when CMG helicase unloading was impaired (**Figure 3.1**).

DNA LIGATION AND GENOMIC INSTABILITY

Defects in ligation at the sites of DNA replication fork termination would most likely resemble defects in Okazaki fragment ligation – creating thousands of single-strand breaks (SSBs). The repair and impact on genomic stability and disease of SSBs are discussed in Chapter 5.

The one known case of human DNA ligase I deficiency was caused by inherited mutant *lig1* alleles: *LIG1 R771W*, which retained about 10% of ligase activity, and *LIG1 E566K*, which is presumed inactive. Cell lines obtained from this individual showed about a 10- to 20-fold reduced LIG1 activity and a strong defect in Okazaki fragment maturation but surprisingly not in cell proliferation. The symptoms resulted in unexplained immunodeficiencies, sensitivity to sunlight, growth retardation and delayed development (reviewed in Ellenberger and Tomkinson 2008). A mouse model has also been established to recapitulate *LIG1 R771W* mutation. The resultant mice recapitulated most of the human phenotypes, but also exhibited increased spontaneous genomic instabilities in the spleen and an increased incidence of spontaneous tumours (Bentley et al. 2002).

REPLISOME DISASSEMBLY

As mentioned before, disassembly of the Mcm2-7 helicase complex during replication termination is the best-characterised step of eukaryotic termination to date. Using both budding yeast and *X. laevis* systems, Maric et al. (2014) and Moreno et al. (2014) showed that disassembly of the replication machinery at the point of termination requires polyubiquitylation of the Mcm2-7 complex. Indeed, this mechanism appears to be widely conserved, as it has now also been confirmed in *C. elegans* (Sonneville et al. 2017). More specifically, it has been found that Mcm7 is the only CMG helicase component to be polyubiquitylated and that K48-linked ubiquitin chains are generated by the SCF^{Dia2} ubiquitin ligase in budding yeast and the E3 ligase $CUL2^{LRR1}$ in *C. elegans* and in *X. laevis* (Maric et al. 2014, Dewar and Walter 2017, Sonneville et al. 2017). The consequence of this polyubiquitylation is that ubiquitin-binding cofactors recruit the p97/VCP/Cdc48 protein remodeler, which disassembles the CMG complex from chromatin using its ATPase activity (Maric et al. 2014, 2017, Moreno et al. 2014). The different factors responsible for this stage of termination will next be discussed in more detail.

Disassembly of the replisome starts with the ubiquitylation of Mcm7 within the terminating CMG. In all eukaryotes studied so far, this is completed by a cullin-type ubiquitin ligase. Cullin RING ligases (CRLs) work through the cullin scaffold, which brings the substrate and ubiquitin conjugating enzyme (E2) into close proximity (**Figure 3.2A**). This family of ubiquitin ligases is activated by neddylation [(N) in **Figure 3.2**], a reversible post-translational modification, which involves covalent attachment of the ubiquitin-like protein Nedd8. This PTM is believed to activate the ligase by bridging the gap between the ubiquitin-charged E2 and the substrate (Saha and Deshaies 2008). Importantly, we can take advantage of this activation route and block cullin ligase activity with the small molecule inhibitor (MLN4924) of the neddylation enzyme Nedd8-activating enzyme E1 (NAE).

The *S. cerevisiae* Mcm7 is ubiquitylated by SCF^{Dia2}; the SCF (Skp1/Cullin/F-box) ubiquitin ligase complex is composed of the Cullin 1 scaffold protein Cdc53, the RING finger protein (E3) Hrt1, the substrate adaptor/linker protein Skp1 and a substrate receptor Dia2 (an F-box protein) (Maric et al. 2014) (**Figure 3.2B**). Deletion of Dia2 leads to CMG persistence on chromatin until the next cell cycle. Interestingly, yeast cells without Dia2 are viable but exhibit masses of chromosomal instabilities (Blake et al. 2006, Koepp et al. 2006). Dia2 itself contains an N-terminal tetratricopeptide repeat (TPR) domain and leucine-rich repeat (LRR) domain. TPR promotes direct interactions with components of the elongating replisome: Mrc1/Claspin and Ctf4/And-1, and as a result Dia2 interacts with the replication fork even when its

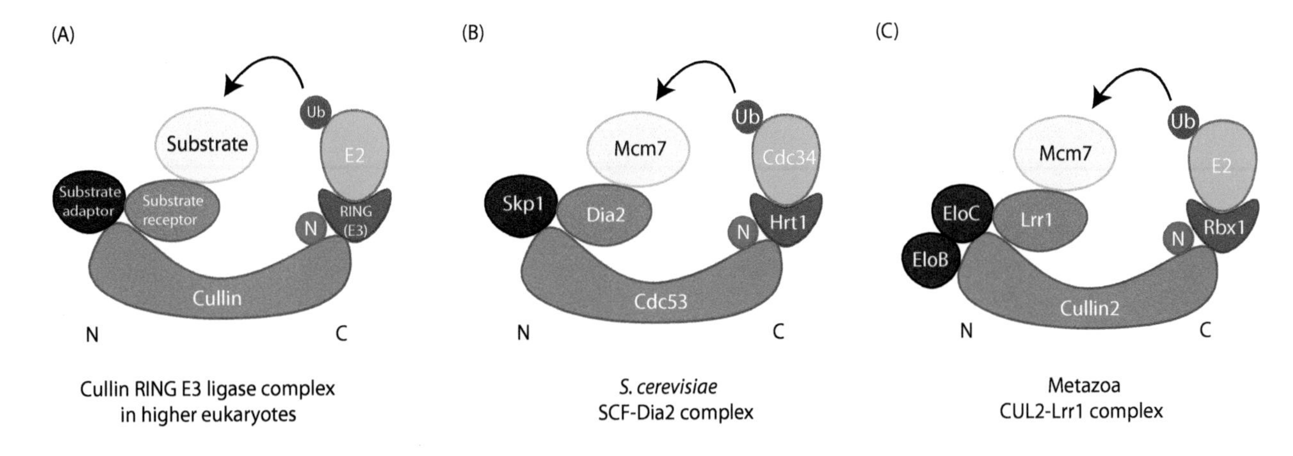

Figure 3.2 Structure of the cullin RING ligases. (A) Cullin RING ligases perform ubiquitylation of specific substrates. They are efficient at this process as they consist of a cullin scaffold, which brings together the substrate and ubiquitin-conjugating enzyme (E2). Cullin ligases become active once they have been modified by neddylation (N). (B) Within *S. cerevisiae* Mcm7 is polyubiquitylated by the SCF^{Dia2} complex. (C) Within metazoan, Mcm7 is polyubiquitylated by the $CUL2^{LRR1}$ complex.

progression is blocked with HU (Morohashi et al. 2009). The relevance of these interactions is not yet fully understood although it is clear that they facilitate polyubiquitylation of Mcm7 even if the TPR domain becomes important exclusively in conditions when Rrm3 helicase (promotes fork passage through protein-DNA barriers) is not present (Maculins et al. 2015).

The ubiquitin ligase ubiquitylating Mcm7 in metazoa (*C. elegans* and *X. laevis*) has also been identified and constitutes the scaffold protein CUL2, the RING finger protein RBX1, the substrate adaptor Elongin B/C and the substrate receptor Lrr1 (Dewar and Walter 2017, Sonneville et al. 2017) (**Figure 3.2C**). Depletion of $CUL2^{LRR1}$ leads to an accumulation of CMGs on chromatin past S phase. Lrr1 protein, similarly to Dia2, contains an LRR domain. It remains to be determined if this domain is essential for recognition of terminated CMG. Interestingly, however, in contrast to *S. cerevisiae*, depletion of the substrate receptor LRR-1 in *C. elegans* does not completely block CMG disassembly. In fact, Sonneville et al. (2017) found evidence to suggest there is a backup pathway for CMG disassembly in the absence of LRR-1, which occurs in mitosis (prophase). Nevertheless, *C. elegans* LRR-1 is an essential gene (Piano et al. 2002).

Once Mcm7 has been polyubiquitylated, the CMG complex becomes a target for the p97/Cdc48/VCP segregase, which promotes protein unfolding and disassembly of the complex from chromatin. This abundant protein, which forms a ring-shaped homo-hexamer, is a central player in regulating the ubiquitin–proteasome system and has been implicated in a large number of cellular processes, including DNA replication (Ramadan et al. 2017). Its ability to function within so many pathways relies upon a large number of cofactors, which are essential for recruiting the p97 ATPase to its substrates. Upon recognising its substrate, ATP is hydrolysed and the hexamer undergoes a conformational change, allowing it to pull the substrate (plus some of the ubiquitin chain) through its narrow central pore, causing it to unfold (Tonddast-Navaei and Stan 2013, Bodnar and Rapoport 2017). The unfolded substrate is thus removed from its cellular location, for example, from the chromatin, and is either degraded by the proteasome or deubiquitylated and recycled. As mentioned, p97/Cdc48/VCP requires cofactors to detect specific substrates. In all organisms studied so far, the major p97 cofactors required for CMG disassembly have been identified as Npl4 and Ufd1 (Dewar and Walter 2017, Maric et al. 2017, Sonneville et al. 2017). When investigating the backup pathway of CMG disassembly in *C. elegans*, Sonneville et al. found that p97 was responsible for removal of CMG helicase from chromatin in mitosis. However, to act in this backup pathway, p97 requires yet another co-factor, UBXN-3, to be involved – the homologue of human Fas-associated factor FAF1 (**Figure 3.3**). FAF1 is an essential gene (Adham et al. 2008) and an established modulator of apoptosis; it regulates NFκB and is involved in ubiquitin-mediated protein turnover (reviewed in Menges et al. 2009). This pathway also requires the activity of the SUMO protease ULP-4 (homologue of SENP6/7) (Sonneville et al. 2017).

Important questions remain about how terminating replisomes are specifically targeted for polyubiquitylation over elongating replisomes. As mentioned earlier, Dewar et al. propose that the terminating replisome complex jumps onto dsDNA following convergence, and so their hypothesis is that this will cause a conformational change in the CMG complex, thus rendering it detectable by the ubiquitin ligase (Dewar et al. 2015, Dewar and Walter 2017). It is possible that an as-yet-unidentified post-translational modification of the CMG complex first occurs at termination, followed by ubiquitylation. It has been proposed that SUMO removal could be such a mark. It has been found that the MCM2-7 complex is SUMOylated at the beginning of S phase, but all Mcms except Mcm7 lose this modification as S phase progresses (Wei and Zhao 2016). Also of note is the observation that the deubiquitylating (DUB) enzyme, Usp7, can specifically deubiquitylate SUMOylated proteins (Lecona et al. 2016). With this in mind, one theory worth investigating is that Usp7 removes ubiquitin from SUMOylated Mcm7 during replication elongation. Once the replication forks terminate, the SUMO tags are removed with the result that Usp7 no longer functions and the competing ubiquitin ligase can then polyubiquitylate Mcm7. Interestingly, as mentioned earlier, a SUMO protease ULP-4 has been identified in *C. elegans*, which seems to be needed for CMG disassembly, albeit in the mitotic backup pathway (Sonneville et al. 2017).

What happens to Mcm7 after it has dissociated from chromatin? Although ubiquitin chains on Mcm7 are K48-linked, which often leads to degradation of a protein by the proteasome, the Gambus group found that proteasomal activity was not needed for chromatin removal of CMG helicase (Moreno et al. 2014). Indeed, Fullbright et al. (2016) suggested that when Mcm7 is ubiquitylated during termination, it is likely to be deubiquitylated, in contrast to ubiquitylated Mcm7 at sites of interstrand cross-link, where it is degraded. Moreover, work published from the Mechali group showed that in *Xenopus* and human cells, a protein called

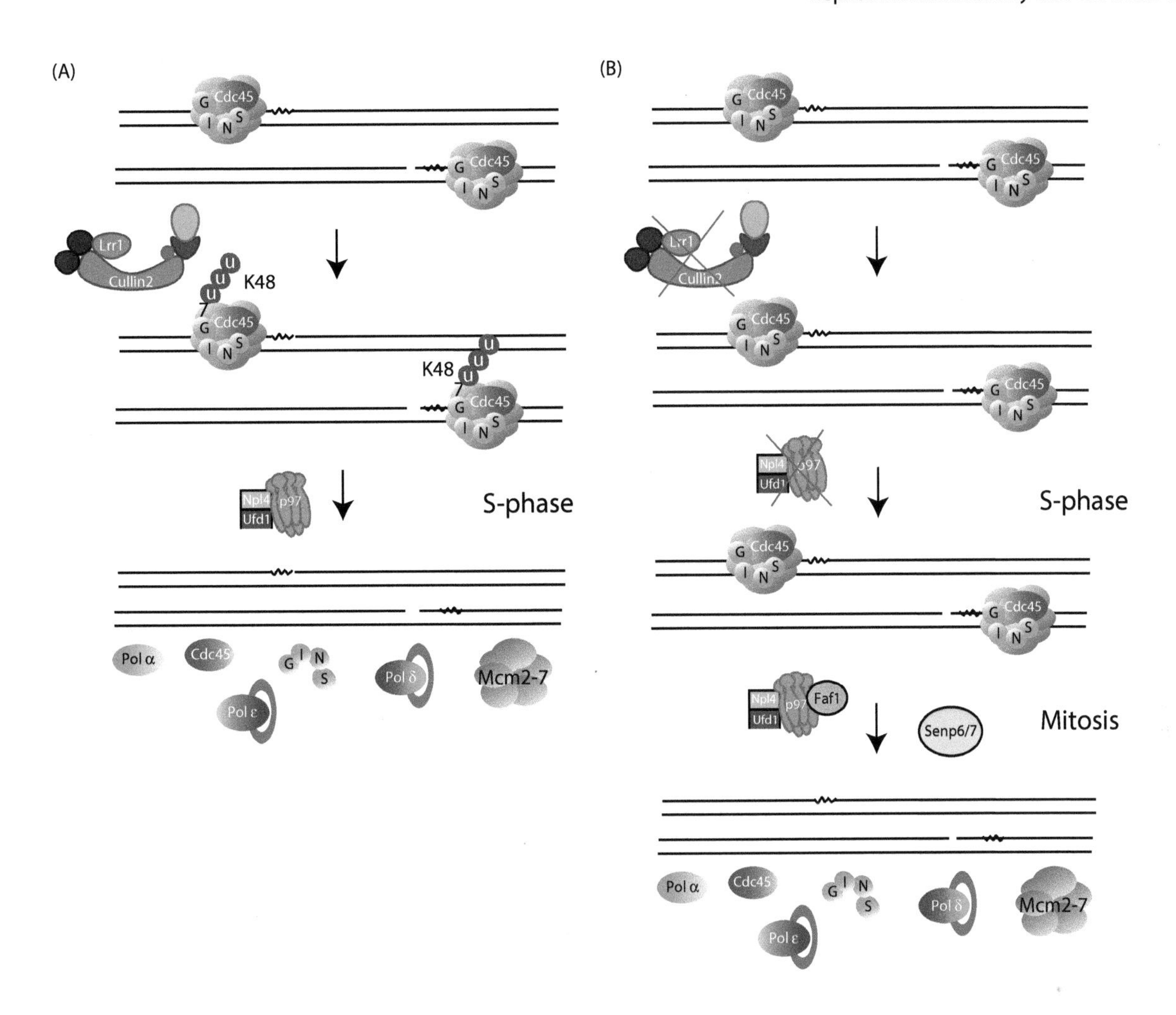

Figure 3.3 **Backup pathway for CMG disassembly.** (A) DNA replication termination occurs throughout S phase, whenever two replication forks from neighbouring origins converge. At termination, the approaching replisomes pass one another, and Mcm7 on each of the CMG complexes is ubiquitylated by the CUL2-Lrr1 cullin RING ligase within metazoa. This polyubiquitylated substrate is recognised by the p97/VCP/Cdc48 protein remodeler, whose major cofactors are Npl4 and Ufd1. With their assistance, p97 promotes disassembly of the entire CMG complex from the chromatin. (B) In the absence of efficient cullin RING ligase activity, that is, when the substrate receptor is non-functional, CMG complexes are retained on the chromatin. Evidence suggests, however, that a backup pathway exists within mitosis to ensure disassembly prior to chromosome separation. This backup pathway involves p97 with its major cofactors Npl4 and Ufd1 but also requires help from an extra cofactor, FAF1, and the SUMO protease SENP6/7.

MCM-BP (MCM-binding protein) can interact with Mcm7 within the nucleus towards the end of S phase, although not whilst it is held within the Mcm2-7 complex on chromatin (Nishiyama et al. 2011). These data suggest that once Mcm7 has been released from chromatin, it could bind MCM-BP, potentially until it is recycled in the next G1 phase. An alternative hypothesis, however, is that MCM-BP is involved in the removal of inactive/dormant MCM proteins from chromatin (Lengronne and Pasero 2014).

Although this step of termination is the best understood at present, it is still unclear whether replisome disassembly occurs prior to, or following, DNA end ligation. Dewar et al. (2015) found evidence for the latter, but DNA fibre analyses by Moreno et al. (2014) suggest otherwise. Of note, helicase disassembly occurs prior to ligation during prokaryotic replication termination, but further work is required to clarify what occurs in eukaryotes (reviewed in Dewar and Walter 2017).

REPLISOME DISASSEMBLY AND GENOMIC INSTABILITY

The consequences of blocking removal of the replisome from chromatin at termination sites depend on when the disassembly occurs. Should it be the case that replisome disassembly is the ultimate stage of termination, as proposed by Dewar et al. (2015), then blocking it will not generate an accumulation of under-replicated or un-ligated DNA. The consequence would, in fact, be DNA-bound CMG complexes, which could in itself lead to several problems. First, the retention of these protein complexes on the DNA could negatively impact chromatin organisation and affect DNA processes downstream, that is, transcription or DNA replication in the next cell cycle. Second, if the persisting helicases have indeed jumped onto the dsDNA by this point and are still able to slide along the DNA, then there is a chance it could encounter an ssDNA–dsDNA junction, for example, an immature Okazaki fragment with a 3′ flap, which would be recognised as a replication fork structure, and the helicase could start unwinding the DNA, promoting re-replication (Kang et al. 2012). Problems with re-replication at replication termination sites have been observed in bacteria by Rudolph et al. (2013), who demonstrated that active mechanisms are in place to avoid this. Finally, with CMG complexes essentially trapped on chromatin, individual components, that is, Cdc45 and GINS, which are recycled through each S phase, will become limiting and cause delays in the subsequent S phase. Such an observation has been made with *dia2Δ* mutants in budding yeast (Blake et al. 2006).

If, on the other hand, disassembly of the CMG complexes occurs prior to DNA ligation, then blocking this stage will result in un-ligated DNA, presented as single-strand breaks at termination sites. Further investigation is required to determine the reality of the situation. Whichever is the true mechanism, there are indications that inhibiting the unloading of CMG helicases leads to genomic instabilities.

At the moment, the evidence suggesting that defective CMG disassembly could be deleterious for genome integrity comes from studies in yeast and *C. elegans*. In budding yeast, lack of the ubiquitin ligase SCF^{Dia2} that ubiquitylates Mcm7 and drives CMG disassembly (Maric et al. 2014) causes high rates of DNA damage and genomic instability, and makes cells sensitive to DNA-damaging agents that affect replication fork progression (Blake et al. 2006, Koepp et al. 2006). Moreover, this is supported by the fact that *dia2Δ* is synthetic lethal with genes implicated in DNA replication, recombination, checkpoints and chromatin remodelling (Blake et al. 2006).

Phenotypes caused by loss of function of the ubiquitin ligase driving Mcm7 ubiquitylation and CMG disassembly in higher eukaryotes – $CUL2^{LRR1}$ – have been studied in *C. elegans*. *lrr-1* is an essential gene for embryonic development in *C. elegans* (Piano et al. 2002), but maternal rescue allows the development of homozygous *lrr-1* mutants until adulthood. Interestingly, *lrr-1* mutants were sterile due to cell proliferation defects in the germline. Importantly, such a sterility phenotype was rescued by inactivation of *atl-1* (worm homologue of ATR gene) or *chk-1* (worm homologue of Chk1 gene) factors of the replication checkpoint, indicating that the defect in germ cell proliferation in *lrr-1* mutants was caused by checkpoint activation. Moreover, downregulation of *lrr-1* by RNAi resulted in a severe embryonic lethal phenotype, with more than 95% of the embryos derived from RNAi-treated worms failing to develop. Similarly to germline cells, such an embryonic lethality phenotype was rescued in *atl-1* mutant embryos, indicating that the replication checkpoint is activated upon *lrr-1* depletion causing embryonic lethality. Importantly, *lrr-1* mutant germline cells accumulated RPA–ssDNA nuclear foci, which were increased upon double mutation of *lrr-1* and *atl-1*. This result suggested that *lrr-1* loss of function generates DNA damage leading to activation of the replication checkpoint and cell cycle arrest. Interestingly, *lrr-1* and *atl-1* mutant germ cells accumulated DNA content greater than 4N, suggesting that some regions of the genome might undergo re-replication in this genetic background (Merlet et al. 2010). Similar phenotypes of embryonic lethality and sterility caused by activation of the replication checkpoint were obtained with a *C. elegans cul-2* temperature-sensitive mutant (Burger et al. 2013). Although the molecular basis of the DNA damage generated in *C. elegans* germ cells upon *lrr-1* downregulation is not well understood, it has recently been reported that the sterility phenotype of *lrr-1* mutants can be rescued by low-dose RNAi against CMG components. Moreover, such RNAi treatment also rescued the embryonic lethality associated with a *C. elegans cul-2* temperature-sensitive mutant (Ossareh-Nazari et al. 2016). These findings suggest that the DNA damage caused by loss of $CUL\text{-}2^{LRR\text{-}1}$ could, at least in part, be due to the resultant defect in CMG extraction from chromatin, as reducing levels of CMG alleviates the sterility and embryonic lethal phenotypes. However, these severe phenotypes caused by loss of function of $CUL\text{-}2^{LRR\text{-}1}$ must result from a synergistic effect of impairing other processes as well, as CMG should be disassembled in *lrr-1* mutants by the UBXN-3 and ULP-4 dependent mitotic backup pathway. However, partial co-downregulation of *lrr-1* and *ubxn-3* or *ulp-4* led to synthetic lethality, suggesting that inhibition of CMG disassembly by partially compromising both S phase and mitotic CMG disassembly pathways resulted in non-viable worms (Sonneville et al. 2017). Moreover, Franz et al. (2016) found that downregulation of *FAF1* by siRNA in human cells causes a pronounced replication stress phenotype: defective fork progression, fork stalling, dormant origin firing and activation of both the S phase checkpoint (ATR-Chk1) and DNA damage checkpoint (ATM-Chk2) pathways.

DECATENATION OF SISTER CHROMATIDS

As mentioned earlier, the enzymes critical for fork progression and convergence in the face of torsional stresses are topoisomerases. Top1 enzymes create single-strand nicks in the DNA, while Top2 enzymes create temporary double-strand cuts. As such, Top2 is the only enzyme able to decatenate intertwined sister chromatids produced after DNA replication (catenanes) (Baxter and Diffley 2008). Dewar et al. (2015) reported that site-specific termination plasmids (described earlier) require Top2 for decatenation of daughter plasmids even though Top2 activity was not required for fork convergence and DNA ligation. Supporting the idea that Top2 works at termination, Sonneville et al. (2017) found Top2 to associate with post-termination replisomes in both *C. elegans* and *X. laevis.*

DECATENATION OF SISTER CHROMATIDS AND GENOMIC INSTABILITY

Topoisomerase function can be studied through different methods. The protein can be depleted from cells, mutations can be made to produce catalytic-dead protein, which is unable to cleave DNA, or finally the enzyme can be trapped on the DNA by a drug, for example, etoposide, which traps Top2. These different methods produce subtly different problems in the cells. Top2 is an essential protein in all organisms. In the absence of Top2 upon conditional depletion of both Top2 isoforms in human cells (Gonzalez et al. 2011), conditional depletion of Top2 in yeast (Baxter and Diffley 2008) and mutation of yeast Top2 so that it is conditionally catalytically dead (*top2-1*) (Brill et al. 1987, Bermejo et al. 2007), the synthesis of DNA can be completed during S phase. The same is true for Top1 deletions (Bermejo et al. 2007, Kegel et al. 2011). However, when both Top1 and Top2 proteins are faulty (in the *top1Δ top2-1* double mutant of yeast), bulk DNA synthesis becomes defective and there are signs of replication fork collapse indicating that Top1 and Top2 work redundantly to promote replication fork progression and stability (Brill et al. 1987, Bermejo et al. 2007).

Although loss of Top2 protein or catalytic activity does not perturb completion of DNA synthesis in the first cell cycle, Baxter and Diffley (2008) found significant differences in later stages of the cell cycle depending on whether the protein was depleted or mutated. Conditional depletion of Top2 protein in yeast caused cells to suffer defects in chromosome segregation at the end of mitosis, which led to extensive DNA damage and activation of the DNA damage checkpoint. Interestingly and conversely, catalytically dead Top2 mutant cells (*top2Y-F*) arrested cells in G2/M-phase of the cell cycle before they could enter mitosis and it was this stage at which the DNA damage checkpoint was activated. Their work revealed that in the absence of Top2 protein, DNA synthesis can be completed but catenanes are not resolved and so chromosome segregation upon mitotic exit becomes disastrous. A catalytic dead Top2 protein, however, which binds DNA as a closed clamp but does not perform cleavage reactions, prevents full completion of DNA replication, and, indeed, gaps (i.e. regions of under-replicated DNA) are detected in the synthesised DNA using more sensitive techniques than flow cytometry. Incomplete decatenation has been shown to create ultra-fine bridges, as Top2 inhibition greatly increases the number of observed UFBs in mitosis (Chan and Hickson 2011).

As mentioned earlier, topoisomerase activity can also be blocked with drugs such as etoposide or ICRF-193. The idea that blocking inactive Top2 on DNA causes incomplete DNA synthesis is supported by the fact that treating *Xenopus* egg extract with ICRF-193 blocks replication termination (Cuvier et al. 2008), and human cells treated with the same arrest in G2 phase (Downes et al. 1994). Although these data indicate that topoisomerase activity itself is required for effective fork convergence and termination, further research is required, as there is some evidence that ICRF-193 treatment actually causes changes to nucleosome spacing and chromatin structure (Germe and Hyrien 2005, Gaggioli et al. 2013) and it could be these alterations that impact termination indirectly. Of note also is that when ICRF-193 is added to *Xenopus* egg extract prior to nuclear assembly, a slowing of replication forks is observed (Lucas et al. 2001).

TARGETS FOR CLINICAL THERAPY

The best-understood part of replication termination thus far is the process of CMG helicase disassembly. Interestingly from the clinical point of view, the mechanism of CMG disassembly is mainly governed, as far as we know, by two targetable enzymes: p97 and a CRL-type ubiquitin ligase ($CUL2^{LRR1}$). Importantly, small molecule inhibitors that compromise the activity of p97 and CRL ligases are being currently tested as anticancer drugs. However, because the role of p97 and CUL2^{LRR1} in CMG disassembly has been discovered only recently, the effect of these drugs on cancer cell death has been attributed to the impairment of the ubiquitin proteasome system.

Oncogene-targeted therapies have successfully contributed to the treatment of many cancers (Ramos and Bentires-Alj 2015). However, these therapies often fail to kill the tumour effectively due to the high plasticity and heterogeneity of the cancerous cell population that result in tumour resistance (Meacham and Morrison 2013). Alternatively, other targets for cancer therapy are represented by cellular processes which are non-oncogenic as such, but become essential in certain types of cancer to deal with the hostile conditions generated within the tumour environment. In other words, in order to survive, tumour cells become addicted to specific cellular processes such as DNA repair pathways (due to accumulation of high levels of genomic instability), mitosis (in order to accomplish the requirement of rapid cancer cell proliferation) and protein homeostasis.

Cancer cells often exhibit a fast rate of protein synthesis to be able to achieve rapid cell proliferation. Therefore, in order to maintain appropriate protein levels, cancer cells are particularly addicted to processes governing protein homeostasis, such as the ubiquitin proteasome system (Van Drie 2011). Drugs targeting the proteasome, such as bortezomib, have been shown to be effective and have been clinically approved for treating haematological cancers. Cancer cells experience a dramatic build-up of misfolded proteins upon treatment with proteasome inhibitors that ultimately leads to apoptotic cell death. However, problems with proteasome therapy include side effects due to the drug attacking normal cells (Zhao and Sun 2013) and the fact that the drug is not effective in solid tumours (Milano et al. 2009, Johnson 2015). Therefore, targeting other factors upstream of the proteasome, such as p97 and CRL ubiquitin ligases, to make therapy more specific has been the focus of hard work resulting in the generation of small molecule inhibitors against these enzymes (Zhao and Sun 2013, Zhou et al. 2015). Studies with the p97 inhibitor CB-5083 showed that the primary mechanism of cancer cell death was through creating high levels of protein cytotoxicity as a consequence of impairment of the ubiquitin proteasome system. Importantly, this inhibitor showed efficiency towards solid tumour models, in contrast to bortezomib (Zhou et al. 2015). Likewise, studies with the small molecule inhibitor of NEDD8-activating enzyme (NAE), whose major effect is the inactivation of CRL ligases, were shown to suppress cancer cell proliferation both in in vitro cell culture as well as in in vivo animal models (Soucy et al. 2009a, 2009b). MLN4924 caused cell death via apoptosis, senescence and autophagy as a result of inhibiting activity of CRL1 and CRL4 ubiquitin ligases, which are amplified in cancer in part due to their role in promoting cell cycle progression and degradation of tumour suppressor proteins. However, although targeting CRLs is less aggressive than targeting the proteasome (CRLs are only responsible for 20% of cellular proteasomal degradation), the strategy still presents specificity problems. For instance, some components of the CRL family are known to be tumour suppressors, such as the CUL2-VHL ligase that promotes degradation of Hif1α during normoxia. Therefore, although it would have a positive impact in cancer cells, targeting all CRLs would still cause side effects, as tumour suppressor proteins in normal cells would be affected (Zhao and Sun 2013).

Could p97 and CRL inhibitors be causing cancer cell death in part by targeting the CMG disassembly pathway? Answering this question still requires much work. First of all, it is essential to confirm that the CMG disassembly pathway discovered in yeast (Maric et al. 2014), frogs (Moreno et al. 2014) and worms (Sonneville et al. 2017) is also recapitulated in mammalian cells. If it is, then it will be important to study this pathway in detail in order to characterise the involvement of other targetable enzymes, as well as the impact on the maintenance of genome integrity of inhibiting CMG disassembly. Factors involved in the assembly of the CMG complex during initiation of DNA replication, such as Cdc7 and TopBP1, are currently being investigated as potential targets for cancer therapy in tumours presenting defects in DNA replication (Montagnoli et al. 2010, Chowdhury et al. 2014). Similarly, inhibiting CMG disassembly in specific genetic backgrounds might be a new avenue for future cancer therapies. Crucially, data obtained by Sonneville et al. (2017) showed that partial depletion of factors involved in both S phase and mitotic pathways of CMG disassembly in *C. elegans* early embryos caused synthetic lethality. This, therefore, provides a potential opportunity for selective cancer therapy. One of the factors required for the mitotic pathway of CMG disassembly – Faf1 (UBXN-3 in worms) – is, in fact, downregulated or mutated in many cancers (Menges et al. 2009). If these pathways are conserved in mammalian cells, it will be very interesting to generate small molecule inhibitors against CUL2^{LRR1} and study their effect in cells from patients with Faf1-deficient tumours. It might turn out that this treatment strategy would be analogous to inhibitors of poly(ADP)ribose polymerase (PARP), which are used against BRCA1-deficient tumours (Konecny and Kristeleit 2016).

CONCLUSION

The purpose of this chapter was to provide an overview of the mechanism of DNA replication termination, which is the least well-characterised event of DNA replication, and to discuss implications in human disease should the process be defective.

With regard to the mechanism, seminal works in the last three years have revealed the first elements of a well-conserved and sophisticated mechanism that regulates termination of DNA replication forks. It was shown that on simple DNA templates, converging forks do not slow down and CMG complexes on the leading strands pass one another. As discussed, further studies are needed to reveal if such a simple scenario will also work in the context of more complex chromatin structures.

Following fork convergence, the leading strand is replicated up to a few bases away from the last Okazaki fragment on the lagging strand. At this point, it is still unclear as to how the final RNA–DNA primer is processed, which polymerase performs synthesis of the last DNA fragment and how the ends are ligated. Should there be a problem in completing this DNA synthesis or ligation, however, cells will certainly accumulate under-replicated DNA or SSBs and suffer genomic instability, driving the development of cancer and other diseases.

The best-understood event of the termination process so far is disassembly of the CMG complex, which requires cullin-type ubiquitin ligases, polyubiquitylation of Mcm7 and the protein remodeler p97/VCP/Cdc48. Inhibiting the cullin ligase can trap the CMG on chromatin past S phase, though evidence suggests there is a second p97-dependent backup pathway for disassembly in mitosis, at least in *C. elegans*.

Importantly, the precise order of events is still under review; more specifically whether CMG disassembly occurs prior to or following end ligation. Knowing this will help us understand the consequences of blocking CMG disassembly, which is important especially because inhibitors against p97 and cullin ligase are already being used as anticancer strategies. Either way, trapping replisome on the chromatin could pose a problem to DNA-driven processes. We can envisage this being a problem for DNA replication and transcription, for example, in the subsequent cell cycle.

Finally, the question remains as to whether the current model of termination is indeed true for human cells and whether targeting this pathway in cancer cells is a feasible strategy. The next few years of research will be telling and exciting.

REFERENCES

Adham, I. M., Khulan, J., Held, T., Schmidt, B., Meyer, B. I., Meinhardt, A., Engel, W. 2008. Fas-associated factor (FAF1) is required for the early cleavage-stages of mouse embryo. *Mol Hum Reprod* 14(4): 207–213.

Ahuja, A. K., Jodkowska, K., Teloni, F., Bizard, A. H., Zellweger, R., Herrador, R., Ortega, S. et al. 2016. A short G1 phase imposes constitutive replication stress and fork remodelling in mouse embryonic stem cells. *Nat Commun* 7: 10660.

Baumann, C., Korner, R., Hofmann, K., Nigg, E. A. 2007. PICH, a centromere-associated SNF2 family ATPase, is regulated by Plk1 and required for the spindle checkpoint. *Cell* 128(1): 101–114.

Baxter, J., Diffley, J. F. 2008. Topoisomerase II inactivation prevents the completion of DNA replication in budding yeast. *Mol Cell* 30(6): 790–802.

Bentley, D. J., Harrison, C., Ketchen, A. M., Redhead, N. J., Samuel, K., Waterfall, M., Ansell, J. D., Melton, D. W. 2002. DNA ligase I null mouse cells show normal DNA repair activity but altered DNA replication and reduced genome stability. *J Cell Sci* 115(Pt 2): 1551–1561.

Bergoglio, V., Boyer, A. S., Walsh, E., Naim, V., Legube, G., Lee, M. Y., Rey, L. et al. 2013. DNA synthesis by Pol eta promotes fragile site stability by preventing under-replicated DNA in mitosis. *J Cell Biol* 201(3): 395–408.

Bermejo, R., Doksani, Y., Capra, T., Katou, Y. M., Tanaka, H., Shirahige, K., Foiani, M. 2007. Top1- and Top2-mediated topological transitions at replication forks ensure fork progression and stability and prevent DNA damage checkpoint activation. *Genes Dev* 21(15): 1921–1936.

Biebricher, A., Hirano, S., Enzlin, J. H., Wiechens, N., Streicher, W. W., Huttner, D., Wang, L. H. et al. 2013. PICH: A DNA translocase specially adapted for processing anaphase bridge DNA. *Mol Cell* 51(5): 691–701.

Blake, D., Luke, B., Kanellis, P., Jorgensen, P., Goh, T., Penfold, S., Breitkreutz, B. J., Durocher, D., Peter, M., Tyers, M. 2006. The F-box protein Dia2 overcomes replication impedance to promote genome stability in *Saccharomyces cerevisiae*. *Genetics* 174(4): 1709–1727.

Bodnar, N. O., Rapoport, T. A. 2017. Molecular mechanism of substrate processing by the Cdc48 ATPase complex. *Cell* 169(4): 722–735 e729.

Brill, S. J., DiNardo, S., Voelkel-Meiman, K., Sternglanz, R. 1987. DNA topoisomerase activity is required as a swivel for DNA replication and for ribosomal RNA transcription. *NCI Monogr* 4: 11–15.

Burger, J., Merlet, J., Tavernier, N., Richaudeau, B., Arnold, A., Ciosk, R., Bowerman, B., Pintard, L. 2013. CRL2(LRR-1) E3-ligase regulates proliferation and progression through meiosis in the *Caenorhabditis elegans* germline. *PLoS Genet* 9(3): e1003375.

Chan, K. L., Hickson, I. D. 2011. New insights into the formation and resolution of ultra-fine anaphase bridges. *Semin Cell Dev Biol* 22(8): 906–912.

Chowdhury, P., Lin, G. E., Liu, K., Song, Y., Lin, F. T., Lin, W. C. 2014. Targeting TopBP1 at a convergent point of multiple oncogenic pathways for cancer therapy. *Nat Commun* 5: 5476.

Costa, A., Ilves, I., Tamberg, N., Petojevic, T., Nogales, E., Botchan, M. R., Berger, J. M. 2011. The structural basis for MCM2-7 helicase activation by GINS and Cdc45. *Nat Struct Mol Biol* 18(4): 471–477.
Cuvier, O., Stanojcic, S., Lemaitre, J. M., Mechali, M. 2008. A topoisomerase II-dependent mechanism for resetting replicons at the S-M-phase transition. *Genes Dev* 22(7): 860–865.
Daigaku, Y., Keszthelyi, A., Muller, C. A., Miyabe, I., Brooks, T., Retkute, R., Hubank, M., Nieduszynski, C. A., Carr, A. M. 2015. A global profile of replicative polymerase usage. *Nat Struct Mol Biol* 22(3): 192–198.
Devbhandari, S., Jiang, J., Kumar, C., Whitehouse, I., Remus, D. 2017. Chromatin constrains the initiation and elongation of DNA replication. *Mol Cell* 65(1): 131–141.
Dewar, J. M., Budzowska, M., Walter, J. C. 2015. The mechanism of DNA replication termination in vertebrates. *Nature* 525(7569): 345–350.
Dewar, J. M., Walter, J. C. 2017. Mechanisms of DNA replication termination. *Nat Rev Mol Cell Biol* 18(8): 507–516.
Downes, C. S., Clarke, D. J., Mullinger, A. M., Gimenez-Abian, J. F., Creighton, A. M., Johnson, R. T. 1994. A topoisomerase II-dependent G2 cycle checkpoint in mammalian cells. *Nature* 372(6505): 467–470.
Du, W. W., Yang, W., Liu, E., Yang, Z., Dhaliwal, P., Yang, B. B. 2016. Foxo3 circular RNA retards cell cycle progression via forming ternary complexes with p21 and CDK2. *Nucleic Acids Res* 44(6): 2846–2858.
Duxin, J. P., Dewar, J. M., Yardimci, H., Walter, J. C. 2014. Repair of a DNA-protein crosslink by replication-coupled proteolysis. *Cell* 159(2): 346–357.
Ellenberger, T., Tomkinson, A. E. 2008. Eukaryotic DNA ligases: Structural and functional insights. *Annu Rev Biochem* 77: 313–338.
Franz, A., Pirson, P. A., Pilger, D., Halder, S., Achuthankutty, D., Kashkar, H., Ramadan, K., Hoppe, T. 2016. Chromatin-associated degradation is defined by UBXN-3/FAF1 to safeguard DNA replication fork progression. *Nat Commun* 7: 10612.
Fullbright, G., Rycenga, H. B., Gruber J. D., Long, D. T. 2016. p97 promotes a conserved mechanism of helicase unloading during DNA cross-link repair. *Mol Cell Biol* 36(23): 2983–2994.
Gaggioli, V., Le Viet, B., Germe, T., Hyrien, O. 2013. DNA topoisomerase IIalpha controls replication origin cluster licensing and firing time in *Xenopus* egg extracts. *Nucleic Acids Res* 41(15): 7313–7331.
Gambus, A., Jones, R. C., Sanchez-Diaz, A., Kanemaki, M., van Deursen, F., Edmondson, R. D., Labib, K. 2006. GINS maintains association of Cdc45 with MCM in replisome progression complexes at eukaryotic DNA replication forks. *Nat Cell Biol* 8(4): 358–366.
Ganai, R. A., Zhang, X. P., Heyer W. D., Johansson, E. 2016. Strand displacement synthesis by yeast DNA polymerase epsilon. *Nucleic Acids Res* 44(17): 8229–8240.
Georgescu, R. E., Langston, L., Yao, N. Y., Yurieva, O., Zhang, D., Finkelstein, J., Agarwal, T., O'Donnell, M. E. 2014. Mechanism of asymmetric polymerase assembly at the eukaryotic replication fork. *Nat Struct Mol Biol* 21(8): 664–670.
Germann, S. M., Schramke, V., Pedersen, R. T., Gallina, I., Eckert-Boulet, N., Oestergaard, V. H., Lisby, M. 2014. TopBP1/Dpb11 binds DNA anaphase bridges to prevent genome instability. *J Cell Biol* 204(1): 45–59.
Germe, T., Hyrien, O. 2005. Topoisomerase II-DNA complexes trapped by ICRF-193 perturb chromatin structure. *EMBO Rep* 6(8): 729–735.
Glover, T. W., Wilson T. E., Arlt, M. F. 2017. Fragile sites in cancer: More than meets the eye. *Nat Rev Cancer* 17(8): 489–501.
Gonzalez, R. E., Lim, C. U., Cole, K., Bianchini, C. H., Schools, G. P., Davis, B. E., Wada, I., Roninson, I. B., Broude, E. V. 2011. Effects of conditional depletion of topoisomerase II on cell cycle progression in mammalian cells. *Cell Cycle* 10(20): 3505–3514.
Hawkins, M., Retkute, R., Muller, C. A., Saner, N., Tanaka, T. U., de Moura, A. P., Nieduszynski, C. A. 2013. High-resolution replication profiles define the stochastic nature of genome replication initiation and termination. *Cell Rep* 5(4): 1132–1141.
Howes, T. R., Tomkinson, A. E. 2012. DNA ligase I, the replicative DNA ligase. *Subcell Biochem* 62: 327–341.
Ivessa, A. S., Lenzmeier, B. A., Bessler, J. B., Goudsouzian, L. K., Schnakenberg S. L., Zakian, V. A. 2003. The *Saccharomyces cerevisiae* helicase Rrm3p facilitates replication past nonhistone protein-DNA complexes. *Mol Cell* 12(6): 1525–1536.
Ivessa, A. S., Zhou J. Q., Zakian, V. A. 2000. The *Saccharomyces* Pif1p DNA helicase and the highly related Rrm3p have opposite effects on replication fork progression in ribosomal DNA. *Cell* 100(4): 479–489.
Johnson, D. E. 2015. The ubiquitin-proteasome system: Opportunities for therapeutic intervention in solid tumors. *Endocr Relat Cancer* 22(1): T1–17.
Johnston, L. H., Nasmyth, K. A. 1978. *Saccharomyces cerevisiae* cell cycle mutant cdc9 is defective in DNA ligase. *Nature* 274(5674): 891–893.
Kang, Y. H., Galal, W. C., Farina, A., Tappin, I., Hurwitz, J. 2012. Properties of the human Cdc45/Mcm2-7/GINS helicase complex and its action with DNA polymerase epsilon in rolling circle DNA synthesis. *Proc Natl Acad Sci USA* 109(16): 6042–6047.
Kegel, A., Betts-Lindroos, H., Kanno, T., Jeppsson, K., Strom, L., Katou, Y., Itoh, T., Shirahige, K., Sjogren, C. 2011. Chromosome length influences replication-induced topological stress. *Nature* 471(7338): 392–396.
Keszthelyi, A., Minchell N. E., Baxter, J. 2016. The causes and consequences of topological stress during DNA replication. *Genes (Basel)* 7(12): 134.
Koepp, D. M., Kile, A. C., Swaminathan, S., Rodriguez-Rivera, V. 2006. The F-box protein Dia2 regulates DNA replication. *Mol Biol Cell* 17(4): 1540–1548.
Konecny, G. E., Kristeleit, R. S. 2016. PARP inhibitors for BRCA1/2-mutated and sporadic ovarian cancer: Current practice and future directions. *Br J Cancer* 115(10): 1157–1173.
Lai, X., Broderick, R., Bergoglio, V., Zimmer, J., Badie, S., Niedzwiedz, W., Hoffmann, J. S., Tarsounas, M. 2017. MUS81 nuclease activity is essential for replication stress tolerance and chromosome segregation in BRCA2-deficient cells. *Nat Commun* 8: 15983.
Langston, L., O'Donnell, M. 2017. Action of CMG with strand-specific DNA blocks supports an internal unwinding mode for the eukaryotic replicative helicase. *Elife* 6: e23449.
Le Beau, M. M., Rassool, F. V., Neilly, M. E., Espinosa, 3rd, R., Glover, T. W., Smith, D. I., McKeithan, T. W. 1998. Replication of a common fragile site, FRA3B, occurs late in S phase and is delayed further upon induction: Implications for the mechanism of fragile site induction. *Hum Mol Genet* 7(4): 755–761.
Lecona, E., Rodriguez-Acebes, S., Specks, J., Lopez-Contreras, A. J., Ruppen, I., Murga, M., Munoz, J., Mendez, J., Fernandez-Capetillo, O. 2016. USP7 is a SUMO deubiquitinase essential for DNA replication. *Nat Struct Mol Biol* 23(4): 270–277.
Lengronne, A., Pasero, P. 2014. Closing the MCM cycle at replication termination sites. *EMBO Rep* 15(12): 1226–1227.
Letessier, A., Millot, G. A., Koundrioukoff, S., Lachages, A. M., Vogt, N., Hansen, R. S., Malfoy, B., O. Brison, Debatisse, M. 2011. Cell-type-specific replication initiation programs set fragility of the FRA3B fragile site. *Nature* 470(7332): 120–123.
Lucas, I., Germe, T., Chevrier-Miller, M., Hyrien, O. 2001. Topoisomerase II can unlink replicating DNA by precatenane removal. *EMBO J* 20(22): 6509–6519.
Lukas, C., Savic, V., Bekker-Jensen, S., Doil, C., Neumann, B., Pedersen, R. S., Grofte, M. et al. 2011. 53BP1 nuclear bodies form around DNA lesions generated by mitotic transmission of chromosomes under replication stress. *Nat Cell Biol* 13(3): 243–253.
Maculins, T., Nkosi, P. J., Nishikawa, H., Labib, K. 2015. Tethering of SCF(Dia2) to the replisome promotes efficient ubiquitylation and disassembly of the CMG helicase. *Curr Biol* 25(17): 2254–2259.
Maric, M., Maculins, T., De Piccoli, G., Labib, K. 2014. Cdc48 and a ubiquitin ligase drive disassembly of the CMG helicase at the end of DNA replication. *Science* 346(6208): 1253596.
Maric, M., Mukherjee, P., Tatham, M. H., Hay, R., Labib, K. 2017. Ufd1-Npl4 recruit Cdc48 for disassembly of ubiquitylated CMG helicase at the end of chromosome replication. *Cell Rep* 18(13): 3033–3042.
McGuffee, S. R., Smith, D. J., Whitehouse, I. 2013. Quantitative, genome-wide analysis of eukaryotic replication initiation and termination. *Mol Cell* 50(1): 123–135.

Meacham, C. E., Morrison, S. J. 2013. Tumour heterogeneity and cancer cell plasticity. *Nature* 501(7467): 328–337.

Menges, C. W., Altomare, D. A., Testa, J. R. 2009. FAS-associated factor 1 (FAF1): Diverse functions and implications for oncogenesis. *Cell Cycle* 8(16): 2528–2534.

Merlet, J., Burger, J., Tavernier, N., Richaudeau, B., Gomes, J. E., Pintard, L. 2010. The CRL2LRR-1 ubiquitin ligase regulates cell cycle progression during *C. elegans* development. *Development* 137(22): 3857–3866.

Milano, A., Perri, F., Caponigro, F. 2009. The ubiquitin-proteasome system as a molecular target in solid tumors: An update on bortezomib. *Onco Targets Ther* 2: 171–178.

Minocherhomji, S., Ying, S., Bjerregaard, V. A., Bursomanno, S., Aleliunaite, A., Wu, W., Mankouri, H. W., Shen, H., Liu, Y., Hickson, I. D. 2015. Replication stress activates DNA repair synthesis in mitosis. *Nature* 528(7581): 286–290.

Montagnoli, A., Moll, J., Colotta, F. 2010. Targeting cell division cycle 7 kinase: A new approach for cancer therapy. *Clin Cancer Res* 16(18): 4503–4508.

Moreno, A., Carrington, J. T., Albergante, L., Al Mamun, M., Haagensen, E. J., Komseli, E. S., Gorgoulis, V. G., Newman, T. J., Blow, J. J. 2016. Unreplicated DNA remaining from unperturbed S phases passes through mitosis for resolution in daughter cells. *Proc Natl Acad Sci USA* 113(39): E5757–5764.

Moreno, S. P., Bailey, R., Campion, N., Herron, S., Gambus, A. 2014. Polyubiquitylation drives replisome disassembly at the termination of DNA replication. *Science* 346(6208): 477–481.

Morohashi, H., Maculins, T., Labib, K. 2009. The amino-terminal TPR domain of Dia2 tethers SCF(Dia2) to the replisome progression complex. *Curr Biol* 19(22): 1943–1949.

Nasmyth, K. A. 1977. Temperature-sensitive lethal mutants in the structural gene for DNA ligase in the yeast *Schizosaccharomyces pombe*. *Cell* 12(4): 1109–1120.

Neelsen, K. J., Lopes, M. 2015. Replication fork reversal in eukaryotes: From dead end to dynamic response. *Nat Rev Mol Cell Biol* 16(4): 207–220.

Nishiyama, A., Frappier, L., Mechali, M. 2011. MCM-BP regulates unloading of the MCM2-7 helicase in late S phase. *Genes Dev* 25(2): 165–175.

Ossareh-Nazari, B., Katsiarimpa, A., Merlet, J., Pintard, L. 2016. RNAi-based suppressor screens reveal genetic interactions between the CRL2LRR-1 E3-ligase and the DNA replication machinery in *Caenorhabditis elegans*. *G3 (Bethesda)* 6(10): 3431–3442.

Petryk, N., Kahli, M., d'Aubenton-Carafa, Y., Jaszczyszyn, Y., Shen, Y., Silvain, M., Thermes, C., Chen, C. L., Hyrien, O. 2016. Replication landscape of the human genome. *Nat Commun* 7: 10208.

Piano, F., Schetter, A. J., Morton, D. G., Gunsalus, K. C., Reinke, V., Kim, S. K., Kemphues, K. J. 2002. Gene clustering based on RNAi phenotypes of ovary-enriched genes in *C. elegans*. *Curr Biol* 12(22): 1959–1964.

Postow, L., Crisona, N. J., Peter, B. J., Hardy, C. D., Cozzarelli, N. R. 2001. Topological challenges to DNA replication: Conformations at the fork. *Proc Natl Acad Sci USA* 98(15): 8219–8226.

Powell, S. K., MacAlpine, H. K., Prinz, J. A., Y. Li, Belsky, J. A., MacAlpine, D. M. 2015. Dynamic loading and redistribution of the Mcm2-7 helicase complex through the cell cycle. *EMBO J* 34(4): 531–543.

Ramadan, K., Halder, S., Wiseman, K., Vaz, B. 2017. Strategic role of the ubiquitin-dependent segregase p97 (VCP or Cdc48) in DNA replication. *Chromosoma* 126(1): 17–32.

Ramos, P., Bentires-Alj, M. 2015. Mechanism-based cancer therapy: Resistance to therapy, therapy for resistance. *Oncogene* 34(28): 3617–3626.

Rivera-Mulia, J. C., Gilbert, D. M. 2016. Replicating large genomes: Divide and conquer. *Mol Cell* 62(5): 756–765.

Rudolph, C. J., Upton, A. L., Stockum, A., Nieduszynski, C. A., Lloyd, R. G. 2013. Avoiding chromosome pathology when replication forks collide. *Nature* 500(7464): 608–611.

Ryba, T., Hiratani, I., Lu, J., Itoh, M., Kulik, M., Zhang, J., Schulz, T. C., Robins, A. J., Dalton, S., Gilbert, D. M. 2010. Evolutionarily conserved replication timing profiles predict long-range chromatin interactions and distinguish closely related cell types. *Genome Res* 20(6): 761–770.

Saha, A., Deshaies, R. J. 2008. Multimodal activation of the ubiquitin ligase SCF by Nedd8 conjugation. *Mol Cell* 32(1): 21–31.

Schalbetter, S. A., Mansoubi, S., Chambers, A. L., Downs, J. A., Baxter, J. 2015. Fork rotation and DNA precatenation are restricted during DNA replication to prevent chromosomal instability. *Proc Natl Acad Sci USA* 112(33): E4565–4570.

Smith, D. J., Whitehouse, I. 2012. Intrinsic coupling of lagging-strand synthesis to chromatin assembly. *Nature* 483(7390): 434–438.

Sonneville, R., Moreno, S. P., Knebel, A., Johnson, C., Hastie, C. J., Gartner, A., Gambus, A., Labib, K. 2017. CUL-2LRR-1 and UBXN-3 drive replisome disassembly during DNA replication termination and mitosis. *Nat Cell Biol* 19(5): 468–479.

Soucy, T. A., Smith, P. G., Milhollen, M. A., Berger, A. J., Gavin, J. M., S. Adhikari, Brownell, J. E., Burke, K. E. et al. 2009a. An inhibitor of NEDD8–activating enzyme as a new approach to treat cancer. *Nature* 458(7239): 732–736.

Soucy, T. A., Smith, P. G., Rolfe, M. 2009b. Targeting NEDD8-activated cullin-RING ligases for the treatment of cancer. *Clin Cancer Res* 15(12): 3912–3916.

Sundin, O., Varshavsky, A. 1980. Terminal stages of SV40 DNA replication proceed via multiply intertwined catenated dimers. *Cell* 21(1): 103–114.

Tonddast-Navaei, S., Stan, G. 2013. Mechanism of transient binding and release of substrate protein during the allosteric cycle of the p97 nanomachine. *J Am Chem Soc* 135(39): 14627–14636.

Van Drie, J. H. 2011. Protein folding, protein homeostasis, and cancer. *Chin J Cancer* 30(2): 124–137.

Wei, L., Zhao, X. 2016. A new MCM modification cycle regulates DNA replication initiation. *Nat Struct Mol Biol* 23(3): 209–216.

Wendel, B. M., Courcelle, C. T., Courcelle, J. 2014. Completion of DNA replication in *Escherichia coli*. *Proc Natl Acad Sci USA* 111(46): 16454–16459.

Yeeles, J. T., Janska, A., Early, A., Diffley, J. F. 2017. How the eukaryotic replisome achieves rapid and efficient DNA replication. *Mol Cell* 65(1): 105–116.

Zhao, Y., Sun, Y. 2013. Cullin-RING ligases as attractive anti-cancer targets. *Curr Pharm Des* 19(18): 3215–3225.

Zheng, L., Shen, B. 2011. Okazaki fragment maturation: Nucleases take centre stage. *J Mol Cell Biol* 3(1): 23–30.

Zhou, H. J., Wang, J., Yao, B., Wong, S., Djakovic, S., Kumar, B., Rice, J. et al. 2015. Discovery of a first-in-class, potent, selective, and orally bioavailable inhibitor of the p97 AAA ATPase (CB-5083). *J Med Chem* 58(24): 9480–9497.

Mechanisms of DNA Damage Tolerance

4

Cyrus Vaziri and Anastasia Zlatanou

INTRODUCTION

Accurate and efficient DNA replication is crucial for the health and survival of all living organisms. Under optimal conditions, the replicative DNA polymerases ε, δ and α can work in concert to ensure that the genome is replicated efficiently with high accuracy in every cell cycle (Burgers 2009). However, DNA is constantly threatened by exogenous and endogenous genotoxic insults, such as solar ultraviolet radiation and reactive oxygen species, generated as by-products of cellular metabolism. Even with multiple repair mechanisms, the normal DNA replication machinery still encounters lesions that have evaded repair, potentially causing replication arrest and genome instability. The most deleterious are DNA lesions that arise during DNA replication, because damaged DNA can act as a steric block to replicative polymerases, leading to incomplete DNA replication or the formation of secondary DNA strand breaks at the sites of replication stalling. Incomplete DNA synthesis and strand breaks are both potential sources of genomic instability.

The very efficient and faithful replicative polymerases guarantee fast and accurate DNA duplication during S phase; however, any distortion in the DNA structure can hinder this process and cause the replication fork to stall. If the stalling is prolonged, the replication fork can collapse, generating DNA double-strand breaks that lead to genome instability. It is therefore crucial to resume replication even in the face of persistent DNA damage (Lopes et al. 2006, Branzei 2011). To survive fork-stalling DNA impediments, cells are equipped with mechanisms of tolerating DNA damage during replication to ensure completion of DNA synthesis. Post-replication repair (PRR) allows for temporal acceptance of the presence of DNA lesions in the genome when there is a risk of cell death. The main role of PRR is to fill in ssDNA gaps in the daughter strand and to restore DNA to its double-stranded state for subsequent DNA repair via other mechanisms. Based on genetic studies in budding yeast, PRR utilizes two major mechanisms that allow DNA damage tolerance (DDT) and facilitate resumption of blocked replication. The first one involves a damage avoidance mechanism and the second one is DNA translesion synthesis (TLS). The process of DNA damage avoidance is a form of indirect bypass that is thought to involve a template switching mechanism whereby the undamaged sister chromatid is used as a temporary replication template for homologous recombination (Xiao et al. 2000). During TLS, the highly precise and efficient replicative DNA polymerases that are blocked by a DNA lesion are replaced by specialized, low-processivity TLS polymerases that are able to directly replicate across the damaged site. These TLS DNA polymerases can either bypass the lesion unassisted or with the help of another TLS polymerase in a two-step process (Woodgate 2001).

PCNA AND REGULATION OF DNA DAMAGE TOLERANCE BY POST-TRANSLATIONAL MODIFICATIONS

The vast majority of studies on the mechanisms of DDT over the last 30 years have mostly come from genetic and mechanistic studies of the *RAD6* epistasis group of genes from yeast (Prakash 1981, Sale 2012). These genes encode a set of low-fidelity DNA polymerases (Yang and Woodgate 2007, Guo et al. 2009, Waters et al. 2009, Sale et al. 2012, Yang 2014), as well as ubiquitin-conjugating E2 enzymes and ubiquitin E3 ligases that ubiquitinate proliferating

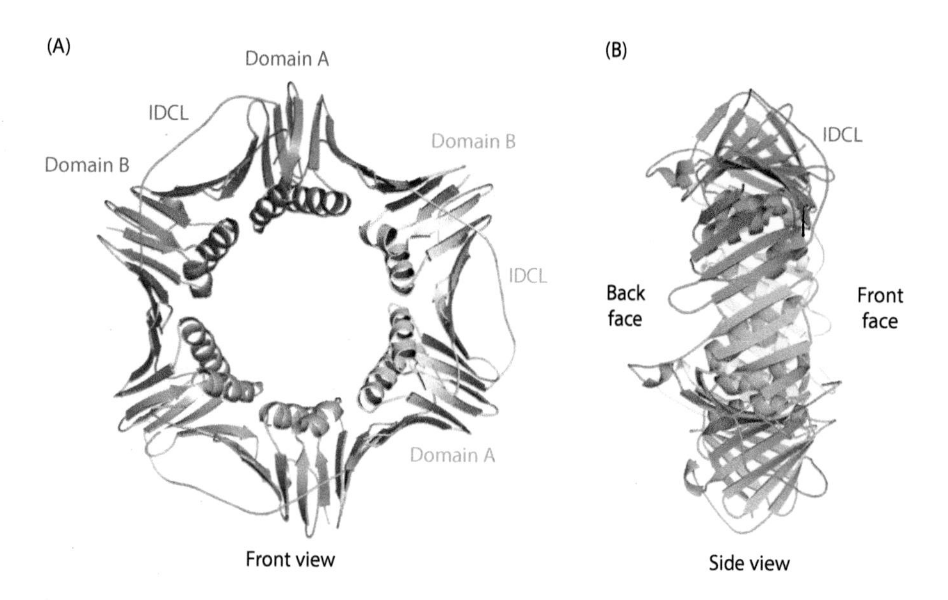

Figure 4.1 Structure of eukaryotic PCNA. Ribbon diagram of the PCNA trimer shown from the front (A) and the side (B) with the individual PCNA subunits coloured red, yellow and blue. The interdomain connector loop (IDCL) is indicated. (From Dieckman, L. M. et al. 2012, *Subcell Biochem* 62: 281–299.)

cell nuclear antigen (PCNA) (Hoege et al. 2002, Friedberg et al. 2005), a sliding clamp protein that encircles DNA and functions as a polymerase processivity factor (Moldovan et al. 2007). Ubiquitin is a small protein that is present exclusively in eukaryotes and that may be covalently attached to a target protein through an enzyme cascade. The E1 ubiquitin-activating enzyme activates ubiquitin and transfers it to an E2 ubiquitin-conjugating enzyme (Ubc), which catalyses formation of an isopeptide bond between ubiquitin and a lysine residue on a target protein. An E3 ubiquitin ligase can bind an E2 enzyme and a target protein simultaneously, mediating the specificity of this process.

PCNA, a replication processivity factor, is a ring-shaped homotrimeric complex which encircles double-stranded DNA and slides along the DNA (**Figure 4.1**) (Krishna et al. 1994). The monomers, each comprising two structurally similar domains, are linked in head-to-tail mode. PCNA monomers interact with DNA through their DNA binding motifs (61–80 residues) located on an internal surface. On the outer surface, the N- and C-terminal halves of PCNA are linked by the interdomain-connecting loop (IDCL) positioned above a hydrophobic pocket that provides a docking site for the PCNA-interacting peptide (PIP) motif of proteins that interact with PCNA (Dieckman et al. 2012). PCNA interacts with multiple proteins involved in replication, cell cycle regulation and DNA repair, and coordinates their access to replication forks (reviewed in Moldovan et al. 2007).

In response to replication fork stalling, PCNA undergoes monoubiquitination at K164 by Rad6 and Rad18 (E2-ubiquitin conjugation and E3-ubiquitin ligase enzyme, respectively) (Hoege et al. 2002, Parker and Ulrich 2009). Monoubiquitinated PCNA interacts with TLS polymerases via their ubiquitin binding domains (UBDs), thereby activating the TLS pathway (**Figure 4.3**). The crystal structure of ubiquitin-modified PCNA has been resolved and showed that the ubiquitin moiety occupies a position on the back face of the PCNA ring without significantly altering the conformation of PCNA (**Figure 4.2**) (Freudenthal et al. 2010). The ubiquitin moiety on PCNA does not act as an allosteric modifier in order to increase the affinity of PCNA for TLS polymerases but instead provides an additional binding surface to which the TLS pols can attach. Rad6-Rad18 is the main source of PCNA monoubiquitination, though some residual, conditional ubiquitination can be observed in yeast and chicken cells lacking Rad6 or Rad18 (Arakawa et al. 2006, Simpson et al. 2006, Kats et al. 2009). There are also reports that human PCNA can be monoubiquitinated by $CRL4^{Cdt2}$ or RNF8 ubiquitin ligases (Zhang et al. 2008, Terai et al. 2010). PCNA ubiquitination is reversible, and modified PCNA can be deubiquitinated via USP1 or BPLF1 (only in human cells) (Huang et al. 2006, Brun et al. 2010, Whitehurst et al. 2012).

PCNA that is monoubiquitinated at K164 can undergo further ubiquitination. In budding yeast, K63-linked polyubiquitin chains, catalysed by the Ubc13-Mms2-Rad5 E2-E3 enzymes, promote template switching (Parker and Ulrich 2009) (**Figure 4.3**). In human cells, there are two Rad5 homologues, HLTF and SHPRH (Motegi et al. 2006, Unk et al. 2006, Motegi et al. 2008, Unk et al. 2008), serving as the E3 ligases for K63-chain formation. Like yeast Rad5, they both interact with RAD6-RAD18 and MMS2-UBC13 complexes (Motegi et al. 2006, 2008); however, their role is not fully understood and their function in damage

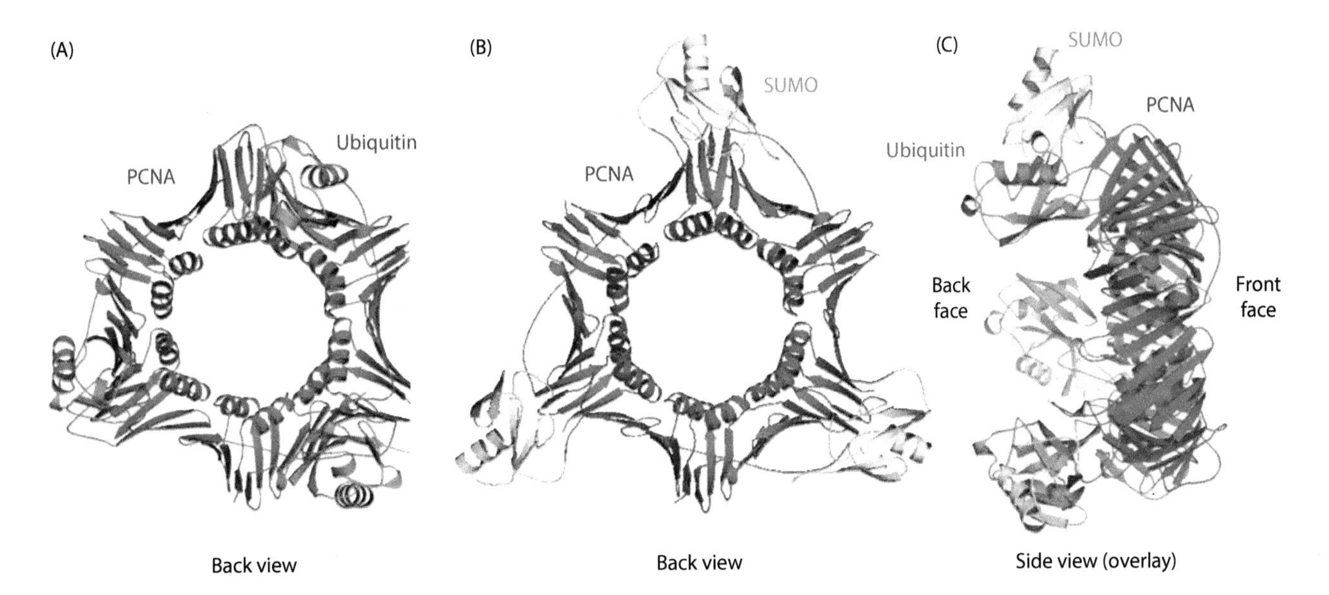

Figure 4.2 **Structures of ubiquitinated and SUMOylated PCNA.** (A) Ribbon diagram of the ubiquitin-modified PCNA trimer depicted from the back with the PCNA ring in blue and the ubiquitin moieties shown in red. (B) Ribbon diagram of the SUMO-modified PCNA trimer shown from the back with the PCNA ring shown in blue and the SUMO moieties shown in yellow. (C) Overlay of the structures of ubiquitin-modified PCNA and SUMO-modified PCNA shown from the side. (From Dieckman, L. M. et al. 2012, *Subcell Biochem* 62: 281–299.)

avoidance and TLS sub-pathways of PRR have been suggested (Lin et al. 2011). Furthermore, the existence of yet another E3 ligase has been suggested based on the observation that a fraction of PCNA was polyubiquitinated even in HLTF/SHPRH double mutant mice (Krijger et al. 2011). An additional difference in PCNA polyubiquitination between lower and higher eukaryotes is also the requirement for MMS2, as this protein seems to be dispensable for this process in mammalian cells (Brun et al. 2008). PCNA polyubiquitination, similar to PCNA monoubiquitination, is negatively regulated by USP1 (Motegi et al. 2008, Brun et al. 2010).

Yeast PCNA can also be ubiquitinated at K107 in response to replication stress caused by the presence of unprocessed Okazaki fragments in ligase-deficient cells. K107 monoubiquitination signals checkpoint activation and both mono- and K29-linked polyubiquitination on K107 involves Mms2, Ubc4 and Rad5 (Das-Bradoo Nguyen et al. 2010). Human PCNA is also ubiquitinated in DNA ligase I–depleted cells, but the modified residue has not yet been identified (Das-Bradoo et al. 2010).

In addition to ubiquitination, PCNA can be modified by SUMO. Initially, this modification was identified in yeast, followed by a handful of other species, including *Xenopus* and chicken DT40 cells (Hoege et al. 2002, Leach and Michael 2005, Arakawa et al. 2006). Recently, a low level of SUMOylation of human PCNA has also been reported (Gali, Juhasz et al. 2012, Moldovan et al.

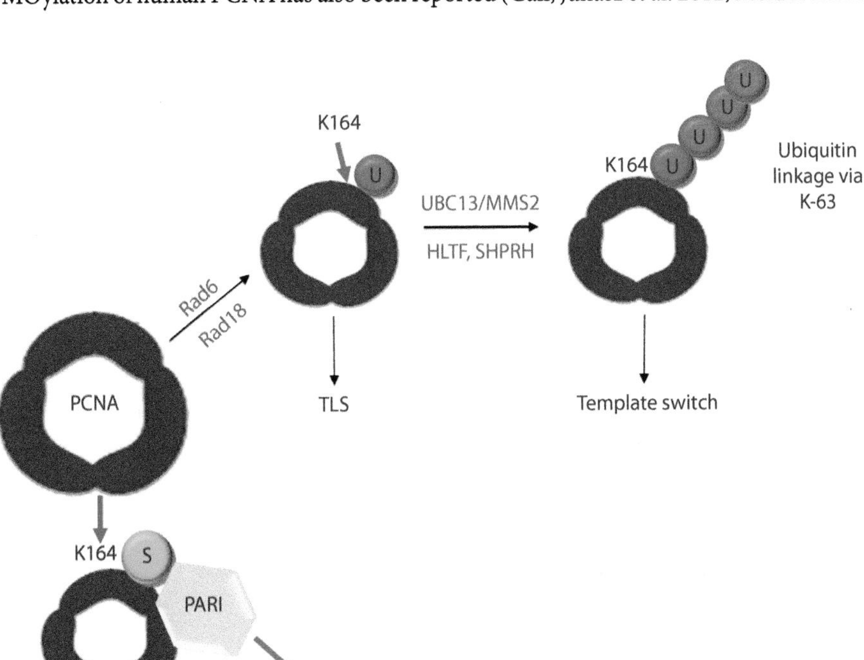

Figure 4.3 **Schematic of ubiquitin and SUMO modifications of PCNA.** Monoubiquitination on Lysine K164 of PCNA promotes TLS, whereas further polyubiquitination of PCNA by HLTF/SHPRH promotes error-free DNA damage avoidance bypass termed template switch (TS). SUMOylation of PCNA enhances binding of the anti-recombination protein PARI to prevent homologous recombination.

2012). SUMOylation is a reversible process and Ulp1 hydrolase removes SUMO from PCNA (Panse et al. 2003, Stelter and Ulrich 2003).

In budding yeast, PCNA SUMOylation occurs mostly during S phase progression and, by influencing PCNA interactions with various partners, controls DNA replication and repair. PCNA SUMOylation promotes the binding of the Srs2 helicase, an inhibitor of recombination, and thereby prevents unwanted recombination events at the replication fork (Papouli et al. 2005). A non-canonical PIP box found in Srs2 has relatively low affinity for unmodified PCNA, but the interaction is significantly strengthened upon PCNA SUMOylation (Kim et al. 2012b). Another interaction involving Elg1, the large subunit of an alternative clamp loading complex, implies the involvement of SUMOylation of PCNA in its unloading from DNA in yeast cells (Parnas et al. 2010, Kubota et al. 2013). Conversely, ATAD5, a human homologue of yeast Elg1, does not seem to have a preference for unloading SUMOylated PCNA despite possessing a SUMO interacting motif (Lee et al. 2010, Lee et al. 2013). On the other hand, it has been shown to be involved in deubiquitinating PCNA by recruiting the USP1 complex to ubiquitinated PCNA (Papouli et al. 2005). PCNA SUMOylation in yeast cells has also been shown to inhibit Eco1-PCNA–dependent sister chromatid cohesion (Moldovan et al. 2006).

Yeast PCNA can be SUMOylated on K164 and to a lesser extent on K127, with the involvement of SUMO conjugating and ligating E2-E3 enzymes, Ubc9 and Siz1 (K164), or just Ubc9 itself (K127) (Hoege et al. 2002). Human PCNA is also SUMOylated on K164 and K254 in order to prevent DSB formation and inappropriate recombination in response to replication fork arrest by DNA lesions (Gali et al. 2012). In another study, the human analogue of Srs2, PARI, was shown to promote the interaction with SUMOylated PCNA, correspondingly obstructing homologous recombination (Moldovan et al. 2012).

Monoubiquitination of PCNA on K164 enhances the interaction with TLS polymerases, whilst another modification, ISGylation, promotes release of the TLS polymerase η (polη) from PCNA (Park et al. 2014). Upon UV irradiation, either K164 or K168 is assumed to undergo ISG15 modification that induces PCNA de-ubiquitination and polη discharge followed by PCNA de-ISGylation and resumption of normal replication. UV-induced PCNA acetylation at K14 causes dissociation of PCNA from a complex with MTH2, and as a consequence, shortens its half-life, as PCNA is more easily degraded by the proteasome (Yu et al. 2009).

ERROR-FREE MODE OF DNA DAMAGE TOLERANCE

Damage avoidance via template switch

The template switching (TS) branch of DDT enables the stalled replication fork to use the newly synthesized daughter strand as the template to avoid damaged DNA. TS is mediated by PCNA post-translational modifications, specifically poly-ubiquitination and SUMOylation. The extension of Rad18-induced K164 mono-ubiquitination to poly-ubiquitination by Ubc13-Mms2 and Rad5 channels DDT to TS.

Most of our understanding of TS was generated from a series of elegant studies in yeast, which provided the basis for the existence of an error-free form of DDT (Prakash 1981, Prakash et al. 1993). This error-free mechanism requires Rad5 (Xiao et al. 2000, Minca and Kowalski 2010), Ubc13-Mms2 (Broomfield et al. 1998), DNA pol δ (Torres-Ramos et al. 1997), a subset of the RAD52 epistasis group (Gangavarapu et al. 2007) and involves recombination between partially replicated sister strands. A study combining 2D gel electrophoresis of DNA replication intermediates and yeast genetics allowed the identification of TS intermediates and defined their relationship with the previously mentioned TS factors (Branzei et al. 2008).

Template switching (Ciccia et al. 2012, Weston et al. 2012, Achar et al. 2015, Kile et al. 2015) has been shown to contribute to DNA damage avoidance by promoting the formation of an X-shaped recombination intermediate-like structure consisting of sister chromatic junctions (SCJs) close to the stalled replication fork (**Figure 4.4**, left). SCJ formation requires Rad51 and the resolution of SCJs structures depends on the Sgs1 helicase or BLM protein in humans (Branzei et al. 2008). The SCJs generated during TS resemble the properties of a DNA crossover intermediate in homologous recombination. A recent study visualized the recombination intermediates using electron microscopy and proved that the undamaged sister chromatid is used as template using a recombination-based mechanism (Giannattasio et al. 2014). During TS, the ssDNA template containing the DNA lesion anneals with the newly synthesized double-stranded sister chromatid to form a three-strand duplex (**Figure 4.4**). This intermediate then

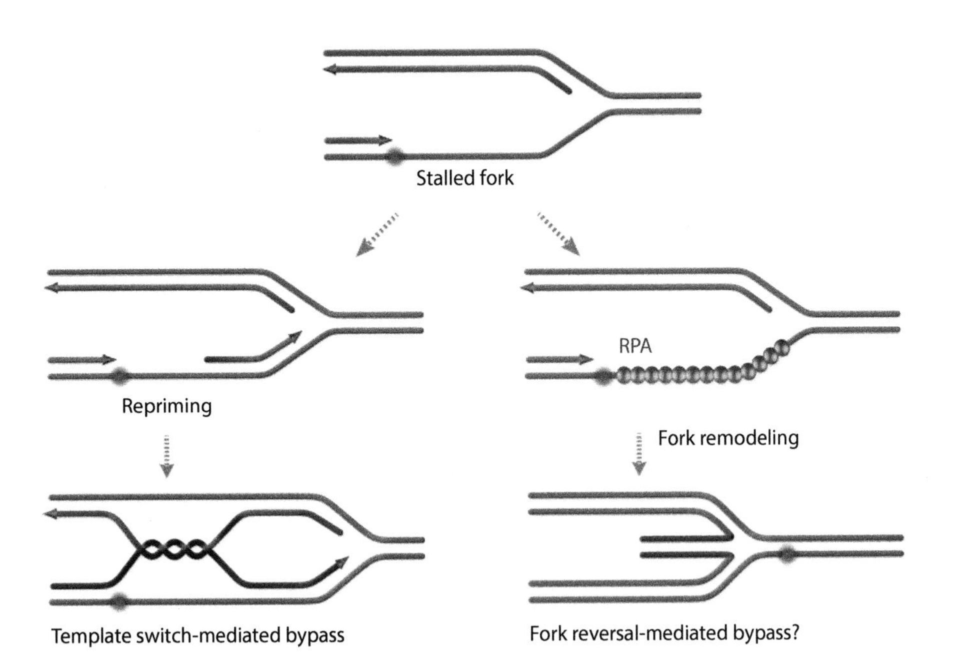

Figure 4.4 **Schematic representation of two main modes of damage bypass upon stalling of the leading polymerase.** Repriming is a crucial step for post-replicative DDT by template switch (represented) or TLS. Fork reversal (on the right) is mediated by various fork remodelling activities in vitro and can promote in certain conditions replication fork stabilization and DDT. (Adapted from Branzei, D., Szakal, B., 2016, *DNA Repair (Amst)* 44: 68–75.)

releases the newly synthesized, undamaged daughter strand from the parental strand so that it can be used as template for damage avoidance synthesis. This causes the formation of a D-loop structure that is subsequently extended to a double Holliday junction (HJ)-like intermediate (pseudo-double HJ), in which the newly synthesized strands are paired by plectonemic pairing induced by DNA synthesis and the parental strands are annealed to each other. The pseudo-double HJ intermediates are then dissolved to hemicatenanes by the Sgs1-Top3-Rmi1 complex before full resolution takes place, likely by Top3 (Giannattasio et al. 2014). In addition to the core TS participants such as Rad5, Rad51 and Sgs1, there is a growing body of evidence that other DNA replication and damage repair factors are also involved in TS. For example, the 9-1-1 complex and Exo1 nuclease are essential for the initiation of TS (Karras et al. 2013), and Ctf4 was found to establish the connection between Pol α/primase and the MCM helicase to protect the replication fork structure that favours TS (Fumasoni et al. 2015). Another model of template switching proposes that a stalled replication fork can be remodelled into a reversed fork structure for stabilization or to facilitate damage bypass (Neelsen and Lopes 2015). The formation of the regressed fork may involve coordinated annealing of the two newly synthesized strands (Figure 4.4, right column). The mechanistic understanding of this process is still very limited but in vitro, several DNA translocases, such as RAD54 (Bugreev et al. 2011), SMARCAL1 (Betous et al. 2012), Mph1 (Xue et al. 2014) and its mammalian homolog FANCM (Gari et al. 2008), ZRANB3 (Ciccia et al. 2012, Yuan et al. 2012, Vujanovic et al. 2017) and Rad5 (Blastyak et al. 2007) and its mammalian homologue HLTF (Kile et al. 2015), can all promote fork remodelling that leads to fork reversal. Helicases such as FBH1 (Fugger et al. 2015), BLM and WRN (Machwe et al. 2006) can also promote the same in vitro reaction. However, in vivo, fork reversal functions have been attributed so far to FBH1 (Fugger et al. 2015), ZRANB3 (Weston et al. 2012) and for SMARCAL1 (Couch et al. 2013).

Although it is clear that Ubc13-Mms2 and Rad5-mediated poly-ubiquitination of PCNA at K164 is a crucial event in TS in budding yeast, the function of the PCNA poly-ubiquitin chain formed during TS remains elusive. In contrast, the significance of PCNA SUMOylation is slightly better understood. PCNA SUMOylation occurs both during normal, unperturbed DNA replication and in response to DNA damage (Hoege et al. 2002, Stelter and Ulrich 2003). SUMOylated PCNA provides an interaction platform for the recruitment of the Srs2 helicase in yeast (Papouli et al. 2005, Pfander et al. 2005). Similar to other PCNA-interacting proteins, Srs2 contains a non-canonical PIP box motif that mediates PCNA binding. However, the interaction between the Srs2 PIP box and PCNA is fairly weak until a second interaction is established between SUMOylated PCNA and a SUMO-interacting motif at the C terminus of Srs2 (Kim et al. 2012a). PCNA-bound Srs2 helicase functions as a safeguard that limits unscheduled recombination at the replication fork by disrupting Rad51 filament formation on ssDNA during normal replication (Krejci et al. 2003, Veaute et al. 2003). A small controversy still exists regarding why PCNA SUMOylation is required for the TS pathway; Srs2 actively removes Rad51 from the replication fork, while template switching requires Rad51 activity. For this reason, it is generally believed that SUMOylation antagonizes the effect of PCNA ubiquitination and inhibits the TS pathway (Hoege et al. 2002, Stelter and Ulrich 2003). However, recently it was shown that a SUMO-like domain protein, Esc2, was found to be recruited to stalled replication forks and was able to displace Srs2, thereby creating a microenvironment that is permissive for Rad51 chromatin-binding (Urulangodi et al. 2015). Therefore, PCNA SUMOylation promotes the usage of DDT on a challenged replication fork by suppressing homologous recombination. When TS is initiated by PCNA poly-ubiquitination, replication fork binding factors such as Esc2 alleviate the inhibition of recombination by PCNA SUMOylation and allow DNA damage avoidance (Pfander et al. 2005).

Error-free DNA damage avoidance is a conserved PRR mechanism in metazoans (Chiu et al. 2006, Izhar et al. 2013). Although the identity of the human Srs2 orthologue is still being debated, human PCNA SUMOylation has also been shown to suppress unscheduled DNA recombination via PARI (PCNA-associated recombination inhibitor), suggesting a conserved mechanism of regulating HR at the replication fork (Gali et al. 2012, Moldovan et al. 2012, Burkovics et al. 2016). PCNA is also poly-ubiquitinated in human cells in response to DNA damage. Blocking K63-linked poly-ubiquitination chain formation sensitises cells to DNA damage, increases UV-induced mutagenesis, and increases the reliance of cells on TLS for DNA damage tolerance (Chiu et al. 2006). Rad5 has evolved into two orthologues, SHPRH (Motegi et al. 2006, Unk et al. 2006) and HLTF, in higher organisms (Motegi et al. 2008, Unk et al. 2008). Both SHPRH and HLTF can poly-ubiquitinate PCNA in vitro but via distinct mechanisms. SHPRH extends RAD18–mediated PCNA mono-ubiquitination, while HLTF transfers the pre-assembled poly-ubiquitin chain to RAD6-RAD18 and eventually onto unmodified PCNA (Motegi et al. 2006, Unk et al. 2006, Masuda et al. 2012). Depletion of SHPRH and HLTF sensitises the cell to DNA damaging agents and reduces PCNA poly-ubiquitination; however, *SHPRH*$^{-/-}$*HLTF*$^{-/-}$ double knockout mouse embryonic fibroblasts are still able to poly-ubiquitinate PCNA, suggesting that other Rad5 orthologues might exist in higher organisms (Krijger et al. 2011a). In addition to poly-ubiquitination of PCNA, HLTF has acquired additional functions in DNA damage tolerance. In response to UV damage, HLTF is able to mono-ubiquitinate PCNA and promote Polη recruitment (Lin et al. 2011). Furthermore, HLTF can also facilitate DNA strand invasion and D-loop formation in a Rad51-independent manner (Burkovics et al. 2014).

Post-replicative DNA damage tolerance

DDT mechanisms are utilised by the cells in order to cope with replication fork barriers that halt the replicative DNA polymerases. Because S phase is a time when the DNA is most vulnerable to injuries, DDT has to be finely tuned to ensure efficient bypass and minimal stalling. One way of achieving this is through limited accessibility or expression of crucial DDT factors until after the end of DNA replication (Waters and Walker 2006, Ortiz-Bazan et al. 2014). In fact, in *S. cerevisiae*, DDT can also be functional during G2/M without significantly delaying the progression of S phase (Daigaku et al. 2010, Karras and Jentsch 2010). Furthermore, delaying the onset of DDT using temporally controlled expression of Rad18 or Polη does not significantly impact cell viability either. These studies suggest that it is possible to detach the DDT from bulk DNA synthesis in the S phase without compromising its function.

Nevertheless, the delayed onset of DDT during S phase could potentially lead to the accumulation of long and fragile ssDNA stretches, especially on the leading strand. Exposed ssDNA in cells is frequently observed when the replicative polymerase is blocked. However, these ssDNA gaps are usually small in size and are located inside a single replicon, regardless of whether they are on the leading or the lagging strand. However, extremely long ssDNA gaps (>3 kb) are rarely observed. This suggests that the leading strand is also synthesised discontinuously when replicating a damaged DNA template, similar to the discontinuous synthesis of the lagging strand (Lopes et al. 2006).

Restart of replication requires a de novo re-priming mechanism downstream (3′) of the stalled leading strand DNA polymerase. This repriming activity is carried out by DnaG in *E. coli* (Yeeles and Marians 2011), and by a specialized polymerase PrimPol in higher organisms (Bianchi et al. 2013, Garcia-Gomez et al. 2013, Mouron et al. 2013, Wan et al. 2013) (see section on PrimPol later in this chapter). This repriming mechanism of post-replicative synthesis explains why UV-induced lesions only cause a slight reduction in fork speed even when polη is mutated in human cells (Elvers et al. 2011). The ability of DDT to function distal (5′) to a newly primed leading strand may provide the necessary time to select the optimal DNA damage bypass mechanism.

Translesion synthesis at stalled replication forks

This mode of DDT is primarily carried out by low-fidelity Y-family (REV1, polη, polι, polκ) and B-family (polζ) DNA polymerases (Yang and Woodgate 2007, Guo et al. 2009, Waters et al. 2009, Sale et al. 2012, Yang 2014). These TLS polymerases have been implicated in filling lesion-containing single-stranded DNA gaps left after replication to prevent the formation of double-stranded breaks (Daigaku et al. 2010, Karras and Jentsch 2010, Diamant et al. 2012). TLS polymerases can insert nucleotides across a range of DNA lesions; however, their accommodating active sites make TLS enzymes highly mutagenic with the rates of nucleotide misincorporation on undamaged DNA being 10^{-1}–10^{-4} orders of magnitude higher than 10^{-6}–10^{-8} for the replicative DNA polymerases polδ and polε (Kunkel 2004, McCulloch and Kunkel 2008). Human translesion DNA polymerases include all members of the Y-family of DNA polymerases (polη, polι, polκ, REV1), as well as the B-family DNA polymerase polζ (Sale et al. 2012, Makarova and Burgers 2015, Vaisman and Woodgate 2017). They also include some A- and X-family DNA polymerases, such as polν, polθ, polβ, polλ and polμ (**Figure 4.5**), but they will not be discussed in this review.

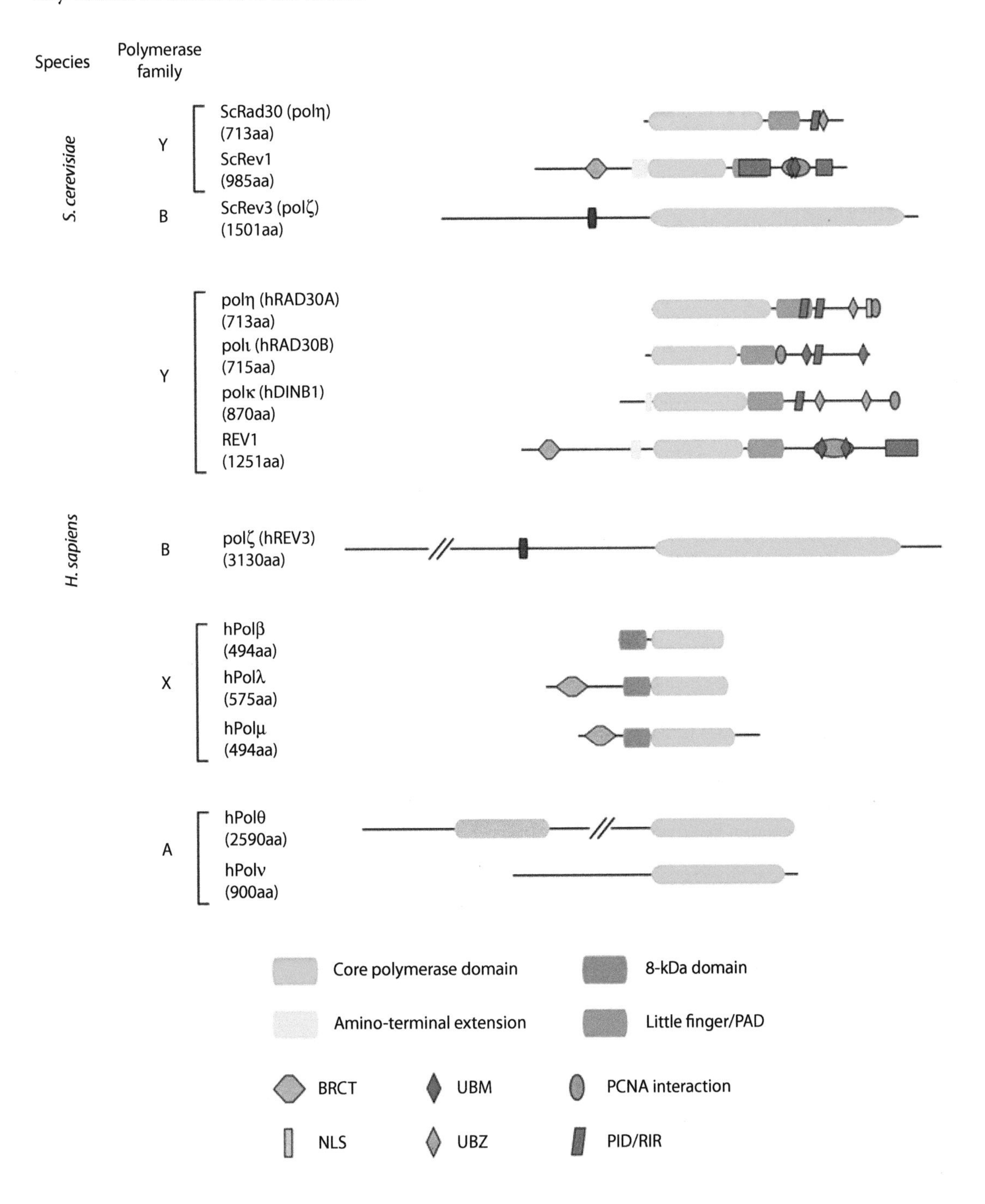

Figure 4.5 **Domain organization of TLS polymerases from *S. cerevisiae* and *Homo sapiens*.** The diagrams are to scale except from hREV3 and hPolθ which are truncated, as indicated by parallel diagonal lines. The REV7 subunit of polζ is not shown. PAD, polymerase-associated domain (also known as the little finger); BRCT, a domain with homology to the BRCA1 carboxyl terminus; UBM and UBZ, ubiquitin-binding domains; PID, polymerase-interacting domain (of REV1); RIR, REV1-interacting region (of other Y-family polymerases). (From Sale, J. E., 2013, *Cold Spring Harb Perspect Biol* 5(3): a012708.)

Extensive investigations during the first decade following the discovery of TLS DNA polymerases from the Y-family as well as the B-family polymerase ζ mainly focused on their biochemical properties (Vaisman et al. 2004, Yang and Woodgate 2007, Guo et al. 2009, Waters et al. 2009, Choi et al. 2010, Goodman and Woodgate 2013, Makarova and Burgers 2015). In general, translesion polymerases are characterized by low catalytic efficiency, fidelity and processivity. This set of properties is not surprising taking into account that the main task of TLS polymerases is to assist the replication fork in overcoming DNA impediments and to ensure continuation of DNA synthesis. The low catalytic activity and distributive mode of DNA synthesis of TLS polymerases prevents these highly inaccurate polymerases from replicating long stretches of DNA and generating a large number of mutations. The low fidelity of these polymerases stems not only from the intrinsic reduced capacity to discriminate between right and wrong nucleotides, but also from the lack of 3′-5′ exonucleolytic proofreading that allows these enzymes to avoid an additional kinetic barrier to translesion synthesis (Khare and Eckert 2002). Yet low fidelity can actually be beneficial for living organisms, such as in the case of somatic hypermutation, a mechanism by which the immune system adapts to foreign antigens. During this process, targeted mutations in the variable regions of immunoglobulin genes must be generated and in some cases this involves DNA synthesis by the error-prone TLS DNA polymerases. While one way or another, all TLS polymerases have been considered as potential suspects responsible for the mutagenesis of antibody genes, only Rev1 and polη have been unequivocally proven to play a central role in somatic hypermutation (Yavuz et al. 2002, Jansen et al. 2006, Krijger et al. 2013, Zhao et al. 2013).

Two modes of TLS: Direct or two-step bypass

One way through which DNA lesion bypass can be effected is as a single step by a single TLS polymerase, for example, the bypass of cyclobutane pyrimidine dimers (CPDs) by polymerase η. However, many in vitro and structural studies have highlighted constraints that polymerases face in solving both problems of incorporation opposite a lesion and extension from the resulting mismatched terminus. Certain polymerases seem to be particularly good at the insertion step, whereas others are more proficient at extension. These findings led to the proposal that lesion bypass is often a two-step process (Johnson et al. 2000). Indeed, the two-step bypass seems to be a central mechanism of TLS for several important lesions (Shachar et al. 2009).

Apart from the structural features of individual TLS polymerases, successful TLS also depends on protein–protein interactions between these polymerases and other cellular proteins that target and regulate their activity. Rev1 functions as a principal scaffolding protein, which recruits other TLS polymerases to first insert a nucleotide opposite the DNA lesion and then eventually help extend the distorted primer-template terminus in what is recognized as the two-step mechanism of TLS (Yang and Woodgate 2007, Waters et al. 2009, Sale et al. 2012). For the insertion step, a particular interface of the Rev1 CTD interacts with the REV1-interacting region (RIR) of the inserter polymerases (polη, polι, polκ). Mutations that disrupt the RIR interface in the Rev1 CTD prevent interaction with the inserter polymerase in yeast-2 hybrid (Y2H) screens (Pozhidaeva et al. 2012, Wojtaszek et al. 2012a, 2012b, Pustovalova et al. 2016). Insertion across from the damaged base can also be less frequently carried out by REV1 and polζ (Waters et al. 2009). In the second step, an extender TLS enzyme, a role most frequently fulfilled by polζ (Rev3-Rev7-POLD2-POLD3) and to a lesser extent by polκ, replaces the inserter and extends the primer-template termini (Pustovalova et al. 2016). For the polζ-mediated extension step, a different interface in REV1 CTD makes contact with specific amino acids located on Rev7. Mutating residues in the Rev7-interface of the REV1 CTD inhibit REV1-REV7 interaction in Y2H studies and sensitise chicken DT40 cells to cisplatin (Wojtaszek et al. 2012a). In addition to bypassing DNA damage at stalled replication forks, TLS polymerases are also involved in filling single-stranded DNA (ssDNA) gaps left behind by replicative polymerases via the less well understood gap-filling mechanism (Sale et al. 2009).

Y-family DNA polymerase modifications (reviewed in McIntyre and Woodgate 2015)

Whilst many of the characterized DNA polymerases have the ability to carry out TLS past certain DNA lesions, the best characterised are the four Y-family DNA polymerases (pols), polη, polι, polκ and REV1 and the B-family DNA polymerase polζ (Sale et al. 2012). Each of these polymerases can bypass certain DNA lesions with differing efficiency and accuracy. Their catalytic domain generally is within the N-terminus, while protein-protein interaction domains and protein-binding motifs are at the C-terminus. Polymerases η, ι and κ possess non-canonical PCNA-binding motifs (PIP box) and a REV1 interacting region (Waters et al. 2009). REV1 interacts with PCNA via a BRCT domain localized at its N-terminus, and the C-terminus of REV1 interacts with polymerases η, ι and κ. All Y-family DNA polymerases have ubiquitin binding domains that bind non-covalently to ubiquitin or to ubiquitinated proteins. Polymerases ι and REV1 possess two UBMs (ubiquitin binding motif), while polymerases η and κ contain UBZs–UBDs that additionally bind a zinc atom (polη has one UBZ, whereas polκ has two UBZs) (**Figure 4.5**) (Bienko et al. 2005, Waters et al. 2009, Bomar et al. 2010).

Besides possessing UBDs that facilitate the interaction with monoubiquitinated PCNA, human polη, polι, mouse polκ and REV1 as well as yeast and nematode polη have been shown to be targets of ubiquitination themselves (Bienko et al. 2005, Guo et al. 2006b, Parker et al. 2007, Skoneczna et al. 2007, Guo et al. 2008, Kim and Michael 2008, Pabla et al. 2008, Bienko et al. 2010). In general, most of the proteins that can non-covalently bind ubiquitin via different types of UBDs are themselves targets of monoubiquitination in a process called coupled monoubiquitination (Hoeller et al. 2007). In this process, ubiquitin attached to an E2 ubiquitin conjugating enzyme or E3 ubiquitin ligase is recruited to the UBD-containing substrate which becomes ubiquitinated in an E3-dependent or independent mode (Haglund and Stenmark 2006).

Despite the lack of primary amino acid sequence homology between DNA polymerases from different families, they all share a common structural topology of a catalytic core often described as a right hand with finger, palm and thumb subdomains holding DNA and positioning the incoming dNTP for incorporation (**Figure 4.6**).

The active centre of the polymerases reside in the palm domain where the essential metal ions required for the nucleotidyl transferase reaction are coordinated. The thumb subdomain plays an important role in binding to DNA and determining the processivity and translocation of the enzyme. The finger domain is largely responsible for the positioning of the DNA template required for optimal nucleotide pairing. Apart from the conserved polymerase domains, DNA polymerases often have additional domains that have evolved to fulfil specific functions. For example, most replicative polymerases possess a 3′-5′ exonuclease domain that proofreads newly synthesised DNA and corrects mismatched base pairs. In contrast, the majority of specialised polymerases lack a 3′-5′ exonuclease domain, with the exception of polζ that contains sequences corresponding to a 3′-5′ exonuclease domain (due to the fact that the catalytic core of polζ is homologous to the B-family replicative polymerases) even though this domain in polζ is non-functional. All REV proteins (REV1, REV3 and REV7) have a mitochondrial localisation domain located at their N-terminus which allows them to assist polγ in the replication of a damaged mitochondrial genome (Zhang et al. 2006). The domain that separates Y-family enzymes from other polymerases is at the C-terminus and is connected to the catalytic core by a flexible linker. This domain mediates the contact of the polymerase with DNA and has been referred to as the 'little finger' (LF) (Ling et al. 2001), 'polymerase-associated domain' (PAD) (Trincao et al. 2001) or 'wrist' (Silvian et al. 2001). Out of all Y-family polymerase domains (palm, thumb and finger) that are highly conserved, the LF domain is unique for each family member and plays a significant role in determining both the biological role and biochemical properties of the enzyme (Boudsocq et al. 2004, Wilson et al. 2013). The general structural organization separating this group from

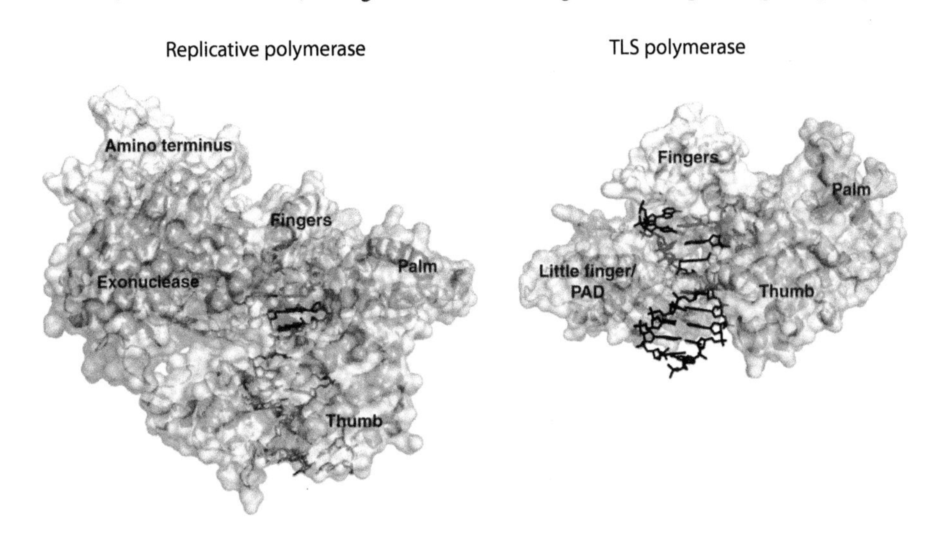

Figure 4.6 **Comparison of the anatomy of a replicative polymerase and a TLS polymerase.** On the left, *S. cerevisiae* polymerase δ, PDB 3IAY (Swan et al. 2009a, 2009b), a replicative B-family polymerase. The domains are shaded: palm, pink; thumb, green; fingers, cyan; exonuclease, purple. The DNA is in black, and the active site in the palm and incoming nucleotide triphosphate is in red. On the right side, *H. sapiens* polymerase η, PDB 3SI8 (Biertumpfel et al. 2010), a Y-family TLS polymerase with a T-T CPD in the +2 position. The domains in common with polδ are shaded the same. The little-finger domain/PAD is shaded in light brown. The DNA is in black, except the CPD, which is pink. The active site and incoming nucleotide triphosphate pairing with the first base after the CPD are in red. The β-strand splint in the little finger/PAD domain that constrains the CPD is highlighted in yellow. (From Sale, J. E., 2013, *Cold Spring Harb Perspect Biol* 5(3): a012708.)

polymerases belonging to other families is that the Y-family polymerases possess a flexible and accessible active site pocket capable of tolerating a variety of damaged template bases. Due to the relaxed constraints of their active sites, Y-family polymerases were expected to be less discriminatory with regard to sugar selection. This is not the case, however, as most DNA polymerases, including those from the Y-family, utilize the same major mechanism of sugar discrimination which relies on a specific amino acid within the nucleotide-binding pocket in the polymerase-active site.

Polη modifications

Polη is possibly the best-characterized Y-family DNA polymerase and is mainly known for efficient replication past cyclobutane pyrimidine dimers, which are the main DNA lesions induced after UV irradiation. As a consequence, a dysfunction in human polη results in the variant form of xeroderma pigmentosum, which is characterized by sunlight sensitivity and a high incidence of skin cancer (Johnson et al. 1999, Masutani et al. 1999b).

It has previously been shown that human polη can be ubiquitinated in vivo in its nuclear localization signal (NLS) motif. K682 was identified as the main ubiquitination site; however, when this residue is unavailable, three other close by lysines (K686, K694 or K709) can serve as a target (Bienko et al. 2005, 2010). Pirh2, an E3 ligase, was discovered to interact with human polη and monoubiquitinate it at one of the four lysine residues at the C-terminus (Jung et al. 2010, 2011). Attaching a ubiquitin moiety to the C-terminus of polη prevents its interaction with PCNA, inhibits its ability to bypass UV-induced lesions, thereby causing an increased sensitivity to UV radiation (Bienko et al. 2010, Jung et al. 2011). Therefore, monoubiquitinated polη needs to be actively de-ubiquitinated prior to interacting with PCNA and its recruitment to a stalled replication fork (Bienko et al. 2010). Additionally, polη is a subject of polyubiquitination by another E3 ligase, Mdm2, that targets polη for proteasomal degradation and controls its stability in response to UV-induced DNA damage (Jung et al. 2012). Wallace et al. (2014) showed that human polη can also be polyubiquitinated by a RING E3 ligase, TRIP (TRAF-interacting protein) (tumour necrosis factor receptor [TNFR]-associated factor) and TRIP promote its localization in nuclear foci. The TRIP homolog in *Drosophila melanogaster*, NOPO, enhances ubiquitination of polη during insect embryogenesis (Wallace et al. 2014). Most probably, NOPO promotes non-proteolytic polyubiquitination, as its overexpression does not cause polη destabilization, and additionally, NOPO interacts with Bendless (Ben), the *Drosophila* homolog of Ubc13, suggesting the formation of K63-linked polyubiquitin chains (Merkle et al. 2009).

Interestingly, in response to DNA damaging agents (MMS and UV), nematode polη becomes SUMOylated by GEI-17 SUMO E3 ligase at K85 and K260, and protects it from degradation mediated by CRL4-CDT-2-dependent ubiquitination (Kim and Michael 2008, Roerink et al. 2012). The SUMOylated lysine residues are conserved in human polη (K86 and K261, respectively), and very recently, human polη has been reported to be SUMOylated at K163 by PIAS1 at replications forks during unchallenged DNA synthesis (Despras et al. 2016). This modification of polη is important for its role in regions of under-replicated DNA such as common fragile sites.

In addition to ubiquitination and SUMOylation, human polη can also be phosphorylated by both ATR kinase and protein kinase C (PKC) in response to UV radiation. Potential phosphorylation sites were reported at S587, T617 and S601 by two independent groups (Chen et al. 2008, Gohler et al. 2011). The phosphorylation of polη seems to be required for cell survival after UV radiation and provides a link between DNA damage-induced checkpoint control and translesion synthesis.

Polι modifications

The C-terminus of polι contains motifs characteristic of other Y-family polymerases and includes PCNA-interacting, Rev1-interacting and ubiquitin-interacting (UBM1 and UBM2) motifs/domains. To date, the only reported post-translational modification in human polι is ubiquitination (Bienko et al. 2005), and the cellular function of ubiquitinated polι is not fully understood. However, our previous studies implied that ubiquitination of either polη or polι is required for the two polymerases to physically interact (McIntyre et al. 2013).

Polκ modifications

Similar to other Y-family DNA polymerases, polκ possesses a catalytic domain at its N-terminus, while the C-terminal half of the protein contains domains involved in protein–protein interactions, including an RIR, two UBZ domains, and at the extreme C-terminus, a PIP box

(Okada et al. 2002, Sale et al. 2012). The UBZ domains mediate the enhanced interaction with PCNA, and some studies report that they are essential for UV-induced nuclear foci formation by polκ, but do not affect the protein's half-life (Guo et al. 2008). On the other hand, Okada et al. (2002) suggested that polκ can function independently of PCNA modification. Recent studies of Wit et al. (2015) concluded that dependent on the type of DNA damage, PCNA ubiquitination may, or may not, be required for polκ activation.

To date, several reports show evidence for post-translational modifications of polκ in eukaryotic cells. Guo et al. (2008) observed monoubiquitination of mouse polκ and confirmed that, similar to other Y-family polymerases, monoubiquitination depends on UBDs. Endogenously expressed murine polκ is a stable protein with a half-life estimated to be ~5.4 h (Guo et al. 2008). The turnover of exogenously expressed polκ is somewhat faster (3.7–4.2 h) but does not change in polκ UBZ mutants. The suggestion that human polκ might also be ubiquitinated comes from the publication of Wallace et al. (2014), which reports an interaction between the C-terminus of polκ and the RING E3 ligase, TRIP. On the other hand, the possibility of polκ E3-independent monoubiquitination has also been reported in vitro (Hoeller et al. 2007). The biological function of polκ modification is not yet known. However, by influencing polκ interactions with other proteins, it may regulate its presence in replication factories.

Regarding other types of polκ modification, Roerink et al. (2012) suggested that GEI-17, which is known to SUMOylate and consequently protect nematode polη from proteasomal degradation, most likely also acts on polκ, so it is entirely possible that polκ may also be SUMOylated in vivo.

REV1 modifications

REV1 is a unique enzyme among Y-family polymerases, as it is only able to incorporate dCMP nucleotides opposite undamaged or damaged template G and some DNA lesions, including an abasic site (Nelson et al. 1996a, 1996b; Lin et al. 1999, Nelson et al. 2000). Besides its distinctive enzymatic activity, REV1 also plays a non-catalytic role in TLS as a scaffolding protein that coordinates the other TLS polymerases. The extreme C-terminus of REV1 in higher eukaryotes is devoted to the interaction with polymerases η, ι, κ and ζ (Guo et al. 2003, Ohashi et al. 2004, Tissier et al. 2004, Ross et al. 2005, Ohashi et al. 2009, Pozhidaeva et al. 2012). Additionally, the two UBMs that are located close to the C-terminal domain permit interaction with ubiquitin, ubiquitinated PCNA or other ubiquitinated proteins (Wood et al. 2007, Bomar et al. 2010). Interestingly, REV1 lacks a well-conserved PIP box that is characteristic of the other Y-family polymerases, and the interaction with PCNA is, instead, through the N-terminal BRCT domain of REV1 (Guo et al. 2006a).

In addition to non-covalent interactions with ubiquitin and ubiquitinated proteins, at least two studies have shown that mouse and human REV1 can also be directly conjugated to ubiquitin (Guo et al. 2006b, Kim et al. 2012a). However, the mechanism and sites of the ubiquitination remain unknown. Kim et al. reported that ubiquitinated human REV1 can be recruited to nuclear foci by the Fanconi anaemia core complex, as it binds directly to the UBZ4 domain of FAAP20 protein (Kim et al. 2012a).

SUMO2 modification of human REV1 at K99 has also been reported (Hendriks et al. 2014). In another proteome-wide analysis, *Drosophila melanogaster* Rev1 was shown to be acetylated at K136 (Weinert et al. 2011). However, in human REV1, this residue is replaced by arginine (R149). These results suggest that REV1 modifications (most probably at the N-terminus), occur either rarely or under highly specific conditions not employed in global proteomic studies.

Biochemical properties and unique features of Y-family polymerases

Polη

The signature feature of the active site of polη is its particularly large size, allowing it to accommodate the linked thymine bases of a cyclobutane pyrimidine dimer and to stabilize them for the correct pairing with an incoming dA (Biertumpfel et al. 2010, Yang 2014). The fascinating feature of polη is that it is not only able to efficiently and accurately 'read' information encoded by the damaged Ts, but also ensures that the reading frame is properly maintained. This is possible because unlike polκ, there is no gap between the catalytic core and the LF domain. As a result, extensive interactions between these domains stabilize the entire polymerase region and act as a 'molecular splint' to keep the newly synthesised duplex in a stable B-form (Biertumpfel et al. 2010). Furthermore, polη processively inserts three more bases past the lesion, at which point it switches back to a distributive mode of DNA synthesis. Further extension is prevented by steric clashes displacing DNA from the enzyme and thus ensuring re-engagement of a high-fidelity DNA polymerase. Interestingly, for the accurate bypass of CPDs, polη relies on two uniquely conserved residues (R61 and Q38 in humans) located near the active site and synergistically contributing to the efficiency and fidelity of the enzyme (Biertumpfel et al. 2010, Su et al. 2015). However, these two residues also play a key role in the efficient misincorporation of dG opposite template T, especially when the template T is immediately preceded by an AT primer/template base pair, known as the WA motif (Acharya et al. 2008). Indeed, such a misincorporation pattern on undamaged DNA correlates with the well-known mutation signature of polη generated during somatic hypermutation of immunoglobulin genes.

Polι

DNA polι is a paralogue of polη (McDonald et al. 1999) and is thought to bypass a number of lesions in vivo, especially when polη is absent (e.g. in XPV cells), and due to its reduced accuracy in synthesising across photoproducts, polι-dependent TLS results in elevated mutagenesis (Wang et al. 2007, Ziv et al. 2009, Jansen et al. 2014). Extensive biochemical studies performed in vitro with the highly purified enzyme suggest that polι is able to bypass a wide range of DNA lesions (Tissier et al. 2000a, Frank and Woodgate 2007, Pence et al. 2009). Interestingly, when copying an undamaged template, its accuracy varies 10,000-fold depending on the template base copied (Johnson et al. 2000, Tissier et al. 2000b, Zhang et al. 2000). Its unusual preference of incorporating G opposite T (3–10-fold greater than the correct base, A) gives polι a distinctive signature (Zhang et al. 2000, Tissier et al. 2001) and is a result of the specific structure of its active site (Nair et al. 2005a, Kirouac and Ling 2011). Polι N-terminus contains two partially overlapping catalytic domains: one with DNA polymerase activity and one with dRP lyase activity (Bebenek et al. 2001, Prasad et al. 2003). Significant insight into the unusual base selectivity of polι that is characterised by a 10^5-fold difference in fidelity depending upon the template base has been gained from the crystal structure of its ternary complex with DNA substrates and an incoming nucleotide (Nair et al. 2004; Nair et al. 2006a, 2006b). Several large aliphatic residues in the active cleft of polι prevent the DNA template and dNTP from binding in its normal position and forming a canonical Watson–Crick base pair. Instead, the active site holds the template A fixed in the *syn* conformation limiting hydrogen-bonding opportunities with any incoming nucleotide other than dT in an anti-conformation and promoting Hoogsteen base pairing that results in the accurate replication of template A (Nair et al. 2004). However, the same active site configuration is also responsible for the extreme infidelity of polι. The unique base selectivity opposite template T is explained by the fact that it is always held by polι in an anti-conformation, irrespective of the identity of the incoming dNTP. Since an incoming dA adopts a *syn* conformation and exhibits reduced base stacking, polι gives preference to dG remaining in an anti-conformation that is stabilized via hydrogen bonding with glutamine (Q59) in the finger subdomain (Nair et al. 2006a, 2006b). At the same time, the narrow groove width imposition of polι on the nascent base pair makes it more accurate than most polymerases when copying DNA containing the oxidative lesion 8-oxoG (Kirouac and Ling 2011). In this case, polι holds the damaged base in the *syn* conformation that would normally promote pairing with the incorrect dA, but its restrictive active site prevents this from occurring and instead results in stable pairing with the smaller and correct dC (Kirouac and Ling 2011).

Polκ

Polκ is able to bypass multiple types of DNA lesions, including abasic sites and bulky adducts, but with rather low efficiency (Ohashi et al. 2000), and due to a constricted active site, cannot incorporate a base opposite a pyrimidine dimer (Lone et al. 2007). When copying an undamaged template, polκ is quite accurate compared to other Y-family polymerases, but can also extend mispaired primer termini (Washington et al. 2002). Furthermore, after UV irradiation, the activity of polκ has been implicated in the gap-filling step of nucleotide excision repair (Ogi and Lehmann 2006). Polκ differs from other family members by an N-terminal extension (N-clasp) involved in DNA binding. A large structural gap divides the catalytic core and the 'little finger' in polκ because of limited interactions between the domains, even in ternary complexes with DNA and dNTPs. The N-clasp stabilizes the polymerase by holding the catalytic core and LF together to encircle DNA (Uljon et al. 2004, Lone et al. 2007, Yang 2014). It has been hypothesised that these features of polκ make it specifically suitable for the extension of a diverse range of primer termini distorted by base mismatches, or lesions in the template, even though in vivo findings suggest that the main 'extender' in cells is polζ. Structural studies also suggested that polκ has a more restrictive active-site cleft compared to other Y-family polymerases and is only able to accommodate only a single Watson–Crick base pair (Lone et al. 2007). Such a configuration explains the relatively higher fidelity of polκ on undamaged DNA, the limited selection of lesions it can tolerate and the inability to bypass dinucleotide lesions (Bavoux et al. 2005). At the same time, however, the structural gap between the catalytic core and LF allows polκ to accommodate bulky polycyclic aromatic hydrocarbons during lesion bypass (Liu et al. 2014).

REV1

REV1 possesses an extended N-terminal region containing breast cancer-associated protein-1 carboxy-terminal (BRCT) and mitochondrial targeting signal (MTS) domains and a C-terminal protein-binding domain. Similar to polκ, the three-dimensional structure of REV1 has a large gap that divides the LF domain and the ring-shaped catalytic core encircling DNA (Yang 2014). This gap is filled by the N-terminal extension from the catalytic core (Nair et al. 2005a, Swan et al. 2009). Two large inserts in the finger and palm domains specific for the catalytic core of human REV1 are thought to play an important role in supporting the peculiar properties and unique functions of the protein in TLS (Swan et al. 2009). One insert (I1) has been proposed to serve as a platform for the protein–protein interactions needed for a specific non-catalytic role of human REV1 in TLS, while the second insert (I2) is important for the limiting selectivity of REV1 in the choice of nucleotide and template substrates in human cells.

Unlike other polymerases that depend on DNA to identify the nucleotide best suited for incorporation, REV1 itself dictates the identity of both the templating base it uses as a substrate (mostly dG and a limited number of N2-adducted dG) and the incoming nucleotide insertion of dC. To achieve this, REV1 swings the template base (dG) out of the helix, replacing it with an arginine residue (R324), which forms a hydrogen bond with the incoming dC instead of the direct pairing of the DNA template with the incoming dC. The displaced template dG is temporarily accommodated within the hydrophobic pocket in the little finger domain (Nair et al. 2005a, 2005b, 2008, Swan et al. 2009).

B-family polymerase ζ in TLS

Polζ (Rev3/Rev7)

REV3 was identified in a screen for reversionless mutants in *S. cerevisiae* as *REV1*. *REV7* was isolated by a similar strategy a few years later (Lawrence et al. 1985). Like *rev1* mutants, *rev3* and *rev7* mutants are severely defective for spontaneous mutagenesis, as well as for mutagenesis induced by a wide variety of DNA-damaging agents, and for mutations induced in various DNA repair and tolerance pathway mutant backgrounds (Lawrence 2004). *REV1*, *REV3* and *REV7* are considered to be in the same branch of the *RAD6* epistasis group based on phenotypic similarity and limited epistasis analysis (Lawrence 2004). Like REV1, DNA polζ plays a key role in most mutagenesis from yeast to humans (Gibbs et al. 1998, McNally et al. 2008) as well as in cisplatin resistance in human cancer cells (Lin et al. 2006a). Together, REV1 and DNA polζ are thought to mediate the vast majority of the mutagenic class of DNA damage tolerance in vivo.

The complex of catalytic REV3 and accessory REV7 subunits constitutes the minimal unit required for polζ function (Nelson et al. 1996a, 1996b). REV7 (Mad2B) is a 24 kDa HORMA (Hop1, Rev7 and Mad2) domain (Aravind and Koonin 1998, Cahill et al. 1999, Hara et al. 2010) that interacts with a short REV7-binding (Rev7BD) motif of REV3 (Nelson et al. 1996a, 1996b, Hara et al. 2010). REV7 undergoes significant conformational change upon REV3 binding, which locks REV7 in a closed conformation and creates a REV7 binding site for the REV1–CT domain (Hara et al. 2010). Recent discovery of a second REV7BD motif in human REV3 (Tomida et al. 2015) and observation that human REV7 undergoes dimerization in solution (Hara et al. 2009) suggests a possibility that the REV3-REV7 assembly may contain two copies of REV7 associated with one another. Beyond mediating interactions with REV7 via the two REV7BD motifs, (Nelson et al. 1996a, 1996b, Hara et al. 2010, Tomida et al. 2015) the functional role of the nearly 2300 residue long N-terminal region preceding the catalytic domain of human REV3 (Gibbs et al. 1998) remains elusive. Recently, several groups have shown that a four-subunit polζ (polζ4), which consists of REV3-REV7 and a complex of two additional polD2 (p50) and polD3 (p66) subunits (that are also accessory subunits of the replicative DNA polymerase polδ), represents a more efficient and processive form of polζ that exhibits polymerase activity an order of magnitude greater than the REV3-REV7 complex (Baranovskiy et al. 2012, Johnson et al. 2012, Makarova et al. 2012, Lee et al. 2014, Makarova and Burgers 2015). In addition to REV3-REV7 interactions, the four-subunit polζ complex is stabilised by interactions between polD2 and the C-terminal region of REV3, which contains an iron–sulphur 4Fe–4S cluster and a zinc-finger domain (Netz et al. 2011, Baranovskiy et al. 2012) and interactions between polD2 and the N-terminal domain of polD3 (Baranovskiy et al. 2008). The four-subunit polζ complex acquires PIP-box and RIR motifs in the C-terminal part of polD3 subunit that enhance polζ interactions with PCNA and REV1, helping displace an 'inserter' TLS enzyme with the 'extender' polζ upon REV1/polζ-dependent TLS (Pustovalova et al. 2016). Overall, these studies suggest that PPIs of the accessory subunits of polζ play important roles in both modulating the enzymatic activity of polζ and mediating polymerase switching events.

Polζ catalytic activity

Unlike most of the TLS polymerases, which are Y family DNA polymerases, REV3 is a member of the B family, which includes the highly accurate replicative DNA polymerases DNA pols δ, ε and α (Morrison et al. 1989, Lawrence 2004). In contrast to most other B family replicative polymerases, DNA polζ lacks functional 3′-to-5′ exonuclease activity (Lawrence 2004). Although it can bypass certain lesions such as a *cis-syn* TT dimer and perform both the insertion and extension steps opposite a thymine glycol lesion in an error-free manner (Nelson et al. 1996a, 1996b, Johnson et al. 2003), polζ appears to be particularly specialized to extend distorted base pairs, such as mismatches that might result from inaccurate base insertion by a TLS polymerase or a base pair involving a bulky DNA lesion (Lawrence 2004, Prakash et al. 2005). In combination with a relatively high error rate for base substitutions, the ability to extend mismatches is what allows polζ to contribute significantly to mutagenesis (Lawrence 2004, Zhong et al. 2006). Despite the lack of conserved PCNA interaction motifs, polζ exhibits increased lesion bypass activity in the presence of PCNA (Garg et al. 2005). However, stimulation of polζ activity is not observed with either monoubiquitinated PCNA or the alternative 9-1-1 processivity clamp (Garg and Burgers 2005, Northam et al. 2006). In vitro studies using damaged DNA templates suggested that among the large variety of DNA lesions that polζ is able to bypass, some are replicated accurately (Johnson et al. 2003, Lin et al. 2014a, 2014b), while TLS past others is highly error-prone (Lin et al. 2014a, 2014b). In agreement with in vitro data, genetic studies have demonstrated that yeast polζ is not only necessary for damage-induced nuclear and mitochondrial genome mutagenesis, it also promotes accurate bypass of some DNA lesions (Baynton et al. 1999, Bresson and Fuchs 2002, Zhang et al. 2006, Kalifa and Sia 2007). On all damaged templates tested, polζ was far less efficient at incorporating a nucleotide opposite the damaged bases than extending the resulting primer terminus, consistent with the idea that its primary role in lesion bypass is confined to the elongation step, while the incorporation step depends on other polymerases. However, polζ has the capacity to catalyse all steps of TLS by itself, acting as both the inserter and an extender (Stone et al. 2011). It should be noted that studies related to the fidelity and TLS activity of yeast and human polζ4 are very limited at the present time, but the available data suggest that in the presence of all accessory subunits, polζ4 catalyses TLS much more efficiently compared to the polζ2 complex (Makarova et al. 2012, Lee et al. 2014).

In humans, the low fidelity of polζ is manifested by chromosome instability and carcinogenesis caused by alterations in its expression (Diaz et al. 2003, Wittschieben et al. 2006, Gan et al. 2008, Lange et al. 2011). It was originally believed that polζ-dependent mutagenesis resulted primarily from the exceptional ability of the enzyme to extend mispairs made by other polymerases. Indeed, steady-state kinetic studies of yeast polζ using primed single-stranded oligonucleotide templates revealed about 100-fold higher discrimination against nucleotide misincorporation (10^3–10^4 fold) compared to mismatched primer extension (10–100 fold) (Johnson et al. 2000). However, subsequent studies using undamaged gapped plasmid DNA substrates have shown that yeast polζ is also able to directly generate its own mismatches. The mutational spectra determined in these studies correlate well with the in vivo findings. They indicate that polζ-dependent errors are restricted to specific sequence contexts and are often clustered in short patches, at a rate that is unprecedented in comparison with other polymerases. The average base substitution fidelity of yeast polζ in gap-filling reactions was significantly lower than that of related pols α, δ and ε (Zhong et al. 2006). However, the overall error rate of polζ was still lower than that of Y-family polymerases.

Polζ interactions

Although it is a very large protein, REV3 does not contain any known protein–protein interaction modules or other regulatory motifs. In *S. cerevisiae*, Rev3 interacts with the C-terminal 100 amino acids of Revl in vitro, and this interaction stimulates the ability of polζ to extend mismatches and bypass specific lesions (Guo et al. 2004, Acharya et al. 2006). However, the majority of the regulation of polζ activity appears to occur through the accessory factor of Rev7. In yeast, Rev7 binds to the 9-1-1 alternative DNA processivity clamp, which participates in DNA damage signalling and checkpoint, and this interaction may recruit DNA polζ to sites of DNA damage (Sabbioneda et al. 2005). Additionally, REV7 interacts with REV1 (Guo et al. 2003, Ohashi et al. 2004, Tissier et al. 2004), which seems likely to promote localization of DNA polζ to DNA lesions.

The physical and genetic interactions of Rev1 with DNA polζ are complex. Despite the fact that each of the three proteins interacts with the others (see earlier), a heterotrimer of Rev1, Rev3 and Rev7 does not appear to be formed between purified proteins, as binding of Rev1 to Rev7 inhibits interaction of purified Rev1 and Rev3 in vitro (Acharya et al. 2006). These findings indicate that the architecture of the Rev1-polζ complex is intricate and that several complexes may exist, possibly in a regulated manner. It is also possible that the post-translational modifications of Rev1 mentioned earlier may influence the nature of Rev1 interaction with DNA polζ in vivo.

Loss of polζ causes embryonic lethality in mice (Bemark et al. 2000, Wittschieben et al. 2000, Van Sloun et al. 2002, Collins et al. 2007), indicating that during rapid proliferation, mammalian cells require a function of polζ. The inability to study *rev3* mutant cell lines in mammalian systems has hampered understanding of polζ function. However, studies with the chicken DT40 line have provided insight into the role of polζ in vivo, in particular, the contribution of *REV1*, *REV3* and *REV7* to chromosomal rearrangements during recombination and interstrand cross-link repair (Okada et al. 2005, Shen et al. 2006). In *S. cerevisiae*, an organism in which *rev3* mutants are viable, Rev3 has also been shown to participate in homologous recombination by mediating the mutagenesis observed in the break-induced replication subpathway of homologous recombination (Rattray et al. 2002).

Despite being a relatively small protein, Rev7 participates in many protein–protein interactions apart from its interactions with Rev1 and Rev3. Many of these additional Rev7 interactions are with cell cycle proteins, indicating a potential link between TLS and regulation of cell growth. In higher eukaryotes, REV7 has been shown to interact with the specificity factors CDH1 and/or CDC20 of the anaphase-promoting complex/cyclosome, as well as the spindle checkpoint protein MAD2, both key regulators of mitotic progression (Chen and Fang 2001, Pfleger et al. 2001). Interaction with REV7 inhibits the ubiquitin ligase activity of the APC/C and prevents the onset of mitotic anaphase (Chen and Fang 2001, Pfleger et al. 2001).

PrimPol (reviewed in Boldinova et al. 2017)

Prim1 belongs to the archaea-eukaryotic primase (AEP) superfamily. Many members of this superfamily possess both primase and DNA polymerase activities and not only play an essential role in initiation of DNA replication, but also undertake a wide variety of cellular roles in DNA replication, damage tolerance and repair, in addition to primer synthesis (Guilliam et al. 2015b). In 2005, Iyer et al. *in silico* predicted the existence of a new hypothetical single subunit human primase encoded by the gene CCDC111 on chromosome 4q35.1. The protein encoded by CCDC111 belongs to the NCLDV-herpesvirus clade of the AEP primases. In 2013, three groups published research articles describing the new enzyme (Bianchi et al. 2013, Garcia-Gomez et al. 2013, Wan et al. 2013). Like other TLS polymerases, Primpol lacks a 3′-5′-exonuclease activity and exhibits low fidelity of DNA synthesis. In the presence of Mg^{2+} ions as a cofactor of DNA polymerization, PrimPol makes one error per 10^2–10^5 nucleotides on undamaged DNA templates (Zafar et al. 2014, Guilliam et al. 2015a), an error rate comparable to the fidelity of the error-prone Y-family, human polη, polι and polκ (Vaisman and Woodgate 2017). However, the error specificity of PrimPol is different from other human DNA polymerases. In particular, PrimPol has a preference to generate base insertions and deletions over base misincorporations (Guilliam et al. 2015a, 2015b, Martinez-Jimenez et al. 2015). The high rate of insertions/deletions potentially leads to frame-shift mutagenesis and has a deleterious effect on cells. PrimPol has low processivity of DNA synthesis, as it incorporates only a few nucleotides per binding event (Keen et al. 2014). The poor processivity can be explained by the lack of contacts with DNA and the low affinity of PrimPol to DNA (Zafar et al. 2014, Rechkoblit et al. 2016). It also preferentially incorporates non-complementary nucleotides opposite the templating bases C and G and efficiently extends from primers with terminal mismatched base pairs contributing to mutation fixation (Guilliam et al. 2015a, 2015b). Human PrimPol replicates through photoproducts, including the highly distorting (6–4) photoproduct lesion, but, in contrast with polη, PrimPol has a constrained active-site cleft with respect to the templating base (Rechkoblit et al. 2016).

Mammalian and avian cells deficient in PrimPol display sensitivity to UV irradiation (Bianchi et al. 2013, Mouron et al. 2013, Wan et al. 2013, Keen et al. 2014, Pilzecker et al. 2016), indicating that PrimPol is important for recovery from UV damage. Because the loss of PrimPol in human xeroderma pigmentosum variant (XPV) cells led to an increase in UV sensitivity, the contribution to tolerance of UV photoproducts by PrimPol is likely to involve a pathway that is independent of polη (Bianchi et al. 2013). Furthermore, chicken PrimPol$^{-/-}$ DT40 cells are also hypersensitive to cisplatin and methylmethane sulphonate; this effect is not epistatic to the polη- and polζ-dependent pathways (Kobayashi et al. 2016).

Domain organization and structure of primpol

In terms of its structure, PrimPol has an N-terminal AEP-like catalytic domain and a C-terminal zinc finger (ZnF) domain that forms contacts to the DNA template (Iyer et al. 2005, Garcia-Gomez et al. 2013, Keen et al. 2014). The conserved I, II, III-motifs in the AEP-like domain are required for both the DNA polymerase and primase activities (Bianchi et al. 2013, Garcia-Gomez et al. 2013). The catalytic core contains the N-helix and two modules called Module N (ModN) and Module C (ModC). The N-helix is connected to ModN via a long flexible linker (Rechkoblit et al. 2016). The ModC module harbours functions of both the finger and palm domains and contains key active site residues interacting with the nascent T-dATP base pair. In PrimPol, ModN together with ModC functions as the finger domain and these modules are in contact with the template DNA strand and the templating base (T). The N-helix interacts with the template strand and resembles the little-finger domain in Y-family polymerases but makes far fewer contacts in the major groove. Moreover, PrimPol does not have an analogue to the thumb domain to grip the template-primer. As a result, PrimPol demonstrates an almost complete lack of contacts to the DNA primer strand. This feature can play a key role in the primase activity of PrimPol, as the lack of contacts to the DNA primer strand eliminates the need for a pre-existing primer and leaves room for a dNTP at the initiation site during de novo DNA synthesis. The contacts of the N-helix with the template DNA likely play an important role in the DNA polymerase activity of PrimPol (Rechkoblit et al. 2016).

The C-terminal domain of PrimPol contains a conserved ZnF motif that is indispensable for de novo synthesis by the enzyme but not necessary for primer elongation (Wan et al. 2013, Keen et al. 2014). In particular, mutations of the residues C419 and H426 abolish the primase activity of PrimPol but retain its DNA polymerase activity (Mouron et al. 2013, Wan et al. 2013, Keen et al. 2014). Nevertheless, the deletion and mutations of the ZnF modulate the processivity and fidelity of DNA synthesis by PrimPol as well as abrogating template-independent dNTP incorporation by the enzyme (Keen et al. 2014). This is because the ZnF reduces the processivity of the enzyme and allows a slower, higher-fidelity incorporation of complementary nucleotides. The C-terminal domain is required for the binding to single-stranded DNA downstream of the primer-template junction (Keen et al. 2014) and likely is responsible for template recognition during repriming. Moreover, the C-terminal domain was shown to be involved in protein interactions with replication protein A (Wan et al. 2013, Guilliam et al. 2015a, 2015b).

Primase or TLS polymerase activity of primpol?

PrimPol is crucial for recovery of stalled replication forks in HeLa and DT40 cells after treatment with the dNTP depleting agent hydroxyurea and chain-terminating nucleoside analogues (Mouron et al. 2013, Wan et al. 2013, Kobayashi et al. 2016). In the absence of DNA damage, disruption of PrimPol function in mammalian cells slows down replication and induces replicative stress, thereby leading to the accumulation of DNA breaks and chromosome instability (Bianchi et al. 2013, Mouron et al. 2013). Therefore, PrimPol may play an important role in replication even in unperturbed cells. In fact, it was shown that PrimPol may facilitate the replication across non-B DNA (Schiavone et al. 2016). Two other studies, however, reported only modest or no effect of PrimPol defects on the replication speed during unperturbed replication in human cells (Wan et al. 2013, Pilzecker et al. 2016).

PrimPol may restart a stalled replication fork by acting as either a translesion DNA polymerase or by repriming DNA synthesis downstream of the lesion. Biochemical studies have indicated that PrimPol is capable of TLS synthesis in vitro. However, recent in vivo studies with a zinc-finger (ZnF) primase-null PrimPol mutant in human and avian DT40 cells suggested that the primary function of PrimPol in nuclear replication is repriming at sites of DNA damage and at stalled replication forks on the leading strand. It was shown that PrimPol reprimes efficiently downstream of UV-induced DNA lesions, AP-sites and cisplatin lesions (Mouron et al. 2013, Keen et al. 2014, Kobayashi et al. 2016, Pilzecker et al. 2016).

It is likely that the primase activity of PrimPol also plays a pivotal role in the re-initiation of DNA synthesis after dNTP depletion by hydroxyurea and chain termination with nucleoside analogues (Mouron et al. 2013, Kobayashi et al. 2016). It was also suggested that PrimPol contributes to replication across G-quadruplexes using close-coupled downstream repriming mechanisms on the leading DNA strand (Schiavone et al. 2016). Taken together, the most likely role of PrimPol is to initiate de novo DNA synthesis downstream of not only DNA lesions, but also non-B DNA structures during normal chromosomal duplication. The discovery of the repriming role of PrimPol in cells explains the previous reports on chromosomal single-stranded gaps formed on the leading strand in S phase during replication and gap filling during TLS in G2 phase of the cell cycle (Elvers et al. 2011, Diamant et al. 2012). The mechanism of repriming is especially important on the leading strand, which is replicated continuously. The remaining gaps can be filled by translesion DNA polymerases or restored by homology-directed repair. More information about repriming function of PrimPol in cells can be found in a recent review (Guilliam and Doherty 2017).

However, a TLS function of PrimPol in cells is also quite possible. In agreement with this, a study using avian PrimPol$^{-/-}$ DT40 cells expressing human PrimPol indicated that its primase activity is required to restore wild-type replication fork rates after UV irradiation, while its DNA polymerase activity is sufficient to maintain regular replisome progression in unperturbed cells (Keen et al. 2014).

TLS polymerases involved in other cellular functions

TLS polymerases have been implicated in various cellular processes. One example is during interstrand cross-link repair in replicating cells where REV1, polι, polκ and polν were shown to be required for DNA synthesis over the ICL on the newly exposed leading strand (Minko et al. 2008, Raschle et al. 2008, Ho et al. 2011, Klug et al. 2012). Similarly, in non-replicating cells, ICL repair depended on the REV1-polζ TLS polymerases to fill the ssDNA-gaps (Clauson et al. 2013). In a similar fashion, both nucleotide excision repair and base excision repair pathways, respectively, can employ polκ and polη to fill the ssDNA gaps left behind after the excising step (Krokan and Bjoras 2013, Scharer 2013). Additionally, polη was recently shown to drive microhomology-mediated break-induced replication (MMBIR) that causes complex genomic rearrangements in yeast and has an important role in homologous recombination in DT40 cells (Kawamoto et al. 2005, Sakofsky et al. 2015). Finally, REV1 was shown to be required for replication of G-quadruplex structures, thereby influencing epigenetic stability (Sarkies et al. 2010). Independent of its role in TLS, REV7 promotes nonhomologous end joining at double-strand breaks and at telomeres by inhibiting CtIP-mediated end resection (Boersma et al. 2015). Additionally, REV7 plays a supporting role in cell cycle regulation by sequestering CDH1, which prevents premature activation of the anaphase-promoting complex, thereby inhibiting an exit from mitosis (Chun et al. 2013). Taken together, all these studies demonstrate the important contribution of TLS polymerases in DNA damage tolerance, DNA repair and epigenetic stability.

PCNA ubiquitination-independent TLS

Some studies showed that the ubiquitination of PCNA was not essential for TLS by polη (Schmutz et al. 2010, Hendel et al. 2011, Krijger et al. 2011b). For example, PCNA ubiquitination was dispensable for polη-catalysed A/T mutagenesis during somatic hypermutation (SHM) (Krijger et al. 2011b). In addition, bypass of DNA alkylation damage is performed by polκ in the absence of PCNA ubiquitination. Experiments in the avian cell line DT40 revealed that replication fork progression in response to UV or 4-nitroquinoline (4-NQO) was unaffected in either *rad18*-/- or *pcna*K164-/- cells that are deficient in damage-induced PCNA ubiquitination (Edmunds et al. 2008). PCNA ubiquitination-independent TLS was also observed for Rev1-mediated bypass of G-quadruplex DNA in these cells (Sarkies et al. 2010, Sarkies et al. 2012). Bypass of UV-induced (6–4) photoproducts was primarily carried out by a PCNA ubiquitination–independent mechanism involving polζ and Rev1 (Szuts et al. 2008). The relevance of PCNA ubiquitination in the activation of polκ is less clear, as some reports showed that polκ critically needs its UBD for activation (Guo et al. 2008), while others have shown that polκ can participate in RAD18-independent TLS of UV lesions (Okada et al. 2002). In support of the latter, a more recent study indicated that polκ can function independently of PCNA modification during TLS of lesions induced by the monofunctional alkylating agent methyl methanesulphonate (MMS) (Wit et al. 2015).

TLS in cancer development and progression

TLS is essential for the completion of DNA synthesis and the maintenance of genomic stability (Waters et al. 2009, Lange et al. 2011, Sale et al. 2012). Defects in this pathway lead to genomic instability and predisposition to cancer (reviewed in Korzhnev and Hadden 2016). The biggest example is sunlight sensitive xeroderma pigmentosum-variant (XP-V) syndrome, where DNA polη function is compromised (Johnson et al. 1999, Masutani et al. 1999a, 1999b). In addition, conditional loss of Rev3 in mice leads to enhanced spontaneous tumorigenesis (Wittschieben et al. 2010), and Rev3-deficient cells display increased frequency of chromosomal aberrations (Sonoda et al. 2003, Wittschieben et al. 2006, Schenten et al. 2009). However, due to its intrinsic error propensity, TLS has been linked to both spontaneous and DNA damage-induced mutations that are mostly carried out in a Rev1/polζ-dependent manner (Lawrence 2002, Li et al. 2002, Lawrence 2004, Wu et al. 2004, Okuda et al. 2005, Hashimoto et al. 2012). Given its role in DNA adduct bypass and resolution of replication stress, it is not surprising that TLS polymerases have been identified as factors in the aetiology of cancer and may fuel carcinogenesis (Futreal et al. 2004, Velculescu 2008, Dumstorf et al. 2009). Furthermore, TLS has also been identified as a mechanism of conferring resistance of cancer cells to genotoxic chemotherapy (Wu et al. 2004, Nojima et al. 2005, Okuda et al. 2005, Lin and Howell 2006, Lin et al. 2006b, Xie et al. 2010). Thus, TLS contributes to the mutator phenotype of cancer cells that exhibit a much higher mutation frequency than that of normal cells. Treatment of cancers with genotoxic chemotherapeutic agents further aggravates the mutator phenotype and gives rise to an accumulation of random mutations, resulting in tumour heterogeneity. The high mutation frequency within the cancer cell population causes tumours to evolve, and cancer cell subpopulations with new mutations may acquire a growth advantage which includes the ability to resist genotoxic treatments. Selection of subpopulations of chemoresistant cancer cells that evolved from increased mutagenesis could be the basis of relapsed tumours (Nojima et al. 2005, Xie et al. 2010).

Many studies have examined the contribution of TLS in the resistance of cancer cells to chemotherapy (reviewed in Korzhnev and Hadden 2016). Others have focused on REV1 and the polζ subunits REV3 and REV7, and in particular on the mechanisms through which they contribute to cisplatin-induced mutagenesis and confer resistance in different cancers (Wu et al. 2004, Nojima et al. 2005, Okuda et al. 2005, Lin and Howell 2006, Lin et al. 2006b, Xie et al. 2010). Reduction in polymerase ζ function through reduced expression of REV3 mRNA in human fibroblasts rendered cells more sensitive to the cytotoxic effect of cisplatin and markedly decreased its mutagenicity. Importantly, it significantly reduced the rate at which cells acquired resistance to cisplatin (Wu et al. 2004). Studies in ovarian cancer cell lines demonstrated that depletion of REV1 with shRNA sensitised these cells to cisplatin and significantly reduced the rate at which the whole population of cells acquired resistance to treatment with this drug (Okuda et al. 2005). Furthermore, reduced expression of REV1 resulted in a decrease in the incidence of DNA mutations and acquired resistance to the chemotherapeutic agent cyclophosphamide in a murine model of Burkitt's lymphoma (Xie et al. 2010). Knockdown of either Rev3 or Rev1 (via shRNA) sensitised B-cell lymphomas to cisplatin treatment in vitro and in vivo (Xie et al. 2010). Rev3-deficient non-small cell lung cancer (NSCLC) transplants treated with cisplatin exhibited pronounced sensitivity to the treatment, leading to a significant extension in the overall survival of treated recipient mice (Doles et al. 2010). Finally, nanoparticle technology to co-deliver cisplatin and a specific siRNA against Rev1/Rev3 encapsulated together completely abolished tumour growth in a murine xenograft model of prostate cancer (Xu et al. 2013). In addition, REV7 depletion has been shown to sensitise ovarian cancer to cisplatin and reduce tumour volumes in nude mice (Niimi et al. 2014). These studies support the hypothesis that TLS inhibition can suppress at least some classes of chemoresistance. Similarly, depletion of: REV3 in cervical cancer cells (Yang et al. 2015); REV1, polζ, polη in HeLa cells (Bartz et al. 2006); and polη in ovarian cancer stem cells (Srivastava et al. 2015) all sensitise cells to cisplatin. Taken together, these data highlight the potential therapeutic effect associated with disruption of TLS in cancer cells and provide important context for developing small molecules that target this process.

Small molecule inhibitors of TLS pols

Researchers at the National Institutes of Health have reported the development and implementation of a novel, broadly applicable assay to identify inhibitors of DNA polymerase enzymatic activity in real time (Dorjsuren et al. 2009). This assay is a primer extension assay based on a tripartite synthetic nucleotide containing the following: (a) a template strand labelled at the 5′-end with a BHQ-2 quencher, (b) a 3′-end primer strand complementary to the template and (c) a reporter strand complementary to the 5′-end of the template labelled at the 3′-end with a rhodamine dye. Incorporation of a nucleotide by an active DNA polymerase at the 3′-end of the primer sequence displaces the fluorescent reporter, which reduces its proximity to the quencher and results in a fluorescent signal. The assay was miniaturised and validated in a 1536-well format, and several small molecules known to inhibit DNA polymerase β or other DNA-processing enzymes were evaluated for their activity against polι and polη. Aurintricarboxylic acid and ellagic acid were identified as potent, nonselective inhibitors of both TLS DNA polymerases. Finally, a small molecule inhibitor that binds to REV7 and inhibits its interaction with REV3 was shown to partially suppress ICL repair (Actis et al. 2016). Whether the same drug could also suppress TLS is worth investigating. Inhibiting TLS polymerases is a promising approach to improve chemotherapy, as it could increase killing of cancer cells, while at the same time reducing the possibility of relapse and acquired drug resistance by reducing chemotherapy-induced mutagenesis. Even cancers known to be intrinsically drug resistant could potentially be sensitised by this approach. Also, a better understanding of synthetic lethal partners of TLS polymerases would provide insights into which tumours might be most susceptible to chemotherapy treatments involving small molecule inhibitors of TLS polymerases. Finally, the effectiveness of small molecule inhibitors of TLS polymerases could be further improved by delivery systems that could target these drugs to specific tumours in cancer patients. Because protein–protein interactions are so important for TLS, drug targets for these interaction interfaces could be promising candidates for cancer therapeutics.

CONCLUSIONS

DDT mechanisms are necessary for the prevention of fork breakage and for DNA replication completion, by promoting damage bypass DNA synthesis either at the fork or post-replicatively. The temporal coordination of DDT pathways that ensure preferential usage early during replication, to error-free post-replicative template switching is still not well understood, but likely involves regulation of replication factors by post-translational modifications and fork stabilization mechanisms triggered by local chromatin and genomic features that block replication fork progression. Understanding how these DDT pathways are deployed and interrelate to each other is crucial for our knowledge of genome integrity mechanisms and replication stress responses. Given that cancer cells treated with chemotherapeutic agents crucially depend on DDT for their proliferation, understanding DDT mechanisms may inform therapeutic approaches by providing targeted inhibition. While many challenging questions remain, improved tools, approaches and working models hold promise to bring insight into the mechanism and intermediate products through which DDT affects genome integrity.

REFERENCES

Achar, Y. J., Balogh, D., Neculai, D., Juhasz, S., Morocz, M., Gali, H., Dhe-Paganon, S., Venclovas, C., Haracska, L. 2015. Human HLTF mediates postreplication repair by its HIRAN domain-dependent replication fork remodelling. *Nucleic Acids Res* 43(21): 10277–10291.

Acharya, N., Johnson, R. E., Prakash, S., Prakash, L. 2006. Complex formation with Rev1 enhances the proficiency of *Saccharomyces cerevisiae* DNA polymerase zeta for mismatch extension and for extension opposite from DNA lesions. *Mol Cell Biol* 26(24): 9555–9563.

Acharya, N., Yoon, J. H., Gali, H., Unk, I., Haracska, L., Johnson, R. E., Hurwitz, J., Prakash, L., Prakash, S. 2008. Roles of PCNA-binding and ubiquitin-binding domains in human DNA polymerase eta in translesion DNA synthesis. *Proc Natl Acad Sci USA* 105(46): 17724–17729.

Actis, M. L., Ambaye, N. D., Evison, B. J., Shao, Y., Vanarotti, M., Inoue, A., McDonald, E. T. et al. 2016. Identification of the first small-molecule inhibitor of the REV7 DNA repair protein interaction. *Bioorg Med Chem* 24(18): 4339–4346.

Arakawa, H., Moldovan, G. L., Saribasak, H., Saribasak, N. N., Jentsch, S., Buerstedde, J. M. 2006. A role for PCNA ubiquitination in immunoglobulin hypermutation. *PLoS Biol* 4(11): e366.

Aravind, L., Koonin, E. V. 1998. The HORMA domain: A common structural denominator in mitotic checkpoints, chromosome synapsis and DNA repair. *Trends Biochem Sci* 23(8): 284–286.

Baranovskiy, A. G., Babayeva, N. D., Liston, V. G., Rogozin, I. B., Koonin, E. V., Pavlov, Y. I., Vassylyev, D. G., Tahirov, T. H. 2008. X-ray structure of the complex of regulatory subunits of human DNA polymerase delta. *Cell Cycle* 7(19): 3026–3036.

Baranovskiy, A. G., Lada, A. G., Siebler, H. M., Zhang, Y., Pavlov, Y. I., Tahirov, T. H. 2012. DNA polymerase delta and zeta switch by sharing accessory subunits of DNA polymerase delta. *J Biol Chem* 287(21): 17281–17287.

Bartz, S. R., Zhang, Z., Burchard, J., Imakura, M., Martin, M., Palmieri, A., Needham, R. et al. 2006. Small interfering RNA screens reveal enhanced cisplatin cytotoxicity in tumor cells having both BRCA network and TP53 disruptions. *Mol Cell Biol* 26(24): 9377–9386.

Bavoux, C., Hoffmann, J. S., Cazaux, C. 2005. Adaptation to DNA damage and stimulation of genetic instability: The double-edged sword mammalian DNA polymerase kappa. *Biochimie* 87(7): 637–646.

Baynton, K., Bresson-Roy, A., Fuchs, R. P. 1999. Distinct roles for Rev1p and Rev7p during translesion synthesis in *Saccharomyces cerevisiae. Mol Microbiol* 34(1): 124–133.

Bebenek, K., Tissier, A., Frank, E. G., McDonald, J. P., Prasad, R., Wilson, S. H., Woodgate, R., Kunkel, T. A. 2001. 5′-Deoxyribose phosphate lyase activity of human DNA polymerase iota *in vitro*. *Science* 291(5511): 2156–2159.

Bemark, M., Khamlichi, A. A., Davies, S. L., Neuberger, M. S. 2000. Disruption of mouse polymerase zeta (Rev3) leads to embryonic lethality and impairs blastocyst development *in vitro*. *Curr Biol* 10(19): 1213–1216.

Betous, R., Mason, A. C., Rambo, R. P., Bansbach, C. E., Badu-Nkansah, A., Sirbu, B. M., Eichman, B. F., Cortez, D. 2012. SMARCAL1 catalyzes fork regression and Holliday junction migration to maintain genome stability during DNA replication. *Genes Dev* 26(2): 151–162.

Bianchi, J., Rudd, S. G., Jozwiakowski, S. K., Bailey, L. J., Soura, V., Taylor, E., Stevanovic, I. et al. 2013. PrimPol bypasses UV photoproducts during eukaryotic chromosomal DNA replication. *Mol Cell* 52(4): 566–573.

Bienko, M., Green, C. M., Crosetto, N., Rudolf, F., Zapart, G., Coull, B., Kannouche, P. et al. 2005. Ubiquitin-binding domains in Y-family polymerases regulate translesion synthesis. *Science* 310(5755): 1821–1824.

Bienko, M., Green, C. M., Sabbioneda, S., Crosetto, N., Matic, I., Hibbert, R. G., Begovic, T., Niimi, A., Mann, M., Lehmann, A. R., Dikic, I. 2010. Regulation of translesion synthesis DNA polymerase eta by monoubiquitination. *Mol Cell* 37(3): 396–407.

Biertumpfel, C., Zhao, Y., Kondo, Y., Ramon-Maiques, S., Gregory, M., Lee, J. Y., Masutani, C., Lehmann, A. R., Hanaoka F., Yang, W. 2010. Structure and mechanism of human DNA polymerase eta. *Nature* 465(7301): 1044–1048.

Blastyak, A., Pinter, L., Unk, I., Prakash, L., Prakash, S., Haracska, L. 2007. Yeast Rad5 protein required for postreplication repair has a DNA helicase activity specific for replication fork regression. *Mol Cell* 28(1): 167–175.

Boersma, V., Moatti, N., Segura-Bayona, S., Peuscher, M. H., van der Torre, J., Wevers, B. A., Orthwein, A., Durocher, D., Jacobs, J. J. L. 2015. MAD2L2 controls DNA repair at telomeres and DNA breaks by inhibiting 5′ end resection. *Nature* 521(7553): 537–540.

Boldinova, E. O., Wanrooij, P. H., Shilkin, E. S., Wanrooij, S., Makarova, A. V. 2017. DNA damage tolerance by eukaryotic DNA polymerase and primase PrimPol. *Int J Mol Sci* 18(7): E1584.

Bomar, M. G., D'Souza, S., Bienko, M., Dikic, I., Walker, G. C., Zhou, P. 2010. Unconventional ubiquitin recognition by the ubiquitin-binding motif within the Y family DNA polymerases iota and Rev1. *Mol Cell* 37(3): 408–417.

Boudsocq, F., Kokoska, R. J., Plosky, B. S., Vaisman, A., Ling, H., Kunkel, T. A., Yang, W., Woodgate, R. 2004. Investigating the role of the little finger domain of Y-family DNA polymerases in low fidelity synthesis and translesion replication. *J Biol Chem* 279(31): 32932–32940.

Branzei, D. 2011. Ubiquitin family modifications and template switching. *FEBS Lett* 585(18): 2810–2817.

Branzei, D., Vanoli, F., Foiani, M. 2008. SUMOylation regulates Rad18-mediated template switch. *Nature* 456(7224): 915–920.

Bresson, A., Fuchs, R. P. 2002. Lesion bypass in yeast cells: Pol eta participates in a multi-DNA polymerase process. *EMBO J* 21(14): 3881–3887.

Broomfield, S., Chow, B. L., Xiao, W. 1998. MMS2, encoding a ubiquitin-conjugating-enzyme-like protein, is a member of the yeast error-free postreplication repair pathway. *Proc Natl Acad Sci USA* 95(10): 5678–5683.

Brun, J., Chiu, R., Lockhart, K., Xiao, W., Wouters, B. G., Gray, D. A. 2008. hMMS2 serves a redundant role in human PCNA polyubiquitination. *BMC Mol Biol* 9: 24.

Brun, J., Chiu, R. K., Wouters, B. G., Gray, D. A. 2010. Regulation of PCNA polyubiquitination in human cells. *BMC Res Notes* 3: 85.

Bugreev, D. V., Rossi, M. J., Mazin, A. V. 2011. Cooperation of RAD51 and RAD54 in regression of a model replication fork. *Nucleic Acids Res* 39(6): 2153–2164.

Burgers, P. M. 2009. Polymerase dynamics at the eukaryotic DNA replication fork. *J Biol Chem* 284(7): 4041–4045.

Burkovics, P., Dome, L., Juhasz, S., Altmannova, V., Sebesta, M., Pacesa, M., Fugger, K. et al. 2016. The PCNA-associated protein PARI negatively regulates homologous recombination via the inhibition of DNA repair synthesis. *Nucleic Acids Res* 44(7): 3176–3189.

Burkovics, P., Sebesta, M., Balogh, D., Haracska, L., Krejci, L. 2014. Strand invasion by HLTF as a mechanism for template switch in fork rescue. *Nucleic Acids Res* 42(3): 1711–1720.

Cahill, D. P., da Costa, L. T., Carson-Walter, E. B., Kinzler, K. W., Vogelstein, B., Lengauer, C. 1999. Characterization of MAD2B and other mitotic spindle checkpoint genes. *Genomics* 58(2): 181–187.

Chen, J., Fang, G. 2001. MAD2B is an inhibitor of the anaphase-promoting complex. *Genes Dev* 15(14): 1765–1770.

Chen, Y. W., Cleaver, J. E., Hatahet, Z., Honkanen, R. E., Chang, J. Y., Yen, Y., Chou, K. M. 2008. Human DNA polymerase eta activity and translocation is regulated by phosphorylation. *Proc Natl Acad Sci USA* 105(43): 16578–16583.

Chiu, R. K., Brun, J., Ramaekers, C., Theys, J., Weng, L., Lambin, P., Gray, D. A., Wouters, B. G. 2006. Lysine 63-polyubiquitination guards against translesion synthesis-induced mutations. *PLoS Genet* 2(7): e116.

Choi, J. Y., Lim, S., Kim, E. J., Jo, A., Guengerich, F. P. 2010. Translesion synthesis across abasic lesions by human B-family and Y-family DNA polymerases alpha, delta, eta, iota, kappa, and REV1. *J Mol Biol* 404(1): 34–44.

Chun, A. C., Kok, K. H., Jin, D. Y. 2013. REV7 is required for anaphase-promoting complex-dependent ubiquitination and degradation of translesion DNA polymerase REV1. *Cell Cycle* 12(2): 365–378.

Ciccia, A., Nimonkar, A. V., Hu, Y., Hajdu, I., Achar, Y. J., Izhar, L., Petit, S. A. et al. 2012. Polyubiquitinated PCNA recruits the ZRANB3 translocase to maintain genomic integrity after replication stress. *Mol Cell* 47(3): 396–409.

Clauson, C., Scharer, O. D., Niedernhofer, L. 2013. Advances in understanding the complex mechanisms of DNA interstrand cross-link repair. *Cold Spring Harb Perspect Biol* 5(10): a012732.

Collins, N. S., Bhattacharyya, S., Lahue, R. S. 2007. Rev1 enhances CAG. CTG repeat stability in *Saccharomyces cerevisiae*. *DNA Repair (Amst)* 6(1): 38–44.

Couch, F. B., Bansbach, C. E., Driscoll, R., Luzwick, J. W., Glick, G. G., Betous, R., Carroll, C. M. et al. 2013. ATR phosphorylates SMARCAL1 to prevent replication fork collapse. *Genes Dev* 27(14): 1610–1623.

Daigaku, Y., Davies, A. A., Ulrich, H. D. 2010. Ubiquitin-dependent DNA damage bypass is separable from genome replication. *Nature* 465(7300): 951–955.

Das-Bradoo, S., Nguyen, H. D., Wood, J. L., Ricke, R. M., Haworth, J. C., Bielinsky, A. K. 2010. Defects in DNA ligase I trigger PCNA ubiquitylation at Lys 107. *Nat Cell Biol* 12(1): 74–79; sup pp 71–20.

Despras, E., Sittewelle, M., Pouvelle, C., Delrieu, N., Cordonnier, A. M., Kannouche, P. L. 2016. Rad18-dependent SUMOylation of human specialized DNA polymerase eta is required to prevent under-replicated DNA. *Nat Commun* 7: 13326.

Diamant, N., Hendel, A., Vered, I., Carell, T., Reissner, T., de Wind, N., Geacinov, N., Livneh, Z. 2012. DNA damage bypass operates in the S and G2 phases of the cell cycle and exhibits differential mutagenicity. *Nucleic Acids Res* 40(1): 170–180.

Diaz, M., Watson, N. B., Turkington, G., Verkoczy, L. K., Klinman, N. R., McGregor, W. G. 2003. Decreased frequency and highly aberrant spectrum of ultraviolet-induced mutations in the hprt gene of mouse fibroblasts expressing antisense RNA to DNA polymerase zeta. *Mol Cancer Res* 1(11): 836–847.

Dieckman, L. M., Freudenthal, B. D., Washington, M. T. 2012. PCNA structure and function: Insights from structures of PCNA complexes and post-translationally modified PCNA. *Subcell Biochem* 62: 281–299.

Doles, J., Oliver, T. G., Cameron, E. R., Hsu, G., Jacks, T., Walker, G. C., Hemann, M. T. 2010. Suppression of Rev3, the catalytic subunit of Pol{zeta}, sensitizes drug-resistant lung tumors to chemotherapy. *Proc Natl Acad Sci USA* 107(48): 20786–20791.

Dorjsuren, D., Wilson, D. M. 3rd, Beard, W. A., McDonald, J. P., Austin, C. P., Woodgate, R., Wilson, S. H., Simeonov, A. 2009. A real-time fluorescence method for enzymatic characterization of specialized human DNA polymerases. *Nucleic Acids Res* 37(19): e128.

Dumstorf, C. A., Mukhopadhyay, S., Krishnan, E., Haribabu, B., McGregor, W. G. 2009. REV1 is implicated in the development of carcinogen-induced lung cancer. *Mol Cancer Res* 7(2): 247–254.

Edmunds, C. E., Simpson, L. J., Sale, J. E. 2008. PCNA ubiquitination and REV1 define temporally distinct mechanisms for controlling translesion synthesis in the avian cell line DT40. *Mol Cell* 30(4): 519–529.

Elvers, I., Johansson, F., Groth, P., Erixon, K., Helleday, T. 2011. UV stalled replication forks restart by re-priming in human fibroblasts. *Nucleic Acids Res* 39(16): 7049–7057.

Frank, E. G., Woodgate, R. 2007. Increased catalytic activity and altered fidelity of human DNA polymerase iota in the presence of manganese. *J Biol Chem* 282(34): 24689–24696.

Freudenthal, B. D., Gakhar, L., Ramaswamy, S., Washington, M. T. 2010. Structure of monoubiquitinated PCNA and implications for translesion synthesis and DNA polymerase exchange. *Nat Struct Mol Biol* 17(4): 479–484.

Friedberg, E. C., Lehmann, A. R., Fuchs, R. P. 2005. Trading places: How do DNA polymerases switch during translesion DNA synthesis? *Mol Cell* 18(5): 499–505.

Fugger, K., Mistrik, M., Neelsen, K. J., Yao, Q., Zellweger, R., Kousholt, A. N., Haahr, P. et al. 2015. FBH1 catalyzes regression of stalled replication forks. *Cell Rep*. doi:10.1016/j.celrep.2015.02.028.

Fumasoni, M., Zwicky, K., Vanoli, F., Lopes, M., Branzei, D. 2015. Error-free DNA damage tolerance and sister chromatid proximity during DNA replication rely on the Polalpha/Primase/Ctf4 Complex. *Mol Cell* 57(5): 812–823.

Futreal, P. A., Coin, L., Marshall, M., Down, T., Hubbard, T., Wooster, R., Rahman, N., Stratton, M. R. 2004. A census of human cancer genes. *Nat Rev Cancer* 4(3): 177–183.

Gali, H., Juhasz, S., Morocz, M., Hajdu, I., Fatyol, K., Szukacsov, V., Burkovics, P., Haracska, L. 2012. Role of SUMO modification of human PCNA at stalled replication fork. *Nucleic Acids Res* 40(13): 6049–6059.

Gan, G. N., Wittschieben, J. P., Wittschieben, B. O., Wood, R. D. 2008. DNA polymerase zeta (pol zeta) in higher eukaryotes. *Cell Res* 18(1): 174–183.

Gangavarapu, V., Prakash, S., Prakash, L. 2007. Requirement of RAD52 group genes for postreplication repair of UV-damaged DNA in *Saccharomyces cerevisiae*. *Mol Cell Biol* 27(21): 7758–7764.

Garcia-Gomez, S., Reyes, A., Martinez-Jimenez, M. I., Chocron, E. S., Mouron, S., Terrados, G., Powell, C. et al. 2013. PrimPol, an archaic primase/polymerase operating in human cells. *Mol Cell* 52(4): 541–553.

Garg, P., Burgers, P. M. 2005. Ubiquitinated proliferating cell nuclear antigen activates translesion DNA polymerases eta and REV1. *Proc Natl Acad Sci USA* 102(51): 18361–18366.

Garg, P., Stith, C. M., Majka, J., Burgers, P. M. 2005. Proliferating cell nuclear antigen promotes translesion synthesis by DNA polymerase zeta. *J Biol Chem* 280(25): 23446–23450.

Gari, K., Decaillet, C., Delannoy, M., Wu, L., Constantinou, A. 2008. Remodeling of DNA replication structures by the branch point translocase FANCM. *Proc Natl Acad Sci USA* 105(42): 16107–16112.

Giannattasio, M., Zwicky, K., Follonier, C., Foiani, M., Lopes, M., Branzei, D. 2014. Visualization of recombination-mediated damage bypass by template switching. *Nat Struct Mol Biol* 21(10): 884–892.

Gibbs, P. E., McGregor, W. G., Maher, V. M., Nisson, P., Lawrence, C. W. 1998. A human homolog of the *Saccharomyces cerevisiae* REV3 gene, which encodes the catalytic subunit of DNA polymerase zeta. *Proc Natl Acad Sci USA* 95(12): 6876–6880.

Gohler, T., Sabbioneda, S., Green, C. M., Lehmann, A. R. 2011. ATR-mediated phosphorylation of DNA polymerase eta is needed for efficient recovery from UV damage. *J Cell Biol* 192(2): 219–227.

Goodman, M. F., Woodgate, R. 2013. Translesion DNA polymerases. *Cold Spring Harb Perspect Biol* 5(10): a010363.

Guilliam, T. A., Doherty, A. J. 2017. PrimPol-Prime time to reprime. *Genes (Basel)* 8(1): 20.

Guilliam, T. A., Jozwiakowski, S. K., Ehlinger, A., Barnes, R. P., Rudd, S. G., Bailey, L. J., Skehel, J. M., Eckert, K. A., Chazin, W. J., Doherty, A. J. 2015a. Human PrimPol is a highly error-prone polymerase regulated by single-stranded DNA binding proteins. *Nucleic Acids Res* 43(2): 1056–1068.

Guilliam, T. A., Keen, B. A., Brissett, N. C., Doherty, A. J. 2015b. Primase-polymerases are a functionally diverse superfamily of replication and repair enzymes. *Nucleic Acids Res* 43(14): 6651–6664.

Guo, C., Fischhaber, P. L., Luk-Paszyc, M. J., Masuda, Y., Zhou, J., Kamiya, K., Kisker, C., Friedberg, E. C. 2003. Mouse Rev1 protein interacts with multiple DNA polymerases involved in translesion DNA synthesis. *EMBO J* 22(24): 6621–6630.

Guo, C., Kosarek-Stancel, J. N., Tang, T. S., Friedberg, E. C. 2009. Y-family DNA polymerases in mammalian cells. *Cell Mol Life Sci* 66(14): 2363–2381.

Guo, C., Sonoda, E., Tang, T. S., Parker, J. L., Bielen, A. B., Takeda, S., Ulrich, H. D., Friedberg, E. C. 2006a. REV1 protein interacts with PCNA: Significance of the REV1 BRCT domain *in vitro* and *in vivo*. *Mol Cell* 23(2): 265–271.

Guo, C., Tang, T. S., Bienko, M., Dikic, I., Friedberg, E. C. 2008. Requirements for the interaction of mouse Polkappa with ubiquitin and its biological significance. *J Biol Chem* 283(8): 4658–4664.

Guo, C., Tang, T. S., Bienko, M., Parker, J. L., Bielen, A. B., Sonoda, E., Takeda, S., Ulrich, H. D., Dikic, I., Friedberg, E. C. 2006b. Ubiquitin-binding motifs in REV1 protein are required for its role in the tolerance of DNA damage. *Mol Cell Biol* 26(23): 8892–8900.

Guo, D., Xie, Z., Shen, H., Zhao, B., Wang, Z. 2004. Translesion synthesis of acetylaminofluorene-dG adducts by DNA polymerase zeta is stimulated by yeast Rev1 protein. *Nucleic Acids Res* 32(3): 1122–1130.

Haglund, K., Stenmark, H. 2006. Working out coupled monoubiquitination. *Nat Cell Biol* 8(11): 1218–1219.

Hara, K., Hashimoto, H., Murakumo, Y., Kobayashi, S., Kogame, T., Unzai, S., Akashi, S., Takeda, S., Shimizu, T., Sato, M. 2010. Crystal structure of human REV7 in complex with a human REV3 fragment and structural implication of the interaction between DNA polymerase zeta and REV1. *J Biol Chem* 285(16): 12299–12307.

Hara, K., Shimizu, T., Unzai, S., Akashi, S., Sato, M., Hashimoto, H. 2009. Purification, crystallization and initial X-ray diffraction study of human REV7 in complex with a REV3 fragment. *Acta Crystallogr Sect F Struct Biol Cryst Commun* 65(Pt 12): 1302–1305.

Hashimoto, K., Cho, Y., Yang, I. Y., Akagi, J., Ohashi, E., Tateishi, S., de Wind, N., Hanaoka, F., Ohmori, H., Moriya, M. 2012. The vital role of polymerase zeta and REV1 in mutagenic, but not correct, DNA synthesis across benzo[a]pyrene-dG and recruitment of polymerase zeta by REV1 to replication-stalled site. *J Biol Chem* 287(12): 9613–9622.

Hendel, A., Krijger, P. H., Diamant, N., Goren, Z., Langerak, P., Kim, J., Reissner, T. et al. 2011. PCNA ubiquitination is important, but not essential for translesion DNA synthesis in mammalian cells. *PLoS Genet* 7(9): e1002262.

Hendriks, I. A., D'Souza, R. C., Yang, B., Verlaan-de Vries, M., Mann, M., Vertegaal, A. C. 2014. Uncovering global SUMOylation signaling networks in a site-specific manner. *Nat Struct Mol Biol* 21(10): 927–936.

Ho, T. V., Guainazzi, A., Derkunt, S. B., Enoiu, M., Scharer, O. D. 2011. Structure-dependent bypass of DNA interstrand crosslinks by translesion synthesis polymerases. *Nucleic Acids Res* 39(17): 7455–7464.

Hoege, C., Pfander, B., Moldovan, G. L., Pyrowolakis, G., Jentsch, S. 2002. RAD6-dependent DNA repair is linked to modification of PCNA by ubiquitin and SUMO. *Nature* 419(6903): 135–141.

Hoeller, D., Hecker, C. M., Wagner, S., Rogov, V., Dotsch, V., Dikic, I. 2007. E3-independent monoubiquitination of ubiquitin-binding proteins. *Mol Cell* 26(6): 891–898.

Huang, T. T., Nijman, S. M., Mirchandani, K. D., Galardy, P. J., Cohn, M. A., Haas, W., Gygi, S. P., Ploegh, H. L., Bernards, R., D'Andrea, A. D. 2006. Regulation of monoubiquitinated PCNA by DUB autocleavage. *Nat Cell Biol* 8(4): 339–347.

Iyer, L. M., Koonin, E. V., Leipe, D. D., Aravind, L. 2005. Origin and evolution of the archaeo-eukaryotic primase superfamily and related palm-domain proteins: Structural insights and new members. *Nucleic Acids Res* 33(12): 3875–3896.

Izhar, L., Ziv, O., Cohen, I. S., Geacintov, N. E., Livneh, Z. 2013. Genomic assay reveals tolerance of DNA damage by both translesion DNA synthesis and homology-dependent repair in mammalian cells. *Proc Natl Acad Sci USA* 110(16): E1462–1469.

Jansen, J. G., Langerak, P., Tsaalbi-Shtylik, A., van den Berk, P., Jacobs, H., de Wind, N. 2006. Strand-biased defect in C/G transversions in hypermutating immunoglobulin genes in Rev1-deficient mice. *J Exp Med* 203(2): 319–323.

Jansen, J. G., Temviriyanukul, P., Wit, N., Delbos, F., Reynaud, C. A., Jacobs, H., de Wind, N. 2014. Redundancy of mammalian Y family DNA polymerases in cellular responses to genomic DNA lesions induced by ultraviolet light. *Nucleic Acids Res* 42(17): 11071–11082.

Johnson, R. E., Kondratick, C. M., Prakash, S., Prakash, L. 1999. hRAD30 mutations in the variant form of xeroderma pigmentosum. *Science* 285(5425): 263–265.

Johnson, R. E., Prakash, L., Prakash, S. 2012. Pol31 and Pol32 subunits of yeast DNA polymerase delta are also essential subunits of DNA polymerase zeta. *Proc Natl Acad Sci U S A* 109(31): 12455–12460.

Johnson, R. E., Washington, M. T., Haracska, L., Prakash, S., Prakash, L. 2000. Eukaryotic polymerases iota and zeta act sequentially to bypass DNA lesions. *Nature* 406(6799): 1015–1019.

Johnson, R. E., Yu, S. L., Prakash, S., Prakash, L. 2003. Yeast DNA polymerase zeta (zeta) is essential for error-free replication past thymine glycol. *Genes Dev* 17(1): 77–87.
Jung, Y. S., Hakem, A., Hakem, R., Chen, X. 2011. Pirh2 E3 ubiquitin ligase monoubiquitinates DNA polymerase eta to suppress translesion DNA synthesis. *Mol Cell Biol* 31(19): 3997–4006.
Jung, Y. S., Liu, G., Chen, X. 2010. Pirh2 E3 ubiquitin ligase targets DNA polymerase eta for 20S proteasomal degradation. *Mol Cell Biol* 30(4): 1041–1048.
Jung, Y. S., Qian, Y., Chen, X. 2012. DNA polymerase eta is targeted by Mdm2 for polyubiquitination and proteasomal degradation in response to ultraviolet irradiation. *DNA Repair (Amst)* 11(2): 177–184.
Kalifa, L., Sia, E. A. 2007. Analysis of Rev1p and Pol zeta in mitochondrial mutagenesis suggests an alternative pathway of damage tolerance. *DNA Repair (Amst)* 6(12): 1732–1739.
Karras, G. I., Fumasoni, M., Sienski, G., Vanoli, F., Branzei, D., Jentsch, S. 2013. Noncanonical role of the 9-1-1 clamp in the error-free DNA damage tolerance pathway. *Mol Cell* 49(3): 536–546.
Karras, G. I., Jentsch, S. 2010. The RAD6 DNA damage tolerance pathway operates uncoupled from the replication fork and is functional beyond S phase. *Cell* 141(2): 255–267.
Kats, E. S., Enserink, J. M., Martinez, S., Kolodner, R. D. 2009. The saccharomyces cerevisiae Rad6 postreplication repair and Siz1/Srs2 homologous recombination-inhibiting pathways process DNA damage that arises in asf1 mutants. *Mol Cell Biol* 29(19): 5226–5237.
Kawamoto, T., Araki, K., Sonoda, E., Yamashita, Y. M., Harada, K., Kikuchi, K., Masutani, C. et al. 2005. Dual roles for DNA polymerase eta in homologous DNA recombination and translesion DNA synthesis. *Mol Cell* 20(5): 793–799.
Keen, B. A., Jozwiakowski, S. K., Bailey, L. J., Bianchi, J., Doherty, A. J. 2014. Molecular dissection of the domain architecture and catalytic activities of human PrimPol. *Nucleic Acids Res* 42(9): 5830–5845.
Khare, V., Eckert, K. A. 2002. The proofreading 3′-->5′ exonuclease activity of DNA polymerases: A kinetic barrier to translesion DNA synthesis. *Mutat Res* 510(1-2): 45–54.
Kile, A. C., Chavez, D. A., Bacal, J., Eldirany, S., Korzhnev, D. M., Bezsonova, I., Eichman, B. F., Cimprich, K. A. 2015. HLTF's ancient HIRAN domain binds 3′ DNA ends to drive replication fork reversal. *Mol Cell* 58(6): 1090–1100.
Kim, H., Yang, K., Dejsuphong, D., D'Andrea, A. D. 2012a. Regulation of Rev1 by the fanconi anemia core complex. *Nat Struct Mol Biol* 19(2): 164–170.
Kim, S. H., Michael, W. M. 2008. Regulated proteolysis of DNA polymerase eta during the DNA-damage response in *C. elegans*. *Mol Cell* 32(6): 757–766.
Kim, S. O., Yoon, H., Park, S. O., Lee, M., Shin, J. S., Ryu, K. S., Lee, J. O., Seo, Y. S., Jung, H. S., Choi, B. S. 2012b. Srs2 possesses a non-canonical PIP box in front of its SBM for precise recognition of SUMOylated PCNA. *J Mol Cell Biol* 4(4): 258–261.
Kirouac, K. N., Ling, H. 2011. Unique active site promotes error-free replication opposite an 8-oxo-guanine lesion by human DNA polymerase iota. *Proc Natl Acad Sci USA* 108(8): 3210–3215.
Klug, A. R., Harbut, M. B., Lloyd, R. S., Minko, I. G. 2012. Replication bypass of N2-deoxyguanosine interstrand cross-links by human DNA polymerases eta and iota. *Chem Res Toxicol* 25(3): 755–762.
Kobayashi, K., Guilliam, T. A., Tsuda, M., Yamamoto, J., Bailey, L. J., Iwai, S., Takeda, S., Doherty, A. J., Hirota, K. 2016. Repriming by PrimPol is critical for DNA replication restart downstream of lesions and chain-terminating nucleosides. *Cell Cycle* 15(15): 1997–2008.
Korzhnev, D. M., Hadden, M. K. 2016. Targeting the translesion synthesis pathway for the development of anti-cancer chemotherapeutics. *J Med Chem* 59(20): 9321–9336.
Krejci, L., Van Komen, S., Li, Y., Villemain, J., Reddy, M. S., Klein, H., Ellenberger, T., Sung, P. 2003. DNA helicase Srs2 disrupts the Rad51 presynaptic filament. *Nature* 423(6937): 305–309.
Krijger, P. H., Lee, K. Y., Wit, N., van den Berk, P. C., Wu, X., Roest, H. P., Maas, A., Ding, H., Hoeijmakers, J. H., Myung, K., Jacobs, H. 2011a. HLTF and SHPRH are not essential for PCNA polyubiquitination, survival and somatic hypermutation: Existence of an alternative E3 ligase. *DNA Repair (Amst)* 10(4): 438–444.
Krijger, P. H., Tsaalbi-Shtylik, A., Wit, N., van den Berk, P. C., de Wind, N., Jacobs, H. 2013. Rev1 is essential in generating G to C transversions downstream of the Ung2 pathway but not the Msh2+Ung2 hybrid pathway. *Eur J Immunol* 43(10): 2765–2770.
Krijger, P. H., van den Berk, P. C., Wit, N., Langerak, P., Jansen, J. G., Reynaud, C. A., de Wind, N., Jacobs, H. 2011b. PCNA ubiquitination-independent activation of polymerase eta during somatic hypermutation and DNA damage tolerance. *DNA Repair (Amst)* 10(10): 1051–1059.
Krishna, T. S., Kong, X. P., Gary, S., Burgers, P. M., Kuriyan, J. 1994. Crystal structure of the eukaryotic DNA polymerase processivity factor PCNA. *Cell* 79(7): 1233–1243.
Krokan, H. E., Bjoras, M. 2013. Base excision repair. *Cold Spring Harb Perspect Biol* 5(4): a012583.
Kubota, T., Nishimura, K., Kanemaki, M. T., Donaldson, A. D. 2013. The Elg1 replication factor C-like complex functions in PCNA unloading during DNA replication. *Mol Cell* 50(2): 273–280.
Kunkel, T. A. 2004. DNA replication fidelity. *J Biol Chem* 279(17): 16895–16898.
Lange, S. S., Takata, K., Wood, R. D. 2011. DNA polymerases and cancer. *Nat Rev Cancer* 11(2): 96–110.
Lawrence, C. W. 2002. Cellular roles of DNA polymerase zeta and Rev1 protein. *DNA Repair (Amst)* 1(6): 425–435.
Lawrence, C. W. 2004. Cellular functions of DNA polymerase zeta and Rev1 protein. *Adv Protein Chem* 69: 167–203.
Lawrence, C. W., Das, G., Christensen, R. B. 1985. REV7, a new gene concerned with UV mutagenesis in yeast. *Mol Gen Genet* 200(1): 80–85.
Leach, C. A., Michael, W. M. 2005. Ubiquitin/SUMO modification of PCNA promotes replication fork progression in *Xenopus laevis* egg extracts. *J Cell Biol* 171(6): 947–954.
Lee, K. Y., Fu, H., Aladjem, M. I., Myung, K. 2013. ATAD5 regulates the lifespan of DNA replication factories by modulating PCNA level on the chromatin. *J Cell Biol* 200(1): 31–44.
Lee, K. Y., Yang, K., Cohn, M. A., Sikdar, N., D'Andrea, A. D., Myung, K. 2010. Human ELG1 regulates the level of ubiquitinated proliferating cell nuclear antigen (PCNA) through Its interactions with PCNA and USP1. *J Biol Chem* 285(14): 10362–10369.
Lee, Y. S., Gregory, M. T., Yang, W. 2014. Human Pol zeta purified with accessory subunits is active in translesion DNA synthesis and complements Pol eta in cisplatin bypass. *Proc Natl Acad Sci USA* 111(8): 2954–2959.
Li, Z., Zhang, H., McManus, T. P., McCormick, J. J., Lawrence, C. W., Maher, V. M. 2002. hREV3 is essential for error-prone translesion synthesis past UV or benzo[a]pyrene diol epoxide-induced DNA lesions in human fibroblasts. *Mutat Res* 510(1–2): 71–80.
Lin, J. R., Zeman, M. K., Chen, J. Y., Yee, M. C., Cimprich, K. A. 2011. SHPRH and HLTF act in a damage-specific manner to coordinate different forms of postreplication repair and prevent mutagenesis. *Mol Cell* 42(2): 237–249.
Lin, W., Xin, H., Zhang, Y., Wu, X., Yuan, F., Wang, Z. 1999. The human REV1 gene codes for a DNA template-dependent dCMP transferase. *Nucleic Acids Res* 27(22): 4468–4475.
Lin, X., Howell, S. B. 2006. DNA mismatch repair and p53 function are major determinants of the rate of development of cisplatin resistance. *Mol Cancer Ther* 5(5): 1239–1247.
Lin, X., Okuda, T., Trang, J., Howell, S. B. 2006b. Human REV1 modulates the cytotoxicity and mutagenicity of cisplatin in human ovarian carcinoma cells. *Mol Pharmacol* 69(5): 1748–1754.
Lin, X., Trang, J., Okuda, T., Howell, S. B. 2006a. DNA polymerase zeta accounts for the reduced cytotoxicity and enhanced mutagenicity of cisplatin in human colon carcinoma cells that have lost DNA mismatch repair. *Clin Cancer Res* 12(2): 563–568.
Lin, Y. C., Li, L., Makarova, A. V., Burgers, P. M., Stone, M. P., Lloyd, R. S. 2014a. Error-prone replication bypass of the primary aflatoxin B1 DNA adduct, AFB1-N7-Gua. *J Biol Chem* 289(26): 18497–18506.

Lin, Y. C., Li, L., Makarova, A. V., Burgers, P. M., Stone, M. P., Lloyd, R. S. 2014b. Molecular basis of aflatoxin-induced mutagenesis-role of the aflatoxin B1-formamidopyrimidine adduct. *Carcinogenesis* 35(7): 1461–1468.

Ling, H., Boudsocq, F., Woodgate, R., Yang, W. 2001. Crystal structure of a Y-family DNA polymerase in action: A mechanism for error-prone and lesion-bypass replication. *Cell* 107(1): 91–102.

Liu, Y., Yang, Y., Tang, T. S., Zhang, H., Wang, Z., Friedberg, E., Yang, W., Guo, C. 2014. Variants of mouse DNA polymerase kappa reveal a mechanism of efficient and accurate translesion synthesis past a benzo[a]pyrene dG adduct. *Proc Natl Acad Sci USA* 111(5): 1789–1794.

Lone, S., Townson, S. A., Uljon, S. N., Johnson, R. E., Brahma, A., Nair, D. T., Prakash, S., Prakash, L., Aggarwal, A. K. 2007. Human DNA polymerase kappa encircles DNA: Implications for mismatch extension and lesion bypass. *Mol Cell* 25(4): 601–614.

Lopes, M., Foiani, M., Sogo, J. M. 2006. Multiple mechanisms control chromosome integrity after replication fork uncoupling and restart at irreparable UV lesions. *Mol Cell* 21(1): 15–27.

Machwe, A., Xiao, L., Groden, J., Orren, D. K. 2006. The Werner and Bloom syndrome proteins catalyze regression of a model replication fork. *Biochemistry* 45(47): 13939–13946.

Makarova, A. V., Burgers, P. M. 2015. Eukaryotic DNA polymerase zeta. *DNA Repair (Amst)* 29: 47–55.

Makarova, A. V., Stodola, J. L., Burgers, P. M. 2012. A four-subunit DNA polymerase zeta complex containing Pol delta accessory subunits is essential for PCNA-mediated mutagenesis. *Nucleic Acids Res* 40(22): 11618–11626.

Martinez-Jimenez, M. I., Garcia-Gomez, S., Bebenek, K., Sastre-Moreno, G., Calvo, P. A., Diaz-Talavera, A., Kunkel, T. A., Blanco, L. 2015. Alternative solutions and new scenarios for translesion DNA synthesis by human PrimPol. *DNA Repair (Amst)* 29: 127–138.

Masuda, Y., Suzuki, M., Kawai, H., Hishiki, A., Hashimoto, H., Masutani, C., Hishida, T., Suzuki, F., Kamiya, K. 2012. En bloc transfer of polyubiquitin chains to PCNA *in vitro* is mediated by two different human E2-E3 pairs. *Nucleic Acids Res* 40(20): 10394–10407.

Masutani, C., Araki, M., Yamada, A., Kusumoto, R., Nogimori, T., Maekawa, T., Iwai, S., Hanaoka, F. 1999a. Xeroderma pigmentosum variant (XP-V) correcting protein from HeLa cells has a thymine dimer bypass DNA polymerase activity. *EMBO J* 18(12): 3491–3501.

Masutani, C., Kusumoto, R., Yamada, A., Dohmae, N., Yokoi, M., Yuasa, M., Araki, M., Iwai, S., Takio, K., Hanaoka, F. 1999b. The XPV (xeroderma pigmentosum variant) gene encodes human DNA polymerase eta. *Nature* 399(6737): 700–704.

McCulloch, S. D., Kunkel, T. A. 2008. The fidelity of DNA synthesis by eukaryotic replicative and translesion synthesis polymerases. *Cell Res* 18(1): 148–161.

McDonald, J. P., Rapic-Otrin, V., Epstein, J. A., Broughton, B. C., Wang, X., Lehmann, A. R., Wolgemuth, D. J., Woodgate, R. 1999. Novel human and mouse homologs of *Saccharomyces cerevisiae* DNA polymerase eta. *Genomics* 60(1): 20–30.

McIntyre, J., Vidal, A. E., McLenigan, M. P., Bomar, M. G., Curti, E., McDonald, J. P., Plosky, B. S., Ohashi, E., Woodgate, R. 2013. Ubiquitin mediates the physical and functional interaction between human DNA polymerases eta and iota. *Nucleic Acids Res* 41(3): 1649–1660.

McIntyre, J., Woodgate, R. 2015. Regulation of translesion DNA synthesis: Posttranslational modification of lysine residues in key proteins. *DNA Repair (Amst)* 29: 166–179.

McNally, K., Neal, J. A., McManus, T. P., McCormick, J. J., Maher, V. M. 2008. hRev7, putative subunit of hPolzeta, plays a critical role in survival, induction of mutations, and progression through S-phase, of UV((254 nm))-irradiated human fibroblasts. *DNA Repair (Amst)* 7(4): 597–604.

Merkle, J. A., Rickmyre, J. L., Garg, A., Loggins, E. B., Jodoin, J. N., Lee, E., Wu, L. P., Lee, L. A. 2009. *no poles* encodes a predicted E3 ubiquitin ligase required for early embryonic development of *Drosophila*. *Development* 136(3): 449–459.

Minca, E. C., Kowalski, D. 2010. Multiple Rad5 activities mediate sister chromatid recombination to bypass DNA damage at stalled replication forks. *Mol Cell* 38(5): 649–661.

Minko, I. G., Harbut, M. B., Kozekov, I. D., Kozekova, A., Jakobs, P. M., Olson, S. B., Moses, R. E., Harris, T. M., Rizzo, C. J., Lloyd, R. S. 2008. Role for DNA polymerase kappa in the processing of N2-N2-guanine interstrand cross-links. *J Biol Chem* 283(25): 17075–17082.

Moldovan, G. L., Pfander, B., Jentsch, S. 2006. PCNA controls establishment of sister chromatid cohesion during S phase. *Mol Cell* 23(5): 723–732.

Moldovan, G. L., Dejsuphong, D., Petalcorin, M. I., Hofmann, K., Takeda, S., Boulton, S. J., D'Andrea, A. D. 2012. Inhibition of homologous recombination by the PCNA-interacting protein PARI. *Mol Cell* 45(1): 75–86.

Moldovan, G. L., Pfander, B., Jentsch, S. 2007. PCNA, the maestro of the replication fork. *Cell* 129(4): 665–679.

Morrison, A., Christensen, R. B., Alley, J., Beck, A. K., Bernstine, E. G., Lemontt, J. F., Lawrence, C. W. 1989. REV3, a *Saccharomyces cerevisiae* gene whose function is required for induced mutagenesis, is predicted to encode a nonessential DNA polymerase. *J Bacteriol* 171(10): 5659–5667.

Motegi, A., Liaw, H. J., Lee, K. Y., Roest, H. P., Maas, A., Wu, X., Moinova, H. et al. 2008. Polyubiquitination of proliferating cell nuclear antigen by HLTF and SHPRH prevents genomic instability from stalled replication forks. *Proc Natl Acad Sci USA* 105(34): 12411–12416.

Motegi, A., Sood, R., Moinova, H., Markowitz, S. D., Liu, P. P., Myung, K. 2006. Human SHPRH suppresses genomic instability through proliferating cell nuclear antigen polyubiquitination. *J Cell Biol* 175(5): 703–708.

Mouron, S., Rodriguez-Acebes, S., Martinez-Jimenez, M. I., Garcia-Gomez, S., Chocron, S., Blanco, L., Mendez, J. 2013. Repriming of DNA synthesis at stalled replication forks by human PrimPol. *Nat Struct Mol Biol* 20(12): 1383–1389.

Nair, D. T., Johnson, R. E., Prakash, S., Prakash, L., Aggarwal, A. K. 2004. Replication by human DNA polymerase-iota occurs by Hoogsteen base-pairing. *Nature* 430(6997): 377–380.

Nair, D. T., Johnson, R. E., Prakash, L., Prakash, S., Aggarwal, A. K. 2005a. Human DNA polymerase iota incorporates dCTP opposite template G via a G.C + Hoogsteen base pair. *Structure* 13(10): 1569–1577.

Nair, D. T., Johnson, R. E., Prakash, L., Prakash, S., Aggarwal, A. K. 2005b. Rev1 employs a novel mechanism of DNA synthesis using a protein template. *Science* 309(5744): 2219–2222.

Nair, D. T., Johnson, R. E., Prakash, L., Prakash, S., Aggarwal, A. K. 2006a. An incoming nucleotide imposes an anti to syn conformational change on the templating purine in the human DNA polymerase-iota active site. *Structure* 14(4): 749–755.

Nair, D. T., Johnson, R. E., Prakash, L., Prakash, S., Aggarwal, A. K. 2006b. Hoogsteen base pair formation promotes synthesis opposite the 1,N6-ethenodeoxyadenosine lesion by human DNA polymerase iota. *Nat Struct Mol Biol* 13(7): 619–625.

Nair, D. T., Johnson, R. E., Prakash, L., Prakash, S., Aggarwal, A. K. 2008. Protein-template-directed synthesis across an acrolein-derived DNA adduct by yeast Rev1 DNA polymerase. *Structure* 16(2): 239–245.

Neelsen, K. J., Lopes, M. 2015. Replication fork reversal in eukaryotes: From dead end to dynamic response. *Nat Rev Mol Cell Biol* 16(4): 207–220.

Nelson, J. R., Gibbs, P. E., Nowicka, A. M., Hinkle, D. C., Lawrence, C. W. 2000. Evidence for a second function for *Saccharomyces cerevisiae* Rev1p. *Mol Microbiol* 37(3): 549–554.

Nelson, J. R., Lawrence, C. W., Hinkle, D. C. 1996a. Deoxycytidyl transferase activity of yeast REV1 protein. *Nature* 382(6593): 729–731.

Nelson, J. R., Lawrence, C. W., Hinkle, D. C. 1996b. Thymine-thymine dimer bypass by yeast DNA polymerase zeta. *Science* 272(5268): 1646–1649.

Netz, D. J., Stith, C. M., Stumpfig, M., Kopf, G., Vogel, D., Genau, H. M., Stodola, J. L., Lill, R., Burgers, P. M., Pierik, A. J. 2011. Eukaryotic DNA polymerases require an iron-sulfur cluster for the formation of active complexes. *Nat Chem Biol* 8(1): 125–132.

Niimi, K., Murakumo, Y., Watanabe, N., Kato, T., Mii, S., Enomoto, A., Asai, M. et al. 2014. Suppression of REV7 enhances cisplatin sensitivity in ovarian clear cell carcinoma cells. *Cancer Sci* 105(5): 545–552.

Nojima, K., Hochegger, H., Saberi, A., Fukushima, T., Kikuchi, K., Yoshimura, M., Orelli, B. J. et al. 2005. Multiple repair pathways mediate tolerance to chemotherapeutic cross-linking agents in vertebrate cells. *Cancer Res* 65(24): 11704–11711.

Northam, M. R., Garg, P., Baitin, D. M., Burgers, P. M., Shcherbakova, P. V. 2006. A novel function of DNA polymerase zeta regulated by PCNA. *EMBO J* 25(18): 4316–4325.

Ogi, T., Lehmann, A. R. 2006. The Y-family DNA polymerase kappa (pol kappa) functions in mammalian nucleotide-excision repair. *Nat Cell Biol* 8(6): 640–642.

Ohashi, E., Hanafusa, T., Kamei, K., Song, I., Tomida, J., Hashimoto, H., Vaziri, C., Ohmori, H. 2009. Identification of a novel REV1-interacting motif necessary for DNA polymerase kappa function. *Genes Cells* 14(2): 101–111.

Ohashi, E., Murakumo, Y., Kanjo, N., Akagi, J., Masutani, C., Hanaoka, F., Ohmori, H. 2004. Interaction of hREV1 with three human Y-family DNA polymerases. *Genes Cells* 9(6): 523–531.

Ohashi, E., Ogi, T., Kusumoto, R., Iwai, S., Masutani, C., Hanaoka, F., Ohmori, H. 2000. Error-prone bypass of certain DNA lesions by the human DNA polymerase kappa. *Genes Dev* 14(13): 1589–1594.

Okada, T., Sonoda, E., Yamashita, Y. M., Koyoshi, S., Tateishi, S., Yamaizumi, M., Takata, M., Ogawa, O., Takeda, S. 2002. Involvement of vertebrate polkappa in Rad18-independent postreplication repair of UV damage. *J Biol Chem* 277(50): 48690–48695.

Okada, T., Sonoda, E., Yoshimura, M., Kawano, Y., Saya, H., Kohzaki, M., Takeda, S. 2005. Multiple roles of vertebrate REV genes in DNA repair and recombination. *Mol Cell Biol* 25(14): 6103–6111.

Okuda, T., Lin, X., Trang, J., Howell, S. B. 2005. Suppression of hREV1 expression reduces the rate at which human ovarian carcinoma cells acquire resistance to cisplatin. *Mol Pharmacol* 67(6): 1852–1860.

Ortiz-Bazan, M. A., Gallo-Fernandez, M., Saugar, I., Jimenez-Martin, A., Vazquez, M. V., Tercero, J. A. 2014. Rad5 plays a major role in the cellular response to DNA damage during chromosome replication. *Cell Rep* 9(2): 460–468.

Pabla, R., Rozario, D., Siede, W. 2008. Regulation of *Saccharomyces cerevisiae* DNA polymerase eta transcript and protein. *Radiat Environ Biophys* 47(1): 157–168.

Panse, V. G., Kuster, B., Gerstberger, T., Hurt, E. 2003. Unconventional tethering of Ulp1 to the transport channel of the nuclear pore complex by karyopherins. *Nat Cell Biol* 5(1): 21–27.

Papouli, E., Chen, S., Davies, A. A., Huttner, D., Krejci, L., Sung, P., Ulrich, H. D. 2005. Crosstalk between SUMO and ubiquitin on PCNA is mediated by recruitment of the helicase Srs2p. *Mol Cell* 19(1): 123–133.

Park, J. M., Yang, S. W., Yu, K. R., Ka, S. H., Lee, S. W., Seol, J. H., Jeon, Y. J., Chung, C. H. 2014. Modification of PCNA by ISG15 plays a crucial role in termination of error-prone translesion DNA synthesis. *Mol Cell* 54(4): 626–638.

Parker, J. L., Bielen, A. B., Dikic, I., Ulrich, H. D. 2007. Contributions of ubiquitin- and PCNA-binding domains to the activity of Polymerase eta in *Saccharomyces cerevisiae*. *Nucleic Acids Res* 35(3): 881–889.

Parker, J. L., Ulrich, H. D. 2009. Mechanistic analysis of PCNA polyubiquitylation by the ubiquitin protein ligases Rad18 and Rad5. *EMBO J* 28(23): 3657–3666.

Parnas, O., Zipin-Roitman, A., Pfander, B., Liefshitz, B., Mazor, Y., Ben-Aroya, S., Jentsch, S., Kupiec, M. 2010. Elg1, an alternative subunit of the RFC clamp loader, preferentially interacts with SUMOylated PCNA. *EMBO J* 29(15): 2611–2622.

Pence, M. G., Blans, P., Zink, C. N., Hollis, T., Fishbein, J. C., Perrino, F. W. 2009. Lesion bypass of N2-ethylguanine by human DNA polymerase iota. *J Biol Chem* 284(3): 1732–1740.

Pfander, B., Moldovan, G. L., Sacher, M., Hoege, C., Jentsch, S. 2005. SUMO-modified PCNA recruits Srs2 to prevent recombination during S phase. *Nature* 436(7049): 428–433.

Pfleger, C. M., Salic, A., Lee, E., Kirschner, M. W. 2001. Inhibition of Cdh1-APC by the MAD2-related protein MAD2L2: A novel mechanism for regulating Cdh1. *Genes Dev* 15(14): 1759–1764.

Pilzecker, B., Buoninfante, O. A., Pritchard, C., Blomberg, O. S., Huijbers, I. J., van den Berk, P. C., Jacobs, H. 2016. PrimPol prevents APOBEC/AID family mediated DNA mutagenesis. *Nucleic Acids Res* 44(10): 4734–4744.

Pozhidaeva, A., Pustovalova, Y., D'Souza, S., Bezsonova, I., Walker, G. C., Korzhnev, D. M. 2012. NMR structure and dynamics of the C-terminal domain from human Rev1 and its complex with Rev1 interacting region of DNA polymerase eta. *Biochemistry* 51(27): 5506–5520.

Prakash, L. 1981. Characterization of postreplication repair in *Saccharomyces cerevisiae* and effects of rad6, rad18, rev3 and rad52 mutations. *Mol Gen Genet* 184(3): 471–478.

Prakash, S., Johnson, R. E., Prakash, L. 2005. Eukaryotic translesion synthesis DNA polymerases: Specificity of structure and function. *Annu Rev Biochem* 74: 317–353.

Prakash, S., Sung, P., Prakash, L. 1993. DNA repair genes and proteins of *Saccharomyces cerevisiae*. *Annu Rev Genet* 27: 33–70.

Prasad, R., Bebenek, K., Hou, E., Shock, D. D., Beard, W. A., Woodgate, R., Kunkel, T. A., Wilson, S. H. 2003. Localization of the deoxyribose phosphate lyase active site in human DNA polymerase iota by controlled proteolysis. *J Biol Chem* 278(32): 29649–29654.

Pustovalova, Y., Magalhaes, M. T., D'Souza, S., Rizzo, A. A., Korza, G., Walker, G. C., Korzhnev, D. M. 2016. Interaction between the Rev1 C-terminal domain and the PolD3 subunit of polzeta suggests a mechanism of polymerase exchange upon Rev1/polzeta-dependent translesion synthesis. *Biochemistry* 55(13): 2043–2053.

Raschle, M., Knipscheer, P., Enoiu, M., Angelov, T., Sun, J., Griffith, J. D., Ellenberger, T. E., Scharer, O. D., Walter, J. C. 2008. Mechanism of replication-coupled DNA interstrand crosslink repair. *Cell* 134(6): 969–980.

Rattray, A. J., Shafer, B. K., McGill, C. B., Strathern, J. N. 2002. The roles of REV3 and RAD57 in double-strand-break-repair-induced mutagenesis of *Saccharomyces cerevisiae*. *Genetics* 162(3): 1063–1077.

Rechkoblit, O., Gupta, Y. K., Malik, R., Rajashankar, K. R., Johnson, R. E., Prakash, L., Prakash, S., Aggarwal, A. K. 2016. Structure and mechanism of human PrimPol, a DNA polymerase with primase activity. *Sci Adv* 2(10): e1601317.

Roerink, S. F., Koole, W., Stapel, L. C., Romeijn, R. J., Tijsterman, M. 2012. A broad requirement for TLS polymerases eta and kappa, and interacting sumoylation and nuclear pore proteins, in lesion bypass during *C. elegans* embryogenesis. *PLoS Genet* 8(6): e1002800.

Ross, A. L., Simpson, L. J., Sale, J. E. 2005. Vertebrate DNA damage tolerance requires the C-terminus but not BRCT or transferase domains of REV1. *Nucleic Acids Res* 33(4): 1280–1289.

Sabbioneda, S., Minesinger, B. K., Giannattasio, M., Plevani, P., Muzi-Falconi, M., Jinks-Robertson, S. 2005. The 9-1-1 checkpoint clamp physically interacts with polzeta and is partially required for spontaneous polzeta-dependent mutagenesis in *Saccharomyces cerevisiae*. *J Biol Chem* 280(46): 38657–38665.

Sakofsky, C. J., Ayyar, S., Deem, A. K., Chung, W. H., Ira, G., Malkova, A. 2015. Translesion polymerases drive microhomology-mediated break-induced replication leading to complex chromosomal rearrangements. *Mol Cell* 60(6): 860–872.

Sale, J. E. 2012. Competition, collaboration and coordination: Determining how cells bypass DNA damage. *J Cell Sci* 125(Pt 7): 1633–1643.

Sale, J. E., Batters, C., Edmunds, C. E., Phillips, L. G., Simpson, L. J., Szuts, D. 2009. Timing matters: Error-prone gap filling and translesion synthesis in immunoglobulin gene hypermutation. *Philos Trans R Soc Lond B Biol Sci* 364(1517): 595–603.

Sale, J. E., Lehmann, A. R., Woodgate, R. 2012. Y-family DNA polymerases and their role in tolerance of cellular DNA damage. *Nat Rev Mol Cell Biol* 13(3): 141–152.

Sarkies, P., Murat, P., Phillips, L. G., Patel, K. J., Balasubramanian, S., Sale, J. E. 2012. FANCJ coordinates two pathways that maintain epigenetic stability at G-quadruplex DNA. *Nucleic Acids Res* 40(4): 1485–1498.

Sarkies, P., Reams, C., Simpson, L. J., Sale, J. E. 2010. Epigenetic instability due to defective replication of structured DNA. *Mol Cell* 40(5): 703–713.

Scharer, O. D. 2013. Nucleotide excision repair in eukaryotes. *Cold Spring Harb Perspect Biol* 5(10): a012609.
Schenten, D., Kracker, S., Esposito, G., Franco, S., Klein, U., Murphy, M., Alt, F. W., Rajewsky, K. 2009. Pol zeta ablation in B cells impairs the germinal center reaction, class switch recombination, DNA break repair, and genome stability. *J Exp Med* 206(2): 477–490.
Schiavone, D., Jozwiakowski, S. K., Romanello, M., Guilbaud, G., Guilliam, T. A., Bailey, L. J., Sale, J. E., Doherty, A. J. 2016. PrimPol is required for replicative tolerance of G quadruplexes in vertebrate cells. *Mol Cell* 61(1): 161–169.
Schmutz, V., Janel-Bintz, R., Wagner, J., Biard, D., Shiomi, N., Fuchs, R. P., Cordonnier, A. M. 2010. Role of the ubiquitin-binding domain of Poleta in Rad18-independent translesion DNA synthesis in human cell extracts. *Nucleic Acids Res* 38(19): 6456–6465.
Shachar, S., Ziv, O., Avkin, S., Adar, S., Wittschieben, J., Reissner, T., Chaney, S. et al. 2009. Two-polymerase mechanisms dictate error-free and error-prone translesion DNA synthesis in mammals. *EMBO J* 28(4): 383–393.
Shen, X., Jun, S., O'Neal, L. E., Sonoda, E., Bemark, M., Sale, J. E., Li, L. 2006. REV3 and REV1 play major roles in recombination-independent repair of DNA interstrand cross-links mediated by monoubiquitinated proliferating cell nuclear antigen (PCNA). *J Biol Chem* 281(20): 13869–13872.
Silvian, L. F., Toth, E. A., Pham, P., Goodman, M. F., Ellenberger, T. 2001. Crystal structure of a DinB family error-prone DNA polymerase from *Sulfolobus solfataricus. Nat Struct Biol* 8(11): 984–989.
Simpson, L. J., Ross, A. L., Szuts, D., Alviani, C. A., Oestergaard, V. H., Patel, K. J., Sale, J. E. 2006. RAD18-independent ubiquitination of proliferating-cell nuclear antigen in the avian cell line DT40. *EMBO Rep* 7(9): 927–932.
Skoneczna, A., McIntyre, J., Skoneczny, M., Policinska, Z., Sledziewska-Gojska, E. 2007. Polymerase eta is a short-lived, proteasomally degraded protein that is temporarily stabilized following UV irradiation in *Saccharomyces cerevisiae. J Mol Biol* 366(4): 1074–1086.
Sonoda, E., Okada, T., Zhao, G. Y., Tateishi, S., Araki, K., Yamaizumi, M., Yagi, T. et al. 2003. Multiple roles of Rev3, the catalytic subunit of polzeta in maintaining genome stability in vertebrates. *EMBO J* 22(12): 3188–3197.
Srivastava, A. K., Han, C., Zhao, R., Cui, T., Dai, Y., Mao, C., Zhao, W., Zhang, X., Yu, J., Wang, Q. E. 2015. Enhanced expression of DNA polymerase eta contributes to cisplatin resistance of ovarian cancer stem cells. *Proc Natl Acad Sci USA* 112(14): 4411–4416.
Stelter, P., Ulrich, H. D. 2003. Control of spontaneous and damage-induced mutagenesis by SUMO and ubiquitin conjugation. *Nature* 425(6954): 188–191.
Stone, J. E., Kumar, D., Binz, S. K., Inase, A., Iwai, S., Chabes, A., Burgers, P. M., Kunkel, T. A. 2011. Lesion bypass by *S. cerevisiae* Pol zeta alone. *DNA Repair (Amst)* 10(8): 826–834.
Su, Y., Patra, A., Harp, J. M., Egli, M., Guengerich, F. P. 2015. Roles of residues Arg-61 and Gln-38 of human DNA polymerase eta in bypass of deoxyguanosine and 7,8-Dihydro-8-oxo-2′-deoxyguanosine. *J Biol Chem* 290(26): 15921–15933.
Swan, M. K., Johnson, R. E., Prakash, L., Prakash, S., Aggarwal, A. K. 2009. Structure of the human Rev1-DNA-dNTP ternary complex. *J Mol Biol* 390(4): 699–709.
Szuts, D., Marcus, A. P., Himoto, M., Iwai, S., Sale, J. E. 2008. REV1 restrains DNA polymerase zeta to ensure frame fidelity during translesion synthesis of UV photoproducts *in vivo*. *Nucleic Acids Res* 36(21): 6767–6780.
Terai, K., Abbas, T., Jazaeri, A. A., Dutta, A. 2010. CRL4(Cdt2) E3 ubiquitin ligase monoubiquitinates PCNA to promote translesion DNA synthesis. *Mol Cell* 37(1): 143–149.
Tissier, A., Frank, E. G., McDonald, J. P., Iwai, S., Hanaoka, F., Woodgate, R. 2000a. Misinsertion and bypass of thymine-thymine dimers by human DNA polymerase iota. *EMBO J* 19(19): 5259–5266.
Tissier, A., Frank, E. G., McDonald, J. P., Vaisman, A., Fernandez de Henestrosa, A. R., Boudsocq, F., McLenigan, M. P., Woodgate, R. 2001. Biochemical characterization of human DNA polymerase iota provides clues to its biological function. *Biochem Soc Trans* 29(Pt 2): 183–187.
Tissier, A., Kannouche, P., Reck, M. P., Lehmann, A. R., Fuchs, R. P., Cordonnier, A. 2004. Co-localization in replication foci and interaction of human Y-family members, DNA polymerase pol eta and REVl protein. *DNA Repair (Amst)* 3(11): 1503–1514.
Tissier, A., McDonald, J. P., Frank, E. G., Woodgate, R. 2000b. Poliota, a remarkably error-prone human DNA polymerase. *Genes Dev* 14(13): 1642–1650.
Tomida, J., Takata, K., Lange, S. S., Schibler, A. C., Yousefzadeh, M. J., Bhetawal, S., Dent, S. Y., Wood, R. D. 2015. REV7 is essential for DNA damage tolerance via two REV3L binding sites in mammalian DNA polymerase zeta. *Nucleic Acids Res* 43(2): 1000–1011.
Torres-Ramos, C. A., Prakash, S., Prakash, L. 1997. Requirement of yeast DNA polymerase delta in post-replicational repair of UV-damaged DNA. *J Biol Chem* 272(41): 25445–25448.
Trincao, J., Johnson, R. E., Escalante, C. R., Prakash, S., Prakash, L., Aggarwal, A. K. 2001. Structure of the catalytic core of *S. cerevisiae* DNA polymerase eta: Implications for translesion DNA synthesis. *Mol Cell* 8(2): 417–426.
Uljon, S. N., Johnson, R. E., Edwards, T. A., Prakash, S., Prakash, L., Aggarwal, A. K. 2004. Crystal structure of the catalytic core of human DNA polymerase kappa. *Structure* 12(8): 1395–1404.
Unk, I., Hajdu, I., Fatyol, K., Hurwitz, J., Yoon, J. H., Prakash, L., Prakash, S., Haracska, L. 2008. Human HLTF functions as a ubiquitin ligase for proliferating cell nuclear antigen polyubiquitination. *Proc Natl Acad Sci USA* 105(10): 3768–3773.
Unk, I., Hajdu, I., Fatyol, K., Szakal, B., Blastyak, A., Bermudez, V., Hurwitz, J., Prakash, L., Prakash, S., Haracska, L. 2006. Human SHPRH is a ubiquitin ligase for Mms2-Ubc13-dependent polyubiquitylation of proliferating cell nuclear antigen. *Proc Natl Acad Sci USA* 103(48): 18107–18112.
Urulangodi, M., Sebesta, M., Menolfi, D., Szakal, B., Sollier, J., Sisakova, A., Krejci, L., Branzei, D. 2015. Local regulation of the Srs2 helicase by the SUMO-like domain protein Esc2 promotes recombination at sites of stalled replication. *Genes Dev* 29(19): 2067–2080.
Vaisman, A., Lehmann, A. R., Woodgate, R. 2004. DNA polymerases eta and iota. *Adv Protein Chem* 69: 205–228.
Vaisman, A., Woodgate, R. 2017. Translesion DNA polymerases in eukaryotes: What makes them tick? *Crit Rev Biochem Mol Biol* 52(3): 274–303.
Van Sloun, P. P., Varlet, I., Sonneveld, E., Boei, J. J., Romeijn, R. J., Eeken, J. C., De Wind, N. 2002. Involvement of mouse Rev3 in tolerance of endogenous and exogenous DNA damage. *Mol Cell Biol* 22(7): 2159–2169.
Veaute, X., Jeusset, J., Soustelle, C., Kowalczykowski, S. C., Le Cam, E., Fabre, F. 2003. The Srs2 helicase prevents recombination by disrupting Rad51 nucleoprotein filaments. *Nature* 423(6937): 309–312.
Velculescu, V. E. 2008. Defining the blueprint of the cancer genome. *Carcinogenesis* 29(6): 1087–1091.
Vujanovic, M., Krietsch, J., Raso, M. C., Terraneo, N., Zellweger, R., Schmid, J. A., Taglialatela, A. et al. 2017. Replication fork slowing and reversal upon DNA damage require PCNA polyubiquitination and ZRANB3 DNA translocase activity. *Mol Cell* 67(5): 882–890. e885.
Wallace, H. A., Merkle, J. A., Yu, M. C., Berg, T. G., Lee, E., Bosco, G., Lee, L. A. 2014. TRIP/NOPO E3 ubiquitin ligase promotes ubiquitylation of DNA polymerase eta. *Development* 141(6): 1332–1341.
Wan, L., Lou, J., Xia, Y., Su, B., Liu, T., Cui, J., Sun, Y., Lou, H., Huang, J. 2013. hPrimpol1/CCDC111 is a human DNA primase-polymerase required for the maintenance of genome integrity. *EMBO Rep* 14(12): 1104–1112.
Wang, Y., Woodgate, R., McManus, T. P., Mead, S., McCormick, J. J., Maher, V. M. 2007. Evidence that in xeroderma pigmentosum variant cells, which lack DNA polymerase eta, DNA polymerase iota causes the very high frequency and unique spectrum of UV-induced mutations. *Cancer Res* 67(7): 3018–3026.
Washington, M. T., Johnson, R. E., Prakash, L., Prakash, S. 2002. Human DINB1-encoded DNA polymerase kappa is a promiscuous extender of mispaired primer termini. *Proc Natl Acad Sci USA* 99(4): 1910–1914.

Waters, L. S., Minesinger, B. K., Wiltrout, M. E., D'Souza, S., Woodruff, R. V., Walker, G. C. 2009. Eukaryotic translesion polymerases and their roles and regulation in DNA damage tolerance. *Microbiol Mol Biol Rev* 73(1): 134–154.

Waters, L. S., Walker, G. C. 2006. The critical mutagenic translesion DNA polymerase Rev1 is highly expressed during G(2)/M phase rather than S phase. *Proc Natl Acad Sci USA* 103(24): 8971–8976.

Weinert, B. T., Wagner, S. A., Horn, H., Henriksen, P., Liu, W. R., Olsen, J. V., Jensen, L. J., Choudhary, C. 2011. Proteome-wide mapping of the *Drosophila* acetylome demonstrates a high degree of conservation of lysine acetylation. *Sci Signal* 4(183): ra48.

Weston, R., Peeters, H., Ahel, D. 2012. ZRANB3 is a structure-specific ATP-dependent endonuclease involved in replication stress response. *Genes Dev* 26(14): 1558–1572.

Whitehurst, C. B., Vaziri, C., Shackelford, J., Pagano, J. S. 2012. Epstein-Barr virus BPLF1 deubiquitinates PCNA and attenuates polymerase eta recruitment to DNA damage sites. *J Virol* 86(15): 8097–8106.

Wilson, R. C., Jackson, M. A., Pata, J. D. 2013. Y-family polymerase conformation is a major determinant of fidelity and translesion specificity. *Structure* 21(1): 20–31.

Wit, N., Buoninfante, O. A., van den Berk, P. C., Jansen, J. G., Hogenbirk, M. A., de Wind, N., Jacobs, H. 2015. Roles of PCNA ubiquitination and TLS polymerases kappa and eta in the bypass of methyl methanesulfonate-induced DNA damage. *Nucleic Acids Res* 43(1): 282–294.

Wittschieben, J. P., Patil, V., Glushets, V., Robinson, L. J., Kusewitt, D. F., Wood, R. D. 2010. Loss of DNA polymerase zeta enhances spontaneous tumorigenesis. *Cancer Res* 70(7): 2770–2778.

Wittschieben, J. P., Reshmi, S. C., Gollin, S. M., Wood, R. D. 2006. Loss of DNA polymerase zeta causes chromosomal instability in mammalian cells. *Cancer Res* 66(1): 134–142.

Wittschieben, J., Shivji, M. K., Lalani, E., Jacobs, M. A., Marini, F., Gearhart, P. J., Rosewell, I., Stamp, G., Wood, R. D. 2000. Disruption of the developmentally regulated Rev3 l gene causes embryonic lethality. *Curr Biol* 10(19): 1217–1220.

Wojtaszek, J., Lee, C. J., D'Souza, S., Minesinger, B., Kim, H., D'Andrea, A. D., Walker, G. C., Zhou, P. 2012a. Structural basis of Rev1-mediated assembly of a quaternary vertebrate translesion polymerase complex consisting of Rev1, heterodimeric polymerase (Pol) zeta, and Pol kappa. *J Biol Chem* 287(40): 33836–33846.

Wojtaszek, J., Liu, J., D'Souza, S., Wang, S., Xue, Y., Walker, G. C., Zhou, P. 2012b. Multifaceted recognition of vertebrate Rev1 by translesion polymerases zeta and kappa. *J Biol Chem* 287(31): 26400–26408.

Wood, A., Garg, P., Burgers, P. M. 2007. A ubiquitin-binding motif in the translesion DNA polymerase Rev1 mediates its essential functional interaction with ubiquitinated proliferating cell nuclear antigen in response to DNA damage. *J Biol Chem* 282(28): 20256–20263.

Woodgate, R. 2001. Evolution of the two-step model for UV-mutagenesis. *Mutat Res* 485(1): 83–92.

Wu, F., Lin, X., Okuda, T., Howell, S. B. 2004. DNA polymerase zeta regulates cisplatin cytotoxicity, mutagenicity, and the rate of development of cisplatin resistance. *Cancer Res* 64(21): 8029–8035.

Xiao, W., Chow, B. L., Broomfield, S., Hanna, M. 2000. The *Saccharomyces cerevisiae* RAD6 group is composed of an error-prone and two error-free postreplication repair pathways. *Genetics* 155(4): 1633–1641.

Xie, K., Doles, J., Hemann, M. T., Walker, G. C. 2010. Error-prone translesion synthesis mediates acquired chemoresistance. *Proc Natl Acad Sci USA* 107(48): 20792–20797.

Xu, X., Xie, K., Zhang, X. Q., Pridgen, E. M., Park, G. Y., Cui, D. S., Shi, J. et al. 2013. Enhancing tumor cell response to chemotherapy through nanoparticle-mediated codelivery of siRNA and cisplatin prodrug. *Proc Natl Acad Sci USA* 110(46): 18638–18643.

Xue, X., Choi, K., Bonner, J., Chiba, T., Kwon, Y., Xu, Y., Sanchez, H., Wyman, C., Niu, H., Zhao, X., Sung, P. 2014. Restriction of replication fork regression activities by a conserved SMC complex. *Mol Cell* 56(3): 436–445.

Yang, L., Shi, T., Liu, F., Ren, C., Wang, Z., Li, Y., Tu, X., Yang, G., Cheng, X. 2015. REV3L, a promising target in regulating the chemosensitivity of cervical cancer cells. *PLOS ONE* 10(3): e0120334.

Yang, W. 2014. An overview of Y-Family DNA polymerases and a case study of human DNA polymerase eta. *Biochemistry* 53(17): 2793–2803.

Yang, W., Woodgate, R. 2007. What a difference a decade makes: Insights into translesion DNA synthesis. *Proc Natl Acad Sci USA* 104(40): 15591–15598.

Yavuz, S., Yavuz, A. S., Kraemer, K. H., Lipsky, P. E. 2002. The role of polymerase eta in somatic hypermutation determined by analysis of mutations in a patient with xeroderma pigmentosum variant. *J Immunol* 169(7): 3825–3830.

Yeeles, J. T., Marians, K. J. 2011. The *Escherichia coli* replisome is inherently DNA damage tolerant. *Science* 334(6053): 235–238.

Yu, Y., Cai, J. P., Tu, B., Wu, L., Zhao, Y., Liu, X., Li, L. et al. 2009. Proliferating cell nuclear antigen is protected from degradation by forming a complex with MutT Homolog2. *J Biol Chem* 284(29): 19310–19320.

Yuan, J., Ghosal, G., Chen, J. 2012. The HARP-like domain-containing protein AH2/ZRANB3 binds to PCNA and participates in cellular response to replication stress. *Mol Cell* 47(3): 410–421.

Zafar, M. K., Ketkar, A., Lodeiro, M. F., Cameron, C. E., Eoff, R. L. 2014. Kinetic analysis of human PrimPol DNA polymerase activity reveals a generally error-prone enzyme capable of accurately bypassing 7,8-dihydro-8-oxo-2′-deoxyguanosine. *Biochemistry* 53(41): 6584–6594.

Zhang, H., Chatterjee, A., Singh, K. K. 2006. *Saccharomyces cerevisiae* polymerase zeta functions in mitochondria. *Genetics* 172(4): 2683–2688.

Zhang, S., Chea, J., Meng, X., Zhou, Y., Lee, E. Y., Lee, M. Y. 2008. PCNA is ubiquitinated by RNF8. *Cell Cycle* 7(21): 3399–3404.

Zhang, Y., Yuan, F., Wu, X., Wang, Z. 2000. Preferential incorporation of G opposite template T by the low-fidelity human DNA polymerase iota. *Mol Cell Biol* 20(19): 7099–7108.

Zhao, Y., Gregory, M. T., Biertumpfel, C., Hua, Y. J., Hanaoka, F., Yang, W. 2013. Mechanism of somatic hypermutation at the WA motif by human DNA polymerase eta. *Proc Natl Acad Sci USA* 110(20): 8146–8151.

Zhong, X., Garg, P., Stith, C. M., Nick McElhinny, S. A., Kissling, G. E., Burgers, P. M., Kunkel, T. A. 2006. The fidelity of DNA synthesis by yeast DNA polymerase zeta alone and with accessory proteins. *Nucleic Acids Res* 34(17): 4731–4742.

Ziv, O., Geacintov, N., Nakajima, S., Yasui, A., Livneh, Z. 2009. DNA polymerase zeta cooperates with polymerases kappa and iota in translesion DNA synthesis across pyrimidine photodimers in cells from XPV patients. *Proc Natl Acad Sci USA* 106(28): 11552–11557.

The Repair of DNA Single-Strand Breaks and DNA Adducts: Mechanisms and Links to Human Disease

5

Alicja Winczura and John J. Reynolds

INTRODUCTION

The DNA in our cells is under constant attack from a multitude of DNA-damaging agents that constitute a threat to genomic integrity (Lindahl 1993). A large variety of DNA lesions can occur from both endogenous and environmental sources, including DNA breaks, base adducts and DNA cross-links. To prevent genome instability, many DNA repair factors operate within overlapping cellular DNA repair pathways to detect, signal and repair DNA damage as it arises. Without these repair pathways, the cell would cease to function due to the accumulation of lethal DNA aberrations.

The importance of DNA repair pathways is highlighted by the fact that loss or mutation of DNA repair factors is strongly linked to human disease. Whilst there is great phenotypic variety amongst these diseases, there are several features that are common to many of them, in particular cancer predisposition, neurological dysfunction and growth retardation (Hoeijmakers 2001, Klingseisen and Jackson 2011).

Neurological dysfunction caused by mutations in DNA repair factors can manifest in different ways, such as impacting either the developing brain or the mature nervous system. Microcephaly is defined as a significant decrease in the head circumference of an individual, resulting from a reduction in the size of the brain (Woods and Parker 2013). The development of the human central nervous system is a complex process involving highly coordinated periods of neuronal proliferation, migration and differentiation. Impaired proliferation of neural progenitor cells during neurodevelopment, resulting in reduced numbers of neurons in the mature brain, is thought to underlie primary microcephaly (microcephaly presenting at birth). A type of neurological dysfunction manifesting in the adult brain that is linked to defective DNA repair is progressive neurodegeneration. Due to the long-lived nature of post-mitotic neurons, combined with the limited regenerative capacity of the adult brain, a failure to maintain neuronal structure and function can lead to neurodegeneration (McKinnon 2017). Genome instability is also one of the driving forces behind the process of tumorigenesis, and as such cancer predisposition is often seen in individuals with mutations in DNA repair genes (Hoeijmakers 2001).

In this chapter, we will be focusing on two DNA repair pathways, and discussing how defects in these pathways impact human health. The first repair pathway is DNA single-strand break repair, which detects and repairs DNA single-strand breaks and small non-helix distorting DNA base modifications. The second repair pathway is nucleotide excision repair, which functions to repair large bulky helix-distorting DNA adducts.

DNA SINGLE-STRAND BREAK REPAIR

Origins and types of DNA single-strand breaks (SSBs)

DNA single-strand breaks are discontinuities in one strand of the DNA sugar phosphate backbone. DNA SSBs typically possess 'damaged' termini and are often associated with a single nucleotide gap. They are believed to be one of the most common types of DNA lesions that occur, and it has been estimated that ten thousands of SSBs can arise in a single cell per day (Lindahl 1993, Caldecott 2001). SSBs can arise in one of several ways: directly via the disintegration of the DNA sugar phosphate backbone, indirectly via the enzymatic removal of damaged DNA bases or due to the abortive action of certain DNA repair enzymes.

Direct SSBs

The most common source of direct SSBs is the disintegration of oxidised sugars following attack of DNA sugar residues by reactive oxygen species. ROS are an abundant class of free radicals, which are highly reactive molecules that contain one or more unpaired electrons. ROS are a significant source of endogenous SSBs, as they are generated as by-products of cellular metabolism (Valko et al. 2007). In addition, ROS can be generated by exogenous sources such as ionising radiation, in the form of either γ-rays or X-rays, or UV light within sunlight (Ward 1998, Schuch et al. 2017). A significant environmental source of ionising radiation is radon, a naturally occurring radioactive gas that emanates from uranium-containing soil and rock (Bissett and McLaughlin 2010).

At low concentrations, ROS have important cellular roles, functioning within intracellular signalling and signal transduction cascades, and are also produced during the inflammatory response in defence against pathogens (Valko et al. 2007). However, due to the potential for free radicals to be harmful at high concentrations by reacting with cellular components, the cell employs numerous antioxidant factors that have the capacity to remove cellular ROS (Valko et al. 2007). Oxidative stress is a cellular state in which the overproduction of oxygen species, and/or an inability to remove excess free radicals, leads to high cellular concentrations of ROS, and consequently causes damage to cellular components.

Types and sources of DNA base damage

Oxidative stress can induce oxidation of DNA bases. It has been estimated that oxidised DNA bases constitute up to 20% of the total DNA damage induced by ionising radiation, and over 70 types of radiation-induced oxidative base and sugar products have been identified and characterised (Demple and DeMott 2002, Sander et al. 2005). Guanine is particularly susceptible to oxidation due to its low redox potential, and 8-oxoG (7,8-dihydro-8-oxodeoxyguanine) is the most abundant type of spontaneously occurring oxidative DNA lesion in the cell (Kalam et al. 2006). Upon oxidative stress, other types of oxidised guanine, such as FapyG (2,6-diamino-4-oxo-5-formamidopyrimidine), can be detected at comparable levels to 8-oxoG (Kalam et al. 2006).

Cytosine, thymine and adenine are also susceptible to oxidative DNA damage, and oxidised forms of these bases can be detected in cellular DNA. In particular, reaction of hydroxyl radicals with cytosine and thymine generates cytosine glycol and thymine glycol, respectively. Further reaction with cytosine glycol can also generate other species of oxidised cytosine, such as uracil glycol, 5-hydroxycytosine or 5-hydroxyuracil (Bjelland and Seeberg 2003). In addition, FapyA (4,6-diamino-5-formamidopyrimidine) and 8-oxoA (7,8-dihydro-8-oxo-adenine) are two oxidised derivatives of adenine that are formed by the attack of ROS on different carbon atoms (Tudek 2003, Kalam et al. 2006).

Methylated DNA bases are another form of base damage that can arise in the cell. S-Adenosyl methionine (SAM) is the cellular methyl-group donor involved in enzymatic methylation events and is a significant source of endogenous DNA base methylation due to its high transfer potential (Sedgwick et al. 2007). The most frequently arising forms of methylated base are 7meG (7-methylguanine) and 3me-A (3-methyladenine), which make up 65%–85% and ~18% of all methylated bases, respectively (Lindahl 1993, Sedgwick et al. 2007). O6-meG (O6-methylguanine) and O4-meT (O4-methylthymine) are much rarer forms of methylated DNA damage, but are of physiological significance due to their mutagenic potential (Lindahl 1993, Sedgwick et al. 2007). An important exogenous source of DNA base methylation is methyl methanosulfonate (MMS), a monofunctional DNA methylating agent that is frequently used in the laboratory and has been shown to induce the same types of base adducts as SAM (Sedgwick et al. 2007).

Another form of spontaneous DNA damage is deamination, a process in which a nitrogen atom is replaced by an oxygen atom within exocyclic amines. Both cytosine and 5-methylcytosine (5mC) can undergo deamination to produce uracil or thymine, respectively. Although spontaneous deamination is a relatively rare event and has been estimated to occur 60 to 500 times per cell per day, it does occur more frequently in ssDNA than dsDNA and can also be generated by ionising radiation-induced oxidative DNA damage (Krokan et al. 2002, Bjelland and Seeberg 2003).

Finally, apurinic and apyramidinic (AP) sites are lesions in which the DNA base has been lost but the DNA sugar phosphate backbone is still intact. AP sites can be generated spontaneously due to the intrinsic instability of DNA in aqueous solution, but can also arise from DNA damage that destabilises the bond between the deoxyribose and the DNA base (Lindahl 1993). AP sites are a major form of endogenous DNA damage, and it is believed that up to 10,000 depurination events can occur in each cell per day.

Indirect SSBs arising from the repair of DNA damage

Apart from the direct disintegration of sugar residues in the DNA backbone, SSBs can also arise indirectly as the result of the enzymatic processing of damaged bases. Oxidation and alkylation of DNA bases result in small DNA lesions that cause very little distortion of the DNA backbone. These small non-helix-distorting DNA base lesions are removed within the base excision repair pathway (BER) by specialised enzymes called DNA glycosylases (Table 5.1) (Krokan and Bjoras 2013). Each DNA glycosylase recognises and removes a small number of specific DNA base modifications. In addition, these enzymes have different preferences for the type of damaged DNA substrate they can accommodate. For example, some DNA glycosylases can remove DNA lesions within dsDNA, ssDNA and DNA mismatches, whilst others can only act on dsDNA (Krokan and Bjoras 2013).

The ability of DNA glycosylases to efficiently recognise and remove small non-helix-distorting DNA lesions within the huge expanse of the human genome allows the BER pathway to efficiently repair approximately 80% of damaged bases that arise in the cell (Dianov and Parsons 2007). The underlying mechanism by which DNA glycosylases find damaged bases is not completely understood, although it appears that they search for DNA lesions using both sliding and distributive methods (Lee and Wallace 2017). Once the glycosylase has found the DNA lesion, it flips out the damaged base and removes it.

There are currently 11 known human nuclear DNA glycosylases that are divided into two classifications depending on their reaction mechanism (Table 5.1). Monofunctional DNA glycosylases excise the damaged base in a single step, leaving an AP site which is further cleaved by AP-endonuclease 1 (APE1) to produce an SSB. Bifunctional DNA glycosylases, which tend to recognise oxidative lesions, remove the damaged base in a multi-step reaction, first excising the damaged base, and then cleaving the DNA backbone in either a β-elimination or a βδ-elimination reaction (Krokan and Bjoras 2013).

Recently, the TET family of DNA dioxygenases, which play an important role in the regulation of DNA methylation, has been proposed as another potential source of endogenous SSBs. 5-methylcytosine (5mC) within CpG dinucleotides is an epigenetic DNA modification that has fundamental roles in gene expression, development and other important cellular processes (Smith and Meissner 2013). Retaining the appropriate pattern of DNA methylation is crucial for cellular homeostasis and DNA methyltransferase enzyme function to both induce de novo methylation and maintain the pattern when appropriate. The TET proteins have been shown to promote DNA demethylation in an active, replication-independent manner. First, the TET proteins convert 5mC into the oxidative cytosine derivatives 5hmC (5-hydroxymethylcytosine), 5fC (5-formylcytosine) or 5caC (5-carboxylcytosine). DNA demethylation is then achieved via the activity of the BER pathway (Ito and Kuraoka 2015). 5fc and 5caC are recognised and removed by the DNA glycosylase TDG, and the resulting AP site is converted to an SSB and repaired (Ito and Kuraoka 2015).

Another cellular DNA repair process that involves the production of an SSB intermediate is ribonucleotide excision repair (RER) (Williams et al. 2016). DNA polymerases possess innate mechanisms that function to prevent ribonucleotides being incorporated into DNA. However, no enzymatic process is completely error free, and it is estimated that during DNA replication, approximately 1 ribonucleotide is incorporated in every 7000 deoxyribonucleotides (Reijns et al. 2012). The consequences of not removing these misincorporated ribonucleotides can include increased mutagenesis and genome instability. Therefore, the cell employs the RER pathway to recognise and remove ribonucleotides within DNA. RER is initiated by RNase-H2, which incises the DNA backbone 5′ to the ribonucleotide, resulting in an SSB which is repaired (Williams et al. 2016).

In all these cases, SSBs are generated as scheduled DNA repair intermediates, either during the repair of different types of DNA lesions, or during cellular processes such as DNA demethylation, and are usually rapidly repaired. However, if these indirect SSBs are mishandled or are unable to be repaired, they have the potential to be deleterious to the cell.

Table 5.1 Human nuclear DNA glycosylases involved in base excision repair.

DNA glycosylase	Classification and reaction mechanism	Preferred substrate specificity	Preference for ssDNA or dsDNA
Monofunctional DNA glycosylases			
Methyl-CpG-binding domain protein 4 (MBD4)	Monofunctional	U, T and fluorouracil (5-FU) paired with G within methylated and unmethylated CpG sequences	dsDNA
DNA-3-methyladenine glycosylase (MPG)	Monofunctional	3-methylguanine (3-meG); 7-methylguanine (7-meG); 3-methyladenine (3-meA); 7-methyladenine (7-meA); 1,N^6-ethenoadenine (εA); hypoxanthine (Hx); 1,N^2-ethenoguanine (1,N^2-εG)	ssDNA and dsDNA
MutY homologue DNA glycosylase (MUTYH)	Monofunctional	A opposite 7,8-dihydro-8-oxodeoxyguanine (8-oxoG), C or G	dsDNA
Single-strand-selective monofunctional uracil-DNA glycosylase (SMUG1)	Monofunctional	U; 5-hydroxymethylcytosine (5-hmU); fluorouracil (5-FU); 3,N^4-ethenocytosine (εC)	ssDNA and dsDNA
Thymine-DNA glycosylase (TDG)	Monofunctional	T, U, fluorouracil (5-FU), 3,N^4-ethenocytosine (εC), 5-formylcytosine (5fC), or 5-carboxylcytosine (5caC) opposite G	dsDNA
Uracil-DNA glycosylase (UNG)	Monofunctional	U; fluorouracil (5-FU)	ssDNA and dsDNA
Bifunctional DNA glycosylases			
8-OxoG DNA glycosylase 1 (OGG1)	Bifunctional (β elimination)	7,8-dihydro-8-oxodeoxyguanine (8-oxoG) opposite C; formamidopyrimidine (FaPy) opposite C	dsDNA
Endonuclease III-like 1 (NTHL1)	Bifunctional (β elimination)	Thymine glycol (Tg); 2,6-diamino-4-hydroxy-5-formamidopyrimidine (FapyG); 5-hydroxycytosine (5-hC); 5-hydroxyuracil (5-hU)	dsDNA
Endonuclease VIII-like glycosylase 1 (NEIL1)	Bifunctional (βδ elimination)	Thymine glycol (Tg); 2,6-diamino-4-hydroxy-5-formamidopyrimidine (FaPyG); 4,6-diamino-5-formamidopyrimidine (FaPyA); 7,8-dihydro-8-oxodeoxyguanine (8-oxoG); 5-hydroxycytosine (5-hC); 5-hydroxyuracil (5-hU)	ssDNA and dsDNA
Endonuclease VIII-like glycosylase 2 (NEIL2)	Bifunctional (βδ elimination)	Thymine glycol (Tg); 2,6-diamino-4-hydroxy-5-formamidopyrimidine (FaPyG); 4,6-diamino-5-formamidopyrimidine (FaPyA); 7,8-dihydro-8-oxodeoxyguanine (8-oxoG); 5-hydroxycytosine (5-hC); 5-hydroxyuracil (5-hU)	ssDNA and dsDNA
Endonuclease VIII-like glycosylase 3 (NEIL3)	Bifunctional (βδ elimination)	2,6-diamino-4-hydroxy-5-formamidopyrimidine (FaPyG); 4,6-diamino-5-formamidopyrimidine (FaPyA)	ssDNA

SSBs arising from abortive enzymatic activity

The imbalanced or abortive activity of certain DNA processing enzymes also represents a significant source of endogenous SSBs. Topoisomerase 1 (TOP1) is an enzyme that acts to relieve superhelical torsional stress generated during transcription and DNA replication (Wang 2002). TOP1 achieves this by nicking one strand of the DNA double helix, allowing the DNA to unwind before the nick is ligated. This process generates a transient TOP1 cleavage complex in which TOP1 is covalently linked via a 3′ phosphotyrosyl bond to the SSB. These TOP1 cleavage complexes are very short-lived and are rapidly resealed under normal conditions. However, in certain situations, an abortive TOP1-SSB complex can form, which cannot be ligated. For example, the presence of pre-existing DNA damage close to the TOP1 cleavage complex, such as an AP-site, or oxidative DNA lesion, can cause misalignment of the 5′-OH within the TOP1 active site, preventing re-ligation and forming an abortive TOP1-SSB (Pourquier et al. 1997, Pommier et al. 2003). Additionally, collisions between DNA or RNA polymerases and TOP1 cleavage complexes are a significant source of abortive TOP1-DNA breaks (Pommier et al. 2003). Furthermore, reversible TOP1 poisons, such as camptothecin (CPT), greatly increase the half-life of the TOP1 cleavage and therefore can lead to increased levels of abortive TOP1-SSBs (Pommier 2009).

Consequences of unrepaired damaged bases and SSBs

Although individual DNA SSBs and damaged base lesions are considered less deleterious to the cell than other types of DNA damage, due to the large quantities in which they arise in each cell per day, unrepaired SSBs and base lesions pose a threat to genome stability. In particular, unrepaired DNA SSBs can be converted into DNA DSBs upon collision with a replication fork (Saleh-Gohari et al. 2005). DSBs are one of the most genotoxic lesions that can arise, and a single DSB may be sufficient to trigger cell cycle arrest and/or cell death (Huang et al. 1996). Therefore, delayed or defective repair of DNA SSBs is a serious risk to genome stability due to the potential to generate large numbers of DSBs during DNA replication.

Unrepaired oxidative and methylated DNA base lesions also need to be repaired, as they have the potential to cause mutations and interfere with DNA replication. 8-oxoG is highly mutagenic, as adenine can be inserted opposite this lesion with the same frequency as cytosine during DNA replication, resulting in G → T point mutations in subsequent rounds of the cell cycle (Shibutani et al. 1991). In addition, T → G conversions can occur when 8-oxodGTP is incorporated into DNA opposite adenine from the free nucleotide pool (Shibutani et al. 1991). The oxidative derivative of guanine, FapyG can also induce G → T point mutations at a higher frequency than 8-oxoG (Kalam et al. 2006). Both FapyA and 8-oxoA are also oxidative lesions that can cause A → C and substitutions; however, their mutagenic potential was found to be much lower than either FapyG or 8-oxoG (Kalam et al. 2006). Although 7meG, one of the most abundant types of methylated base damage, has no mutagenic potential itself, it can be converted into an AP-site or the highly mutagenic 7me-FapyG lesion (Tudek 2003). Furthermore, the presence of unrepaired O6-meG and O4-meT in DNA can result in G → A and T → C conversions, respectively (Swann 1990, Sedgwick et al. 2007), and spontaneous deamination of cytosine and 5mC to uracil or thymine, respectively, can result in C → T point mutations (Lindahl 1993). In addition to their mutagenic potential, certain types of oxidative and methylated DNA lesions can also act as obstacles to the progression of DNA polymerases. Both FapyG and 3me-A have been shown to efficiently block DNA polymerases, and in particular the persistence of 3me-A in the genome results in S-phase arrest, increased levels of chromosomal breaks and apoptosis (Engelward et al. 1998).

DNA single-strand break repair

Although DNA SSBs can arise in a variety of ways, they are all repaired in overlapping DNA repair sub-pathways within a global repair pathway collectively termed single-strand break repair (SSBR) (Caldecott 2014a). SSBR is very rapid and consists of four coordinated phases (**Figure 5.1**): detection of the SSB, end processing of the 'damaged' DNA ends to restore conventional 3′-OH and 5′-P chemistries, DNA synthesis to fill the gap within the SSB and DNA ligation to restore continuity of the DNA strand.

Detection of the DNA SSB

SSB detection is a crucial stage of SSBR, as it is required for the rapid recruitment of the downstream repair machinery. Detection of direct SSBs is predominantly carried out by members of the poly(ADP)-ribose polymerase family of ADP-ribotrasferase enzymes which use NAD+ to modify their protein targets with either single ADP-ribose units or poly(ADP-ribose) chains (PAR) (Caldecott 2014b). PARP1, PARP2 and PARP3 are three members of the PARP family that function as DNA break sensors, and poly(ADP-ribosylate) themselves and other target proteins upon induction of DNA breaks.

PARP1 is a very abundant repair factor and was the first PARP involved in DNA repair to be identified. PARP1 activity is stimulated more than 500 times upon binding to DNA strand breaks, and has been shown to be responsible for over 90% of detectable PAR synthesis following DNA damage (Ame et al. 1999). Once PARP1 is bound to an SSB, it rapidly autoribosylates itself with long-branching PAR chains, and then dissociates from the DNA (Ferro and Olivera 1982).

One major function of this damage-induced ribosylation signal is to facilitate the rapid recruitment of DNA repair factors, such as XRCC1, to sites of DNA SSBs, and loss of PARP1 results in defective recruitment of SSBR factors, delayed SSBR and hypersensitivity to DNA SSBs (El-Khamisy et al. 2003, Hanzlikova et al. 2017). XRCC1 is a molecular scaffold protein that directly interacts with poly(ADP-ribosylated) PARP1 and multiple SSBR components (including POLβ, PNKP, APTX, APE1, PCNA and DNA LIGIII), and plays a critical role in their recruitment and assembly at DNA SSBs (Vidal et al. 2001, Whitehouse et al. 2001,

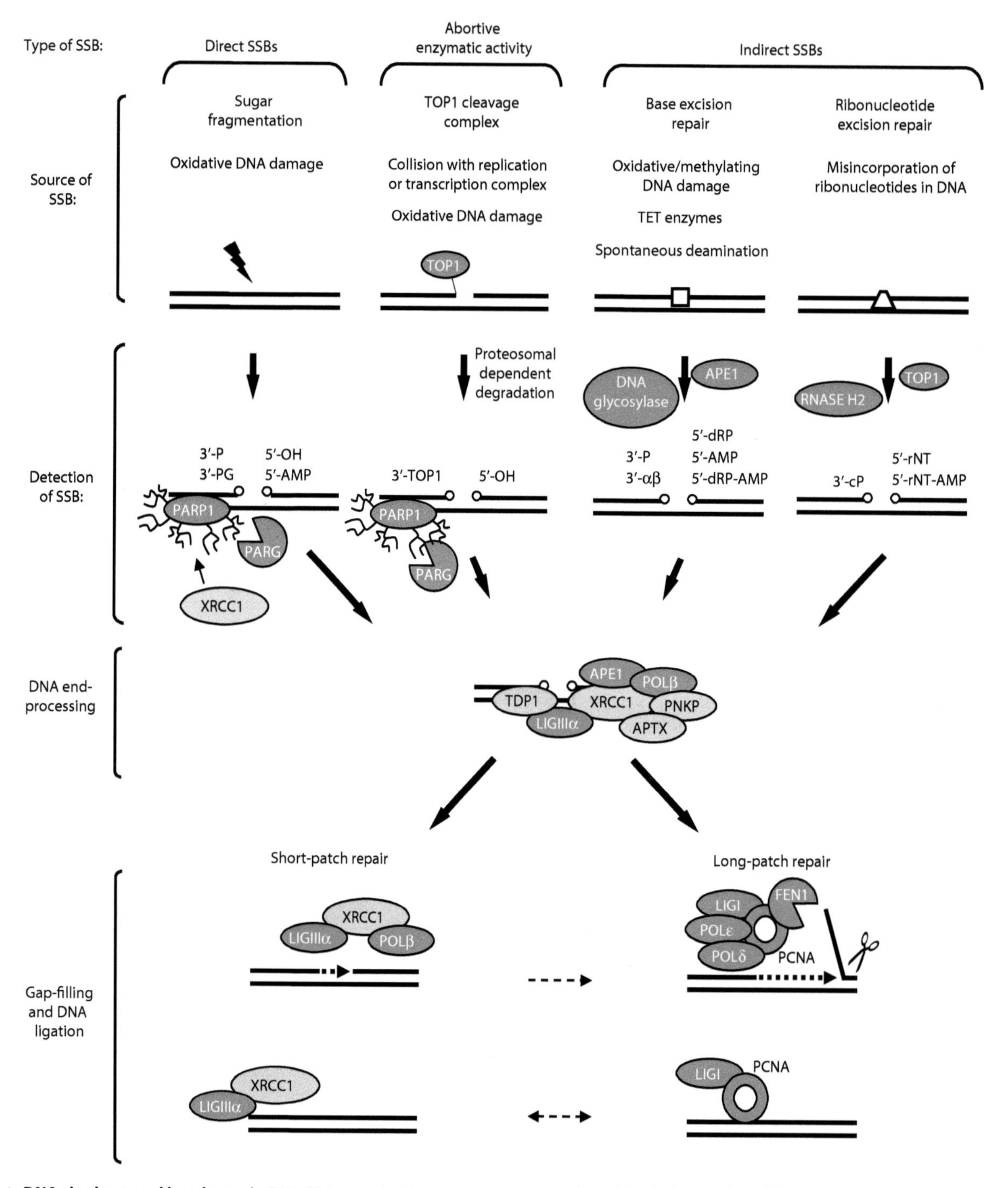

Figure 5.1 DNA single-strand break repair. DNA SSBs can arise in one of several ways: directly by disintegration of the DNA sugar phosphate backbone, from the abortive enzymatic activity of TOP1 or indirectly during either the repair of damages bases by DNA glycosylases (base excision repair) or the removal of misincorporated ribonucleotides from DNA by RNase H2 (ribonucleotide excision repair). Irrespective of how they arise, DNA SSBs are rapidly repaired by overlapping repair pathways collectively called single-strand break repair. **Detection of SSB:** Direct SSBs are recognised by PARP1, which rapidly modifies itself and other target proteins with long branching chains of poly(ADP-ribose) that act as a signal for the rapid recruitment of XRCC1. PARP1 also becomes activated at abortive TOP1 cleavage complexes and is required for the rapid repair of TOP1-SSBs. Indirect SSBs are scheduled repair intermediates within the base excision repair or ribonucleotide excision repair pathways and are thought to be 'handed off' to the next stage of repair. Therefore, it is thought that they do not need to be detected by PARP1, unless they become uncoupled from their respective repair pathways. **DNA end-processing:** The majority of DNA SSBs possess 'damaged' DNA termini and need to be processed to restore the conventional 3′-OH and 5′-P termini. Due to the variety of different chemistries that can arise, DNA end-processing is the most enzymatically diverse stage of SSBR. XRCC1 is crucial for this stage of repair, as although it has no enzymatic activity itself, it functions to recruit, and stimulate, the end-processing factors required to repair the DNA ends. **Gap-filling and DNA ligation:** Most SSBs possess a one-nucleotide gap which needs to be filled once the damaged termini have been repaired. At this stage SSBR splits into two separate repair sub-pathways. Within short-patch repair, a single nucleotide is inserted into the gap and the resulting DNA nick is ligated. It is thought that this is predominantly carried out by POLβ and LIGIIIα. In contrast, long-patch repair involves the synthesis of a longer stretch of DNA (2-12 nucleotides) by POLβ, POLδ and/or POLε, and the displacement of a 5′ single-stranded flap which is removed by FEN1. Repair is then completed by LIGI. (Adapted from Caldecott, K. W., 2014, *Exp Cell Res* 329: 2–8.)

Plo et al. 2003, Clements et al. 2004). In addition, XRCC1 has also been shown to stimulate the activities of its interacting partners to facilitate rapid SSBR (Vidal et al. 2001, Whitehouse et al. 2001). Therefore, even though XRCC1 has no enzymatic activity itself, its role in the recruitment and stimulation of SSBR proteins is critical for the rapid repair of DNA SSBs, and loss of XRCC1 confers global defects in SSBR, hypersensitivity to DNA-damaging agents and embryonic lethality (Tebbs et al. 2003, Breslin and Caldecott 2009).

The half-life of PAR chains in cells is very short, and typically is only detectable for a few minutes (Fisher et al. 2007). This is due to poly(ADP-ribose) glycohydrolase (PARG) rapidly degrading PAR chains and restoring PARP1 to its 'naked' unmodified form (Cortes et al. 2004, Fisher et al. 2007). This cycle of PARP1 modification and demodification is thought to allow PARP1 to be recycled so it is always ready to detect further SSBs. However, the rapid deribosylation of PARP1 may also play other roles, such as preventing SSBR repair factors from being sequestered at SSBs at higher levels of DNA damage, and allowing them to be released and recruited to other DNA breaks (Fisher et al. 2007). The importance of PARG in facilitating SSBR is demonstrated by the observation that depletion of PARG results in an SSBR defect that is similar to, and epistatic with, the repair defect observed in PARP1-depleted cells (Fisher et al. 2007).

PARP1 has also been suggested to play a role in promoting the repair of TOP1-SSBs. PARP1-dependent poly(ADP-ribosylation) of TDP1, the end-processing enzyme responsible for the repair of TOP1-SSBs, has been shown to promote its recruitment to abortive TOP1 cleavage complexes, and combined loss of both TDP1 and PARP1 leads to levels of cellular hypersensitivity to TOP1 inhibition which is similar to individual loss of TDP1 or PARP1 (Das et al. 2014).

PARP2 and PARP3 are two additional DNA-break-induced PARP enzymes that have been implicated in SSBR. It has been suggested that PARP1 and PARP2 have overlapping roles in the recruitment of XRCC1 and PNKP, and indeed combined loss of both PARP1 and PARP2 is required to completely suppress oxidative stress–induced poly(ADP-ribosylation) and recruitment of XRCC1 and PNKP (Hanzlikova et al. 2017). Additionally, PARP3 has also been shown to bind to SSBs within chromatin and is required for efficient SSBR in DT40 cells (Grundy et al. 2016).

Although PARP1 seems to be activated by the presence of indirect SSBs generated during BER, the role it plays in BER is unclear. Damaged DNA bases are detected by specific DNA glycosylases, and BER is then thought to occur via a 'handoff' mechanism by which the SSB is passed directly to the required downstream SSBR factors (Prasad et al. 2010). Indeed, following removal of the damaged base, monofunctional DNA glycosylases remain bound to the resulting AP site until they are displaced by APE1. Furthermore, multiple interactions between SSBR factors and DNA glycosylases exist, and it has been suggested that DNA glycosylases exist in pre-assembled repair complexes that can facilitate complete repair of damaged DNA bases (Parlanti et al. 2007). Therefore, it is possible that PARP1 is only required for the repair of SSBs that become uncoupled from BER.

DNA end processing

The majority of DNA SSB DNA termini possess 'damaged' termini and need to be processed to restore the conventional 5′-phosphate (5′-P) and 3′-hydroxyl (3′-OH) moieties required for DNA ligation. Depending on the origin of the SSB, a broad range of different chemistries can be present at damaged termini, and consequently a large variety of DNA end-processing factors are employed during SSBR (Table 5.2). Processing of damaged 3′ termini is thought to be particularly important, as a failure to repair broken 3′ DNA ends would block both DNA repair synthesis and DNA ligation, leading to persisting unrepaired SSBs. In contrast, unprocessed damaged 5′ termini can be repaired via long-patch repair during downstream repair events (Sung and Demple 2006).

An end-processing factor that repairs SSBs arising from a variety of sources is polynucleotide kinase 3′-phosphatase (PNKP), a bifunctional enzyme with 3′-phosphatase and 5′-kinase activities (Whitehouse et al. 2001, Bernstein et al. 2005). The importance of PNKP is highlighted by the fact that mutations in PNKP are associated with neurodevelopmental defects in humans and mice, and complete loss of PNKP is embryonically lethal (Shen et al. 2010, Shimada et al. 2015). The severity of the defects observed upon loss of PNKP can potentially be explained by the large volume of DNA SSBs arising in the cell that requires PNKP activity to repair. Indeed, DNA SSBs with 3′-P termini are a particularly abundant type of DNA lesion, and are present at approximately 70% of SSBs arising from ROS-induced

Table 5.2 Common types of damaged DNA termini at DNA SSBs and the end-processing factors that repair them.

Type of damaged DNA termini	Source	Major end-processing factor[a]
Types of damaged 3′ termini		
3′-phosphate (3′-P)	3′-phosphate termini can arise from: direct sugar fragmentation following oxidative DNA damage; cleavage of an AP site by a bifunctional DNA glycosylase (βδ elimination) following removal of an oxidative DNA base lesion; cleavage of a 3′-phosphotyrosyl bond within a TOP1-SSB by TDP1	Polynucleotide kinase 3′-phosphatase (PNKP)
3′-phosphoglycolate (3′-PG)	3′-phosphoglycolate termini arise from direct sugar fragmentation following oxidative DNA damage	AP endonuclease 1 (APE1)
3′-αβ unsaturated aldehyde (3′-PUA)	3′-αβ unsaturated aldehyde termini result from the cleavage of an AP site by a bifunctional DNA glycosylase (β elimination) following removal of an oxidative DNA base lesion	AP endonuclease 1 (APE1)
3′-TOP1 peptide/3′-phosphotyrosyl	3′-TOP1 peptides are generated by proteolytic degradation of an abortive TOP1-SSB cleavage complex	Tyrosyl-DNA phosphodiesterase 1 (TDP1)
2′,3′-cyclic phosphate (3′-cP)	2′,3′-cyclic phosphate termini arise when TOP1 cleaves DNA containing a ribonucleotide at the TOP1 cleavage site	?
Types of damaged 5′ termini		
5′-hydroxyl (5′-OH)	3′-hydroxyl termini can result from: direct sugar fragmentation following oxidative DNA damage; repair of a TOP1-SSB by TDP1	Polynucleotide kinase 3′-phosphatase (PNKP)
5′-aldehyde	5′-aldehyde termini result from direct sugar fragmentation following oxidative DNA damage	?
5′-deoxyribose phosphate (5′-dRP)	5′-deoxyribose phosphate termini arise from the cleavage of an AP site by APE1	DNA polymerase β (POLβ)
5′-ribonucleotide (5′-rNT)	Cleavage of ribonucleotides within DNA by RNase H2	NA
5′-adenosine monophosphate (5′-AMP)	5′-adenosine monophosphate termini result from abortive ligation reactions	Aprataxin (APTX)
5′-adenosine monophosphate ribonucleotide (5′-AMP-rNT)	5′-adenosine monophosphate ribonucleotide termini result from the abortive ligation of SSBs containing a 5′-ribonucleotide	Aprataxin (APTX)
5′-adenosine monophosphate-deoxyribose phosphate (5′-AMP-dRP)	5′-adenosine monophosphate deoxyribose phosphate termini result from the abortive ligation of SSBs containing a 5′-deoxyribose phosphate	Aprataxin (APTX)

[a]Does not include minor activities of DNA end-processing enzymes.

sugar fragmentation. They are also generated indirectly via the excision and cleavage of base damage in a βδ elimination reaction by bifunctional DNA glycosylases and are produced following the TDP1-dependent repair of TOP1-SSBs (Ward 1998, Caldecott 2001, Interthal et al. 2001). Furthermore, 5′-OH termini are also present at a small portion of SSBs caused by ROS-induced sugar disintegration (Ward 1998, Caldecott 2001). Loss/mutation of PNKP results in defective SSBR and cellular hypersensitivity to ionising radiation, hydroxide peroxide, MMS and camptothecin (Breslin and Caldecott 2009, Reynolds et al. 2012).

APE1 is another important end-processing factor in SSBR. In addition to its role in repairing AP sites, APE1 can resolve 3′-PG termini (Winters et al. 1994), which arise at approximately 30% of direct SSBs induced by ROS (Ward 1998, Caldecott 2001), and to a lesser extent can repair 3′P termini (Izumi et al. 2000). Conditional inactivation of APE1 leads to an accumulation of unrepaired AP sites and SSBs, and like PNKP, complete loss of APE1 confers embryonic lethality (Izumi et al. 2000, 2005). Therefore, PNKP and APE1 are both crucial end-processing factors and together are responsible for the repair of the majority of direct SSBs induced by oxidative stress.

An end-processing factor that has a vital role in the processing of indirect SSBs, rather than direct SSBs, is DNA polymerase β (POLβ), a member of the X-family of polymerases that possess both a polymerase domain and a small N-terminal dRP lyase domain (Beard and Wilson 2000). Cleavage of AP sites by APE1 generates undamaged 3′-OH termini and damaged 5′-dRP termini, which are substrates for the AP lyase activity of POLβ (Sobol et al. 2000). Interestingly, the AP lyase activity of POLβ seems to play a more important role than its polymerase activity, and expression of the N-terminal dRP lyase domain without the polymerase domain can rescue the hypersensitivity of $POL\beta^{-/-}$ MEFs to DNA methylating agents (Sobol et al. 2000). However, it is worth noting that 5′-dRP termini are potentially reactive and under some conditions may be further modified (e.g. becoming oxidised), in which case they become refractory to repair by POLβ and are repaired by long-patch SSBR as described later (Sung and Demple 2006).

Damaged DNA termini can also arise from abortive enzymatic activity. Unsuccessful TOP1 activity can result in an abortive TOP1-SSB in which the tyrosine residue within the active site of TOP1 is linked to the 3′ termini of the SSB via a phosphotyrosyl linkage (Interthal et al. 2001). Before repair of the TOP1-SSB can proceed, full-length TOP1 undergoes proteasomal degradation to a TOP1 small peptide (Lin et al. 2008). Tyrosyl-DNA phosphodiesterase 1 (TDP1) can then hydrolyse the phosphotyrosyl linkage to remove the TOP1 peptide, leaving a nick with 3′-P and 5′-OH termini, which requires further processing by PNKP (Interthal et al. 2001, El-Khamisy et al. 2005).

Another example of damaged DNA termini resulting from abortive enzymatic activity are 5′-AMP termini, which arise from abortive ligation reactions (Ahel et al. 2006). DNA ligases catalyse phosphodiseter bond formation in a universal three-step reaction mechanism, in which the ligase first reacts with ATP to form a ligase-adenylate intermediate, before transferring the AMP group from its active site to the 5′-P of the DNA, forming an SSB with a 5′-AMP (Tomkinson et al. 2006). This transient AMP-DNA intermediate then undergoes nucleophilic attack by the 3′OH to form a phosphodiester bond, displacing the AMP molecule and completing the ligation reaction (Tomkinson et al. 2006). Under certain situations, such as a premature attempt to ligate a SSB with either a damaged 3′ termini (e.g. 3′-P) or unligatable 5′ termini (e.g. 5′-dRP or 5′-rNT), the ligation reaction will stall and a stable abortive ligation intermediate will persist (Ahel et al. 2006, Tumbale et al. 2014, Caglayan et al. 2015). Adenylated SSBs are repaired by APTX, which releases the AMP group from the DNA to reset the SSB to the previous state (Ahel et al. 2006).

There are several types of damaged termini for which an end-processing factor has not yet been discovered. For example, cleavage of an AP site in a β elimination reaction by bifunctional DNA glycosylases will generate a 3′-αβ unsaturated aldehyde (Krokan and Bjoras 2013) and the presence of a ribonucleotide within DNA at the site of TOP1 cleavage will result in 2′,3′-cyclic phosphate termini (Williams et al. 2013). In addition, 5′-aldehyde termini can arise at a small fraction of ROS-induced sugar fragmentation events (Ward 1998), and nicks with 5′-ribonucleotide termini are generated by the RNase H2–dependent incision of DNA containing ribonucleotides during RER. None of these damaged termini can act as substrates for any known processing enzymes; however, damaged 5′-aldehyde and 5′-ribonucleotide termini can still be repaired by long-patch SSBR pathways (Williams et al. 2016).

XRCC1 also plays a crucial role during DNA end processing. It is present within complexes with many end-processing factors, including PNKP, APE1, APTX, POLβ and TDP1, and is required for their rapid assembly at sites of SSBs (Vidal et al. 2001, Whitehouse et al. 2001, Plo et al. 2003, Clements et al. 2004). In addition, XRCC1 has also been shown to stimulate the activities of several of its interacting partners, including PNKP, APE1 and TDP1, to facilitate rapid repair of SSBs (Vidal et al. 2001, Whitehouse et al. 2001, Plo et al. 2003).

Gap filling and DNA ligation

Once the canonical 3′-OH and 5′-P groups have been restored to the DNA ends, the DNA can then be ligated to complete repair. However, as most SSBs possess a one-nucleotide gap, a gap-filling step must occur before ligation can proceed. SSBR splits into two sub-pathways at this stage (Kim and Wilson 2012). Short-patch repair involves the incorporation of a single nucleotide into the gap, whilst long-patch repair involves the synthesis of a longer patch of 2–12 nucleotides, displacing the 5′ nucleotides of the SSB to form a 5′ flap. This 5′ flap is then removed by Flap endonuclease 1 (FEN1).

Several DNA polymerases have been implicated in both the short-patch and long-patch repair pathways, although the exact roles of each polymerase are still unclear. POLβ is thought to be the main polymerase involved in short-patch repair (Kim and Wilson 2012), although POLλ has been suggested to be able to compensate for loss of POLβ (Braithwaite et al. 2010, Brown et al. 2011). POLβ and POLλ are also thought to have a role in long-patch repair alongside the replicative polymerases (POLδ and POLε) (Balakrishnan et al. 2009, Brown et al. 2011). In addition, PARP1 and FEN1 have also been found to stimulate long-patch repair synthesis (Caldecott 2014a).

It is also not currently clear what influences the choice between short-patch and long-patch SSBR, and it is likely to be a combination of factors. It is known, for example, that long-patch repair is the pathway of choice in the event that a blocked 5′ DNA terminus cannot be processed (Sung and Demple 2006). In addition, following RNASH2-dependent incision, ribonucleotides in DNA are removed and repaired by RER in a process very similar to long-patch repair (Williams et al. 2016). It has also been observed that the identity of the glycosylase that initiates BER seems to influence which pathway is used (Fortini et al. 1999). Moreover, it has been proposed that the levels of ATP local to the SSB can determine the pathway choice via a XRCC1-Ligase IIIα–dependent mechanism (Petermann et al. 2006).

Finally, once the gap is filled, the remaining nick is ligated by either DNA Ligase IIIα (LIGIIIα) and XRCC1, or DNA Ligase I (LIGI) and PCNA (Tomkinson et al. 2006). Traditionally, XRCC1-LIGIIIα has been thought to predominantly function during short-patch repair, whilst PCNA-LIGI was typically associated with long-patch (Caldecott 2008). However, it is becoming clear that this model of different ligases functioning separately in short-patch repair and long-patch repair is overly simplistic, and that the roles of XRCC1-LIGIIIα and PCNA-LIGI in SSBR are likely to be more overlapping/redundant than previously thought.

DNA SSBs and human disease

Due to the high quantities of SSBs that can arise in the cell, and the deleterious consequences of persisting unrepaired SSBs, it is unsurprising that mutations in SSBR factors are associated with human disease. In particular, the human central nervous system seems to be especially sensitive to defective SSBR, as demonstrated by the discovery of five neuropathological disorders associated with defective SSBR: (1) spinocerebellar ataxia with axonal neuropathy 1 (SCAN1) caused by mutation of TDP1; (2) ataxia oculomotor apraxia 1 (AOA1) caused by mutation or loss of APTX; (3) microcephaly, early onset, intractable seizures and developmental delay (MCSZ) and (4) ataxia oculomotor apraxia 4 (AOA4), which are both caused by PNKP mutations; and (5) spinocerebellar ataxia 26 (SCAR26) caused by mutation of XRCC1 (Table 5.3) (Date et al. 2001, Moreira et al. 2001, Takashima et al. 2002, Shen et al. 2010, Bras et al. 2015, Hoch et al. 2017). Interestingly, defective DNA end processing seems to underlie all the SSBR disorders. TDP1, APTX and PNKP are all critical end-processing factors involved in repairing damaged termini within SSBs (Reynolds and Stewart 2013, Jiang et al. 2017). Furthermore, XRCC1 interacts with multiple end-processing factors, and functions to facilitate their rapid recruitment to SSBs and to stimulate their activities. This emphasises the importance of repairing damaged DNA termini present within DNA breaks. As there are many end-processing factors involved in SSBR other than PNKP, APTX and TDP1, it is also possible that mutations in some of these other factors are responsible for yet uncharacterised cerebellar ataxias or microcephalic disorders.

Last, deficiencies in the DNA glycosylases UNG and MUTYH have been found to cause immunodeficiency and familial colorectal cancer, respectively (Table 5.3) (Al-Tassan et al. 2002, Imai et al. 2003). In this section, we will discuss the clinical phenotypes and underlying mechanistic defects associated with these SSBR/BER human disorders.

DNA end processing and human neurological disease

Although neurological dysfunction is a common symptom of many DNA repair defective human diseases, a striking feature of the SSBR disorders is the lack of extraneurological symptoms. This raises two important related questions. Why are post-mitotic neurons in the adult brain so sensitive to SSBR defects? And why can other tissues tolerate loss of SSBR factors?

There are several attributes of post-mitotic neurons that may contribute to their sensitivity to unrepaired SSBs. Neurons are highly metabolically active and the brain has a very high oxygen demand (Barzilai 2007). In addition, despite the fact that neurons exist in an environment full of reactive metabolic by-products, they are less able to neutralise free radicals than other tissue types (Barzilai 2007). As neurons are very long-lived, have a limited regenerative capacity and are exposed to high levels of oxidative DNA damage, they are therefore highly dependent on DNA repair pathways to maintain their functionality (Chen et al. 2007, Steward et al. 2013). One of the consequences of a failure to repair DNA lesions is that DNA damage can interfere with transcription, and the progressive accumulation of unrepaired DNA lesions

Table 5.3 **Human diseases associated with defective single-strand break repair.**

Gene	Gene function	Disease	MOI	Clinical presentation	OMIM #	References
APTX	SSBR end-processing factor that removes DNA breaks with 5′AMP termini	Ataxia oculomotor apraxia 1 (AOA1)	AR	Progressive cerebellar atrophy, ataxia, oculomotor apraxia, muscle weakness, peripheral axonal motor and sensory neuropathy	208920	Moreira et al. (2001), Date et al. (2001)
TDP1	SSBR end-processing factor that removes 3′ TOP1 peptides from DNA breaks	Spinocerebellar ataxia, autosomal recessive with axonal neuropathy (SCAN1)	AR	Cerebellar atrophy, ataxia, peripheral axonal motor and sensory neuropathy, gait disturbance	607250	Takashima et al. (2002)
PNKP[a]	Essential SSBR DNA end-processing factor with 3′ phosphatase and 5′ kinase activities	Microcephaly, seizures, and developmental delay (MCSZ)	AR	Progressive microcephaly, developmental delay, mental retardation, muscular atrophy, seizures, speech difficulties	613402	Shen et al. (2010)
		Microcephaly, seizures, and developmental delay/ ataxia oculomotor apraxia (MCSZ-AOA)	AR	Progressive microcephaly, short stature, progressive cerebellar atrophy, ataxia, developmental delay, muscular atrophy, seizures	613402	Poulton et al. (2013)
		Ataxia oculomotor apraxia 1 (AOA4)	AR	Progressive cerebellar atrophy, ataxia, oculomotor apraxia, muscle weakness, peripheral neuropathy, muscle weakness	616267	Bras et al. (2015)
XRCC1	Essential factor required for the recruitment of downstream SSBR proteins	Spinocerebellar ataxia 26 (SCAR26)	?[b]	Cerebellar atrophy, ataxia, oculomotor apraxia, peripheral axonal motor and sensory neuropathy	617633	Hoch et al. (2017)
UNG	DNA glycosylase involved in BER that removes uracil from DNA	Immunodeficiency with hyper IgM, type 5 (HIGM5)	AR	Recurrent respiratory tract infections, immunodeficiency, elevated IgM	608106	Imai et al. (2003)
MUTYH	DNA glycosylase involved in BER that removes adenine mispaired with 8-oxo-G, cytosine or guanine	Familial adenomatous polyposis 2 (FAP2)	AR	Predisposition to colorectal carcinoma, development of colorectal and colonic adenomatous polyps	608456	Al-Tassan et al. (2002)

Abbreviations: AR, autosomal recessive; MOI, mode of inheritance; OMIM, online mendelian inheritance in man.

[a]Mutations in this gene are associated with multiple distinct clinical presentations.

[b]Insufficient data to determine mode of inheritance.

could potentially result in increasing levels of stalled transcription complexes. This could lead to increased cell death, changes in mRNA levels and/or aberrant cellular function over time, ultimately resulting in neurodegeneration (Bendixen et al. 1990, Ljungman and Lane 2004, Jiang et al. 2017).

In contrast, primary microcephaly reflects a problem occurring within the developing brain. To generate the huge numbers of neurons that make up the adult brain, mammalian neurodevelopment involves a prolonged period of rapid stem cell expansion and neuronal proliferation. Neurons in the developing brain have been shown to have a much lower threshold of apoptosis for DNA damage than post-mitotic neurons (Gatz et al. 2011), potentially explaining why neurodevelopment is particularly sensitive to the persistence of unrepaired DNA damage. Defective DNA repair during this delicate period could lead to increased neuronal cell death, and impact the total numbers of neurons produced during neurodevelopment, therefore resulting in microcephaly.

The existence of redundant repair mechanisms that can compensate for the loss of SSBR factors in cycling cells can potentially explain the lack of extraneurological pathologies in SSBR-defective patients. For example, SSBs that collide with a replication fork will be converted into a DSB and can be repaired by homologous recombination (discussed in Chapter 6 in detail), an S- and G2-phase specific pathway that does not operate in non-cycling cells (Saleh-Gohari et al. 2005). In addition, as homologous recombination is a predominantly error-free repair pathway that has critical roles in the prevention of tumourigenesis, this could explain the lack of cancer predisposition in individuals with SSBR defects (Katsuki and Takata 2016). Specific compensatory repair mechanisms may also be relevant to the different diseases and will be discussed in more detail.

Ataxia oculomotor apraxia 1

AOA1 is one of the most common autosomal recessive ataxias and is characterised by early onset ataxia, oculomotor apraxia (inability to control eye movements), cerebellar atrophy and axonal motor neuropathy (Date et al. 2001, Moreira et al. 2001). Loss of Purkinje cells has also been observed post-mortem in AOA1 patients (Sugawara et al. 2008). Gait ataxia is usually the first symptom to present and can be the most debilitating, as patients can be severely disabled and wheelchair-bound by later life.

APTX possesses three distinct domains: a C-terminal zinc finger motif (DNA-binding domain), a catalytic histidine triad (HIT) domain which is critical for its deadenylase activity, and an N-terminal fork-head-associated (FHA) domain (phosphopeptide binding) that mediates its interaction with CK2 phosphorylated XRCC1 (Moreira et al. 2001, Clements et al. 2004, Ahel et al. 2006). Currently, 25 different mutations in APTX, including both missense and frameshift mutations, as well as whole genome deletions, have been found in AOA1 individuals (Yoon et al. 2008, Schellenberg et al. 2015). The majority of APTX mutations are mutations that reside within, and/or inactivate, the catalytic HIT domain, and indeed cell-free extracts from AOA1 cell lines are completely unable to repair abortive ligation intermediates in vitro (Seidle et al. 2005, Ahel et al. 2006, Reynolds et al. 2009). The remaining mutations are present in the N-terminal FHA domain and are predicted to disrupt DNA binding, disturb important interdomain interactions or destabilise the protein (Tumbale et al. 2011). There is some variation in the severity of clinical presentations within AOA1 individuals, and some patients may not develop oculomotor apraxia (Le Ber et al. 2003, Yokoseki et al. 2011).

It is not clear how common 5′-adenylated SSBs are in human cells. Although abortive ligation intermediates can arise at direct and indirect SSBs induced by oxidative damage, or at SSBs generated from APE1-dependent cleavage of AP sites, it is possible that these are relatively rare occurrences, and they have not been detected in vivo (Ahel et al. 2006, Caglayan et al. 2015). However, abortive ligation intermediates seem to commonly occur at SSBs with 5′-rNT termini resulting from the RNase H2-dependent cleavage of misincorporated ribonucleotides (Tumbale et al. 2014).

Conflicting evidence from different studies exists on whether AOA1 cell lines are sensitive to DNA-damaging agents or have delayed global levels of SSBR. Various studies have shown that some AOA1 cell lines are hypersensitive to SSB-induced agents and display delayed rates of SSBR, whilst other studies have shown normal levels of cellular sensitivity and normal SSBR rates (Reynolds and Stewart 2013, Jiang et al. 2017). It is possible that this reflects the nature and position of the APTX mutations present in the AOA1 cell lines used in the different studies, or perhaps is a result of the different experimental setups used, although this remains unclear.

It has also been proposed that alternative repair mechanisms can compensate for the loss of APTX. Indeed, factors involved in long-patch SSBR, such as the replicative DNA polymerase repair and FEN1, seem to compensate for loss of APTX in *Aptx*$^{-/-}$ mouse astrocytes, ATPX null DT40 cells and cell-free extracts from AOA1 lymphoblastoid cell lines (Reynolds et al. 2009, Caglayan et al. 2015). Additionally, yeast strains deleted for Hnt3 (*Saccharomyces cerevisiae* homologue of APTX), become sensitive to oxidative and methylation DNA damage upon additional loss of rad27 (homologue of FEN1) (Daley et al. 2010). Furthermore, disrupting repair of damaged 3′ termini also results in increased cellular sensitivity to SSB-inducing agents in Hnt3Δ yeast cells, and reduced rates of SSBR in *Aptx*$^{-/-}$ mice, possibly by preventing extension from the 3′ termini of SSBs during long-patch repair (El-Khamisy et al. 2009, Daley et al. 2010). Moreover, POLβ has also been shown to remove 5′-AMP-dRP termini using its lyase activity (Caglayan et al. 2015).

The fact that long-patch repair can compensate for loss of APTX may account for the lack of extraneurological symptoms in patients with AOA1. Indeed, DNA replicative factors, such as DNA POLδ, POLε, PCNA and FEN1, which function during long-patch repair, would be significantly attenuated in non-cycling cells (Caldecott 2008, Kim and Wilson 2012). Therefore, it is possible that post-mitotic neurons are more dependent on APTX-dependent SSBR than cycling cells, which would possess redundant repair mechanisms, particularly S-phase and G2-associated repair pathways.

Although the mechanism underlying AOA1 still remains unclear, at least in part due to the lack of any obvious phenotypes in the *Aptx*$^{-/-}$ mice (Ahel et al. 2006), the progressive accumulation of unrepaired DNA damage is considered to be the most likely cause of neurodegeneration in AOA1 patients (Rass et al. 2007). In addition, it is informative that increasing the levels of oxidative stress in *Aptx*$^{-/-}$ mice by crossing them with mice possessing a mutant form of the antioxidant enzyme superoxide dismutase (*Sod1*^{G93A}) results in defective survival of motor neurons and premature ageing in the *Aptx*$^{-/-}$*Sod1*^{G93A} mice (Carroll et al. 2015). *Aptx*$^{-/-}$ *Sod1*^{G93A} MEFs also exhibited delayed rates of SSR following oxidative DNA damage (Carroll et al. 2015). This supports the hypothesis that APTX functions to repair abortive ligation intermediates that arise at oxidative SSBs. APTX has also been reported to repair oxidative SSBs with 3′-P and 3′-PG termini, which may contribute to the accumulation of oxidative SSBs in the absence of APTX (Takahashi et al. 2007). Although SSBs possessing a 5′-rNT seem to be particularly susceptible to the formation of abortive ligation intermediates (Tumbale et al. 2014), ribonucleotides are predominantly misincorporated during DNA replication, and therefore it is unclear whether SSBs with 5′-AMP-rNT termini are a relevant lesion in non-cycling post-mitotic neurons (Nick McElhinny et al. 2010).

Spinocerebellar ataxia with axonal neuropathy 1

SCAN1 is an adult-onset autosomal recessive ataxia characterised by progressive cerebellar ataxia, dysarthia (difficulties with speech) and progressive atrophy of the muscles of the fingers and feet (Takashima et al. 2002). SCAN1 typically has a later age of onset and a milder clinical presentation than AOA1, and has not been associated with oculomotor apraxia or cognitive decline. To date, mutations in TDP1 have only been identified in nine individuals from a single Saudi Arabian family. All SCAN1 patients are homozygous for a single amino-acid substitution of histidine 493 (H493R), which is a critical residue within the catalytic motif of the phosphodiesterase domain (Takashima et al. 2002). The H493R mutation causes a 25-fold reduction in TDP1 activity in vitro and also results in abortive TDP1 activity, consequently leaving TDP1 bound at the 3′ terminus of the DNA strand break and leading to an accumulation of covalent TDP1-DNA intermediates (Interthal et al. 2005).

Consistent with the role of TDP1 in the repair of abortive TOP1-SSBs, SCAN1 cells have been shown to accumulate more DNA breaks in the presence of camptothecin (El-Khamisy et al. 2005). Several lines of evidence also show that the repair of both replication-dependent and transcription-dependent TOP1-SSBs is defective in SCAN1 cells (El-Khamisy and Caldecott 2006, Miao et al. 2006). Furthermore, SCAN1 cell lines also exhibit delayed SSBR of oxidative DNA damage (El-Khamisy et al. 2005, Katyal et al. 2007). As the close proximity of DNA lesions to the topoisomerase cleavage complex increases the likelihood of an abortive reaction (Pourquier et al. 1997), this could reflect an increase in oxidative stress–induced TOP1-SSBs. Alternatively, it could also reflect defective processing of oxidative DNA breaks and base damage, as several studies have shown that TDP1 can repair different types of damaged 3′ termini and can hydrolyse AP sites (Zhou et al. 2005, Lebedeva et al. 2012). It has been proposed that accumulating levels of transcription arrest caused by unrepaired DNA lesions, most likely TOP1-SSBs, underlie the neurodegeneration seen in SCAN1 individuals.

Several *Tdp*$^{-/-}$ mouse models have been generated, and, importantly, cells derived from these mice recapitulate the cellular defects seen in SCAN1 cells (Katyal et al. 2007, Hawkins et al. 2009). However, none of the studies found any overt neurological abnormalities or ataxic phenotypes in the *Tdp*$^{-/-}$ mice compared to their littermate controls. One study did observe a progressive decrease in cerebellar size in the *Tdp*$^{-/-}$ (Katyal et al. 2007), but this decrease was not found in another study (Hawkins et al. 2009). It is not known why the *Tdp*$^{-/-}$ mice fail to phenocopy SCAN1 patients, although a complete loss of TDP1 is not the same as possessing the H493R mutation, and it is possible that the accumulation of the TDP1-DNA intermediates in the brains of SCAN1 individuals contributes to the disease phenotype.

The existence of dedicated DNA replication–associated repair pathways in cycling cells that operate in parallel to TDP1 to repair TOP1-SSBs and TOP1-DSBs may explain why mutation of TDP1 predominantly impacts on post-mitotic neurons. When an ongoing replication fork collides with a TOP1-SSB, this generates a TOP1-DSB which is repaired by the MRE11-RAD50-NBS complex during homologous recombination (discussed further in Chapter 12) (Aparicio et al. 2016). Furthermore, TOP1-SSBs can also be repaired in cycling cells by a DNA replication-associated pathway that is dedicated to removing DNA-protein cross-links that collide with the replisome (detailed in Chapter 12) (Vaz et al. 2016). This pathway involves the proteolytic degradation of DNA-protein cross-links by the SPRTN metallopeptidase. As neither of these pathways would operate within non-cycling cells, post-mitotic neurons would therefore be more dependent on TDP1 to repair abortive TOP1 complexes.

Spinocerebellar ataxia 26

Biallelic mutations in XRCC1 were recently identified in a patient that presented with cerebellar atrophy, ataxia, oculomotor apraxia and peripheral neuropathy (Hoch et al. 2017).

Patient-derived fibroblasts exhibited greatly reduced levels of XRCC1 and its obligate partner LIGIIIα. Consistent with the critical role for XRCC1 in SSBR, SCAR26 patient-derived fibroblasts exhibited a significant delay in the repair of oxidative stress-induced SSBs, but did not show delayed repair of oxidative DSBs (Hoch et al. 2017). Furthermore, a substantial increase in the levels of spontaneous sister chromatid exchanges were observed in patient-derived lymphoblastoid cell lines, reflecting homologous recombination-dependent repair of collapsed replication forks that have collided with unrepaired SSBs (Saleh-Gohari et al. 2005). Interestingly, the study also found that excessive levels of PARP1-dependent poly(ADP-ribosylation) persisted in SCAR26 patient cell lines following treatment with hydrogen peroxide and camptothecin (Hoch et al. 2017).

Importantly, the conditional deletion of XRCC1 from the mouse central nervous system during neurodevelopment (*Xrcc1*$^{Nes\text{-}cre}$) also resulted in neurological dysfunction (Lee et al. 2009). Loss of XRCC1 in the central nervous system led to loss of cerebellar interneurons, hippocampal dysfunction, progressive accumulation of DNA breaks within the brain and increased levels of apoptosis in cerebellar granule neurons (Lee et al. 2009, Hoch et al. 2017). Furthermore, the *Xrcc1*$^{Nes\text{-}cre}$ mice also developed profound ataxia. Elevated levels of poly(ADP-ribosylation) were also observed in the cerebellum of the *Xrcc1*$^{Nes\text{-}cre}$ mice, and, crucially, deletion of PARP1 in the *Xrcc1*$^{Nes\text{-}cre}$ mice reduced the excessive levels of ADP-ribose, restored neuronal density to normal levels and reduced the ataxic behaviours observed in the mice (Hoch et al. 2017). Therefore, this supports the model that the neuronal cell death observed in the absence of XRCC1 is due to PARP1 hyperactivity and NAD+ depletion resulting from persisting unrepaired SSBs.

This is consistent with several previous studies that have demonstrated the toxicity of excessive PARP1 activation upon DNA damage. It has been reported that neuronal damage following cerebral ischaemia is due to the excessive activation of PARP1, and subsequent cell death due to depletion of NAD+ levels (Eliasson et al. 1997). Indeed, another study showed that deletion of PARP1 suppressed MMS-induced tissue toxicity and reduced MMS-induced motor dysfunction in *Mpg* $^{-/-}$ mice (Calvo et al. 2013). Therefore, it seems that neuronal cell death caused by PARP1 hyperactivity and NAD+ depletion may be a common mechanism underlying the neurodegeneration seen in DNA repair defective disorders.

Microcephaly, early onset, intractable seizures and developmental delay, and ataxia oculomotor apraxia 4

PNKP dysfunction results in several distinct disorders that can be characterised by microcephaly as well as neurodegeneration (Dumitrache and McKinnon 2017). The first PNKP disorder to be discovered was MCSZ, which is characterised by primary microcephaly, early onset seizures, developmental delay and hyperactivity (Shen et al. 2010). MCSZ patients retain normal brain structures and do not suffer from neurodegeneration or ataxia. In contrast, AOA4 is a disorder characterised by early-onset progressive neurodegeneration, oculomotor apraxia and neuropathy, but patients lack any sign of microcephaly (Bras et al. 2015, Paucar et al. 2016, Tzoulis et al. 2017). The clinical spectrum of PNKP disorders was expanded further by the discovery of individuals with PNKP mutations presenting with microcephaly, seizures, progressive cerebellar degeneration and polyneuropathy, thereby constituting a third disorder that is associated with a combination of symptoms characteristic of both MCSZ and AOA4 (MCSZ-AOA) (Poulton et al. 2013). Furthermore, one MCSZ patient presented with cerebellar atrophy and primordial dwarfism in addition to microcephaly, seizures and developmental delay (Nair et al. 2016).

PNKP is a bifunctional DNA end-processing factor with an N-terminal FHA domain (which mediates its interaction with XRCC1), a phosphatase domain and C-terminal kinase domain (Bernstein et al. 2005). To date, 13 out of 14 disease-associated PNKP mutations reside in either the phosphatase or kinase domains (Jiang et al. 2017). A common truncating mutation (T424Gfs48X) was found to be homozygous in MCSZ and MCSZ-AOA patients, and was found in one allele of an AOA4 individual (Shen et al. 2010, Poulton et al. 2013, Bras et al. 2015). It is not clear why the same homozygous mutation can result in two different clinical phenotypes, although it is likely that other factors, such as genetic modifiers, contribute to the clinical variability. Interestingly, although PNKP mutations associated with MCSZ are found in both the phosphastase and kinase domain, AOA4 associated mutations have so far been restricted to the kinase domain (Bras et al. 2015, Paucar et al. 2016, Tzoulis et al. 2017).

To date, all of the disease-causing mutations have proved to be hypomorphic and retain varying levels of PNKP activity (Reynolds et al. 2012). Indeed, it is striking that all of the PNKP mutations tested retained 3′-phosphatase activity in vitro at 30°C, and only a single mutant (L176F) exhibited significantly reduced repair of 3′-P termini (Reynolds et al. 2012). In contrast, three out of four mutations (L176F, T424Gfs48X and exon15Δfs4X) greatly reduced or completely ablated 5′ kinase activity. Furthermore, the E326K mutation, which is homozygous in four MCSZ individuals, retained WT levels of phosphatase and kinase activities at permissive temperatures (Reynolds et al. 2012). Complete loss of PNKP is embryonically lethal (Shimada et al. 2015), and it is likely that it is the phosphatase activity, rather than the kinase activity of PNKP, which is essential for life. This perhaps reflects the large quantity of SSBs with a 3′-P that can arise in a cell, and a lack of redundant mechanisms that can repair damaged 3′ termini, in comparison to damaged 5′-OH termini, which are much less prevalent and can be repaired by long-patch SSBR. This may also explain why mutations in the phosphatase domain are exclusively associated with the more severe MCSZ and MCSZ-AOA1 diseases rather than AOA4. However, a common feature of the PNKP mutations tested was that they all exhibited significantly reduced stability in vitro at physiological temperatures, and cell lines from MCSZ or AOA4 individuals analysed by Western blot have shown significant reductions in cellular levels of PNKP (Shen et al. 2010, Reynolds et al. 2012, Poulton et al. 2013, Paucar et al. 2016). Furthermore, all MCSZ cell lines studied so far exhibited defective repair of ROS-induced SSBs (Reynolds et al. 2012).

In addition to complete loss of *Pnkp*, homozygosity of the *T424Gfs* mutation is also embryonically lethal in mice. However, a viable PNKP knockout mouse model, in which cellular *Pnkp* levels were reduced by 90%, exhibited a neurological phenotype reminiscent of MCSZ patients (Shimada et al. 2015). Detailed analysis of the developing mouse brain showed that loss of PNKP caused an accumulation of DNA damage and increased levels of apoptosis in neural progenitor cells, leading to microcephaly and seizures (Shimada et al. 2015). In addition, the absence of PNKP also caused loss of certain neural cell types in the adult brain, such as glial cells and oligodendrocytes (Shimada et al. 2015). The phenotypes of the PNKP knockout mice therefore reflect both the neurodevelopmental and neurodegenerative defects seen in MCSZ and AOA4.

Together these data support the hypothesis that defective repair of DNA breaks, resulting in increased levels of p53-dependent apoptosis of neural progenitor cells, underlies the neurodevelopmental defects associated with PNKP mutations. In addition, the similarities between the clinical phenotypes of AOA4 and SCAR26 patients, combined with the overlapping neurological abnormalities in *Pnkp* and *Xrcc1* knockout mice suggest that common mechanisms lie beneath both diseases (Lee et al. 2009, Shimada et al. 2015, Hoch et al. 2017). Indeed, similar to SCAR26 patient fibroblasts, cells from AOA4 patients also exhibited elevated levels of PARylation following DNA damage, suggesting that PARP1 hyperactivation and NAD+ depletion may also contribute to neurodegeneration upon PNKP dysfunction (Hoch et al. 2017).

Base excision repair deficiency and human disease

There are two known autosomal recessive human diseases caused by mutations in DNA glycosylases. The first is a colorectal adenoma and carcinoma predisposition syndrome called familial adenomatous polyposis 2 (FAP2), which is caused by biallelic mutations in MUTYH (Al-Tassan et al. 2002). MUTYH repairs adenine bases that have been misincorporated opposite 8-oxo-G, and analysis of patient tumour samples possessed characteristic G → T somatic mutations (Al-Tassan et al. 2002). Therefore, increased mutagenesis due to misincorporation of adenine opposite 8-oxo-G was deemed to the underlying cause of the predisposition to cancer in the FAP2 patients. It is unknown why colorectal tissues seem to be particularly sensitive to the increased mutagenesis caused by loss of MUTYH, although this is reminiscent of the types of cancer caused by mutations of mismatch repair factors (discussed in Chapter 12) (Sijmons and Hofstra 2016).

The second autosomal human disease is an immunodeficiency syndrome caused by mutations in UNG (Imai et al. 2003). In addition to its role in the repair of damaged DNA bases, the BER pathway also functions within immunoglobulin (Ig) class-switch recombination (CSR), a cellular pathway in B lymphocytes that promotes Ig gene alterations in response to foreign antigens (see also Chapter 7) (Methot and Di Noia 2017). CSR involves the generation of programmed DSBs via the activity of activation-induced deaminase (AID) and UNG (Methot and Di Noia 2017). The first step in the generation of the DSB is the AID-dependent deamination of cytosine to uracil within class-switch regions, followed by generation of AP sites by UNG. Further processing of the AP sites by APE1 and POLβ generate SSBs in close proximity to each other, which form the precursors to DSBs (Methot and Di Noia 2017). The importance of BER in CSR is highlighted by the existence of three patients with an immunodeficiency syndrome called hyper IgM syndrome type 5 (HIGM5) who possess mutations in UNG. HIGM5 patient-derived cells are completely defective in the ability to undergo immunoglobulin CSR, resulting in recurrent bacterial infections, lymphoid hyperplasia, increased serum IgM concentrations and decreased serum IgG and IgA concentrations (Imai et al. 2003). The lack of cancer predisposition in individuals with UNG mutations likely reflects the activity of other DNA glycoslyases, such as MBD4, SMUG1 and TDG, that can remove uracil from DNA. It has also been shown that MBD4-deficient mice do not exhibit immunodeficiency, and it has been proposed that MBD4, SMUG1 and TDG cannot compensate for loss of UNG within CSR (Bardwell et al. 2003, Imai et al. 2003).

Therefore, even though DNA repair pathways are critical to prevent genome instability and therefore protect against neurological dysfunction and cancer, the HIGM5 syndrome highlights the fact that DNA repair pathways can also have specific roles in other cellular processes, for example, in the production of antibodies in the immune system.

Conclusions and future perspectives

Significant progress has been made into understanding the molecular defects behind the neuropathologies associated with defective SSBR. However, there are several important questions that remain answered. APTX, TDP1 and PNKP have all been implicated in double-strand break repair (Clements et al. 2004, Karimi-Busheri et al. 2007, Heo et al. 2015). Therefore, it is possible that the accumulation of a combination of both SSBs and DSBs underlies the neurological dysfunction associated with the diseases discussed earlier. Furthermore, it is also possible that defective repair of mitochondrial DNA also contributes to the disease phenotypes. APTX, TDP1 and PNKP have all been found in the mitochondria, and, importantly, delayed repair of mitochondrial 5′-AMP SSBs was observed in $APTX^{-/-}$ human cells, which could not be compensated for by alternative repair mechanisms (Das et al. 2010, Tahbaz et al. 2012, Akbari et al. 2015).

As described earlier, it has been proposed that transcription arrest caused by unrepaired DNA damage is a significant contributor to the neurodegeneration seen in SSBR-defective individuals. Furthermore, the lack of extraneurological symptoms, such as cancer predisposition, has also been suggested to be due to the existence of redundant repair pathways in other tissue types. However, these hypotheses need to be explored in further research. Due to the failure of many mouse models to phenocopy human disease, this may require alternative model systems, such as neuronal organotypic cultures (Tzur-Gilat et al. 2013).

NUCLEOTIDE EXCISION REPAIR

Nucleotide excision repair is arguably one of the most studied DNA repair pathways within human cells. Indeed, seminal research in the 1960s demonstrated that cells derived from patients with xeroderma pigmentosum, an autosomal recessive syndrome associated with cancer predisposition and extreme sun sensitivity, exhibited defective repair of DNA lesions induced by UV light (Cleaver 1969). This was the first time that DNA repair defects were linked to inherited cancer predisposition (Cleaver 1969). Subsequent research focused on understanding the molecular pathology underlying this disease then paved the way to defining the biochemistry behind the NER pathway.

One of the key features of the NER pathway is its ability to recognise and repair a wide range of structurally unrelated, bulky DNA lesions. This versatility is partly due to the fact that the NER machinery detects distortions of the DNA double-helix, which is a feature common to bulky DNA adducts, rather than recognising specific DNA lesions (Spivak 2015). Bulky adducts can arise from both environmental and endogenous sources. A major environmental source of bulky DNA lesions is UV light, which induces many types of pyrimidine dimers, of which 6-4 pyrimidine–pyrimidone photoproducts (6-4PPs) and cyclobutane–pyrimidine dimers are the most common (Chatterjee and Walker 2017). Another environmental source of bulky DNA adducts is polycyclic aromatic hydrocarbons, which are organic compounds composed of multiple aromatic rings and are commonly found in cigarette smoke, cooked foods, vehicle exhaust and smoke from the incomplete combustion of fossil fuels (Chatterjee and Walker 2017). Polycyclic aromatic hydrocarbons, such as benzo(*a*)-pyrene and dibenzo[*a,l*]pyrene, readily react with DNA to form cytotoxic bulky adducts (Chatterjee and Walker 2017). A significant source of endogenous bulky adducts are ROS, which can induce the formation of cyclopurines within DNA (Shafirovich and Geacintov 2017). Finally, NER can also repair intrastrand cross-links, which involve a covalent linkage between two bases within the same DNA strand and can be induced by cross-linking chemotherapeutic therapies used to treat cancer (Kelland 2007).

Mammalian NER is divided into two sub-pathways which differ in their mode of DNA damage recognition, but converge into one pathway for downstream processing and repair events. Global genome NER functions to recognize and repair DNA lesions throughout the genome and is most efficient at repairing DNA adducts that result in a large distortion of the DNA double-helix, such as 6-4PPs (Sugasawa 2016). The presence of ssDNA caused by disrupted base pairing is common to all bulky lesions repaired by GG-NER. Adducts which cause only a minor destabilisation of the DNA, such as CPDs, can go undetected by the GG-NER surveillance machinery (Sugasawa 2016). These lesions are predominantly repaired via the second NER sub-pathway, namely transcription-coupled NER. TC-NER is triggered when an elongating RNA polymerase II (RNA POLII) molecule stalls upon collision with a DNA lesion during transcription (Pani and Nudler 2017). Following recognition of the DNA lesion, the two NER sub-pathways converge and then proceed through the same stages of repair (Figure 5.2). First, the DNA helix surrounding the DNA lesion is unwound to verify the presence of DNA damage, the damaged DNA strand is then cleaved and excised and finally the resulting DNA gap is filled and ligated.

DNA damage recognition

Global genome nucleotide excision repair

As mentioned earlier, GG-NER functions to efficiently repair bulky DNA adducts throughout the genome. Within this pathway, DNA damage recognition is carried out by XPC, which constantly probes the genomic DNA for the presence of helix-distortions caused by DNA adducts. When it encounters a DNA lesion, XPC interacts with the ssDNA that is exposed by the disrupted base pairing within the helix-distortion (Sugasawa et al. 2001). XPC then recruits the TFIIH complex, which is a large multi-subunit complex that possesses two DNA helicase subunits with opposing polarities (Volker et al. 2001). XPD is a DNA helicase with a 5′-3′ polarity, whilst XPB unwinds DNA in a 3′-5′ direction (Coin et al. 1998, Coin et al. 2007). TFIIH proceeds to unwind the DNA to generate a 20- to 30-nucleotide bubble surrounding the lesion (Tapias et al. 2004, Compe and Egly 2012). This allows the presence of DNA damage to be verified by XPD (Sugasawa et al. 2009). If a bulky lesion is present within the bubble, this will act as a block to XPD activity, thereby confirming the presence of DNA damage and allowing repair to proceed to the next stage (Mathieu et al. 2010). However, if the XPD helicase does not detect any lesions within the DNA, then repair is aborted to prevent undamaged DNA from being cleaved unnecessarily (Mathieu et al. 2013).

There are several other factors that function to support DNA damage recognition and verification within NER. The first is XPA, a central NER factor that interacts with many proteins involved in the repair of bulky adducts and binds to altered nucleotides within ssDNA (Sugitani et al. 2016). XPA is recruited to the unwound DNA bubble, stimulates the helicase activity of TFIIH and helps XPD to recognise the presence of bulky lesions (Li et al. 2015). At this stage, a single RPA molecule interacts with the undamaged strand of the DNA bubble, and together XPA and RPA function to direct the later stages of repair towards the damaged DNA strand (de Laat et al. 1998).

Another important DNA damage recognition factor is the UV–DDB complex, consisting of XPE and DDB1, which functions to stimulate the binding of XPC to CPDs. As CPDs only cause a minor distortion of the DNA, they are poorly recognised by XPC (Sugasawa et al. 2001). To compensate for this, UV–DDB binds directly to UV-induced DNA lesions and kinks the DNA, exposing ssDNA that allows XPC to bind the CPD (Scrima et al. 2008). Loss or mutation of XPE results in defective NER-dependent repair of CPDs, but not 6-4PPs (Fitch et al. 2003). Once XPC detects the presence of DNA damage, the repair reaction proceeds to the next stage.

Transcription-coupled nucleotide excision repair

Even with the presence of the UV–DDB complex, the repair of CPDs is still much less efficient than that of 6-4PPs, and the persistence of unrepaired CPDs can cause problems during replication and transcription. Replication forks that stall upon encountering unrepaired minor helix distorting lesions have alternative mechanisms to repair the lesion and promote replication restart. For example, translesion synthesis polymerases can be engaged to bypass the lesion

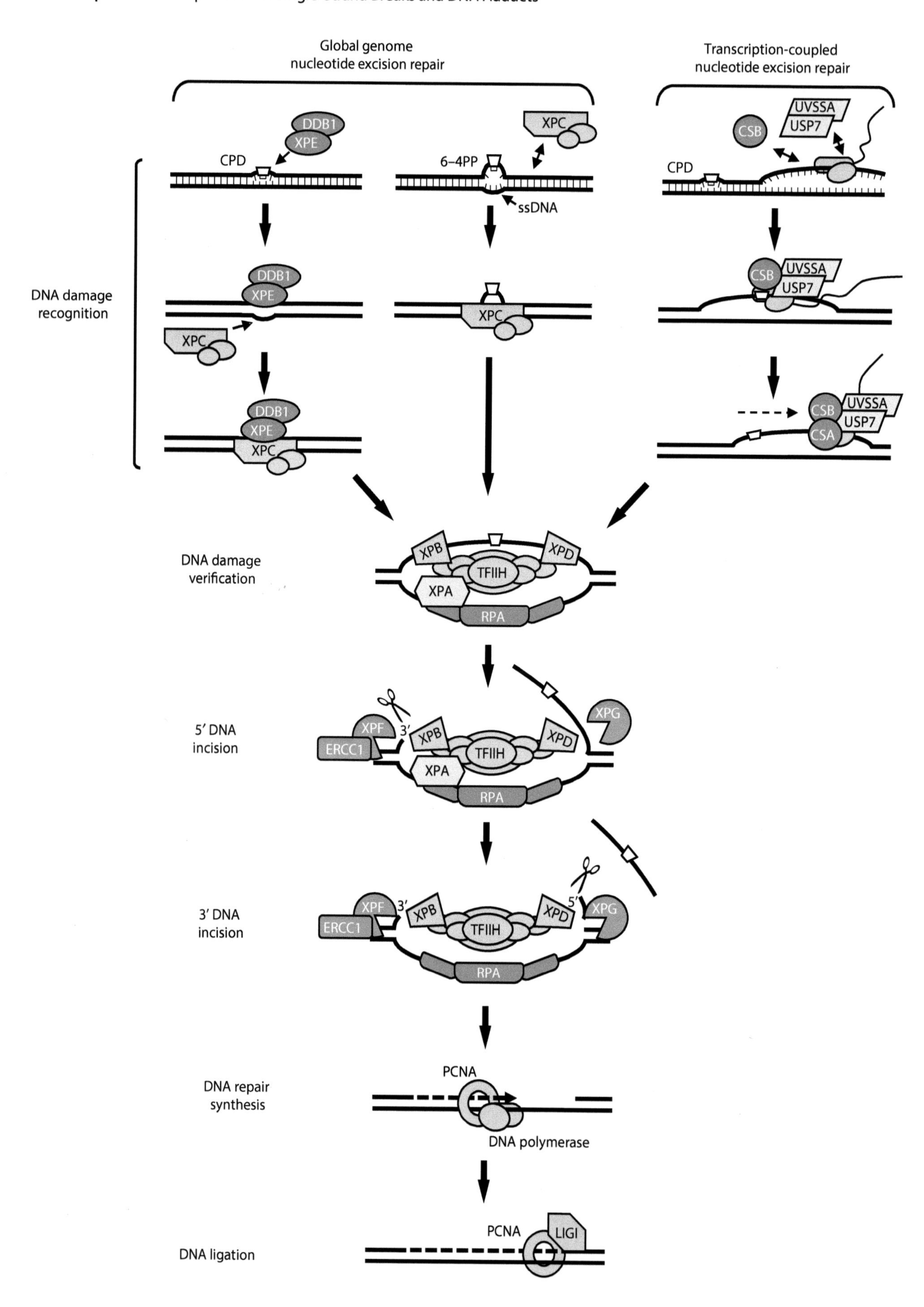

Global genome nucleotide excision repair
Transcription-coupled nucleotide excision repair
DNA damage recognition
DDB1
XPE
XPC
CPD
6–4PP
ssDNA
CSB
UVSSA
USP7
CSA
DNA damage verification
XPB
XPD
TFIIH
XPA
RPA
5′ DNA incision
XPF
ERCC1
3′
XPG
3′ DNA incision
5′
DNA repair synthesis
PCNA
DNA polymerase
DNA ligation
LIGI

(discussed in Chapter 4), or various mechanisms, such as fork reversal, can be engaged to avoid the damage and resume DNA replication (detailed in Chapters 4 and 6) (Lopes et al. 2006, Berti and Vindigni 2016). DNA lesions that block RNA polymerases are repaired by TC-NER, a dedicated sub-pathway that exclusively repairs transcription-blocking DNA adducts (Pani and Nudler 2017). Apart from CPDs, a range of DNA lesions can interfere with transcription and are repaired by TC-NER, and loss of TC-NER renders cells hypersensitive to oxidative stress (de Waard et al. 2004).

TC-NER is initiated by the recruitment of CSB to a blocked RNA POLII molecule upon collision with a DNA lesion (Fousteri et al. 2006). UVSSA and the deubiquitylating enzyme USP7 are also recruited to the stalled transcription complex, following which USP7 deubiquinates CSB, preventing it from being degraded and strengthening the association between CSB and RNA POLII (Nakazawa et al. 2012, Schwertman et al. 2012). CSB is also responsible for recruiting CSA, and together they recruit downstream NER repair factors (Fousteri et al. 2006). CSA is part of an E3 ubiquitin ligase complex with CUL4A and ROC1, which has several roles in the regulation of TC-NER, such as targeting CSB for degradation once repair is complete to allow resumption of transcription (Groisman et al. 2006). Another important event in TC-NER is the removal of the stalled RNA POLII in order for the downstream NER repair factors to gain access to the DNA adduct. The most likely mechanism for the removal of the RNA polymerase from the CPD is CSB-mediated backtracking of RNA POLII (Cheung and Cramer 2011). Finally, the TC-NER and GG-NER pathways converge once TFIIH is recruited to the CPD and performs DNA unwinding and DNA damage verification.

Excision of damaged DNA, repair synthesis and ligation

Once the presence of a DNA adduct within the DNA has been verified, the next stage of NER involves the excision of the damaged DNA strand by nucleolytic incision within the DNA bubble on either side of the lesion. This is achieved by the action of two structure-specific nucleases, XPG and XPF-ERCC1. XPF-ERCC1 is responsible for the incision 5′ from the lesion, whilst XPG incises the DNA bubble 3′ from the lesion. Due to the potential dangers of allowing ssDNA gaps to persist within the genome, this stage of repair is tightly coordinated to ensure that the DNA is only incised when all the required NER factors are present and repair can proceed rapidly to completion (Godon et al. 2012). XPA and RPA also have a critical role in correctly orientating the two nucleases to ensure they cleave to the DNA in the appropriate locations (de Laat et al. 1998). Once the 5′ incision has been made by XPF-ERCC1, DNA synthesis can occur within the ssDNA gap before the second incision (Fagbemi et al. 2011). The endonuclease activity of XPG is then triggered by the presence of XPF-ERCC1 and the 5′ incision, resulting in the 3′ incision and excision of the damaged strand (Tapias et al. 2004, Fagbemi et al. 2011). Finally, the ssDNA gap is filled and ligated by the coordinated action of

Figure 5.2 **Nucleotide excision repair (previous page).** Nucleotide excision repair is split into two sub-pathways that differ in their modes of DNA damage recognition. Global-genome nucleotide excision repair (GG-NER) recognises and repairs bulky helix-disorting DNA adducts as they arise within the genome, whilst transcription-coupled nucleotide excision repair (TC-NER) solely repairs DNA lesions that collide with, and stall, an ongoing transcription complex. **DNA damage detection:** XPC is the major factor that recognises DNA damage within the GG-NER pathway. To achieve this, XPC complex constantly interrogates the DNA for the presence of ssDNA, a structure that is common to many helix-distorting DNA adducts (e.g. 6-4PPs). In addition, the XPE-DDB1 complex binds to DNA lesions that cause only minor disruption of the DNA double helix, and which are poorly recognised by XPC (such as CPDs). This functions to help XPC to detect minor helix distorting lesions. Once XPC detects a DNA adduct (with or without the aid of XPE-DDB1), it binds to the ssDNA within the distortion and recruits TFIIH. TC-NER functions to recognise and repair DNA lesions within actively transcribed genes that have evaded detection by GG-NER and have collided with a RNA POLII molecule. During normal transcription, CSB and UVSSA-USP7 transiently interact with RNA POLII. Upon collision with a DNA adduct, RNA POLII stalls and strongly interacts with CSB and UVSSA-USP7. CSA is then recruited and the transcription complex backtracks to reveal the DNA lesion. At this stage, TFIIH is recruited to the site of DNA damage. **DNA damage verification:** Once TFIIH is recruited to the DNA adduct, the two NER pathways converge and follow the same stages of repair. Upon recruitment, the XPB and XPD helicase subunits within TFIIH unwind the DNA to form a 20–30 bp bubble. With the help of XPA, XPD verifies the presence of a DNA adduct. If no DNA damage is detected, the repair reaction can be aborted. **5′ DNA incision:** Once XPD has verified the presence of a DNA adduct within the bubble, XPA and RPA direct the XPF-ERCC1 endonuclease to incise the DNA 5′ to the DNA lesion. The activity of XPF-ERCC1 requires the presence of XPG. **3′ DNA incision:** The XPG endonuclease then incises the DNA bubble 3′ to the DNA lesion and excises the damaged strand. **DNA repair synthesis and ligation:** Once the damaged DNA is excised, the gap is filled and the resulting DNA nick is ligated to complete repair.

PCNA, POLε and DNA Ligase I in replicating cells, and POLδ/POLκ and XRCC1-LIGIIIα in non-dividing cells (Spivak 2015).

Nucleotide excision repair and human disease

Mutations in numerous NER genes have been identified in patients with a diverse range of clinical phenotypes (Table 5.4) (Menck and Munford 2014). Depending on which gene is mutated, patients can present with severe cancer predisposition and sun sensitivity (xeroderma pigmentosum), sun sensitivity without an increased risk of cancer (UV-sensitive syndrome) or neurodevelopmental defects and premature ageing without cancer predisposition (Cockayne syndrome, cerebro-oculo-facio skeletal syndrome or trichothiodystrophy). The differing clinical presentations likely reflect which of the two NER sub-pathways are impacted, as well as the nature of the DNA repair defect.

Xeroderma pigmentosum

Xeroderma pigmentosum is a rare autosomal recessive disease characterised by high sensitivity to sunlight, a 10,000-fold increased risk of sunlight-induced non-melanoma skin cancer, and a 10-fold increased risk of developing internal tumours (Kraemer et al. 1994, DiGiovanna and Kraemer 2012). XP is caused by mutations in factors that are involved in the recognition of DNA lesions within the GG-NER pathway (XPC and XPE) or in factors involved in NER following convergence of the two sub-pathways (XPB, XPD, XPG and XPF). Therefore, it seems that defective GG-NER, and the persistence of unrepaired UV-induced bulky DNA adducts, is the underlying cause of XP.

The increased cancer predisposition can be explained by the increased levels of mutagenesis generated by repair pathways that operate in the absence of GG-NER. To prevent prolonged fork stalling by unrepaired DNA adducts, error-prone TLS polymerases will bypass the DNA lesion and allow replication to proceed (discussed in Chapter 4) (Yang and Woodgate 2007). Whilst this prevents cell death from incomplete replication, the use of error-prone polymerases leads to increased mutagenesis, which is a very strong driver of tumourigenesis. The increased risk of skin cancer is most likely due to the accumulation of UV-induced DNA adducts, whilst the predisposition to internal tumours likely reflects an inability to repair endogenous DNA lesions, perhaps such as those induced by oxidative stress (Brooks 2008).

In addition to cancer and sun-induced skin abnormalities, 25% of XP patients will also develop progressive neurodegeneration (DiGiovanna and Kraemer 2012). It is noteworthy that abnormalities of the nervous system have not been found in individuals with mutations in factors that solely function within GG-NER (XPC and XPE), but instead are associated with mutations in core NER factors that impact both GG-NER and TC-NER (XPA, XPB, XPD, XPF, XPG), which suggests a link between defects in TC-NER and neurodegeneration (DiGiovanna and Kraemer 2012). XP can also be caused by mutations in the TLS polymerase POLη (Masutani et al. 1999). This form of XP is associated with sun sensitivity and increased risk of skin cancer, but due to a lack of a cellular NER defect, is referred to as an XP variant.

Cockayne syndrome

Cockayne syndrome (CS) is another rare autosomal disease associated with mutations in NER factors (Laugel 2013). Similar to XP patients, CS patients exhibit sun sensitivity and patient-derived cells are hypersensitive to UV irradiation. However unlike XP, CS is not associated with an increased risk of cancer (DiGiovanna and Kraemer 2012). Additionally, CS patients present with neurological abnormalities, such as microcephaly, neurodegeneration, progressive ataxia, delayed psychomotor development and mental retardation, as well as features of premature ageing, such as cachexia, kyphosis, thin hair, loss of subcutaneous fat and osteoporosis. CS is caused by mutations in the TC-NER factors, CSA and CSB. Interestingly, some individuals with mutations in XPB, XPD, XPG and XPF suffer from XP/CS, a disorder associated with a combination of features characteristic of both XP and CS (DiGiovanna and Kraemer 2012).

Defects of TC-NER are therefore associated with neurological abnormalities and premature ageing, rather than cancer predisposition. Indeed, cell lines from individuals with CS or XP/CS are unable to recover RNA synthesis following UV irradiation, an observation which is used for the diagnosis of these diseases (Mayne and Lehmann 1982). It is therefore proposed that the underlying cause of the neurological defects and premature ageing seen in CS and XP/CS individuals is defective transcription due to the persistence of RNA POLII blocking DNA lesions, leading to increased cell dysfunction or death (Balajee et al. 1997, Ljungman and Lane 2004).

Table 5.4 Human diseases associated with defective nucleotide excision repair.

Gene	Gene function	Disease	MOI	Clinical presentation	OMIM #	References
XPA	XPA has a central role in NER	Xeroderma pigmentosum type A (XPA)	AR	Microcephaly, cerebellar atrophy, ataxia, photophobia, increased risk of sunlight-induced skin cancer, UV-induced freckles, UV sensitivity, telangiectasia and other skin abnormalities	278700	Tanaka et al. (1990)
XPB/ ERCC3[a]	3′-5′ DNA helicase subunit of the TFIIH complex that functions within NER	Xeroderma pigmentosum type B (XPB)	AR	UV sensitivity, photophobia, increased risk of sunlight-induced skin cancer, sun-induced freckles	610651	Oh et al. (2006)
		Xeroderma pigmentosum type B/Cockayne syndrome (XPB/CS)	AR	Short stature, microcephaly, features of premature ageing, UV sensitivity, cerebellar atrophy, ataxia, increased risk of sunlight-induced skin cancer	610651	Weeda et al. (1990)
		Trichothiodystrophy 2, photosensitive (TTD2)	AR	Photosensitivity; skin abnormalities; coarse, brittle hair	616390	Weeda et al. (1997)
XPC	NER factor involved in recognising and verifying the presence of DNA damage	Xeroderma pigmentosum type C (XPC)	AR	UV sensitivity, photophobia, increased risk of sunlight-induced skin cancer, skin atrophy, telangiectasia, sun-induced freckles	278720	Li et al. (1993)
XPD/ ERCC2[a]	5′-3′ DNA helicase subunit of the TFIIH complex that functions within NER	Xeroderma pigmentosum type D (XPD)	AR	Microcephaly, cerebellar atrophy, ataxia, photophobia, increased risk of sunlight-induced skin cancer, UV-induced freckles, UV sensitivity, telangiectasia and other skin abnormalities	278730	Frederick et al. (1994)
		Xeroderma pigmentosum type D/Cockayne syndrome (XPD/CS)	AR	Short stature, microcephaly, features of premature ageing, UV sensitivity, cerebellar atrophy, ataxia, increased risk of sunlight-induced skin cancer	NA[b]	Broughton et al. (1995)
		Trichothiodystrophy 1, photosensitive (TTD1)	AR	Microcephaly; intellectual disability; growth retardation; UV sensitivity; photosensitivity; skin abnormalities; coarse, brittle hair	601675	Takayama et al. (1996)
		Cerebro-oculo-facio-skeletal syndrome 2 (COFS2)	AR	Severe microcephaly, intrauterine and postnatal growth retardation, congenital cataracts, severe developmental delay, arthrogryposis, congenital kyphosis, profound intellectual disability, facial dysmorphia, features of premature ageing	610756	Graham et al. (2001)
XPE/ DDB2	NER factor involved in the recognition of DNA damage	Xeroderma pigmentosum type E (XPE)	AR	UV sensitivity, photophobia, increased risk of sunlight-induced skin cancer, skin atrophy, telangiectasia, sun-induced freckles	278740	Nichols et al. (1996)
XPG/ ERCC5[a]	Structure-specific nuclease involved in the repair of DNA adducts	Xeroderma pigmentosum, type G (XPG)	AR	High sensitivity to sunlight, increased risk of skin cancer	278780	Lalle et al. (2002)
		Xeroderma pigmentosum, type G/Cockayne syndrome (XPG/CS)	AR	Short stature, microcephaly, features of premature ageing, UV sensitivity, cerebellar atrophy, ataxia, increased risk of sunlight-induced skin cancer	278780	Zafeiriou et al. (2001)
		Cerebro-oculo-facio-skeletal syndrome 3 (COFS3)	AR	Severe microcephaly, intrauterine and postnatal growth retardation, congenital cataracts, severe developmental delay, arthrogryposis, congenital kyphosis, profound intellectual disability, facial dysmorphia, features of premature ageing	616570	Zafeiriou et al. (2001), Drury et al. (2014)

Table 5.4 (*Continued*) Human diseases associated with defective nucleotide excision repair.

Gene	Gene function	Disease	MOI	Clinical presentation	OMIM #	References
XPF/ ERCC4/ FANCQ[a]	Structure-specific nuclease with multiple roles in DNA repair, including NER, homologous recombination and ICL repair	Xeroderma pigmentosum type F (XPF)	AR	High sensitivity to sunlight, increased risk of skin cancer	278760	Sijbers et al. (1996)
		Xeroderma pigmentosum, type F/Cockayne syndrome (XPF/CS)	AR	High sensitivity to sunlight, increased risk of skin cancer, intellectual disability, cerebral atrophy, cerebellar ataxia, microcephaly, short stature	278760	Kashiyama et al. (2013)
		Fanconi anaemia type Q (FANCQ)	AR	Growth retardation, microcephaly, bone marrow failure, abnormalities of the heart and skin	615272	Bogliolo et al. (2013), Kashiyama et al. (2013)
ERCC1	Binding partner of the structure-specific nuclease XPF with roles in DNA repair	Cerebro-oculo-facio-skeletal syndrome 4 (COFS4)	AR	Severe microcephaly, intrauterine and postnatal growth retardation, congenital cataracts, severe developmental delay, arthrogryposis, congenital kyphosis, profound intellectual disability, facial dysmorphia, features of premature ageing	610758	Jaspers et al. (2007), Kashiyama et al. (2013)
CSA/ ERCC8[a]	Plays an important role in TC-NER	Cockayne syndrome group A (CSA)	AR	Intrauterine and postnatal growth retardation; microcephaly; intellectual disability; progressive cerebellar atrophy; ataxia; features of premature ageing; multiple facial abnormalities; multiple abnormalities of the skeleton, heart, skin and kidneys; skin UV sensitivity	216400	Henning et al. (1995)
		UV-sensitive syndrome 2 (UVSS2)	AR	Skin photosensitivity, freckling	614621	Nardo et al. (2009)
CSB/ ERCC6[a]	Plays an important role in TC-NER	Cockayne syndrome B (CSB)	AR	Intrauterine and postnatal growth retardation; microcephaly; intellectual disability; progressive cerebellar atrophy; ataxia; features of premature ageing; multiple facial abnormalities; multiple abnormalities of the skeleton, heart, skin and kidneys; skin UV sensitivity	133540	Mallery et al. (1998)
		Cerebro-oculo-facio-skeletal syndrome 1 (COFS1)	AR	Severe microcephaly, intrauterine and postnatal growth retardation, congenital cataracts, severe developmental delay, arthrogryposis, congenital kyphosis, profound intellectual disability, facial dysmorphia, features of premature ageing	214150	Meira et al. (2000)
		De Sanctis-Cacchione syndrome	AR	Short stature, microcephaly, intellectual disability, cerebral and cerebellar atrophy, ataxia, skin UV sensitivity, sun-induced freckles, predisposition to developing skin cancer, sensorineural deafness	278800	Greenhaw et al. (1992)
		UV-sensitive syndrome 1 (UVSS1)	AR	Skin UV sensitivity, freckling, telangiectasia	600630	Horibata et al. (2004)
		Premature ovarian failure 11 (POF11)	AD	Premature ovarian failure	616946	Qin et al. (2015)
UVSSA	Plays an important role in TC-NER	UV-sensitive syndrome 3 (UVSS3)	AR	Skin UV sensitivity, freckling, telangiectasia	614640	Nakazawa et al. (2012)
TFB5	Subunit of the TFIIH complex with a critical role in NER	Trichothiodystrophy 3, photosensitive (TTD3)	?[c]	Short stature; intellectual disability; UV sensitivity; intellectual disability; coarse, brittle hair	616395	Giglia-Mari et al. (2004)

Table 5.4 (*Continued*) Human diseases associated with defective nucleotide excision repair.

Gene	Gene function	Disease	MOI	Clinical presentation	OMIM #	References
PCNA	A critical replisome factor that has multiple roles in DNA replication and DNA repair, including NER	Ataxia-telangiectasia-like disorder 2 (ATLD2)	AR	Neurodegeneration, spinocerebellar ataxia, developmental delay, postnatal growth retardation, premature ageing, telangiectasia, photosensitivity, predisposition to UV-induced malignancies	615919	Baple et al. (2014)
POLH/ Pol eta	Member of the Y family of DNA polymerases that function within TLS	Xeroderma pigmentosum, variant type (XPV)	AR	Increased sun sensitivity, increased risk of skin cancer	278750	Masutani et al. (1999)

Abbreviations: AR, autosomal recessive; MOI, mode of inheritance; NA, not applicable; OMIM, online mendelian inheritance in man.

[a]Mutations in these genes cause multiple distinct clinical presentations.

[b]OMIM designation has not yet been given.

[c]Insufficient data to determine mode of inheritance.

The nature of the endogenous transcription-blocking lesions that cause increased cell death in TC-NER–defective individuals is still not clear. However, oxidative lesions induced by ROS are thought to contribute to the pathology of CS. Indeed, CSA and CSB patient–derived fibroblasts are defective in the repair of 8-oxoG, a finding which is recapitulated in CSB-defective mouse models (D'Errico et al. 2007, Kirkali et al. 2009). However, as loss of XPC also results in cellular hypersensitivity to oxidative DNA damage, but is not associated with either neurological or progeroid symptoms, it is likely that there are other contributing factors involved (D'Errico et al. 2006).

Other disorders associated with defective NER

Apart from XP and CS, there are other disorders associated with defective NER. For example, UV-sensitive syndrome (UVSS) is a disorder caused by mutations in CSA, CSB and UVSSA. UVSS is characterised by sun sensitivity, but lacks any form of neurological defects, premature ageing or cancer predisposition that is seen in XP or CS. However, in a similar situation to CS, UVSSA patient–derived cell lines also exhibit a failure to recover RNA synthesis following UV-induced DNA damage (Spivak 2005). Therefore, it seems that defects in TC-NER are associated with a range of clinical presentations, including severe disorders (CS, cerebro-oculo-facio skeletal syndrome and trichothiodystrophy), and a milder disease (UVSS). It is unclear why this is the case, especially as both severe and mild clinical presentations can be caused by specific mutations in CSA and CSB. One hypothesis that has been suggested is that defects in TC-NER which lead to a stalled transcription complex stably associated with the DNA, thus blocking the DNA lesion from being accessed by other repair processes and preventing transcription of the damaged gene, will result in the much more severe clinical presentation seen in CS individuals (Marteijn et al. 2014). However, if the RNA POL II complex can be degraded, then the DNA lesion can still be repaired by alternative pathways, and a milder clinical phenotype may result.

Furthermore, mutations in XPB, XPD and TFB5 are found in individuals with trichothiodystrophy (TTD), a disease with similar neurological and progeroid symptoms to CS, but also associated with fragile hair and nails (Stefanini et al. 2010). XPB, XPD and TFB5 are all subunits within the TFIIH complex, and TTD-associated mutations result in decreased stability of the entire complex (Vermeulen et al. 2001, de Boer et al. 2002, Theil et al. 2013). Therefore, it is highly likely that these mutations would impact upon both transcription and DNA repair, which could potentially explain the additional symptoms seen in TTD individuals.

Another NER disorder is cerebro-oculo-facio skeletal syndrome (COFS), which is a much more severe form of CS. COFS individuals present with microcephaly, underdevelopment of the cerebellum, profound intellectual disability and delayed/lack of motor and speech development, multiple skeletal abnormalities and premature ageing. These symptoms are typically present before birth, and individuals will not often survive past their second year of life (Laugel 2013). COFS is caused by mutations in XPD, XPG, ERCC1 and CSB. XPG and XPF-ERCC1 are both structure-specific nucleases that have multiple roles in DNA repair outside of NER. For example, XPF-ERCC1 has critical roles in the repair of DNA interstrand cross-links and in the resolution of replication intermediates arising from replication stress (Bhagwat et al. 2009, Naim et al. 2013). Furthermore, mutations in XPF also result in Fanconi anaemia, a multigenic disease characterised by bone marrow failure, multiple congenital anomalies and cancer predisposition (discussed in Chapter 12) (Kashiyama et al. 2013). The FA pathway is predominantly focused on detecting and repairing DNA ICLs during DNA replication (Ceccaldi et al. 2016). This suggests that defects in other DNA replication and repair pathways may contribute to the severe abnormalities seen in some COFS individuals.

The large variety of clinical phenotypes associated with defective NER is noteworthy. However, despite the striking differences in severity between UVSSA, CS and COFS, they can be considered to be the same disease at opposite ends of a continuous spectrum of severity, rather than distinct entities (Laugel 2013). Obvious genotype–phenotype correlations do not exist within the NER defective disorders; although it is possible that the severity of the clinical phenotypes associated with CS, UVSSA, COFS and/or TTD patients may correlate with how much the underlying defects impact transcription. Additionally, the presence of combined defects in GG-NER, TC-NER and/or other repair pathways may also contribute to the severity disease phenotype.

A final disorder associated with defective NER is ataxia-telangiectasia-like disorder 2 (ATLD2), which is caused by hypomorphic mutation of PCNA (Baple et al. 2014). ATLD2 patients presented with short stature, premature ageing, photosensitivity, telangiectasia, sensorineural hearing loss, neurodegeneration, cerebellar atrophy, ataxia and UV-induced skin cancer (Baple et al. 2014). A striking aspect to ATLD2 was that the patients exhibited a combination of clinical phenotypes typically found in ataxia telangiectasia (Chapter 13), Cockayne syndrome and xeroderma pigmentosum. In addition, despite the critical role for PCNA during DNA replication, patient-derived fibroblasts did not display any gross DNA replication abnormalities, but instead exhibited defective repair of UV-induced DNA damage (Baple et al. 2014, Green et al. 2014). Indeed, both GG-NER and TC-NER pathways were found to be affected. Although defects in other DNA repair pathways were not reported, it was also considered possible that the clinical phenotypes of the PCNA patients were caused by the combined dysfunction of different DNA repair pathways (Baple et al. 2014).

Conclusions and future perspectives

We are gaining a greater understanding of how GG-NER functions to protect against tumorigenesis and how TC-NER prevents neurological dysfunction and premature ageing. The loss of GG-NER seems to cause a greater impact on cycling cells, due to error-prone TLS polymerases functioning to promote cell survival by bypassing unrepaired bulky DNA lesions during DNA replication, leading to increased mutagenesis. In contrast, non-proliferating, or slowly cycling, tissues seem to be particularly prone to accelerated ageing in CS individuals, suggesting that non-cycling cells are heavily dependent on TC-NER to prevent transcription stress–induced cell death. Furthermore, the lack of cancer predisposition in CS has been proposed to be due to increased levels of apoptosis killing cells with unrepaired DNA damage, preventing them from becoming cancerous (Ljungman and Lane 2004, Caputo et al. 2013).

However, despite our progress, there are still many questions that remain unanswered. For example, the nature of the DNA lesions that underlies the increased risk of developing internal tumours in XP patients is unclear, and it is unknown which RNA polymerase–blocking DNA adducts accumulate in TC-NER–defective individuals. Furthermore, it has been suggested that defective repair of mitochondrial DNA may also contribute to the disease pathology in individuals with CSB mutations (Stevnsner et al. 2002). The answers to these and other questions will be forthcoming with future research.

CONCLUSIONS

As discussed in this chapter, the processes of SSBR and NER are distinct DNA repair pathways that function to repair DNA SSBs and base damage in the case of the former, or bulky DNA adducts in the latter. However, whilst the two pathways have significant differences, there are some intriguing overlaps and similarities between them which may allow understanding of their roles in the maintenance of human health in a broader context.

On the face of it, SSBR and GG-NER seem to be more similar as they both function to constantly survey the genome for the presence of DNA damage, and once a DNA lesion is detected, they immediately engage in the process of repair. In contrast, TC-NER is only initiated under specific circumstances, specifically the collision of an RNA POLII transcription complex with an unrepaired DNA lesion.

However, whilst GG-NER is crucial for the repair of DNA lesions arising from external sources (e.g. UV-induced DNA adducts), SSBR and TC-NER both appear to be vital for the repair of endogenously arising DNA lesions, and indeed both pathways function to repair oxidative DNA damage. This may provide an explanation as to why the primary clinical phenotype associated with GG-NER defects is a predisposition to UV-induced skin cancers, whilst deficiencies of SSBR- and TC-NER pathways have a large impact internally on the central nervous system. Indeed, the neurodevelopmental and neurodegenerative pathologies observed in both SSBR- and TC-NER-deficient patients are proposed to arise from cellular dysfunction and cell death, at least in part due to the impact of unrepaired DNA damage on transcription. In addition, mutation of TC-NER specific factors and SSBR factors are not associated with cancer predisposition and it has been suggested that the lack of cancer is due to the presence of alternative 'error-free' repair pathways. For example, GG-NER is still functional in CSA- and CSB-mutated individuals, and replication-associated HR mechanisms function to repair collapsed replication forks within SSBR defective patients.

Nevertheless, despite our progress, questions remain unanswered. It is not clear why mutations in the same gene, and sometimes the same specific mutation, can give rise to such phenotypic variability. It is possible that this reflects the nature of the mutation or may be the result of unknown genetic modifiers. Also, in some cases, the exact nature of the deleterious unrepaired DNA lesions eludes us, such as in the case of AOA1 or CS, and it is hoped that further research will shed some light on this. Furthermore, there is much inconsistency involving mouse models of SSBR and NER diseases, and as often as not, mouse models fail to recapitulate the disease phenotype. Why this is the case is not known, but research involving patient cell lines can only take us so far and it is clear that we need to obtain more useful disease models to help us understand the clinical pathologies. Additionally, there are many inherited human diseases that have yet to be linked with a disease-causing mutation, and it is highly likely that mutations in other DNA repair genes are responsible for yet uncharacterised cerebellar neurological, developmental and/or cancer predisposition disorders. The discovery of novel disorders will allow a greater understanding of the mechanisms and pathways important for the maintenance of neuronal homeostasis and prevention of tumourigenesis, and further study of both novel and existing disorders will hopefully yield potential therapeutic strategies in the future.

REFERENCES

Ahel, I., Rass, U., El-Khamisy, S. F., Katyal, S., Clements, P. M., McKinnon, P. J., Caldecott, K. W., West, S. C. 2006. The neurodegenerative disease protein aprataxin resolves abortive DNA ligation intermediates. *Nature* 443: 713–716.

Akbari, M., Sykora, P., Bohr, V. A. 2015. Slow mitochondrial repair of 5′-AMP renders mtDNA susceptible to damage in APTX deficient cells. *Sci Rep* 5: 12876.

Al-Tassan, N., Chmiel, N. H., Maynard, J., Fleming, N., Livingston, A. L., Williams, G. T., Hodges, A. K. et al. 2002. Inherited variants of MYH associated with somatic G:C → T:A mutations in colorectal tumors. *Nat Genet* 30: 227–12832.

Ame, J. C., Rolli, V., Schreiber, V., Niedergang, C., Apiou, F., Decker, P., Muller, S., Hoger, T., Menissier-De Murcia, J., De Murcia, G. 1999. PARP-2, A novel mammalian DNA damage-dependent poly(ADP-ribose) polymerase. *J Biol Chem* 274: 17860–8.

Aparicio, T., Baer, R., Gottesman, M., Gautier, J. 2016. MRN, CtIP, and BRCA1 mediate repair of topoisomerase II-DNA adducts. *J Cell Biol* 212: 399–408.

Balajee, A. S., May, A., Dianov, G. L., Friedberg, E. C., Bohr, V. A. 1997. Reduced RNA polymerase II transcription in intact and permeabilized Cockayne syndrome group B cells. *Proc Natl Acad Sci USA* 94: 4306–4311.

Balakrishnan, L., Brandt, P. D., Lindsey-Boltz, L. A., Sancar, A., Bambara, R. A. 2009. Long patch base excision repair proceeds via coordinated stimulation of the multienzyme DNA repair complex. *J Biol Chem* 284: 15158–15172.

Baple, E. L., Chambers, H., Cross, H. E., Fawcett, H., Nakazawa, Y., Chioza, B. A., Harlalka, G. V. et al. 2014. Hypomorphic PCNA mutation underlies a human DNA repair disorder. *J Clin Invest* 124: 3137–3146.

Bardwell, P. D., Martin, A., Wong, E., Li, Z., Edelmann, W., Scharff, M. D. 2003. Cutting edge: The G-U mismatch glycosylase methyl-CpG binding domain 4 is dispensable for somatic hypermutation and class switch recombination. *J Immunol* 170: 1620–1624.

Barzilai, A. 2007. The contribution of the DNA damage response to neuronal viability. *Antioxid Redox Signal* 9: 211–218.

Beard, W. A., Wilson, S. H. 2000. Structural design of a eukaryotic DNA repair polymerase: DNA polymerase beta. *Mutat Res* 460: 231–244.

Bendixen, C., Thomsen, B., Alsner, J., Westergaard, O. 1990. Camptothecin-stabilized topoisomerase I-DNA adducts cause premature termination of transcription. *Biochemistry* 29: 5613–5619.

Bernstein, N. K., Williams, R. S., Rakovszky, M. L., Cui, D., Green, R., Karimi-Busheri, F., Mani, R. S. et al. 2005. The molecular architecture of the mammalian DNA repair enzyme, polynucleotide kinase. *Mol Cell* 17: 657–670.

Berti, M., Vindigni, A. 2016. Replication stress: Getting back on track. *Nat Struct Mol Biol* 23: 103–109.

Bhagwat, N., Olsen, A. L., Wang, A. T., Hanada, K., Stuckert, P., Kanaar, R., D'andrea, A., Niedernhofer, L. J., Mchugh, P. J. 2009. XPF-ERCC1 participates in the Fanconi anemia pathway of cross-link repair. *Mol Cell Biol* 29: 6427–6437.

Bissett, R. J., Mclaughlin, J. R. 2010. Radon. *Chronic Dis Can* 29(Suppl. 1): 38–50.

Bjelland, S., Seeberg, E. 2003. Mutagenicity, toxicity and repair of DNA base damage induced by oxidation. *Mutat Res* 531: 37–80.

Bogliolo, M., Schuster, B., Stoepker, C., Derkunt, B., Su, Y., Raams, A., Trujillo, J. P. et al. 2013. Mutations in ERCC4, encoding the DNA-repair endonuclease XPF, cause Fanconi anemia. *Am J Hum Genet* 92: 800–806.

Braithwaite, E. K., Kedar, P. S., Stumpo, D. J., Bertocci, B., Freedman, J. H., Samson, L. D., Wilson, S. H. 2010. DNA polymerases beta and lambda mediate overlapping and independent roles in base excision repair in mouse embryonic fibroblasts. *PLOS ONE* 5: e12229.

Bras, J., Alonso, I., Barbot, C., Costa, M. M., Darwent, L., Orme, T., Sequeiros, J., Hardy, J., Coutinho, P., Guerreiro, R. 2015. Mutations in PNKP cause recessive ataxia with oculomotor apraxia type 4. *Am J Hum Genet* 96: 474–9.

Breslin, C., Caldecott, K. W. 2009. DNA 3′-phosphatase activity is critical for rapid global rates of single-strand break repair following oxidative stress. *Mol Cell Biol* 29: 4653–4662.

Brooks, P. J. 2008. The 8,5′-cyclopurine-2′-deoxynucleosides: Candidate neurodegenerative DNA lesions in xeroderma pigmentosum, and unique probes of transcription and nucleotide excision repair. *DNA Repair (Amst)* 7: 1168–1179.

Broughton, B. C., Thompson, A. F., Harcourt, S. A., Vermeulen, W., Hoeijmakers, J. H., Botta, E., Stefanini, M. et al. 1995. Molecular and cellular analysis of the DNA repair defect in a patient in xeroderma pigmentosum complementation group D who has the clinical features of xeroderma pigmentosum and Cockayne syndrome. *Am J Hum Genet* 56: 167–174.

Brown, J. A., Pack, L. R., Sanman, L. E., Suo, Z. 2011. Efficiency and fidelity of human DNA polymerases lambda and beta during gap-filling DNA synthesis. *DNA Repair (Amst)* 10: 24–33.

Caglayan, M., Horton, J. K., Prasad, R., Wilson, S. H. 2015. Complementation of aprataxin deficiency by base excision repair enzymes. *Nucleic Acids Res* 43: 2271–2281.

Caldecott, K. W. 2001. Mammalian DNA single-strand break repair: An X-ra(y)ted affair. *BioEssays* 23: 447–455.

Caldecott, K. W. 2008. Single-strand break repair and genetic disease. *Nat Rev Genet* 9: 619–631.

Caldecott, K. W. 2014a. DNA single-strand break repair. *Exp Cell Res* 329: 2–8.

Caldecott, K. W. 2014b. Protein ADP-ribosylation and the cellular response to DNA strand breaks. *DNA Repair (Amst)* 19: 108–113.

Calvo, J. A., Moroski-Erkul, C. A., Lake, A., Eichinger, L. W., Shah, D., Jhun, I., Limsirichai, P. et al. 2013. Aag DNA glycosylase promotes alkylation-induced tissue damage mediated by Parp1. *PLoS Genet* 9: e1003413.

Caputo, M., Frontini, M., Velez-Cruz, R., Nicolai, S., Prantera, G., Proietti-De-Santis, L. 2013. The CSB repair factor is overexpressed in cancer cells, increases apoptotic resistance, and promotes tumor growth. *DNA Repair (Amst)* 12: 293–9.

Carroll, J., Page, T. K., Chiang, S. C., Kalmar, B., Bode, D., Greensmith, L., McKinnon, P. J., Thorpe, J. R., Hafezparast, M., El-Khamisy, S. F. 2015. Expression of a pathogenic mutation of SOD1 sensitizes aprataxin-deficient cells and mice to oxidative stress and triggers hallmarks of premature ageing. *Hum Mol Genet* 24: 828–840.

Ceccaldi, R., Sarangi, P., D'andrea, A. D. 2016. The Fanconi anaemia pathway: New players and new functions. *Nat Rev Mol Cell Biol* 17: 337–349.

Chatterjee, N., Walker, G. C. 2017. Mechanisms of DNA damage, repair, and mutagenesis. *Environ Mol Mutagen* 58: 235–263.

Chen, L., Lee, H. M., Greeley, G. H., Jr., Englander, E. W. 2007. Accumulation of oxidatively generated DNA damage in the brain: A mechanism of neurotoxicity. *Free Radic Biol Med* 42: 385–393.

Cheung, A. C., Cramer, P. 2011. Structural basis of RNA polymerase II backtracking, arrest and reactivation. *Nature* 471: 249–253.

Cleaver, J. E. 1969. Xeroderma pigmentosum: A human disease in which an initial stage of DNA repair is defective. *Proc Natl Acad Sci USA* 63: 428–435.

Clements, P. M., Breslin, C., Deeks, E. D., Byrd, P. J., Ju, L., Bieganowski, P., Brenner, C., Moreira, M. C., Taylor, A. M., Caldecott, K. W. 2004. The ataxia-oculomotor apraxia 1 gene product has a role distinct from ATM and interacts with the DNA strand break repair proteins XRCC1 and XRCC4. *DNA Repair (Amst)* 3: 1493–1502.

Coin, F., Marinoni, J. C., Rodolfo, C., Fribourg, S., Pedrini, A. M., Egly, J. M. 1998. Mutations in the XPD helicase gene result in XP and TTD phenotypes, preventing interaction between XPD and the p44 subunit of TFIIH. *Nat Genet* 20: 184–188.

Coin, F., Oksenych, V., Egly, J. M. 2007. Distinct roles for the XPB/p52 and XPD/p44 subcomplexes of TFIIH in damaged DNA opening during nucleotide excision repair. *Mol Cell* 26: 245–256.

Compe, E., Egly, J. M. 2012. TFIIH: When transcription met DNA repair. *Nat Rev Mol Cell Biol* 13: 343–354.

Cortes, U., Tong, W. M., Coyle, D. L., Meyer-Ficca, M. L., Meyer, R. G., Petrilli, V., Herceg, Z., Jacobson, E. L., Jacobson, M. K., Wang, Z. Q. 2004. Depletion of the 110-kilodalton isoform of poly(ADP-ribose) glycohydrolase increases sensitivity to genotoxic and endotoxic stress in mice. *Mol Cell Biol* 24: 7163–7178.

Daley, J. M., Wilson, T. E., Ramotar, D. 2010. Genetic interactions between HNT3/Aprataxin and RAD27/FEN1 suggest parallel pathways for 5′ end processing during base excision repair. *DNA Repair (Amst)* 9: 690–699.

Das, B. B., Dexheimer, T. S., Maddali, K., Pommier, Y. 2010. Role of tyrosyl-DNA phosphodiesterase (TDP1) in mitochondria. *Proc Natl Acad Sci USA* 107: 19790–5.

Das, B. B., Huang, S. Y., Murai, J., Rehman, I., Ame, J. C., Sengupta, S., Das, S. K. et al. 2014. PARP1-TDP1 coupling for the repair of topoisomerase I-induced DNA damage. *Nucleic Acids Res* 42: 4435–4449.

Date, H., Onodera, O., Tanaka, H., Iwabuchi, K., Uekawa, K., Igarashi, S., Koike, R. et al. 2001. Early-onset ataxia with ocular motor apraxia and hypoalbuminemia is caused by mutations in a new HIT superfamily gene. *Nat Genet* 29: 184–188.

de Boer, J., Andressoo, J. O., De Wit, J., Huijmans, J., Beems, R. B., Van Steeg, H., Weeda, G. et al. 2002. Premature aging in mice deficient in DNA repair and transcription. *Science* 296: 1276–1279.

de Laat, W. L., Appeldoorn, E., Sugasawa, K., Weterings, E., Jaspers, N. G., Hoeijmakers, J. H. 1998. DNA-binding polarity of human replication protein A positions nucleases in nucleotide excision repair. *Genes Dev* 12: 2598–2609.

Demple, B., Demott, M. S. 2002. Dynamics and diversions in base excision DNA repair of oxidized abasic lesions. *Oncogene* 21: 8926–8934.

D'errico, M., Parlanti, E., Teson, M., Degan, P., Lemma, T., Calcagnile, A., Iavarone, I. et al. 2007. The role of CSA in the response to oxidative DNA damage in human cells. *Oncogene* 26: 4336–4343.

D'errico, M., Parlanti, E., Teson, M., De Jesus, B. M., Degan, P., Calcagnile, A., Jaruga, P. et al. 2006. New functions of XPC in the protection of human skin cells from oxidative damage. *EMBO J* 25: 4305–4315.

de Waard, H., De Wit, J., Andressoo, J. O., Van Oostrom, C. T., Riis, B., Weimann, A., Poulsen, H. E., Van Steeg, H., Hoeijmakers, J. H., Van Der Horst, G. T. 2004. Different effects of CSA and CSB deficiency on sensitivity to oxidative DNA damage. *Mol Cell Biol* 24: 7941–7948.

Dianov, G. L., Parsons, J. L. 2007. Co-ordination of DNA single strand break repair. *DNA Repair (Amst)* 6: 454–460.

Digiovanna, J. J., Kraemer, K. H. 2012. Shining a light on xeroderma pigmentosum. *J Invest Dermatol* 132: 785–796.

Drury, S., Boustred, C., Tekman, M., Stanescu, H., Kleta, R., Lench, N., Chitty, L. S., Scott, R. H. 2014. A novel homozygous ERCC5 truncating mutation in a family with prenatal arthrogryposis – Further evidence of genotype-phenotype correlation. *Am J Med Genet A* 164A: 1777–1783.

Dumitrache, L. C., Mckinnon, P. J. 2017. Polynucleotide kinase-phosphatase (PNKP) mutations and neurologic disease. *Mech Ageing Dev* 161: 121–129.

Eliasson, M. J., Sampei, K., Mandir, A. S., Hurn, P. D., Traystman, R. J., Bao, J., Pieper, A. et al. 1997. Poly(ADP-ribose) polymerase gene disruption renders mice resistant to cerebral ischemia. *Nat Med* 3: 1089–1095.

El-Khamisy, S. F., Caldecott, K. W. 2006. TDP1-dependent DNA single-strand break repair and neurodegeneration. *Mutagenesis* 21: 219–224.

El-Khamisy, S. F., Katyal, S., Patel, P., Ju, L., Mckinnon, P. J., Caldecott, K. W. 2009. Synergistic decrease of DNA single-strand break repair rates in mouse neural cells lacking both Tdp1 and aprataxin. *DNA Repair (Amst)* 8: 760–766.

El-Khamisy, S. F., Masutani, M., Suzuki, H., Caldecott, K. W. 2003. A requirement for PARP-1 for the assembly or stability of XRCC1 nuclear foci at sites of oxidative DNA damage. *Nucleic Acids Res* 31: 5526–5533.

El-Khamisy, S. F., Saifi, G. M., Weinfeld, M., Johansson, F., Helleday, T., Lupski, J. R., Caldecott, K. W. 2005. Defective DNA single-strand break repair in spinocerebellar ataxia with axonal neuropathy-1. *Nature* 434: 108–113.

Engelward, B. P., Allan, J. M., Dreslin, A. J., Kelly, J. D., Wu, M. M., Gold, B., Samson, L. D. 1998. A chemical and genetic approach together define the biological consequences of 3-methyladenine lesions in the mammalian genome. *J Biol Chem* 273: 5412–5418.

Fagbemi, A. F., Orelli, B., Scharer, O. D. 2011. Regulation of endonuclease activity in human nucleotide excision repair. *DNA Repair (Amst)* 10: 722–729.

Ferro, A. M., Olivera, B. M. 1982. Poly(ADP-ribosylation) *in vitro*. Reaction parameters and enzyme mechanism. *J Biol Chem* 257: 7808–7813.

Fisher, A. E., Hochegger, H., Takeda, S., Caldecott, K. W. 2007. Poly(ADP-ribose) polymerase 1 accelerates single-strand break repair in concert with poly(ADP-ribose) glycohydrolase. *Mol Cell Biol* 27: 5597–5605.

Fitch, M. E., Nakajima, S., Yasui, A., Ford, J. M. 2003. *In vivo* recruitment of XPC to UV-induced cyclobutane pyrimidine dimers by the DDB2 gene product. *J Biol Chem* 278: 46906–46910.

Fortini, P., Parlanti, E., Sidorkina, O. M., Laval, J., Dogliotti, E. 1999. The type of DNA glycosylase determines the base excision repair pathway in mammalian cells. *J Biol Chem* 274: 15230–6.

Fousteri, M., Vermeulen, W., Van Zeeland, A. A., Mullenders, L. H. 2006. Cockayne syndrome A and B proteins differentially regulate recruitment of chromatin remodeling and repair factors to stalled RNA polymerase II *in vivo*. *Mol Cell* 23: 471–482.

Frederick, G. D., Amirkhan, R. H., Schultz, R. A., Friedberg, E. C. 1994. Structural and mutational analysis of the xeroderma pigmentosum group D (XPD) gene. *Hum Mol Genet* 3: 1783–1788.

Gatz, S. A., Ju, L., Gruber, R., Hoffmann, E., Carr, A. M., Wang, Z. Q., Liu, C., Jeggo, P. A. 2011. Requirement for DNA ligase IV during embryonic neuronal development. *J Neurosci* 31: 10088–10100.

Giglia-Mari, G., Coin, F., Ranish, J. A., Hoogstraten, D., Theil, A., Wijgers, N., Jaspers, N. G. et al. 2004. A new, tenth subunit of TFIIH is responsible for the DNA repair syndrome trichothiodystrophy group A. *Nat Genet* 36: 714–719.

Godon, C., Mourgues, S., Nonnekens, J., Mourcet, A., Coin, F., Vermeulen, W., Mari, P. O., Giglia-Mari, G. 2012. Generation of DNA single-strand displacement by compromised nucleotide excision repair. *EMBO J* 31: 3550–3563.

Graham, J. M., Jr., Anyane-Yeboa, K., Raams, A., Appeldoorn, E., Kleijer, W. J., Garritsen, V. H., Busch, D., Edersheim, T. G., Jaspers, N. G. 2001. Cerebro-oculo-facio-skeletal syndrome with a nucleotide excision-repair defect and a mutated XPD gene, with prenatal diagnosis in a triplet pregnancy. *Am J Hum Genet* 69: 291–300.

Green, C. M., Baple, E. L., Crosby, A. H. 2014. PCNA mutation affects DNA repair not replication. *Cell Cycle* 13: 3157–3158.

Greenhaw, G. A., Hebert, A., Duke-Woodside, M. E., Butler, I. J., Hecht, J. T., Cleaver, J. E., Thomas, G. H., Horton, W. A. 1992. Xeroderma pigmentosum and Cockayne syndrome: Overlapping clinical and biochemical phenotypes. *Am J Hum Genet* 50: 677–689.

Groisman, R., Kuraoka, I., Chevallier, O., Gaye, N., Magnaldo, T., Tanaka, K., Kisselev, A. F., Harel-Bellan, A., Nakatani, Y. 2006. CSA-dependent degradation of CSB by the ubiquitin-proteasome pathway establishes a link between complementation factors of the Cockayne syndrome. *Genes Dev* 20: 1429–1434.

Grundy, G. J., Polo, L. M., Zeng, Z., Rulten, S. L., Hoch, N. C., Paomephan, P., Xu, Y. et al. 2016. PARP3 is a sensor of nicked nucleosomes and monoribosylates histone H2B(Glu2). *Nat Commun* 7: 12404.

Hanzlikova, H., Gittens, W., Krejcikova, K., Zeng, Z., Caldecott, K. W. 2017. Overlapping roles for PARP1 and PARP2 in the recruitment of endogenous XRCC1 and PNKP into oxidized chromatin. *Nucleic Acids Res* 45: 2546–12557.

Hawkins, A. J., Subler, M. A., Akopiants, K., Wiley, J. L., Taylor, S. M., Rice, A. C., Windle, J. J., Valerie, K., Povirk, L. F. 2009. *In vitro* complementation of Tdp1 deficiency indicates a stabilized enzyme-DNA adduct from tyrosyl but not glycolate lesions as a consequence of the SCAN1 mutation. *DNA Repair (Amst)* 8: 654–663.

Henning, K. A., Li, L., Iyer, N., Mcdaniel, L. D., Reagan, M. S., Legerski, R., Schultz, R. A. et al. 1995. The Cockayne syndrome group A gene encodes a WD repeat protein that interacts with CSB protein and a subunit of RNA polymerase II TFIIH. *Cell* 82: 555–564.

Heo, J., Li, J., Summerlin, M., Hays, A., Katyal, S., Mckinnon, P. J., Nitiss, K. C., Nitiss, J. L., Hanakahi, L. A. 2015. TDP1 promotes assembly of non-homologous end joining protein complexes on DNA. *DNA Repair (Amst)* 30: 28–37.

Hoch, N. C., Hanzlikova, H., Rulten, S. L., Tetreault, M., Komulainen, E., Ju, L., Hornyak, P. et al. 2017. XRCC1 mutation is associated with PARP1 hyperactivation and cerebellar ataxia. *Nature* 541: 87–91.

Hoeijmakers, J. H. 2001. Genome maintenance mechanisms for preventing cancer. *Nature* 411: 366–374.

Horibata, K., Iwamoto, Y., Kuraoka, I., Jaspers, N. G., Kurimasa, A., Oshimura, M., Ichihashi, M., Tanaka, K. 2004. Complete absence of Cockayne syndrome group B gene product gives rise to UV-sensitive syndrome but not Cockayne syndrome. *Proc Natl Acad Sci USA* 101: 15410–5.

Huang, L. C., Clarkin, K. C., Wahl, G. M. 1996. Sensitivity and selectivity of the DNA damage sensor responsible for activating p53-dependent G1 arrest. *Proc Natl Acad Sci USA* 93: 4827–4832.

Imai, K., Slupphaug, G., Lee, W. I., Revy, P., Nonoyama, S., Catalan, N., Yel, L. et al. 2003. Human uracil-DNA glycosylase deficiency associated with profoundly impaired immunoglobulin class-switch recombination. *Nat Immunol* 4: 1023–1028.

Interthal, H., Chen, H. J., Kehl-Fie, T. E., Zotzmann, J., Leppard, J. B., Champoux, J. J. 2005. SCAN1 mutant Tdp1 accumulates the enzyme-DNA intermediate and causes camptothecin hypersensitivity. *EMBO J* 24: 2224–2233.

Interthal, H., Pouliot, J. J., Champoux, J. J. 2001. The tyrosyl-DNA phosphodiesterase Tdp1 is a member of the phospholipase D superfamily. *Proc Natl Acad Sci USA* 98: 12009–12014.

Ito, S., Kuraoka, I. 2015. Epigenetic modifications in DNA could mimic oxidative DNA damage: A double-edged sword. *DNA Repair (Amst)* 32: 52–57.

Izumi, T., Brown, D. B., Naidu, C. V., Bhakat, K. K., Macinnes, M. A., Saito, H., Chen, D. J., Mitra, S. 2005. Two essential but distinct functions of the mammalian abasic endonuclease. *Proc Natl Acad Sci USA* 102: 5739–5743.

Izumi, T., Hazra, T. K., Boldogh, I., Tomkinson, A. E., Park, M. S., Ikeda, S., Mitra, S. 2000. Requirement for human AP endonuclease 1 for repair of 3′-blocking damage at DNA single-strand breaks induced by reactive oxygen species. *Carcinogenesis* 21: 1329–1334.

Jaspers, N. G., Raams, A., Silengo, M. C., Wijgers, N., Niedernhofer, L. J., Robinson, A. R., Giglia-Mari, G. et al. 2007. First reported patient with human ERCC1 deficiency has cerebro-oculo-facio-skeletal syndrome with a mild defect in nucleotide excision repair and severe developmental failure. *Am J Hum Genet* 80: 457–466.

Jiang, B., Glover, J. N., Weinfeld, M. 2017. Neurological disorders associated with DNA strand-break processing enzymes. *Mech Ageing Dev* 161: 130–140.

Kalam, M. A., Haraguchi, K., Chandani, S., Loechler, E. L., Moriya, M., Greenberg, M. M., Basu, A. K. 2006. Genetic effects of oxidative DNA damages: Comparative mutagenesis of the imidazole ring-opened formamidopyrimidines (Fapy lesions) and 8-oxo-purines in simian kidney cells. *Nucleic Acids Res* 34: 2305–2315.

Karimi-Busheri, F., Rasouli-Nia, A., Allalunis-Turner, J., Weinfeld, M. 2007. Human polynucleotide kinase participates in repair of DNA double-strand breaks by nonhomologous end joining but not homologous recombination. *Cancer Res* 67: 6619–6625.

Kashiyama, K., Nakazawa, Y., Pilz, D. T., Guo, C., Shimada, M., Sasaki, K., Fawcett, H. et al. 2013. Malfunction of nuclease ERCC1-XPF results in diverse clinical manifestations and causes Cockayne syndrome, xeroderma pigmentosum, and Fanconi anemia. *Am J Hum Genet* 92: 807–819.

Katsuki, Y., Takata, M. 2016. Defects in homologous recombination repair behind the human diseases: FA and HBOC. *Endocr Relat Cancer* 23: T19–T37.

Katyal, S., El-Khamisy, S. F., Russell, H. R., Li, Y., Ju, L., Caldecott, K. W., Mckinnon, P. J. 2007. TDP1 facilitates chromosomal single-strand break repair in neurons and is neuroprotective *in vivo*. *EMBO J* 26: 4720–4731.

Kelland, L. 2007. The resurgence of platinum-based cancer chemotherapy. *Nat Rev Cancer* 7: 573–584.

Kim, Y. J., Wilson, D. M. 3rd, 2012. Overview of base excision repair biochemistry. *Curr Mol Pharmacol* 5: 3–13.

Kirkali, G., De Souza-Pinto, N. C., Jaruga, P., Bohr, V. A., Dizdaroglu, M. 2009. Accumulation of (5′S)-8,5′-cyclo-2′-deoxyadenosine in organs of Cockayne syndrome complementation group B gene knockout mice. *DNA Repair (Amst)* 8: 274–278.

Klingseisen, A., Jackson, A. P. 2011. Mechanisms and pathways of growth failure in primordial dwarfism. *Genes Dev* 25: 2011–2024.

Kraemer, K. H., Lee, M. M., Andrews, A. D., Lambert, W. C. 1994. The role of sunlight and DNA repair in melanoma and nonmelanoma skin cancer. The xeroderma pigmentosum paradigm. *Arch Dermatol* 130: 1018–1021.

Krokan, H. E., Bjoras, M. 2013. Base excision repair. *Cold Spring Harb Perspect Biol* 5: a012583.

Krokan, H. E., Drablos, F., Slupphaug, G. 2002. Uracil in DNA—Occurrence, consequences and repair. *Oncogene* 21: 8935–8948.

Lalle, P., Nouspikel, T., Constantinou, A., Thorel, F., Clarkson, S. G. 2002. The founding members of xeroderma pigmentosum group G produce XPG protein with severely impaired endonuclease activity. *J Invest Dermatol* 118: 344–351.

Laugel, V. 2013. Cockayne syndrome: The expanding clinical and mutational spectrum. *Mech Ageing Dev* 134: 161–170.

Lebedeva, N. A., Rechkunova, N. I., El-Khamisy, S. F., Lavrik, O. I. 2012. Tyrosyl-DNA phosphodiesterase 1 initiates repair of apurinic/apyrimidinic sites. *Biochimie* 94: 1749–1753.

Le Ber, I., Moreira, M. C., Rivaud-Pechoux, S., Chamayou, C., Ochsner, F., Kuntzer, T., Tardieu, M. et al. 2003. Cerebellar ataxia with oculomotor apraxia type 1: Clinical and genetic studies. *Brain* 126: 2761–2772.

Lee, A. J., Wallace, S. S. 2017. Hide and seek: How do DNA glycosylases locate oxidatively damaged DNA bases amidst a sea of undamaged bases? *Free Radic Biol Med* 107: 170–178.

Lee, Y., Katyal, S., Li, Y., El-Khamisy, S. F., Russell, H. R., Caldecott, K. W., Mckinnon, P. J. 2009. The genesis of cerebellar interneurons and the prevention of neural DNA damage require XRCC1. *Nat Neurosci* 12: 973–980.

Li, C. L., Golebiowski, F. M., Onishi, Y., Samara, N. L., Sugasawa, K., Yang, W. 2015. Tripartite DNA lesion recognition and verification by XPC, TFIIH, and XPA in nucleotide excision repair. *Mol Cell* 59: 1025–1034.

Li, L., Bales, E. S., Peterson, C. A., Legerski, R. J. 1993. Characterization of molecular defects in xeroderma pigmentosum group C. *Nat Genet* 5: 413–417.

Lin, C. P., Ban, Y., Lyu, Y. L., Desai, S. D., Liu, L. F. 2008. A ubiquitin-proteasome pathway for the repair of topoisomerase I-DNA covalent complexes. *J Biol Chem* 283: 21074–21083.

Lindahl, T. 1993. Instability and decay of the primary structure of DNA. *Nature* 362: 709–715.

Ljungman, M., Lane, D. P. 2004. Transcription—Guarding the genome by sensing DNA damage. *Nat Rev Cancer* 4: 727–737.

Lopes, M., Foiani, M., Sogo, J. M. 2006. Multiple mechanisms control chromosome integrity after replication fork uncoupling and restart at irreparable UV lesions. *Mol Cell* 21: 15–27.

Mallery, D. L., Tanganelli, B., Colella, S., Steingrimsdottir, H., Van Gool, A. J., Troelstra, C., Stefanini, M., Lehmann, A. R. 1998. Molecular analysis of mutations in the CSB (ERCC6) gene in patients with Cockayne syndrome. *Am J Hum Genet* 62: 77–85.

Marteijn, J. A., Lans, H., Vermeulen, W., Hoeijmakers, J. H. 2014. Understanding nucleotide excision repair and its roles in cancer and ageing. *Nat Rev Mol Cell Biol* 15: 465–481.

Masutani, C., Kusumoto, R., Yamada, A., Dohmae, N., Yokoi, M., Yuasa, M., Araki, M., Iwai, S., Takio, K., Hanaoka, F. 1999. The XPV (xeroderma pigmentosum variant) gene encodes human DNA polymerase eta. *Nature* 399: 700–704.

Mathieu, N., Kaczmarek, N., Naegeli, H. 2010. Strand- and site-specific DNA lesion demarcation by the xeroderma pigmentosum group D helicase. *Proc Natl Acad Sci USA* 107: 17545–17550.

Mathieu, N., Kaczmarek, N., Ruthemann, P., Luch, A., Naegeli, H. 2013. DNA quality control by a lesion sensor pocket of the xeroderma pigmentosum group D helicase subunit of TFIIH. *Curr Biol* 23: 204–212.

Mayne, L. V., Lehmann, A. R. 1982. Failure of RNA synthesis to recover after UV irradiation: An early defect in cells from individuals with Cockayne's syndrome and xeroderma pigmentosum. *Cancer Res* 42: 1473–1478.

McKinnon, P. J. 2017. Genome integrity and disease prevention in the nervous system. *Genes Dev* 31: 1180–1194.

Meira, L. B., Graham, J. M., Jr., Greenberg, C. R., Busch, D. B., Doughty, A. T., Ziffer, D. W., Coleman, D. M., Savre-Train, I., Friedberg, E. C. 2000. Manitoba aboriginal kindred with original cerebro-oculo-facio-skeletal syndrome has a mutation in the Cockayne syndrome group B (CSB) gene. *Am J Hum Genet* 66: 1221–1228.

Menck, C. F., Munford, V. 2014. DNA repair diseases: What do they tell us about cancer and aging? *Genet Mol Biol* 37: 220–233.

Methot, S. P., Di Noia, J. M. 2017. Molecular mechanisms of somatic hypermutation and class switch recombination. *Adv Immunol* 133: 37–87.

Miao, Z. H., Agama, K., Sordet, O., Povirk, L., Kohn, K. W., Pommier, Y. 2006. Hereditary ataxia SCAN1 cells are defective for the repair of transcription-dependent topoisomerase I cleavage complexes. *DNA Repair (Amst)* 5: 1489–1494.

Moreira, M. C., Barbot, C., Tachi, N., Kozuka, N., Uchida, E., Gibson, T., Mendonca, P. et al. 2001. The gene mutated in ataxia-ocular apraxia 1 encodes the new HIT/Zn-finger protein aprataxin. *Nat Genet* 29: 189–193.

Naim, V., Wilhelm, T., Debatisse, M., Rosselli, F. 2013. ERCC1 and MUS81-EME1 promote sister chromatid separation by processing late replication intermediates at common fragile sites during mitosis. *Nat Cell Biol* 15: 1008–1015.

Nair, P., Hamzeh, A. R., Mohamed, M., Saif, F., Tawfiq, N., El Halik, M., Al-Ali, M. T., Bastaki, F. 2016. Microcephalic primordial dwarfism in an Emirati patient with PNKP mutation. *Am J Med Genet A* 170: 2127–2132.

Nakazawa, Y., Sasaki, K., Mitsutake, N., Matsuse, M., Shimada, M., Nardo, T., Takahashi, Y. et al. 2012. Mutations in UVSSA cause UV-sensitive syndrome and impair RNA polymerase IIo processing in transcription-coupled nucleotide-excision repair. *Nat Genet* 44: 586–592.

Nardo, T., Oneda, R., Spivak, G., Vaz, B., Mortier, L., Thomas, P., Orioli, D. et al. 2009. A UV-sensitive syndrome patient with a specific CSA mutation reveals separable roles for CSA in response to UV and oxidative DNA damage. *Proc Natl Acad Sci USA* 106: 6209–6214.

Nichols, A. F., Ong, P., Linn, S. 1996. Mutations specific to the xeroderma pigmentosum group E Ddb- phenotype. *J Biol Chem* 271: 24317–24320.

Nick McElhinny, S. A., Watts, B. E., Kumar, D., Watt, D. L., Lundstrom, E. B., Burgers, P. M., Johansson, E., Chabes, A., Kunkel, T. A. 2010. Abundant ribonucleotide incorporation into DNA by yeast replicative polymerases. *Proc Natl Acad Sci USA* 107: 4949–4954.

Oh, K. S., Khan, S. G., Jaspers, N. G., Raams, A., Ueda, T., Lehmann, A., Friedmann, P. S. et al. 2006. Phenotypic heterogeneity in the XPB DNA helicase gene (ERCC3): Xeroderma pigmentosum without and with Cockayne syndrome. *Hum Mutat* 27: 1092–1103.

Pani, B., Nudler, E. 2017. Mechanistic insights into transcription coupled DNA repair. *DNA Repair (Amst)* 56: 42–50.

Parlanti, E., Locatelli, G., Maga, G., Dogliotti, E. 2007. Human base excision repair complex is physically associated to DNA replication and cell cycle regulatory proteins. *Nucleic Acids Res* 35: 1569–1577.

Paucar, M., Malmgren, H., Taylor, M., Reynolds, J. J., Svenningsson, P., Press, R., Nordgren, A. 2016. Expanding the ataxia with oculomotor apraxia type 4 phenotype. *Neurol Genet* 2: e49.

Petermann, E., Keil, C., Oei, S. L. 2006. Roles of DNA ligase III and XRCC1 in regulating the switch between short patch and long patch BER. *DNA Repair (Amst)* 5: 544–e55.

Plo, I., Liao, Z. Y., Barcelo, J. M., Kohlhagen, G., Caldecott, K. W., Weinfeld, M., Pommier, Y. 2003. Association of XRCC1 and tyrosyl DNA phosphodiesterase (Tdp1) for the repair of topoisomerase I-mediated DNA lesions. *DNA Repair (Amst)* 2: 1087–1100.

Pommier, Y. 2009. DNA topoisomerase I inhibitors: Chemistry, biology, and interfacial inhibition. *Chem Rev* 109: 2894–2902.

Pommier, Y., Redon, C., Rao, V. A., Seiler, J. A., Sordet, O., Takemura, H., Antony, S. et al. 2003. Repair of and checkpoint response to topoisomerase I-mediated DNA damage. *Mutat Res* 532: 173–203.

Poulton, C., Oegema, R., Heijsman, D., Hoogeboom, J., Schot, R., Stroink, H., Willemsen, M. A. et al. 2013. Progressive cerebellar atrophy and polyneuropathy: Expanding the spectrum of PNKP mutations. *Neurogenetics* 14: 43–51.

Pourquier, P., Pilon, A. A., Kohlhagen, G., Mazumder, A., Sharma, A., Pommier, Y. 1997. Trapping of mammalian topoisomerase I and recombinations induced by damaged DNA containing nicks or gaps. Importance of DNA end phosphorylation and camptothecin effects. *J Biol Chem* 272: 26441–7.

Prasad, R., Shock, D. D., Beard, W. A., Wilson, S. H. 2010. Substrate channeling in mammalian base excision repair pathways: Passing the baton. *J Biol Chem* 285: 40479–40488.

Qin, Y., Guo, T., Li, G., Tang, T. S., Zhao, S., Jiao, X., Gong, J. et al. 2015. CSB-PGBD3 mutations cause premature ovarian failure. *PLoS Genet* 11: e1005419.

Rass, U., Ahel, I., West, S. C. 2007. Defective DNA repair and neurodegenerative disease. *Cell* 130: 991–1004.

Reijns, M. A., Rabe, B., Rigby, R. E., Mill, P., Astell, K. R., Lettice, L. A., Boyle, S. et al. 2012. Enzymatic removal of ribonucleotides from DNA is essential for mammalian genome integrity and development. *Cell* 149: 1008–1022.

Reynolds, J. J., El-Khamisy, S. F., Katyal, S., Clements, P., Mckinnon, P. J., Caldecott, K. W. 2009. Defective DNA ligation during short-patch single-strand break repair in ataxia oculomotor apraxia 1. *Mol Cell Biol* 29: 1354–1362.

Reynolds, J. J., Stewart, G. S. 2013. A single strand that links multiple neuropathologies in human disease. *Brain* 136: 14–27.

Reynolds, J. J., Walker, A. K., Gilmore, E. C., Walsh, C. A., Caldecott, K. W. 2012. Impact of PNKP mutations associated with microcephaly, seizures and developmental delay on enzyme activity and DNA strand break repair. *Nucleic Acids Res* 40: 6608–6619.

Saleh-Gohari, N., Bryant, H. E., Schultz, N., Parker, K. M., Cassel, T. N., Helleday, T. 2005. Spontaneous homologous recombination is induced by collapsed replication forks that are caused by endogenous DNA single-strand breaks. *Mol Cell Biol* 25: 7158–7169.

Sander, M., Cadet, J., Casciano, D. A., Galloway, S. M., Marnett, L. J., Novak, R. F., Pettit, S. D. et al. 2005. Proceedings of a workshop on DNA adducts: Biological significance and applications to risk assessment Washington, DC, April 13–14, 2004. *Toxicol Appl Pharmacol* 208: 1–20.

Schellenberg, M. J., Tumbale, P. P., Williams, R. S. 2015. Molecular underpinnings of Aprataxin RNA/DNA deadenylase function and dysfunction in neurological disease. *Prog Biophys Mol Biol* 117: 157–165.

Schuch, A. P., Moreno, N. C., Schuch, N. J., Menck, C. F. M., Garcia, C. C. M. 2017. Sunlight damage to cellular DNA: Focus on oxidatively generated lesions. *Free Radic Biol Med* 107: 110–124.

Schwertman, P., Lagarou, A., Dekkers, D. H., Raams, A., Van Der Hoek, A. C., Laffeber, C., Hoeijmakers, J. H. et al. 2012. UV-sensitive syndrome protein UVSSA recruits USP7 to regulate transcription-coupled repair. *Nat Genet* 44: 598–602.

Scrima, A., Konickova, R., Czyzewski, B. K., Kawasaki, Y., Jeffrey, P. D., Groisman, R., Nakatani, Y., Iwai, S., Pavletich, N. P., Thoma, N. H. 2008. Structural basis of UV DNA-damage recognition by the DDB1-DDB2 complex. *Cell* 135: 1213–1223.

Sedgwick, B., Bates, P. A., Paik, J., Jacobs, S. C., Lindahl, T. 2007. Repair of alkylated DNA: Recent advances. *DNA Repair (Amst)* 6: 429–442.

Seidle, H. F., Bieganowski, P., Brenner, C. 2005. Disease-associated mutations inactivate AMP-lysine hydrolase activity of aprataxin. *J Biol Chem* 280: 20927–20931.

Shafirovich, V., Geacintov, N. E. 2017. Removal of oxidatively generated DNA damage by overlapping repair pathways. *Free Radic Biol Med* 107: 53–61.

Shen, J., Gilmore, E. C., Marshall, C. A., Haddadin, M., Reynolds, J. J., Eyaid, W., Bodell, A. et al. 2010. Mutations in PNKP cause microcephaly, seizures and defects in DNA repair. *Nat Genet* 42: 245–249.

Shibutani, S., Takeshita, M., Grollman, A. P. 1991. Insertion of specific bases during DNA synthesis past the oxidation-damaged base 8-oxodG. *Nature* 349: 431–434.

Shimada, M., Dumitrache, L. C., Russell, H. R., Mckinnon, P. J. 2015. Polynucleotide kinase-phosphatase enables neurogenesis via multiple DNA repair pathways to maintain genome stability. *EMBO J* 34: 2465–2480.

Sijbers, A. M., De Laat, W. L., Ariza, R. R., Biggerstaff, M., Wei, Y. F., Moggs, J. G., Carter, K. C. et al. 1996. Xeroderma pigmentosum group F caused by a defect in a structure-specific DNA repair endonuclease. *Cell* 86: 811–822.

Sijmons, R. H., Hofstra, R. M. 2016. Review: Clinical aspects of hereditary DNA mismatch repair gene mutations. *DNA Repair (Amst)* 38: 155–162.

Smith, Z. D., Meissner, A. 2013. DNA methylation: Roles in mammalian development. *Nat Rev Genet* 14: 204–220.

Sobol, R. W., Prasad, R., Evenski, A., Baker, A., Yang, X. P., Horton, J. K., Wilson, S. H. 2000. The lyase activity of the DNA repair protein beta-polymerase protects from DNA-damage-induced cytotoxicity. *Nature* 405: 807–810.

Spivak, G. 2005. UV-sensitive syndrome. *Mutat Res* 577: 162–169.

Spivak, G. 2015. Nucleotide excision repair in humans. *DNA Repair (Amst)* 36: 13–18.

Stefanini, M., Botta, E., Lanzafame, M., Orioli, D. 2010. Trichothiodystrophy: From basic mechanisms to clinical implications. *DNA Repair (Amst)* 9: 2–10.

Stevnsner, T., Nyaga, S., De Souza-Pinto, N. C., Van Der Horst, G. T., Gorgels, T. G., Hogue, B. A., Thorslund, T., Bohr, V. A. 2002. Mitochondrial repair of 8-oxoguanine is deficient in Cockayne syndrome group B. *Oncogene* 21: 8675–8682.

Steward, M. M., Sridhar, A., Meyer, J. S. 2013. Neural regeneration. *Curr Top Microbiol Immunol* 367: 163–191.

Sugasawa, K. 2016. Molecular mechanisms of DNA damage recognition for mammalian nucleotide excision repair. *DNA Repair (Amst)* 44: 110–117.

Sugasawa, K., Akagi, J., Nishi, R., Iwai, S., Hanaoka, F. 2009. Two-step recognition of DNA damage for mammalian nucleotide excision repair: Directional binding of the XPC complex and DNA strand scanning. *Mol Cell* 36: 642–653.

Sugasawa, K., Okamoto, T., Shimizu, Y., Masutani, C., Iwai, S., Hanaoka, F. 2001. A multistep damage recognition mechanism for global genomic nucleotide excision repair. *Genes Dev* 15: 507–521.

Sugawara, M., Wada, C., Okawa, S., Kobayashi, M., Sageshima, M., Imota, T., Toyoshima, I. 2008. Purkinje cell loss in the cerebellar flocculus in patients with ataxia with ocular motor apraxia type 1/early-onset ataxia with ocular motor apraxia and hypoalbuminemia. *Eur Neurol* 59: 18–23.

Sugitani, N., Sivley, R. M., Perry, K. E., Capra, J. A., Chazin, W. J. 2016. XPA: A key scaffold for human nucleotide excision repair. *DNA Repair (Amst)* 44: 123–135.

Sung, J. S., Demple, B. 2006. Roles of base excision repair subpathways in correcting oxidized abasic sites in DNA. *FEBS J* 273: 1620–1629.

Swann, P. F. 1990. Why do O6-alkylguanine and O4-alkylthymine miscode? The relationship between the structure of DNA containing O6-alkylguanine and O4-alkylthymine and the mutagenic properties of these bases. *Mutat Res* 233: 81–94.

Tahbaz, N., Subedi, S., Weinfeld, M. 2012. Role of polynucleotide kinase/phosphatase in mitochondrial DNA repair. *Nucleic Acids Res* 40: 3484–3495.

Takahashi, T., Tada, M., Igarashi, S., Koyama, A., Date, H., Yokoseki, A., Shiga, A. et al. 2007. Aprataxin, causative gene product for EAOH/AOA1, repairs DNA single-strand breaks with damaged 3′-phosphate and 3′-phosphoglycolate ends. *Nucleic Acids Res* 35: 3797–3809.

Takashima, H., Boerkoel, C. F., John, J., Saifi, G. M., Salih, M. A., Armstrong, D., Mao, Y. et al. 2002. Mutation of TDP1, encoding a topoisomerase I-dependent DNA damage repair enzyme, in spinocerebellar ataxia with axonal neuropathy. *Nat Genet* 32: 267–272.

Takayama, K., Salazar, E. P., Broughton, B. C., Lehmann, A. R., Sarasin, A., Thompson, L. H., Weber, C. A. 1996. Defects in the DNA repair and transcription gene ERCC2(XPD) in trichothiodystrophy. *Am J Hum Genet* 58: 263–270.

Tanaka, K., Miura, N., Satokata, I., Miyamoto, I., Yoshida, M. C., Satoh, Y., Kondo, S., Yasui, A., Okayama, H., Okada, Y. 1990. Analysis of a human DNA excision repair gene involved in group A xeroderma pigmentosum and containing a zinc-finger domain. *Nature* 348: 73–76.

Tapias, A., Auriol, J., Forget, D., Enzlin, J. H., Scharer, O. D., Coin, F., Coulombe, B., Egly, J. M. 2004. Ordered conformational changes in damaged DNA induced by nucleotide excision repair factors. *J Biol Chem* 279: 19074–19083.

Tebbs, R. S., Thompson, L. H., Cleaver, J. E. 2003. Rescue of Xrcc1 knockout mouse embryo lethality by transgene-complementation. *DNA Repair (Amst)* 2: 1405–1417.

Theil, A. F., Nonnekens, J., Steurer, B., Mari, P. O., De Wit, J., Lemaitre, C., Marteijn, J. A. et al. 2013. Disruption of TTDA results in complete nucleotide excision repair deficiency and embryonic lethality. *PLoS Genet* 9: e1003431.

Tomkinson, A. E., Vijayakumar, S., Pascal, J. M., Ellenberger, T. 2006. DNA ligases: Structure, reaction mechanism, and function. *Chem Rev* 106: 687–99.

Tudek, B. 2003. Imidazole ring-opened DNA purines and their biological significance. *J Biochem Mol Biol* 36: 12–19.

Tumbale, P., Appel, C. D., Kraehenbuehl, R., Robertson, P. D., Williams, J. S., Krahn, J., Ahel, I., Williams, R. S. 2011. Structure of an aprataxin-DNA complex with insights into AOA1 neurodegenerative disease. *Nat Struct Mol Biol* 18: 1189–1195.

Tumbale, P., Williams, J. S., Schellenberg, M. J., Kunkel, T. A., Williams, R. S. 2014. Aprataxin resolves adenylated RNA-DNA junctions to maintain genome integrity. *Nature* 506: 111–115.

Tzoulis, C., Sztromwasser, P., Johansson, S., Gjerde, I. O., Knappskog, P., Bindoff, L. A. 2017. PNKP mutations identified by whole-exome sequencing in a Norwegian patient with sporadic ataxia and edema. *Cerebellum* 16: 272–275.

Tzur-Gilat, A., Ziv, Y., Mittelman, L., Barzilai, A., Shiloh, Y. 2013. Studying the cerebellar DNA damage response in the tissue culture dish. *Mech Ageing Dev* 134: 496–505.

Valko, M., Leibfritz, D., Moncol, J., Cronin, M. T., Mazur, M., Telser, J. 2007. Free radicals and antioxidants in normal physiological functions and human disease. *Int J Biochem Cell Biol* 39: 44–84.

Vaz, B., Popovic, M., Newman, J. A., Fielden, J., Aitkenhead, H., Halder, S., Singh, A. N. et al. 2016. Metalloprotease SPRTN/DVC1 orchestrates replication-coupled DNA-protein crosslink repair. *Mol Cell* 64: 704–719.

Vermeulen, W., Rademakers, S., Jaspers, N. G., Appeldoorn, E., Raams, A., Klein, B., Kleijer, W. J., Hansen, L. K., Hoeijmakers, J. H. 2001. A temperature-sensitive disorder in basal transcription and DNA repair in humans. *Nat Genet* 27: 299–303.

Vidal, A. E., Boiteux, S., Hickson, I. D., Radicella, J. P. 2001. XRCC1 coordinates the initial and late stages of DNA abasic site repair through protein-protein interactions. *EMBO J* 20: 6530–6539.

Volker, M., Mone, M. J., Karmakar, P., Van Hoffen, A., Schul, W., Vermeulen, W., Hoeijmakers, J. H., Van Driel, R., Van Zeeland, A. A., Mullenders, L. H. 2001. Sequential assembly of the nucleotide excision repair factors *in vivo*. *Mol Cell* 8: 213–224.

Wang, J. C. 2002. Cellular roles of DNA topoisomerases: A molecular perspective. *Nat Rev Mol Cell Biol* 3: 430–440.

Ward, J. F. 1998. Nature of lesions formed by ionizing radiation. In: Nickoloff, J. A., Hoekstra, M. F. editors. *DNA Damage and Repair: Volume 2: DNA Repair in Higher Eukaryotes*, 65–84. Humana Press, Totowa, NJ.

Weeda, G., Eveno, E., Donker, I., Vermeulen, W., Chevallier-Lagente, O., Taieb, A., Stary, A., Hoeijmakers, J. H., Mezzina, M., Sarasin, A. 1997. A mutation in the XPB/ERCC3 DNA repair transcription gene, associated with trichothiodystrophy. *Am J Hum Genet* 60: 320–329.

Weeda, G., Van Ham, R. C., Vermeulen, W., Bootsma, D., Van Der Eb, A. J., Hoeijmakers, J. H. 1990. A presumed DNA helicase encoded by ERCC-3 is involved in the human repair disorders xeroderma pigmentosum and Cockayne's syndrome. *Cell* 62: 777–791.

Whitehouse, C. J., Taylor, R. M., Thistlethwaite, A., Zhang, H., Karimi-Busheri, F., Lasko, D. D., Weinfeld, M., Caldecott, K. W. 2001. XRCC1 stimulates human polynucleotide kinase activity at damaged DNA termini and accelerates DNA single-strand break repair. *Cell* 104: 107–117.

Williams, J. S., Lujan, S. A., Kunkel, T. A. 2016. Processing ribonucleotides incorporated during eukaryotic DNA replication. *Nat Rev Mol Cell Biol* 17: 350–363.

Williams, J. S., Smith, D. J., Marjavaara, L., Lujan, S. A., Chabes, A., Kunkel, T. A. 2013. Topoisomerase 1-mediated removal of ribonucleotides from nascent leading-strand DNA. *Mol Cell* 49: 1010–1015.

Winters, T. A., Henner, W. D., Russell, P. S., Mccullough, A., Jorgensen, T. J. 1994. Removal of 3′-phosphoglycolate from DNA strand-break damage in an oligonucleotide substrate by recombinant human apurinic/apyrimidinic endonuclease 1. *Nucleic Acids Res* 22: 1866–1873.

Woods, C. G., Parker, A. 2013. Investigating microcephaly. *Arch Dis Child* 98: 707–713.

Yang, W., Woodgate, R. 2007. What a difference a decade makes: Insights into translesion DNA synthesis. *Proc Natl Acad Sci USA* 104: 15591–8.

Yokoseki, A., Ishihara, T., Koyama, A., Shiga, A., Yamada, M., Suzuki, C., Sekijima, Y. et al. 2011. Genotype-phenotype correlations in early onset ataxia with ocular motor apraxia and hypoalbuminaemia. *Brain* 134: 1387–1399.

Yoon, G., Westmacott, R., Macmillan, L., Quercia, N., Koutsou, P., Georghiou, A., Christodoulou, K., Banwell, B. 2008. Complete deletion of the aprataxin gene: Ataxia with oculomotor apraxia type 1 with severe phenotype and cognitive deficit. *J Neurol Neurosurg Psychiatry* 79: 234–236.

Zafeiriou, D. I., Thorel, F., Andreou, A., Kleijer, W. J., Raams, A., Garritsen, V. H., Gombakis, N., Jaspers, N. G., Clarkson, S. G. 2001. Xeroderma pigmentosum group G with severe neurological involvement and features of Cockayne syndrome in infancy. *Pediatr Res* 49: 407–412.

Zhou, T., Lee, J. W., Tatavarthi, H., Lupski, J. R., Valerie, K., Povirk, L. F. 2005. Deficiency in 3′-phosphoglycolate processing in human cells with a hereditary mutation in tyrosyl-DNA phosphodiesterase (TDP1). *Nucleic Acids Res* 33: 289–297.

Homologous Recombination at Replication Forks

6

Eva Petermann

INTRODUCTION

Homologous recombination (HR) is a remarkable genome maintenance pathway that brings together DNA replication and DNA repair. Because of this, it is absolutely central to replication stress and to diseases that are characterized by replication stress or treated with replication stress-inducing agents (see Chapter 15). As a DNA repair pathway, HR mediates double-strand break (DSB) repair in mitotic interphase. As a genetic recombination pathway, it mediates crossover formation in meiosis. At all times, HR uses extensive DNA synthesis to copy the homologous template, and it is becoming increasingly evident that HR processes frequently occur at perturbed DNA replication forks in S phase. HR proteins are rapidly recruited to perturbed replication forks, and HR intermediates are observed at forks even during unperturbed DNA replication. This chapter will first discuss the standard models of DSB repair and replication fork restart by HR repair, before moving on to newly described roles of HR at stressed replication forks and the significance of HR at replication forks in cancer development and therapy. It will show how fundamental insights into DNA repair mechanisms can become relevant for the clinic.

HOMOLOGOUS RECOMBINATION REPAIR OF DOUBLE-STRAND BREAKS

HR is one of several pathways that repair DSBs that have been directly induced by ionizing radiation (gamma rays or x-rays) or restriction enzymes such as I-SceI. In fact the Rad52 epistasis group of HR genes, which includes the central recombinase Rad51, was initially identified as a group of genes that promotes the survival of ionizing radiation in yeast (Game and Cox 1971). At the same time, genetic studies in yeast helped to generate detailed models of the mechanism of HR repair of DSBs (Szostak et al. 1983). However, because HR requires a sister chromatid for faithful repair, it is only used for DSB repair in proliferating cells. The process of recombination also has the potential to induce chromosomal rearrangements, especially in higher eukaryotes where the genomes are rich in repetitive sequences. Because of this, mammalian cells tightly regulate HR repair and repair more DSBs using non-homologous end joining (Rothkamm et al. 2003), which is overall faster and may even be less likely to cause deleterious mutations. However, if DSBs arise at replication forks or in transcriptionally active chromatin (Aymard et al. 2014), then HR is favoured.

The central protein in HR is recombinase, which is called RecA in bacteria and Rad51/RAD51 in eukaryotes (Shinohara et al. 1992). The function of RecA and RAD51 is to bind single-stranded DNA, search the genome for homologous sequences and catalyse strand invasion to initiate repair synthesis. RAD51 loading at DSBs requires a 3′ overhang of single-stranded DNA (ssDNA), and this ssDNA overhang is generated by exonuclease resection of the 5′ ends of the break (**Figure 6.1**). DNA end resection requires high cyclin-dependent kinase (CDK) activity, which restricts it to the S and G2 phase of the cell cycle, when homologous sister chromatids are present (Aylon et al. 2004, Ira et al. 2004). Resection is dependent on the tumour suppressor BRCA1, a large E3 ubiquitin ligase that also functions in cell cycle checkpoints and transcription, on CtIP, the MRE11-RAD50-NBS1 (MRN) complex and on the exonucleases DNA2 and EXO1. All resection depends on BRCA1, MRE11 and CtIP, but

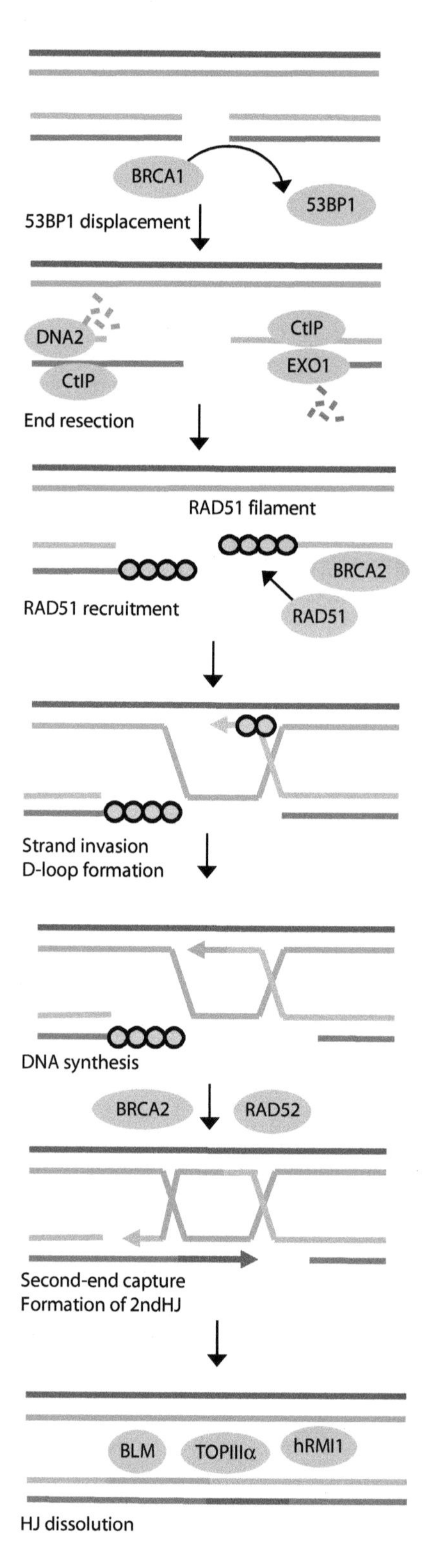

Figure 6.1 Model of double-strand break repair by homologous recombination. BRCA1-mediated removal of 53BP1 from the break enables CtIP-dependent resection of the double-stranded DNA ends by DNA2 and EXO1. Resection creates a 3′ overhang for RAD51 loading, which is promoted by BRCA2 and other mediators. The RAD51 filament catalyzes homology search, strand invasion and D-loop formation. Repair synthesis is established in the D-loop. Second-end capture, mediated by BRCA2 or RAD52, establishes DNA synthesis extension of the second 3′ end. Re-annealing the 3′ end from the D-loop back to the original template generates a second Holliday junction. The double HJ is resolved by the BLM-TOPIIIα-hRMI1 complex. Sister chromatids are color-coded in blue/green and red/yellow, respectively.

the bulk of long-range resection can be carried out by either DNA2 or EXO1, with help from RPA and the DNA helicase BLM (Zhu et al. 2008, Nimonkar et al. 2011). RAD51 competes for ssDNA binding with the single-stranded DNA binding protein RPA, which has a higher affinity to ssDNA. RAD51 loading therefore requires a large number of RAD51 interacting proteins or mediators that promote both the efficient replacement of RPA with RAD51 as well as the stability of the resulting RAD51-ssDNA complex. The most important mediators are BRCA2, a large scaffold protein than can bind multiple RAD51 molecules and stabilize RAD51 multimers and filaments, and the RAD51 paralogues XRCC2, XRCC3, RAD51B, RAD51C and RAD51D (Bishop et al. 1998, O'Regan et al. 2001, Sigurdsson et al. 2001, Takata et al. 2001). These proteins also ensure that RAD51 is specifically loaded onto 3′-ssDNA overhangs, although it can also bind to DNA gaps (Shahid et al. 2014) (Figure 6.1). During the initiation of HR, large numbers of RAD51 molecules are loaded onto the ssDNA overhang to generate a long protein-DNA filament (Ogawa et al. 1993). The filament performs the homology search and strand invasion into the homologous sequence (Baumann et al. 1996). The cell cycle regulation of resection and the presence of cohesin, which holds sister chromatids together after replication, ensures that this homologous sequence is usually the sister chromatid (Sjogren and Nasmyth 2001). Strand invasion generates the D-loop, where one strand of the double helix is displaced by the filament and thus forms a single-stranded loop. The D-loop contains a Holliday junction (HJ). The 3′ end of the D-loop can be used to prime DNA synthesis, which allows copying of information from the undamaged homologous sequence, which is also called repair synthesis. For completion of DNA repair, the second 3′ end anneals to the displaced strand of the D-loop, which provides it with a template for more repair synthesis. This second-end capture step can be mediated by RAD52 or by BRCA2 (McIlwraith and West 2008, Mazloum and Holloman 2009). The now extended 3′ end of the D-loop re-anneals with its original partner, generating a second HJ. DNA ligation then finally results in two completely unbroken DNA molecules (reviewed in Li and Heyer 2008). The remaining two HJ structures can be cleaved by a group of endonucleases called HJ resolvases such as GEN1, but the preferred pathway is cleavage-free HJ dissolution by topoisomerase IIIα (TOPIIIα) in a complex with BLM helicase and hRMI1 (Wu and Hickson 2003, Ip et al. 2008).

Another homology-directed DSB repair pathway, called single-strand annealing (SSA), also uses sequence homology and depends on resection and resection factors such as BRCA1. However, SSA is independent of RAD51 and RAD51 paralogues and instead requires the Rad52 proteins, which catalyse microhomology-dependent annealing of single-stranded DNA overhangs, followed by flap removal and ligation. Human RAD52 has therefore been proposed to act as a backup to RAD51-dependent recombination (Bhowmick et al. 2016), but it can also perform backup functions in RAD51 loading in cells lacking functional BRCA2 (Lok et al. 2013). Such microhomology-mediated pairings have the potential to induce genomic rearrangements (VanHulle et al. 2007).

Cell cycle control of resection means that HR repair is normally limited to S and G2 phase (Rothkamm et al. 2003). One exception is the repair of DSBs in ribosomal DNA (rDNA) repeats in the nucleolus, where HR repair has been observed even in G1 phase (van Sluis and McStay 2015). CDK1/2 activity increases in S/G2 phase and promotes resection, for example, via phosphorylation of CtIP (Huertas and Jackson 2009). BRCA1, whose protein levels increase in S/G2 phase (Dimitrov et al. 2013), also promotes resection. Resection is normally prevented by the DNA damage response protein 53BP1, which is recruited to DSBs by the checkpoint kinase ATM. 53BP1 needs to be displaced by BRCA1 for activation of HR (Bouwman et al. 2010, Bunting et al. 2010).

However, CDK activity also restricts HR, for example, by modulating binding of the C-terminal TR2 domain of BRCA2 to RAD51 multimers (Esashi et al. 2005). This CDK regulation prevents aberrant HR in G2/M phase when CDK activity is high, but not in S phase when CDK activity is lower, and HR is presumably required for repairing replication-dependent DNA damage. When DNA is damaged, this activates the cell cycle checkpoints which downregulate CDK activity and thus help shift the balance towards HR (Esashi et al. 2005). Further, cell cycle checkpoint kinases can directly regulate HR by phosphorylating RAD51, BRCA2 and other HR proteins (Andreassen et al. 2004, Sorensen et al. 2005, Wang et al. 2007, Bahassi et al. 2008) so that inhibiting checkpoint kinases can lead to defects in HR (Sorensen et al. 2005, Bryant and Helleday 2006).

HOMOLOGOUS RECOMBINATION AT REPLICATION FORKS

It is well established that HR is also activated by genotoxic treatment that does not directly induce DSBs, but instead interferes with the progression of DNA replication forks. Although HR-defective mutant cell lines were initially isolated based on sensitivity to ionizing radiation, they are often much more sensitive to agents that interfere with replication forks. Mammalian HR is strongly activated by replication fork–stalling treatments, such as hydroxyurea (HU) or thymidine, and promotes the survival of these treatments (Arnaudeau et al. 2001, Saintigny et al. 2001, Lundin et al. 2002). Replication inhibitors generally stall replication fork progression by inducing DNA polymerase–blocking lesions on the template (e.g. UV, MMS and interstrand cross-linkers such as mitomycin c or platinum compounds), depleting deoxyribonucleotides (HU, thymidine) or acting as catalytic inhibitors of the replicative DNA polymerases (aphidicolin). Campothecin collapses replication forks because they run into unrepaired single-strand breaks. HR activation in response to replication blocks can be assessed in a number of ways. First, HR proteins such as BRCA2 and RAD51 are recruited to stalled replication forks (Petermann et al. 2010, Kolinjivadi et al. 2017). Second, replication inhibitor treatments stimulate recombination in reporter genes (Arnaudeau et al. 2001, Saintigny et al. 2001) and HR is activated at engineered site-specific replication fork barriers in vivo (Lambert and Carr 2005, Willis et al. 2014). Third, RAD51 and BRCA2 prevent fork collapse and accumulation of replication-associated DSBs in response to replication-blocking treatments (Sonoda et al. 1998, Lomonosov et al. 2003, Lundin et al. 2003). Finally, replication fork restart assays, for example, using DNA fibre analysis, have shown that HR proteins such as RAD51 and RecA promote the restart of stalled replication forks across all domains of life (Horiuchi and Fujimura 1995, Lambert et al. 2010, Petermann et al. 2010).

There is also another HR-related DNA repair pathway, the Fanconi anaemia (FA) pathway, that plays roles at perturbed replication forks (also discussed in Chapter 12). The FA pathway was initially described as a DNA repair pathway involved in the repair of DNA interstrand cross-links, such as those induced by mitomycin c (MMC). The pathway depends on at least 21 proteins that are encoded by the genes of the Fanconi anaemia complementation groups, which include the FANC genes FANCA – FANCL and HR genes such as RAD51. While the FA pathway has long been considered a specific ICL repair pathway, it now appears evident that many FA proteins also participate in the protection or repair of replication forks that have been perturbed by other insults such as nucleotide depletion (Pichierri et al. 2004). This occurs in conjunction with canonical HR proteins such as BRCA1, BRCA2 and RAD51, all three themselves encoded by FA genes. The central FA pathway protein FANCD2, recruits the resection protein CtIP to damaged replication forks and is implicated in promoting RAD51 foci formation (Gravells et al. 2013, Yeo et al. 2014, Sato et al. 2016). The FA pathway also includes many proteins that are not just involved in promoting RAD51 loading, such as the DNA helicases FANCJ and FANCM (controversial as a FA protein), but have the ability to remodel perturbed replication forks and promote fork progression through difficult-to-replicate sequences (Gari et al. 2008, Wu et al. 2008). It is also important to note that some FA proteins such as FANCD2 and FANCJ perform specific functions at replication forks without involvement of the FA core complex or the FANCD2 monoubiquitination that is central to the response to ICLs (Chaudhury et al. 2013, Raghunandan et al. 2015, Madireddy et al. 2016).

HR repair of collapsed replication forks: Break-induced replication

What are the lesions that activate HR at stalled forks? First, replication fork stalling can lead to double-strand break formation through cleavage of the fork by the structure-specific endonuclease complexes MUS81-EME1 (Hanada et al. 2007) or MUS81-EME2 (Pepe and

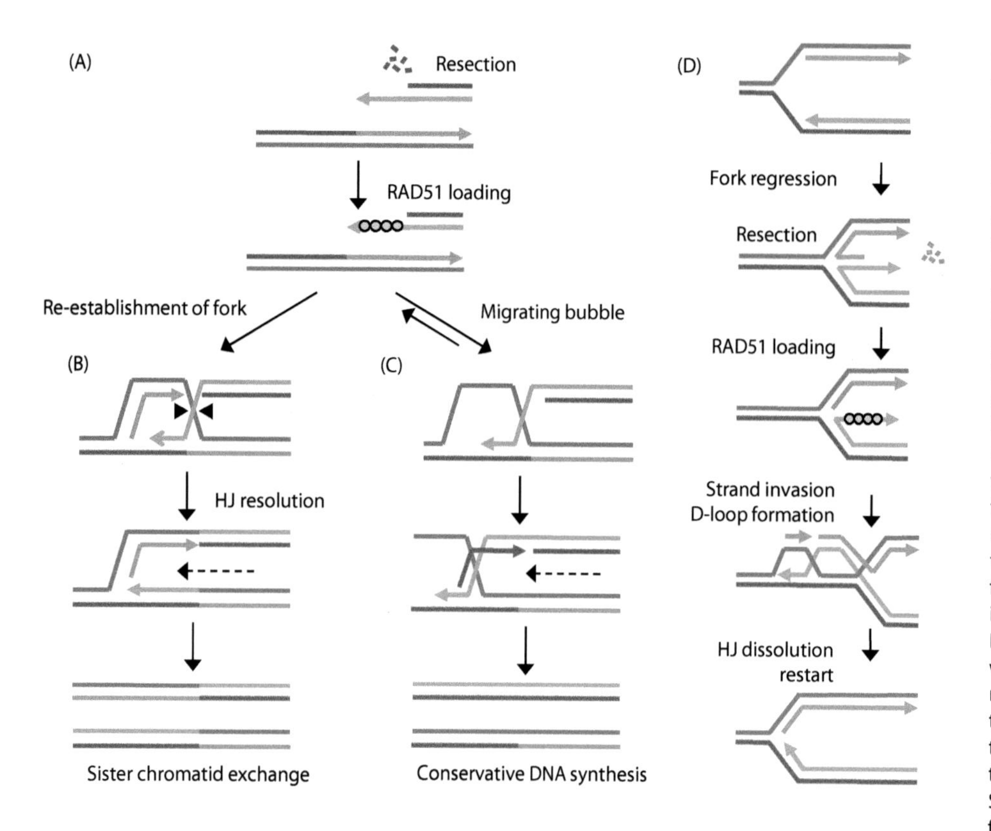

Figure 6.2 **Models for the restart of damaged replication forks by homologous recombination.** (A) If a stalled or collapsed fork has been converted into a DSB, it can be restarted and repaired by break-induced replication (BIR). End resection followed by RAD51 loading allows homology search and strand invasion, which also depends on RAD52. Strand invasion forms a D-loop, which allows re-initiation of DNA synthesis by the regular replication machinery and non-essential DNA polymerase δ subunit POLD3. (B) In the semi-conservative model of BIR, the resulting single HJ is removed by a HJ resolvase activity to re-establish a normal replication fork. This is associated with sister chromatid exchange (SCE). (C) In the conservative model for BIR, the HJ is not resolved and a migrating bubble is maintained. Lagging strand information is copied from the leading strand rather than the leading strand template, resulting in conservative DNA synthesis. This D-loop may be unstable with the 3′ end repeatedly dissociating and re-annealing. (D) For restart of stalled replication forks without a break, forks may regress to form a "chicken foot" structure that contains an HJ and a double-stranded DNA end. Some amount of resection may be required for RAD51 loading. RAD51-mediated strand invasion and D-loop formation allows fork restart. This generates a double HJ which is resolved by the BLM-TOPIIIα-hRMI1 complex to re-establish a normal replication fork without sister chromatid exchange.

West 2014). This process is conserved throughout evolution, even though the responsible endonucleases are not. In bacteria, fork-associated DSBs are generated by the RuvABC protein complex, which has no homologues in eukaryotes (Seigneur et al. 1998). This suggests that even though DSBs are highly toxic and mutagenic lesions, their generation at stalled forks can be beneficial if it allows fork restart via the HR mechanism of break-induced replication (BIR).

Endonuclease cleavage of collapsed forks results in the generation of one-ended DSBs, which lack a second end for repair because that part of the chromosome has not yet been replicated (**Figure 6.2A**). Such one-ended DSBs are also formed spontaneously when replication forks encounter unrepaired single-strand breaks, and they are repaired in the same way. BIR is the only pathway that can faithfully repair one-ended DSBs, in a process that restarts the collapsed replication fork and repairs the DSB in the process. Long known in yeast, we have recently gained much better insights into how this pathway proceeds in mammalian cells (Costantino et al. 2014). Initially, the double-stranded DNA end may need to be resected to form the 3′ ssDNA overhang for RAD51 loading (Chung et al. 2010). RAD51 loading is followed by strand invasion and D-loop formation, as in standard DSB repair. In yeast, and humans, strand invasion and D-loop formation for BIR also require Rad52/RAD52 (Malkova et al. 1996, Davis and Symington 2004, Sotiriou et al. 2016). Strand invasion is followed by recruitment of nearly all factors involved in normal S-phase replication fork progression, such as GINS and DNA polymerase ε (epsilon) (Lydeard et al. 2010, Hashimoto et al. 2012). Interestingly, however, the DNA polymerase δ (delta) subunit Pol32 (yeast) or POLD3 (human) is uniquely required for BIR while dispensable for normal replication (Lydeard et al. 2007, Costantino et al. 2014). Furthermore, the DNA helicase PIF1, which is not a canonical replicative helicase, stimulates DNA synthesis during BIR (Wilson et al. 2013). BIR thus re-establishes a replication fork that can then continue to replicate the genome. The extent of DNA repair synthesis may therefore be greater in BIR than in normal DSB repair by HR.

Two different models have been suggested for the DNA synthesis that follows the establishment of a replication fork during BIR. According to the first model, the D-loop is resolved and replication is completed in a semi-conservative way, just like at a normal DNA replication fork (Lydeard et al. 2010) (**Figure 6.2B**). In the second model, BIR proceeds by a migrating replication bubble that copies DNA in a conservative way (Donnianni and Symington 2013, Saini et al. 2013). Conservative DNA synthesis can result if the invading strand is extended to copy genetic information from the sister chromatid (the migrating D-loop or bubble), and this invading strand, rather than the identical strand that forms the bubble, is then copied again to generate double-stranded DNA (**Figure 6.2C**).

In addition to standard BIR, there is an alternative model termed microhomology-mediated BIR (MMBIR). MMBIR involves microhomology-mediated annealing of the 3′-overhang of the one-ended DSB with another region of single-stranded DNA, such as found at another replication fork. Like in SSA, two single-stranded DNA regions anneal, and therefore this process does not involve strand invasion and does not require RAD51 (Hastings et al. 2009). In yeast, BIR can also be either Rad51-dependent (Davis and Symington 2004, Malkova et al. 2005) or -independent (Malkova et al. 1996).

In mammalian cells, large numbers of DSBs only start to appear after several hours of continuous treatment with replication inhibitor HU (Saintigny et al. 2001, Hanada et al. 2007, Petermann et al. 2010), suggesting that MUS81-mediated fork cleavage is not the first or primary response of mammalian cells to replication fork stalling. Furthermore, RAD51 foci and recombination in reporter sequences are also only observed after longer HU blocks, and they do not seem to be directly connected to fork restart. Instead, they may be required for the post-replicative repair of collapsed forks (Petermann et al. 2010). This is in line with a report that RAD51-dependent mechanisms remove spontaneous DNA lesions preferentially during the G2 phase of the cell cycle, suggesting that HR is temporally separated from DNA replication that creates these lesions (Su et al. 2008). If HR repair of collapsed forks occurs when DNA replication has been mostly completed, it might act on different substrates than would be present during S phase. For example, if a replication fork from an adjacent origin arrived at an un-repaired collapsed fork, this would result in a two-ended DSB instead of a one-ended break. Such a DSB could also be a substrate for NHEJ, which could explain the HU sensitivity of NHEJ-defective cells (Saintigny et al. 2001, Lundin et al. 2002).

HR and replication fork restart without a DSB

If a stalled replication fork does not collapse into a DSB, then it might still require DNA damage response pathways to be able to resume progression once the obstacle on the DNA has been removed or nucleotide pools have been replenished (reviewed in Jones and Petermann 2012). Methods such as site-specific fork barriers and single-molecule analyses of replication fork progression using the DNA fibre technique have demonstrated that replication fork restart without DSB formation also involves the HR machinery (Lambert et al. 2010, Petermann et al. 2010). RAD51 can be recruited to stalled forks independent of DSB formation (Hanada et al. 2007, Hashimoto et al. 2010, Petermann et al. 2010), suggesting that RAD51 might be involved in a different fork restart mechanism avoiding DSB formation. This process does not, however, involve RAD51 foci formation or detectable long patch recombination, arguing that it is different from RAD51-mediated restart of collapsed forks through BIR (Petermann et al. 2010). Both RAD51 and RAD51 paralogues are required for efficient replication fork restart after short HU blocks (Petermann et al. 2010, Somyajit et al. 2015). This mechanism may be especially relevant after shorter replication blocks in mammalian cells and is also a predominant mechanism of fork restart in yeast (Lambert et al. 2010).

The DSB-independent mechanism of RAD51-mediated fork restart likely involves the generation of a HJ despite the absence of a break (**Figure 6.2D**). Electron microscopy of replication intermediates isolated from cells has revealed that replication forks can reverse or 'regress' by re-annealing of the template strands combined with pairing of the displaced daughter strands, resulting in HJ structures that are also called chicken foot structures due to their shape (Sogo et al. 2002). In *E. coli*, fork reversal has been suggested to be a spontaneous process caused by topological strain on the DNA (Postow et al. 2001); in budding yeast, such torsional stress can arise when a replication fork meets an unresolved transcription intermediate (Bermejo et al. 2011). In budding yeast, fork reversal is only observed in S phase checkpoint-deficient cells and is therefore likely to be a pathological process that is associated with fork collapse and is detrimental to genome integrity (Cotta-Ramusino et al. 2005). More recently, however, electron microscopy has been used to show that checkpoint-proficient human cells display high levels of fork reversal, further supporting that fork reversal may be required to protect or restart perturbed forks (Ray Chaudhuri et al. 2012).

RAD51 loading onto the stalled or reversed fork requires ssDNA and might therefore also depend on resection, either to generate an ssDNA gap or a 3′ overhang if RAD51 is loaded after formation of the chicken foot. In line with this, MRE11, CtIP and the alternative ssDNA-binding protein hSSB1, which is involved in HR repair of DSBs, are also required for fork restart (Bryant et al. 2009, Bolderson et al. 2014, Yeo et al. 2014). The model then predicts that RAD51 will restart the fork by catalysing invasion of the leading strand overhang of the chicken foot into the intact double-stranded template, forming a D-loop, which, analogous to BIR, will then be used to facilitate origin-independent replisome loading. It also predicts that a double HJ will be formed during restart, which can be resolved by the BLM-TOPIIIα-hRMI1 complex without crossing over (Wu and Hickson 2003). However in the last few years, this model has been modified, as impressive progress has been made on the role of HR proteins in promoting both stalled fork regression and protection of stalled replication forks from nucleolytic attack, and these new findings will be discussed next.

EMERGING ROLES FOR HOMOLOGOUS RECOMBINATION AT REPLICATION FORKS

Promoting replication fork regression

It has recently become evident that fork regression in mammalian cells does not occur passively but is an active process (Figure 6.3A). The first DNA repair factor shown to promote fork regression in human cells was, however, not RAD51 but the poly(ADP-ribose) polymerase PARP1, which promoted fork regression in response to low-dose camptothecin (Ray Chaudhuri et al. 2012). It was shown that PARP1 actually maintains the reversed chicken

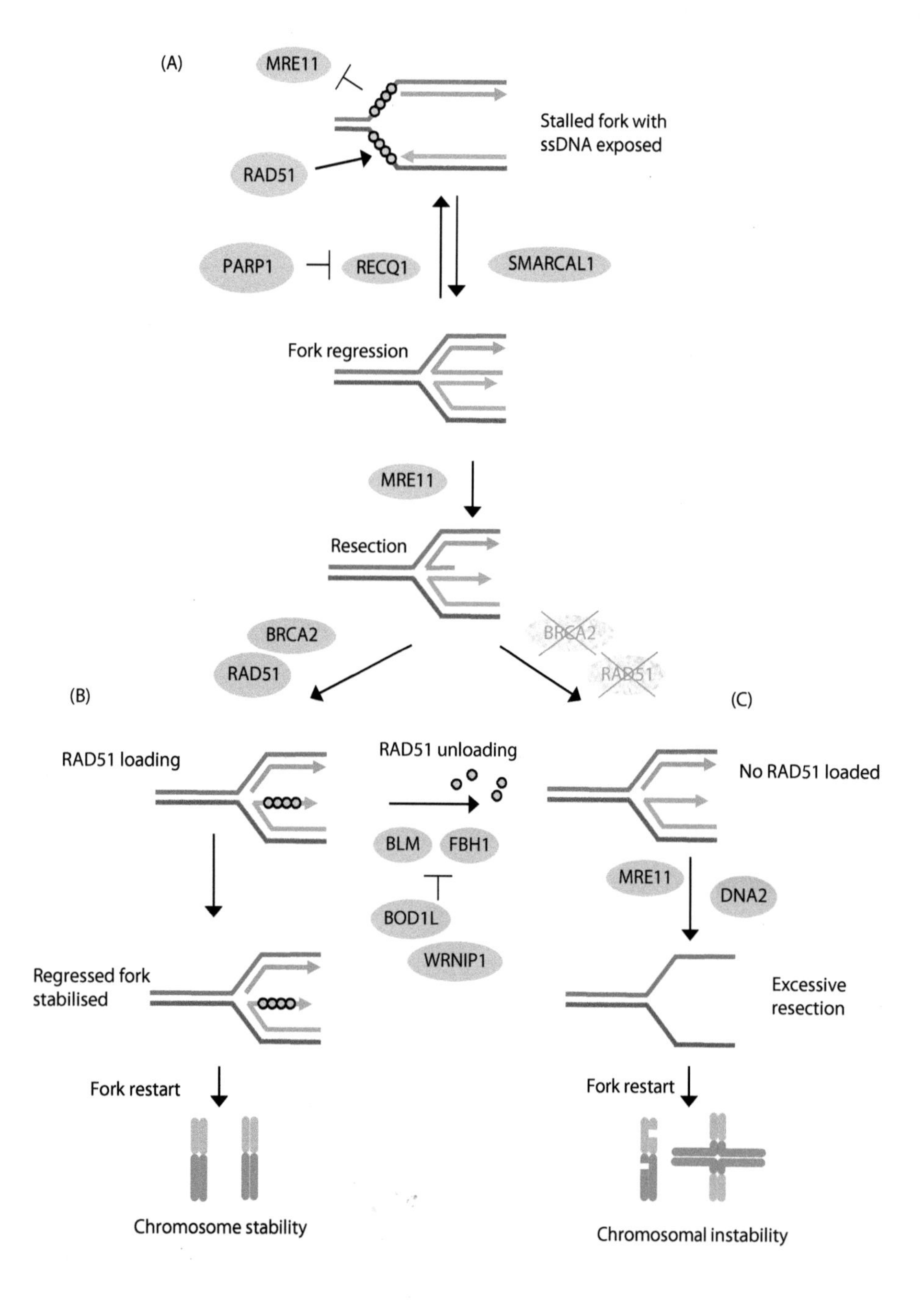

Figure 6.3 Emerging models of regression and excessive resection at stalled replication forks. (A) Replication fork stalling exposes excessive single-stranded DNA (ssDNA). RAD51 loading to ssDNA prevents resection by MRE11. The extent of ssDNA depends on the stalling agents and correlates with increased fork regression. Fork regression is catalyzed by SMARCAL1 and reversed by RECQ1. PARP1 promotes fork regression by inhibiting RECQ1. MRE11 is required for limited resection to allow RAD51 loading at the regressed fork. (B) In the presence of BRCA2 and RAD51, RAD51 loading onto the regressed fork structure stabilizes the structure, which promotes fork restart and genome stability. (C) In absence of BRCA2 or RAD51, the regressed fork is subject to excessive resection by MRE11 and DNA2. Activities that remove RAD51 filaments such as BLM and FBH1 also promote excessive resection, but are counteracted by BOD1L and WRNIP1. Excessive resection allows fork restart, but at the expense of genome stability.

foot structure of the fork by counteracting the RECQ family DNA helicase RECQ1. RECQ1 otherwise facilitates a reversal of the regression back to the regular fork structure, which is proposed to promote fork restart after removal of the damage (Berti et al. 2013). Soon after, it was also reported that fork reversal is promoted by the DNA helicase SMARCAL1 (Kolinjivadi et al. 2017) and also requires RAD51 (Zellweger et al. 2015). Fork regression was observed even at low concentrations of genotoxin that slow forks without stalling them (Zellweger et al. 2015). The different levels of ssDNA accumulation at forks in response to different genotoxins correlate with the percentages of regressed forks under the same conditions, suggesting that ssDNA serves to load RAD51, which then promotes regression (Zellweger et al. 2015). In fact, RecA had been similarly reported to promote HJ formation at stalled replication forks in *E. coli* many years before (Seigneur et al. 2000). Alternatively, it has also been suggested that RAD51, like PARP1, may stabilise chicken foot structures rather than actively driving their formation (**Figure 6.3B**). Though not strictly considered an HR factor, PARP1 has previously been implicated promoting RAD51 function at replication forks in response to camptothecin (Sugimura et al. 2008) and hydroxyurea (Bryant et al. 2009), which may be related to its role in promoting fork regression.

While fork regression can only be visualised by electron microscopy, it is also generally associated with replication fork slowing as measured by DNA fibre assay (Ray Chaudhuri et al. 2012). In fact, it has been known for some time that replication fork slowing in response to a number of replication-blocking treatments is not passive but depends on DNA repair factors – these happen to be the same factors that are reported to promote fork regression. PARP1, RAD51 and the RAD51 paralogues XRCC3 and XRCC2 actively slow fork progression in response to treatments with cisplatin, camptothecin or MMC (Henry-Mowatt et al. 2003, Sugimura et al. 2008, Zellweger et al. 2015). It therefore seems that it is fork regression facilitated by RAD51 and PARP1 that slows fork progression in response to many replication inhibitor treatments (Henry-Mowatt et al. 2003). Consequently, fork slowing by DNA fibre assay can sometimes be used as a proxy for fork regression (Berti et al. 2013). Exceptions are treatment such as with HU, where the depletion of nucleotides directly slows forks and cannot be overcome by inhibiting PARP1 or RAD51 (Zellweger et al. 2015). This fork-slowing function of RAD51 may, however, be different from the requirement for HR for cell survival, for example, of cisplatin damage, which is likely mediated through the DSB repair function of HR (Takata et al. 2001).

What is the benefit of fork regression? A chicken foot may be a more stable structure than a stalled fork and facilitate replication restart, as described earlier. Some data also suggest that fork regression may protect forks from being converted into DSBs (Ray Chaudhuri et al. 2012, Berti et al. 2013). On the other hand, the chicken foot structure presents a potential substrate for structure-specific endonucleases such as MUS81-EME1, making them potentially even more prone to DSB formation than non-regressed stalled-fork structures (Neelsen et al. 2013). As the activities of HJ endonucleases are regulated by CDK activity and thus largely restricted to mitosis, different cell cycle stages and levels of checkpoint activation may ultimately lead to different fates of reversed forks (Neelsen et al. 2013).

Preventing excessive resection of stalled replication forks

A related novel function of HR in mammalian cells is to prevent excessive nuclease resection of stalled replication forks (Hashimoto et al. 2010, Schlacher et al. 2011, 2012) (**Figure 6.3C**). This phenomenon is currently the subject of intensive investigation, partly due to the fact that it can be detected by DNA fibre methodology, which is more accessible technically than electron microscopy of replication intermediates. If replication forks are stalled for more than 4 hours in high concentrations of hydroxyurea or gemcitabine (another drug that depletes dNTPs), then the daughter strands that were synthesized prior to the replication block grow progressively shorter, eventually losing stretches of incorporated nucleotides of more than 10 kb in length (Schlacher et al. 2011). This shortening depends on nuclease activities of MRE11 and DNA2, suggesting it is due to resection and therefore associated with generation of very long stretches of single-stranded DNA (Schlacher et al. 2011, Thangavel et al. 2015). BRCA1, BRCA2, FANCD2 and RAD51 all prevent this excessive resection, together with several other proteins that are implicated in RAD51 function (Somyajit et al. 2015). Based on separation-of-function experiments using a BRCA2 C-terminal mutant, fork protection has been shown to be a distinct function from the HR repair of frank DSBs as induced by the I-SceI restriction enzyme (Schlacher et al. 2011). However, the initial stages of RAD51 loading certainly appear to be involved, with the data suggesting that resection is prevented by promoting RAD51 recruitment and the stability of RAD51 filaments (Hashimoto et al. 2010,

Schlacher et al. 2011). It seems likely that MRE11-dependent resection, initially required to allow RAD51 loading, becomes excessive when RAD51 cannot be efficiently recruited. In line with this, the anti-recombinase FBH1 has been reported to promote excessive resection, at least if its activity is not counteracted by genome maintenance proteins such as BOD1L (Higgs et al. 2015) and WRNIP1 (Leuzzi et al. 2016). One model to explain this is that FBH1 removes RAD51 from stalled forks, thereby promoting resection (**Figure 6.3C**).

Because RAD51 both promotes fork regression and prevents excessive resection, this might be taken to suggest that the former prevents the latter. However, the experimental conditions that have been used to measure fork regression and fork resection are quite different. Excessive fork resection as measured by DNA fibre analyses apparently requires very strong and prolonged inhibition of the ribonucleotide reductase, achieved with high doses of hydroxyurea or gemcitabine (Schlacher et al. 2011). While increased resection has been observed after MMS treatment in RAD51-deficient *Xenopus laevis* egg extracts, the resulting single-stranded regions were very short at only a few hundred nucelotides (Hashimoto et al. 2010). Fork regression, on the other hand, can be observed immediately after treatment with many different drugs and at low doses (Zellweger et al. 2015). A recent study uses in vitro approaches with *Xenopus laevis* egg extracts and reconstituted systems to clarify these points (Kolinjivadi et al. 2017). A likely model emerging from this study is that RAD51 and BRCA2 may not actively promote fork regression, but that RAD51 nucleofilament formation stabilises regressed forks by preventing resection. Regressed forks therefore promote excessive resection, but only if they are not protected by RAD51. Without fork regression, there is no excessive resection (Kolinjivadi et al. 2017) (**Figure 6.3C**). Interestingly, the anti-recombinase FBH1 has also been reported to promote fork regression (Fugger et al. 2015), which might provide an alternative explanation as to why FBH1 is required for excessive resection in cells that lack BOD1L or WRNIP1 (Higgs et al. 2015, Leuzzi et al. 2016).

What is the impact of excessive resection on the long-term fate of stalled forks? Excessive resection correlated with reduced fork restart and reduced cell survival of replication inhibitor treatments in a number of studies (Berti et al. 2013, Leuzzi et al. 2016), and it is clearly connected with increased genomic instability and chromosome rearrangements (Schlacher et al. 2011, Higgs et al. 2015, Leuzzi et al. 2016). It has not yet been linked to DSB formation at forks (Leuzzi et al. 2016). Interestingly, resection could be relevant for the mechanism of action of cytotoxic cancer treatments (see later).

Error-free damage bypass

Replication fork stalling by replication-blocking DNA lesions such as bulky base modifications leads to monoubiquitination of PCNA by RAD6 and RAD18 (Hoege et al. 2002). This activates the non-HR DNA damage tolerance pathway, translesion synthesis (TLS), which allows the lesion to be bypassed by specialised DNA polymerases. But PCNA can also be polyubiquitinated at the same lysine residue (K164), which is catalysed by Rad5 in yeast, and most likely by two different Rad5 homologues, HLTF and SHPRH, in mammalian cells (Motegi et al. 2008, Lin et al. 2011). Polyubiquitination of PCNA activates not TLS but an alternative 'error-free' damage bypass mechanism that involves HR-like activities, in particular template switching to the undamaged sister chromatid. This bypass mechanism was recently described in some more detail in budding yeast (Branzei et al. 2008, Giannattasio et al. 2014) (**Figure 6.4**). Error-free bypass operates behind the replication fork, as the fork itself continues replicating downstream of the blocking lesion by unscheduled repriming, and template switching is activated at an ssDNA gap that is left around the lesion (Fumasoni et al. 2015). Template switching is mediated by Rad51 and allows the copying of genetic information from an undamaged template after pairing of the newly replicated daughter strands, which involves pseudo-double HJ structures in which the daughter strands are interwound to form a hemicatenane and the template strands are unpaired (Branzei et al. 2008, Giannattasio et al. 2014). Repriming, which prevents accumulation of large ssDNA gaps at the fork, also suppresses fork regression in both budding yeast and vertebrates (Fumasoni et al. 2015, Kolinjivadi et al. 2017). In yeast, repriming and post-replicative bypass are much favoured over fork regression; in vertebrates, where fork regression seems to be much more common, the balance may be different.

Promoting normal replication fork progression

While the connection to fork regression implicates HR in slowing replication, there is also evidence that HR factors, at least those involved in RAD51 loading, promote normal speeds of replication fork progression in unperturbed S phase, as mammalian cells deficient in

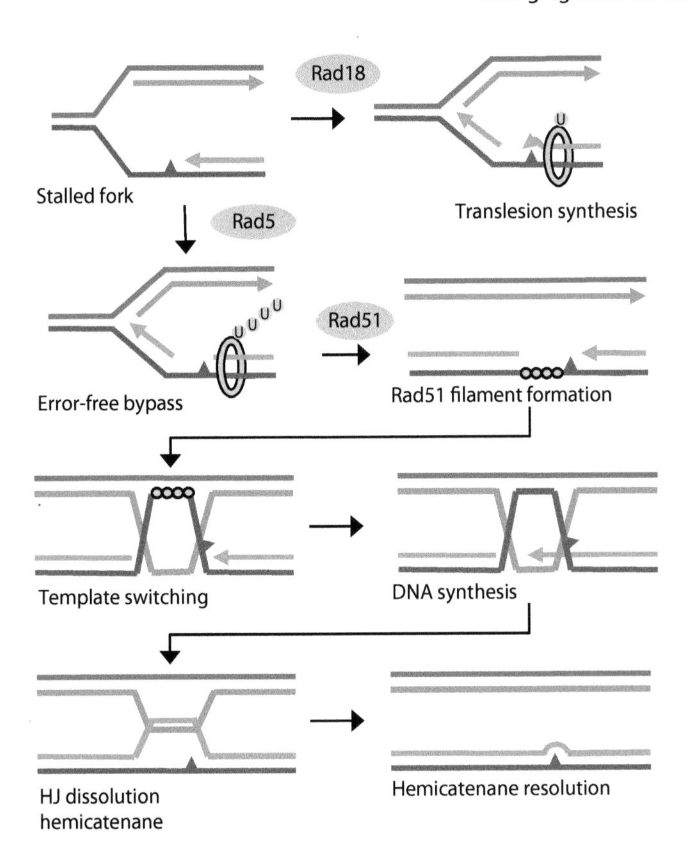

Figure 6.4 **Proposed mechanism for error-free bypass in budding yeast.** If a replication fork stalls at a blocking lesion (triangle), then replication can resume by re-repriming downstream of the lesion. This leaves a single-stranded gap, which requires post-replicative gap filling or repair. The exposed single-stranded DNA (ssDNA) acts as a signal for PCNA monoubiquitination by Rad18, promoting error-prone translesion synthesis. Alternatively, poly-ubiquitination of PCNA by Rad5 promotes error-free bypass involving template switching. In this model of error-free bypass, Rad51 is loaded onto the single-stranded template strand. Template switching allows genetic information to be copied from the undamaged sister. The double HJ is dissolved by Sgs1-Top3-Rmi1 (mammalian BLM-TOPIIIα-hRMI1) followed by resolution of a hemicatenane formed by the daughter strands. (Adapted from Branzei, D. and B. Szakal (2016). DNA Repair (Amst) 44: 68–75.)

XRCC2, BRCA2 or RAD51 can display mildly reduced basal fork speeds (Daboussi et al. 2008, Schlacher et al. 2011, Wilhelm et al. 2014, 2016). Spontaneous fork slowing has not been systematically investigated in cells defective in other FA pathway components or BRCA1. Nevertheless, the available data may support mild (5%–10%) fork slowing in BRCA1 and FA core complex, but not FANCJ- or FANCM-defective cells (Schwab et al. 2010, 2013, Schlacher et al. 2012). This could suggest a requirement for RAD51 loading to stabilise or restart forks that stall when encountering endogenous obstacles, such as spontaneous DNA damage. A completely new angle to this is provided by recent reports that FA proteins protect replication forks from conflicts with transcription and collisions with RNA–DNA hybrids (R-loops). Replication and transcription share the same template leading to inevitable collisions despite some spatial segregation (French 1992). Replication–transcription collisions can slow, stall or collapse replication forks, and therefore activate homologous recombination; this is also called transcription-associated recombination (TAR) (Gottipati and Helleday 2009). One particular source of such collisions and TAR are R-loops, DNA–RNA hybrids in which the nascent RNA remains associated to the template DNA and can impair replication fork progression (Gan et al. 2011). FANCD2, BRCA2 and FANCM, through its translocase activity, have been reported to counteract R-loop accumulation (Schwab et al. 2015, Madireddy et al. 2016). Because R-loops can slow replication fork progression, this suggests that fork slowing in HR-defective cells could be due to conflicts with transcription and especially R-loops.

Oxidative stress has also been made responsible for replication fork slowing in HR-deficient cells. Oxidative stress is a common cause of endogenous DNA damage that may interfere with replication fork progression if not properly dealt with (Hegde et al. 2013). Reducing the levels of reactive oxygen species by treating cells with antioxidants could rescue the defect (Wilhelm et al. 2016). However, the interpretation of these results was complicated by the observation that antioxidant treatment also, surprisingly, increased cellular deoxyribonucleotide (dNTP) pools, which could easily underlie that improvement in fork progression (Wilhelm et al. 2016). Nevertheless, HR-deficient cells also displayed higher levels of endogenous ROS compared to wild type, and reducing oxygen levels to a more physiological concentration of 3% O_2 abolished the requirement for HR proteins for replication fork progression (Wilhelm et al. 2016). This suggests that fork slowing in HR-deficient cells is due to a reduced ability to cope with an ROS-damaged DNA template, as well as an increased initial damage load. It also suggests that cells may only require HR factors for normal replication fork progression if they are exposed to high levels of oxygen, but not under physiological conditions.

HR AT REPLICATION FORKS AND GENOMIC INSTABILITY

Recombination-mediated replication fork restart can promote genomic instability

HR is usually considered an 'error-free' repair pathway that does not cause mutations; however, this is not entirely correct because recombination can induce genomic rearrangements. Replication restart by HR can cause chromosomal rearrangements if the wrong template is used (non-allelic HR), which contributes to chromosomal instability (CIN). This was first shown using a site-specific replication fork barrier in fission yeast, where HR-mediated restart can lead to ectopic recombination between chromosomes, and gross chromosomal rearrangements including translocations and deletions (Lambert et al. 2005, 2010). Using this system, it was also shown that the DNA synthesis by replication forks that have been restarted by recombination is more error-prone than DNA synthesis by regular replication forks, regardless of whether they were restarted from a DSB using BIR or restarted without a break. The restarted forks can perform template switching, such as U-turns at palindromes, and are more prone to generating microhomology-mediated insertions or deletions by replication slippage (Iraqui et al. 2012, Mizuno et al. 2013). At least in fission yeast, the complex chromosomal rearrangements that can be induced by HR at forks are not detected by the cell cycle checkpoints, leading to chromosome bridging and breakage, which are powerful drivers of CIN (Mohebi et al. 2015).

At least in the case of BIR, the low fidelity of DNA synthesis has been ascribed to the conservative or migrating bubble mode of replication of the restarted fork. Because the migrating D-loop or bubble is unstable, DNA synthesis during BIR and MMBIR involves repeated rounds of dissociation and strand invasion by the extending 3′ end (**Figure 6.2**). This increases the chance of pairing with different repeated sequences across the genome and chromosomal rearrangements (Smith et al. 2007, Hastings et al. 2009) (**Figure 6.5**). DNA synthesis during BIR is also prone to small 1-base pair deletions which cause frameshift mutations (Deem et al. 2011).

If recombination activity at forks can be detrimental to genome stability, it is not surprising that several activities can counteract recombination at forks. Specialized DNA helicases or 'anti-recombinases' counteract HR activities at forks. The RECQ helicases BLM and WRN are the best examples, although there are many more, such as FBH1 (**Figure 6.3**). Anti-recombinases can disrupt recombination intermediate structures such as the D-loop and the HJ, thus preventing spontaneous HR and genomic instability at perturbed replication forks (Constantinou et al. 2000, Karow et al. 2000). BLM dislodges RAD51 from ssDNA (Bugreev et al. 2007) and suppresses recombination via its HJ dissolution activity (Wu and Hickson 2003). Lack of functional BLM or WRN leads to genomic instability; in the rare human disorders Bloom and Werner syndrome, this is associated with increased cancer predisposition (Chu and Hickson 2009, Rossi et al. 2010) (also discussed in Chapter 12).

Break-induced replication prevents genomic instability at common fragile sites

Nevertheless, DNA repair by HR is of course important for genome stability, and one newly described function of BIR is to maintain genome stability at common fragile sites in mitosis. Common fragile sites are regions in the genome that are particularly prone to incomplete replication at mitotic entry, which has been attributed to low replication origin density and transcription–replication conflicts (Helmrich et al. 2011, Letessier et al. 2011). CFS can be detected as breaks or gaps in DAPI-stained metaphase chromosomes, which is why they have long been interpreted as sites of DNA breakage. The appearance of these DAPI-negative areas is also called 'CFS expression'. CFS display instability in Fanconi anaemia and cancer cells and in cells treated with low-dose replication inhibitors such as aphidicolin (Chan et al. 2009, Naim and Rosselli 2009, Bignell et al. 2010). In addition, aphidicolin treatment also increases the expression of rare fragile sites, which can be anywhere in the genome.

Without completion of replication termination and proper resolution of catenated DNA structures at termination sites, sister chromatids cannot be correctly separated during mitosis. Incomplete sister chromatid separation at CFS and other under-replicated regions can be observed as DNA bridges that connect the daughter nuclei during anaphase and are marked by FANCD2 and FANCI foci, suggesting that the FA pathway, and possibly HR, are activated

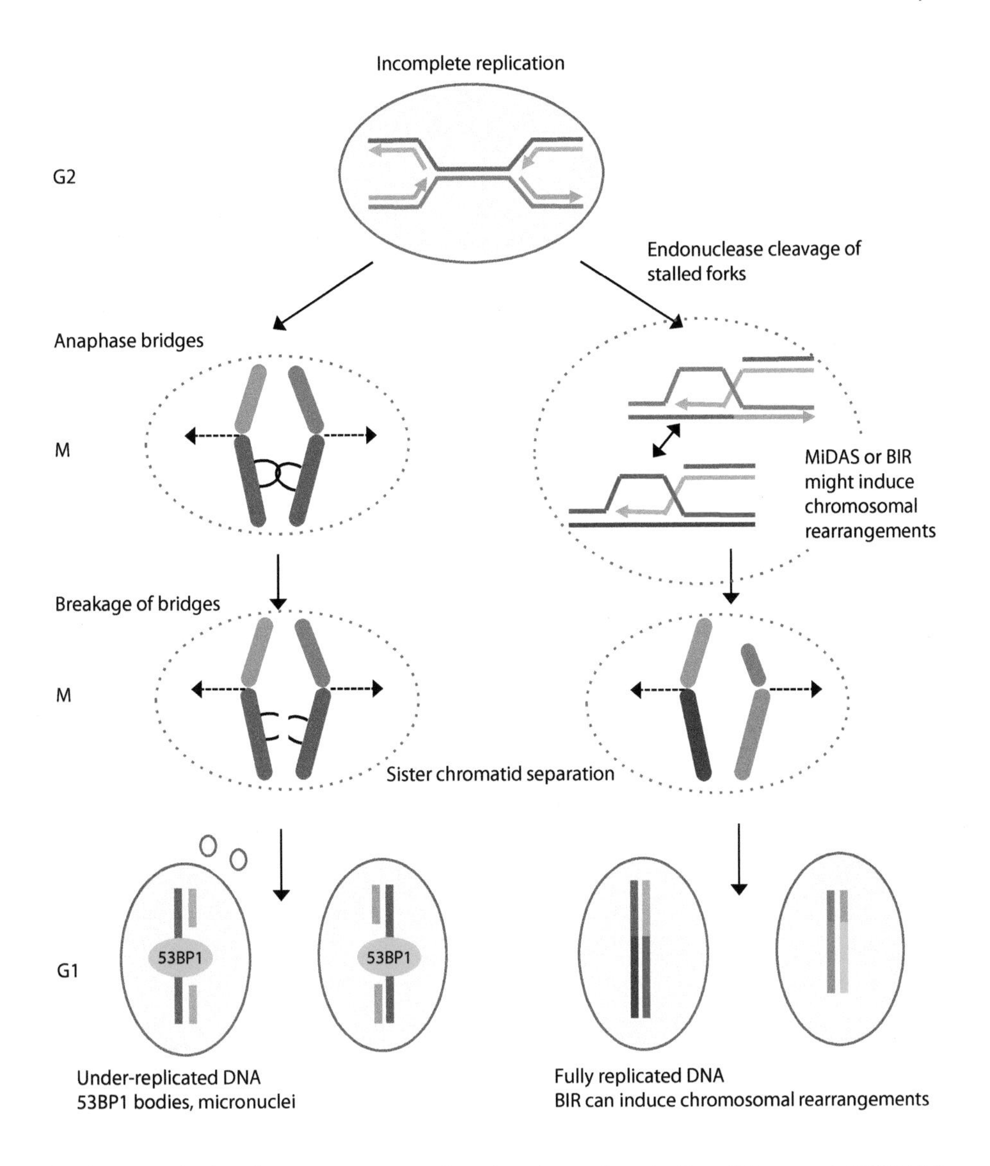

Figure 6.5 **Emerging impacts of BIR on genome stability. Far left:** G2, M, G1 indicates cell cycle phase. **Left:** Without completion of replication, sister chromatids cannot be correctly separated during mitosis, leading to DNA bridges during anaphase. Breakage of these anaphase bridges can lead to loss or rearrangement of genetic material. 53BP1 bodies mark any remaining under-replicated DNA in the following G1 phase. **Right:** Anaphase bridge cleavage by MUS81-EME1 and SLX4, followed by mitotic DNA synthesis (MiDAS), completes DNA replication in mitosis. MiDAS requires fork cleavage and depends on RAD52 and POLD3, and thus resembles BIR. BIR, however, can also induce chromosomal rearrangements if using the wrong template. Whether MiDAS is also prone to such rearrangements is not known.

by the perturbed replication structures that give rise to these bridges (Chan et al. 2009, Naim and Rosselli 2009). MUS81-EME1 and SLX4 are required to resolve these structures during mitosis (Ying et al. 2013) (Figure 6.5). This involves mitotic DNA synthesis (dubbed MiDAS by the authors), which requires MUS81-EME1 and POLD3, and so could be identical to BIR (Minocherhomji et al. 2015). It is frequently observed on only one sister chromatid, suggesting it might occur through conservative replication (Bhowmick et al. 2016). MUS81 is required for loading of PCNA, POLD1 and POLD3. MiDAS appeared to be dependent on DNA condensation in prophase, leading the authors to speculate that chromosome condensation exposes under-replicated DNA (Minocherhomji et al. 2015). The metaphase chromatid breaks and gaps that are the hallmark of CFS could therefore represent sites of BIR DNA synthesis rather than DNA breaks. This idea is supported by the observation that CFS expression itself depends on MUS81 (Ying et al. 2013). The same group then showed that MiDAS also depends on RAD52, but not RAD51 (Bhowmick et al. 2016). Altogether these data suggest that MiDAS resembles the current model for BIR.

If under-replicated regions and CFS are not properly resolved by MUS81 and MiDAS, this generates DNA damage foci (called 53BP1 bodies) during the following G1 phase (Lukas et al. 2011) and can also promote genomic instability. Breakage of anaphase bridges can lead to the exclusion of chromosome fragments from the daughter nuclei to form micronuclei, resulting in loss or rearrangement of genetic material and even shattering of chromosomes (chromothripsis) (Naim and Rosselli 2009, Zhang et al. 2015). Failure to complete DNA replication, followed by mitotic problems such as anaphase bridging and common fragile site expression, has been associated with the gross chromosomal instability that is the most common form of genomic instability in sporadic cancers (Dereli-Oz et al. 2011, Burrell et al. 2013).

HR AND CANCER THERAPY

BRCA1 and BRCA2 as tumour suppressors

HR function is crucial for cell survival, organismal development and health. Depending on the severity of the resulting HR defect, biallelic mutations in HR genes are either embryonic lethal or cause the autosomal recessive genetic disorder Fanconi anaemia. Fanconi anaemia is cancer-prone, and heterozygous germline mutations in HR genes such as BRCA1, BRCA2, PALB2 or BRIP1 also cause familial predisposition to breast, ovarian or pancreatic cancer in humans (Hruban et al. 2010, Tischkowitz and Xia 2010, van der Groep et al. 2011). Finally, mutations in or epigenetic silencing of BRCA1, BRCA2 and other HR genes also frequently arise in sporadic cancers (Turner et al. 2004, Muggia and Safra 2014). BRCA1 and BRCA2 are the two major cancer susceptibility genes of the HR pathway. Germline mutations in BRCA1 confer average risks of approximately 60% for breast cancer development and 40% for ovarian cancer development by the age of 70 in women. Germline mutations in BRCA2 confer average risks of approximately 50% for breast cancer development and 15% for ovarian cancer development by the age of 70 in women (Nielsen et al. 2016).

Non-cancer cells with heterozygous BRCA1 or PALB2 mutations, obtained from normal tissue of mutation carriers, displayed increased replication stress (Nikkila et al. 2013). PALB2 mutant cell lines displayed increased density of replication initiation, an established feature of replication stress that is nevertheless difficult to explain using current models of HR function (Nikkila et al. 2013). Primary cells with heterozygous BRCA1 mutations displayed defects in replication fork repair, while DSB repair, cell cycle checkpoint and non-DNA repair functions of BRCA1 were unaffected (Pathania et al. 2014). BRCA1 heterozygotes displayed defective CtIP recruitment, ssDNA generation and RPA recruitment at stalled forks, while the same processes were functional at DSBs. This suggests that BRCA1 is haploinsufficient for replication fork maintenance functions and that preventing replication stress is an important tumour suppressor function of HR proteins (Pathania et al. 2014).

Conventional cancer therapies

In line with the tumour suppressor function of HR, HR defects are frequent in cancers. However, increased expression of RAD51 or other HR genes has also been reported in prostate cancer cells, pancreatic cancers and a subset of non-small cell lung cancer (Maacke et al. 2000, Fan et al. 2004, Qiao et al. 2005). While this does not always correlate with more efficient DNA repair, it is important not to simply expect all cancers to be HR-defective.

Nevertheless, cancers are routinely treated with cytotoxic therapies that induce DNA lesions that require HR for repair. Many conventional chemotherapies either induce DSBs or block replication. HR-deficient cells are generally more sensitive to these treatments than HR-proficient cells, but they are especially sensitive to treatments that effectively induce replication-dependent DSBs, namely interstrand cross-linkers (cisplatin, carboplatin, oxaliplatin and mitomycin c) (Clark et al. 2012, Muggia and Safra 2014) and topoisomerase I inhibitors (topotecan and irinotecan). Data also suggest that defects in HR or HR inhibitors sensitise tumours to radiotherapy (Barker and Powell 2010).

When it comes to replication inhibitors that are less certain to effectively induce DSBs, then HR-defective cells can sometimes be similar, or even less sensitive, than HR-proficient cells. This applies to anti-metabolites such as 5-fluorouracil (5-FU), fludarabine and gemcitabine (Crul et al. 2003, van der Heijden et al. 2005, de Campos-Nebel et al. 2008, Choudhury et al. 2009, Issaeva et al. 2010, Tsai et al. 2010). Active metabolites of 5-FU and gemcitabine deplete nucleotide pools through inhibition of thymidylate synthase (TS) and ribonucleotide reductase (RNR), respectively, and also interfere with DNA replication after incorporation into nascent DNA (Ewald et al. 2008, Wilson et al. 2014). The DNA damage response to 5-FU and gemcitabine remains poorly understood and their ability to induce fork collapse is not well understood. Our recent observations suggest that BRCA2 and RAD51 activities are actually required to promote DSB formation at gemcitabine-stalled forks. RAD51 loading promotes DSB generation by the structure-specific endonucleases MUS81 and XPF, possibly because it is required for the formation of HR intermediates, such as regressed forks, that are preferentially cleaved by these structure-specific endonucleases (Jones et al. 2014).

A challenge in exploiting HR defects for personalized cancer therapy is to find reliable biomarkers, other than BRCA1/BRCA2 mutations and platinum sensitivity, that can detect

HR defects in a clinical setting (Stover et al. 2016). This can potentially be circumvented by using HR inhibitors to generate HR defects regardless of the genetic makeup of the tumours. Several studies have tested or are testing combinations of conventional therapies with small molecule inhibitors that directly or indirectly inhibit HR; indirect inhibitors include a number of agents that target cell signalling such as erlotinib and gefitinib, which inhibit signalling from the epidermal growth factor receptor (EGFR), and histone deacetylase (HDAC) inhibitors that relax chromatin structure (Adimoolam et al. 2007). Both types of treatments can downregulate expression of RAD51, BRCA1 or BRCA2 and impair HR activity. Catalytic inhibitors of RAD51 itself are also being developed (Budke et al. 2013). Such combination approaches aim to sensitize tumours to DNA-damaging treatments by reducing repair. The disadvantage of this approach is that HR inhibition targets both cancer and normal tissues, thus potentially reducing cancer selectivity of the treatment.

PARP inhibitors and cancer therapy by synthetic lethality

Homologous recombination-deficient cancers have also provided the first real-life example of cancer therapy by synthetic lethality. The idea to treat BRCA1- or BRCA2-deficient cells with small molecule PARP inhibitors was initially born out of basic science studies investigating a potential role for PARP1 in HR. These revealed that PARP1 activity had no effect on HR repair of DSBs, but that PARP1-deficient cell displayed increased spontaneous HR activity, evidenced by elevated sister chromatid exchanges (Schultz et al. 2003). This led to the hypothesis that HR is essential for DNA repair in the absence of PARP activity. The proposed model stated that PARP inhibition decreases the repair of spontaneous single-strand breaks (SSBs) by PARP1-dependent SSB repair. Larger numbers of unrepaired SSBs lead to increased fork collapse and formation of one-ended DSBs, which require HR for repair by BIR. HR thus acts as a backup repair pathway for PARP-dependent SSBR and is therefore required for survival of PARP-deficient cells, and vice versa (**Figure 6.6**). Failure of HR results in aberrant repair of collapsed forks by NHEJ, chromosomal rearrangements and cell death (Farmer et al. 2005). The principle was first demonstrated in BRCA-mutant rodent cell lines, which were shown to be

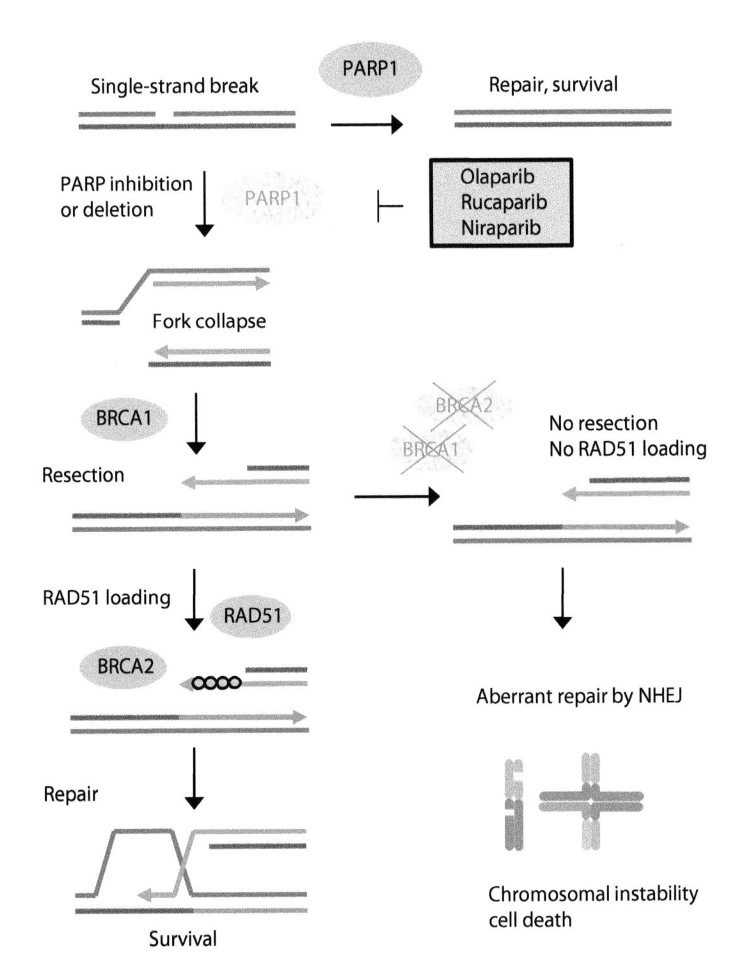

Figure 6.6 **Model for the mechanism of synthetic lethality between BRCA1/2 deficiency and PARP inhibitors.** PARP inhibition decreases the repair of spontaneous single-strand breaks (SSBs), which lead to increased fork collapse and formation of one-ended DSBs. BRCA1 and BRCA2 are required for resection and RAD51 loading for BIR repair of collapsed forks. HR thus acts as a backup repair pathway for PARP-dependent SSB repair and is therefore required for survival of PARP inhibitor treatments. Failure of HR results in aberrant repair of collapsed forks by non-homologous end joining (NHEJ), chromosomal rearrangements and cell death. Yellow box: Clinical PARP inhibitors.

exquisitely sensitive to loss of PARP activity (Bryant et al. 2005, Farmer et al. 2005), and then quickly progressed into clinical trials. As of 2017, the PARP inhibitors olaparib and rucaparib have been approved for treatment of BRCA-mutated advanced ovarian cancer, and the PARP inhibitor niraparib was approved for maintenance treatment of epithelial ovarian, fallopian tube or primary peritoneal cancers that respond at least partially to platinum compounds. Several more are in advanced clinical trials.

The success of synthetic lethality using PARP inhibitors has prompted a huge research effort to find other approaches to specifically target HR-deficient cancers. Most of these are still focusing on BRCA1 or BRCA2-deficient cells. However, the concept of synthetic lethality with PARP inhibitors also works in other cancer-relevant HR-defective contexts, as has been demonstrated for ATM-deficient chronic lymphocytic leukaemia (CLL) (Weston et al. 2010). Often, the new therapeutic angles are discovered during fundamental basic research because any process that impacts HR can potentially be exploited. Mateos-Gomez et al. (2015) investigated the role of DNA polymerase θ (theta) during aberrant NHEJ at dysfunctional telomeres when they discovered that Polθ-depleted cells displayed increased HR activity, suggesting that HR is activated to compensate for loss of Polθ. Indeed, BRCA1-deficient cells were more sensitive to loss of Polθ. Mechanistically, Polθ acts downstream of PARP1 in the alt-NHEJ pathway of DSB repair.

A major drawback of single-agent therapies such as the synthetic lethal PARP inhibitor treatment is that cancers can easily develop treatment resistance. PARP inhibitor resistance can develop through upregulation of drug efflux pumps, but also via genetic reversal of the HR defect. BRCA2-mutant cells can acquire resistance to PARP inhibitors and platinum compounds through secondary mutations in the BRCA2 gene that restore a functional protein (Edwards et al. 2008). BRCA1-mutant cells can develop resistance through loss of 53BP1, which is the principal antagonist of BRCA1 in the regulation of resection (Bouwman et al. 2010, Bunting et al. 2010).

It has been demonstrated that it is possible to find drugs that can kill even PARP inhibitor–resistant cells. The first one was 6-thioguanine, which is active against BRCA1- and BRCA2-deficient tumours or cells that have acquired resistance to PARP inhibitors (Issaeva et al. 2010). Zimmer et al. (2016) reported that compounds that interact with and therefore stabilize DNA G4-quadruplex structures can be used to target BRCA- or RAD51-deficient cells. G4 quadruplexes are naturally occurring secondary DNA structures that pose blocks to replication forks. Stabilising these structures with compounds such as pyridostatin (PDS) will block more forks, cause replication stress and likely activate HR. What is interesting is that G4-stabilizing drugs were also able to kill cells from BRCA-mutant tumours with acquired PARP inhibitor resistance. This suggests that blocking a replication fork with G4 quadruplex is not the same as damaging it with PARP inhibitors, as far as the role of HR is concerned.

A rational search for treatments to target acquired resistance in HR mutant cancers is currently hampered by the fact that it is still not entirely clear which of the many roles of HR at replication forks are involved in protecting cells from PARP inhibitors. There is good evidence that PARP inhibitor sensitivity correlates very well with sensitivity to platinum-based chemotherapy. Cancers that become resistant to PARP inhibitors are usually also resistant to platinum compounds, and vice versa. However, differences are beginning to emerge, at least for BRCA1-defective cells. These revolve around the roles of BRCA1 in the protection of stalled replication forks versus its role in resection during DSB repair. A provocative study recently proposed that it is loss of the fork protection function, rather than loss of the DSB repair function, that sensitizes BRCA1-defective cells to PARP inhibitors and platinum compounds (Ray Chaudhuri et al. 2016) (**Figure 6.7**). This is, however, contradicted by previous findings that patient cells with heterozygous BRCA1 mutations display fork protection defects but normal DSB repair, and are only sensitive to platinum compounds but not to PARP inhibitors (Pathania et al. 2014). Additionally, resistance to PARP inhibitors and ionizing radiation requires the BRCA1 ubiquitin ligase function, which also promotes resection at DSBs, while resistance to platinum compounds is independent of this function (Densham et al. 2016). Finally, rescuing resection in BRCA1-deficient cells by co-deleting 53BP1 restores PARP inhibitor resistance, but not cisplatin resistance, at least in some cell types (Bunting et al. 2012). It is therefore likely that PARP inhibitor resistance relies more on DSB resection and platinum compound resistance relies more on fork protection. This would make sense considering the lesions involved, as cisplatin adducts will initially stall replication forks, whereas PARP inhibition is thought to cause fork collapse into DSBs straight away. The former is likely to involve loading of RAD51

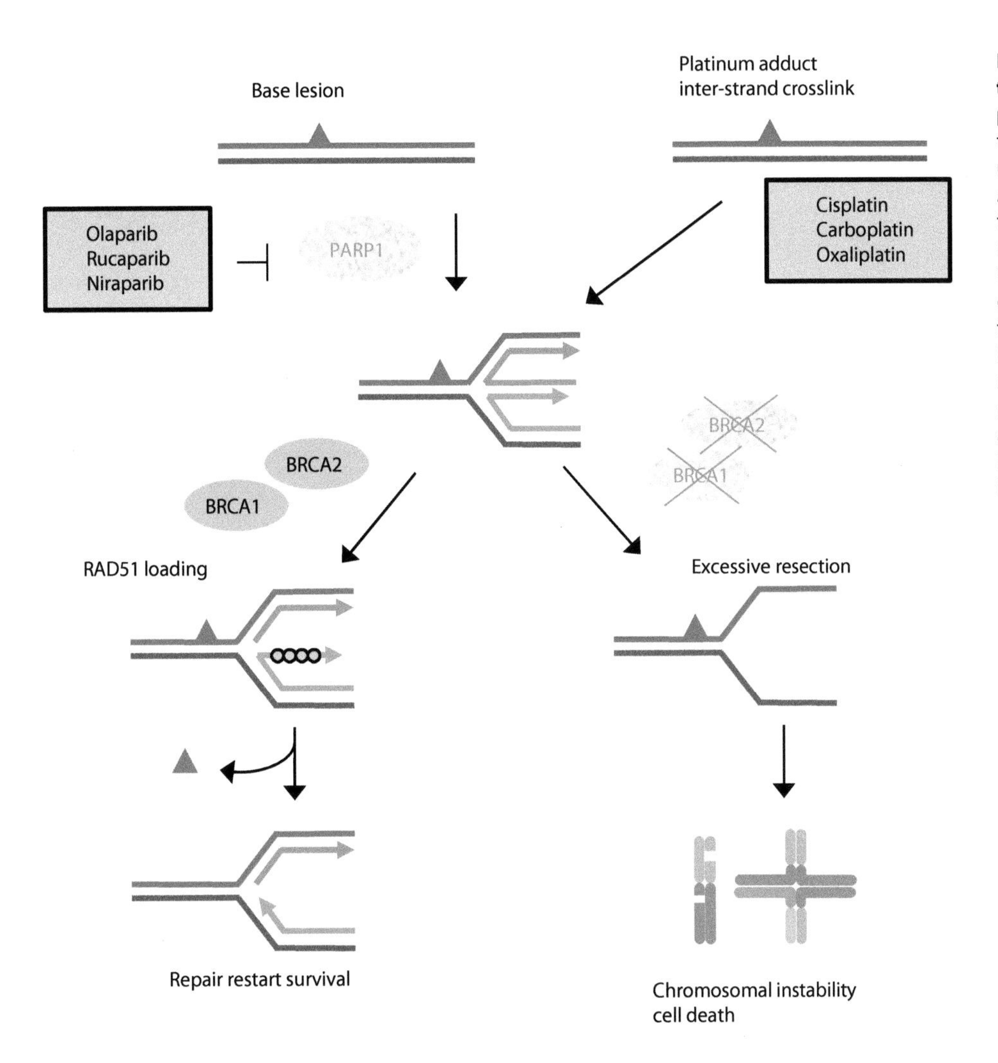

Figure 6.7 **Potential role of fork protection in the response to PARP inhibitors or platinum compounds.** Loss of fork protection by BRCA1/BRCA2 has been proposed to underlie PARP inhibitor synthetic lethality. A similar mechanism may be relevant to the action of platinum compounds. In this model, increased levels of replication blocking DNA lesions, either from PARP inhibition or platinum compound treatment, lead to fork stalling and fork regression. BRCA1 and BRCA2 prevent cell death by protecting stalled forks from excessive resection. This mechanism does not involve DSB formation and is distinct from the BIR repair of a collapsed fork. Yellow boxes: Clinical PARP inhibitors and platinum compounds.

onto a stalled fork structure, the same process as for fork protection. The latter would require BIR for repair, which involves extensive resection.

CONCLUSIONS

Research into HR at replication forks continues to provide new conceptual advances in our understanding of how genome stability is maintained during DNA replication and how this could be exploited for cancer therapy. Future challenges in basic research will include gaining better understanding of HR factors and pathways that have been relatively neglected in higher eukaryotes, often for technical reasons. These include RAD52 and the molecular mechanisms of error-free damage bypass.

It has long been a pressing question to understand why HR defects predispose to cancer specifically in breast and ovarian tissues, and not to cancers in other tissues. This question remains as relevant as ever. It will also be important to describe the roles of HR in response to under-researched replication fork–blocking agents such as 5-FU and to environmental genotoxins. It will be necessary to decipher the precise significance of replication fork protection, and to understand the impact of fork regression and excessive resection on DNA damage and survival in mammalian cells. While progress is now being made on this question in BRCA1-deficient cells, it will also be necessary to better understand the mechanisms determining PARP inhibitor and platinum compound sensitivity in BRCA2-deficient cells.

Finally, another future challenge will consist of developing tools to study the molecular processes described here in animal models that are also suitable for modelling human disease. The observed impact of oxygen concentration of HR functions at forks (Wilhelm et al. 2016) highlights the need for more physiological conditions to study replication stress and DNA repair and their impact on health and disease.

REFERENCES

Adimoolam, S., Sirisawad, M., Chen, J., Thiemann, P., Ford, J. M., Buggy, J. J. 2007. HDAC inhibitor PCI-24781 decreases RAD51 expression and inhibits homologous recombination. *Proc Natl Acad Sci USA* 104: 19482–19487.

Andreassen, P. R., D'Andrea, A. D., Taniguchi, T. 2004. ATR couples FANCD2 monoubiquitination to the DNA-damage response. *Genes Dev* 18: 1958–1963.

Arnaudeau, C., Lundin, C., Helleday, T. 2001. DNA double-strand breaks associated with replication forks are predominantly repaired by homologous recombination involving an exchange mechanism in mammalian cells. *J Mol Biol* 307: 1235–1245.

Aylon, Y., Liefshitz, B., Kupiec, M. 2004. The CDK regulates repair of double-strand breaks by homologous recombination during the cell cycle. *Embo J* 23: 4868–4875.

Aymard, F., Bugler, B., Schmidt, C. K., Guillou, E., Caron, P., Briois, S., Iacovoni, J. S. et al. 2014. Transcriptionally active chromatin recruits homologous recombination at DNA double-strand breaks. *Nat Struct Mol Biol* 21: 366–374.

Bahassi, E. M., Ovesen, J. L., Riesenberg, A. L., Bernstein, W. Z., Hasty, P. E., Stambrook, P. J. 2008. The checkpoint kinases Chk1 and Chk2 regulate the functional associations between hBRCA2 and Rad51 in response to DNA damage. *Oncogene* 27: 3977–3985.

Barker, C. A., Powell, S. N. 2010. Enhancing radiotherapy through a greater understanding of homologous recombination. *Semin Radiat Oncol* 20: 267–273. e263.

Baumann, P., Benson, F. E., West, S. C. 1996. Human Rad51 protein promotes ATP-dependent homologous pairing and strand transfer reactions *in vitro*. *Cell* 87: 757–766.

Bermejo, R., Capra, T., Jossen, R., Colosio, A., Frattini, C., Carotenuto, W., Cocito, A. et al. 2011. The replication checkpoint protects fork stability by releasing transcribed genes from nuclear pores. *Cell* 146: 233–246.

Berti, M., Chaudhuri, A. R., Thangavel, S., Gomathinayagam, S., Kenig, S., Vujanovic, M., Odreman, F. et al. 2013. Human RECQ1 promotes restart of replication forks reversed by DNA topoisomerase I inhibition. *Nat Struct Mol Biol* 20: 347–354.

Bhowmick, R., Minocherhomji, S., Hickson, I. D. 2016. RAD52 facilitates mitotic DNA synthesis following replication stress. *Mol Cell* 64: 1117–1126.

Bignell, G. R., Greenman, C. D., Davies, H., Butler, A. P., Edkins, S., Andrews, J. M., Buck, G. et al. 2010. Signatures of mutation and selection in the cancer genome. *Nature* 463: 893–898.

Bishop, D. K., Ear, U., Bhattacharyya, A., Calderone, C., Beckett, M., Weichselbaum, R. R., Shinohara, A. 1998. Xrcc3 is required for assembly of Rad51 complexes *in vivo*. *J Biol Chem* 273: 21482–21488.

Bolderson, E., Petermann, E., Croft, L., Suraweera, A., Pandita, R. K., Pandita, T. K., Helleday, T., Khanna, K. K., Richard, D. J. 2014. Human single-stranded DNA binding protein 1 (hSSB1/NABP2) is required for the stability and repair of stalled replication forks. *Nucleic Acids Res* 42: 6326-6336.

Bouwman, P., Aly, A., Escandell, J. M., Pieterse, M., Bartkova, J., van der Gulden, H., Hiddingh, S., Thanasoula, M. et al. 2010. 53BP1 loss rescues BRCA1 deficiency and is associated with triple-negative and BRCA-mutated breast cancers. *Nat Struct Mol Biol* 17: 688–695.

Branzei, D., Szakal, B. 2016. DNA damage tolerance by recombination: Molecular pathways and DNA structures. *DNA Repair (Amst)* 44: 68–75.

Branzei, D., Vanoli, F., Foiani, M. 2008. SUMOylation regulates Rad18-mediated template switch. *Nature* 456: 915–920.

Bryant, H. E., Helleday, T. 2006. Inhibition of poly (ADP-ribose) polymerase activates ATM which is required for subsequent homologous recombination repair. *Nucleic Acids Res* 34: 1685–1691.

Bryant, H. E., Petermann, E., Schultz, N., Jemth, A. S., Loseva, O., Issaeva, N., Johansson, F., Fernandez, S., McGlynn, P., Helleday, T. 2009. PARP is activated at stalled forks to mediate Mre11-dependent replication restart and recombination. *Embo J* 28: 2601–2615.

Bryant, H. E., Schultz, N., Thomas, H. D., Parker, K. M., Flower, D., Lopez, E., Kyle, S., Meuth, M., Curtin, N. J., Helleday, T. 2005. Specific killing of BRCA2-deficient tumours with inhibitors of poly(ADP-ribose) polymerase. *Nature* 434: 913–917.

Budke, B., Kalin, J. H., Pawlowski, M., Zelivianskaia, A. S., Wu, M., Kozikowski, A. P., Connell, P. P. 2013. An optimized RAD51 inhibitor that disrupts homologous recombination without requiring Michael acceptor reactivity. *J Med Chem* 56: 254–263.

Bugreev, D. V., Yu, X., Egelman, E. H., Mazin, A. V. 2007. Novel pro- and anti-recombination activities of the Bloom's syndrome helicase. *Genes Dev* 21: 3085–3094.

Bunting, S. F., Callen, E., Kozak, M. L., Kim, J. M., Wong, N., Lopez-Contreras, A. J., Ludwig, T. et al. 2012. BRCA1 functions independently of homologous recombination in DNA interstrand crosslink repair. *Mol Cell* 46: 125–135.

Bunting, S. F., Callen, E., Wong, N., Chen, H. T., Polato, F., Gunn, A., Bothmer, A. et al. 2010. 53BP1 inhibits homologous recombination in BRCA1-deficient cells by blocking resection of DNA breaks. *Cell* 141: 243–254.

Burrell, R. A., McClelland, S. E., Endesfelder, D., Groth, P., Weller, M. C., Shaikh, N., Domingo, E. et al. 2013. Replication stress links structural and numerical cancer chromosomal instability. *Nature* 494: 492–496.

Chan, K. L., Palmai-Pallag, T., Ying, S., Hickson, I. D. 2009. Replication stress induces sister-chromatid bridging at fragile site loci in mitosis. *Nat Cell Biol* 11: 753–760.

Chaudhury, I., Sareen, A., Raghunandan, M., Sobeck, A. 2013. FANCD2 regulates BLM complex functions independently of FANCI to promote replication fork recovery. *Nucleic Acids Res* 41, 6444–6459.

Choudhury, A., Zhao, H., Jalali, F., Al Rashid, S., Ran, J., Supiot, S., Kiltie, A. E., Bristow, R. G. 2009. Targeting homologous recombination using imatinib results in enhanced tumor cell chemosensitivity and radiosensitivity. *Mol Cancer Ther* 8: 203–213.

Chu, W. K., Hickson, I. D. 2009. RecQ helicases: Multifunctional genome caretakers. *Nature Reviews Cancer* 9: 644–654.

Chung, W. H., Zhu, Z., Papusha, A., Malkova, A., Ira, G. 2010. Defective resection at DNA double-strand breaks leads to *de novo* telomere formation and enhances gene targeting. *PLoS Genet* 6: e1000948.

Clark, C. C., Weitzel, J. N., O'Connor, T. R. 2012. Enhancement of synthetic lethality via combinations of ABT-888, a PARP inhibitor, and carboplatin *in vitro* and *in vivo* using BRCA1 and BRCA2 isogenic models. *Mol Cancer Ther* 11: 1948–1958.

Constantinou, A., Tarsounas, M., Karow, J. K., Brosh, R. M., Bohr, V. A., Hickson, I. D., West, S. C. 2000. Werner's syndrome protein (WRN) migrates Holliday junctions and co-localizes with RPA upon replication arrest. *EMBO Rep* 1: 80–84.

Costantino, L., Sotiriou, S. K., Rantala, J. K., Magin, S., Mladenov, E., Helleday, T., Haber, J. E., Iliakis, G., Kallioniemi, O. P., Halazonetis, T. D. 2014. Break-induced replication repair of damaged forks induces genomic duplications in human cells. *Science* 343: 88–91.

Cotta-Ramusino, C., Fachinetti, D., Lucca, C., Doksani, Y., Lopes, M., Sogo, J., Foiani, M. 2005. Exo1 processes stalled replication forks and counteracts fork reversal in checkpoint-defective cells. *Mol Cell* 17: 153–159.

Crul, M., van Waardenburg, R. C., Bocxe, S., van Eijndhoven, M. A., Pluim, D., Beijnen, J. H., Schellens, J. H. 2003. DNA repair mechanisms involved in gemcitabine cytotoxicity and in the interaction between gemcitabine and cisplatin. *Biochem Pharmacol* 65: 275–282.

Daboussi, F., Courbet, S., Benhamou, S., Kannouche, P., Zdzienicka, M. Z., Debatisse, M., Lopez, B. S. 2008. A homologous recombination defect affects replication-fork progression in mammalian cells. *J Cell Sci* 121: 162–166.

Davis, A. P., Symington, L. S. 2004. RAD51-dependent break-induced replication in yeast. *Mol Cell Biol* 24: 2344–2351.

de Campos-Nebel, M., Larripa, I., Gonzalez-Cid, M. 2008. Non-homologous end joining is the responsible pathway for the repair of fludarabine-induced DNA double strand breaks in mammalian cells. *Mutat Res* 646: 8–16.

Deem, A., Keszthelyi, A., Blackgrove, T., Vayl, A., Coffey, B., Mathur, R., Chabes, A., Malkova, A. 2011. Break-induced replication is highly inaccurate. *PLoS Biol* 9: e1000594.

Densham, R. M., Garvin, A. J., Stone, H. R., Strachan, J., Baldock, R. A., Daza-Martin, M., Fletcher, A. et al. 2016. Human BRCA1-BARD1 ubiquitin ligase activity counteracts chromatin barriers to DNA resection. *Nat Struct Mol Biol* 23: 647–655.

Dereli-Oz, A., Versini, G., Halazonetis, T. D. 2011. Studies of genomic copy number changes in human cancers reveal signatures of DNA replication stress. *Mol Oncol* 5: 308–314.

Dimitrov, S. D., Lu, D., Naetar, N., Hu, Y., Pathania, S., Kanellopoulou, C., Livingston, D. M. 2013. Physiological modulation of endogenous BRCA1 p220 abundance suppresses DNA damage during the cell cycle. *Genes Dev* 27: 2274–2291.

Donnianni, R. A., Symington, L. S. 2013. Break-induced replication occurs by conservative DNA synthesis. *Proc Natl Acad Sci USA* 110: 13475–13480.

Edwards, S. L., Brough, R., Lord, C. J., Natrajan, R., Vatcheva, R., Levine, D. A., Boyd, J., Reis-Filho, J. S., Ashworth, A. 2008. Resistance to therapy caused by intragenic deletion in BRCA2. *Nature* 451: 1111–1115.

Esashi, F., Christ, N., Gannon, J., Liu, Y., Hunt, T., Jasin, M., West, S. C. 2005. CDK-dependent phosphorylation of BRCA2 as a regulatory mechanism for recombinational repair. *Nature* 434: 598–604.

Ewald, B., Sampath, D., Plunkett, W. 2008. Nucleoside analogs: Molecular mechanisms signaling cell death. *Oncogene* 27: 6522–6537.

Fan, R., Kumaravel, T. S., Jalali, F., Marrano, P., Squire, J. A., Bristow, R. G. 2004. Defective DNA strand break repair after DNA damage in prostate cancer cells: Implications for genetic instability and prostate cancer progression. *Cancer Res* 64: 8526–8533.

Farmer, H., McCabe, N., Lord, C. J., Tutt, A. N., Johnson, D. A., Richardson, T. B., Santarosa, M. et al. 2005. Targeting the DNA repair defect in BRCA mutant cells as a therapeutic strategy. *Nature* 434: 917–921.

French, S. 1992. Consequences of replication fork movement through transcription units *in vivo*. *Science* 258: 1362–1365.

Fugger, K., Mistrik, M., Neelsen, K. J., Yao, Q., Zellweger, R., Kousholt, A. N., Haahr, P. et al. 2015. FBH1 catalyzes regression of stalled replication forks. *Cell Reports* 10: 1749–1757.

Fumasoni, M., Zwicky, K., Vanoli, F., Lopes, M., Branzei, D. 2015. Error-free DNA damage tolerance and sister chromatid proximity during DNA replication rely on the Polalpha/Primase/Ctf4 Complex. *Mol Cell* 57: 812–823.

Game, J. C., Cox, B. S. 1971. Allelism tests of mutants affecting sensitivity to radiation in yeast and a proposed nomenclature. *Mutat Res* 12: 328–331.

Gan, W., Guan, Z., Liu, J., Gui, T., Shen, K., Manley, J. L., Li, X. 2011. R-loop-mediated genomic instability is caused by impairment of replication fork progression. *Genes Dev* 25: 2041–2056.

Gari, K., Decaillet, C., Stasiak, A. Z., Stasiak, A., Constantinou, A. 2008. The Fanconi anemia protein FANCM can promote branch migration of Holliday junctions and replication forks. *Mol Cell* 29: 141–148.

Giannattasio, M., Zwicky, K., Follonier, C., Foiani, M., Lopes, M., Branzei, D. 2014. Visualization of recombination-mediated damage bypass by template switching. *Nat Struct Mol Biol* 21: 884–892.

Gottipati, P., Helleday, T. 2009. Transcription-associated recombination in eukaryotes: Link between transcription, replication and recombination. *Mutagenesis* 24: 203–210.

Gravells, P., Hoh, L., Solovieva, S., Patil, A., Dudziec, E., Rennie, I. G., Sisley, K., Bryant, H. E. 2013. Reduced FANCD2 influences spontaneous SCE and RAD51 foci formation in uveal melanoma and Fanconi anaemia. *Oncogene* 32: 5338–5346.

Hanada, K., Budzowska, M., Davies, S. L., van Drunen, E., Onizawa, H., Beverloo, H. B., Maas, A., Essers, J., Hickson, I. D., Kanaar, R. 2007. The structure-specific endonuclease Mus81 contributes to replication restart by generating double-strand DNA breaks. *Nat Struct Mol Biol* 14: 1096–1104.

Hashimoto, Y., Chaudhuri, A. R., Lopes, M., Costanzo, V. 2010. Rad51 protects nascent DNA from Mre11-dependent degradation and promotes continuous DNA synthesis. *Nat Struct Mol Biol* 17: 1305–1311.

Hashimoto, Y., Puddu, F., Costanzo, V. 2012. RAD51- and MRE11-dependent reassembly of uncoupled CMG helicase complex at collapsed replication forks. *Nat Struct Mol Biol* 19: 17–24.

Hastings, P. J., Ira, G., Lupski, J. R. 2009. A microhomology-mediated break-induced replication model for the origin of human copy number variation. *PLoS Genet* 5: e1000327.

Hegde, M. L., Hegde, P. M., Bellot, L. J., Mandal, S. M., Hazra, T. K., Li, G. M., Boldogh, I., Tomkinson, A. E., Mitra, S. 2013. Prereplicative repair of oxidized bases in the human genome is mediated by NEIL1 DNA glycosylase together with replication proteins. *Proc Natl Acad Sci USA* 110: E3090–3099.

Helmrich, A., Ballarino, M., Tora, L. 2011. Collisions between replication and transcription complexes cause common fragile site instability at the longest human genes. *Mol Cell* 44: 966–977.

Henry-Mowatt, J., Jackson, D., Masson, J. Y., Johnson, P. A., Clements, P. M., Benson, F. E., Thompson, L. H., Takeda, S., West, S. C., Caldecott, K. W. 2003. XRCC3 and Rad51 modulate replication fork progression on damaged vertebrate chromosomes. *Mol Cell* 11: 1109–1117.

Higgs, M. R., Reynolds, J. J., Winczura, A., Blackford, A. N., Borel, V., Miller, E. S., Zlatanou, A. et al. 2015. BOD1L is required to suppress deleterious resection of stressed replication forks. *Mol Cell* 59: 462–477.

Hoege, C., Pfander, B., Moldovan, G. L., Pyrowolakis, G., Jentsch, S. 2002. RAD6-dependent DNA repair is linked to modification of PCNA by ubiquitin and SUMO. *Nature* 419: 135–141.

Horiuchi, T., Fujimura, Y. 1995. Recombinational rescue of the stalled DNA replication fork: A model based on analysis of an *Escherichia coli* strain with a chromosome region difficult to replicate. *J Bacteriol* 177: 783–791.

Hruban, R. H., Canto, M. I., Goggins, M., Schulick, R., Klein, A. P. 2010. Update on familial pancreatic cancer. *Adv Surg* 44, 293–311.

Huertas, P., Jackson, S. P. 2009. Human CtIP mediates cell cycle control of DNA end resection and double strand break repair. *J Biol Chem* 284: 9558–9565.

Ip, S. C., Rass, U., Blanco, M. G., Flynn, H. R., Skehel, J. M., West, S. C. 2008. Identification of Holliday junction resolvases from humans and yeast. *Nature* 456: 357–361.

Ira, G., Pellicioli, A., Balijja, A., Wang, X., Fiorani, S., Carotenuto, W., Liberi, G. et al. 2004. DNA end resection, homologous recombination and DNA damage checkpoint activation require CDK1. *Nature* 431: 1011–1017.

Iraqui, I., Chekkal, Y., Jmari, N., Pietrobon, V., Freon, K., Costes, A., Lambert, S. A. 2012. Recovery of arrested replication forks by homologous recombination is error-prone. *PLoS Genet* 8: e1002976.

Issaeva, N., Thomas, H. D., Djureinovic, T., Jaspers, J. E., Stoimenov, I., Kyle, S., Pedley, N. et al. 2010. 6-thioguanine selectively kills BRCA2-defective tumors and overcomes PARP inhibitor resistance. *Cancer Res* 70: 6268–6276.

Jones, R. M., Kotsantis, P., Stewart, G. S., Groth, P., Petermann, E. 2014. BRCA2 and RAD51 promote double-strand break formation and cell death in response to Gemcitabine. *Mol Cancer Ther* 13: 2412–2421.

Jones, R. M., Petermann, E. 2012. Replication fork dynamics and the DNA damage response. *Biochem J* 443: 13–26.

Karow, J. K., Constantinou, A., Li, J. L., West, S. C., Hickson, I. D. 2000. The Bloom's syndrome gene product promotes branch migration of holliday junctions. *Proc Natl Acad Sci USA* 97: 6504–6508.

Kolinjivadi, A. M., Sannino, V., De Antoni, A., Zadorozhny, K., Kilkenny, M., Techer, H., Baldi, G. et al. 2017. Smarcal1-mediated fork reversal triggers Mre11-dependent degradation of nascent DNA in the absence of Brca2 and stable Rad51 nucleofilaments. *Mol Cell* 67: 867–881.e7.

Lambert, S., Carr, A. M. 2005. Checkpoint responses to replication fork barriers. *Biochimie* 87: 591–602.

Lambert, S., Mizuno, K., Blaisonneau, J., Martineau, S., Chanet, R., Freon, K., Murray, J. M., Carr, A. M., Baldacci, G. 2010. Homologous recombination restarts blocked replication forks at the expense of genome rearrangements by template exchange. *Mol Cell* 39: 346–359.

Lambert, S., Watson, A., Sheedy, D. M., Martin, B., Carr, A. M. 2005. Gross chromosomal rearrangements and elevated recombination at an inducible site-specific replication fork barrier. *Cell* 121: 689–702.

Letessier, A., Millot, G. A., Koundrioukoff, S., Lachages, A. M., Vogt, N., Hansen, R. S., Malfoy, B., Brison, O., Debatisse, M. 2011. Cell-type-specific replication initiation programs set fragility of the FRA3B fragile site. *Nature* 470: 120–123.
Leuzzi, G., Marabitti, V., Pichierri, P., Franchitto, A. 2016. WRNIP1 protects stalled forks from degradation and promotes fork restart after replication stress. *EMBO J* 35: 1437–1451.
Li, X., Heyer, W. D. 2008. Homologous recombination in DNA repair and DNA damage tolerance. *Cell Res* 18: 99–113.
Lin, J. R., Zeman, M. K., Chen, J. Y., Yee, M. C., Cimprich, K. A. 2011. SHPRH and HLTF act in a damage-specific manner to coordinate different forms of postreplication repair and prevent mutagenesis. *Mol Cell* 42: 237–249.
Lok, B. H., Carley, A. C., Tchang, B., Powell, S. N. 2013. RAD52 inactivation is synthetically lethal with deficiencies in BRCA1 and PALB2 in addition to BRCA2 through RAD51-mediated homologous recombination. *Oncogene* 32: 3552–3558.
Lomonosov, M., Anand, S., Sangrithi, M., Davies, R., Venkitaraman, A. R. 2003. Stabilization of stalled DNA replication forks by the BRCA2 breast cancer susceptibility protein. *Genes Dev* 17: 3017–3022.
Lukas, C., Savic, V., Bekker-Jensen, S., Doil, C., Neumann, B., Pedersen, R. S., Grofte, M. et al. 2011. 53BP1 nuclear bodies form around DNA lesions generated by mitotic transmission of chromosomes under replication stress. *Nat Cell Biol* 13: 243–253.
Lundin, C., Erixon, K., Arnaudeau, C., Schultz, N., Jenssen, D., Meuth, M., Helleday, T. 2002. Different roles for nonhomologous end joining and homologous recombination following replication arrest in mammalian cells. *Mol Cell Biol* 22: 5869–5878.
Lundin, C., Schultz, N., Arnaudeau, C., Mohindra, A., Hansen, L. T., Helleday, T. 2003. RAD51 is involved in repair of damage associated with DNA replication in mammalian cells. *J Mol Biol* 328: 521–535.
Lydeard, J. R., Jain, S., Yamaguchi, M., Haber, J. E. 2007. Break-induced replication and telomerase-independent telomere maintenance require Pol32. *Nature* 448: 820–823.
Lydeard, J. R., Lipkin-Moore, Z., Sheu, Y. J., Stillman, B., Burgers, P. M., Haber, J. E. 2010. Break-induced replication requires all essential DNA replication factors except those specific for pre-RC assembly. *Genes Dev* 24: 1133–1144.
Maacke, H., Jost, K., Opitz, S., Miska, S., Yuan, Y., Hasselbach, L., Luttges, J., Kalthoff, H., Sturzbecher, H. W. 2000. DNA repair and recombination factor Rad51 is over-expressed in human pancreatic adenocarcinoma. *Oncogene* 19: 2791–2795.
Madireddy, A., Kosiyatrakul, S. T., Boisvert, R. A., Herrera-Moyano, E., Garcia-Rubio, M. L., Gerhardt, J., Vuono, E. A. et al. 2016. FANCD2 facilitates replication through common fragile sites. *Mol Cell* 64: 388–404.
Malkova, A., Ivanov, E. L., Haber, J. E. 1996. Double-strand break repair in the absence of RAD51 in yeast: A possible role for break-induced DNA replication. *Proc Natl Acad Sci USA* 93: 7131–7136.
Malkova, A., Naylor, M. L., Yamaguchi, M., Ira, G., Haber, J. E. 2005. RAD51-dependent break-induced replication differs in kinetics and checkpoint responses from RAD51-mediated gene conversion. *Mol Cell Biol* 25: 933–944.
Mateos-Gomez, P. A., Gong, F., Nair, N., Miller, K. M., Lazzerini-Denchi, E., Sfeir, A. 2015. Mammalian polymerase theta promotes alternative NHEJ and suppresses recombination. *Nature* 518: 254–257.
Mazloum, N., Holloman, W. K. 2009. Second-end capture in DNA double-strand break repair promoted by Brh2 protein of Ustilago maydis. *Mol Cell* 33: 160–170.
McIlwraith, M. J., West, S. C. 2008. DNA repair synthesis facilitates RAD52-mediated second-end capture during DSB repair. *Mol Cell* 29: 510–516.
Minocherhomji, S., Ying, S., Bjerregaard, V. A., Bursomanno, S., Aleliunaite, A., Wu, W., Mankouri, H. W., Shen, H., Liu, Y., Hickson, I. D. 2015. Replication stress activates DNA repair synthesis in mitosis. *Nature* 528: 286–290.
Mizuno, K., Miyabe, I., Schalbetter, S. A., Carr, A. M., Murray, J. M. 2013. Recombination-restarted replication makes inverted chromosome fusions at inverted repeats. *Nature* 493: 246–249.
Mohebi, S., Mizuno, K., Watson, A., Carr, A. M., Murray, J. M. 2015. Checkpoints are blind to replication restart and recombination intermediates that result in gross chromosomal rearrangements. *Nat Commun* 6: 6357.
Motegi, A., Liaw, H. J., Lee, K. Y., Roest, H. P., Maas, A., Wu, X., Moinova, H. et al. 2008. Polyubiquitination of proliferating cell nuclear antigen by HLTF and SHPRH prevents genomic instability from stalled replication forks. *Proc Natl Acad Sci USA* 105: 12411–12416.
Muggia, F., Safra, T. 2014. 'BRCAness' and its implications for platinum action in gynecologic cancer. *Anticancer Res* 34: 551–556.
Naim, V., Rosselli, F. 2009. The FANC pathway and BLM collaborate during mitosis to prevent micro-nucleation and chromosome abnormalities. *Nat Cell Biol* 11: 761–768.
Neelsen, K. J., Zanini, I. M., Herrador, R., Lopes, M. 2013. Oncogenes induce genotoxic stress by mitotic processing of unusual replication intermediates. *J Cell Biol* 200: 699–708.
Nielsen, F. C., van Overeem Hansen, T., Sorensen, C. S. 2016. Hereditary breast and ovarian cancer: New genes in confined pathways. *Nat Rev Cancer* 16: 599–612.
Nikkila, J., Parplys, A. C., Pylkas, K., Bose, M., Huo, Y. Y., Borgmann, K., Rapakko, K. et al. 2013. Heterozygous mutations in PALB2 cause DNA replication and damage response defects. *Nat Commun* 4: 2578.
Nimonkar, A. V., Genschel, J., Kinoshita, E., Polaczek, P., Campbell, J. L., Wyman, C., Modrich, P., Kowalczykowski, S. C. 2011. BLM-DNA2-RPA-MRN and EXO1-BLM-RPA-MRN constitute two DNA end resection machineries for human DNA break repair. *Genes Dev* 25: 350–362.
Ogawa, T., Yu, X., Shinohara, A., Egelman, E. H. 1993. Similarity of the yeast RAD51 filament to the bacterial RecA filament. *Science* 259: 1896–1899.
O'Regan, P., Wilson, C., Townsend, S., Thacker, J. 2001. XRCC2 is a nuclear RAD51-like protein required for damage-dependent RAD51 focus formation without the need for ATP binding. *J Biol Chem* 276: 22148–22153.
Pathania, S., Bade, S., Le Guillou, M., Burke, K., Reed, R., Bowman-Colin, C., Su, Y. et al. 2014. BRCA1 haploinsufficiency for replication stress suppression in primary cells. *Nat Commun* 5: 5496.
Pepe, A., West, S. C. 2014. MUS81-EME2 promotes replication fork restart. *Cell Reports* 7: 1048–1055.
Petermann, E., Orta, M. L., Issaeva, N., Schultz, N., Helleday, T. 2010. Hydroxyurea-stalled replication forks become progressively inactivated and require two different RAD51-mediated pathways for restart and repair. *Mol Cell* 37: 492–502.
Pichierri, P., Franchitto, A., Rosselli, F. 2004. BLM and the FANC proteins collaborate in a common pathway in response to stalled replication forks. *Embo J* 23: 3154–3163.
Postow, L., Ullsperger, C., Keller, R. W., Bustamante, C., Vologodskii, A. V., Cozzarelli, N. R. 2001. Positive torsional strain causes the formation of a four-way junction at replication forks. *J Biol Chem* 276: 2790–2796.
Qiao, G. B., Wu, Y. L., Yang, X. N., Zhong, W. Z., Xie, D., Guan, X. Y., Fischer, D., Kolberg, H. C., Kruger, S., Stuerzbecher, H. W. 2005. High-level expression of Rad51 is an independent prognostic marker of survival in non-small-cell lung cancer patients. *Br J Cancer* 93: 137–143.
Raghunandan, M., Chaudhury, I., Kelich, S. L., Hanenberg, H., Sobeck, A. 2015. FANCD2, FANCJ and BRCA2 cooperate to promote replication fork recovery independently of the Fanconi anemia core complex. *Cell Cycle* 14: 342–353.
Ray Chaudhuri, A., Callen, E., Ding, X., Gogola, E., Duarte, A. A., Lee, J. E., Wong, N. et al. 2016. Replication fork stability confers chemoresistance in BRCA-deficient cells. *Nature* 535: 382–387.
Ray Chaudhuri, A., Hashimoto, Y., Herrador, R., Neelsen, K. J., Fachinetti, D., Bermejo, R., Cocito, A., Costanzo, V., Lopes, M. 2012. Topoisomerase I poisoning results in PARP-mediated replication fork reversal. *Nat Struct Mol Biol* 19: 417–423.
Rossi, M. L., Ghosh, A. K., Bohr, V. A. 2010. Roles of Werner syndrome protein in protection of genome integrity. *DNA Repair (Amst)* 9: 331–344.
Rothkamm, K., Kruger, I., Thompson, L. H., Lobrich, M. 2003. Pathways of DNA double-strand break repair during the mammalian cell cycle. *Mol Cell Biol* 23: 5706–5715.

Saini, N., Ramakrishnan, S., Elango, R., Ayyar, S., Zhang, Y., Deem, A., Ira, G., Haber, J. E., Lobachev, K. S., Malkova, A. 2013. Migrating bubble during break-induced replication drives conservative DNA synthesis. *Nature* 502: 389–392.

Saintigny, Y., Delacote, F., Vares, G., Petitot, F., Lambert, S., Averbeck, D., Lopez, B. S. 2001. Characterization of homologous recombination induced by replication inhibition in mammalian cells. *Embo J* 20: 3861–3870.

Sato, K., Shimomuki, M., Katsuki, Y., Takahashi, D., Kobayashi, W., Ishiai, M., Miyoshi, H., Takata, M., Kurumizaka, H. 2016. FANCI-FANCD2 stabilizes the RAD51-DNA complex by binding RAD51 and protects the 5′-DNA end. *Nucleic Acids Res* 44: 10758–10771.

Schlacher, K., Christ, N., Siaud, N., Egashira, A., Wu, H., Jasin, M. 2011. Double-strand break repair-independent role for BRCA2 in blocking stalled replication fork degradation by MRE11. *Cell* 145: 529–542.

Schlacher, K., Wu, H., Jasin, M. 2012. A distinct replication fork protection pathway connects Fanconi anemia tumor suppressors to RAD51-BRCA1/2. *Cancer Cell* 22: 106–116.

Schultz, N., Lopez, E., Saleh-Gohari, N., Helleday, T. 2003. Poly(ADP-ribose) polymerase (PARP-1) has a controlling role in homologous recombination. *Nucleic Acids Res* 31: 4959–4964.

Schwab, R. A., Blackford, A. N., Niedzwiedz, W. 2010. ATR activation and replication fork restart are defective in FANCM-deficient cells. *EMBO J* 29: 806–818.

Schwab, R. A., Nieminuszczy, J., Shah, F., Langton, J., Lopez Martinez, D., Liang, C. C., Cohn, M. A., Gibbons, R. J., Deans, A. J., Niedzwiedz, W. 2015. The Fanconi anemia pathway maintains genome stability by coordinating replication and transcription. *Mol Cell* 60: 351–361.

Schwab, R. A., Nieminuszczy, J., Shin-ya, K., Niedzwiedz, W. 2013. FANCJ couples replication past natural fork barriers with maintenance of chromatin structure. *J Cell Biol* 201: 33–48.

Seigneur, M., Bidnenko, V., Ehrlich, S. D., Michel, B. 1998. RuvAB acts at arrested replication forks. *Cell* 95: 419–430.

Seigneur, M., Ehrlich, S. D., Michel, B. 2000. RuvABC-dependent double-strand breaks in dnaBts mutants require recA. *Mol Microbiol* 38: 565–574.

Shahid, T., Soroka, J., Kong, E., Malivert, L., McIlwraith, M. J., Pape, T., West, S. C., Zhang, X. 2014. Structure and mechanism of action of the BRCA2 breast cancer tumor suppressor. *Nat Struct Mol Biol* 21: 962–968.

Shinohara, A., Ogawa, H., Ogawa, T. 1992. Rad51 protein involved in repair and recombination in *S. cerevisiae* is a RecA-like protein. *Cell* 69: 457–470.

Sigurdsson, S., Van Komen, S., Bussen, W., Schild, D., Albala, J. S., Sung, P. 2001. Mediator function of the human Rad51B-Rad51C complex in Rad51/RPA-catalyzed DNA strand exchange. *Genes Dev* 15: 3308–3318.

Sjogren, C., Nasmyth, K. 2001. Sister chromatid cohesion is required for postreplicative double-strand break repair in Saccharomyces cerevisiae. *Curr Biol* 11: 991–995.

Smith, C. E., Llorente, B., Symington, L. S. 2007. Template switching during break-induced replication. *Nature* 447: 102–105.

Sogo, J. M., Lopes, M., Foiani, M. 2002. Fork reversal and ssDNA accumulation at stalled replication forks owing to checkpoint defects. *Science* 297: 599–602.

Somyajit, K., Saxena, S., Babu, S., Mishra, A., Nagaraju, G. 2015. Mammalian RAD51 paralogs protect nascent DNA at stalled forks and mediate replication restart. *Nucleic Acids Res* 43: 9835–9855.

Sonoda, E., Sasaki, M. S., Buerstedde, J. M., Bezzubova, O., Shinohara, A., Ogawa, H., Takata, M., Yamaguchi-Iwai, Y., Takeda, S. 1998. Rad51-deficient vertebrate cells accumulate chromosomal breaks prior to cell death. *Embo J* 17: 598–608.

Sorensen, C. S., Hansen, L. T., Dziegielewski, J., Syljuasen, R. G., Lundin, C., Bartek, J., Helleday, T. 2005. The cell-cycle checkpoint kinase Chk1 is required for mammalian homologous recombination repair. *Nat Cell Biol* 7: 195–201.

Sotiriou, S. K., Kamileri, I., Lugli, N., Evangelou, K., Da-Re, C., Huber, F., Padayachy, L. 2016. Mammalian RAD52 functions in break-induced replication repair of collapsed DNA replication forks. *Mol Cell* 64: 1127–1134.

Stover, E. H., Konstantinopoulos, P. A., Matulonis, U. A., Swisher, E. M. 2016. Biomarkers of response and resistance to DNA repair targeted therapies. *Clin Cancer Res* 22: 5651–5660.

Su, X., Bernal, J. A., Venkitaraman, A. R. 2008. Cell-cycle coordination between DNA replication and recombination revealed by a vertebrate N-end rule degron-Rad51. *Nat Struct Mol Biol* 15: 1049–1058.

Sugimura, K., Takebayashi, S., Taguchi, H., Takeda, S., Okumura, K. 2008. PARP-1 ensures regulation of replication fork progression by homologous recombination on damaged DNA. *J Cell Biol* 183: 1203–1212.

Szostak, J. W., Orr-Weaver, T. L., Rothstein, R. J., Stahl, F. W. 1983. The double-strand-break repair model for recombination. *Cell* 33: 25–35.

Takata, M., Sasaki, M. S., Tachiiri, S., Fukushima, T., Sonoda, E., Schild, D., Thompson, L. H., Takeda, S. 2001. Chromosome instability and defective recombinational repair in knockout mutants of the five Rad51 paralogs. *Mol Cell Biol* 21: 2858–2866.

Thangavel, S., Berti, M., Levikova, M., Pinto, C., Gomathinayagam, S., Vujanovic, M., Zellweger, R. et al. 2015. DNA2 drives processing and restart of reversed replication forks in human cells. *J Cell Biol* 208: 545–562.

Tischkowitz, M., Xia, B. 2010. PALB2/FANCN: Recombining cancer and Fanconi anemia. *Cancer Res* 70: 7353–7359.

Tsai, M. S., Kuo, Y. H., Chiu, Y. F., Su, Y. C., Lin, Y. W. 2010. Down-regulation of Rad51 expression overcomes drug resistance to gemcitabine in human non-small-cell lung cancer cells. *J Pharmacol Exp Ther* 335: 830–840.

Turner, N., Tutt, A., Ashworth, A. 2004. Hallmarks of 'BRCAness' in sporadic cancers. Nature reviews. *Cancer* 4: 814–819.

van der Groep, P., van der Wall, E., van Diest, P. J. 2011. Pathology of hereditary breast cancer. *Cell Oncol (Dordr)* 34: 71–88.

van der Heijden, M. S., Brody, J. R., Dezentje, D. A., Gallmeier, E., Cunningham, S. C., Swartz, M. J., DeMarzo, A. M. et al. 2005. *In vivo* therapeutic responses contingent on Fanconi anemia/BRCA2 status of the tumor. *Clin Cancer Res* 11: 7508–7515.

VanHulle, K., Lemoine, F. J., Narayanan, V., Downing, B., Hull, K., McCullough, C., Bellinger, M., Lobachev, K., Petes, T. D., Malkova, A. 2007. Inverted DNA repeats channel repair of distant double-strand breaks into chromatid fusions and chromosomal rearrangements. *Mol Cell Biol* 27: 2601–2614.

van Sluis, M., McStay, B. 2015. A localized nucleolar DNA damage response facilitates recruitment of the homology-directed repair machinery independent of cell cycle stage. *Genes Dev* 29: 1151–1163.

Wang, X., Kennedy, R. D., Ray, K., Stuckert, P., Ellenberger, T., D'Andrea, A. D. 2007. Chk1-mediated phosphorylation of FANCE is required for the Fanconi anemia/BRCA pathway. *Mol Cell Biol* 27: 3098–3108.

Weston, V. J., Oldreive, C. E., Skowronska, A., Oscier, D. G., Pratt, G., Dyer, M. J., Smith, G. et al. 2010. The PARP inhibitor olaparib induces significant killing of ATM-deficient lymphoid tumor cells *in vitro* and *in vivo*. *Blood* 116: 4578–4587.

Wilhelm, T., Magdalou, I., Barascu, A., Techer, H., Debatisse, M., Lopez, B. S. 2014. Spontaneous slow replication fork progression elicits mitosis alterations in homologous recombination-deficient mammalian cells. *Proc Natl Acad Sci USA* 111: 763–768.

Wilhelm, T., Ragu, S., Magdalou, I., Machon, C., Dardillac, E., Techer, H., Guitton, J., Debatisse, M., Lopez, B. S. 2016. Slow replication fork velocity of homologous recombination-defective cells results from endogenous oxidative stress. *PLoS Genet* 12: e1006007.

Willis, N. A., Chandramouly, G., Huang, B., Kwok, A., Follonier, C., Deng, C., Scully, R. 2014. BRCA1 controls homologous recombination at Tus/Ter-stalled mammalian replication forks. *Nature* 510: 556–559.

Wilson, M. A., Kwon, Y., Xu, Y., Chung, W. H., Chi, P., Niu, H., Mayle, R. et al. 2013. Pif1 helicase and Poldelta promote recombination-coupled DNA synthesis via bubble migration. *Nature* 502: 393–396.

Wilson, P. M., Danenberg, P. V., Johnston, P. G., Lenz, H. J., Ladner, R. D. 2014. Standing the test of time: Targeting thymidylate biosynthesis in cancer therapy. Nature reviews. *Clin Oncol* 11: 282–298.

Wu, L., Hickson, I. D. 2003. The Bloom's syndrome helicase suppresses crossing over during homologous recombination. *Nature* 426: 870–874.

Wu, Y., Shin-ya, K., Brosh, R. M., Jr. 2008. FANCJ helicase defective in Fanconia anemia and breast cancer unwinds G-quadruplex DNA to defend genomic stability. *Mol Cell Biol* 28: 4116–4128.
Yeo, J. E., Lee, E. H., Hendrickson, E. A., Sobeck, A. 2014. CtIP mediates replication fork recovery in a FANCD2-regulated manner. *Hum Mol Genet* 23: 3695–3705.
Ying, S., Minocherhomji, S., Chan, K. L., Palmai-Pallag, T., Chu, W. K., Wass, T., Mankouri, H. W., Liu, Y., Hickson, I. D. 2013. MUS81 promotes common fragile site expression. *Nat Cell Biol* 15: 1001–1007.
Zellweger, R., Dalcher, D., Mutreja, K., Berti, M., Schmid, J. A., Herrador, R., Vindigni, A., Lopes, M. 2015. Rad51-mediated replication fork reversal is a global response to genotoxic treatments in human cells. *J Cell Biol* 208: 563–579.
Zhang, C. Z., Spektor, A., Cornils, H., Francis, J. M., Jackson, E. K., Liu, S., Meyerson, M., Pellman, D. 2015. Chromothripsis from DNA damage in micronuclei. *Nature* 522: 179–184.
Zhu, Z., Chung, W. H., Shim, E. Y., Lee, S. E., Ira, G. 2008. Sgs1 helicase and two nucleases Dna2 and Exo1 resect DNA double-strand break ends. *Cell* 134: 981–994.
Zimmer, J., Tacconi, E. M., Folio, C., Badie, S., Porru, M., Klare, K., Tumiati, M. et al. 2016. Targeting BRCA1 and BRCA2 deficiencies with G-quadruplex-interacting compounds. *Mol Cell* 61: 449–460.

Mechanism of Double-Strand Break Repair by Non-Homologous End Joining

7

Michal Malewicz

INTRODUCTION

Cellular DNA is under constant threat of damage by reactive oxygen species (ROS) derived from cellular metabolism, exogenous high-energy radiation and aberrant activity of nuclear enzymes. Although DNA can be damaged in a number of different ways, double-strand breaks (DSBs) are among the most cytotoxic DNA lesions, as they interfere with essential cellular processes that involve DNA, such as transcription and replication. Furthermore persistent DSBs activate DNA damage response pathways, potentially leading to cell cycle arrest, cellular senescence or programmed cell death. Mammalian cells possess several distinct DSB repair pathways, most DSBs are repaired, via a pathway termed non-homologous end joining (NHEJ). NHEJ is activated to join two DNA ends directly to restore the linear unperturbed structure of DNA. NHEJ evolved to rapidly eliminate DSBs without the help of an undamaged template DNA and it is thus somewhat error prone. Defective NHEJ results in hypersensitivity to DNA-damaging agents, impaired development of lymphocytes and neurons, and predisposition to tumours. These observations highlight an essential nature of this DNA repair pathway in humans. This chapter highlights recent progress in understanding the mechanism and regulation of NHEJ in mammalian cells.

INTRODUCTION TO DNA END JOINING

DNA is under a constant threat from a variety of sources, such as endogenous reactive oxygen species, replication errors, action of cellular nucleases, abortive activity of topoisomerases and exogenous ionising radiation (IR) (Ciccia and Elledge 2010, Chang et al. 2017). Although DNA can be damaged in a variety of different ways, double-strand breaks are considered most cytotoxic (Jackson and Bartek 2009). This is partly due to a dramatic inhibition of replication and transcription that DSBs impose (Ceccaldi et al. 2016). Persistent unrepaired DSBs engage the cellular DNA damage response (DDR) machinery, leading to adaptive changes in cellular physiology, cell cycle and gene expression, potentially leading to an induction of apoptosis or senescence (Lukas et al. 2011). Measurements of metaphase chromosome and chromatid breaks performed in cell culture estimate the frequency of DSBs at around 10 per day per cell (Martin et al. 1985). Mammalian cells are dependent on two general strategies to resolve DNA double-strand breaks: homology-directed repair (HDR) (also known as homologous recombination [HR]) and non-homologous end joining (Waters et al. 2014). Both HDR and NHEJ encompass a number of sub-pathways. HDR relies on extensive homologies to direct the repair. NHEJ sub-pathways differ in the extent of terminal DNA microhomology needed for repair (Chang et al. 2017). Microhomology-dependent end joining refers to a situation in which DNA end ligation is assisted by Watson-Crick DNA end annealing of terminal complementary bases of less than 20 bp in length (**Figure 7.1**). In so-called classical NHEJ (c-NHEJ), no microhomology is required for repair, and the process is dependent on the Ku70/Ku86 DNA damage sensor that can inhibit alternative DSB repair pathways (Bennardo et al. 2008). In the absence of Ku loading, the repair can proceed via so-called alternative NHEJ (alt-NHEJ), which typically requires microhomology for efficient end joining, although direct joining can also occur in alt-NHEJ (Boboila et al. 2012). Homology of greater that 20 bp would typically be associated with HDR. Gene conversion is the most common HDR reaction and is typically referred to as homologous recombination (Prakash et al. 2015). HR preferentially

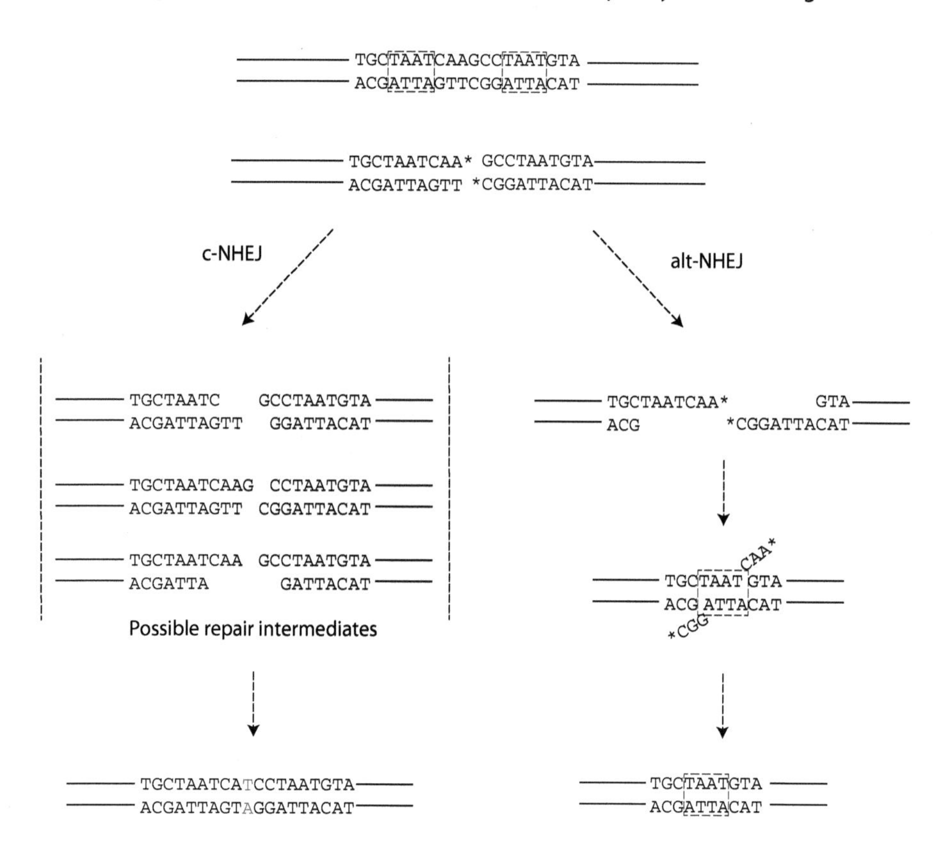

Figure 7.1 Comparison of classical (c-NHEJ) and alternative (alt-NHEJ) non-homologous end joining (NHEJ) pathways of DNA double strand break repair. DNA double-strand breaks can be repaired either via c-NHEJ or alt-NHEJ. C-NHEJ machinery processes DNA ends to remove ligation-blocking groups (*), with each DNA end undergoing individual gain or loss of nucleotides (possible repair intermediates). As the result of c-NHEJ repair, the final repair product shown in the example has lost its original AG sequence and gained T insertion. The repair through alt-NHEJ proceeds via initial short-range resection to expose terminal microhomology (dashed boxes) present at the vicinity of the DSB prior to repair. DNA end annealing through exposed microhomology serves as means of stabilising the end synapsis prior to flap removal and ligation. Note that alt-NHEJ typically results in more extensive information loss/gain at the break owing to the need for uncovering microhomology.

utilises an undamaged sister chromatid as a template for repair and is thus error free. HR can only operate in S/G2 phases of cell cycle once the chromosomes have replicated. Therefore, cells in G1 are dependent on NHEJ for DSB repair. However, NHEJ is also active outside G1, and cells have developed intricate regulatory mechanisms to facilitate utilisation of the most suitable repair strategy to avoid extensive competition between alternative pathways (Ceccaldi et al. 2016). Unlike any other DNA repair pathway, NHEJ has evolved substantial flexibility and multifunctionality in order to facilitate repair of a range of substrate DNA ends at DSBs (Lieber 2010). NHEJ's main function is to re-join chromosome ends and, because it does not take advantage of homology to repair the existing damage, NHEJ is inherently error prone. Of note, this propensity of NHEJ is utilised in current CRISPR-based approaches of genome editing. Accordingly, chromosomal cuts directed by the Cas9/gRNA complex when repaired through the NHEJ pathway typically acquire altered sequences at the cut site (indels) (Harrison et al. 2014). NHEJ, however, is capable of achieving accurate repair, but only of those DNA ends that can be directly ligated and do not require processing.

Most DSBs in multicellular eukaryotes are pathological and occur at random, with the notable exception for so-called programmed DSBs induced by specialised enzymes at defined chromosomal locations in developing lymphocytes (see VDJ recombination and class-switch recombination [CSR]). These programmed DSBs are physiological and are required for proper lymphocyte development. Of note, the enzymes responsible for induction of programmed DSBs (RAG1/2 complex for VDJ; AID enzyme for CSR) produce a certain number of off-target DSBs (Hasham et al. 2010). These breaks are considered pathological and are a frequent cause of translocations in developing lymphocytes.

The majority of DSBs that occur in dividing cells are a result of replication fork collision with a DNA single-strand break. These S phase–specific one-ended DSBs are not suitable for repair through NHEJ and, therefore, are typically repaired by HR (Saleh-Gohari and Helleday 2004). ROS-derived DSBs are another common DSB type and arise when mitochondria-derived hydroxyl radicals react with DNA to cause single-strand breaks. These so-called staggered DSBs form after the appearance of two SSBs on opposite DNA strands in close proximity (Woodbine et al. 2011). Further sources of DSBs include environmental ionising radiation (cosmic rays, nuclear accidents and cancer radiotherapy), which causes DNA breaks directly by interacting with DNA and indirectly through radiolysis of water and subsequent generation of ROS (Lomax et al. 2013). Furthermore, activity of certain nuclear enzymes that act on DNA can lead to DSB generation. Examples include failure of topoisomerase I (Topo1/TOP1), which functions to relieve DNA torsional stress by transiently breaking one DNA strand. Collision of the Topo1-DNA complex with the replication machinery can lead to a DSB generation (Strumberg et al. 2000). As mentioned earlier, RAG recombinase and activation-induced deaminase acting in lymphoid cells can generate off-target DSBs located outside the antigen receptor gene loci.

LIGATABLE VERSUS BLOCKED DNA END CONFIGURATIONS

ROS-derived DNA breaks can have several end configurations, which require cleaning/processing prior to re-ligation (Jena 2012). For example, 3′ phosphoglycoaldehyde is one of such products arising as a result of the oxidation of the 3′ position of deoxyribose. Other examples of oxidative DNA end damage include 3′ phosphoglycolate (Figure 7.2B) and 5′ aldehyde (Menon and Povirk 2016). Most such modifications are unstable (except 3′ phosphoglycolate [3′ PG]) and spontaneously break down, giving rise to 3′ and 5′ phosphates. DNA ends can also be blocked by covalent protein adducts. Most typical examples include those generated by the failure of topoisomerases to complete their reaction – a process that can occur spontaneously at a low rate in normal cells. Protein-DNA cross-links (also termed Topo-DNA cleavage complexes) arising from failed topoisomerase activity can be dramatically up-regulated after treatment of cells with several common cancer chemotherapy drugs, which function by trapping topoisomerases on DNA (Froelich-Ammon and Osheroff 1995). Examples include camptothecin (topoisomerase I inhibitor) and etoposide (topoisomerase II inhibitor). Prolonged treatment of proliferating cells with these compounds leads to apoptosis due to catastrophic levels of DNA damage resulting from replication fork collapse (Avemann et al. 1988). Topoisomerase I can be trapped on a SSB with a 5′-hydroxyl end with the enzyme covalently linked to the 3′ DNA terminus (Pommier 2006). A DSB can form at such an adduct after collision with replication or transcription machinery. Topoisomerase II-type enzymes (TOP2/Topo2, TOP2b and Spo11) form a DSB as a reaction intermediate with 3′-hydroxyl and 4bp 5′-overhang with a tyrosyl-linked enzyme moiety (Figure 7.2C). Again a DSB results from a failure of Topo2 to complete the reaction (Pommier et al. 2010).

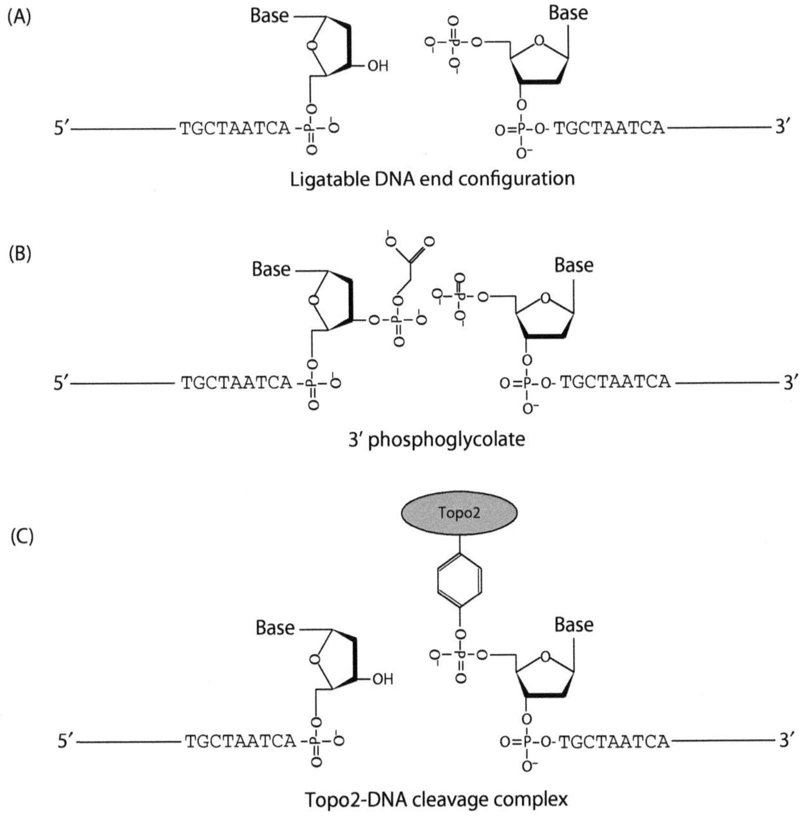

(A) Ligatable DNA end configuration

(B) 3′ phosphoglycolate

(C) Topo2-DNA cleavage complex

Figure 7.2 **Examples of common DNA end ligation-blocking lesions.** (A) 3′ OH and 5′ phosphate groups constitute the default ligatable DNA end configuration and are the desired products of DNA end processing prior to ligation. (B) 3′ phosphoglycolate (3′PG) is an example of a common DNA lesion derived from the oxidative breakdown of deoxyribose. (C) Topo2-DNA cleavage complex is an example of common DNA-protein cross-link (DPC) arising through abortive activity of mammalian DNA type 2 topoisomerases (Topo2 transiently breaks both DNA strands during its catalytic cycle; here only one strand is shown for simplicity.)

REGULATION OF DSB REPAIR PATHWAY CHOICE

Given that a number of DNA repair pathways are available to resolve DSBs, a question arises of the choice of DSB repair pathway. The main point of regulation is the extent of resection occurring at a DSB (Symington and Gautier 2011). Resection is a process of 3′ single-stranded DNA (ssDNA) generation at DSBs by specialised nucleases, which digest away the 5′ DNA strand (Symington 2016). Classical NHEJ does not require DNA end resection. Moreover, Ku binding is inhibited by ssDNA generated at the junction, so the more resection is allowed to progress, the less likely the participation of c-NHEJ in the repair reaction (Huertas 2010). Short-range resection (up to several hundred base pairs from the beak) catalysed by CtIP and MRN promote repair via alt-NHEJ through the use of uncovered microhomologies. Resection is controlled by a variety of mechanisms at the level of chromatin and throughout the cell cycle (Hustedt and Durocher 2016). Regulation at the level of chromatin involves the 53BP1 protein that acts as an antagonist of the resection by blocking access of the initial resection factors (CtIP and MRN) to DSBs (Panier and Boulton 2014). In S phase, the action of 53BP1 is antagonised by BRCA1 protein, which favours resection (Daley et al. 2015).

53BP1 is recruited to chromatin-flanking DSBs by a mechanism dependent on ATM-mediated histone H2A.X phosphorylation and RNF8/168-dependent chromatin ubiquitylation (Dantuma and van Attikum 2016). Interestingly, proteins constituting the ATM-H2A.X-53BP1 axis (including the MRN complex) are capable of spreading along the chromatin flanking DSBs into megabase domains (in yeast, these domains occupy up to 100,000 bp from the break) (Rogakou et al. 1999, Kim et al. 2007). It is not clear why such an extensive loading of these DDR mediators is required, as involvement of H2A.X and 53BP1 in pathological DSB resolution is limited (Fernandez-Capetillo et al. 2003, Morales et al. 2003). A more likely function of this extensive chromatin build-up of DDR mediators is to amplify signalling from DSBs to enforce checkpoint induction (Stewart et al. 2003). Importantly, however, DSBs generated during CSR somatic recombination do require ATM-H2A.X-53BP1 for repair (Petersen et al. 2001, Manis et al. 2004). In summary, although 53BP1does not participate in the actual DNA repair at the break, it has a strong stimulatory effect on NHEJ, presumably by establishing a physical barrier to resection in chromatin in the DSB's vicinity (Panier and Boulton 2014). Once resection has reached a certain threshold, additional proteins responsible for long-range resection become involved (e.g. BLM, EXO1 and DNA2 nucleases) and the repair becomes committed to HDR (Liu and Huang 2016). The cell cycle exerts a strong control on the resection efficacy by CDK-mediated phosphorylation of resection factors (e.g. CtIP) in such a way that resection is inhibited in G1 but promoted in S/G2 phases of the cell cycle (Huertas and Jackson 2009).

THE CLASSICAL END JOINING MECHANISM (C-NHEJ)

To resolve DSBs, NHEJ employs a set of specialised proteins termed NHEJ core factors that function by sensing DSBs, aligning broken DNA ends and acting as a scaffold for a variety of end cleaning/processing enzymes. The concerted action of these proteins results in the generation of ligatable DNA ends and sealing of the damage. Restoration of the linear unperturbed DNA structure is the primary role of NHEJ reactions and is typically achieved at the expense of accuracy. This is well tolerated in higher eukaryotes due to poor gene density of their genome (Deriano and Roth 2013). The mechanism of NHEJ reaction will be presented based on current understanding of vertebrate NHEJ. Classical end joining (c-NHEJ) is responsible for the bulk (typically more than 90%) of all NHEJ actions in mammalian cells, with the rest being achieved through microhomology dependent alt-NHEJ (Lieber 2010). Of note, haploid yeast cells preferentially utilise HR over NHEJ, a choice that probably reflects their high gene density and ongoing DNA replication, thus requiring faithful repair (Daley et al. 2005). When repair of pathological IR-induced DSBs is followed in G1-arrested mammalian cells, biphasic repair kinetics are observed (Riballo et al. 2004). Most DSBs are rapidly repaired within the first few hours after irradiation. However, around 20% to 30% of these DSBs are repaired with slower kinetics over a period of 24–48 hours. Several repair models have been invoked to explain these observations. It is believed that repair of breaks bearing simple ligatable DNA ends proceeds relatively quickly, whereas ends requiring extensive processing require more time to resolve. This interpretation is consistent with a specific requirement for Artemis nuclease activity for the slowly repaired DSBs (Riballo et al. 2004). Another possible explanation is that the slowly repaired DSBs are positioned within non-permissive chromatin repair domains such as heterochromatin (Goodarzi et al. 2008). Substantial chromatin relaxation is necessary for access of repair machinery to these lesions. Requirement for ATM-dependent chromatin remodelling is consistent with this model (Ziv et al. 2006).

During c-NHEJ-mediated repair, DSBs are initially sensed by the Ku heterodimer (Walker et al. 2001). The Ku-DNA complex protects DNA ends from degradation (and from resection initiated by competing MRN and CtIP) and serves as a platform for recruitment of other NHEJ proteins (Downs and Jackson 2004). It is presumed that each DNA end is bound by at least one Ku dimer and processing of both ends occurs independently. It is believed that this flexibility is responsible for the vast array of possible repair outcomes (Figure 7.1). Although ligation of cohesive ends is more efficient than blunt ends, c-NHEJ joints show no preference for microhomologies (average microhomology is 0 bp). Such a model is consistent with the so-called iterative end-processing scheme where generation of a 5′ phosphate end serves as a recognition element for the ligase and terminates processing (Reid et al. 2017). Ku dimer loaded on DSBs undergoes an allosteric transition exposing a flexible C-terminal end of Ku86, which recruits DNA-PKcs protein (Davis and Chen 2013). DNA-PKcs is responsible for synapsis of DNA ends and recruitment of a specialised c-NHEJ nuclease, Artemis (Ma et al. 2002). Furthermore, DNA-PKcs facilitates ordered access of enzymes responsible for end cleaning/processing to DNA ends (Figure 7.3). DNA end synapsis is a crucial process in NHEJ and ensures that DNA ends are held in close proximity (Reid et al. 2015). It is currently believed that synapsis occurs in several steps and begins with long-range end bridging mediated by the DNA-PK complex, followed by short-range synapsis facilitated by XRCC4 protein and its paralogue XLF (Graham et al. 2016). Besides synaptic functions, XRCC4 and two of its

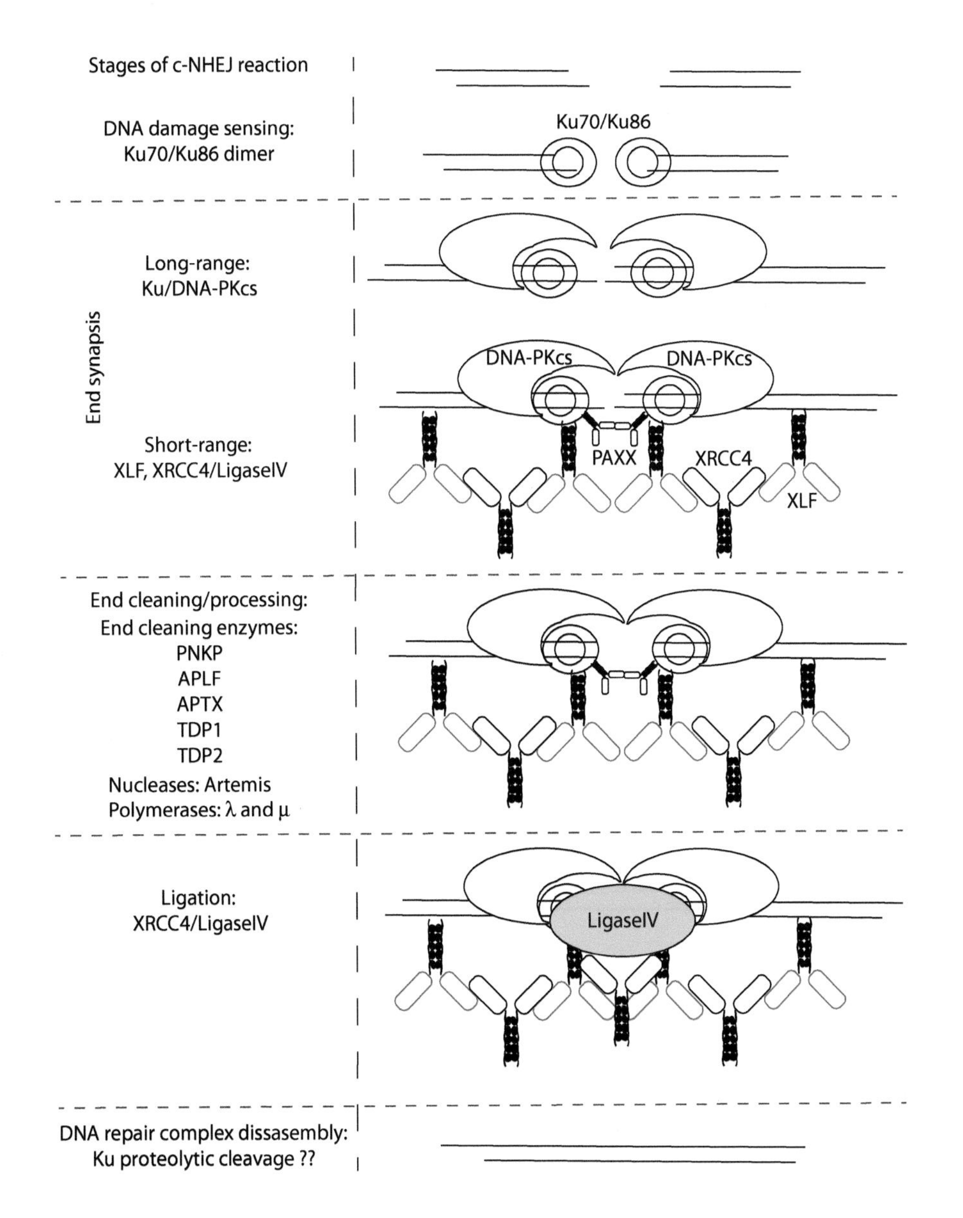

Figure 7.3 **Mechanism of c-NHEJ reaction.** Double-strand DNA breaks (DSBs) are initially recognised by the Ku70/Ku86 DNA damage sensor. Upon DSB binding, Ku translocates inward on DNA and recruits c-NHEJ core proteins: DNA-PKcs, PAXX and XLF-XRCC4. DNA ends are initially tethered (end synapsis) by DNA-PKcs; XLF-XRCC4 filaments that make contact with DNA further facilitate close proximity alignment. This assembly is stabilised by PAXX that contacts Ku70/Ku86. Once the DNA ends have been stabilised and sheltered from access of undesired enzymatic activities, controlled DNA end processing occurs. DNA processing enzymes are able to access DNA ends as permitted by the NHEJ core complex to remove DNA end blocking groups and achieve ligatable DNA end configuration. Processing is followed by the recruitment of the XRCC4/ligase IV ligation complex that can sense the generation of the 5′ phosphate group after processing. DNA ends are ligated and the NHEJ core complex is disassembled. Ku70/86 might require removal by proteolysis after successful repair, as it can become trapped on DNA after ligation.

paralogues (XLF and PAXX) participate in c-NHEJ by providing a range of specific functions (see later). XRCC4, for example, serves as a chaperone for ligase IV, which is responsible for end ligation. As mentioned earlier, prior to ligation, DNA ends are subject to extensive end cleaning (to remove ligation blocking lesions) and processing (to fill in gaps or trim flaps to achieve a ligatable configuration) (Menon and Povirk 2016). Finally, the DNA break is sealed by the XRCC4-ligase IV complex (Chang et al. 2017). In vivo DSBs arise in the context of chromatin, which constitutes a barrier to repair. Therefore, limited chromatin remodelling associated with NHEJ and involving nucleosome removal or remodelling has been described in mammalian cells (Berkovich et al. 2007) and yeast (Tsukuda et al. 2005). Furthermore, Ku itself is able to bind nucleosome-associated DNA and peel away up to 50 bp from the histone octamer exposing free DNA ends for processing. There are also a number of poly-ADP-dependent mechanisms that facilitate opening of chromatin at DSBs to allow for access to repair machinery (Rulten et al. 2011, Luijsterburg et al. 2016). The following section and Figure 7.4 describe c-NHEJ factors in greater detail.

The core c-NHEJ factors

Current literature divides protein components involved in NHEJ into so-called core NHEJ factors and accessory NHEJ factors. This distinction is mostly based on the severity of the phenotype associated with the loss of a given protein. Thus, deficiency of a core NHEJ factor typically produces a strong defect in c-NHEJ execution, whereas depletion of NHEJ accessory factors results in a mild phenotype (e.g. DSB repair delay). A common example of an NHEJ

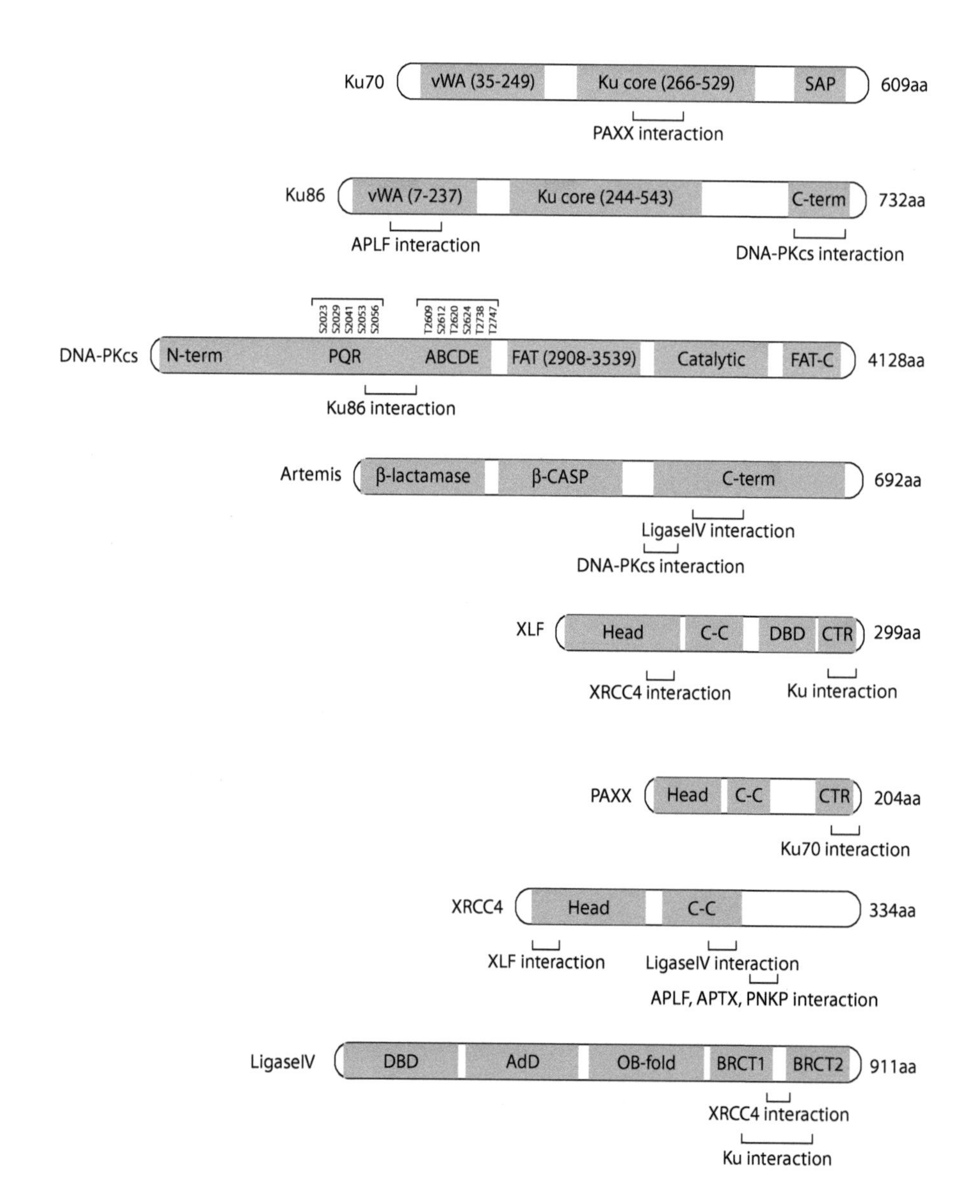

Figure 7.4 Core c-NHEJ proteins: primary structures and key interactions. Schematic description of the primary structures of core c-NHEJ proteins. Functional domains are depicted as brown boxes. Major interactions among NHEJ components are shown. In DNA-PKcs, protein amino acids phosphorylated within PQR and ABCDE clusters are indicated.

accessory factor is the APLF protein, which contributes to NHEJ complex stability in chromatin without being absolutely essential (Rulten et al. 2011). There is, however, a caveat with such NHEJ component classification in that both XLF and PAXX proteins, although considered core factors, produce a mild phenotype when individually depleted. Their core factor nature is clearly evident from the phenotype of double XLF/PAXX knockout mice showing severe c-NHEJ deficiency resembling XRCC4 and ligase IV knockout animals (Balmus et al. 2016). Furthermore, although Artemis is considered a core NHEJ factor, it is only required for repair of a subset of DSBs – those requiring specific Artemis nuclease-mediated processing. Because of substantial genetic redundancy between different proteins acting in DNA repair, a possibility exists that many factors currently perceived as accessory might be reconsidered as core factors when analysed in the context of a redundant counterpart deficiency. For the purpose of this review, Ku70/Ku86, DNA-PKcs, Artemis, XLF, PAXX and XRCC4-ligase IV will be referred to as core NHEJ factors.

Ku

Ku serves as a DSB sensor during c-NHEJ and has very high affinity for dsDNA. Ku is a heterodimer composed of two subunits: Ku70 and Ku86 (numbers indicate respective sizes in kilodaltons [kDa]; Ku86 is sometimes referred to as Ku80). Structurally, Ku70 and Ku86 share an N-terminal von Willebrand type A-like (vWA) domain and central core domains (containing dimerisation interfaces); however, their C-termini differ substantially (**Figure 7.4**). Ku proteins form a tight heterodimer supporting each other's stability (Walker et al. 2001). The Ku complex has a cradle-shaped structure and upon initial DNA binding translocates inward onto a DNA duplex. DNA is threaded through Ku86 so that the Ku70 core and vWA domains position close to free DNA ends and make contact with the DNA phosphate backbone. This binding mode explains Ku's high specificity for DNA ends and lack of preference for any specific DNA sequence. Ku can only load and unload at DNA ends and can become trapped on DNA upon circularisation. Proteolytic degradation has been proposed as the main mechanism of removal of Ku after DSB repair (Postow 2011). As explained earlier, stable binding of Ku to DNA serves as a critical regulatory step in c-NHEJ initiation and commitment. Early in vitro studies of the Ku/DNA binding mechanism suggested a simple 1:1 binding stoichiometry (Gottlieb and Jackson 1993). Recently, this model has been validated in vivo by super-resolution microscopy by demonstrating the formation of small Ku foci at DSBs (Britton et al. 2013). DSB processing requires removal of 5′ deoxyribose phosphates (dRPs), which are typical breakdown products of abasic sites. Although polymerases involved in NHEJ appear to possess the required lyase activity, Ku70 is the main 5′ dRP/AP lyase at DNA ends. Ku70 lyase activity cleaves abasic sites positioned at DNA overhangs to facilitate further processing and ligation (Roberts et al. 2010).

DNA-PKcs

Together, the Ku/DNA-PKcs complex is termed DNA-PK kinase with DNA-PKcs serving as the catalytic component of the enzyme (Gottlieb and Jackson 1993). The high-resolution crystal structure of DNA-PKcs has recently been determined (Sibanda et al. 2017). The protein can be structurally divided into three subdomains: the N-terminal region, the central circular cradle and a C-terminal head region. Each subdomain contains a large number of so-called HEAT (Huntingtin, elongation factor 3, PP2 A and TOR1) repeats distributed throughout. DNA-PKcs is a member of the PI3K kinase sub-family, and historically several distinct regions of homology shared by this class of enzymes have been described. These include the FAT (FRAP, ATM and TRRRAP) domain, the catalytic domain resembling other PI3K kinases and the C-terminal FAT-C region (**Figure 7.4**) (Blackford and Jackson 2017). DNA-PKcs serine/threonine kinase activity is activated at DSBs by the C-terminal fragment of Ku86 when bound to DNA and is essential for efficient end-joining (Sibanda et al. 2017). Although most c-NHEJ factors can be phosphorylated by DNA-PK, it is DNA-PKcs itself that is the main functional substrate of phosphorylation (Chan et al. 2002). DNA-PKcs is autophosphorylated on several residues grouped into distinct clusters (**Figure 7.4**). DNA-PKcs undergoes several structural transitions during the execution of NHEJ repair reactions, which correlate with distinct autophosphorylation patterns (Hammel et al. 2010). For example, phosphorylation of the PQR cluster results in a closed DNA-PKcs conformation, whereas ABCDE cluster phosphorylation is compatible with more efficient access by end modifying enzymes to the DNA lesion (Cui et al. 2005). Furthermore, DNA-PKcs autophosphorylation is essential to allow the release of the DNA-PKcs from DNA after repair (Dobbs et al. 2010). The importance of ordered DNA-PKcs release from DSBs is highlighted by the phenotype of knockin mice, in which several DNA-PKcs autophosphorylation sites have been mutated into alanine. These animals show a bone marrow failure suggestive of dysfunction of other DNA repair pathways such as HDR or the Fanconi anaemia (FA) DNA repair pathway (Zhang et al. 2011). As mentioned earlier, one of the key functions of DNA-PKcs is recruitment and activation of the Artemis nuclease, which requires a specific autophosphorylated form of DNA-PKcs to become active at DSBs (Goodarzi et al. 2006). DNA-PKcs is also capable of stimulating the ligase IV activity. Therefore, DNA-PKcs performs key functions in NHEJ, specifically facilitating end tethering, end processing and ligation (Jette and Lees-Miller 2015).

Artemis

Artemis is the main NHEJ nuclease and its activity is required for repair of pathological (radiation-derived) and programmed (RAG1/2-dependent) DSBs (Chang and Lieber 2016). The Artemis N-terminal region contains the catalytic metallo-β-lactamase (MBL) domain, similar to that found in other nucleases (Allerston et al. 2015), whereas its C-terminus contains a DNA-PKcs interaction interface and ligase IV binding site (**Figure 7.4**). In response to DNA damage, Artemis is extensively phosphorylated on its C-terminal region by both DNA-PKcs and the related ATM kinase (Poinsignon et al. 2004). The Artemis/DNA-PKcs complex has endonuclease activity, which preferentially cleaves at single-strand to double-strand DNA transitions found in DNA hairpins, overhangs, flaps and so on. Such substrate preference is likely a reflection of a way by which the Artemis/DNA-PKcs complex binds to its DNA substrate by localising on 4bp of ssDNA adjacent to a single-/double-strand transition (Chang and Lieber 2016). Artemis is crucial for facilitating correct hairpin opening at coding DNA ends during VDJ reactions after RAG1/2–mediated DNA cleavage (Ma et al. 2002). Failure of Artemis at this stage results in greatly diminished efficiency of VDJ recombination (**Figure 7.5**) (Rooney et al. 2002). Rare coding joints formed independently of Artemis-mediated hairpin opening show characteristic nucleotide insertions (P-nucleotides), which can be used as a surrogate readout for Artemis activity when screening immunodeficient patients for underlying biochemical defects (van der Burg et al. 2009).

XRCC4-ligase IV complex

XRCC4 is a 335aa protein composed of an N-terminal globular head domain followed by a long coiled coil (Li et al. 1995, Junop et al. 2000). XRCC4 functions as a dimer with a portion of its coiled coil serving as the dimerisation interface. XRCC4 performs important scaffolding functions in NHEJ complex assembly (see later in the section describing XLF for additional details). The XRCC4 C-terminal region contains a short motif phosphorylated by casein kinase II (CK2), which mediates phospho-dependent interaction with APLF, PNKP and aprataxin/APTX (all of which are accessory c-NHEJ factors) through their N-terminal FHA domains (Waters et al. 2014). The XRCC4 protein is capable of sequence-independent binding to DNA, although it requires at least a 100-bp fragment for stable interaction (Modesti et al. 1999). One of the essential functions of XRCC4 in c-NHEJ is to provide chaperone activity to ligase IV (in XRCC4-deficient cells ligase IV protein is unstable) (Wu et al. 2009).

Ligase IV N-terminal and central regions contain DNA-binding and catalytic domains, respectively, whereas its C-terminal region comprises two BRCT domains (**Figure 7.4**). Ligase IV binds to the C-terminal end of the XRCC4 coiled coil via a short stretch of amino acids separating its BRCT domains (Sibanda et al. 2001). The ligase IV catalytic domain is responsible for enzyme adenylation (Ochi et al. 2013). Most of the cellular ligase IV exists in an adenylated form (Robins and Lindahl 1996). The ligase IV catalytic mechanism involves initial enzyme adenylation with a subsequent transfer of the AMP group to the 5′ phosphate of one DNA end. This is followed by a nucleophilic attack by the 3′ hydroxyl group of a second DNA end and release of AMP to yield the final ligation product. Ligase IV is mechanistically versatile and can perform a variety of ligation reactions including ligation across gaps of incompatible DNA ends and even a single strand across a DSB (Lieber 2010). Such activity is important, as ligation of a single DNA strand converts a DSB into am SSB (single-strand DNA break), which requires a mechanistically simpler repair reaction.

XLF/Cernunnos

XLF (also called Cernunnos) was first found as a gene mutated in a patient with immunodeficiency and, independently, as a protein physically associated with XRCC4 (Ahnesorg et al. 2006, Buck et al. 2006a). XLF is considered an XRCC4 paralogue as it shares its overall domain organisation. Interestingly, the primary sequence of XLF does not show any substantial similarity to XRCC4. The XLF N-terminal domain forms a globular head, its central coiled coil contains a dimerisation interface but, in contrast to XRCC4, is appreciably shorter and folds back on itself (Y. Li et al. 2008). The XLF C-terminal domain (CTR) contains a Ku interaction site (Yano et al. 2011). XLF binds to XRCC4 in a head-to-head manner (Hammel et al. 2011)

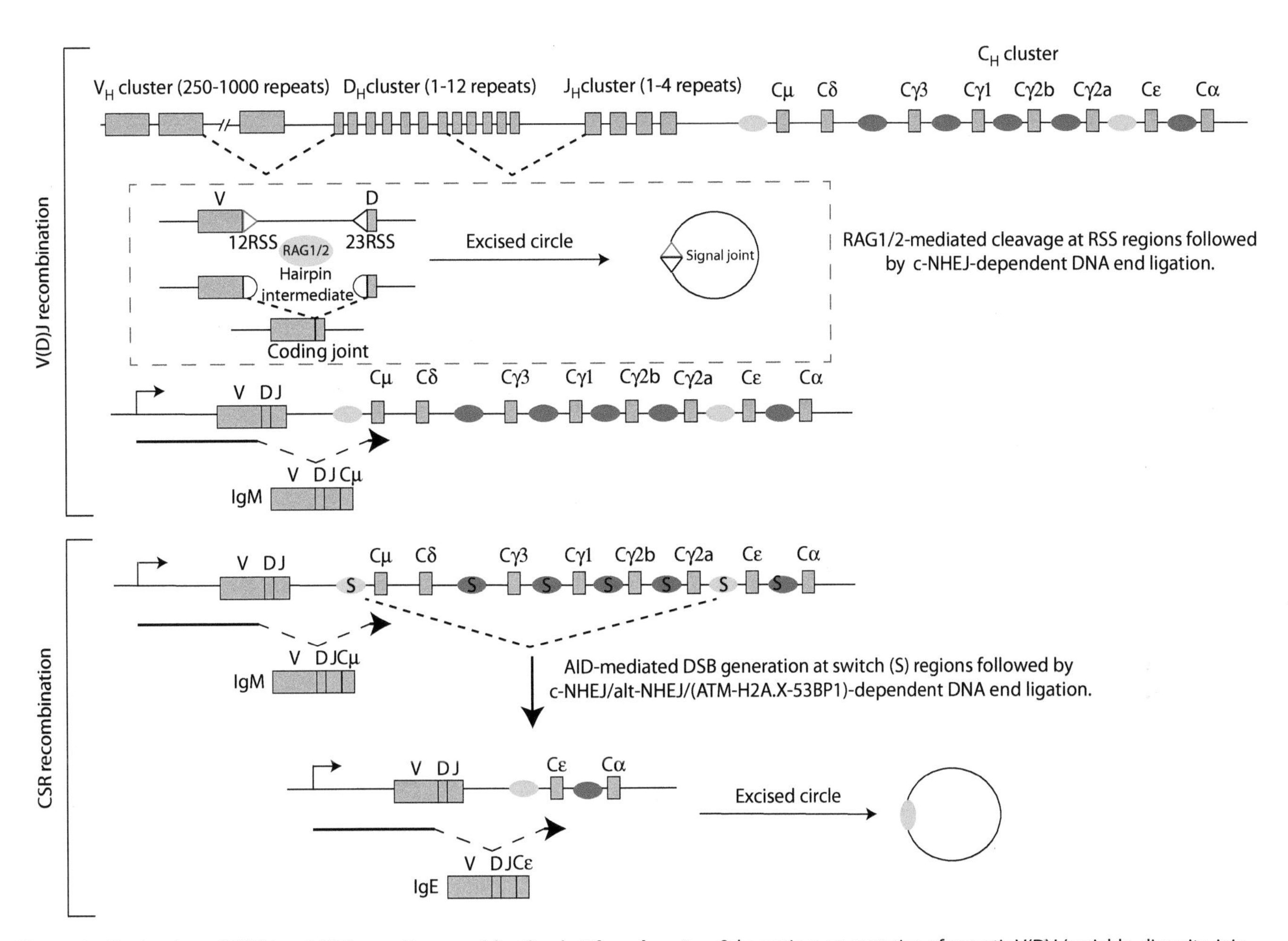

Figure 7.5 **Mechanism of V(D)J and CSR somatic recombination in B lymphocytes.** Schematic representation of somatic V(D)J (variable, diversity, joining) recombination process followed by class-switch recombination (CSR) at a prototypical immunoglobulin heavy chain (IGH) B-cell locus. The top panel shows the structure of the locus in germ line configuration. V (variable), D (diversity), J (joining), C (constant) clusters are arranged from the 5′ end as repetitive units of variable numbers as indicated. The process of V(D)J recombination occurs in pre-B cell stage and is required for lymphocyte survival. The result of successful VDJ recombination is the expression of the assembled VDJ segment as a part of a functional immunoglobulin molecule. After accomplishing VDJ recombination, the B cell will express IgM on the cell surface (transcriptional start site is indicated with an arrow, primary transcript structure is depicted under the locus structure; note that constant region encoding exon is spliced into VDJ pre-arranged transcript during pre-mRNA processing). The VDJ recombination reaction involves RAG1/2 complex–dependent cleavage of DNA at two RSS sites brought to close proximity. The resolution of this recombination reaction results in the creation of a coding joint between a V and D or J segment and excision of the chromosomal segment containing the RSS sites. This segment possesses the signal joint created by the c-NHEJ machinery (excised circle). In subsequent CSR, extracellular signalling triggered by B-cell activation results in AID-enzyme-dependent DSB generation at switch regions (S blue and red ovals). Resolution of CSR intermediates will result in linking the remote switch regions (S blue ovals) with subsequent elimination of the intervening DNA segment as a circle (excised circle). CSR will result in a switch in expression from IgM immunoglobulin to IgG, IgA or IgE (the specificity of the immunoglobulin remains unaltered).

and the XLF-XRCC4 dimers form extended filaments that warp around DNA and contribute to end alignment/synapsis (Brouwer et al. 2016). Although the existence of such higher-order XLF-XRCC4 assemblies was initially inferred from structural data, recently super-resolution microscopy has confirmed their presence in cells (Reid et al. 2015). XLF can bind DNA though a stretch of amino acids present in its C-terminal region, although it requires at least 60 bp of DNA for stable binding (Lu et al. 2007). XLF's DNA-binding ability is likely important in the function of XLF-XRCC4 filaments. XLF is also important in stimulating the ligation of non-compatible DNA ends (Tsai et al. 2007). In the context of VDJ recombination in humans, XLF has been shown to promote TdT-dependent nucleotide addition (N-nucleotides). Impairment of this step of VDJ recombination contributes to a limited repertoire of VDJ rearrangements and is one of the causes of immune system dysfunction in XLF-deficient patients (Ijspeert et al. 2016).

PAXX

The most recently discovered NHEJ factor is PAXX (Craxton et al. 2015, Ochi et al. 2015, Xing et al. 2015). It also is considered an XRCC4 paralogue, although it is of substantially lower molecular weight (204aa versus 336aa). Unlike XRCC4, PAXX is not associated with XLF but rather co-purifies with the DNA-PK complex. A limited evolutionary presence of PAXX (which is absent from fungi, worms and flies) resembles that of DNA-PKcs and coincides with the emergence of VDJ recombination. A handful of simple unicellular eukaryotes possess a protein resembling PAXX, although its primary sequence conservation is limited to the extreme C-terminal end. Strikingly, this conserved region (CTR) serves as the Ku70 binding domain in mammalian PAXX (Tadi et al. 2016) and is important in exerting its activity in NHEJ (Ochi et al. 2015, Xing et al. 2015). Interestingly, molecular modelling of the coral PAXX homologue predicts a high degree of structural similarity to the mammalian counterpart (Craxton et al. 2015). In contrast to XRCC4 and XLF, PAXX does not seem to be able to bind DNA or form filamentous structures at DSBs. A number of studies have found a role for PAXX in facilitating NHEJ complex assembly. PAXX dimers presumably act by direct interaction with two Ku molecules across a DSB facilitating end bridging (Figure 7.3). An additional role for PAXX appears to be the stimulation of ligation efficiency of non-compatible DNA ends. Genetic and biochemical evidence suggests a strong functional overlap between XLF and PAXX (Kumar et al. 2016, Lescale et al. 2016).

The main c-NHEJ processing and end cleaning factors

The main polymerases specifically involved in NHEJ are pol mu (μ), lambda (λ) and TdT (terminal deoxynucleotidyl transferase) (Waters et al. 2014). These enzymes belong to the X polymerase family and are responsible for nucleotide insertions at processed DSBs, although they differ in the requirement for complementary sequences opposite the primer DNA. Pol lambda requires at least one base complementarity, whereas pol mu does not require any. TdT can add nucleotides to ssDNA overhangs in a template-independent fashion (Waters et al. 2014, Pryor et al. 2015). Pol mu and lambda associate with NHEJ complexes through their N-terminal BRCT domains and contribute essential DNA polymerase activity to 3′ overhangs at DSBs (Mueller et al. 2008). Pol mu can also show TdT-like activity potentially converting incompatible DSB ends into compatible overhang ends (polymerase-derived microhomology). TdT is primarily expressed in developing lymphocytes and contributes to joint diversity during VDJ reactions. However, currently there is a lack of evidence for strong contribution of TdT to radiation-related DSB resolution. This may possibly be explained by restricted tissue-specific expression of TdT and the ability of other polymerases involved in NHEJ to exhibit a TdT-like activity (Dudley et al. 2005).

PNKP

Mammalian PNKP is a multifunctional enzyme involved in NHEJ possessing both kinase and phosphatase activities (Weinfeld et al. 2011). The kinase domain of PNKP has a preference for double-stranded DNA ends with recessed 5′ termini. PNKP uses a separate centrally located phosphatase domain to remove 3′ phosphates. These activities act in concert to convert 5′ hydroxyl and 3′ phosphate DNA ends to a ligatable configuration. Such blocked DNA ends are generated by both ionising radiation and the abortive action of Topo1. PNKP docks to the NHEJ complex via N-terminal FHA domains and directly interacts with CK2-phosphorylated XRCC4 (Aceytuno et al. 2017).

APTX

As mentioned earlier, DNA end ligation proceeds with 5′ phosphate groups being transiently adenylated. If the ligase aborts at this step, then the adenylated 5′ phosphate becomes a ligation-blocking group, which needs to be removed prior to another ligation attempt. This reaction is accomplished by the aprataxin (APTX) protein that resets the 5′ phosphate. APTX possesses an FHA domain and, similar to PNKP, is capable of docking to XRCC4 in a phospho-dependent fashion (Ahel et al. 2006).

APLF

APLF (aprataxin and PNKP-like factor) has been shown to possess a 3′ exonuclease activity (this type of nuclease activity is notably lacking in Artemis). However, APLF's primary role in NHEJ appears to be as a scaffolding molecule. APLF can bind to XRCC4 through its N-terminal FHA domain and to Ku86 through its MID domain (Grundy et al. 2013, Cherry et al. 2015). APLF harbours a poly-ADP-ribose binding domain, which facilitates its retention at DNA breaks (Rulten et al. 2011). Furthermore, APLF C-terminus shows histone chaperone activity that is sufficient to partially remodel nucleosomes in vitro and likely contributes to DNA repair by facilitating local chromatin remodelling (Mehrotra et al. 2011).

TDP1 and TDP2

As mentioned earlier, DNA topoisomerases facilitate transcription, replication and DNA repair by altering DNA topology. During the reaction cycle, topoisomerases form a covalent protein-DNA intermediate, in which the catalytic tyrosine of the enzyme is linked to a DNA strand break terminus. Mammalian type I topoisomerases cleave one DNA strand and form a 3′ phosphotyrosine intermediate whereas type II enzymes transiently break both strands, forming 5′ phosphotyrosine intermediates (**Figure 7.2C**) (Pommier et al. 2010). Tyrosyl-DNA phosphodiesterases (TDPs) have evolved to specifically deal with such ligation-blocking protein-DNA cross-links. TDP1 has an enzymatic activity directed towards peptides covalently linked to 3′ DNA termini and can also efficiently remove 3′ phosphoglycolate (**Figure 7.2B**). TDP2, on the other hand, is targeted primarily towards 5′-tyrosyl-DNA cross-links (Marchand et al. 2014). TDP1 leaves behind a 3′ phosphate moiety creating a substrate for PNKP, whereas the action of TDP2 results in the generation of directly ligatable 5′ phosphate termini.

One surprising fact related to the function of DNA end cleaning proteins in DSB repair is the absence of a strong DSB repair defect in cells deficient in the major DNA end cleaning enzymes. This is in striking contrast to phenotypes associated with core NHEJ factor loss. A possible explanation is that DSBs are less sensitive to a loss of any given end cleaning factor, as other mechanisms specific to DSB repair exist to deal with DNA end blocking lesions. For example, Artemis can bypass blocked ends to remove the whole segment of nucleotides, including the lesion. Such mechanisms might be of limited availability to single-strand DNA break repair machinery. Mutations in SSBR components (e.g. XRCC1, APTX, PNKP and TDP1) are associated with the development of neurodegenerative diseases such as ataxias (Takashima et al. 2002, Reynolds and Stewart 2013) (detailed in Chapter 5). Notably, ataxias are not commonly observed in NHEJ deficiency.

Additional NHEJ accessory factors

MRN/MRX

The MRN complex (Mre11-Rad50-Nbs1) is capable of DSB sensing owing to its ability to recognise DNA breaks through its Rad50 ATPase DNA binding domain. Yeast cells have an evolutionarily related complex termed MRX (Mre11-Rad50-Xrs2). MRN DSB sensing is independent of Ku (Britton et al. 2013). MRN is recruited early to DSBs, and is important in sustaining the activation of ATM, a prominent DNA damage-activated kinase. Although initial DSB recognition by MRN is direct, this complex is subsequently maintained at sites of DNA damage by Nbs1-mediated interaction with MDC1, which itself is recruited to DSBs by binding to phospho-H2A.X (an ATM substrate). This mode of recruitment involves an establishment of a positive feedback loop in which the initial ATM activity is sustained by phospho-H2A.X–dependent MRN-assisted activation at the chromatin level. The Mre11 protein has a nuclease activity that is primarily involved in DNA end processing and in the early steps of resection, which channels DSB repair into both alt-NHEJ and HR. Furthermore, MRN has a DNA end bridging ability. Mammalian MRN has been shown to mainly participate in alt-NHEJ and CSR. Although MRN activity had been implicated in V(D)J recombination, its involvement in this process appears limited (Dinkelmann et al. 2009).

CtIP

CtIP (C-terminal binding protein [CtBP]-interacting protein) has a 5′ flap nuclease activity and, in conjunction with the MRN complex, is primarily involved in the initiation of resection prior to repair via HR (Sartori et al. 2007). Additionally, CtIP has been shown to play a role in alt-NHEJ by assisting MRN in resection initiation to expose available microhomologies proximal to the break. Furthermore, CtIP has been involved in resolving Topo1- and Topo2-DNA cleavage complexes in G1 suggesting that CtIP can act on specific DNA blocked ends and could participate in clustered DNA damage repair.

Metnase

Metnase is an enzyme specifically found in primates, which functions as a nuclease and histone methylase. Metnase bears certain similarity to Artemis, as it is capable of interaction with ligase IV and can cleave at ssDNA/dsDNA junctions. Metnase contributes to NHEJ in chromatin by methylating lysine 36 of histone H3 (Fnu et al. 2011).

WRN

WRN is the protein mutated in the rare autosomal, premature ageing disorder Werner syndrome (WS). It belongs to the RecQ family of DNA helicases and acts as both a 3′-5′ helicase and a 3′-5′ exonuclease. WRN protein can physically interact with Ku and the XRCC4-ligase IV complex via XRCC4. Ku is able to stimulate WRN nuclease activity. However, as WRN is primarily capable of degrading 3′ hydroxyl DNA substrates but not 3′ blocked substrates (e.g. 3′ phosphoglycolate) its activity is unlikely to be involved in processing of blocked DNA ends. NHEJ is not dramatically affected by WRN deficiency, although a recent study reported a mild NHEJ delay in WRN-deficient cells (Grundy et al. 2016).

ALTERNATIVE END JOINING PATHWAYS (ALT-NHEJ)

In the absence of DNA ligase IV or XRCC4, chromosomal translocations still occur, pointing to the existence of an alternative NHEJ (alt-NHEJ) pathway (Sfeir and Symington 2015). Moreover, when c-NHEJ–deficient mice (e.g. Ku86, DNA-PKcs or XRCC4 mutant strains) are crossed to p53 knockout animals to permit lymphocyte survival, these mice invariably develop oncogenic translocations catalysed by alt-NHEJ machinery (Gao et al. 2000). Furthermore, although V(D)J recombination is greatly reduced in the absence of functional c-NHEJ, rare junctions are still formed at the receptor loci and these bear features of alt-NHEJ, such as large deletions/insertions. Because of the joint characteristics, alt-NHEJ must involve enzymes generating limited resection, proteins utilising microhomologies to stabilise DNA ends and nucleases capable of removing incompatible 5′ and 3′ overhangs (**Figure 7.1**). It remains uncertain whether alt-NHEJ represents a single biochemical pathway. Currently accepted mammalian alt-NHEJ factors include PARP-1, ligase I and III, MRN and CtIP (Sfeir and Symington 2015). The main difficulty in studying alt-NHEJ is the dominant activity of c-NHEJ, reflected by the fact that under normal circumstances, only a small proportion of DSBs are repaired with the use of alt-NHEJ (Chiruvella et al. 2013). The propensity of alt-NHEJ to promote chromosomal translocations is currently being explained by its low efficiency of end tethering (c-NHEJ has evolved more sophisticated end synapsis/alignment machinery), increased free DNA end movement in the nucleus or slow DNA repair (with an increased chance for linking incorrect ends).

Recently a polymerase specifically involved in alt-NHEJ has been characterised as polymerase theta (pol τ; encoded by the *POLQ* gene) (Wood and Doublie 2016). The pol theta protein belongs to the class A mammalian polymerases and is highly unusual in that it possesses an N-terminal helicase domain. The pol theta C-terminal polymerase domain is separated from the helicase domain by a large central region of unknown function. Unlike most polymerases, pol theta can extend ssDNA substrates. The mechanism by which pol theta can repair DSBs requires two steps: initial microhomology-independent DNA synapse formation and subsequent microhomology-based annealing, which enhances the base pairing between the ends, leading to efficient overhang extension (Kent et al. 2015). Thus, in alt-NHEJ repair, a single protein can both stabilise DNA ends and process them in a way that makes them suitable for subsequent further processing and ligation. Although pol theta is involved in alt-NHEJ, it appears to be required for a specific subset of DSBs repaired through this sub-pathway (Wyatt et al. 2016). In contrast to typical alt-NHEJ, pol theta supresses chromosomal translocations, rather than contributing to their formation. Another remarkable ability of polymerase theta is that it can join ends derived from collapsed replication forks and thus constitutes a backup pathway to HDR (Wyatt et al. 2016). Interestingly, HDR-deficient cancers become strongly dependent on pol theta for survival after DSB induction (Mateos-Gomez et al. 2015).

V(D)J RECOMBINATION

V(D)J recombination is a process of somatic DNA recombination restricted to developing lymphocytes (T and B cells), which serves to create a diverse repertoire of antigen receptors enabling the response of the immune system to virtually any foreign antigen (de Villartay et al. 2003). Immune antigen receptors (TCR and BCR) are encoded in the germ line as highly unusual non-functional separate gene segments. Antigen binding domains are encoded within these receptor loci as arrays of variable (V), diversity (D; not present in light immunoglobulin chains) and joining (J) gene segments. Each class of segments is represented by multiple variant copies (**Figure 7.5**). In the process of V(D)J recombination V, D and J segments are brought into close proximity, and this results in an assembly of a complete receptor gene, where each lymphocyte expresses a receptor of unique specificity. During lymphocyte differentiation, these genes become transcriptionally active and their chromatin undergoes remodelling to an open state to allow access to the recombination machinery. The process of V(D)J recombination is regulated by restricted lymphocyte-specific expression of recombination proteins RAG1 and

RAG2, as well as by accessibility of RAG1/2 to so-called RSS (recombination signal sequences). V(D)J recombination is initiated at the pre-T/pre-B cell stage of differentiation when the RAG1/2 recombinase complex binds in a sequence-specific fashion to RSS present at antigen receptor segment junctions. RSS consists of heptamer or nonamer consensus sequences separated by 12 or 23 bp non-conserved spacer sequences. For the recombination to occur, two RSS sites (one 12 bp and one 23 bp; 12/23 rule) are required. Cleavage at a pair of RSS sequences produces two DSBs (four DNA ends) followed by NHEJ-dependent repair, linking a given V segment to a J segment (or V segment to a prearranged DJ segment). Two DNA coding ends emerging from RAG cleavage are covalently sealed, forming terminal hairpin structures. Two other ends which terminate at RSS are called signal ends and are joined by the NHEJ machinery to create an extra-chromosomal excised circle (Schatz and Ji 2011). Coding ends terminated with hairpins are opened by the DNA-PKcs/Artemis nuclease to leave a 4 bp 3′ overhang and are eventually joined to create a so-called coding join (Ma et al. 2002) (Figure 7.5). Terminal deoxynucleotidyl transferase (TdT) acts on released coding DNA ends after Artemis-dependent cleavage but prior to NHEJ-dependent ligation to insert random non-template nucleotides (N-nucleotides). Furthermore, other NHEJ-dependent polymerases mu and lambda are able to perform fill-in reactions at these ends to increase the junctional diversity of coding joints (Lieber 2006). Most coding joints do not show significant terminal microhomology. Apart from the induction of DSBs, RAG1/2 recombinase is also responsible for tethering of DNA ends to assist the NHEJ machinery in correctly linking DNA ends. Furthermore, RAG1/2 is somehow able to direct the repair to c-NHEJ at the expense of other competing pathways such as alt-NHEJ (Deriano et al. 2011). This ability of RAG1/2 might be facilitated by physical interaction between RAGs and the Ku complex. Despite such an extensive regulation, it has become increasingly clear that errors occurring during execution of V(D)J recombination (e.g. chromosomal translocations) can have severely deleterious consequences such as development of lymphomas and leukaemias (Roth 2014). Most core c-NHEJ factors are essential for V(D)J recombination (DNA-Pkcs and Artemis knockout animals have a selective defect in coding end formation), with the exception of XLF and PAXX (Table 7.1) (Blackford and Jackson 2017). It is currently hypothesised that XLF function can be compensated by PAXX and vice versa (see later). Interestingly, however, the ATM-H2A.X axis and DNA-PKcs are all capable of providing compensatory activities in the absence of XLF, as mice deficient in XLF and ATM or H2A.X are severely compromised in signal end generation (Zha et al. 2011). Curiously, double ATM/PAXX knockout mice do not show a V(D)J recombination defect arguing for both specific and distinct functions provided by XLF and PAXX in c-NHEJ execution (Balmus et al. 2016). Similarly, ATM/DNA-PKcs double knockout mice show defective signal end formation, demonstrating a functional

Table 7.1 Phenotypes associated with c-NHEJ factor loss in mice/humans and the process of V(D)J and CSR somatic recombination.

Protein	Mouse knockout (ko) phenotype	V(D)J recombination (in mice)	Class switch recombination (in mice)	Human patients
Ku70	SCID, growth retardation	Profound defect	Profound defect	Not found
Ku86	SCID, growth retardation	Almost complete abrogation	Profound defect	Not found
DNA-PKcs	SCID	Almost complete abrogation	Mild defect	RS-SCID, microcephaly
Artemis	SCID	Profound defect	Mild defect	RS-SCID
XLF	Mild radiosensitivity (XLF/PAXX double ko: embryonic lethal, neuronal apoptosis)	Mild defect (in XLF/PAXX double ko: profound defect)	Mild defect	Immunodeficiency, radiosensitivity microcephaly
PAXX	Mild radiosensitivity (XLF/PAXX double ko: embryonic lethal, neuronal apoptosis)	No defect (in XLF/PAXX double ko: profound defect)	Not determined	Not found
XRCC4	Embryonic lethal, neuronal apoptosis	Almost complete abrogation	Mild defect	Growth retardation, radiosensitivity microcephaly,
Ligase IV	Embryonic lethal, neuronal apoptosis	Almost complete abrogation	Mild defect	Immunodeficiency, growth retardation, radiosensitivity, microcephaly

Note: Mouse ko phenotypes are graded from least severe (no defect) to intermediate (mild or profound defect) to most severe (almost complete abrogation).

overlap between these DNA-activated kinases in somatic recombination (Gapud and Sleckman 2011). In summary, analysis of c-NHEJ mutant mice has demonstrated specific functions for components of this pathway in V(D)J recombination but also uncovered a substantial level of redundancy, especially at the level of DNA and tethering and alignment.

CLASS-SWITCH RECOMBINATION

Antibodies secreted by activated B-cells are central mediators of humoral immunity and act by neutralising pathogens and facilitating recruitment of cells responsible for cellular immunity (Schroeder and Cavacini 2010). Antibodies exist as several functional classes (IgM, IgD, IgG, IgA and IgE) characterised by distinct constant regions of their heavy chains. Each immunoglobulin class has a specific effector function, tissue distribution and efficacy against pathogens. Naïve B cells express exclusively IgM or IgD isotypes. However, IgMs are pentameric and, owing to their large size, are largely ineffective in passing into the extravascular space. Thus monomeric IgGs, IgEs and monomeric/dimeric IgAs can be systemically distributed to various tissues to mediate specialised biological effector functions. Switching expression from IgM/IgD to IgG/IgE or IgA is achieved through the second lymphocyte-specific somatic DNA recombination process known as class switch recombination (Xu et al. 2012). Constant regions of different immunoglobulin isotypes are encoded by distinct C_H exons, which are present in 3′ position to the variable region (Figure 7.5). CSR results in the replacement of the default C_H encoding exon (e.g. C_m for IgM) for $C\gamma$, $C\alpha$ or $C\varepsilon$ (giving rise to IgG, IgA or IgE) in such a way that the antigen-specific variable region remains unaltered (Figure 7.5). At the molecular level, CSR is initiated by introduction of DSBs in so-called switch regions located between DNA sequences encoding C_H effector cassettes. DSBs in switch regions arise indirectly as a by-product of the action of the AID (activation-induced cytidine deaminase) enzyme. AID deaminates cytidine residues at switch regions leading to an appearance of deoxyuracil in DNA, which as an unnatural base is recognised as a DNA lesion by the base-excision repair machinery. Excision of uracil by BER generates single-stranded DNA breaks. A high density of SSBs occurring in close proximity produces staggered DSBs (Chaudhuri and Alt 2004). AID preferentially acts on ssDNA, which forms opposite to R-loops at switch regions. R-loops occur between the DNA template and the nascent RNA transcript derived from the transcribed C_H locus. Thus, the specific mechanism of DSB generation during CSR differs substantially from the V(D)J recombination reactions reflecting the requirement for transcription through the switch region and a range of additional factors responsible for execution of all of the steps leading to DSB generation. Thus, strictly speaking, DSB induction in CSR is both indirect and not entirely sequence-specific, although it is believed that the majority of the AID activity is directed to switch regions. Similar to the V(D)J process in CSR, there is a need to join distant DNA fragments (Figure 7.5). However, unlike in V(D)J recombination, CSR does not employ a specialised sequence-specific complex for DNA end tethering and thus requires a special type of DNA end alignment. This appears to be facilitated by factors related to the ATM-dependent DDR activation axis at DSBs (such as H2AX, 53BP1 and its effector Rif1), which are prominently involved in CSR, while showing less prominent contribution to V(D) J recombination (Reina-San-Martin et al. 2003, Manis et al. 2004). Furthermore, CSR, unlike V(D)J recombination, is substantially dependent on the MRN complex (Dinkelmann et al. 2009) and the alt-NHEJ machinery (Frit et al. 2014). In summary, two lymphocyte-specific somatic recombination processes commonly involve programmed DSB generation in defined chromosomal locations. Both processes utilise cellular generic DNA repair machinery for the repair of physiological DSBs, although they show a distinct preference for repair factor utilisation, reflecting different mechanisms of DSB induction and repair.

PHENOTYPES ASSOCIATED WITH NHEJ DEFICIENCY IN MICE

Our knowledge of the physiological role of NHEJ proteins has been greatly enhanced by the generation and characterisation of mouse mutants bearing mutations in NHEJ factor encoding genes. Given the important functions of NHEJ proteins in somatic recombination of developing lymphocytes, it is not surprising that deficient lymphocyte development and function are hallmarks of NHEJ mutant mice (Table 7.1). Knockout mice for Ku subunits, DNA-PKcs and Artemis all succumb to severe-combined immunodeficiency (SCID), characterised by an almost complete absence of mature T and B cells (Zhu et al. 1996, Gu et al. 1997, Gao et al. 1998a, Rooney et al. 2002). Ku knockout mice additionally show growth retardation. Interestingly, while Ku-deficient lymphocytes fail to generate both signal and coding joints during V(D)J recombination, DNA-PKcs and Artemis mutants are specifically defective in coding joint formation. This presumably reflects a specific hairpin opening activity of the DNA-PKcs/Artemis complex, which is not compensated by any other nuclease. Residual hairpin opening can be detected in the absence of Artemis. However, joints formed in this way contain so-called P-nucleotides (palindromic) caused by aberrant opening of the hairpin away from the apex. As mentioned earlier, frequency of P-nucleotide insertions can be used as a surrogate measure of the contribution of Artemis to hairpin opening on a given genetic background (van der Burg et al. 2009). p53 transcription factor is essential for apoptosis induction by persistent DSBs. Thus, lymphocyte survival (even after failed V(D)J recombination) can be rescued in Ku or DNA-PKcs knockout mice when crossed with p53 mutant animals. Although substantial numbers of mature lymphocytes can be detected in double knockouts, these animals invariably suffer from various lymphoid tumours such as lymphomas (Rooney et al. 2004). It is assumed that persistent DSBs in such animals give rise to genomic instability leading to tumorigenesis. XRCC4 and ligase IV knockouts show the most dramatic phenotype of all c-NHEJ knockout animals. These mice are not viable and present a growth defect and apoptosis of developing neurons in the embryonic brain (Frank et al. 1998, Gao et al. 1998b). It is hypothesised that the severe phenotype of XRCC4 and ligase IV knockout animals is a consequence of Ku being trapped on DSBs obstructing repair by alternative pathways. Consistent with this scenario, embryonic lethality of ligase IV mutant mice can be rescued by crossing to a Ku-deficient strain (Karanjawala et al. 2002). Furthermore, XRCC4 (or ligase IV) knockout animals survive when crossed to p53 mutant mice; however, they develop a SCID phenotype and lymphoid tumours resembling the phenotype observed in Ku and DNA-PKcs mutants (Gao et al. 2000). In contrast to XRCC4/ligase IV deficiency, XLF-deficient animals have a surprisingly mild phenotype (G. Li et al. 2008). Although cells derived from XLF mutant animals are radiosensitive, the development of lymphocytes proceeds almost normally (there is a mild reduction in the overall numbers of lymphocytes in these mice). An initial explanation for this phenotype was that XLF function could be partially taken over by other factors, such as components of the H2A.X-ATM axis (Zha et al. 2011). Strikingly, soon after the discovery of PAXX in 2015, the mystery of this mild phenotype of XLF knockout mice was resolved by showing that combined XLF/PAXX double knockouts developed a phenotype resembling XRCC4 or ligase IV–deficient animals (Balmus et al. 2016). Thus XLF/PAXX double knockout animals show embryonic lethality, growth retardation, an absence of lymphocytes and induction of apoptosis in the brain (PAXX single mutant mice show mild radiation sensitivity but are otherwise normal). Notably, fibroblasts derived from any of the available NHEJ knockout animals uniformly show enhanced radiosensitivity. Collectively, data obtained from the analysis of NHEJ-deficient animals have proven a central role for this pathway in repair of both pathological and physiological DNA double-strand breaks.

PHENOTYPES ASSOCIATED WITH NHEJ DEFICIENCY IN HUMANS

One of the most successful routes to the identification of individuals carrying NHEJ mutations has been the effort to characterise genes mutated in patients suffering from severe immunodeficiencies (Woodbine et al. 2014). Patients deficient in NHEJ display both radiosensitivity (RS) and severe combined immunodeficiency, a phenotype that has been described as RS-SCID (Noordzij et al. 2003). This is in contrast to individuals suffering from SCID, unrelated to NHEJ, such as patients with RAG mutations, who are not radiosensitive (Tasher and Dalal 2012). Given the severe consequences of NHEJ deficiency, it is not surprising that mutations in NHEJ factor-encoding genes found in humans typically present a partial loss of function (hypomorphic mutations). In many cases, the severity of the phenotype is directly linked to the impact of the mutation on DSB repair efficiency, with a mild DSB repair defect being mirrored by relatively minor clinical phenotypical manifestations (Woodbine et al. 2014). Mutations in the Artemis-encoding gene are among the most common causes of RS-SCID in humans (Moshous et al. 2001). Most of these patients show genomic deletions, resulting in the absence of transcription of Artemis-encoding mRNA. These Artemis-null patients present an SCID phenotype, necessitating bone marrow transplantation for survival (Pannicke et al. 2010). Another class of patients with Artemis mutations are those displaying a so-called leaky Artemis deficiency characterised by various point mutations affecting Artemis function to a variable degree. Such patients frequently display progressive immunodeficiency as well as EBV-associated lymphomas (Moshous et al. 2003).

DNA-PKcs loss in mice is similar to Artemis knockouts in that both animal models do not show any additional phenotype beyond the SCID. Therefore, it was surprising that for many years no DNA-PKcs mutant patients had been described. In 2009, however, the first individual with a DNA-PKcs mutation was reported with features similar to Artemis deficiency (van der Burg et al. 2009). That patient carried a missense DNA-PKcs mutation (L3062R), which did not impair the expression or activity of the protein. However, sequencing of bone marrow–derived V(D)J coding joints revealed an increased number of P-nucleotide insertions, characteristic for impaired Artemis function. More recently, another patient with DNA-PKcs deficiency was identified, showing defective neuronal development. In this case two heterozygous mutations in DNA-PKcs–encoding genes were found to result in markedly reduced expression of DNA-PKcs protein. Thus, perhaps surprisingly, DNA-PKcs appears to have a major role in human neuronal development (Woodbine et al. 2013).

LigIV syndrome, caused by mutations in DNA ligase IV, was among the first RS-SCID disorders to be reported (O'Driscoll et al. 2001). All ligase IV mutations found to date in humans are hypomorphic (giving rise to a partial loss of function). Common features of ligase IV deficiency include immunodeficiency, characteristic dimorphic facial features, microcephaly (small head size) and developmental delay. Interestingly, microcephaly observed in ligase IV syndrome patients is present at birth but is not progressive (Buck et al. 2006b, van der Burg et al. 2006).

Similarly to DNA-PKcs, the search for XRCC4 mutant individuals went on for a number of years until in 2015 several XRCC4 mutant patients were identified. Although ligase IV, Artemis and DNA-PKcs mutations in humans are associated with SCID, XRCC4 mutations mostly produce growth retardation (primordial dwarfism), microcephaly and radiosensitivity (Murray et al. 2015, Rosin et al. 2015, Saito et al. 2016). Lack of obvious immunodeficiency in XRCC4 mutant patients is surprising (de Villartay 2015). It is possible that as identified XRCC4 mutations show diminished, but not absent protein expression, the residual level of XRCC4 protein is sufficient to permit its contribution to somatic recombination reactions.

XLF was originally discovered as a gene mutated in a cohort of patients displaying growth retardation, microcephaly, immunodeficiency and cellular radiosensitivity (Revy et al. 2006). Most mutations found in XLF-deficient patients appear to strongly compromise its activity. XLF in general appears not to be essential for NHEJ reactions, although its absence profoundly affects the efficiency of NHEJ (Riballo et al. 2009).

It is worth mentioning that, unlike rodents, in human cells, Ku proteins are essential and DNA-PKcs loss is associated with severe proliferative defects (Li et al. 2002, Ruis et al. 2008). Furthermore, expression levels of DNA-PK in human cells are substantially elevated in comparison to rodent cells (Raschella et al. 2017). These observations probably reflect additional important functions which these proteins have acquired during the evolution of primates, such as the involvement of human Ku in telomere maintenance (Wang et al. 2009, Huang et al. 2014) and the contribution of DNA-PKcs to progression through mitosis (Huang et al. 2014). These additional critical functions acquired by Ku and DNA-PKcs in human cells do not appear to be specifically related to c-NHEJ, however.

Collectively, data obtained from the analysis of human patients with NHEJ mutations have revealed a strong contribution of this pathway to radiosensitivity and the development and functioning of the human immune system and the brain (Table 7.1). These observations are consistent with data derived from studies on mutant mice, although the contribution of c-NHEJ core factors to human development is more pronounced in comparison to mice. To date, no mutation in Ku or PAXX-encoding genes has been found in humans.

NHEJ IN YEAST AND BACTERIA

Unicellular organisms such as yeast and bacteria are capable of performing NHEJ-type reactions, although their NHEJ machinery is simplified compared to vertebrates. *S. cerevisiae* possesses homologues of mammalian c-NHEJ core factors such as Ku (Yku70/80) and XRCC4/XLF/ligase IV (Dnl4/Nej1/Lif1) (Daley et al. 2005, Callebaut et al. 2006). These proteins constitute the basic set of c-NHEJ components in yeast, as DNA-PKcs, Artemis and PAXX homologues are not present. Yeast cells also employ one type of X polymerase enzyme Pol4, which is a homologue of the mammalian polymerases lambda and mu (Pardo et al. 2006). In yeast, the MRX complex, which is the functional counterpart of mammalian MRN, is required for NHEJ (Boulton and Jackson 1998). Interestingly, yeast can also perform alt-NHEJ reactions; however, yeast alt-NHEJ requires longer microhomology tracks (>6 bp)

(Daley and Wilson 2005). This notable difference in comparison to vertebrate alt-NHEJ is believed to reflect a lack of DNA end annealing factors such as PARP-1 and polymerase theta in yeast. As a result, the efficiency of alt-NHEJ in yeast strongly depends on the stability of base pairing between the DNA ends destined for ligation. The bacterial NHEJ process is further simplified, as such cells typically only have one Ku-like molecule (acting as a homodimer) and a multifunctional ligase enzyme capable of end processing and ligation (Weller et al. 2002, Shuman and Glickman 2007).

CONCLUSIONS

Over the last 30 years, extensive research into mechanisms of NHEJ has greatly improved our understanding of the way this DSB repair pathway operates. Although the main proteins involved in NHEJ have already been characterised, there is a need for better understanding of the regulation of NHEJ throughout the cell cycle, during organismal ageing and across different specialised tissues. In addition, several post-translational modifications (e.g. phosphorylation, ubiquitination and poly-ADP-ribosylation) of key NHEJ proteins have been described; however, their functional consequences are unclear and the enzymes responsible for these modifications are, for the most part, unknown. Furthermore, although NHEJ components are, in principle, excellent targets for cancer therapies, few successful attempts have been described involving pharmacological modulation of NHEJ factors. Some promising specific NHEJ inhibitors failed due to their in vivo toxicity (e.g. inhibitors of DNA-PKcs). Clearly more work is needed to realise the full potential of NHEJ modulation in cancer therapy. Given that multiple DNA damage sensors are capable of DSB recognition in a competitive and cooperative fashion, another area for further investigation is the question of repair pathway choice. For example, the determinants of DSB pathway choice and how the choice of particular repair pathway is regulated have not been fully clarified and will be a subject of intensive future investigations.

REFERENCES

Aceytuno, R. D., Piett, C. G., Havali-Shahriari, Z., Edwards, R. A., Rey, M., Ye, R., Javed, F. et al. 2017. Structural and functional characterization of the PNKP-XRCC4-LigIV DNA repair complex. *Nucleic Acids Res* 45(10): 6238–6251.

Ahel, I., Rass, U., El-Khamisy, S. F., Katyal, S., Clements, P. M., McKinnon, P. J., Caldecott, K. W., West, S. C. 2006. The neurodegenerative disease protein aprataxin resolves abortive DNA ligation intermediates. *Nature* 443(7112): 713–716.

Ahnesorg, P., Smith, P., Jackson, S. P. 2006. XLF interacts with the XRCC4-DNA ligase IV complex to promote DNA nonhomologous end-joining. *Cell* 124(2): 301–313.

Allerston, C. K., Lee, S. Y., Newman, J. A., Schofield, C. J., McHugh, P. J., Gileadi, O. 2015. The structures of the SNM1A and SNM1B/Apollo nuclease domains reveal a potential basis for their distinct DNA processing activities. *Nucleic Acids Res* 43(22): 11047–11060.

Avemann, K., Knippers, R., Koller, T., Sogo, J. M. 1988. Camptothecin, a specific inhibitor of type I DNA topoisomerase, induces DNA breakage at replication forks. *Mol Cell Biol* 8(8): 3026–3034.

Balmus, G., Barros, A. C., Wijnhoven, P. W., Lescale, C., Hasse, H. L., Boroviak, K., le Sage, C. et al. 2016. Synthetic lethality between PAXX and XLF in mammalian development. *Genes Dev* 30(19): 2152–2157.

Bennardo, N., Cheng, A., Huang, N., Stark, J. M. 2008. Alternative-NHEJ is a mechanistically distinct pathway of mammalian chromosome break repair. *PLoS Genet* 4(6): e1000110.

Berkovich, E., Monnat, R. J. Jr., Kastan, M. B. 2007. Roles of ATM and NBS1 in chromatin structure modulation and DNA double-strand break repair. *Nat Cell Biol* 9(6): 683–690.

Blackford, A. N., Jackson, S. P. 2017. ATM, ATR, and DNA-PK: The trinity at the heart of the DNA damage response. *Mol Cell* 66(6): 801–817.

Boboila, C., Alt, F. W., Schwer, B. 2012. Classical and alternative end-joining pathways for repair of lymphocyte-specific and general DNA double-strand breaks. *Adv Immunol* 116: 1–49.

Boulton, S. J., Jackson, S. P. 1998. Components of the Ku-dependent non-homologous end-joining pathway are involved in telomeric length maintenance and telomeric silencing. *EMBO J* 17(6): 1819–1828.

Britton, S., Coates, J., Jackson, S. P. 2013. A new method for high-resolution imaging of Ku foci to decipher mechanisms of DNA double-strand break repair. *J Cell Biol* 202(3): 579–595.

Brouwer, I., Sitters, G., Candelli, A., Heerema, S. J., Heller, I., de Melo, A. J., Zhang, H. et al. 2016. Sliding sleeves of XRCC4-XLF bridge DNA and connect fragments of broken DNA. *Nature* 535(7613): 566–569.

Buck, D., Malivert, L., de Chasseval, R., Barraud, A., Fondaneche, M. C., Sanal, O., Plebani, A. et al. 2006a. Cernunnos, a novel nonhomologous end-joining factor, is mutated in human immunodeficiency with microcephaly. *Cell* 124(2): 287–299.

Buck, D., Moshous, D., de Chasseval, R., Ma, Y., le Deist, F., Cavazzana-Calvo, M., Fischer, A., Casanova, J. L., Lieber, M. R., de Villartay, J. P. 2006b. Severe combined immunodeficiency and microcephaly in siblings with hypomorphic mutations in DNA ligase IV. *Eur J Immunol* 36(1): 224–235.

Callebaut, I., Malivert, L., Fischer, A., Mornon, J. P., Revy, P., de Villartay, J. P. 2006. Cernunnos interacts with the XRCC4 × DNA-ligase IV complex and is homologous to the yeast nonhomologous end-joining factor Nej1. *J Biol Chem* 281(20): 13857–13860.

Ceccaldi, R., Rondinelli, B., D'Andrea, A. D. 2016. Repair pathway choices and consequences at the double-strand break. *Trends Cell Biol* 26(1): 52–64.

Chan, D. W., Chen, B. P., Prithivirajsingh, S., Kurimasa, A., Story, M. D., Qin, J., Chen, D. J. 2002. Autophosphorylation of the DNA-dependent protein kinase catalytic subunit is required for rejoining of DNA double-strand breaks. *Genes Dev* 16(18): 2333–2338.

Chang, H. H., Lieber, M. R. 2016. Structure-specific nuclease activities of Artemis and the Artemis: DNA-PKcs complex. *Nucleic Acids Res* 44(11): 4991–4997.

Chang, H. H. Y., Pannunzio, N. R., Adachi, N., Lieber, M. R. 2017. Non-homologous DNA end joining and alternative pathways to double-strand break repair. *Nat Rev Mol Cell Biol* 18(8): 495–506.

Chaudhuri, J., Alt, F. W. 2004. Class-switch recombination: Interplay of transcription, DNA deamination and DNA repair. *Nat Rev Immunol* 4(7): 541–552.

Cherry, A. L., Nott, T. J., Kelly, G., Rulten, S. L., Caldecott, K. W., Smerdon, S. J. 2015. Versatility in phospho-dependent molecular recognition of the XRCC1 and XRCC4 DNA-damage scaffolds by aprataxin-family FHA domains. *DNA Repair (Amst)* 35: 116–125.

Chiruvella, K. K., Liang, Z., Wilson, T. E. 2013. Repair of double-strand breaks by end joining. *Cold Spring Harb Perspect Biol* 5(5): a012757.

Ciccia, A., Elledge, S. J. 2010. The DNA damage response: Making it safe to play with knives. *Mol Cell* 40(2): 179–204.

Craxton, A., Somers, J., Munnur, D., Jukes-Jones, R., Cain, K., Malewicz, M. 2015. XLS (c9orf142) is a new component of mammalian DNA double-stranded break repair. *Cell Death Differ* 22(6): 890–897.

Cui, X., Yu, Y., Gupta, S., Cho, Y. M., Lees-Miller, S. P., Meek, K. 2005. Autophosphorylation of DNA-dependent protein kinase regulates DNA end processing and may also alter double-strand break repair pathway choice. *Mol Cell Biol* 25(24): 10842–10852.

Daley, J. M., Niu, H., Miller, A. S., Sung, P. 2015. Biochemical mechanism of DSB end resection and its regulation. *DNA Repair (Amst)* 32: 66–74.

Daley, J. M., Palmbos, P. L., Wu, D., Wilson, T. E. 2005. Nonhomologous end joining in yeast. *Annu Rev Genet* 39: 431–451.

Daley, J. M., Wilson, T. E. 2005. Rejoining of DNA double-strand breaks as a function of overhang length. *Mol Cell Biol* 25(3): 896–906.

Dantuma, N. P., van Attikum, H. 2016. Spatiotemporal regulation of posttranslational modifications in the DNA damage response. *EMBO J* 35(1): 6–23.

Davis, A. J., Chen, D. J. 2013. DNA double strand break repair via non-homologous end-joining. *Transl Cancer Res* 2(3): 130–143.

Deriano, L., Chaumeil, J., Coussens, M., Multani, A., Chou, Y., Alekseyenko, A. V., Chang, S., Skok, J. A., Roth, D. B. 2011. The RAG2 C terminus suppresses genomic instability and lymphomagenesis. *Nature* 471(7336): 119–123.

Deriano, L., Roth, D. B. 2013. Modernizing the nonhomologous end-joining repertoire: Alternative and classical NHEJ share the stage. *Annu Rev Genet* 47: 433–455.

de Villartay, J. P. 2015. When natural mutants do not fit our expectations: The intriguing case of patients with XRCC4 mutations revealed by whole-exome sequencing. *EMBO Mol Med* 7(7): 862–864.

de Villartay, J. P., Fischer, A., Durandy, A. 2003. The mechanisms of immune diversification and their disorders. *Nat Rev Immunol* 3(12): 962–972.

Dinkelmann, M., Spehalski, E., Stoneham, T., Buis, J., Wu, Y., Sekiguchi, J. M., Ferguson, D. O. 2009. Multiple functions of MRN in end-joining pathways during isotype class switching. *Nat Struct Mol Biol* 16(8): 808–813.

Dobbs, T. A., Tainer, J. A., Lees-Miller, S. P. 2010. A structural model for regulation of NHEJ by DNA-PKcs autophosphorylation. *DNA Repair (Amst)* 9(12): 1307–1314.

Downs, J. A., Jackson, S. P. 2004. A means to a DNA end: The many roles of Ku. *Nat Rev Mol Cell Biol* 5(5): 367–378.

Dudley, D. D., Chaudhuri, J., Bassing, C. H., Alt, F. W. 2005. Mechanism and control of V(D)J recombination versus class switch recombination: Similarities and differences. *Adv Immunol* 86: 43–112.

Fernandez-Capetillo, O., Celeste, A., Nussenzweig, A. 2003. Focusing on foci: H2AX and the recruitment of DNA-damage response factors. *Cell Cycle* 2(5): 426–427.

Fnu, S., Williamson, E. A., De Haro, L. P., Brenneman, M., Wray, J., Shaheen, M., Radhakrishnan, K., Lee, S. H., Nickoloff, J. A., Hromas, R. 2011. Methylation of histone H3 lysine 36 enhances DNA repair by nonhomologous end-joining. *Proc Natl Acad Sci USA* 108(2): 540–545.

Frank, K. M., Sekiguchi, J. M., Seidl, K. J., Swat, W., Rathbun, G. A., Cheng, H. L., Davidson, L., Kangaloo, L., Alt, F. W. 1998. Late embryonic lethality and impaired V(D)J recombination in mice lacking DNA ligase IV. *Nature* 396(6707): 173–177.

Frit, P., Barboule, N., Yuan, Y., Gomez, D., Calsou, P. 2014. Alternative end-joining pathway(s): Bricolage at DNA breaks. *DNA Repair (Amst)* 17: 81–97.

Froelich-Ammon, S. J., Osheroff, N. 1995. Topoisomerase poisons: Harnessing the dark side of enzyme mechanism. *J Biol Chem* 270(37): 21429–21432.

Gao, Y., Chaudhuri, J., Zhu, C., Davidson, L., Weaver, D. T., Alt, F. W. 1998a. A targeted DNA-PKcs-null mutation reveals DNA-PK-independent functions for KU in V(D)J recombination. *Immunity* 9(3): 367–376.

Gao, Y., Ferguson, D. O., Xie, W., Manis, J. P., Sekiguchi, J., Frank, K. M., Chaudhuri, J., Horner, J., DePinho, R. A., Alt, F. W. 2000. Interplay of p53 and DNA-repair protein XRCC4 in tumorigenesis, genomic stability and development. *Nature* 404(6780): 897–900.

Gao, Y., Sun, Y., Frank, K. M., Dikkes, P., Fujiwara, Y., Seidl, K. J., Sekiguchi, J. M. et al. 1998b. A critical role for DNA end-joining proteins in both lymphogenesis and neurogenesis. *Cell* 95(7): 891–902.

Gapud, E. J., Sleckman, B. P. 2011. Unique and redundant functions of ATM and DNA-PKcs during V(D)J recombination. *Cell Cycle* 10(12): 1928–1935.

Goodarzi, A. A., Noon, A. T., Deckbar, D., Ziv, Y., Shiloh, Y., Lobrich, M., Jeggo, P. A. 2008. ATM signaling facilitates repair of DNA double-strand breaks associated with heterochromatin. *Mol Cell* 31(2): 167–177.

Goodarzi, A. A., Yu, Y., Riballo, E., Douglas, P., Walker, S. A., Ye, R., Harer, C. et al. 2006. DNA-PK autophosphorylation facilitates Artemis endonuclease activity. *EMBO J* 25(16): 3880–3889.

Gottlieb, T. M., Jackson, S. P. 1993. The DNA-dependent protein kinase: Requirement for DNA ends and association with Ku antigen. *Cell* 72(1): 131–142.

Graham, T. G., Walter, J. C., Loparo, J. J. 2016. Two-stage synapsis of DNA ends during non-homologous end joining. *Mol Cell* 61(6): 850–858.

Grundy, G. J., Rulten, S. L., Arribas-Bosacoma, R., Davidson, K., Kozik, Z., Oliver, A. W., Pearl, L. H., Caldecott, K. W. 2016. The Ku-binding motif is a conserved module for recruitment and stimulation of non-homologous end-joining proteins. *Nat Commun* 7: 11242.

Grundy, G. J., Rulten, S. L., Zeng, Z., Arribas-Bosacoma, R., Iles, N., Manley, K., Oliver, A., Caldecott, K. W. 2013. APLF promotes the assembly and activity of non-homologous end joining protein complexes. *EMBO J* 32(1): 112–11125.

Gu, Y., Seidl, K. J., Rathbun, G. A., Zhu, C., Manis, J. P., van der Stoep, N., Davidson, L. et al. 1997. Growth retardation and leaky SCID phenotype of Ku70-deficient mice. *Immunity* 7(5): 653–665.

Hammel, M., Rey, M., Yu, Y., Mani, R. S., Classen, S., Liu, M., Pique, M. E. et al. 2011. XRCC4 protein interactions with XRCC4-like factor (XLF) create an extended grooved scaffold for DNA ligation and double strand break repair. *J Biol Chem* 286(37): 32638–32650.

Hammel, M., Yu, Y., Mahaney, B. L., Cai, B., Ye, R., Phipps, B. M., Rambo, R. P. et al. 2010. Ku and DNA-dependent protein kinase dynamic conformations and assembly regulate DNA binding and the initial non-homologous end joining complex. *J Biol Chem* 285(2): 1414–1423.

Harrison, M. M., Jenkins, B. V., O'Connor-Giles, K. M., Wildonger, J. 2014. A CRISPR view of development. *Genes Dev* 28(17): 1859–1872.

Hasham, M. G., Donghia, N. M., Coffey, E., Maynard, J., Snow, K. J., Ames, J., Wilpan, R. Y., He, Y., King, B. L., Mills, K. D. 2010. Widespread genomic breaks generated by activation-induced cytidine deaminase are prevented by homologous recombination. *Nat Immunol* 11(9): 820–826.

Huang, B., Shang, Z. F., Li, B., Wang, Y., Liu, X. D., Zhang, S. M., Guan, H., Rang, W. Q., Hu, J. A., Zhou, P. K. 2014. DNA-PKcs associates with PLK1 and is involved in proper chromosome segregation and cytokinesis. *J Cell Biochem* 115(6): 1077–1088.

Huertas, P. 2010. DNA resection in eukaryotes: Deciding how to fix the break. *Nat Struct Mol Biol* 17(1): 11–16.

Huertas, P., Jackson, S. P. 2009. Human CtIP mediates cell cycle control of DNA end resection and double strand break repair. *J Biol Chem* 284(14): 9558–9565.

Hustedt, N., Durocher, D. 2016. The control of DNA repair by the cell cycle. *Nat Cell Biol* 19(1): 1–9.

Ijspeert, H., Rozmus, J., Schwarz, K., Warren, R. L., van Zessen, D., Holt, R. A., Pico-Knijnenburg, I. et al. 2016. XLF deficiency results in reduced N-nucleotide addition during V(D)J recombination. *Blood* 128(5): 650–659.

Jackson, S. P., Bartek, J. 2009. The DNA-damage response in human biology and disease. *Nature* 461(7267): 1071–1078.

Jena, N. R. 2012. DNA damage by reactive species: Mechanisms, mutation and repair. *J Biosci* 37(3): 503–517.

Jette, N., Lees-Miller, S. P. 2015. The DNA-dependent protein kinase: A multifunctional protein kinase with roles in DNA double strand break repair and mitosis. *Prog Biophys Mol Biol* 117(2–3): 194–205.

Junop, M. S., Modesti, M., Guarne, A., Ghirlando, R., Gellert, M., Yang, W. 2000. Crystal structure of the Xrcc4 DNA repair protein and implications for end joining. *EMBO J* 19(22): 5962–5970.

Karanjawala, Z. E., Adachi, N., Irvine, R. A., Oh, E. K., Shibata, D., Schwarz, K., Hsieh, C. L., Lieber, M. R. 2002. The embryonic lethality in DNA ligase IV-deficient mice is rescued by deletion of Ku: Implications for unifying the heterogeneous phenotypes of NHEJ mutants. *DNA Repair (Amst)* 1(12): 1017–1026.

Kent, T., Chandramouly, G., McDevitt, S. M., Ozdemir, A. Y., Pomerantz, R. T. 2015. Mechanism of microhomology-mediated end-joining promoted by human DNA polymerase theta. *Nat Struct Mol Biol* 22(3): 230–237.

Kim, J. A., Kruhlak, M., Dotiwala, F., Nussenzweig, A., Haber, J. E. 2007. Heterochromatin is refractory to gamma-H2AX modification in yeast and mammals. *J Cell Biol* 178(2): 209–218.

Kumar, V., Alt, F. W., Frock, R. L. 2016. PAXX and XLF DNA repair factors are functionally redundant in joining DNA breaks in a G1-arrested progenitor B-cell line. *Proc Natl Acad Sci USA* 113(38): 10619–10624.

Lescale, C., Lenden Hasse, H., Blackford, A. N., Balmus, G., Bianchi, J. J., Yu, W., Bacoccina, L. et al. 2016. Specific roles of XRCC4 paralogs PAXX and XLF during V(D)J recombination. *Cell Rep* 16(11): 2967–2979.

Li, G., Alt, F. W., Cheng, H. L., Brush, J. W., Goff, P. H., Murphy, M. M., Franco, S., Zhang, Y., Zha, S. 2008. Lymphocyte-specific compensation for XLF/cernunnos end-joining functions in V(D)J recombination. *Mol Cell* 31(5): 631–640.

Li, G., Nelsen, C., Hendrickson, E. A. 2002. Ku86 is essential in human somatic cells. *Proc Natl Acad Sci USA* 99(2): 832–837.

Li, Y., Chirgadze, D. Y., Bolanos-Garcia, V. M., Sibanda, B. L., Davies, O. R., Ahnesorg, P., Jackson, S. P., Blundell, T. L. 2008. Crystal structure of human XLF/Cernunnos reveals unexpected differences from XRCC4 with implications for NHEJ. *EMBO J* 27(1): 290–300.

Li, Z., Otevrel, T., Gao, Y., Cheng, H. L., Seed, B., Stamato, T. D., Taccioli, G. E., Alt, F. W. 1995. The XRCC4 gene encodes a novel protein involved in DNA double-strand break repair and V(D)J recombination. *Cell* 83(7): 1079–1089.

Lieber, M. R. 2006. The polymerases for V(D)J recombination. *Immunity* 25(1): 7–9.

Lieber, M. R. 2010. The mechanism of double-strand DNA break repair by the nonhomologous DNA end-joining pathway. *Annu Rev Biochem* 79: 181–211.

Liu, T., Huang, J. 2016. DNA end resection: Facts and mechanisms. *Genomics Proteomics Bioinformatics* 14(3): 126–130.

Lomax, M. E., Folkes, L. K., O'Neill, P. 2013. Biological consequences of radiation-induced DNA damage: Relevance to radiotherapy. *Clin Oncol (R Coll Radiol)* 25(10): 578–585.

Lu, H., Pannicke, U., Schwarz, K., Lieber, M. R. 2007. Length-dependent binding of human XLF to DNA and stimulation of XRCC4.DNA ligase IV activity. *J Biol Chem* 282(15): 11155–11162.

Luijsterburg, M. S., de Krijger, I., Wiegant, W. W., Shah, R. G., Smeenk, G., de Groot, A. J., Pines, A. et al. 2016. PARP1 links CHD2-mediated chromatin expansion and H3.3 deposition to DNA repair by non-homologous end-joining. *Mol Cell* 61(4): 547–562.

Lukas, J., Lukas, C., Bartek, J. 2011. More than just a focus: The chromatin response to DNA damage and its role in genome integrity maintenance. *Nat Cell Biol* 13(10): 1161–1169.

Ma, Y., Pannicke, U., Schwarz, K., Lieber, M. R. 2002. Hairpin opening and overhang processing by an Artemis/DNA-dependent protein kinase complex in nonhomologous end joining and V(D)J recombination. *Cell* 108(6): 781–794.

Manis, J. P., Morales, J. C., Xia, Z., Kutok, J. L., Alt, F. W., Carpenter, P. B. 2004. 53BP1 links DNA damage-response pathways to immunoglobulin heavy chain class-switch recombination. *Nat Immunol* 5(5): 481–487.

Marchand, C., Huang, S. Y., Dexheimer, T. S., Lea, W. A., Mott, B. T., Chergui, A., Naumova, A. et al. 2014. Biochemical assays for the discovery of TDP1 inhibitors. *Mol Cancer Ther* 13(8): 2116–2126.

Martin, G. M., Smith, A. C., Ketterer, D. J., Ogburn, C. E., Disteche, C. M. 1985. Increased chromosomal aberrations in first metaphases of cells isolated from the kidneys of aged mice. *Isr J Med Sci* 21(3): 296–301.

Mateos-Gomez, P. A., Gong, F., Nair, N., Miller, K. M., Lazzerini-Denchi, E., Sfeir, A. 2015. Mammalian polymerase theta promotes alternative NHEJ and suppresses recombination. *Nature* 518(7538): 254–257.

Mehrotra, P. V., Ahel, D., Ryan, D. P., Weston, R., Wiechens, N., Kraehenbuehl, R., Owen-Hughes, T., Ahel, I. 2011. DNA repair factor APLF is a histone chaperone. *Mol Cell* 41(1): 46–55.

Menon, V., Povirk, L. F. 2016. End-processing nucleases and phosphodiesterases: An elite supporting cast for the non-homologous end joining pathway of DNA double-strand break repair. *DNA Repair (Amst)* 43: 57–68.

Modesti, M., Hesse, J. E., Gellert, M. 1999. DNA binding of Xrcc4 protein is associated with V(D)J recombination but not with stimulation of DNA ligase IV activity. *EMBO J* 18(7): 2008–2018.

Morales, J. C., Xia, Z., Lu, T., Aldrich, M. B., Wang, B., Rosales, C., Kellems, R. E., Hittelman, W. N., Elledge, S. J., Carpenter, P. B. 2003. Role for the BRCA1 C-terminal repeats (BRCT) protein 53BP1 in maintaining genomic stability. *J Biol Chem* 278(17): 14971–14977.

Moshous, D., Callebaut, I., de Chasseval, R., Corneo, B., Cavazzana-Calvo, M., Le Deist, F., Tezcan, I. et al. 2001. Artemis, a novel DNA double-strand break repair/V(D)J recombination protein, is mutated in human severe combined immune deficiency. *Cell* 105(2): 177–186.

Moshous, D., Pannetier, C., Chasseval Rd, R., Deist Fl, F., Cavazzana-Calvo, M., Romana, S., Macintyre, E. et al. 2003. Partial T and B lymphocyte immunodeficiency and predisposition to lymphoma in patients with hypomorphic mutations in Artemis. *J Clin Invest* 111(3): 381–387.

Mueller, G. A., Moon, A. F., Derose, E. F., Havener, J. M., Ramsden, D. A., Pedersen, L. C., London, R. E. 2008. A comparison of BRCT domains involved in nonhomologous end-joining: Introducing the solution structure of the BRCT domain of polymerase lambda. *DNA Repair (Amst)* 7(8): 1340–1351.

Murray, J. E., van der Burg, M., Ijspeert, H., Carroll, P., Wu, Q., Ochi, T., Leitch, A. et al. 2015. Mutations in the NHEJ component XRCC4 cause primordial dwarfism. *Am J Hum Genet* 96(3): 412–424.

Noordzij, J. G., Verkaik, N. S., van der Burg, M., van Veelen, L. R., de Bruin-Versteeg, S., Wiegant, W., Vossen, J. M. et al. 2003. Radiosensitive SCID patients with Artemis gene mutations show a complete B-cell differentiation arrest at the pre-B-cell receptor checkpoint in bone marrow. *Blood* 101(4): 1446–1452.

Ochi, T., Blackford, A. N., Coates, J., Jhujh, S., Mehmood, S., Tamura, N., Travers, J. et al. 2015. PAXX, a paralog of XRCC4 and XLF, interacts with Ku to promote DNA double-strand break repair. *Science* 347(6218): 185–188.

Ochi, T., Gu, X., Blundell, T. L. 2013. Structure of the catalytic region of DNA ligase IV in complex with an Artemis fragment sheds light on double-strand break repair. *Structure* 21(4): 672–679.

O'Driscoll, M., Cerosaletti, K. M., Girard, P. M., Dai, Y., Stumm, M., Kysela, B., Hirsch, B. et al. 2001. DNA ligase IV mutations identified in patients exhibiting developmental delay and immunodeficiency. *Mol Cell* 8(6): 1175–1185.

Panier, S., Boulton, S. J. 2014. Double-strand break repair: 53BP1 comes into focus. *Nat Rev Mol Cell Biol* 15(1): 7–18.

Pannicke, U., Honig, M., Schulze, I., Rohr, J., Heinz, G. A., Braun, S., Janz, I. et al. 2010. The most frequent DCLRE1C (ARTEMIS) mutations are based on homologous recombination events. *Hum Mutat* 31(2): 197–207.

Pardo, B., Ma, E., Marcand, S. 2006. Mismatch tolerance by DNA polymerase Pol4 in the course of nonhomologous end joining in *Saccharomyces cerevisiae*. *Genetics* 172(4): 2689–2694.

Petersen, S., Casellas, R., Reina-San-Martin, B., Chen, H. T., Difilippantonio, M. J., Wilson, P. C., Hanitsch, L. et al. 2001. AID is required to initiate Nbs1/gamma-H2AX focus formation and mutations at sites of class switching. *Nature* 414(6864): 660–665.

Poinsignon, C., de Chasseval, R., Soubeyrand, S., Moshous, D., Fischer, A., Hache, R. J., de Villartay, J. P. 2004. Phosphorylation of Artemis following irradiation-induced DNA damage. *Eur J Immunol* 34(11): 3146–3155.

Pommier, Y. 2006. Topoisomerase I inhibitors: Camptothecins and beyond. *Nat Rev Cancer* 6(10): 789–802.

Pommier, Y., Leo, E., Zhang, H., Marchand, C. 2010. DNA topoisomerases and their poisoning by anticancer and antibacterial drugs. *Chem Biol* 17(5): 421–433.

Postow, L. 2011. Destroying the ring: Freeing DNA from Ku with ubiquitin. *FEBS Lett* 585(18): 2876–2882.

Prakash, R., Zhang, Y., Feng, W., Jasin, M. 2015. Homologous recombination and human health: The roles of BRCA1, BRCA2, and associated proteins. *Cold Spring Harb Perspect Biol* 7(4): a016600.

Pryor, J. M., Waters, C. A., Aza, A., Asagoshi, K., Strom, C., Mieczkowski, P. A., Blanco, L., Ramsden, D. A. 2015. Essential role for polymerase specialization in cellular nonhomologous end joining. *Proc Natl Acad Sci USA* 112(33): E4537–E4545.

Raschella, G., Melino, G., Malewicz, M. 2017. New factors in mammalian DNA repair-the chromatin connection. *Oncogene* 36(33): 4673–4681.

Reid, D. A., Conlin, M. P., Yin, Y., Chang, H. H., Watanabe, G., Lieber, M. R., Ramsden, D. A., Rothenberg, E. 2017. Bridging of double-stranded breaks by the nonhomologous end-joining ligation complex is modulated by DNA end chemistry. *Nucleic Acids Res* 45(4): 1872–1878.

Reid, D. A., Keegan, S., Leo-Macias, A., Watanabe, G., Strande, N. T., Chang, H. H., Oksuz, B. A. et al. 2015. Organization and dynamics of the nonhomologous end-joining machinery during DNA double-strand break repair. *Proc Natl Acad Sci USA* 112(20): E2575–E2584.

Reina-San-Martin, B., Difilippantonio, S., Hanitsch, L., Masilamani, R. F., Nussenzweig, A., Nussenzweig, M. C. 2003. H2AX is required for recombination between immunoglobulin switch regions but not for intra-switch region recombination or somatic hypermutation. *J Exp Med* 197(12): 1767–1778.

Revy, P., Malivert, L., de Villartay, J. P. 2006. Cernunnos-XLF, a recently identified non-homologous end-joining factor required for the development of the immune system. *Curr Opin Allergy Clin Immunol* 6(6): 416–420.

Reynolds, J. J., Stewart, G. S. 2013. A single strand that links multiple neuropathologies in human disease. *Brain* 136(Pt 1): 14–27.

Riballo, E., Kuhne, M., Rief, N., Doherty, A., Smith, G. C., Recio, M. J., Reis, C. et al. 2004. A pathway of double-strand break rejoining dependent upon ATM, Artemis, and proteins locating to gamma-H2AX foci. *Mol Cell* 16(5): 715–724.

Riballo, E., Woodbine, L., Stiff, T., Walker, S. A., Goodarzi, A. A., Jeggo, P. A. 2009. XLF-Cernunnos promotes DNA ligase IV-XRCC4 re-adenylation following ligation. *Nucleic Acids Res* 37(2): 482–492.

Roberts, S. A., Strande, N., Burkhalter, M. D., Strom, C., Havener, J. M., Hasty, P., Ramsden, D. A. 2010. Ku is a 5′-dRP/AP lyase that excises nucleotide damage near broken ends. *Nature* 464(7292): 1214–1217.

Robins, P., Lindahl, T. 1996. DNA ligase IV from HeLa cell nuclei. *J Biol Chem* 271(39): 24257–24261.

Rogakou, E. P., Boon, C., Redon, C., Bonner, W. M. 1999. Megabase chromatin domains involved in DNA double-strand breaks *in vivo*. *J Cell Biol* 146(5): 905–916.

Rooney, S., Sekiguchi, J., Whitlow, S., Eckersdorff, M., Manis, J. P., Lee, C., Ferguson, D. O., Alt, F. W. 2004. Artemis and p53 cooperate to suppress oncogenic N-myc amplification in progenitor B cells. *Proc Natl Acad Sci USA* 101(8): 2410–2415.

Rooney, S., Sekiguchi, J., Zhu, C., Cheng, H. L., Manis, J., Whitlow, S., DeVido, J. et al. 2002. Leaky SCID phenotype associated with defective V(D)J coding end processing in Artemis-deficient mice. *Mol Cell* 10(6): 1379–1390.

Rosin, N., Elcioglu, N. H., Beleggia, F., Isguven, P., Altmuller, J., Thiele, H., Steindl, K. et al. 2015. Mutations in XRCC4 cause primary microcephaly, short stature and increased genomic instability. *Hum Mol Genet* 24(13): 3708–3717.

Roth, D. B. 2014. V(D)J recombination: Mechanism, errors, and fidelity. *Microbiol Spectr* 2(6). doi: 10.1128/microbiolspec.MDNA3-0041-2014.

Ruis, B. L., Fattah, K. R., Hendrickson, E. A. 2008. The catalytic subunit of DNA-dependent protein kinase regulates proliferation, telomere length, and genomic stability in human somatic cells. *Mol Cell Biol* 28(20): 6182–6195.

Rulten, S. L., Fisher, A. E., Robert, I., Zuma, M. C., Rouleau, M., Ju, L., Poirier, G., Reina-San-Martin, B., Caldecott, K. W. 2011. PARP-3 and APLF function together to accelerate nonhomologous end-joining. *Mol Cell* 41(1): 33–45.

Saito, S., Kurosawa, A., Adachi, N. 2016. Mutations in XRCC4 cause primordial dwarfism without causing immunodeficiency. *J Hum Genet* 61(8): 679–685.

Saleh-Gohari, N., Helleday, T. 2004. Conservative homologous recombination preferentially repairs DNA double-strand breaks in the S phase of the cell cycle in human cells. *Nucleic Acids Res* 32(12): 3683–3688.

Sartori, A. A., Lukas, C., Coates, J., Mistrik, M., Fu, S., Bartek, J., Baer, R., Lukas, J., Jackson, S. P. 2007. Human CtIP promotes DNA end resection. *Nature* 450(7169): 509–514.

Schatz, D. G., Ji, Y. 2011. Recombination centres and the orchestration of V(D)J recombination. *Nat Rev Immunol* 11(4): 251–263.

Schroeder, H. W. Jr., Cavacini, L. 2010. Structure and function of immunoglobulins. *J Allergy Clin Immunol* 125(Suppl. 2): S41–S52.

Sfeir, A., Symington, L. S. 2015. Microhomology-mediated end joining: A back-up survival mechanism or dedicated pathway? *Trends Biochem Sci* 40(11): 701–714.

Shuman, S., Glickman, M. S. 2007. Bacterial DNA repair by non-homologous end joining. *Nat Rev Microbiol* 5(11): 852–861.

Sibanda, B. L., Chirgadze, D. Y., Ascher, D. B., Blundell, T. L. 2017. DNA-PKcs structure suggests an allosteric mechanism modulating DNA double-strand break repair. *Science* 355(6324): 520–524.

Sibanda, B. L., Critchlow, S. E., Begun, J., Pei, X. Y., Jackson, S. P., Blundell, T. L., Pellegrini, L. 2001. Crystal structure of an Xrcc4-DNA ligase IV complex. *Nat Struct Biol* 8(12): 1015–1019.

Stewart, G. S., Wang, B., Bignell, C. R., Taylor, A. M., Elledge, S. J. 2003. MDC1 is a mediator of the mammalian DNA damage checkpoint. *Nature* 421(6926): 961–966.

Strumberg, D., Pilon, A. A., Smith, M., Hickey, R., Malkas, L., Pommier, Y. 2000. Conversion of topoisomerase I cleavage complexes on the leading strand of ribosomal DNA into 5′-phosphorylated DNA double-strand breaks by replication runoff. *Mol Cell Biol* 20(11): 3977–3987.

Symington, L. S. 2016. Mechanism and regulation of DNA end resection in eukaryotes. *Crit Rev Biochem Mol Biol* 51(3): 195–212.

Symington, L. S., Gautier, J. 2011. Double-strand break end resection and repair pathway choice. *Annu Rev Genet* 45: 247–271.

Tadi, S. K., Tellier-Lebegue, C., Nemoz, C., Drevet, P., Audebert, S., Roy, S., Meek, K., Charbonnier, J. B., Modesti, M. 2016. PAXX is an accessory c-NHEJ factor that associates with Ku70 and has overlapping functions with XLF. *Cell Rep* 17(2): 541–555.

Takashima, H., Boerkoel, C. F., John, J., Saifi, G. M., Salih, M. A., Armstrong, D., Mao, Y. et al. 2002. Mutation of TDP1, encoding a topoisomerase I-dependent DNA damage repair enzyme, in spinocerebellar ataxia with axonal neuropathy. *Nat Genet* 32(2): 267–272.

Tasher, D., Dalal, I. 2012. The genetic basis of severe combined immunodeficiency and its variants. *Appl Clin Genet* 5: 67–80.

Tsai, C. J., Kim, S. A., Chu, G. 2007. Cernunnos/XLF promotes the ligation of mismatched and noncohesive DNA ends. *Proc Natl Acad Sci USA* 104(19): 7851–7856.

Tsukuda, T., Fleming, A. B., Nickoloff, J. A., Osley, M. A. 2005. Chromatin remodelling at a DNA double-strand break site in *Saccharomyces cerevisiae*. *Nature* 438(7066): 379–383.

van der Burg, M., Ijspeert, H., Verkaik, N. S., Turul, T., Wiegant, W. W., Morotomi-Yano, K., Mari, P. O. et al. 2009. A DNA-PKcs mutation in a radiosensitive T-B- SCID patient inhibits Artemis activation and nonhomologous end-joining. *J Clin Invest* 119(1): 91–98.

van der Burg, M., van Veelen, L. R., Verkaik, N. S., Wiegant, W. W., Hartwig, N. G., Barendregt, B. H., Brugmans, L. et al. 2006. A new type of radiosensitive T-B-NK+ severe combined immunodeficiency caused by a LIG4 mutation. *J Clin Invest* 116(1): 137–145.

Walker, J. R., Corpina, R. A., Goldberg, J. 2001. Structure of the Ku heterodimer bound to DNA and its implications for double-strand break repair. *Nature* 412(6847): 607–614.

Wang, Y., Ghosh, G., Hendrickson, E. A. 2009. Ku86 represses lethal telomere deletion events in human somatic cells. *Proc Natl Acad Sci USA* 106(30): 12430–12435.

Waters, C. A., Strande, N. T., Wyatt, D. W., Pryor, J. M., Ramsden, D. A. 2014. Nonhomologous end joining: A good solution for bad ends. *DNA Repair (Amst)* 17: 39–51.

Weinfeld, M., Mani, R. S., Abdou, I., Aceytuno, R. D., Glover, J. N. 2011. Tidying up loose ends: The role of polynucleotide kinase/phosphatase in DNA strand break repair. *Trends Biochem Sci* 36(5): 262–271.

Weller, G. R., Kysela, B., Roy, R., Tonkin, L. M., Scanlan, E., Della, M., Devine, S. K. et al. 2002. Identification of a DNA nonhomologous end-joining complex in bacteria. *Science* 297(5587): 1686–1689.

Wood, R. D., Doublie, S. 2016. DNA polymerase theta (POLQ), double-strand break repair, and cancer. *DNA Repair (Amst)* 44: 22–32.

Woodbine, L., Brunton, H., Goodarzi, A. A., Shibata, A., Jeggo, P. A. 2011. Endogenously induced DNA double strand breaks arise in heterochromatic DNA regions and require ataxia telangiectasia mutated and Artemis for their repair. *Nucleic Acids Res* 39(16): 6986–6997.

Woodbine, L., Gennery, A. R., Jeggo, P. A. 2014. The clinical impact of deficiency in DNA non-homologous end-joining. *DNA Repair (Amst)* 16: 84–96.

Woodbine, L., Neal, J. A., Sasi, N. K., Shimada, M., Deem, K., Coleman, H., Dobyns, W. B. et al. 2013. PRKDC mutations in a SCID patient with profound neurological abnormalities. *J Clin Invest* 123(7): 2969–2980.

Wu, P. Y., Frit, P., Meesala, S., Dauvillier, S., Modesti, M., Andres, S. N., Huang, Y. et al. 2009. Structural and functional interaction between the human DNA repair proteins DNA ligase IV and XRCC4. *Mol Cell Biol* 29(11): 3163–3172.

Wyatt, D. W., Feng, W., Conlin, M. P., Yousefzadeh, M. J., Roberts, S. A., Mieczkowski, P., Wood, R. D., Gupta, G. P., Ramsden, D. A. 2016. Essential roles for polymerase theta-mediated end joining in the repair of chromosome breaks. *Mol Cell* 63(4): 662–673.

Xing, M., Yang, M., Huo, W., Feng, F., Wei, L., Jiang, W., Ning, S. et al. 2015. Interactome analysis identifies a new paralogue of XRCC4 in non-homologous end joining DNA repair pathway. *Nat Commun* 6: 6233.

Xu, Z., Zan, H., Pone, E. J., Mai, T., Casali, P. 2012. Immunoglobulin class-switch DNA recombination: Induction, targeting and beyond. *Nat Rev Immunol* 12(7): 517–6531.

Yano, K., Morotomi-Yano, K., Lee, K. J., Chen, D. J. 2011. Functional significance of the interaction with Ku in DNA double-strand break recognition of XLF. *FEBS Lett* 585(6): 841–846.

Zha, S., Guo, C., Boboila, C., Oksenych, V., Cheng, H. L., Zhang, Y., Wesemann, D. R. et al. 2011. ATM damage response and XLF repair factor are functionally redundant in joining DNA breaks. *Nature* 469(7329): 250–254.

Zhang, S., Yajima, H., Huynh, H., Zheng, J., Callen, E., Chen, H. T., Wong, N. et al. 2011. Congenital bone marrow failure in DNA-PKcs mutant mice associated with deficiencies in DNA repair. *J Cell Biol* 193(2): 295–305.

Zhu, C., Bogue, M. A., Lim, D. S., Hasty, P., Roth, D. B. 1996. Ku86-deficient mice exhibit severe combined immunodeficiency and defective processing of V(D)J recombination intermediates. *Cell* 86(3): 379–389.

Ziv, Y., Bielopolski, D., Galanty, Y., Lukas, C., Taya, Y., Schultz, D. C., Lukas, J., Bekker-Jensen, S., Bartek, J., Shiloh, Y. 2006. Chromatin relaxation in response to DNA double-strand breaks is modulated by a novel ATM- and KAP-1 dependent pathway. *Nat Cell Biol* 8(8): 870–876.

Protein Methylation and the DNA Damage Response

8

Martin R. Higgs and Clare Davies

INTRODUCTION

As already described elsewhere in this volume, various post-translational modifications play a vital role in regulating the DNA damage response (DDR). In this chapter, we discuss how methylation of lysine and arginine residues on both histone and non-histone proteins plays a crucial function in controlling DNA repair and maintaining genome integrity.

LYSINE METHYLATION AND THE DNA DAMAGE RESPONSE

Lysine methylation exists as one of three methyl states – mono-, di- or tri-methylation – and is catalysed by the activity of lysine methyltransferases (KMTs). Much of our knowledge surrounding lysine (Lys) methylation and KMTs originates from the study of histone methylation, which was first identified in the late 1950s. These original studies surmised that since the half-lives of methylated lysine and of the methylated histone were similar, lysine methylation was irreversible. Identification of the first lysine demethylase (KDM), however, challenged this dogma, and it is now clear that protein lysine methylation is dynamic, reversible and wide-ranging. In addition, since many methyl-binding proteins, or 'methyl readers', display methylation state-specific recruitment, it is clear that the three methyl-lysine states can have distinct functional outcomes.

In addition, the importance of non-histone lysine methylation is becoming increasingly apparent, especially given the recent advances in proteomics that allow detection of methyl-lysine. To date, thousands of methylated Lys residues on hundreds of non-histone substrates have been identified. It is now clear that non-histone lysine methylation plays an important role in a wide range of important cellular processes, including the DDR, and is deregulated in a variety of diseases. Indeed, whilst many KMTs were once considered to be histone specific, many have been found to catalyse lysine methylation of other substrates. This chapter therefore builds on this evidence, and focuses on the effectors and dynamics of lysine methylation and their roles in maintaining genome integrity.

Effectors, erasers and readers of lysine methylation

As mentioned earlier, protein lysine methylation is a dynamic process, comprising KMTs that catalyse methylation, methyl readers that bind methylated residues, and KDMs to remove methylated marks.

Lysine methyltransferases (KMTs)

Lysine methylation 'writers', or KMTs, catalyse the methylation (mono-, di- or tri-) of the ε-group of lysine residues on target proteins. Lysine methylation does not modify the side chain's positive charge and causes only a small change in mass (14, 28 or 42 Da). The majority of these enzymes contain a SET domain, a $\sim$140 amino acid sequence motif that was initially characterised in the *Drosophila* proteins Su(var)3-9, Enhancer-of-zeste and Trithorax (giving rise to the SET acronym). These enzymes usually recognise short linear peptides containing the target amino acid. Although this SET domain contains all the catalytically important residues, these KMTs also contain either I-SET (immunoglobulin-SET) or post-SET domains, which contribute to peptide or cofactor binding. However, several non-SET-domain KMTs have also been characterised, which are postulated to distinguish their substrates by recognising tertiary protein structures. Even though these enzymes differ in their catalytic mechanisms, both use S-adenosyl-L-methionine as a methyl donor. KMTs also demonstrate specificity both in terms of the degree of methylation catalysed, but also the target residue within a protein. Moreover, interacting partners of various KMTs can modify their activity, or even their target lysine.

Several KMTs have been shown to play an integral role in the DDR (see Table 8.1), and it is likely that many more perform as-yet-uncharacterised functions in the response to genotoxic stress. These known roles are detailed more thoroughly in subsequent sections.

Methyl-lysine readers

Several protein domains confer the ability to bind or 'read' methylated lysine residues. This can occur in a methyl-dependent or sequence-specific fashion, providing specificity and functional flexibility to many biological systems. Methyl-lysine domains fall into several protein motif families, including Bromo, MBT, PHD and PWWP motifs, and the 'Royal' family of reader modules: Tudor, chromodomain and double chromodomain.

Despite the wide variety of folds that bind methyl-lysine, they share similar structural features that allow binding to methylation marks (Taverna et al. 2007). In particular, members of the Royal superfamily of folds and PHD domains both bind methyl-lysine through conserved aromatic cages consisting of two to four aromatic residues. Although lysine methylation does not alter the charge of the residue, it will alter the hydrogen bonding potential of the methyl-lysine; incremental addition of methyl groups increases hydrophobicity and decreases the ability to donate hydrogen bonds. Hydrogen bonding between the methyl-lysine and the aromatic residues within this 'cage' provides the basis for recognition of mono-, di- or tri-methylated residues by methyl-binding domains, allowing a state-specific readout by different reader molecules. In addition, the study of unmodified lysine residues within histone tails has also highlighted a further level of specificity within this system: readers can also preferentially bind unmodified lysine residues, and lysine methylation can therefore function to block/weaken molecular interactions.

Several well-characterised proteins containing methyl-lysine reader domains play an important role in the DDR by binding both histone and non-histone proteins (see Table 8.2). As with KMTs, the role of these proteins in the DDR is still poorly characterised. Furthermore, since many proteins containing methyl-binding modules also have other functions, it is likely that lysine methylation acts as a vital regulatory step to recruit enzymes and co-factors containing these domains to promote DNA repair. In this regard, it is important to note that several KMT enzymes themselves contain methyl-lysine reader domains, including SUV39H1, KMT2A, KMT5B/C and SETDB1 (KMT1E), KDM2, KDM5 and KDM4, allowing them to bind methylated lysines. For example, SUV39H1 (KMT1A) binds to H3K9me3 through its chromodomain (Wang et al. 2012b), allowing methylation of neighbouring histones from a mono-methylated to a tri-methylated state.

Lysine demethylases (KDMs)

Lysine demethylases, or KDMs, act as 'erasers' to remove methyl-lysine marks. Mechanistically, KDMs can be subdivided into two categories: (1) flavin adenine dinucleotide (FAD)-dependent monoamine oxygenases or (2) oxygenases such as JmjC domain-containing KDMs that use oxygen, Fe(II) and α-ketoglutarate as cofactors. Due to their differing mechanisms of action, KDMs of these two classes have different demethylase capacities: FAD-dependent oxygenases can demethylate mono- and di-methylated lysines, whereas JmjC enzymes can demethylate all three methylation states. There is substantial debate over the precise classification of these enzymes, as the KDM denomination does not accurately reflect the oxygenase activity of some of these enzymes towards other substrates. Furthermore, structural differences between enzymes within the same family also dictate the degree of demethylation performed.

As well as specificity for the degree of methylation of a lysine moiety, KDMs also demonstrate specificity for particular target residues within methylated proteins. As is the case with KMTs and methyl-readers, both these factors can be altered by interacting partners, controlling access of KDMs to methylated lysines, and providing further specificity to the system.

Several KDMs play important roles in the DDR, acting to demethylate either histone or non-histone substrates (Table 8.3). Interestingly, formaldehyde is commonly produced as a

Table 8.1 **Lysine methyltransferase (KMT) gene family.**

Approved gene symbol	Approved name	Widely used synonyms	Role in DDR
SUV39H1	Suppressor of variegation 39 homologue 1	KMT1A	Regulates H3K9 methylation and KAT5 recruitment in DSBR
SUV39H2	Suppressor of variegation 39 homologue 2	KMT1B	
EHMT2	Euchromatic histone lysine methyltransferase 2	KMT1C, G9A, NG36/G9a,	
EHMT1	Euchromatic histone lysine methyltransferase 1	KMT1D, Eu-HMTase1	
SETDB1	SET domain bifurcated 1	KMT1E, ESET, KG1T	Promotes HR via H3K9me
SETDB2	SET domain bifurcated 2	KMT1F, C13orf4, CLLL8	
KMT2A	Lysine methyltransferase 2A	MLL, MLL1A, ALL-1, TRX1	Methylates H3K4, regulates origin firing
KMT2B	Lysine methyltransferase 2B	MLL4, TRX2, WBP7 (Mll2 in *M. musculus*)	Methylates H3K4, role in replication fork protection via PTIP
KMT2C	Lysine methyltransferase 2C	MLL3, HALR	
KMT2D	Lysine methyltransferase 2D	MLL2, TNRC21, ALR (Mll4 in *M. musculus*)	Regulates transcription-associated DNA damage
KMT2E	Lysine methyltransferase 2E	MLL5	
SETD1A	SET domain containing 1A	KMT2F	May have role in replication stress
SETD1B	SET domain containing 1B	KMT2G	
ASH1L	ASH1-like histone methyltransferase	KMT2H, ASH1, ASH1L1	
SETD2	SET domain containing 2	KMT3A, HIF-1	Regulates H3K36 methylation and HR
NSD1	Nuclear receptor binding SET domain protein 1	KMT3B, STO	
SMYD2	SET and MYND domain containing 2	KMT3C, HSKM-B	Methylates PARP1
SMYD1	SET and MYND domain containing 1	KMT3D, BOP	
SMYD3	SET and MYND domain containing 3	KMT3E, ZMYND1, ZNF3A1	Transcribes HR gene products
NSD3	Nuclear receptor binding SET domain protein 3	KMT3F, WHSC1L1, WHISTLE	
NSD2	Nuclear receptor binding SET domain protein 2	KMT3G, WHSC1, MMSET	Involved in CSR, NHEJ and HR
DOT1L	DOT1-like histone lysine methyltransferase	KMT4, DOT1	
KMT5A	Lysine methyltransferase 5A	SETD8, SET8, PR-Set7	Regulates H4K20 methylation; impacts 53BP1 recruitment during DSBR and origin licencing during G1 phase
KMT5B	Lysine methyltransferase 5B	SUV420H1	
KMT5C	Lysine methyltransferase 5C	SUV420H2	
EZH2	Enhancer of zeste 2 polycomb repressive complex 2 subunit	KMT6A, EZH1, KMT6	Regulates DDR gene expression
EZH1	Enhancer of zeste 1 polycomb repressive complex 2 subunit	KMT6B	
SETD7	SET domain containing lysine methyltransferase 7	KMT7, SET7, SET7/9, Set9	Methylates KMT1A during DSBR
PRDM2	PR/SET domain 2	KMT8A, KMT8, RIZ1, RIZ2	

Table 8.1 **(*Continued*) Lysine methyltransferase (KMT) gene family.**

Approved gene symbol	Approved name	Widely used synonyms	Role in DDR
PRDM9	PR/SET domain 9	KMT8B, MSBP3, PFM6	
PRDM6	PR/SET domain 6	KMT8C, PRISM	
PRDM8	PR/SET domain 8	KMT8D	
MECOM	MDS1 and EVI1 complex locus	KMT8E, MDS1, EVI1, PRDM3	
PRDM16	PR/SET domain 16	KMT8F, MEL1	

Note: The HUGO gene nomenclature and name are given for each KMT, along with widely used synonyms from the literature to prevent confusion. A brief summary of the role of each KMT in the DDR is also given.

Table 8.2 **Methyl-lysine readers important in the DDR.**

Approved gene symbol	Approved name	Widely used synonyms	Role in DDR
KDM4A	Lysine demethylase 4A	JMJD2, JMJD2A	Tudor domain binds H4K20me2/3, blocking 53BP1 binding. Required for DSB repair
TP53BP1	Tumour suppressor p53-binding protein 1	53BP1	Tandem TUDOR binds H4K20me2; promotes DSB repair. Plays a prominent role in replication stress via formation of G1 53BP1 bodies. Also binds methylated p53 and Rb
VCP	Valosin-containing protein	p97, TER	Masks H4K20me2 to prevent 53BP1 binding. Removed by RNF8/RNF168
L3MBTL1	Lethal(3)malignant brain tumour-like 1	L3MBT	Masks H4K20me2 to prevent 53BP1 binding. Removed by RNF8/RNF168
TONSL	Tonsoku-like, DNA repair protein	NFKBIL2	Binds non-methylated H4K20. Plays a key role in replication stress response alongside MMS22L
CHD1	Chromodomain helicase DNA binding protein 1	–	Binds methylated H3K4. Stimulates recruitment of CtIP and promotes end resection during DSB repair
CHD3	Chromodomain helicase DNA binding protein 3	Mi2-ALPHA	Binds H3K36, important role in repair of heterochromatic DSBs
CHD4	Chromodomain Helicase DNA binding protein 4	Mi2-BETA	Binds methylated H3K9, repelled by H3K4me, implicated in HR
MAD2L2	Mitotic arrest deficient 2-like protein 2	REV7, hREV7	Regulates resection and replication stress. HORMA domain binds methylated H3K4

Note: The HUGO gene nomenclature and name are given for each reader, along with widely used synonyms from the literature to prevent confusion. A brief summary of the role of each KMT in the DDR is also given.

by-product during demethylase reactions. Given that formaldehyde causes interstrand and protein-DNA cross-links, requiring the combined actions of proteases and components of the Fanconi anaemia pathway to resolve, it is interesting to speculate that the activity of KDMs might also act as a potential source of genome instability.

Roles for lysine methylation and demethylation in double-strand break repair

The study of Lys methylation in the DNA damage response, and characterisation of the roles for KMTs, KDMs and Lys readers, is an emerging and fast-growing field. Here, we discuss the growing body of evidence for both histone and non-histone methylation in the maintenance of genome integrity, focusing first on its role in double-strand break (DSB) repair, and speculating on future areas of research. The importance of Lys methylation in DSB repair is perhaps the most thoroughly characterised facet of this PTM in the DNA damage response. Methylation of both histone and non-histone residues helps to modulate chromatin dynamics around

Table 8.3 Lysine demethylase (KDM) gene family.

Approved gene symbol	Approved name	Widely used synonyms	Role in DDR
KDM1A	Lysine demethylase 1A	LSD1, KDM1, AOF2	
KDM1B	Lysine demethylase 1B	LSD2, AOF1	
KDM2A	Lysine demethylase 2A	FBXL11, JHDM1A, CXXC8	
KDM2B	Lysine demethylase 2B	FBXL10, JHDM1B,CXXC2	
KDM3A	Lysine demethylase 3A	JMJD1A, JMJD1	
KDM3B	Lysine demethylase 3B	JMJD1B, NET22	
JMJD1C	Jumonji domain–containing 1C	TRIP8, KDM3C	Demethylates MDC1 after DSB formation, promoting MDC1-RNF8 interaction
KDM4A	Lysine demethylase 4A	JMJD2, JMJD2A	Binds H4K20me2/3; evicted by RNF6/168-dependent Ub to allow 53BP1 binding
KDM4B	Lysine demethylase 4B	JMJD2B	Involved in DSB repair via H3K9 demethylation
KDM4C	Lysine demethylase 4C	JMJD2C	
KDM4D	Lysine demethylase 4D	JMJD2D	Required for CDC45 binding to pre-RC and origin firing
KDM4E	Lysine demethylase 4E	JMJD2E, KDM5E, KDM4DL	
KDM4F	Lysine demethylase 4F	JMJD2F	
KDM5A	Lysine demethylase 5A	JARID1A, RBBP2	
KDM5B	Lysine demethylase 5B	JARID1B	Required for DSB repair and recruitment of BRCA1/KU70 to DSBs; perhaps via H3K4 demethylation
KDM5C	Lysine demethylase 5C	JARID1C, SMCX	Demethylates H3K4me2/3 to promote origin firing and PCNA/CDC45 binding
KDM5D	Lysine demethylase 5D	JARID1D, SMCY	
KDM6A	Lysine demethylase 6A	UTX	
KDM6B	Lysine demethylase 6B	JMJD3	
UTY	Ubiquitously transcribed tetratricopeptide repeat containing, Y-linked	KDM6C	
KDM7A	Lysine demethylase 7A	JHDM1D	
PHF8	PHD finger protein 8	KDM7B, JHDM1F	Crucial for DSB repair and recruitment of DDR factors BLM and KU70 to DSBs
PHF2	PHD finger protein 2	KDM7C, JHDM1E	
KDM8	Lysine demethylase 8	JMJD5[a]	Regulates RAD51 and HR; perhaps via H3K36 demethylation

Note: The HUGO gene nomenclature and name are given for each KDM, along with widely used synonyms from the literature to prevent confusion. A brief summary of the role of each KDM in the DDR is also given.

[a]The functional assignment of the KDM8/JMJD5 hydroxylase as a KDM is controversial (see Markolovic et al. 2016).

DSBs, control recruitment of repair and checkpoint proteins and modify their function. Unsurprisingly, the modification of several Lys residues within histones is central to this.

Lysine methylation and the early stages of double-strand break repair

The apical ATM kinase phosphorylates a multitude of substrates involved in DSB repair, including several proteins involved in Lys methylation. However, complete activation of ATM requires the acetyltransferase activity of the methyl-Lys reader KAT5 (TIP60), which is recruited and activated by histone H3 Lys 9 methylation at the break site. This methylation is itself mediated by the rapid recruitment of the methyltransferase SUV39H1 to DSBs, which catalyses waves of H3K9 methylation. Subsequent cycles of H3K9 methylation bring in additional SUV39H1 via interaction with the methyl-Lys reader HP1 and its recruitment to nascent H3K9me3 (Ayrapetov et al. 2014). As well as being a marker for compact silent heterochromatin, thus potentially suppressing transcription at sites of DSBs, H3K9me3 also acts to recruit the repair factors BRCA1 and BARD1 via their interactions with HP1 and KAP1. However, compact heterochromatin may also limit DNA repair; therefore, activated ATM rapidly phosphorylates KAP1, removing SUV39H1/HP1/KAP1 complexes from the DSB and providing a powerful feedback loop to allow efficient DSB repair. Thus, SUV39H1-mediated H3K9 methylation plays a vital role in the earlier stages of the DSBR signalling cascade by promoting a transient upregulation of H3K9 methylation at DSBs to ensure efficient ATM activation. Furthermore, SUV39H1 is itself regulated by Lys methylation in response to genotoxic stress, which negatively regulates its KMT activity (Wang et al. 2013). This regulation is mediated by SETD7 (KMT7), and both the interaction between these KMTs and the methylation of SUV39H1 increase after DNA damage. Precisely how and why SUV39H1 Lys methylation occurs, and whether this regulates ATM activation, is not known.

In addition to histone methylation, methylation of non-histone proteins also plays a vital stimulatory role in the early stages of DSB recognition and repair (Figure 8.1). Poly(ADP-ribose) polymerase-1 (PARP1) plays a vital role in the early stages of many types of DNA damage recognition and repair, including DSB repair, where it acts to promote MRE11-RAD50-NBS1 complex recruitment by catalysing poly-ADP-ribosylation of target proteins. Interestingly, PARP1 itself is Lys methylated by the KMT SMYD2 (KMT3C) in response to DNA damage (Piao et al. 2014). Although this stimulates its activity upon oxidative stress, the impact on the ability of PARP1 to promote DSB repair has yet to be analysed. As well as PARP1, there is also limited evidence that components of the MRN complex itself are Lys methylated,

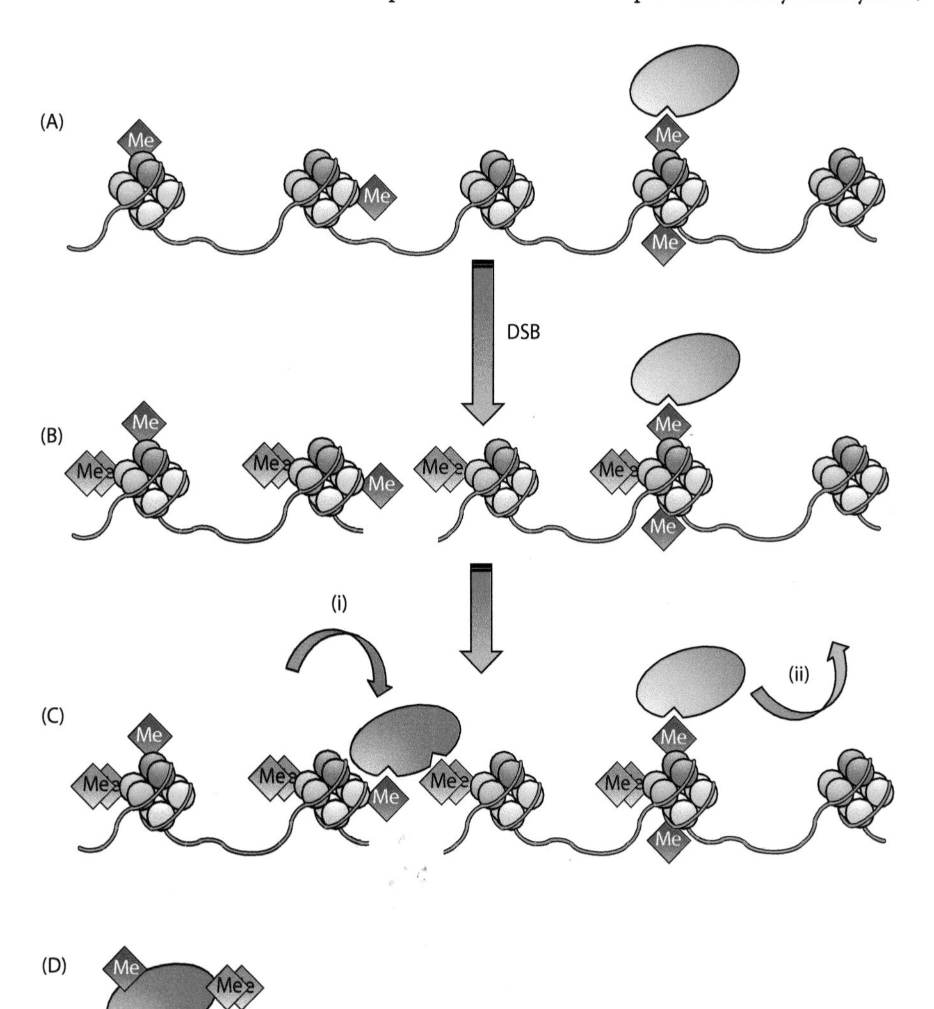

Figure 8.1 Histone Lys methylation and DSB repair. Repair of DSBs is in part governed by the methylation of Lys residues of both histone and non-histone proteins. (A) Pre-existing methylation marks such as K4K20 (purple) act as binding sites for methyl-readers (blue), which are chromatin bound before DSB formation and mask the methyl mark. (B) Upon DSB formation, additional Lys methylation events (orange) are catalysed on nucleosomes surrounding the break. (C) These new methyl marks (i), combined with the eviction of 'masking' methyl-binding proteins (ii), expose binding sites for repair factors such as 53BP1. (D) These factors may also be methylated/demethylated in response to DNA damage (e.g. PARP1, MDC1).

including MRE11 and RAD50 (Kish et al. 2016). These findings are supported by the results of large-scale Lys methylation proteomics analyses, but as yet the functional significance of this methylation is unknown. As with many of the Lys modifications covered in this chapter, it is likely that Lys methylation of the MRN provides a binding platform for methyl Lys readers to be recruited to damaged chromatin, although it may be that these modifications also alter the function of the MRN complex.

Lys methylation and DSB repair pathway choice

As outlined elsewhere in this volume, DSB repair occurs via two main pathways: non-homologous end joining (NHEJ) and homologous recombination (HR). Perhaps unsurprisingly, Lys methylation of both histones and non-histone proteins also plays a key role in determining the choice between these DSB repair pathways, mainly by governing the recruitment of different repair proteins to methylated chromatin. A prime example is the recruitment and binding of 53BP1 to DSBs, which is a pivotal determinant of NHEJ. This is governed in part by binding of the tandem Tudor domain of 53BP1 to di-methylated Lys20 of histone H4. Since H4K20 di-methylation is highly abundant and is present on over 80% of nucleosomes, the current model for recruitment of the tandem Tudor domain of 53BP1 to methylated Lys residues represents a bivalent process: the exposure of pre-existing H4K20me2 to permit 53BP1 recruitment, and the catalysis of de novo H4K20me2 at DSB sites.

H4K20 methylation is mediated by members of the KMT5 family of methyltransferases: KMT5A (SETD8) catalyses mono-methylation of H4K20, whereas KMT5B/C (SUV420H1/2) catalyses di-methylation. In addition to 53BP1, several other methyl-lysine readers bind to H4K20me2, including L3MBTL1 and KDM4A. Several papers have demonstrated that RNF8 and RNF168 stimulate ubiquitination of these reader proteins, leading to the eviction or degradation of these factors at DSBs in a manner dependent on the chaperone Valosin-containing protein (VCP) (Acs et al. 2011, Mallette et al. 2012). This exposes di-methylated H4K20 and permits 53BP1 recruitment, since the affinity of 53BP1 for H4K20 is lower than that of either reader protein.

In addition, it is proposed that H4K20me2 is also induced locally at DSBs; accordingly, H4K20me increases at site-specific DSBs. The mono-methyltransferase activity of KMT5A is necessary for 53BP1 recruitment, and KMT5A is recruited to DSBs through interaction with KU70 and DNA-PKcs. However, H4K20me1 at DSB sites is insufficient for 53BP1 nucleation, which also requires the di-methyltransferase activity of KMT5B/C. It is currently unclear how the two facets of this model overlap, and whether de novo H4K20me catalysed by KMT5A/B/C at DSBs is also coated with L3MBTL1 and KDM4A, which needs to be removed by RNF8 or RNF168. Further studies must also address whether KDM4A and KMT5 act in opposition to regulate H4K20me.

Furthermore, Lys methylation of DNA-PKcs itself also contributes to NHEJ. Dynamic methylation of DNA-PKcs on three Lys residues, K1150, K2746 and K3248, is important to promote chromatin recruitment after DSB formation, although the KMT has yet to be identified. These methylated residues act as a binding platform for the heterochromatic factor HP1β, promoting recruitment of DNA-PKcs to chromatin. This then acts as a platform for KMT5A recruitment, and ultimately permits 53BP1 binding to H4K20me2 (Liu et al. 2013). It is an intriguing notion that, as is the case with H4K20me, these methylated residues may be masked by further adaptor proteins to regulate NHEJ. In addition, whilst DNA-PKcs Lys methylation increases after DSB induction, Lys methylation of its binding partner, KU80, concomitantly decreases (Liu et al. 2013). Thus, Lys methylation of proteins within the same DNA repair complex is tightly regulated in a spatiotemporal fashion.

H3K36 methylation also plays a role in promoting both NHEJ and HR, depending upon the context of methylation. Di-methylation of H3K36 occurs rapidly upon DSB induction through the recruitment of the KMT fusion protein SETMAR, and both the enzyme and modification are required for NHEJ. In a similar fashion to H3K9me, KMT2A is displaced from H3K36me2 in an ATM-dependent manner, exposing binding sites for methyl-lysine readers such as PHRF1, which enhances MRN complex recruitment. In addition to these early roles in DSB signalling, H3K36me3 also plays an integral role in HR repair, especially of actively transcribed genes. This is in agreement with the known role for the only known H3K36me3 methyltransferase, SETD2 (KMT3A), in HR (Aymard et al. 2014). Once again, this occurs via the recruitment of a methyl reader, LEDGF, via interaction of its PWWP domain and the methyl Lys residue. LEDGF promotes recruitment of the pro-resection factor CtIP (RBBP8), leading to resection of the break and consequent repair by HR. Unlike di-methylated H3K36, no increase in the tri-methylated form is seen after DSB induction. It is unclear whether existing H3K36me3 needs to be 'exposed' to allow LEDGF binding, or whether de novo methylation does occur at DSBs in human cells, although it seems likely that the H3K36me mark is pre-established.

Methylation of Lys 4 of histone H3, and in particular the KMT family of methyltransferases that catalyse methylation of this residue, has also been implicated in DSB repair. Members of this family of KMTs accumulate at DSBs induced in both yeast and human cells, along with H3K4me3. In agreement, cells lacking Set1 (the yeast homologue of SETD1A [KMT2F] and SETD1B [KMT2G]) are defective for DSB repair by NHEJ (Faucher and Wellinger 2010). This seems to be independent of transcription/shutdown, but may be linked to chromatin remodelling by the ING family of methyl reader proteins, which is also crucial for other aspects of the DDR. Whether the human SETD1A and SETD1B enzymes play analogous roles in human cells is unknown, and the complexity of H3K4 methylation in DSB repair situations is further complicated by the apparent role for H3K4 demethylation in the efficient repair of DSBs (see later).

Importance of methyl readers in governing DSB repair

As outlined earlier, methyl-Lys readers are clearly vital in governing all steps of DSB repair. It is interesting to speculate that the presence of pre-existing histone modifications such as H4K20me2, H3K9me2 and H3K36me3 in specific genomic regions may govern the type of DSB repair that occurs within that region, simply through the recruitment of pro-NHEJ or pro-HR methyl readers. However, since several of these marks overlap within gene bodies, further investigations must be performed to determine how these methyl-Lys marks regulate the balance between NHEJ and HR. Moreover, since H3K36me2 favours NHEJ, whereas H3K36me3 promotes HR, it is clear that the degree of Lys methylation is also an important factor in DSB repair pathway choice, likely through the affinity of different methyl Lys readers for di- or tri-methylated marks.

Perhaps the best-studied methyl-Lys reader in the DDR is 53BP1. As previously discussed, 53BP1 binds to H4K20me2 marks via its tandem Tudor domains (as well as ubiquitinylated H2A and H2AX) and plays a prominent role in controlling DSB repair (reviewed in detail in Panier and Boulton 2014). It is of interest to note that binding of 53BP1 to methyl-lysine is not restricted to H4K20me: various reports have also demonstrated binding of the tandem Tudor domains to methylated p53, Rb and H3K36me.

Another particularly important reader of methyl-Lys is VCP (also called p97). VCP is an essential and highly conserved ATP-dependent chaperone implicated in a wide range of cellular processes, including DNA repair. As previously described, VCP plays an important regulatory role in recruiting 53BP1 to damaged chromatin, by chaperoning and removing L3MBTL1 and KDM4A from pre-existing H4K20me2 marks, 'unmasking' them to permit 53BP1 binding. Several recent papers have identified a novel family of KMTs, one of which (VCP-KMT or METTL21D) methylates VCP on Lys315 in vitro and in vivo (Kernstock et al. 2012, Cloutier et al. 2013, Fusser et al. 2015). Lys methylation of VCP does not affect basal activity, ATP hydrolysis or stability of VCP, and is not required for survival under unstressed conditions: however, it is entirely conceivable that this modification may regulate other functions of VCP, perhaps including its role in DSB repair. Thus, it is of great interest to the field to examine whether VCP-SET is itself regulated during the DNA damage response. Furthermore, as described later, Lys methylation is not restricted to DSB repair, and the potential role of VCP and its methylation in these other pathways has not been examined.

In addition, the CHD (chromodomain-helicase-DNA binding) family of chromatin re-modellers also acts as a reader of methyl-lysine marks, and a growing body of evidence suggests that they also play a role in regulating the DNA damage response. Peptide-binding studies have revealed that different CHDs demonstrate specificity to different methyl-lysine moieties (Bartke et al. 2010). Of particular interest in regulating the DDR are CHD1, CHD3 and CHD4 (reviewed in Stanley et al. 2013). Recently, a role for CHD1 in HR has been described, and depletion of CHD1 sensitises cells to DSBs (Kari et al. 2016). This is likely due to the ability of CHD1 to stimulate recruitment of CtIP to chromatin to promote end resection, perhaps by opening the chromatin surrounding the DSB to facilitate access. Interestingly, CHD1 binds methylated H3K4 via its tandem chromodomains, suggesting an indirect role for H3K4 methylation in recruiting CtIP. CHD3 also plays a key role in repair of DSBs. Under unperturbed conditions, CHD3 binds chromatin-associated KAP1; this interaction is perturbed upon DSB induction by ATM-mediated phosphorylation of KAP1. This dispersal of CHD3 activity is necessary for repair of heterochromatic DSBs. Finally, the H3K9me3 reader CHD4 also plays a role in HR, although it may have a somewhat paradoxical role. Loss of CHD4 sensitises cells to a variety of genotoxic stress, including DSBs, and has been implicated in HR. However, loss of CHD4 may also engender resistance to cisplatin in the absence of BRCA2, independently of HR.

Recent evidence suggests that the translesion synthesis factor REV7 (MAD2L2) plays an important role in controlling resection, acting to inhibit 5′ end resection downstream of 53BP1 (reviewed by Sale 2015). Of particular interest are reports that the HORMA domain of REV7 and the related protein MAD2 both bind methylated H3K4 (Schibler et al. 2016), suggesting that Lys methylation may directly regulate end resection of DSBs by controlling anti-resection factors.

Although these data portray a convincing role for methyl-lysine readers in regulating DSB repair and pathway choice, clearly more work remains to be performed to understand how these factors function and how they themselves are regulated.

DSB repair and Lys demethylation

In addition to pre-existing and de novo Lys methylation, demethylation of methylated target proteins also plays an important role in the repair of DSBs and in repair pathway choice.

H3K36 methylation, and subsequent repair processes, may also be regulated through demethylation. The JmjC-containing protein KDM8 (JMJD5) has been shown to regulate H3K36me2 in *C. elegans*, and to play a crucial role in regulating HR by promoting RAD51 eviction after strand invasion, and thus in HR-mediated repair (Amendola et al. 2017). However, its catalytic role in histone demethylation in humans is contentious, with several studies unable to demonstrate JMJD5-catalysed demethylation, suggesting that JMJD5 hydroxylates other substrates (Markolovic et al. 2016). Precisely how removal of di-methyl marks from H3K36 regulates HR remains to be determined.

The H3K4 demethylase KDM5B is also required for DSB repair and for maintaining genome stability after ionising radiation-induced DSBs (Li et al. 2014). The demethylase activity of KMD5B is required for the recruitment of KU70 and BRCA1 to sites of DSBs, and loss of KDM5B sensitises cells to genotoxic damage. In agreement, both HR and NHEJ efficiency is decreased upon KDM5B depletion. This has been attributed to demethylation of H3K4me3 at DSB sites, although how this regulates Ku70 and BRCA1 recruitment is unclear. Peptide pull-down data suggest that BRCA1 can bind unmodified H3 peptides more efficiently than tri-methylated sequences, although the mechanisms remain to be investigated.

KDM4B also seems to be involved in DSB resolution after ionising radiation. Resolution of DSBs is retarded after KDM4B depletion, and this is concomitant with decreased H3K9me levels at shorter time points following DSB induction. How this demethylation is temporally controlled, and how this fits into the SUV39H1-mediated methylation of H3K9 after DSB initiation, is unknown.

Demethylation of DDR regulators themselves is also important in DSB repair. The critical adaptor protein MDC1 regulates the DNA damage-induced ubiquitin cascade mediated by RNF8/RNF168, ultimately allowing recruitment of repair factors such as BRCA1 and 53BP1. MDC1 is Lys methylated on Lys45, and this methylation decreases upon DSB formation in a JMJD1C-dependent manner (Watanabe et al. 2013). Mechanistically, this demethylation promotes the interaction between MDC1 and RNF8, allowing recruitment of BRCA1 and subsequent DSB repair. However, questions remain over the KMT responsible for MDC1 methylation and how this methylation is 'protected' from JMJD1C activity until DSB signalling is activated.

As discussed earlier, much of the research on Lys methylation and DSB repair has focused on the interplay between HR and NHEJ, and on the role of chromatin modifications in this process. However, several alternative pathways of DSB repair also exist, including alternative end-joining (Alt-EJ) and single-strand annealing. Although significant mechanistic advances have recently been made in the understanding of these DSB repair pathways, nothing is yet known about the role of Lys methylation and KMTs in these processes, providing an intriguing avenue of future study. Interestingly, the NSD1 (MMSET) methyltransferase has been shown to regulate class switch recombination by modulating H4K20 methylation and thus 53BP1 recruitment (Pei et al. 2013), suggesting that other KMTs may also be vital for this process.

Lysine methylation during DNA replication and replication stress

Lysine methylation also plays an important role in regulating DNA replication and in dealing with replication stress (defined in Chapter 1). As noted for double-strand break repair, this involves Lys methylation of both histone and non-histone proteins, although the majority of

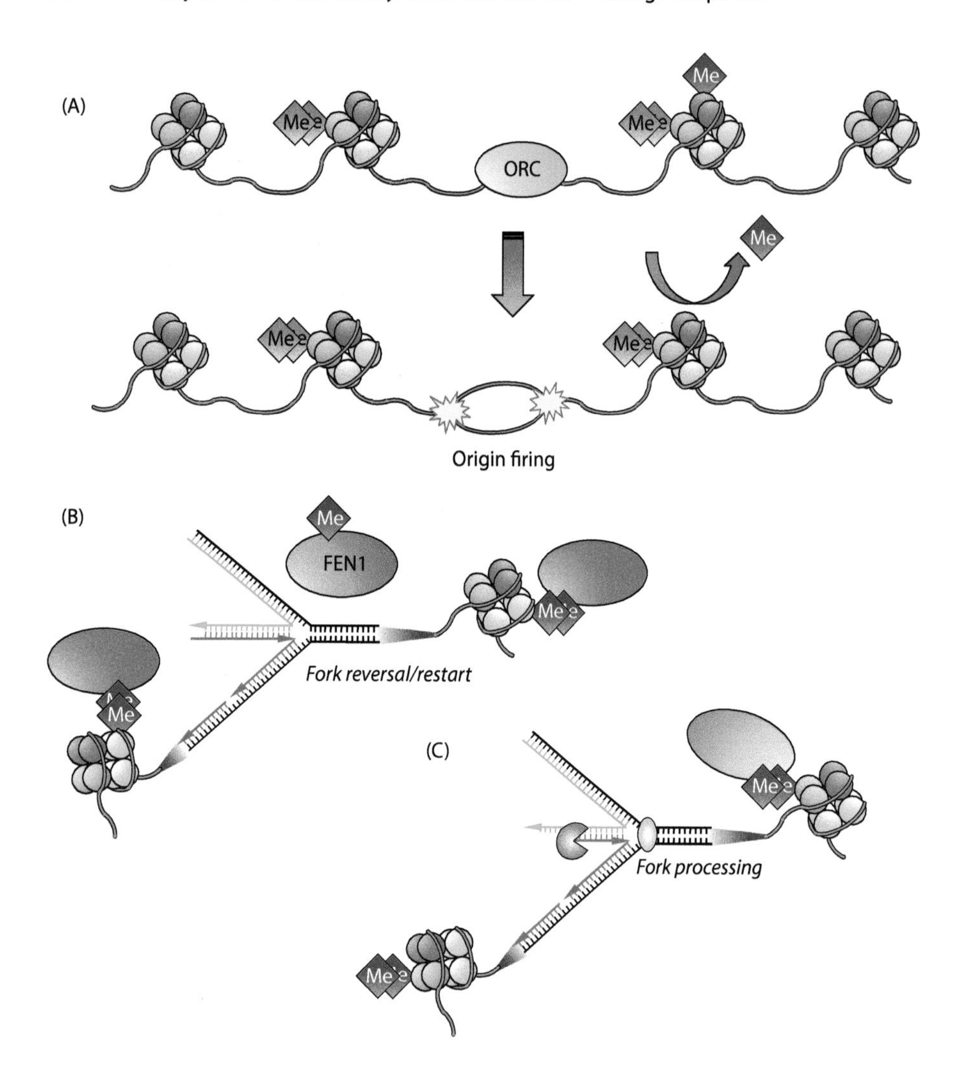

Figure 8.2 **Lys methylation in replication stress.** Both replication and replication stress are regulated by the methylation of Lys residues of both histone and non-histone proteins. In addition, KMTs and KDMs regulate histone methylation on nascent chromatin (not shown). (A) Pre-existing methylation of histones provides areas for ORC binding and regulating origin firing. Demethylation of the H3K4me methyl mark I (lower panel) is required for CDC45 and PCNA recruitment, and for origin firing. Similar mechanisms also regulate origin firing during replication stress. (B) Histone and non-histone (e.g. FEN1) Lys methylation also regulate fork reversal and restart during replication stress. (C) KMTs (such as KMT2B/C; blue oval) regulate nucleolytic processing of stalled (yellow oval) replication forks, especially in the absence of protective factors such as BRCA2.

research to date has focused on histone Lys methylation (Figure 8.2). Unlike Lys methylation and DSB repair, however, there is much less known about this PTM in controlling genome duplication. Moreover, histone Lys methylation in this context must be considered upon the background of chromatin reassembly following DNA replication, as the presence of methyl lysine marks must be tightly controlled on both newly synthesised and parental histones.

Lys methylation and early DNA replication

Methylation of Lys residues on histones H3 and H4 modulates several early stages of DNA replication, including origin recognition, licencing and firing, as well as replication elongation itself. Although the precise mechanisms behind origin recognition are still unclear, it seems that defined origins are enriched for di-methylated H4K20. Several components of the origin recognition complex (ORC), including the methyl-reader ORC1, bind H4K20 at these sites, creating areas of high ORC occupancy. Importantly, depletion of KMT5B/C, the KMTs that catalyse H4K20me2, reduce ORC1 occupancy (Beck et al. 2012). H4K20 methylation also plays a key role in origin licencing, and methylated H4K20 levels are elevated at origins of replication. As previously described, KMT5A mono-methylates H4K20, which is a prerequisite for further methylation by KMT5B/C. To prevent re-replication, KMT5A is degraded in a CRL4-dependent fashion prior to S phase. Tethering of KMT5A to a specific locus leads to elevated H4K20 methylation, which is dependent on KMT5B/C, and loading of the pre-RC complex to that locus. Thus, it is clear that regulation of H4K20 methylation by the KMT5 family is vital in controlling the early stages of genome duplication.

The recruitment of CDC45 to the pre-RC complex is required for origin firing. Interestingly, Lys methylation of H3 on Lys 4, 36 and 79 may also play a role in regulating replication origin firing by controlling this recruitment. Whilst CDC45 binds to unmodified H3 peptides, it cannot interact with those tri-methylated at Lys4 (Liu et al. 2010), suggesting that methylation of this residue may provide a mechanism by which origin firing is regulated.

Although a role for H3K4 methylation has been suggested in regulating origin firing during replication stress (see later), no role has yet been described for H3K4 methylation in CDC45 binding during unperturbed replication. Moreover, it is unclear whether the KMT2 family of methyltransferases that catalyse H3K4 methylation play a role in this process. In addition, since H3K4me3 is enriched at actively transcribed promoters, it may be that this mechanism acts to prevent origin firing near actively transcribed regions. In addition to H3K4me3, studies in yeast suggest that different states of H3K36 methylation are also important in regulating CDC45 origin binding (Pryde et al. 2009): mono-methylation is enriched at sites of CDC45 binding at early replicating regions, whereas demethylation is associated with later-replicating origins. However, it is unclear whether this regulation occurs in humans, or whether NSD1 (KMT3B) and SMYD2 (which catalyse H3K36 methylation in humans) are involved. Finally, di-methylated H3K79 is also enriched at replication origins in human cells, although it seems that this may play a repressive role in preventing deleterious re-replication. Depletion of DOT1L (KMT4), which catalyses H3K79 methylation, leads to re-replication of certain genomic regions (Fu et al. 2013). Thus, different Lys methylation marks on varying histone residues impact replication initiation in different ways, and it is likely that different organisms display differential methylation patterns depending upon conservation of KMTs and methyl-lysine readers.

A small but growing body of evidence suggests that components of the replicative machinery are Lys methylated. In addition to its role in regulating unperturbed origin firing via H4K20 methylation (see earlier), KMT5A also methylates the key replication protein PCNA on Lys 248 both in vitro and in vivo. Perturbation of this methylation by depletion of KMT5A or mutation of the target Lys destabilised PCNA, whilst methylation of this site increased interaction with the FEN1 endonuclease, which is important during Okazaki fragment maturation. Accordingly, loss of PCNA methylation retards Okazaki fragment maturation and impairs replication, causing DNA damage and replication stress (Takawa et al. 2012).

Excitingly, several components of the replicative MCM helicase have also been shown to be Lys methylated using large-scale mass-spectrometry approaches (Moore et al. 2013), including MCM2, MCM3, MCM4, MCM5 and MCM6, as well as ORC2. Although no functional significance has yet been attributed to these modifications, and the KMTs responsible have yet to be identified, it is tempting to speculate that these modifications may influence origin firing, replication progression or perhaps even replication termination, or to act as a binding platform for methyl-reader cofactors necessary for replication.

Thus, methylation of key non-histone replication factors may also play a key role in safeguarding genome stability and promoting unperturbed DNA replication. However, further examination is needed to establish exactly how Lys methylation regulates this key cellular process.

Re-establishment of methyl histone marks during replication

Unlike with DSB repair, the role of histone modifications in replication and replicative stress is complicated by the fact that these modifications must be considered upon the background of chromatin disassembly and reassembly at the replication fork, and the maintenance of epigenetic marks on nascent chromatin. Current data suggest that parental histone tetramers bearing methyl-Lys marks randomly segregate to either of the two nascent strands of DNA during replication. However, this affects the propagation of modifications to the daughter strands, influencing not only epigenetic inheritance but also impacting the DDR. Therefore, the re-establishment of methyl-Lys marks is of utmost importance for cellular and organismal viability. Two major chaperones control the reassembly of chromatin at the replication fork: CAF-1, which deposits H3.1-H4 at a rate coupled to synthesis, and ASF1, which associates with the MCM helicase and co-ordinates recycling and deposition of newly synthesised histones (Burgess and Zhang 2013). Once the newly synthesised and parental nucleosomes have been assembled onto the nascent DNA, pre-existing methyl marks on both parental and newly synthesised nucleosomes recruit KMTs to further modify the newly deposed chromatin. Several KMTs contain methyl-Lys binding domains that recognise and bind methylated histones to recruit them to sites of replication, including SUV39H1, SETDB1, KMT2A and KMT5B/C. Several co-exist in macromolecular complexes that couple methyl-Lys recognition with the enzymatic activity required for further methylation. One example is the EZH2 (KMT6A) binding partner EED, which recruits the PRC2 complex containing EZH2 to sites of H3K27me3. This permits modification of neighbouring histones and the re-establishment of the H3K27 mark. Recent evidence suggests that this is mediated via the transcriptional co-repressor CDYL, which bridges the MCM helicase, CAF-1 and EZH2 to promote Lys methylation of newly deposed histone H3 (Liu et al. 2017).

In addition, KMTs can also act on non-nucleosomal histones before their incorporation into chromatin. This includes both parental histones evicted in front of the replication fork as well as newly synthesised histones. For example, SETDB1 (KMT1E) mono-methylates newly synthesised, soluble H3 prior to incorporation at Lys9 (Loyola et al. 2009). It is unclear whether deficient H3K9 methylation impacts nucleosome formation and/or replication rates, or whether other KMTs also play a role in regulating histone removal/deposition. However, as is observed with defects in histone chaperones, it is likely that loss of KMT activity at the fork results in replication stress as a result of improper nucleosome deposition or defective recruitment of methyl-binding co-factors.

Moreover, protein methylation may also regulate the function of the histone chaperone machinery itself. Methylation of glutamine residues in H2A disrupt the binding to the histone chaperone FACT (Tessarz et al. 2014), leading to reduced histone incorporation. This raises the possibility that lysine methylation may also have an analogous effect on the replicative chaperones CAF1 and ASF1, or may regulate their function directly.

Lys methylation during replication stress

In addition to regulating 'unperturbed' replication, Lys methylation also plays a role in regulating the cellular response to replication stress and in maintaining genome stability at damaged replication forks resulting from unrepaired replication stress.

A crucial aspect of the cellular response to replication stress is activation of the intra-S phase checkpoint, which prevents unscheduled DNA synthesis in the presence of replication stress by downregulating global origin firing upon detection of replication stress, precluding replication of damaged DNA and maintaining genome stability (described in Chapter 1). As mentioned earlier, origin firing is regulated in part by CDC45 binding to the pre-RC complex, and this may be influenced by Lys methylation of Lys4 of histone H3. In line with this, the methyltransferase KMT2A (MLL1) has been shown to regulate origin firing under conditions of replication stress by tri-methylating H3K4 at late origins (Liu et al. 2010). Ultimately, this contributes to activation of the intra-S phase checkpoint and the maintenance of genome integrity. In general agreement, shRNA-mediated depletion of WDR5, which acts as a crucial cofactor of the KMT2 family of KMTs, also perturbs CDC45 chromatin binding (Carugo et al. 2016), although the impact on the intra-S phase checkpoint is unknown. Moreover, RNAi-mediated knockdown of other components of the KMT2 enzyme complexes also perturb this checkpoint after re-replication and lead to genome instability (Lu et al. 2016). However, these data cannot exclude the role of other KMT2 family methyltransferases in this process, such as KMT2B, KMT2D, SETD1A or SETD1B, since WDR5 and other cofactors are required for the KMT activity of all these enzymes.

Intriguingly, data from yeast suggest that H3K4 methylation is also required to maintain genome stability upon replication stress, and that this is due to the activity of the yeast H3K4 KMT Set1 (Faucher and Wellinger 2010). Taken together, these data suggest that H3K4 methylation by the KMT2 enzymes is vital to maintain genome stability upon replication stress, although further work is clearly necessary to delineate which enzymes are involved and to elucidate the underlying mechanisms. Moreover, as is the case with DSB repair, it is unclear which enzymes are involved in this process in human cells.

A growing body of recent literature demonstrates that maintenance of replication fork stability under conditions of replication stress is vital to ensure continued genomic integrity (reviewed by Higgs and Stewart 2016). Protective factors such as BRCA1, BRCA2 and members of the Fanconi anaemia pathway act to promote stabilisation of stalled forks to allow repair by HR. Perhaps unsurprisingly, Lys methylation has also been implicated in this process, although the targets for methylation have yet to be identified. In particular, recent work has demonstrated that the KMT2B and KMT2C enzymes, together with their cofactor PTIP, promote replication fork instability and degradation in the absence of BRCA1/2. Acquisition of chemoresistance in BRCA-deficient cells is associated with loss of PTIP and the resultant increase in fork protection (Ray Chaudhuri et al. 2016). Whether the other members of the KMT2 family play an analogous role in the absence of other fork protection factors is unknown, but can be hypothesised, for example, SETD1B and FANCD2. Moreover, since fork reversal is likely a prerequisite for fork stability, and as PARP1 plays a key role in regulating fork reversal and fork degradation (Ying et al. 2012, Neelsen and Lopes 2015), it may be that Lys methylation of PARP1 by SMYD2 (see earlier) may also regulate replication fork stability. However, whether this involves the KMT2 methyltransferases, or other KMTs, is unclear. Finally, it is unclear whether these KMTs regulate fork stability per se or alter efficiency of other related processes.

In addition to fork stabilisation, another crucial aspect of the response to replication stress is the restart of stalled replication forks. The KMT SETMAR is vital for fork restart after replication stress, as well as other roles in DSB repair (De Haro et al. 2010). This restart capability is dependent on the catalytic SET domain of SETMAR, but not on its ability to catalyse H3K36 methylation (Kim et al. 2015), perhaps because SETMAR cannot catalyse H3K36 methylation in vitro. Together, this suggests that Lys methylation of another target is involved. However, the identification of only a single target (snRNP70) of SETMAR-dependent methylation has failed to shed any light on how this KMT regulates restart. Interestingly, SETMAR interacts with and promotes the loading of the EXO1 exonuclease to ssDNA at stalled replication forks (Kim et al. 2017) and stimulates the activity of TOP2A (Williamson et al. 2008); although whether these functions are mediated via auto-methylation of SETMAR, or via methylation of these substrates, is unclear. Recently, the SMYD3 enzyme (KMT3E) has been shown to regulate expression of HR genes, ensuring that there are sufficient levels of these factors to allow restart of stalled forks upon replication stress (Chen et al. 2017).

Recent studies from the Richard group have demonstrated that the KMT SETD7 methylates the endonuclease FEN1, which plays a vital role in Okazaki fragment maturation. However, methylation by SETD7 on Lys 377 does not affect FEN1 activity, but rather is required for the response to replication stress (Thandapani et al. 2017). Loss of this residue renders cells exquisitely sensitive to replication stress and suggests that Lys methylation of the replicative machinery is vital to maintain genome integrity after replication stress.

As well as their roles in the replication stress response, several KMTs also play well-characterised roles in regulating transcriptional initiation, elongation and termination. These two roles are not necessarily separate: the KMT2D enzyme has been shown to regulate transcription-associated replication stress and maintain genome stability (Kantidakis et al. 2016). In addition, recent studies have suggested that H3K9me2/me3 suppresses transcription-associated replication stress (Zeller et al. 2016), suggesting that SUV39H1/2 may also be involved in the replication stress response (discussed in detail in Chapter 12).

Methyl readers and genome duplication

As is the case for DSB repair, methyl readers are also important in genome duplication and in the cellular response to replication stress. However, the literature surrounding this aspect of the Lys methylation field is much less comprehensive than that for double-strand break repair, and much more remains to be determined. However, several more recent studies have begun to shed light on the role methyl readers play during replication stress.

Several publications have demonstrated a key role for the MMS22L-TONSL complex in maintaining genome stability after replication stress. More recent data now suggest that this complex acts as a methyl reader, 'reading' the absence of methylation (me0) on H4K20 that is present on newly synthesised histones during S phase (Saredi et al. 2016). This therefore acts to recruit the MMS22L-TONSL complex to replication forks and post-replicative chromatin, where it can act to load RAD51 to repair replication stress-associated lesions. Whilst H4K20me0 on newly synthesised histones seems to be protected from methylation, it is unclear whether this is due to MMS22L-TONSL or another methyl reader.

In addition to its already-discussed role in DSB repair, the methyl binder 53BP1 also plays an important role in marking under-replicated DNA as it transits through mitosis and into G1 phase, forming G1-specific 53BP1 'bodies' (Lukas et al. 2011). The prevalence of these bodies increases upon unresolved replication stress, suggesting that, since 53BP1 chromatin binding is regulated by H4K20 methylation as well as H2A ubiquitylation, these marks may well decorate under-replicated DNA to allow protection by 53BP1. However, it is unclear how H4K20 methylation is regulated to allow 53BP1 body formation or how under-replicated DNA is detected to allow histone Lys methylation.

The TLS regulatory factor REV7 (MAD2L2) also plays a vital role in the cellular response to replication stress induced by intrastrand cross-links, as well as its role in controlling end resection during DSB repair. Indeed, recent studies have demonstrated that REV7 is mutated in patients suffering from the replication stress disorder Fanconi anaemia (Bluteau et al. 2016). As mentioned earlier, REV7 contains a HORMA domain that binds to H3K4 methylated on Lys 4. Whilst the function of this domain in REV7 is unknown, it is tempting to speculate that Lys methylation may play some role in regulating the Fanconi anaemia pathway by regulating the activity or recruitment of REV7. Indeed, in line with this, the biallelic mutations of REV7 observed in patients with Fanconi anaemia occur within this methyl-binding HORMA domain.

Another methyl reader involved in genome duplication is L3MBTL1, which plays roles in both DSB repair and replication. L3MBTL1 interacts with several of the replisome components, notably CDC45, MCM2-7 and PCNA, and is required for normal replication fork progression. Depletion of this factor sensitises cells to replication stress and increases genomic instability (Gurvich et al. 2010). This raises the possibility that H4K20me2, the binding site for the MBT domains of L3MBTL1, is crucial to recruit L3MBTL1 and therefore to promote replication. Moreover, since VCP plays a key role in removing ubiquitinylated L3MBTL1 at DSBs, it may be that a similar mechanism exists to negatively regulate origin firing. In addition, since VCP is itself Lys methylated (see earlier), this may also play a role.

Replication stress and Lys demethylation

Unsurprisingly, since Lys methylation plays an important role in both genome duplication and replication stress, KDMs also regulate these processes, although very little is known about the mechanistic details. For example, the H3K9me demethylase, KDM4D, is required for pre-RC complex formation; depletion of this KDM impairs CDC45 recruitment and origin firing, and reduces replication elongation (Wu et al. 2017). This seems to be linked to H3K9 demethylation, as reduction of H3K9 at replicative origins is required for efficient DNA replication. Moreover, since H3K9 methylation is lost on nascent DNA, concomitant with increased H3K9 acetylation, it seems that a dynamic demethylation may be necessary for DNA replication (Aranda et al. 2014). Furthermore, and in agreement with the hypothesis that H3K4me2/3 prevents origin firing (see earlier), demethylation of this mark by KDM5C promotes replication origin firing by driving binding of PCNA and CDC45 (Rondinelli et al. 2015). Whether these or other Lys demethylases also act on non-histone substrates to regulate replication and replication stress remains to be determined.

Regulation of Lys methylation by DDR-induced post-translational modifications: The missing link?

Although our understanding of the function and structure of much of the Lys methylation apparatus involved in the DDR has increased significantly in the last few years, the regulation of these proteins is still poorly understood. Little, if anything, is known about how their ability to methylate, demethylate and bind Lys residues in response to genotoxic damage is governed. Since the DDR is governed by a complex array of overlapping PTMs, and as Lys methylation plays a key role, this is clearly a crucial area of future work. Nevertheless, there are several examples in the literature which serve to underline the importance of this area of study.

The apical kinases that govern much of the DDR, ATM, ATR and DNA-PK, phosphorylate a vast number of target proteins to control DNA repair and the associated checkpoints. Amongst these are several components of the Lys methylation pathway, although only a limited number of KMTs and KDMs have been demonstrated to be phosphorylated. For example, EZH2 interacts with, and is phosphorylated by, DNA-PK at Ser 729. This phosphorylation impairs EZH2 KMT activity and may serve to regulate the balance between apoptosis and repair after DNA damage (Wang et al. 2016b). Another example involves the ATR-dependent phosphorylation of KMT2A on Ser 516, which perturbs its interaction with the SCF E3 ligase, preventing its degradation. Therefore, ATR-dependent phosphorylation of KMT2A plays a key role in the response to replication stress and activation of the intra-S phase checkpoint (Liu et al. 2010). Furthermore, several KMTs are putative targets of ATM/ATR-mediated phosphorylation, including KMT1C and KMT2D (Matsuoka et al. 2007), although the functional consequences are unknown.

In addition to phosphorylation, it is likely that KMTs, KDMs and methyl readers are modified by other DDR-induced PTMs including ubiquitination and SUMOylation, contributing to their activity and/or abundance. In line with this, KDM4A is regulated by the DDR Ub E3 ligases RNF8 and RNF168 during DSB repair, as mentioned earlier (Mallette et al. 2012), which regulates binding of 53BP1 to H4K20me3. Whether other factors involved in the Lys methylation apparatus are ubiquitinated or SUMOylated by DDR Ub/SUMO ligases remains to be examined, although several of these proteins (KDM5C, SETDB1 and KDM1) are indeed SUMOylated (Bruderer et al. 2011). In addition, several KMTs and KDMs are regulated and/or degraded in a cell cycle–dependent fashion (see Black et al. 2012 for a review), which may be significantly altered during checkpoint activation upon DNA damage. For example, KMT5A is regulated by three separate ubiquitin ligase complexes at various stages of the cell cycle, including SCF^{SKP2}, $CRL4^{CTD2}$ and APC^{CHD1}. Deubiquitination also plays a role in stabilising Lys methyltransferase/demethylase enzymes; for example, the DDR-associated DUB USP7 promotes the deubiquitination and stabilisation of PHF8 (KDM7B), which is crucial for DSB repair and recruitment of DDR factors BLM and KU70 to DSBs (Wang et al. 2016a). However, as is the case with phosphorylation, the functional impact of these PTMs on the KDMs/KMTs involved, and the ultimate effect on the DDR, is unknown. This clearly represents a pressing area for future study.

Summary: Lys methylation and the DDR

During the last few years, our knowledge and understanding of Lys methylation and demethylation in the DNA damage response have increased enormously. Studies have mainly focused on modification of histones and chromatin, and a complex picture has evolved, with many different histone methylation marks playing an important role in DNA repair. A growing body of evidence also demonstrates that Lys methylation of non-histone DDR proteins is vital in regulating their function. It is now abundantly clear that the enzymes that catalyse Lys methylation/demethylation, and those readers that bind these moieties, play a crucial role in regulating genome stability, especially during DSB repair and genome duplication.

However, as with many aspects of the DDR, summarising these data into a coherent picture of how Lys methylation regulates the DDR is challenging, and there are many contradictory aspects that remain unresolved. Moreover, how these enzymes are regulated by the various and complex signalling pathways involved in the DDR is barely understood. Further structural and molecular studies on KMTs, KDMs and methyl readers will provide greater insight into the function and regulation of these proteins, and allow us to understand their contribution to maintaining genome stability and safeguarding human health.

Crosstalk between arginine and Lys methylation factors

One intriguing but under-studied area of research is the crosstalk between Arg and Lys methylation, particularly the methylation of KMTs and KDMs by Arg methyltransferases (PRMTs). The next section of this chapter deals with Arg methylation in greater detail. However, it is interesting to note that several components of the KMT2 Lys methyltransferases, including the co-factor ASH2L, are Arg methylated by two different PRMTs (PRMT1 and PRMT5) (Butler et al. 2011). Although the impact of this Arg methylation is unknown, it may be that this regulates recruitment of the KMT2 enzymes to different substrates, affects chromatin interaction or affects formation of the KMT2 'holoenzyme' complex.

ARGININE METHYLATION AND THE DDR

Arginine methylation was identified over 45 years ago (Baldwin and Carnegie 1971, Brostoff and Eylar 1971); however, in comparison to lysine methylation, our knowledge into the precise functional significance of arginine methylation for the DDR is still in its infancy, with only a very small number of DDR proteins having been identified and characterised as bona fide PRMT substrates. Moreover, and in comparison to lysine methylation, no study has yet showed a direct role for histone arginine methylation as a mechanism to recruit protein complexes during the DDR; hence, methylation of non-histone proteins is to date the main mechanism by which arginine methylation influences DNA repair. Many of the identified substrates are considered core to the DNA damage response, including those involved in the replication stress response, DSB and BER. However, it is becoming increasingly apparent that the impact of arginine methylation is far greater than simply the methylation of DDR sensors, mediators, transducers and effectors. Indeed, epigenetic-mediated regulation of gene expression enabling cell fate choices, mRNA splicing and RNA biology after genotoxic insult also appears to be heavily regulated by arginine methylation during the DDR.

In this section of the chapter, we will first introduce the concept of arginine methylation and the enzymes that catalyse this modification, protein arginine methyltransferases (PRMTs). We will then discuss how studies using genetically null cells have unveiled the importance of particular PRMTs for maintaining genome stability. We will then consider how the identification of DDR proteins, as PRMT substrates has provided insight into the potential mechanisms by which PRMTs regulate genome stability and responses to genotoxic insult.

Protein arginine methyltransferases: The 'writers' of arginine methylation

Methylation of arginine residues is an highly abundant modification, with as many as 2% of arginine residues of nuclear proteins being methylated in vivo (Boffa et al. 1977). Arginine is a unique amino acid, as the guanidino group contains two nitrogens and five potential hydrogen bond donors. Methylation of either or both of the guanidino nitrogens removes a potential hydrogen bond donor and induces the formation of a more bulky and hydrophobic residue that can both positively and negatively regulate protein–protein or protein–nucleic acid interactions. Consequently, arginine methylation can have profound influences on protein behaviour and cellular function.

Arginine methylation is catalysed by a family of nine protein arginine methyltransferases (PRMTs) that use S-adenosyl methionine (SAM) as a donor of methyl groups. The predominant feature of all PRMTs is a conserved catalytic methyltransferase core of around 300 residues comprising a Rossmann-fold like seven-β-strand (7BS) methyltransferase domain that binds SAM, a double-E and THW sequence specific to the PRMT subfamily of methyltransferases, and a β-barrel that contributes to substrate binding (Katz et al. 2003, Cheng et al. 2005). In contrast, the N-terminal domains are relatively diverse, facilitating substrate specificity, catalytic activity via cofactor binding and dimerisation.

Enzymatically, PRMTs are classed into three groups: Type I, Type II and Type III. All PRMTs catalyse the transfer of a single methyl group onto a single terminal nitrogen of the guanidino group. Following this, Type I enzymes generate asymmetric dimethylation (ADMA), where two methyl groups are placed onto one of the terminal nitrogen atoms, whilst Type II enzymes generate symmetric dimethylation (SDMA), where one methyl group is placed onto each of the terminal nitrogen groups. PRMT7 is the sole Type III enzyme that, due to the substitution of conserved residues within the double-E and THW loop, can only generate a mono-methylated substrate (Zurita-Lopez et al. 2012).

PRMTs have been proposed to preferentially methylate arginine residues embedded within glycine-arginine rich (GAR) regions, the so-called RG/RGG motif (Thandapani et al. 2013). However, it is becoming increasingly apparent that the tertiary structure of the methyl-acceptor sequence is highly important; hence, many PRMT substrates are actually methylated at non-RG/RGG sequences. Since this means that *in silico* prediction of PRMT substrates is particularly challenging, the development of mass spectrometry approaches such as affinity purification of methylated proteins (Boisvert et al. 2003) and enzyme-substrate trapping interactions (Clarke et al. 2017) has been instrumental in deciphering the biological significance of methylation.

PRMT1 and PRMT5 are the main asymmetric and symmetric di-methyltransferases in mammalian cells, with 85% of all methylation events mediated by PRMT1 (Tang et al. 2000). For the majority of substrates, individual PRMT family members display substrate specificity. Indeed, the genetic deletion of either PRMT1 or PRMT5 results in early embryonic lethality (Yu et al. 2009, Tee et al. 2010), highlighting a lack of functional redundancy between the two main methyltransferases. One prominent exception to this rule is histone H4. Whilst PRMT1 and PRMT5 both di-methylate Arg3 on the histone H4 tail (H4R3me2), the biological effect of the methylation event is completely opposite. PRMT1-mediated asymmetric methylation is associated with histone acetylation and gene expression (Li et al. 2010). In contrast, PRMT5-mediated symmetric demethylation recruits DNMT3A, promoting DNA methylation and gene repression (Zhao et al. 2009). These findings clearly emphasise the biological specificity of asymmetric versus symmetric di-methylation events, and highlight the biochemical differences mediated by the positioning of methyl groups on the guanidino group.

Biochemical effects and 'readers' of arginine methylation

As with other post-translational modification, methylated arginine residues are recognised by specific 'reader' domains. By far the best-characterised is the Tudor domain that can interact with both methyl-arginine and methyl-lysine residues. However, since the accommodation of the dimethyl-arginine requires a narrow aromatic cage, very few Tudor domains can actually 'read' this PTM, and those that can show a strong preference for SDMA over ADMA (Tripsianes et al. 2011, Liu et al. 2012). Recently, Migliori et al. (2012) identified a second SDMA reader, the WD40 β-propeller domain of WDR5. Here, PRMT5-mediated symmetrical dimethylation of histone H3R2 promotes MLL/SETD1A recruitment via the WD40 domain of WDR5 and H3K4 trimethylation. Given that the majority of methylation in the cell is PRMT1-catalysed ADMA, it is speculated that other methyl-arginine reader domains exist, particularly those that recognise ADMA. Alternatively, rather than specific reader domains, methylation-induced increase in Van der Walls interactive forces could be the dominant biochemical mechanism by which alterations in protein interactions are mediated.

Methylation also has the potential to negatively regulate protein–protein/protein–nucleic acid interactions by removing potential hydrogen bonds. For example, ADMA of histone H3R2 by PRMT6 completely blocks the binding of proteins containing chromo, PHD, WD40 and Tudor domains (Iberg et al. 2008). Thus, in contrast to other post-translational modifications, the biochemical significance of a methyl-arginine is challenging to predict.

Regulation of PRMT activity

A key question that still remains largely unanswered is how the activity of PRMTs are directed towards a specific target in a stimulus-specific manner. This is particularly important, as PRMTs are largely ubiquitously expressed and constitutively active enzymes; hence, fine-tuning of activity is expected. One mechanism by which this could occur is via the PTM of PRMTs themselves. Supporting this hypothesis, we and others have demonstrated that methylation of proteins involved in the DDR is increased after DNA damage (Jansson et al. 2008, Guo et al. 2010, Clarke et al. 2017), implying that DNA damage signalling pathways potentially feed into the activation of PRMTs. Whilst the details by which this occurs has yet to be elucidated, there is precedence for PRMTs being targeted for phosphorylation. For example, CARM1 is phosphorylated at Ser217 by an unidentified kinase in a cell cycle–dependent manner, which disrupts hydrogen bond formation and SAM cofactor binding (Feng et al. 2009), whilst phosphorylation of PRMT1 at Tyr291 regulates substrate specificity (Rust et al. 2014). Furthermore, pre-existing post-translational modifications on PRMT target substrates can also influence PRMT activity. For example, CBP/p300-mediated acetylation of histone H3K18 facilitates CARM1-mediated H3R17 methylation (Daujat et al. 2002), whilst H4K5 acetylation preferentially facilitates PRMT5 rather than PRMT1-directed methylation of H4R3 (Feng et al. 2011).

A second mechanism by which the catalytic activity of PRMTs can be controlled is through binding partner interactions. This is exemplified by PRMT5. Structural analysis has demonstrated that PRMT5 is tightly associated with its essential cofactor MEP50, forming a functional PRMT5-MEP50 heter-octomeric complex, with MEP50 functioning by orienting the arginine-containing substrate towards the active site of PRMT5 (Antonysamy et al. 2012, Ho et al. 2013). The importance of MEP50 is highlighted by the findings that phosphorylation of PRMT5 by the constitutively active oncogenic Janus Kinase 2 (JAK2) mutant, JAK2-V617F, decreases MEP50 binding and enzymatic activity (Liu et al. 2011). Whilst MEP50 is the primary regulator of PRMT5, other binding partners can also modulate PRMT5 function, including the histone-binding protein COPR5, which specifically directs H4R3 rather than H3R8 methylation (Lacroix et al. 2008). Understanding the composition of the functional PRMT complex and the extent of PTM after a specific stimulus, for example, DNA damage, will be essential to fully appreciate how specific substrate targeting is achieved in a temporal and spatial manner.

Can arginine methylation be reversed?

As extensively reviewed in this chapter, lysine methylation levels are dynamically regulated through the combined action of lysine methyltransferases and lysine demethylases (KDMs). Since the levels of arginine methylation on a particular substrate appear to be induced by biological events, it would be expected that it should also be dynamically regulated by enzymatic removal. Surprisingly, the identification of a bona fide arginine demethylase has been particularly challenging and is still an unresolved question for the field.

The first indication that arginine demethylases existed was the finding that oestrogen stimulation of MCF7 cells promotes cyclic arginine methylation of histone H3R17 in a time frame that could not be explained by protein turnover (Métivier et al. 2003). Whilst the H3R17me2a demethylase has yet to be identified, this finding kick-started the hunt for arginine demethylases resulting in the proposition that JMJD6 demethylated H3R2me2 and H4R3me2 (Chang et al. 2007). However, structural analysis in combination with detailed mass spectrometry analysis has now conclusively shown that JMJD6 is actually a lysine hydroxylase, rather than an arginine demethylase, and does not catalyse H3R2me2 demethylation (Webby et al. 2009, Mantri et al. 2010). Interestingly, the Schofield group recently described that a select few members of the JmjC enzymes can actually catalyse mono- and di-demethylation (Walport et al. 2016). Using mass spectrometry to detect demethylation of R-methylated histone peptides as a change in mass, they identified KDM4E and KDM5C as effective in vitro demethylases. Remarkably, these enzymes were able to demethylate both ADMA and SDMA on the same arginine on histone H3/H4. Whilst these findings were incredibly encouraging in the discovery of the elusive arginine demethylase, in vivo demethylase activity of KDM4E and KDM5C towards ADMA and SDMA has so far not been demonstrated.

Alternatively, it has been proposed that the peptidylarginine deiminase (PADI) enzymes could function as arginine demethylases. PADIs convert arginine to the non-ribosomally encoded amino acid citruilline through an enzymatic reaction termed citrullination, or deimination. Since many sites of histone arginine methylation overlap with those that are citrullinated, it has been suggested that PADI enzymes may represent a non-conventional arginine demethylase. In this context, citrullination removes the methyl mark but is not a traditional demethylase, as the residue is not reverted back to an unmodified arginine. However, in vitro evidence that PADI enzymes function as arginine demethylases is largely lacking. Peptide substrates containing MMA, ADMA or SDMA are inefficiently, if at all, deiminated by PADI4 (Kearney et al. 2005) implying that the in vivo inverse correlation between methylation and citruillination could be actually be via citrullination antagonising methylation. Indeed, only unmodified or MMA arginine are viable substrates for PAD4 (Cuthbert et al. 2004).

With the seemingly elusive nature of the arginine demethylases, one hypothesis is that arginine methylation is simply reversed through protein turnover or via protein exchange with unmodified species. However, given the importance of this PTM for cellular events, we speculate that this is unlikely to be the whole story. Thus, the quest to identify a true arginine demethylase is continuing.

PRMT1, the main asymmetric methyltransferase, and the DDR

Since PRMT1 is the main arginine methyltransferase in mammalian cells, it is probably not that surprising that a number of studies have investigated the significance of this enzyme for the DDR. Clear indication that PRMT1 is critical for maintaining genome stability came from studies demonstrating that genetic null PRMT1 mouse embryonic fibroblasts (MEFs) exhibit spontaneous DNA damage, polyploidy and chromosomal instability, and checkpoint defects after DNA damage (Yu et al. 2009). Whilst this implicates PRMT1 in DSB repair and the replication stress response, only a handful of DNA damage proteins have been established as PRMT1 substrates: MRE11, hnRNPUL1, hnRNPK, BRCA1, 53BP1 and DNA polymerase β (**Figure 8.3**).

PRMT1 and DSB repair

The Richard group and their studies into MRE11 provided the first indication that PRMT1 regulates the DNA damage response. MRE11 is a component of the MRE11-RAD50-NBS1 (MRN) complex, a key sensor of DSBs and one of the earliest DNA repair complexes recruited to DSB contributing to the activation of ATM (Nelms et al. 1998, Lee and Paull 2005). The importance of MRE11 is underscored by the finding that deleterious mutations leads to the genomic instability disorder ataxia-telangiectasia (A-T)-like disease (A-TLD) (Stewart et al. 1999). MRE11 comprises an N-terminal nuclease domain, a C-terminal DNA binding domain and a GAR region. Since PRMT1 is known to methylate RG/RGG GAR motifs, and the function of the GAR domain was at that time unknown, Boisvert et al. (2005a) reasoned that it could

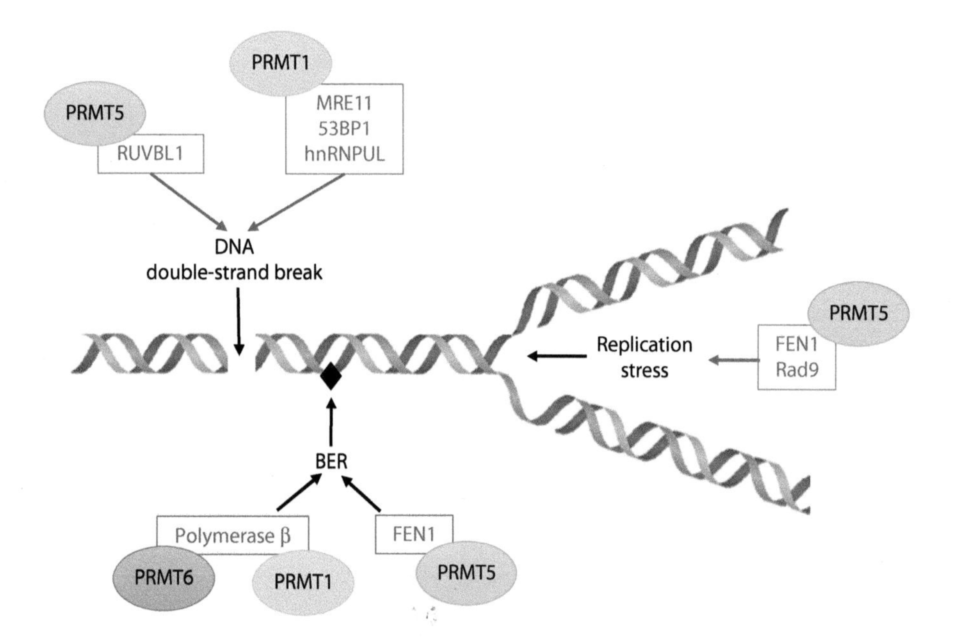

Figure 8.3 DNA repair proteins identified as PRMT substrates, and their involvement in double-strand break repair, base excision repair (BER) or replication stress.

be targeted for methylation. Using ASYM25, a pan-asymmetric dimethyl antibody generated against a peptide comprising of six consecutive asymmetric methylated RGG motifs, they identified MRE11 as an in vivo methylated protein. By using ES cells deficient in PRMT1, they demonstrated that PRMT1 was most likely the main PRMT mediating methylation, because methylation of MRE11 was markedly reduced in PRMT1 null cells. Interestingly, whilst the formation of the MRN complex was not affected by arginine to lysine substitutions within the GAR domain, exonuclease activity was severely impaired, as was the intra-S phase checkpoint response. Consequently, methyl-deficient MRE11 expressing cells aberrantly progressed through the cell cycle after DNA damage (Boisvert et al. 2005a). Moreover, methylation appears to promote the recruitment and anchoring of MRE11 to DSBs enabling DNA repair signal activation, and H2AX phosphorylation (γH2AX) (Boisvert et al. 2005b, Déry et al. 2008). PRMT1 was found localised with MRE11 in PML nuclear bodies, a nuclear structure that is thought to function as a dynamic reservoir of DNA repair proteins enabling local increases in protein concentration after extensive DNA damage (Conlan et al. 2004). PML bodies are also sites by which DDR proteins can acquire PTMs; thus, the finding that methyl-MRE11 is readily detected on nuclear structures suggests that arginine methylation of MRE11 within PML nuclear bodies prepares the MRN complex for further activation and recruitment to sites of DNA damage. Supporting this, PRMT1 interacts with MRE11 but not RAD50 or NBS (Déry et al. 2008).

To understand the physiological relevance of methyl-MRE11, a genetic knock-in mouse was generated where the methyl arginines within the GAR domain were replaced by lysines (the *Mre11*^{RK} mouse) (Yu et al. 2012). Like *Mre11* null mice, these mice are hypersensitive to ionising radiation (IR), genetically validating the importance of PRMT1-mediated methylation of MRE11 for DNA repair. However, not all MRE11-dependent phenotypes were methyl-dependent, as the *Mre11*^{RK} mouse was viable, fertile and able to undergo B cell class switch recombination (CSR). Isolated MEFs also displayed defective genomic stability and a loss of G2/M checkpoint control. Importantly, and in agreement with the previous cell culture–based study by the Richard group (Boisvert et al. 2005a), MEFs isolated from *Mre11*^{RK} embryos were also defective in exonuclease activity. Consequently, DSB formation in these cells failed to promote ATR activation, and RPA and RAD51 foci formation, implying defective end resectioning and ssDNA formation. Taken together, these genetic studies clearly demonstrate that methylation of the MRE11 GAR domain is physiologically relevant, enabling ATR pathway activation and DSB repair (Yu et al. 2012). One question that remains is whether PRMT1-mediated MRE11 methylation is a dynamic process induced by DNA damage signals. BTG2 (TIS21/PC3), a PRMT1 interacting partner that regulates its activity, is upregulated after DNA damage and is required for maintaining genome stability. Whilst MRE11 methylation was found to be enhanced after overexpression of BTG2 (Choi et al. 2012), the physiological relevance of BTG2 regulation is still unclear, as BTG2 overexpression altered ATM rather than ATR signalling (Yu et al. 2012). Studies enabling the genetic dissection of BTG2 for MRE11 methylation are still lacking; thus, the in vivo relevance of this pathways is still unclear.

PRMT1 and the methylation of RNA binding proteins

RNA binding proteins often possess GAR motifs, and are thus potentially methylated by PRMT1 (Thandapani et al. 2013). One such protein, hnRNPUL, possesses a C-terminal GAR motif that is required for association with NBS1 and recruitment of the MRN complex along with BLM helicase to damaged DNA (Polo and Jackson 2011). PRMT1 was shown to methylate hnRNPUL at a number of residues, including those within the C-terminal GAR motif (Gurunathan et al. 2015). Indeed, mutation of the methyl-acceptor sites to lysine resulted in reduced NBS1-hnRNPUL interactions and a failure to recruit hnRNPUL to sites of DNA damage. In contrast, methyl-deficient hnRNPUL-expressing cells were still able to recruit BLM, implying that methyl-hnRNPUL is functionally involved in regulation of upstream rather than downstream components of the DDR. Since hnRNPUL is involved in both RNA metabolism and DNA repair, these studies imply that arginine methylation could be one mechanism by which specific pools of hnRNPUL are directed to a certain biological response.

hnRNPK is also an RNA/DNA binding protein that has been implicated in numerous cellular events, including chromatin remodelling, transcription, mRNA splicing, mRNA stability and the DNA damage response. Phosphorylation of hnRNPK is elevated after DNA damage in an ATM-dependent manner, facilitating p53-mediated gene expression (Moumen et al. 2013). PRMT1 methylates hnRNPK (Chen et al. 2008, Yang et al. 2014). However, the significance of this event for the DNA damage response is unclear. Using the pan-methyltransferase inhibitor, MTA, methylation was shown to enhance hnRNPK-p53 interactions and transcriptional activity (Chen et al. 2008); however, mutation of the methyl-acceptor sites Arg296 and Arg299 to Lys actually promoted rather than inhibited DNA-damage-induced apoptosis in a p53-independent manner (Yang et al. 2014). One explanation for this discrepancy is that MTA will inhibit all methylation events in the cell, and decreases the methylation status on all arginine five sites (not to mention any lysine residues that are yet to be identified). Thus, it appears that Arg296 and Arg299 methylation directs a particular function of hnRNPK that suppresses DNA damage–induced apoptosis. How this occurs is unknown, but given that loss of hnRNPK-Arg296/Arg299 methylation sensitises cancer cells to DNA damage–induced apoptosis implies that targeting the arginine methylation of these residues could have anti-cancer properties.

PRMT1 and the transcriptional regulation of DNA repair

BRCA1 and its orthologue BRCA2 are cancer-susceptible genes; germline loss-of-function mutations give rise to a high predisposition to breast, ovarian and pancreatic cancers. BRCA1, in conjunction with its binding partner BARD1, forms a number of distinct complexes resulting in multiple cellular roles, including transcription and DNA damage repair. BRCA1 is known to regulate cell cycle checkpoints, DNA cross-link repair, replication fork stability and DSB repair. Perhaps the most defined role of BRCA1 is in HR repair. During S/G2 phase of the cell cycle, CDK2-mediated phosphorylation of CtIP promotes the recruitment of BRCA1 to DSBs enabling BRCA1-mediated H2A ubiquitination. This then recruits the chromatin remodeller SMARCAD1, enabling the redistribution of 53BP1 away from DSB ends (Densham et al. 2016). Since retention of 53BP1 at DSB ends is the main block to DNA end resectioning, an essential component of HR repair, BRCA1 loss-of-function mutations result in DSBs being preferentially repaired by NHEJ, a far error-prone DNA repair pathway. BRCA1 may also regulate transcription during the DNA damage response. Gal4-fusion reporter assays have demonstrated that the C-terminus constituting residues 1760–1863 composed of two BRCT domains are sufficient for gene expression (Guendel et al. 2010). BRCA1 and its BRCT domain bind a number of proteins that are known transcriptional factors or regulators, including p53, CBP/p300, STAT1 and oestrogen receptor. Consequently, BRCA1 upregulates the expression of key DNA damage response genes including MDM2, BAX, p21/WAF, p27/KIP1 and GADD45α (Mullan et al. 2006).

It has been suggested that post-translational modification of BRCA1 could be one mechanism by which specific events after DNA damage are controlled. Using pan-methyl antibodies, Guendal et al. (2010) identified BRCA1 as a PRMT1 substrate, and identified a highly conserved DNA binding domain region spanning residues 504–802 as being the methylated domain. Since methylation alters protein–nucleic acid interactions by changing hydrogen bond formation, these findings suggest that BRCA1 methylation has the potential to alter BRCA1–DNA interactions. Supporting this, Guendal et al. (2010) found that treatment of cells with the pan-methyltransferase inhibitor AdOX promotes differential BRCA1 recruitment to target gene promoters. For example, hypomethylated BRCA1 binds more strongly to the APEX, ARHG and GADD45G promoter, but more weakly to the ESR2, SREB2 and FGF9 gene promoters. BRCA1 promoter occupancy after PRMT1 depletion was validated for the APEX and GADD45D promoters (other promoters were not assessed). Whilst this study implicates methylation of BRCA1 as a regulator of its transcriptional properties, no study has yet defined whether methylation regulates other BRCA1-dependent events, especially in the context of HR-mediated repair. Given that methylation occurs within the DNA binding domain and is enhanced in S phase of the cell cycle, it would be interesting to determine if methylation regulates the recruitment of BRCA1 to DSBs or stalled replication forks.

PRMT1 and the methylation of 53BP1: An unknown biological role

Two independent studies have demonstrated that 53BP1 is methylated by PRMT1 within its GAR domain (Adams et al. 2005, Boisvert et al. 2005c); however, the functional significance of this methylation event is still unclear. 53BP1 foci formation after IR was unaffected in PRMT1 null embryonic stem cells or in cells expressing 53BP1 that had been mutated in the methyl acceptor sites. Since 53BP1 methylation was not induced after DNA damage, implying that it is a constitutive rather than regulated event, it is plausible that methylation plays more of a housekeeping role regarding 53BP1 function.

PRMT1 and base excision repair

Base excision repair is primarily responsible for the removal of small, non-helix-distorting base lesions that if left unresolved could cause mutations or chromosomal breakages. DNA polymerase β (Pol β) is a single polypeptide enzyme that has 5′dRP lysase activity and can perform gap-filling synthesis, and has critical roles in BER. Accordingly, Pol β–deficient MEFs display hypersensitivity to alkylating agents (Sobol et al. 1996). Since deficiency in Pol β leads to DNA damage, but overexpression has been correlated with cancer, the activity of Pol β appears to require tight regulation by a number of post-translational modifications, including acetylation, phosphorylation and arginine methylation (El-Andaloussi et al. 2006, 2007). Pol β was found to interact with both PRMT1 and PRMT6. PRMT6-mediated methylation of Arg83/Arg152 was found to strongly stimulate polymerase activity by enhancing DNA binding and processivity. The biological significance of this for DNA repair was confirmed by Arg to Lys mutations of the PRMT6-methyl acceptor site promoting hypersensitivity to DNA alkylating agents (El-Andaloussi et al. 2006). In contrast, PRMT1-mediated methylation of Arg137 did not affect polymerase activity, but abolished the interaction of Pol β with PCNA, implying a negative regulation of BER (El-Andaloussi et al. 2007). Hence, the significance of arginine methylation for Pol β regulation is highly dependent on the nature of the PRMT that methylates the protein.

PRMT5, the main symmetric methyltransferase, and the DDR

PRMT5 is the main mammalian symmetric di-methyltransferase and is a ubiquitously expressed protein. PRMT5 knockout mice are early embryonic lethal owing to aberration of pluripotent cells in the blastocyst (Tee et al. 2010). PRMT5 is a prominent regulator of gene expression; can epigenetically repress expression via the methylation of H2AR3, H3R8 and H4R; activate expression via the methylation of H3R2; or function as a transcriptional coactivator (Zhao et al. 2009, Migliori et al. 2012, Wei et al. 2013, Park et al. 2014, Liu et al. 2016). Whilst many studies have focused on the epigenetic role of PRMT5, it is becoming increasingly clear that PRMT5 methylates a vast number of non-histone proteins, including those that regulate mRNA processing. Indeed, selective deletion of PRMT5 within the murine CNS leads to postnatal lethality due to defects in the core splicing machinery and aberrant p53 pathway activity (Bezzi et al. 2013), whilst methylation of hnRNPA1 facilitates IRES-dependent translational of cyclin D1 and c-Myc (Gao et al. 2017).

The role of PRMT5 in the DNA damage response has been largely unexplored, with only a handful of DNA repair substrates known to be modified (Auclair and Richard 2013). Importantly, whilst these studies addressed the consequence of expressing a methyl-deficient form of these proteins, investigations into the significance of PRMT5 per se during the DNA damage response have remained elusive. Recently, using a combination of shRNA-mediated knockdown and genetic deletion approaches, Clarke et al. (2017) demonstrated for the first time a role for PRMT5 in HR-mediated repair. Depletion of PRMT5 mildly sensitised HeLa cells to ionising radiation, but greatly reduced cell survival after camptothecin- and olaparib-induced DNA damage, agents that require HR for repair. Further supporting an HR defect, depletion of PRMT5 reduced ssDNA formation, ATR checkpoint activation and RAD51 recruitment. Interestingly, MEFs isolated from PRMT5-deficient embryos display spontaneous DNA damage and double-strand breaks (Clarke et al. 2017). Taken together, this study not only highlights the importance of PRMT5 for the HR response, but also demonstrates that PRMT5 is required for maintaining genome stability after endogenous replication stress.

To date, the only identified PRMT5 substrates involved in the DNA damage response are RUVBL1, p53, FEB1 and Rad9 (**Figure 8.3**).

PRMT5 and the regulation of HR-mediated DSB repair

As previously mentioned, the primary function of 53BP1 is to protect DSB ends from over-processing by the DNA end resection machinery, thereby favouring NHEJ over HR-mediated DSB repair (Schultz 2000, Anderson et al. 2001, Mallette and Richard 2012). 53BP1 recruitment and removal from DSB ends is incredibly complex, but an excellent example of how chromatin-bound complexes integrate local chromatin architecture to repair pathway choice. DSB formation induces activation of the ATM-signalling pathway leading to RNF8 and RNF168-mediated polyubiquitylation of H2A and degradation of the H3K20me2 lysine demethylase JMJD2A. Since 53BP1 possesses a Tudor domain in tandem with a UDR domain, it is able to specifically read H4K20me2 marks in combination with ubiquitylated histone H2A (Panier and Boulton 2014). If the damage is too complex and a sister chromatid is present, 53BP1 is displaced, enabling end processing and the generation of long stretches of ssDNA and HR-mediated repair. 53BP1 displacement occurs through two mechanisms. First, CDK2-mediated phosphorylation of CtIP recruits BRCA1 enabling mono-ubiquitylation of H2A, SMARCAD1 recruitment and 53BP1 mobilisation away from break ends (Densham et al. 2016). Second, the TIP60 DNA repair complex, which has core chromatin remodelling and acetyltransferase activity (Ikura et al. 2000), acetylates histone H4K16 which disrupts 53BP1 binding by affecting salt bridge formation between unmodified H4K16 residues and the 53BP1 Tudor domain (Sun et al. 2009, Tang et al. 2013), or acetylates H2AK15 which competes with RNF168-mediated ubiquitylation (Jacquet et al. 2016). Together, these complex mechanisms involving various post-translational modifications maintain the fidelity of 53BP1 recruitment/ mobilisation and genome stability.

During their investigations into the regulation of HR by PRMT5, Clarke et al. (2017) utilised TAP-tag affinity purification with methyl-deficient PRMT5 and identified the AAA+ ATPase RUVBL1 (Pontin/Tip49) as a novel PRMT5 substrate. RUVBL1 and its binding partner RUVBL2 are present in a number of separate high molecular weight nuclear complexes that are known to regulate DNA repair (Jha and Dutta 2009). One of these is the TIP60 complex, with RUVBL1 shown to regulate TIP60 acetyltransferase activity and the assembly of a functional complex (Jha et al. 2008, 2013). Mass spectrometry and mutational analysis identified RUVBL1-R205 as the residue targeted by PRMT5. Expression of methyl-deficient RUVBL1 mimicked that of PRMT5 depletion increasing cellular sensitivity to IR, camptothecin and olaparib, and promoted retention of 53BP1 at sites of damage. Indeed, epistatic analysis indicate that PRMT5 and methyl-RUVBL1 are acting in the same DNA repair pathway, and that RUVBL1 is the dominant PRMT5 substrate in the repair of IR-induced DNA lesions (Clarke et al. 2017). Importantly, depletion of 53BP1 in PRMT5 deficient cells, or in cells expressing methyl-deficient RUVBL1, restored RPA and RAD51 recruitment after IR-induced damage, implying that methylation of RUVBL1 is required for 53BP1 mobilisation and HR-mediated repair. Given that TIP60-mediated acetylation of H4K16 effectively prevents 53BP1 from interacting with lysine methylated histone tails, and that RUVBL1 is an integral component of the TIP60 complex and is required for RAD51 foci formation (Gospodinov et al. 2009), Clarke et al. (2017) hypothesised that PRMT5-dependent methylation of RUVBL1 promotes TIP60 acetyltransferase activity at H4K16, facilitating 53BP1 mobilisation. Using a cell culture–based system in which an enzymatic double-strand break can be induced in a controlled manner at a specific loci (Tang et al. 2013), enabling detection of the chromatin composition by chromatin immunoprecipitation (ChIP), they showed that H4K16Ac levels at DSBs were reduced after PRMT5 depletion or expression of methyl-deficient RUVBL1. Conversely, 53BP1 levels were retained at the same breaks. TIP60 is also known to acetylate and activate ATM leading to ATM pathway activation (Sun et al. 2010). Interestingly, ATM kinase activity, as assessed by Chk2 and Kap1 phosphorylation, was unaffected by PRMT5 depletion or RUVBL1 methylation, implying that methylation of RUVBL1 is specifically required for some, but not all, TIP60-regulated DNA damage repair events. Indeed, ATM activation is a very early event in the DSB response, yet RUVBL1 methylation was only induced 2 hours after the initial damage. Collectively, these findings highlight the importance of arginine methylation as a mechanism to regulate the enzymatic activity of TIP60 in a temporal manner directing substrate specificity and DSB repair pathway choice.

A key question that remains is how Arg205 methylation activates TIP60 enzymatic activity. Arg205 is surface facing and located within Domain II of RUVBL1, an OB fold proposed to act as a nucleotide/protein-binding interface. Since methylation regulates protein binding, it is tempting to speculate that methylation of RUVBL1 could be modulating TIP60 activity by altering the binding of specific cofactors. TIP60 complex composition at sites of DNA damage after PRMT5 depletion should be able to shed light on this matter.

PRMT5 and the replication stress response

FEN1 (flap endonuclease 1) is a multifunctional enzyme that has both endo- and exonuclease activity and is best known for its involvement in DNA replication and repair. The ability of FEN1 to perform multiple roles has been attributed to the formation of specific FEN1-containing complexes. For example, the FEN1–PCNA interaction coordinates the sequential actions of polymerase δ, FEN1 and DNA ligase 1 during Okazaki fragment maturation; whilst in response to stalled replication forks, the FEN1-WRN complex activates the gap endonuclease activity of FEN1 initiating break-induced recombination (Zheng et al. 2005). FEN1 is also heavily phosphorylated and acetylated. UV-induced DNA damage promotes FEN1 phosphorylation at Ser187 by Cdk2-cyclin E and nuclear relocalisation (Guo et al. 2008), whilst acetylation by CBP/p300 reduces DNA binding and nuclease activity (Hasan et al. 2001).

FEN1 is methylated by PRMT5 on four residues, but primarily at Arg192 (Guo et al. 2010). Interestingly, arginine methylation blocks Cdk2-cyclin E phosphorylation of Ser187, facilitating FEN1-PCNA interactions, thus demonstrating how arginine methylation can crosstalk and influence the deposition of other PTMs. Here, PRMT5-mediated methylation of FEN1 in early

S phase prevents FEN1 phosphorylation, thereby maintaining the FEN1-PCNA interaction. This allows substitution of the polymerase for the nuclease, DNA flap cleavage and Okazaki fragment maturation. Consequently, expression of methyl-deficient FEN1 results in defective recruitment of FEN1 to replication foci, a delay in cell cycle progression and increased genome instability, as determined by γH2AX foci formation (Guo et al. 2010). Importantly, this model of methylation regulating FEN1-PCNA recruitment to replication forks implies that arginine methylation needs to be actively removed to enable Cdk-cyclin phosphorylation. If this was not achieved, the constant FEN1-PCNA-DNA interactions would lead to inhibition of DNA ligase I action and incomplete Okazaki fragment maturation. Whilst methylation of FEN1 was markedly reduced in late S phase compared to early S phase, identification of the arginine demethylase responsible remains elusive.

Rad9 is a cell cycle checkpoint protein that, together with Rad1 and HUS1, forms the 9-1-1 clamp complex, which is critically involved in triggering cell cycle checkpoint signalling, cell cycle arrest and DNA damage repair after replication stress. The intra-S phase checkpoint monitors S phase cells for DNA damage and prevents transition of damaged cells from S phase into G2. Upon replication stress, RPA-coated ssDNA recruits DNA polymerase α, which in turn recruits the clamp loader Rad17-replication factor C (Rad17-RFC) that enables 9-1-1 clamp complex loading and TOPBP1 association. In a parallel pathway, the ATR-ATRIP complex is recruited through direct interactions of ATRIP with RPA. Both events are required for optimal ATR signalling pathway activation and cell cycle arrest (Delacroix et al. 2007). Rad9 is heavily phosphorylated, both constitutively and by the ATM and ATR kinases (Furuya et al. 2004), implying that its activity is dynamically regulated by post-translational modifications. PRMT5 was identified as a Rad9 interacting protein by affinity purification and mass spectrometry analysis of Rad9-containing complexes, and demonstrated to be methylated in vivo at a RGRR motif spanning residues 172–175 of human Rad9 (He et al. 2011). Treatment of cells with hydroxyurea (HU), an agent that depletes the nucleotide pool, resulting in stalling and collapse of the replication fork and increased Rad9 methylation, whilst expression of methyl-deficient Rad9, where all three methyl-acceptor sites had been mutated to alanine, resulted in defective Chk1 phosphorylation, failure to activate cell cycle checkpoints and enhanced hypersensitivity to HU. Whilst the precise mechanism by which methylation of Rad9 leads to Chk1 activation has yet to be determined, particularly as 9-1-1 complex formation on chromatin was methyl-independent, the FEN1 and Rad9 studies do implicate PRMT5 as a component of the replication stress response through the methylation of a least two important components.

Regulation of gene expression during the DDR

Surprisingly, the role of histone arginine methylation and the direct regulation of DDR complex recruitment/dissociation to chromatin have yet to explored. It would be very interesting to see if histone arginine methylation is induced after DNA damage, if this is dependent on a specific type of DNA lesion and how transient this modification is (thus indicative of regulating chromatin composition rather than epigenetic-mediated gene expression). Until these experiments have been conducted, the role of PRMT-mediated histone methylation and methylation of chromatin-bound transcription factor complexes has so far been restricted to the transcriptional response to DNA damage (Figure 8.4).

p53 is a major tumour suppressor gene and one of the most frequently mutated genes in cancer. Upon genotoxic stress and protein stabilisation, p53 transcribes a number of genes involved in cell cycle arrest, DNA repair and apoptosis. Since distinct sets of genes regulate cell cycle arrest

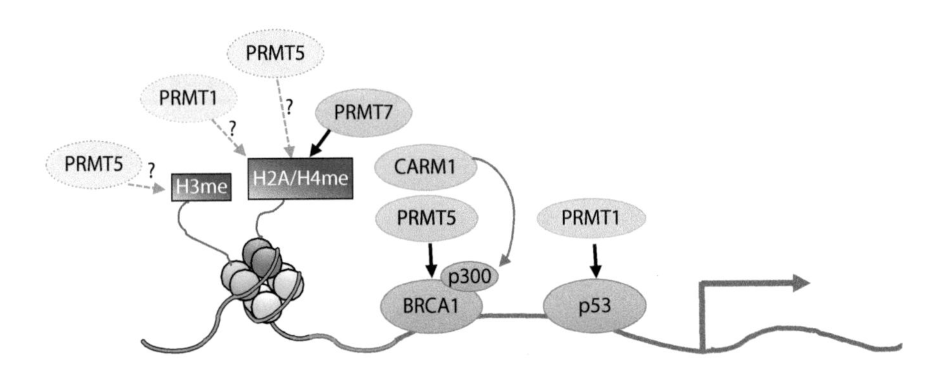

Figure 8.4 **Transcriptional control of the DNA damage response by PRMTs.** PRMT5 and PRMT1 directly methylate and modulate the transcriptional activity of BRCA1 and p53, respectively, leading to cell cycle arrest. CARM1 indirectly regulates BRCA1 transactivation by methylating the BRCA1 coregulator, p300. PRMT7 directly methylates histone H2A/H4 at the promoters of DNA damage genes. PRMT1 regulates p53-dependent cell cycle genes and suppresses pro-apoptotic genes. Epigenetic regulation of gene expression after DNA damage by PRMT1 and PRMT5 has yet to demonstrated, but could potentially be an additional layer of complexity.

versus apoptosis, regulator mechanisms are in place to enable target specific p53-mediated gene expression. Like other transcription factors, recruitment of cofactors and accessory proteins is one mechanism of transcriptional control. One such protein is Strap, a cofactor that bridges the p53 cofactors JMY and the histone acetyltransferases CBP/p300 (Demonacos et al. 2001). CBP/p300 is phosphorylated by ATM after DNA damage, enabling nuclear accumulation and a p53-mediated apoptotic response (Demonacos et al. 2004). In the search for Strap-interacting proteins, Jansson et al. (2008) co-purified PRMT5 and established that PRMT5 methylated p53 at Arg333, Arg225 and Arg337, residues located within the p53 oligomisation and intracellular location domain. Indeed, expression of methyl-deficient p53 resulted in a reduced nuclear localisation and a reduction in p53 oligomisation, implying that methylation affects the biochemical properties of the transcriptionally functional p53 protein complex. Moreover, p53 methylation was induced after etoposide treatment, a chemotherapeutic drug that inhibits topoisomerase II and induces DSB formation. Methylation was shown to be functionally relevant because the transcriptional properties of methylated p53 were distinct from wild-type p53. PRMT5 depletion amplified p53 binding to apoptotic gene promoters, whereas overexpression of PRMT5 facilitated a cell cycle arrest via p21 induction. Since arginine methylation of p53 alters the specificity of p53 to target gene promoters, inhibitors of PRMT5 have the potential to modulate the p53 response. This could be potentially useful in the drug targeting of p53 wild type tumours.

PRMT7 is the only arginine methyltransferase with Type III mono-methylation activity. The first indication that PRMT7 is involved in modulating cellular responses to DNA damaging agents came from studies showing that knockdown of PRMT7 in fibroblast and HeLa cells promoted hypersensitivity to the topoisomerase I inhibitor, camptothecin (Verbiest et al. 2008). However, this effect appears to be cell type– and DNA damage–specific, as depletion of PRMT7 in NIH3T3 fibroblasts promotes resistance rather than hypersensitivity to cisplatin, chlorambucil and mitomycin C (Karkhanis et al. 2012). Whilst limited to only two studies, these results do imply that the contribution of PRMT7 to the DNA damage response is highly context specific. The mechanism by which PRMT7 regulates the DNA damage response remains to be fully investigated. Nevertheless, the finding that PRMT7 was enriched at DNA damage gene promoters, including POLD, APEX2 and ALKBH5, leading to an induction in H2AR3 and H4R3 methylation (Karkhanis et al. 2012) is highly suggestive that epigenetic mechanisms are a major contributing factor. As this is possibly the only study that has investigated the epigenetic contribution of PRMTs to the regulation of gene expression after DNA damage, and that both PRMT1 and PRMT5 are major epigenetic regulators, it would be important to determine the contribution of various PRMTs and their specific histone marks to the epigenetic landscape.

PRMT2 has very weak histone methylation activity, but does regulate gene expression in both an activating and repressive manner by functioning as a transcriptional co-regulator. For example, PRMT2-mediated methylation of steroid hormones/nuclear receptors promotes gene expression, whilst PRMT2-medaited methylation of E2F and NFκB represses gene expression (Yang and Bedford 2013). Very few studies have considered a role for PRMT2 in the DDR; however, it was recently shown that depletion of PRMT2 results in a downregulation of a number of core DNA damage genes, including *BRCA1* and *CDK1* (Oh et al. 2014), implying a positive regulation of DNA repair gene expression. However, and somewhat paradoxical given that PRMT2 appears to be positively regulating HR genes, depletion of PRMT2 appeared to increase HR efficiency, as examined by the I-*Sce*I HR GRP-reporter assay. A number of PRMT2 isoforms exist, which may confuse interpretation of knockdown analysis; thus, studies that determine the sensitivity of PRMT2 genetically null cells to DNA damaging agents are critical to fully understand if PRMT2 is a true transcriptional regulator of the DNA damage response.

CARM1 (PRMT4)-mediated methylation of histone H3R17 and H3R26 is thought to mediate gene expression (Di Lorenzo and Bedford 2011). However, since conclusive ChIP analysis has yet to be conducted, the main mechanisms by which CARM1 is thought to drive gene expression is via the modulation of transcription factor complex activity. Two such coactivators are the highly related histone acetyltransferases p300 and CBP, which are methylated at multiple residues by CARM1. Methylation of one residue, Arg754, located within the KIX domain of p300 and previously identified as methylated in CBP (Chevillard-Briet et al. 2002), was shown to enhance p300 interactions with the C-terminal BRCT domain of BRCA1 and facilitate the recruitment of BRCA1 to the p21 promoter (Lee et al. 2011). Thus, methylation of p300 by CARM1 appears to regulate the transcriptional tumour suppressor functions of BRCA1.

Future directions

Arginine methylation of splicing factors: Novel mechanism of genome stability maintenance and DNA repair?

Genome-wide siRNA screens have revealed an unexpected enrichment for mRNA-processing factors in the maintenance of genome integrity (Paulsen et al. 2009). Whilst this could simply reflect a mechanism to alter protein levels of key DNA repair genes, there has been a growing appreciation that splicing factors/RNA binding proteins (RBPs) play direct roles in DNA damage prevention and repair by integrating transcription and pre-mRNA processing (Naro et al. 2015). Supporting this, some RBPs are directly phosphorylated by DNA damage sensing kinases (e.g. ATM, ATR) which modulate their activity and nucleic acid binding, whilst others are directly recruited to DNA damage sites (Dutertre et al. 2014). This is significant, because RBPs are prominent targets for arginine methylation (Boisvert et al. 2002). Thus, whilst arginine methylation regulates splicing activity (Boisvert et al. 2002), it is also conceivable to suggest that methylation may also contribute to mechanism by which splicing factors directly regulate DNA repair. For instance, SRSF1 (also called SF/ASF), the prototype of SR family of splicing regulator, is methylated at three residues within a GAR domain by PRMT1, with methylation required for nuclear subcellular localisation and correct splicing (Bressan et al. 2009, Sinha et al. 2010). SRSF1-depleted cells accumulate R-loops (Li and Manley 2005), a three-stranded hybrid nucleic acid structure comprising of nascent RNA hybridising to one of the exposed DNA strands. Since R-loops can lead to DSB formation (Aguilera and García-Muse 2012), SRSF1-depleted cells are hypermutagenic. Hence, correct SRSF1 activity is required to prevent RNA-induced genome instability. One mechanism by which this occurs is via the co-operation of DNA topoisomerase I (TOP-1). During DNA replication and RNA transcription, TOP-1 nicks DNA, thereby relieving DNA from the topological tensions induced by these processes (Leppard and Champoux 2005). Since RBPs are highly enriched as TOP-1 binding partners (Czubaty et al. 2005), many of which are PRMT substrates (e.g. SRSF1, hnRNP U, hnRNP A1, hnRNP A2, hnRNP K, HuR), it would be interesting to determine if arginine methylation indirectly regulates TOP-1 activity though RBP methylation.

PRMT5 also appears to regulate the resolution of R-loops via the methylation of the RNA polymerase II subunit POLR2A. Methylation of Arg1810 of POLR2A recruits the Tudor domain–containing splicing factor the survival of motor neuron (SMN) that interacts with senataxin, a helicase required for the resolution of R-loops in transcriptional termination regions (Zhao et al. 2016). These findings are significant because expression of POLR2A-R1810A, or depletion of PRMT5 or SMN resulted in the accumulation of γH2AX indicative of DNA damage.

A third example of a splicing factor that is heavily arginine methylated is Ewing sarcoma protein (EWS) (Belyanskaya et al. 2001). Both PRMT1 and PRMT8 asymmetrically di-methylate EWS in vitro (Kim et al. 2008) promoting nuclear accumulation (Belyanskaya et al. 2003). EWS-deleted embryonic fibroblasts are highly sensitive to IR. Indeed, EWS is required for HR during B cell development (Li et al. 2007). How EWS modulates HR is still unclear, with some studies suggesting direct roles in annealing of homologous DNA or indirectly via the regulation of alternative splicing of DNA damage response genes (Naro et al. 2015). Nevertheless, given that EWS is heavily methylated and PRMT1 is critical for maintaining genome stability, it would be interesting to ascertain the significance of methylated EWS for DNA repair.

A more direct example of arginine methylation regulating the splicing of proteins involved in the DDR was provided by the Guccione laboratory. Using mice conditionally deleted for PRMT5 within the neuronal stem cell populations (Nestin-Cre), they demonstrated that deletion of PRMT5 resulted in reduced methylation of Sm proteins and alternative splicing of mRNA with weak 5′ donor sites. Intriguingly, they showed that the activation of the p53 pathway was the top upregulated gene category, with alternative splicing of *mdm4* resulting in the formation of a shorter, unstable *mdm4* transcript and p53 activation (Bezzi et al. 2013). Whether the correct control of mRNA processing by PRMT5-regulated splicing events during the DNA damage response is an important factor has yet to be determined.

Non-coding RNA and RNA binding proteins: Regulation by arginine methylation?

Non-coding RNA (ncRNAs) are increasingly being recognised as important regulators of the DNA damage response. DICER and DROSHA, components of the RNA interference pathway, promote the formation of small RNAs that are recruited to sites of DNA damage. These DICER and DROSHA-dependent small RNAs (DDRNAs) were found to control DNA damage pathway activation, foci formation and checkpoint enforcement in response to a number of distinct DNA lesions, including oncogene-induced replication stress, IR and site-specific endonucleases (Francia et al. 2012). RNA binding proteins, many of which possess a GAR motif susceptible to PRMT methylation, bind ncRNAs. Given that the role of ncRNAs in the DDR is an emerging field, it would be interesting to determine if PRMT activity and the methylation of RBPs modulates this response. If so, this would add an extra layer of complexity by which PRMTs contribute to its proper regulation.

Small molecule inhibitors of PRMTs and the targeting of the DDR

The development of inhibitor 'tool' compounds, such as those developed by the Structural Genomics Consortium (SGC), are invaluable resources that help decipher the role of enzymes in health and disease. As discussed, PRMTs are extensively linked to the DDR. Given that cancer chemotherapies still largely rely on the induction of DNA damage for cytotoxicity, small molecule drug targeting of PRMTs could have significant clinical benefit as a novel strategy for cancer treatments.

The main challenge has been the development of cell permeable compounds that target a specific PRMT. This has been particularly true for Type I enzyme subgroupings whose cofactor binding sites have high sequence conservation. Indeed, despite the majority of arginine methylation events in the cell being catalysed by PRMT1, a specific inhibitor has yet to be discovered. Compounds that targeted the substrate binding pocket of PRMT1 displayed some effect on suppressing the growth of cancer cell lines, but also inhibit to varying degrees other Type I enzymes (Wang et al. 2012a, 2012b, Yan et al. 2014, Eram et al. 2015, Yu et al. 2015).

Excitingly, the identification of small molecule compounds that specifically target PRMT5 has been more successful, leading to the development of pre-clinical compounds that display a potent tumour suppressive activity in vivo (Chan-Penebre et al. 2015, Duncan et al. 2016). Structural analysis demonstrates that EPZ015666 has an unusual mode of interaction forming cation–pie interactions with cofactor SAM in addition to interacting with the substrate binding site, possibly explaining the high specificity of EPZ01566 for PRMT5 over PRMTs and other lysine methyltransferases (Chan-Penebre et al. 2015). Small molecule PRMT5 inhibitors are set to enter clinical trials for cancer treatment in the near future.

Summary: Arginine methylation and the DDR

With the development of better tools in which to study arginine methylation, including substrate-specific and pan-methyl antibodies, and quantitative methyl-proteomics platforms, it is more than likely that our knowledge into the significance of PRMTs for the DDR and maintenance of genome stability will grow immensely. However, in-depth studies into how methylation of specific residues alters the biological activity of a protein regarding their role in the DDR, and how this is integrated into the complex process of DNA repair is largely missing. Other key questions still remain: Does histone arginine methylation play any role in chromatin composition at break ends or stalled forks? How is the activity of PRMTs, believed to be constitutive enzymes, regulated in a DNA damage-dependent manner? Are PRMTs themselves recruited to damaged chromatin or do all methylation events occur in the nucleoplasm/cytoplasm? Are there any more specific readers of arginine methylation in addition to Tudor domain–containing proteins? Is there a true arginine demethylase in mammalian cells that regulates dynamic arginine methylation after DNA damage? Addressing these questions will allow a much clearer picture of the contribution of this PTM to DNA repair. Moreover, as PRMTs are often overexpressed in cancer, drug targeting of PRMTs could provide novel therapeutic strategies in which to provoke catastrophic DNA damage and cancer cell cytotoxicity.

REFERENCES

Acs, K., Luijsterburg, M. S., Ackermann, L., Salomons, F. A., Hoppe, T., Dantuma, N. P. 2011. The AAA-ATPase VCP/p97 promotes 53BP1 recruitment by removing L3MBTL1 from DNA double-strand breaks. *Nat Struct Mol Biol* 18: 1345–1350.

Adams, M. M., Wang, B., Xia, Z., Morales, J. C., Lu, X., Donehower, L. A., Bochar, D. A., Elledge, S. J., Carpenter, P. B. 2005. 53BP1 oligomerization is independent of its methylation by PRMT1. *Cell Cycle* 4: 1854–1861.

Aguilera, A., García-Muse, T. 2012. R loops: From transcription byproducts to threats to genome stability. *Mol Cell* 46: 115–124.

Amendola, P. G., Zaghet, N., Ramalho, J. J., Vilstrup Johansen, J., Boxem, M., Salcini, A. E. 2017. JMJD-5/KDM8 regulates H3K36me2 and is required for late steps of homologous recombination and genome integrity. *PLoS Genet* 13: e1006632.

Anderson, L., Henderson, C., Adachi, Y. 2001. Phosphorylation and rapid relocalization of 53BP1 to nuclear foci upon DNA damage. *Mol Cell Biol* 21: 1719–1729.

Antonysamy, S., Bonday, Z., Campbell, R. M., Doyle, B., Druzina, Z., Gheyi, T., Han, B. et al. 2012. Crystal structure of the human PRMT5:MEP50 complex. *Proc Natl Acad Sci USA* 109: 17960–17965.

Aranda, S., Rutishauser, D., Ernfors, P. 2014. Identification of a large protein network involved in epigenetic transmission in replicating DNA of embryonic stem cells. *Nucleic Acids Res* 42: 6972–6986.

Auclair, Y., Richard, S. 2013. The role of arginine methylation in the DNA damage response. *DNA Repair (Amst)* 12(7): 459–465.

Aymard, F., Bugler, B., Schmidt, C. K., Guillou, E., Caron, P., Briois, S., Iacovoni, J. S. et al. 2014. Transcriptionally active chromatin recruits homologous recombination at DNA double-strand breaks. *Nat Struct Mol Biol* 21: 366–374.

Ayrapetov, M. K., Gursoy-Yuzugullu, O., Xu, C., Xu, Y., Price, B. D. 2014. DNA double-strand breaks promote methylation of histone H3 on lysine 9 and transient formation of repressive chromatin. *Proc Natl Acad Sci USA* 111: 9169–9174.

Baldwin, G. S., Carnegie, P. R. 1971. Specific enzymic methylation of an arginine in the experimental allergic encephalomyelitis protein from human myelin. *Science* 171: 579–581.

Bartke, T., Vermeulen, M., Xhemalce, B., Robson, S. C., Mann, M., Kouzarides, T. 2010. Nucleosome-interacting proteins regulated by DNA and histone methylation. *Cell* 143: 470–484.

Beck, D. B., Burton, A., Oda, H., Ziegler-Birling, C., Torres-Padilla, M. E., Reinberg, D. 2012. The role of PR-Set7 in replication licensing depends on Suv4-20h. *Genes Dev* 26: 2580–2589.

Belyanskaya, L. L., Delattre, O., Gehring, H. 2003. Expression and subcellular localization of Ewing sarcoma (EWS) protein is affected by the methylation process. *Exp Cell Res* 288: 374–381.

Belyanskaya, L. L., Gehrig, P. M., Gehring, H. 2001. Exposure on cell surface and extensive arginine methylation of Ewing sarcoma (EWS) protein. *J Biol Chem* 276: 18681–18687.

Bezzi, M., Teo, S. X., Muller, J., Mok, W. C., Sahu, S. K., Vardy, L. A., Bonday, Z. Q., Guccione, E. 2013. Regulation of constitutive and alternative splicing by PRMT5 reveals a role for Mdm4 pre-mRNA in sensing defects in the spliceosomal machinery. *Genes Dev* 27: 1903–1916.

Black, J. C., Van Rechem, C., Whetstine, J. R. 2012. Histone lysine methylation dynamics: Establishment, regulation, and biological impact. *Mol Cell* 48: 491–507.

Bluteau, D., Masliah-Planchon, J., Clairmont, C., Rousseau, A., Ceccaldi, R., Dubois d'Enghien, C., Bluteau, O. et al. 2016. Biallelic inactivation of REV7 is associated with Fanconi anemia. *J Clin Invest* 126: 3580–3584.

Boffa, L. C., Karn, J., Vidali, G., Allfrey, V. G. 1977. Distribution of NG, NG,-dimethylarginine in nuclear protein fractions. *Biochem Biophys Res Commun* 74: 969–976.

Boisvert, F.-M., Côté, J., Boulanger, M.-C., Cleroux, P., Bachand, F., Autexier, C., Richard, S. 2002. Symmetrical dimethylarginine methylation is required for the localization of SMN in Cajal bodies and pre-mRNA splicing. *J Cell Biol* 159: 957–969.

Boisvert, F.-M., Côté, J., Boulanger, M.-C., Richard, S. 2003. A proteomic analysis of arginine-methylated protein complexes. *Mol Cell Proteomics* 2: 1319–1330.

Boisvert, F.-M., Déry, U., Masson, J.-Y., Richard, S. 2005a. Arginine methylation of MRE11 by PRMT1 is required for DNA damage checkpoint control. *Genes Dev* 19: 671–676.

Boisvert, F.-M., Hendzel, M. J., Masson, J.-Y., Richard, S. 2005b. Methylation of MRE11 regulates its nuclear compartmentalization. *Cell Cycle* 4: 981–989.

Boisvert, F.-M., Rhie, A., Richard, S., Doherty, A. J. 2005c. The GAR motif of 53BP1 is arginine methylated by PRMT1 and is necessary for 53BP1 DNA binding activity. *Cell Cycle* 4: 1834–1841.

Bressan, G. C., Moraes, E. C., Manfiolli, A. O., Kuniyoshi, T. M., Passos, D. O., Gomes, M. D., Kobarg, J. 2009. Arginine methylation analysis of the splicing-associated SR protein SFRS9/SRP30C. *Cell Mol Biol Lett* 14: 657–669.

Brostoff, S., Eylar, E. H. 1971. Localization of methylated arginine in the A1 protein from myelin. *Proc Natl Acad Sci USA* 68: 765–769.

Bruderer, R., Tatham, M. H., Plechanovova, A., Matic, I., Garg, A. K., Hay, R. T. 2011. Purification and identification of endogenous polySUMO conjugates. *EMBO Rep* 12: 142–148.

Burgess, R. J., Zhang, Z. 2013. Histone chaperones in nucleosome assembly and human disease. *Nat Struct Mol Biol* 20: 14–22.

Butler, J. S., Zurita-Lopez, C. I., Clarke, S. G., Bedford, M. T., Dent, S. Y. 2011. Protein-arginine methyltransferase 1 (PRMT1) methylates Ash2L, a shared component of mammalian histone H3K4 methyltransferase complexes. *J Biol Chem* 286: 12234–12244.

Carugo, A., Genovese, G., Seth, S., Nezi, L., Rose, J. L., Bossi, D., Cicalese, A. et al. 2016. *In vivo* functional platform targeting patient-derived xenografts identifies WDR5-Myc association as a critical determinant of pancreatic cancer. *Cell Rep* 16: 133–147.

Chan-Penebre, E., Kuplast, K. G., Majer, C. R., Boriack-Sjodin, P. A., Wigle, T. J., Johnston, L. D., Rioux, N. et al. 2015. A selective inhibitor of prMt5 with *in vivo* and *in vitro* potency in MCL models. *Nat Chem Biol* 1–10.

Chang, B., Chen, Y., Zhao, Y., Bruick, R. K. 2007. JMJD6 is a histone arginine demethylase. *Science* 318: 444–447.

Chen, Y., Zhou, X., Liu, N., Wang, C., Zhang, L., Mo, W., Hu, G. 2008. Arginine methylation of hnRNP K enhances p53 transcriptional activity. *FEBS Lett* 582: 1761–1765.

Chen, Y. J., Tsai, C. H., Wang, P. Y., Teng, S. C. 2017. SMYD3 promotes homologous recombination via regulation of H3K4-mediated gene expression. *Sci Rep* 7: 3842.

Cheng, X., Collins, R. E., Zhang, X. 2005. Structural and sequence motifs of protein (histone) methylation enzymes. *Annu Rev Biophys Biomol Struct* 34: 267–294.

Chevillard-Briet, M., Trouche, D., Vandel, L. 2002. Control of CBP co-activating activity by arginine methylation. *EMBO J* 21: 5457–5466.

Choi, K.-S., Kim, J. Y., Lim, S.-K., Choi, Y. W., Kim, Y. H., Kang, S. Y., Park, T. J., Lim, I. K. 2012. TIS21(/BTG2/PC3) accelerates the repair of DNA double strand breaks by enhancing Mre11 methylation and blocking damage signal transfer to the Chk2(T68)-p53(S20) pathway. *DNA Repair (Amst)* 11: 965–975.

Clarke, T. L., Sanchez-Bailon, M. P., Chiang, K., Reynolds, J. J., Herrero-Ruiz, J., Bandeiras, T. M., Matias, P. M. et al. 2017. PRMT5-dependent methylation of the TIP60 coactivator RUVBL1 is a key regulator of homologous recombination. *Mol Cell* 65(5): 900–916.e7.

Cloutier, P., Lavallee-Adam, M., Faubert, D., Blanchette, M., Coulombe, B. 2013. A newly uncovered group of distantly related lysine methyltransferases preferentially interact with molecular chaperones to regulate their activity. *PLoS Genet* 9: e1003210.

Conlan, L. A., McNees, C. J., Heierhorst, J. 2004. Proteasome-dependent dispersal of PML nuclear bodies in response to alkylating DNA damage. *Oncogene* 23: 307–310.

Cuthbert, G. L., Daujat, S., Snowden, A. W., Erdjument-Bromage, H., Hagiwara, T., Yamada, M., Schneider, R. et al. 2004. Histone deimination antagonizes arginine methylation. *Cell* 118: 545–553.

Czubaty, A., Girstun, A., Kowalska-Loth, B., Trzci ska, A. M., Purta, E. B., Winczura, A., Grajkowski, W., Staro, K. 2005. Proteomic analysis of complexes formed by human topoisomerase I. *Biochim Biophys Acta (BBA)–Proteins Proteomics* 1749: 133–141.

Daujat, S., Bauer, U.-M., Shah, V., Turner, B., Berger, S., Kouzarides, T. 2002. Crosstalk between CARM1 methylation and CBP acetylation on histone H3. *Curr Biol* 12: 2090–2097.

De Haro, L. P., Wray, J., Williamson, E. A., Durant, S. T., Corwin, L., Gentry, A. C., Osheroff, N., Lee, S. H., Hromas, R., Nickoloff, J. A. 2010. Metnase promotes restart and repair of stalled and collapsed replication forks. *Nucleic Acids Res* 38: 5681–5691.

Delacroix, S., Wagner, J. M., Kobayashi, M., Yamamoto, K.-I., Karnitz, L. M. 2007. The Rad9-Hus1-Rad1 (9-1-1) clamp activates checkpoint signaling via TopBP1. *Genes Dev* 21: 1472–1477.

Demonacos, C., Krstic-Demonacos, M., La Thangue, N. B. 2001. A TPR motif cofactor contributes to p300 activity in the p53 response. *Mol Cell* 8: 71–84.

Demonacos, C., Krstic-Demonacos, M., Smith, L., Xu, D., O'Connor, D. P., Jansson, M., La Thangue, N. B. 2004. A new effector pathway links ATM kinase with the DNA damage response. *Nat Cell Biol* 6: 968–976.

Densham, R. M., Garvin, A. J., Stone, H. R., Strachan, J., Baldock, R. A., Daza-Martin, M., Fletcher, A. et al. 2016. Human BRCA1-BARD1 ubiquitin ligase activity counteracts chromatin barriers to DNA resection. *Nat Struct Mol Biol* 23: 647–655.

Di Lorenzo, A., Bedford, M. T. 2011. Histone arginine methylation. *FEBS Lett* 585: 2024–2031.

Duncan, K. W., Rioux, N., Boriack-Sjodin, P. A., Munchhof, M. J., Reiter, L. A., Majer, C. R., Jin, L. et al. 2016. Structure and property guided design in the identification of PRMT5 tool compound EPZ015666. *ACS Med Chem Lett* 7: 162–166.

Dutertre, M., Lambert, S., Carreira, A., Amor-Guéret, M., Vagner, S. 2014. DNA damage: RNA-binding proteins protect from near and far. *Trends Biochem Sci* 39: 141–149.

Déry, U., Coulombe, Y., Rodrigue, A., Stasiak, A., Richard, S., Masson, J.-Y. 2008. A glycine-arginine domain in control of the human MRE11 DNA repair protein. *Mol Cell Biol* 28: 3058–3069.

El-Andaloussi, N., Valovka, T., Toueille, M., Hassa, P. O., Gehrig, P., Covic, M., Hübscher, U., Hottiger, M. O. 2007. Methylation of DNA polymerase beta by protein arginine methyltransferase 1 regulates its binding to proliferating cell nuclear antigen. *FASEB J* 21: 26–34.

El-Andaloussi, N., Valovka, T., Toueille, M., Steinacher, R., Focke, F., Gehrig, P., Covic, M. et al. 2006. Arginine methylation regulates DNA polymerase beta. *Mol Cell* 22: 51–62.

Eram, M. S., Shen, Y., Szewczyk, M. M., Wu, H., Senisterra, G., Li, F., Butler, K. V. et al. 2015. A potent, selective, and cell-active inhibitor of human type I protein arginine methyltransferases. *ACS Chem Biol* 11(3): 772–781.

Faucher, D., Wellinger, R. J. 2010. Methylated H3K4, a transcription-associated histone modification, is involved in the DNA damage response pathway. *PLoS Genet* 6: e1001082.

Feng, Q., He, B., Jung, S.-Y., Song, Y., Qin, J., Tsai, S. Y., Tsai, M.-J., O'Malley, B. W. 2009. Biochemical control of CARM1 enzymatic activity by phosphorylation. *J Biol Chem* 284: 36167–36174.

Feng, Y., Wang, J., Asher, S., Hoang, L., Guardiani, C., Ivanov, I., Zheng, Y. G. 2011. Histone H4 acetylation differentially modulates arginine methylation by an in Cis mechanism. *J Biol Chem* 286: 20323–20334.

Francia, S., Michelini, F., Saxena, A., Tang, D., de Hoon, M., Anelli, V., Mione, M., Carninci, P., d'Adda di Fagagna, F. 2012. Site-specific DICER and DROSHA RNA products control the DNA-damage response. *Nature* 488: 231–235.

Fu, H., Maunakea, A. K., Martin, M. M., Huang, L., Zhang, Y., Ryan, M., Kim, R., Lin, C. M., Zhao, K., Aladjem, M. I. 2013. Methylation of histone H3 on lysine 79 associates with a group of replication origins and helps limit DNA replication once per cell cycle. *PLoS Genet* 9: e1003542.

Furuya, K., Poitelea, M., Guo, L., Caspari, T., Carr, A. M. 2004. Chk1 activation requires Rad9 S/TQ-site phosphorylation to promote association with C-terminal BRCT domains of Rad4TOPBP1. *Genes Dev* 18: 1154–1164.

Fusser, M., Kernstock, S., Aileni, V. K., Egge-Jacobsen, W., Falnes, P. O., Klungland, A. 2015. Lysine methylation of the valosin-containing protein (VCP) is dispensable for development and survival of mice. *PLOS ONE* 10: e0141472.

Gao, G., Dhar, S., Bedford, M. T. 2017. PRMT5 regulates IRES-dependent translation via methylation of hnRNP A1. *Nucleic Acids Res* 45: 4359–4369.

Gospodinov, A., Tsaneva, I., Anachkova, B. 2009. RAD51 foci formation in response to DNA damage is modulated by TIP49. *Int J Biochem Cell Biol* 41: 925–933.

Guendel, I., Carpio, L., Pedati, C., Schwartz, A., Teal, C., Kashanchi, F., Kehn-Hall, K. 2010. Methylation of the tumor suppressor protein, BRCA1, influences its transcriptional cofactor function. *PLOS ONE* 5: e11379.

Guo, Z., Qian, L., Liu, R., Dai, H., Zhou, M., Zheng, L., Shen, B. 2008. Nucleolar localization and dynamic roles of flap endonuclease 1 in ribosomal DNA replication and damage repair. *Mol Cell Biol* 28: 4310–4319.

Guo, Z., Zheng, L., Xu, H., Dai, H., Zhou, M., Pascua, M. R., Chen, Q. M., Shen, B. 2010. Methylation of Fen1 suppresses nearby phosphorylation and facilitates PCNA binding. *Nat Chem Biol* 6: 766–773.

Gurunathan, G., Yu, Z., Coulombe, Y., Masson, J.-Y., Richard, S. 2015. Arginine methylation of hnRNPUL1 regulates interaction with NBS1 and recruitment to sites of DNA damage. *Sci Rep* 5: 10475.

Gurvich, N., Perna, F., Farina, A., Voza, F., Menendez, S., Hurwitz, J., Nimer, S. D. 2010. L3MBTL1 polycomb protein, a candidate tumor suppressor in del(20q12) myeloid disorders, is essential for genome stability. *Proc Natl Acad Sci USA* 107: 22552–22557.

Hasan, S., Stucki, M., Hassa, P. O., Imhof, R., Gehrig, P., Hunziker, P., Hübscher, U., Hottiger, M. O. 2001. Regulation of human flap endonuclease-1 activity by acetylation through the transcriptional coactivator p300. *Mol Cell* 7: 1221–1231.

He, W., Ma, X., Yang, X., Zhao, Y., Qiu, J., Hang, H. 2011. A role for the arginine methylation of Rad9 in checkpoint control and cellular sensitivity to DNA damage. *Nucleic Acids Res* 39: 4719–4727.

Higgs, M. R., Stewart, G. S. 2016. Protection or resection: BOD1L as a novel replication fork protection factor. *Nucleus* 7: 34–40.

Ho, M.-C., Wilczek, C., Bonanno, J. B., Xing, L., Seznec, J., Matsui, T., Carter, L. G. et al. 2013. Structure of the arginine methyltransferase PRMT5-MEP50 reveals a mechanism for substrate specificity. *PLOS ONE* 8: e57008.

Iberg, A. N., Espejo, A., Cheng, D., Kim, D., Michaud-Levesque, J., Richard, S., Bedford, M. T. 2008. Arginine methylation of the histone H3 tail impedes effector binding. *J Biol Chem* 283: 3006–3010.

Ikura, T., Ogryzko, V. V., Grigoriev, M., Groisman, R., Wang, J., Horikoshi, M., Scully, R., Qin, J., Nakatani, Y. 2000. Involvement of the TIP60 histone acetylase complex in DNA repair and apoptosis. *Cell* 102: 463–473.

Jacquet, K., Fradet-Turcotte, A., Avvakumov, N., Lambert, J.-P., Roques, C., Pandita, R. K., Paquet, E. et al. 2016. The TIP60 complex regulates bivalent chromatin recognition by 53BP1 through direct H4K20me binding and H2AK15 acetylation. *Mol Cell* 62: 409–421.

Jansson, M., Durant, S. T., Cho, E.-C., Sheahan, S., Edelmann, M., Kessler, B., La Thangue, N. B. 2008. Arginine methylation regulates the p53 response. *Nat Cell Biol* 10: 1431–1439.

Jha, S., Dutta, A. 2009. RVB1/RVB2: Running rings around molecular biology. *Mol Cell* 34: 521–533.

Jha, S., Gupta, A., Dar, A., Dutta, A. 2013. RVBs are required for assembling a functional TIP60 complex. *Mol Cell Biol* 33: 1164–1174.

Jha, S., Shibata, E., Dutta, A. 2008. Human Rvb1/Tip49 is required for the histone acetyltransferase activity of Tip60/NuA4 and for the downregulation of phosphorylation on H2AX after DNA damage. *Mol Cell Biol* 28: 2690–2700.

Kantidakis, T., Saponaro, M., Mitter, R., Horswell, S., Kranz, A., Boeing, S., Aygun, O. et al. 2016. Mutation of cancer driver MLL2 results in transcription stress and genome instability. *Genes Dev* 30: 408–420.

Kari, V., Mansour, W. Y., Raul, S. K., Baumgart, S. J., Mund, A., Grade, M., Sirma, H. et al. 2016. Loss of CHD1 causes DNA repair defects and enhances prostate cancer therapeutic responsiveness. *EMBO Rep* 17: 1609–1623.

Karkhanis, V., Wang, L., Tae, S., Hu, Y.-J., Imbalzano, A. N., Sif, S. 2012. Protein arginine methyltransferase 7 regulates cellular response to DNA damage by methylating promoter histones H2A and H4 of the polymerase delta catalytic subunit gene, POLD1. *J Biol Chem* 287(35): 29801–29814.

Katz, J. E., Dlakić, M., Clarke, S. 2003. Automated identification of putative methyltransferases from genomic open reading frames. *Mol Cell Proteomics* 2: 525–540.

Kearney, P. L., Bhatia, M., Jones, N. G., Yuan, L., Glascock, M. C., Catchings, K. L., Yamada, M., Thompson, P. R. 2005. Kinetic characterization of protein arginine deiminase 4: A transcriptional corepressor implicated in the onset and progression of rheumatoid arthritis. *Biochemistry* 44: 10570–10582.

Kernstock, S., Davydova, E., Jakobsson, M., Moen, A., Pettersen, S., Maelandsmo, G. M., Egge-Jacobsen, W., Falnes, P. O. 2012. Lysine methylation of VCP by a member of a novel human protein methyltransferase family. *Nat Commun* 3: 1038.

Kim, H. S., Kim, S. K., Hromas, R., Lee, S. H. 2015. The SET domain is essential for metnase functions in replication restart and the 5′ end of SS-overhang cleavage. *PLOS ONE* 10: e0139418.

Kim, H. S., Williamson, E. A., Nickoloff, J. A., Hromas, R. A., Lee, S. H. 2017. Metnase mediates loading of exonuclease 1 onto single strand overhang DNA for end resection at stalled replication forks. *J Biol Chem* 292: 1414–1425.

Kim, J.-D., Kako, K., Kakiuchi, M., Park, G. G., Fukamizu, A. 2008. EWS is a substrate of type I protein arginine methyltransferase, PRMT8. *Int J Mol Med* 22: 309–315.

Kish, A., Gaillard, J. C., Armengaud, J., Elie, C. 2016. Post-translational methylations of the archaeal Mre11:Rad50 complex throughout the DNA damage response. *Mol Microbiol* 100: 362–378.

Lacroix, M., Messaoudi, E. S., Rodier, G., Le Cam, A., Sardet, C., Fabbrizio, E. 2008. The histone-binding protein COPR5 is required for nuclear functions of the protein arginine methyltransferase PRMT5. *EMBO Rep* 9: 452–458.

Lee, J.-H., Paull, T. T. 2005. ATM activation by DNA double-strand breaks through the Mre11-Rad50-Nbs1 complex. *Science* 308: 551–554.

Lee, Y.-H., Bedford, M. T., Stallcup, M. R. 2011. Regulated recruitment of tumor suppressor BRCA1 to the p21 gene by coactivator methylation. *Genes Dev* 25: 176–188.

Leppard, J. B., Champoux, J. J. 2005. Human DNA topoisomerase I: Relaxation, roles, and damage control. *Chromosoma* 114: 75–85.

Li, H., Watford, W., Li, C., Parmelee, A., Bryant, M. A., Deng, C., O'Shea, J., Lee, S. B. 2007. Ewing sarcoma gene EWS is essential for meiosis and B lymphocyte development. *J Clin Invest* 117: 1314–1323.

Li, X., Hu, X., Patel, B., Zhou, Z., Liang, S., Ybarra, R., Qiu, Y., Felsenfeld, G., Bungert, J., Huang, S. 2010. H4R3 methylation facilitates beta-globin transcription by regulating histone acetyltransferase binding and H3 acetylation. *Blood* 115: 2028–2037.

Li, X., Liu, L., Yang, S., Song, N., Zhou, X., Gao, J., Yu, N. et al. 2014. Histone demethylase KDM5B is a key regulator of genome stability. *Proc Natl Acad Sci USA* 111: 7096–7101.

Li, X., Manley, J. L. 2005. Inactivation of the SR protein splicing factor ASF/SF2 results in genomic instability. *Cell* 122: 365–378.

Liu, F., Zhao, X., Perna, F., Wang, L., Koppikar, P., Abdel-Wahab, O., Harr, M. W. et al. 2011. JAK2V617F-mediated phosphorylation of PRMT5 downregulates its methyltransferase activity and promotes myeloproliferation. *Cancer Cell* 19: 283–294.

Liu, H., Galka, M., Mori, E., Liu, X., Lin, Y. F., Wei, R., Pittock, P. et al. 2013. A method for systematic mapping of protein lysine methylation identifies functions for HP1beta in DNA damage response. *Mol Cell* 50: 723–735.

Liu, H., Takeda, S., Kumar, R., Westergard, T. D., Brown, E. J., Pandita, T. K., Cheng, E. H., Hsieh, J. J. 2010. Phosphorylation of MLL by ATR is required for execution of mammalian S-phase checkpoint. *Nature* 467: 343–346.

Liu, K., Guo, Y., Liu, H., Bian, C., Lam, R., Liu, Y., MacKenzie, F. et al. 2012. Crystal structure of TDRD3 and methyl-arginine binding characterization of TDRD3, SMN and SPF30. *PLOS ONE* 7: e30375.

Liu, L., Zhao, X., Zhao, L., Li, J., Yang, H., Zhu, Z., Liu, J., Huang, G. 2016. Arginine methylation of SREBP1a via PRMT5 promotes *de novo* lipogenesis and tumor growth. *Cancer Res* 76: 1260–1272.

Liu, Y., Liu, S., Yuan, S., Yu, H., Zhang, Y., Yang, X., Xie, G. et al. 2017. Chromodomain protein CDYL is required for transmission/restoration of repressive histone marks. *J Mol Cell Biol* 9: 178–194.

Loyola, A., Tagami, H., Bonaldi, T., Roche, D., Quivy, J. P., Imhof, A., Nakatani, Y., Dent, S. Y., Almouzni, G. 2009. The HP1alpha-CAF1-SetDB1-containing complex provides H3K9me1 for Suv39-mediated K9me3 in pericentric heterochromatin. *EMBO Rep* 10: 769–775.

Lu, F., Wu, X., Yin, F., Chia-Fang Lee, C., Yu, M., Mihaylov, I. S., Yu, J., Sun, H., Zhang, H. 2016. Regulation of DNA replication and chromosomal polyploidy by the MLL-WDR5-RBBP5 methyltransferases. *Biol Open* 5: 1449–1460.

Lukas, C., Savic, V., Bekker-Jensen, S., Doil, C., Neumann, B., Pedersen, R. S., Grofte, M. et al. 2011. 53BP1 nuclear bodies form around DNA lesions generated by mitotic transmission of chromosomes under replication stress. *Nat Cell Biol* 13: 243–253.

Mallette, F. A., Mattiroli, F., Cui, G., Young, L. C., Hendzel, M. J., Mer, G., Sixma, T. K., Richard, S. 2012. RNF8- and RNF168-dependent degradation of KDM4A/JMJD2A triggers 53BP1 recruitment to DNA damage sites. *EMBO J* 31: 1865–1878.

Mallette, F. A., Richard, S. 2012. K48-linked ubiquitination and protein degradation regulate 53BP1 recruitment at DNA damage sites. *Cell Res* 22: 1221–1223.

Mantri, M., Krojer, T., Bagg, E. A., Webby, C. A., Butler, D. S., Kochan, G., Kavanagh, K. L., Oppermann, U., McDonough, M. A., Schofield, C. J. 2010. Crystal structure of the 2-oxoglutarate- and Fe(II)-dependent lysyl hydroxylase JMJD6. *J Mol Biol* 401(2):211–22.

Markolovic, S., Leissing, T. M., Chowdhury, R., Wilkins, S. E., Lu, X., Schofield, C. J. 2016. Structure-function relationships of human JmjC oxygenases-demethylases versus hydroxylases. *Curr Opin Struct Biol* 41: 62–72.

Matsuoka, S., Ballif, B. A., Smogorzewska, A., McDonald, E. R., 3rd, Hurov, K. E., Luo, J., Bakalarski, C. E. et al. 2007. ATM and ATR substrate analysis reveals extensive protein networks responsive to DNA damage. *Science* 316: 1160–1166.

Metivier, R., Penot, G., Hübner, M. R., Reid, G., Brand, H., Kos, M., Gannon, F. 2003. Estrogen receptor-alpha directs ordered, cyclical, and combinatorial recruitment of cofactors on a natural target promoter. *Cell* 115: 751–763.

Migliori, V., Müller, J., Phalke, S., Low, D., Bezzi, M., Mok, W. C., Sahu, S. K. et al. 2012. Symmetric dimethylation of H3R2 is a newly identified histone mark that supports euchromatin maintenance. *Nat Struct Mol Biol* 19: 136–144.

Moore, K. E., Carlson, S. M., Camp, N. D., Cheung, P., James, R. G., Chua, K. F., Wolf-Yadlin, A., Gozani, O. 2013. A general molecular affinity strategy for global detection and proteomic analysis of lysine methylation. *Mol Cell* 50: 444–456.

Moumen, A., Magill, C., Dry, K. L., Jackson, S. P. 2013. ATM-dependent phosphorylation of heterogeneous nuclear ribonucleoprotein K promotes p53 transcriptional activation in response to DNA damage. *Cell Cycle* 12: 698–704.

Mullan, P. B., Quinn, J. E., Harkin, D. P. 2006. The role of BRCA1 in transcriptional regulation and cell cycle control. *Oncogene* 25: 5854–5863.

Naro, C., Bielli, P., Pagliarini, V., Sette, C. 2015. The interplay between DNA damage response and RNA processing: The unexpected role of splicing factors as gatekeepers of genome stability. *Front Genet* 6: 142.

Neelsen, K. J., Lopes, M. 2015. Replication fork reversal in eukaryotes: From dead end to dynamic response. *Nat Rev Mol Cell Biol* 16: 207–220.

Nelms, B. E., Maser, R. S., MacKay, J. F., Lagally, M. G., Petrini, J. H. 1998. *In situ* visualization of DNA double-strand break repair in human fibroblasts. *Science* 280: 590–592.

Oh, T. G., Bailey, P., Dray, E., Smith, A. G., Goode, J., Eriksson, N., Funder, J. W. et al. 2014. PRMT2 and RORγ expression are associated with breast cancer survival outcomes. *Mol Endocrinol* 28: 1166–1185.

Panier, S., Boulton, S. J. 2014. Double-strand break repair: 53BP1 comes into focus. *Nat Rev Mol Cell Biol* 15: 7–18.

Park, J. H., Szemes, M., Vieira, G. C., Melegh, Z., Malik, S., Heesom, K. J., Wallwitz-Freitas, V. L. et al. 2014. Protein arginine methyltransferase 5 is a key regulator of the MYCN oncoprotein in neuroblastoma cells. *Mol Oncol* 9(3): 617–27.

Paulsen, R. D., Soni, D. V., Wollman, R., Hahn, A. T., Yee, M.-C., Guan, A., Hesley, J. A. et al. 2009. A genome-wide siRNA screen reveals diverse cellular processes and pathways that mediate genome stability. *Mol Cell* 35: 228–239.

Pei, H., Wu, X., Liu, T., Yu, K., Jelinek, D. F., Lou, Z. 2013. The histone methyltransferase MMSET regulates class switch recombination. *J Immunol* 190: 756–763.

Piao, L., Kang, D., Suzuki, T., Masuda, A., Dohmae, N., Nakamura, Y., Hamamoto, R. 2014. The histone methyltransferase SMYD2 methylates PARP1 and promotes poly(ADP-ribosyl)ation activity in cancer cells. *Neoplasia* 16: 257–264, 264 e252.

Polo, S. E., Jackson, S. P. 2011. Dynamics of DNA damage response proteins at DNA breaks: A focus on protein modifications. *Genes Dev* 25: 409–433.

Pryde, F., Jain, D., Kerr, A., Curley, R., Mariotti, F. R., Vogelauer, M. 2009. H3 k36 methylation helps determine the timing of cdc45 association with replication origins. *PLOS ONE* 4: e5882.

Ray Chaudhuri, A., Callen, E., Ding, X., Gogola, E., Duarte, A. A., Lee, J. E., Wong, N. et al. 2016. Replication fork stability confers chemoresistance in BRCA-deficient cells. *Nature* 535: 382–387.

Rondinelli, B., Schwerer, H., Antonini, E., Gaviraghi, M., Lupi, A., Frenquelli, M., Cittaro, D., Segalla, S., Lemaitre, J. M., Tonon, G. 2015. H3K4me3 demethylation by the histone demethylase KDM5C/JARID1C promotes DNA replication origin firing. *Nucleic Acids Res* 43: 2560–2574.

Rust, H. L., Subramanian, V., West, G. M., Young, D. D., Schultz, P. G., Thompson, P. R. 2014. Using unnatural amino acid mutagenesis to probe the regulation of PRMT1. *ACS Chem Biol* 9: 649–655.

Sale, J. E. 2015. REV7/MAD2L2: The multitasking maestro emerges as a barrier to recombination. *EMBO J* 34: 1609–1611.

Saredi, G., Huang, H., Hammond, C. M., Alabert, C., Bekker-Jensen, S., Forne, I., Reveron-Gomez, N. et al. 2016. H4K20me0 marks post-replicative chromatin and recruits the TONSL-MMS22L DNA repair complex. *Nature* 534: 714–718.

Schibler, A., Koutelou, E., Tomida, J., Wilson-Pham, M., Wang, L., Lu, Y., Cabrera, A. P. et al. 2016. Histone H3K4 methylation regulates deactivation of the spindle assembly checkpoint through direct binding of Mad2. *Genes Dev* 30: 1187–1197.

Schultz, L. B. 2000. p53 binding protein 1 (53BP1) is an early participant in the cellular response to DNA double-strand breaks. *J Cell Biol* 151: 1381–1390.

Sinha, R., Allemand, E., Zhang, Z., Karni, R., Myers, M. P., Krainer, A. R. 2010. Arginine methylation controls the subcellular localization and functions of the oncoprotein splicing factor SF2/ASF. *Mol Cell Biol* 30: 2762–2774.

Sobol, R. W., Horton, J. K., Kühn, R., Gu, H., Singhal, R. K., Prasad, R., Rajewsky, K., Wilson, S. H. 1996. Requirement of mammalian DNA polymerase-beta in base-excision repair. *Nature* 379: 183–186.

Stanley, F. K., Moore, S., Goodarzi, A. A. 2013. CHD chromatin remodelling enzymes and the DNA damage response. *Mutat Res* 750: 31–44.

Stewart, G. S., Maser, R. S., Stankovic, T., Bressan, D. A., Kaplan, M. I., Jaspers, N. G., Raams, A., Byrd, P. J., Petrini, J. H., Taylor, A. M. 1999. The DNA double-strand break repair gene hMRE11 is mutated in individuals with an ataxia-telangiectasia-like disorder. *Cell* 99: 577–587.

Sun, Y., Jiang, X., Price, B. D. 2010. Tip60: Connecting chromatin to DNA damage signaling. *Cell Cycle* 9: 930–936.

Sun, Y., Jiang, X., Xu, Y., Ayrapetov, M. K., Moreau, L. A., Whetstine, J. R., Price, B. D. 2009. Histone H3 methylation links DNA damage detection to activation of the tumour suppressor Tip60. *Nat Cell Biol* 11: 1376–1382.

Takawa, M., Cho, H. S., Hayami, S., Toyokawa, G., Kogure, M., Yamane, Y., Iwai, Y. et al. 2012. Histone lysine methyltransferase SETD8 promotes carcinogenesis by deregulating PCNA expression. *Cancer Res* 72: 3217–3227.

Tang, J., Cho, N. W., Cui, G., Manion, E. M., Shanbhag, N. M., Botuyan, M. V., Mer, G., Greenberg, R. A. 2013. Acetylation limits 53BP1 association with damaged chromatin to promote homologous recombination. *Nat Struct Mol Biol* 20: 317–325.

Tang, J., Frankel, A., Cook, R. J., Kim, S., Paik, W. K., Williams, K. R., Clarke, S., Herschman, H. R. 2000. PRMT1 is the predominant type I protein arginine methyltransferase in mammalian cells. *J Biol Chem* 275: 7723–7730.

Taverna, S. D., Li, H., Ruthenburg, A. J., Allis, C. D., Patel, D. J. 2007. How chromatin-binding modules interpret histone modifications: Lessons from professional pocket pickers. *Nat Struct Mol Biol* 14: 1025–1040.

Tee, W.-W., Pardo, M., Theunissen, T. W., Yu, L., Choudhary, J. S., Hajkova, P., Surani, M. A. 2010. Prmt5 is essential for early mouse development and acts in the cytoplasm to maintain ES cell pluripotency. *Genes Dev* 24: 2772–2777.

Tessarz, P., Santos-Rosa, H., Robson, S. C., Sylvestersen, K. B., Nelson, C. J., Nielsen, M. L., Kouzarides, T. 2014. Glutamine methylation in histone H2A is an RNA-polymerase-I-dedicated modification. *Nature* 505: 564–568.

Thandapani, P., Couturier, A. M., Yu, Z., Li, X., Couture, J. F., Li, S., Masson, J. Y., Richard, S. 2017. Lysine methylation of FEN1 by SET7 is essential for its cellular response to replicative stress. *Oncotarget* 8(39): 64918–64931.

Thandapani, P., O'Connor, T. R., Bailey, T. L., Richard, S. 2013. Defining the RGG/RG motif. *Mol Cell* 50: 613–623.

Tripsianes, K., Madl, T., Machyna, M., Fessas, D., Englbrecht, C., Fischer, U., Neugebauer, K. M., Sattler, M. 2011. Structural basis for dimethylarginine recognition by the Tudor domains of human SMN and SPF30 proteins. *Nat Struct Mol Biol* 18: 1414–1420.

Verbiest, V., Montaudon, D., Tautu, M. T., Moukarzel, J., Portail, J.-P., Markovits, J., Robert, J., Ichas, F., Pourquier, P. 2008. Protein arginine (N)-methyl transferase 7 (PRMT7) as a potential target for the sensitization of tumor cells to camptothecins. *FEBS Lett* 582: 1483–1489.

Walport, L. J., Hopkinson, R. J., Chowdhury, R., Schiller, R., Ge, W., Kawamura, A., Schofield, C. J. 2016. Arginine demethylation is catalysed by a subset of JmjC histone lysine demethylases. *Nat Commun* 7: 11974.

Wang, D., Zhou, J., Liu, X., Lu, D., Shen, C., Du, Y., Wei, F. Z. et al. 2013. Methylation of SUV39H1 by SET7/9 results in heterochromatin relaxation and genome instability. *Proc Natl Acad Sci USA* 110: 5516–5521.

Wang, J., Chen, L., Sinha, S. H., Liang, Z., Chai, H., Muniyan, S., Chou, Y.-W. et al. 2012a. Pharmacophore-based virtual screening and biological evaluation of small molecule inhibitors for protein arginine methylation. *J Med Chem* 55: 7978–7987.

Wang, Q., Ma, S., Song, N., Li, X., Liu, L., Yang, S., Ding, X. et al. 2016a. Stabilization of histone demethylase PHF8 by USP7 promotes breast carcinogenesis. *J Clin Invest* 126: 2205–2220.

Wang, T., Xu, C., Liu, Y., Fan, K., Li, Z., Sun, X., Ouyang, H. et al. 2012b. Crystal structure of the human SUV39H1 chromodomain and its recognition of histone H3K9me2/3. *PLOS ONE* 7: e52977.

Wang, Y., Sun, H., Wang, J., Wang, H., Meng, L., Xu, C., Jin, M. et al. 2016b. DNA-PK-mediated phosphorylation of EZH2 regulates the DNA damage-induced apoptosis to maintain T-cell genomic integrity. *Cell Death Dis* 7: e2316.

Watanabe, S., Watanabe, K., Akimov, V., Bartkova, J., Blagoev, B., Lukas, J., Bartek, J. 2013. JMJD1C demethylates MDC1 to regulate the RNF8 and BRCA1-mediated chromatin response to DNA breaks. *Nat Struct Mol Biol* 20: 1425–1433.

Webby, C. J., Wolf, A., Gromak, N., Dreger, M., Kramer, H., Kessler, B., Nielsen, M. L. et al. 2009. Jmjd6 catalyses lysyl-hydroxylation of U2AF65, a protein associated with RNA splicing. *Science* 325: 90–93.

Wei, H., Wang, B., Miyagi, M., She, Y., Gopalan, B., Huang, D.-B., Ghosh, G., Stark, G. R., Lu, T. 2013. PRMT5 dimethylates R30 of the p65 subunit to activate NF-κB. *Proc Natl Acad Sci USA* 110: 13516–13521.

Williamson, E. A., Rasila, K. K., Corwin, L. K., Wray, J., Beck, B. D., Severns, V., Mobarak, C., Lee, S. H., Nickoloff, J. A., Hromas, R. 2008. The SET and transposase domain protein Metnase enhances chromosome decatenation: Regulation by automethylation. *Nucleic Acids Res* 36: 5822–5831.

Wu, R., Wang, Z., Zhang, H., Gan, H., Zhang, Z. 2017. H3K9me3 demethylase Kdm4d facilitates the formation of pre-initiative complex and regulates DNA replication. *Nucleic Acids Res* 45: 169–180.

Yan, L., Yan, C., Qian, K., Su, H., Kofsky-Wofford, S. A., Lee, W.-C., Zhao, X., Ho, M.-C., Ivanov, I., Zheng, Y. G. 2014. Diamidine compounds for selective inhibition of protein arginine methyltransferase 1. *J Med Chem* 57: 2611–2622.

Yang, J.-H., Chiou, Y.-Y., Fu, S.-L., Shih, I.-Y., Weng, T.-H., Lin, W.-J., Lin, C.-H. 2014. Arginine methylation of hnRNPK negatively modulates apoptosis upon DNA damage through local regulation of phosphorylation. *Nucleic Acids Res* 42: 9908–9924.

Yang, Y., Bedford, M. T. 2013. Protein arginine methyltransferases and cancer. *Nat Rev Cancer* 13: 37–50.

Ying, S., Hamdy, F. C., Helleday, T. 2012. Mre11-dependent degradation of stalled DNA replication forks is prevented by BRCA2 and PARP1. *Cancer Res* 72: 2814–2821.

Yu, X.-R., Tang, Y., Wang, W.-J., Ji, S., Ma, S., Zhong, L., Zhang, C.-H. et al. 2015. Discovery and structure-activity analysis of 4-((5-nitropyrimidin-4-yl)amino)benzimidamide derivatives as novel protein arginine methyltransferase 1 (PRMT1) inhibitors. *Bioorg Med Chem Lett* 25: 5449–5453.

Yu, Z., Chen, T., Hébert, J., Li, E., Richard, S. 2009. A mouse PRMT1 null allele defines an essential role for arginine methylation in genome maintenance and cell proliferation. *Mol Cell Biol* 29: 2982–2996.

Yu, Z., Vogel, G., Coulombe, Y., Dubeau, D., Spehalski, E., Hébert, J., Ferguson, D. O., Masson, J.-Y., Richard, S. 2012. The MRE11 GAR motif regulates DNA double-strand break processing and ATR activation. *Cell Res* 22: 305–320.

Zeller, P., Padeken, J., van Schendel, R., Kalck, V., Tijsterman, M., Gasser, S. M. 2016. Histone H3K9 methylation is dispensable for *Caenorhabditis elegans* development but suppresses RNA:DNA hybrid-associated repeat instability. *Nat Genet* 48: 1385–1395.

Zhao, D. Y., Gish, G., Braunschweig, U., Li, Y., Ni, Z., Schmitges, F. W., Zhong, G. et al. 2016. SMN and symmetric arginine dimethylation of RNA polymerase II C-terminal domain control termination. *Nature* 529: 48–53.

Zhao, Q., Rank, G., Tan, Y. T., Li, H., Moritz, R. L., Simpson, R. J., Cerruti, L. et al. 2009. PRMT5-mediated methylation of histone H4R3 recruits DNMT3A, coupling histone and DNA methylation in gene silencing. *Nat Struct Mol Biol* 16: 304–311.

Zheng, L., Zhou, M., Chai, Q., Parrish, J., Xue, D., Patrick, S. M., Turchi, J. J., Yannone, S. M., Chen, D., Shen, B. 2005. Novel function of the flap endonuclease 1 complex in processing stalled DNA replication forks. *EMBO Rep* 6: 83–89.

Zurita-Lopez, C. I., Sandberg, T., Kelly, R., Clarke, S. G. 2012. Human protein arginine methyltransferase 7 (PRMT7) is a type III enzyme forming ω-NG-monomethylated arginine residues. *J Biol Chem* 287: 7859–7870.

Ubiquitin, SUMO and the DNA Double-Strand Break Response

9

Ruth M. Densham, Alexander J. Garvin, and Joanna R. Morris

INTRODUCTION

The mammalian DNA double-strand break (DSB) response is highly coordinated through both time and space to elicit rapid and accurate DNA repair and maintain genome integrity. DSBs are efficiently sensed by the cell and signalled through a complex network of post-translational modifications (PTMs), including, but not limited to, phosphorylation, acetylation, methylation and PARylation, and also by modification with small proteins such as ubiquitin (ub) and small ubiquitin-like modifier (SUMO). Here, we review recent insights into the roles of ub and SUMO in the DSB damage response, specifically in controlling signal amplification and timing, and in priming chromatin for repair. Key enzymes in both modification pathways have been implicated in clinical disease and the development of small molecule inhibitors that can modulate the response is an exciting area of current research.

SMALL MODIFIER BIOLOGY

Ubiquitin and SUMO conjugation enzyme architectures

The Ub system

The archetypal small modifier is the 8.5 kDa globular protein ubiquitin (ub) (Goldstein et al. 1975). Since its discovery in 1975, around 20 more ub-like proteins have been identified (reviewed in van der Veen and Ploegh 2012) including SUMO (Matunis et al. 1996). All share the same underlying structural fold and a similar enzymatic conjugation architecture (**Figure 9.1**). In the first instance, both ub and SUMO gene products initially undergo maturation by post-translational processing of an inactive precursor to expose C-terminal di-glycine residues required for conjugation to a substrate lysine. The modification of substrate proteins requires a three-enzyme cascade consisting of the E1 activating enzyme, the E2 conjugating enzyme and an E3 ligase (Hershko et al. 1983). Mechanistically, a cysteine in the E1 active site binds the modifier's C-terminal di-glycine residues. This causes a conformational change in the E1, which promotes the subsequent interaction with the E2 and transfer of the modifier from the E1 to another cysteine in the E2 active site (reviewed in Schulman and Harper 2009). The E2 loaded with the modifier in the active site binds an E3 partner and together this E3-E2-ub complex aids substrate recognition and promotes final transfer of the modifier to the substrate lysine and formation of an isopeptide bond. While the enzymatic architecture of the cascades (E1-E2-E3) are the same, each cascade is specific for either the ub or SUMO pathways with no sharing of enzyme components (Desterro et al. 1997).

In the ub pathway, there are only two ub E1 activating enzymes encoded in the human genome, UBA1 and UBA6, compared to ~40 E2 conjugating enzymes and over 600 E3-ligases (reviewed in Weissman 2001). This vast array of enzymes brings with it specificity, selectivity, complexity and flexibility. As a result, tens of thousands of unique ub modification sites have been identified (Kim et al. 2011, Elia et al. 2015), and it is estimated that every protein could potentially be modified by ub. Specificity is largely thought to be provided by the E3-E2~ub complex. In recent years, work that solved the structure of this catalytic complex (Plechanovova et al. 2012, Pruneda et al. 2012, McGinty et al. 2014) has revealed that contacts between all members (E2, E3 and ub), and between the complex and the substrate can dictate residue

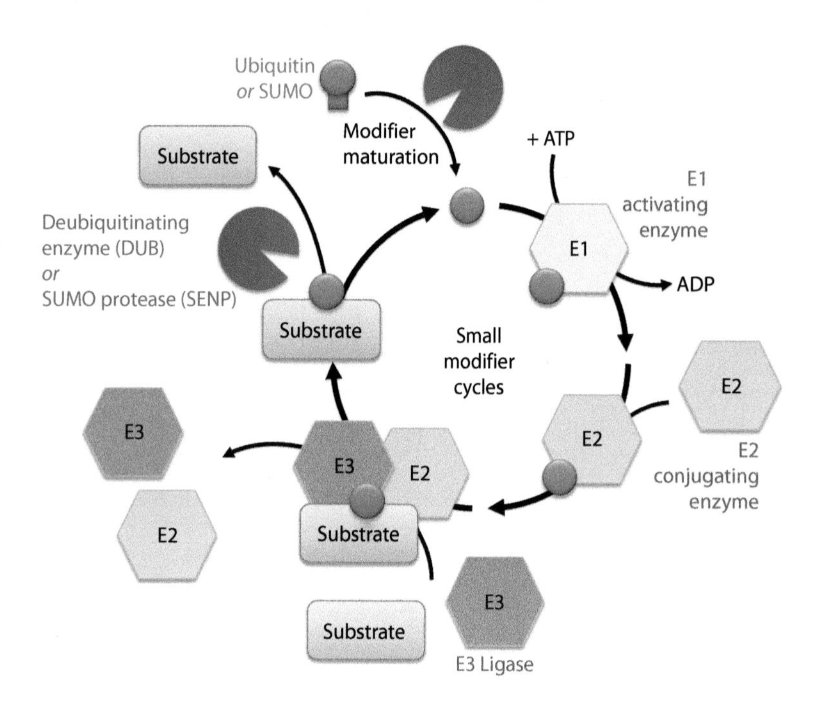

Figure 9.1 Small modifier enzyme cycles. The immature gene products of small modifiers ub or SUMO (blue) are processed by proteases (DUB or SENP (red) to produce mature conjugatable forms. The ub and SUMO modification pathways share the same architectural three-enzyme cascade (E1-E2-E3) (green), although the specific enzymes involved are unique to each pathway. DUB or SENP proteases edit, modify and/or remove the ub or SUMO from substrates in a dynamic process. Release of free small modifiers replenishes intracellular pools ready for another conjugation–deconjugation cycle.

specific modifications. This substrate site specificity is particularly important when ub is acting as a signalling entity rather than as a marker for protein degradation.

The E3-ub ligases are themselves divided into a number of different families, comprising: the homologous E6-AP C-terminus (HECT) ligases, the Really Interesting New Gene (RING) ligases which include the large multi-subunit cullins and the RING-between-RING (RBR) E3 ligases (Weissman 2001). HECT E3 ligases actively remove ub from the E2 before ligation to the substrate lysine, unlike RING E3 ligases which were historically thought to be passive partners of the E2, providing substrate specificity but little else. However, more recent work has demonstrated that binding of the E2 to a conserved patch in the RING domain stimulates an allosteric change in the E2 that primes the E2 active site for ub transfer (Pruneda et al. 2012). In addition, the formation of a tertiary 'closed' E3-E2~ub complex following E3 binding restricts ub motility and opens substrate access to the E2~ub active site (Plechanovova et al. 2012, Pruneda et al. 2012). The RING domain itself contains two zinc ions and the E3s can be either monomeric, for example RNF168 (Campbell et al. 2012); homodimeric, for example, RNF4 (Liew et al. 2010), or heterodimeric, for example, BRCA1-BARD1 (Brzovic et al. 2001).

The RING E3 ligases also include the multi-protein complex cullin family (reviewed in Chen et al. 2015). The minimal complex consists of the E3 ligase RING-box, an adaptor (typically an F-box protein), and one of six cullin scaffolds (Cul1, Cul2, Cul3, Cul4A and B, and Cul7). The final family of E3 ligases is the RBRs, which have a central domain sandwiched between N- and C-terminal RINGs. The N-terminal RING domain has a classical two-zinc ion structure and is essential for ligase activity. In contrast, the C-terminal RING only has one zinc ion, is less well conserved and, in some cases, is dispensable for ub ligase activity (Eisenhaber et al. 2007).

The SUMO system

In turning to the SUMO pathway, the conjugation cascade is much more restricted. The E1 consists of a single heterodimer comprised of SAE1-SAE2 subunits, a single E2 (UBE2I also known as Ubc9) and approximately a dozen E3 ligases (reviewed in Pichler et al. 2017). SUMO conjugation is more specific than ub, as there is a marked preference for substrate lysines occurring within the minimal consensus sequence ψKxE/D, where ψ is a large hydrophobic residue. More than 75% of all known SUMOylation events occur at these and similar consensus sites (Hendriks et al. 2017). In addition, while SUMO E3s enhance final SUMO transfer to the substrate, it has been shown that the E2 UBE2I is sufficient to direct SUMOylation at consensus lysines in the absence of an E3 (Sampson et al. 2001).

There are three classes of SUMO E3 ligases (PIAS, RanBP2 and ZNF451) that catalyse the final transfer of SUMO from the charged UBE2I~SUMO. For a detailed review of SUMO enzymologies, see Pichler et al. (2017). The protein inhibitor of activated STAT (PIAS) family consists of four members (PIAS1–4) and their various splice variants. PIAS proteins share a similar domain structure containing DNA-interacting SAF-A/B, Acinus and PIAS (SAP) domains, Pro-Ile-Asn-Ile-Thr (PINIT) domains that mediate protein–protein interactions and SUMO interaction motifs (SIMs) that bind SUMO and aid catalysis (reviewed in Rabellino et al. 2017). Finally, the RING-like catalytic domains of PIAS members are termed Siz/PIAS-RINGs (SP-RINGs). These are structurally similar to the RINGs found in ub E3 ligases but lack two of the conserved zinc-binding cysteines found in classical RINGs. This SP-RING domain is the only shared domain found in the more distantly related member of the PIAS family, MMS21/Nse2 (Potts and Yu 2005). Recent structural work demonstrates the cooperation between the E3 PIAS family member Siz1-E2~SUMO complex and the substrate PCNA. Here they demonstrate how molecular contacts from the E3 (Siz1) are able to override E2 (UBE2I) specificity to promote SUMO modification of a non-consensus lysine in the substrate PCNA (Streich and Lima 2016). The second class of SUMO E3s is RanBP2, which is a large nuclear pore–associated protein that is not homologous to any other E3 ligase. Ligase function of RanBP2 does not rely on a zinc finger; rather, two repeat elements separated by a linker are responsible for promoting SUMO transfer (Werner et al. 2102). Third, the E3 ligase ZNF451, which was originally characterized as a SUMOylated PML body-associated transcription co-activator (Karvonen et al. 2008), acts via two N-terminal SIMs and a zinc finger to generate SUMO2/3 polymers. It demonstrates a clear preference for SUMO2/3 and is so far unique among the E3s for its SUMO isoform specificity (Cappadocia et al. 2015, Eisenhardt et al. 2015). Other SUMO E3 ligases that are less well characterised in terms of their ligase enzymology include KAP1/TRIM28, which primarily autoSUMOylates itself via a PHD domain (Ivanov et al. 2007), and CBX4/Pc2, a component of polycomb complexes (Kagey et al. 2003). Enzymes from each step of the SUMO cascade (E1-E2-E3) have been detected at sites of DSBs, and roles for SUMO modification in the damage response are emerging (Galanty et al. 2009, Morris et al. 2009, Shima et al. 2013).

Much of the complexity in the SUMO system comes from the fact that, unlike ub where all mature forms are identical, the multiple SUMO genes give rise to distinct protein products, SUMO1/2/3/4. While SUMO fundamentally shares the same tertiary structure with ub, it has only limited sequence identity and, therefore, differs in the distribution of surface charge (Bayer et al. 1998). SUMO1 was first identified as an ub-like modifier of the GTPase RanGAP1 (Matunis et al. 1996). The later identified isoforms, SUMO2 and SUMO3, are more distantly related to SUMO1 but share 97% sequence identity between themselves and are often referred to as SUMO2/3 (Kamitani et al. 1998). Among the differences between SUMO1 and SUMO2/3 is the presence of a ψKxE/D consensus sequence found at lysine 11 in SUMO2/3, but absent in SUMO1, which allows the formation of SUMO2/3 chains. SUMO4 has a proline in close proximity to the C-terminal di-glycine that prevents its maturation and therefore is generally assumed to be inactive due to its inability to be conjugated to substrates (Owerbach et al. 2005).

Substrate modifications

Ub modifications

Substrates can be modified by a single ub at one lysine residue (mono-ub), or at multiple lysines in the same substrate (multi-mono-ub) (**Figure 9.2**). In addition, ub itself has seven conserved lysines (K6, K11, K27, K29, K33, K48, K63) and the N-terminal methionine, all of which can be conjugated to form poly-ub chains (reviewed in Yau and Rape 2016). In a further level of complexity, poly-ub chains can be pure (where all ub–ub linkages are through the same residue), mixed-linkage (where ub–ub linkages in the same chain are through more than one lysine residue), or branched (where a single ub in the chain has multiple linkages) (reviewed in Yau and Rape 2016). The three-dimensional structures of ub chains with different linkages are unique, can be sensed differently by the cell and are used to signal diverse processes. While K48 chains are most famously linked with proteasomal degradation, both K48 and K63 chains are critical components in the DSB response (Wang and Elledge 2007). In addition, there are a few reports emerging of the formation of K11 and K27 chains in response to DNA damage (Gatti et al. 2015, Paul and Wang 2017).

Finally, mass spectrometry site mapping has revealed an added layer of complexity and regulation in small modifier signalling by the finding that ub is modified by PTMs such as phosphorylation and acetylation (Swaney et al. 2013, Ohtake et al. 2015). Whether these additional modifications will affect the roles of small modifiers in the signalling of DNA damage remains to be determined.

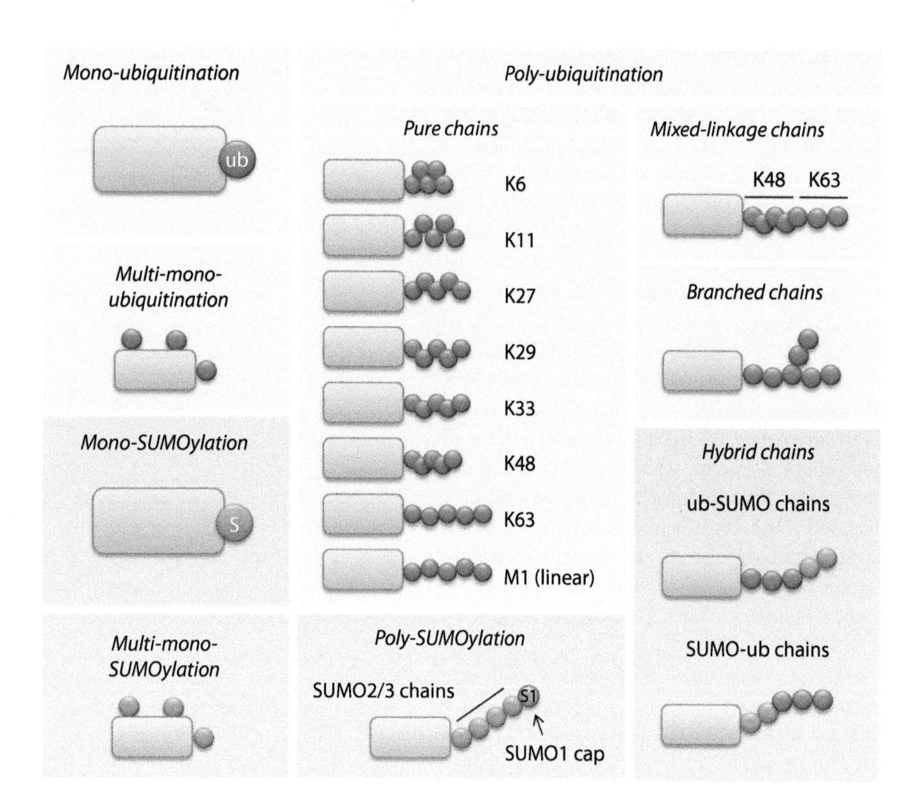

Figure 9.2 The complexity of small modifier modifications. Substrates (grey boxes) can be modified by ub (blue) or SUMO (orange) as both mono- or poly-modifications. Ub can form multiple different types of poly-ub chains by linkages with any one of seven internal lysine residues or the N-terminal methionine. These chains can be pure (all linkages are through the same residue), mixed linkages (different lysine linkages in the same chain) or branched (two or more linkages on a single ub moiety in a chain). SUMO chains are formed between SUMO2/3 isoforms at the internal SUMO consensus at lysine-11. The SUMO1 isoform lacks this consensus so may act to cap SUMO chain lengths. Crossover between the ub and SUMO pathways is seen in the formation of ub-SUMO and SUMO-ub hybrid chains. Each of these modifications has a different topology and as such is sensed differently by the cell.

SUMO modifications

SUMO conjugation can also occur as mono, multi-mono, polymers and branched chains (Figure 9.2). SUMO2/3 contains a SUMO consensus lysine at position 11 through which the majority of chains are formed (Tatham et al. 2001, Hendriks and Vertegaal 2016). Large-scale SUMO site mapping has also shown that SUMO2/3 can form chains on all of their non-consensus lysine residues, and therefore, like ub, can form multiple chain types (Yau and Rape 2016). What roles these non-consensus chains have in cell signalling are currently unknown (Hendriks and Vertegaal 2016, Hendriks et al. 2017). Also, as with ub, SUMO2/3 can form branched chains in which two SUMOs are conjugated at different lysines to a single SUMO (Tammsalu et al. 2014).

SUMO1 is less proficient in forming chains due to a lack of a strong SUMO consensus site. However, SUMO1 can be incorporated into SUMO2/3 polymers where its inability to polymerise further is thought to promote chain capping and, as a result, SUMO1 could have important roles in restricting SUMO2/3 chain length (Tatham et al. 2001). However, SUMO site mapping proteomics have shown that SUMO2/3 is also able to conjugate to SUMO1 at multiple non-consensus lysines (Tammsalu et al. 2014, Hendriks and Vertegaal 2016, Hendriks et al. 2017) arguing that SUMO1 has additional non-capping roles in chain formation.

Hybrid SUMO-Ub modifications

The interconnectivity of ub and SUMO signalling is well highlighted by the discovery of hybrid ub-SUMO/SUMO-ub linkages in cells. SUMO1 and SUMO2/3 are ubiquitinated at several lysine residues, most likely by SUMO targeted ub E3 ligases (STUbLs) such as RNF4 and RNF111 (Sun et al. 2007, Tatham et al. 2008). In addition, ub is extensively modified by SUMO2, suggesting both ubiquitination of SUMO and SUMOylation of ub, although the identity of factors that SUMOylate ub are currently unknown (Hendriks and Vertegaal 2016). Site mapping of both ub and SUMO suggest that most lysines are available to be conjugated by each other. Therefore, the possible number of chain types arising from these hybrid modifications is enormous. The makeup of hybrid chains is also non-random, as SUMO2/3 conjugation switches from K63 on ub to K11 during proteotoxic stress (Hendriks et al. 2017). Incorporation of SUMO2/3 into ub polymers at sites such as ub-K63 would presumably impact the cell's ability to signal through this type of ub linkage and would generate entirely new chain topologies for other binding domains. Therefore, crosstalk between ub and SUMO has the potential to substantially increase the complexity of small modifier recognition in cells (Hendriks et al. 2017).

Small modifiers promote protein–protein interactions

Ub interaction modules

A major function of small modifier conjugation such as ub and SUMO is the additional interaction interfaces they provide to their modified substrates. This allows docking of protein domains that would otherwise not occur on unmodified substrates. Given the complexity in the ub system, it is perhaps unsurprising that the cell has acquired a plethora of diverse ub binding domains (UBDs) that act as readers of the modifications. These have been reviewed elsewhere (Dikic et al. 2009, Husnjak and Dikic 2012). To date, over 20 different families of binding domains have been identified that are all characterised by the non-covalent interaction with ub. They demonstrate a range of differing structural complexity. For example, they can be formed from a simple alpha-helix (e.g. ub interacting motif [UIM] and the related motif interacting with ub [MIU], and UIM and MIU-related UBD [UMI] domains), to folded alpha-helical structures (e.g. ub associated domains [UBA]) or CUE domains (coupling of ub conjugation to endoplasmic reticulum degradation), to zinc fingers (e.g. A20 zinc finger; ub binding zinc finger [UBZ]) and also ub-conjugating enzyme variant [UEV] domains. These UBDs, whether found in isolation or in tandem with another UBD or PTM or protein-binding domain in the same protein, are able to provide exquisite specificity in reading the ub signals. This can be exemplified by examples of UBDs in the double-strand break DNA damage response.

For instance, the scaffold protein RAP80 has tandem UIM domains that specifically bind K63-linked ub and are required for RAP80 recruitment to DNA damage (Kim et al. 2007, Sobhian et al. 2007, Wang et al. 2007). Specific recognition of K63-ub chains is controlled by the finely tuned length of the linker between RAP80's tandem UIMs to orient and position the domains accurately (Sato et al. 2009). This specific recognition of K63-ub signalling is key to regulating recruitment of the BRCA1-A complex to sites of damage. Indeed, a RAP80 variant where amino acid E81 is deleted from UIM1 was identified in patients with familial breast cancer (Nikkila et al. 2009). This variant showed decreased ub binding and impaired recruitment of BRCA1 to double-strand breaks.

The scaffold protein 53BP1, critical in the DNA DSB response, contains a UBD domain required for its recruitment to damage sites (Fradet-Turcotte et al. 2013). Here, recruitment and specificity are achieved through cooperation of both the UBD and the neighbouring methyl binding tandem TUDOR domains in 53BP1. Together they allow the highly specific recognition of the damage-modified nucleosome through TUDOR domain binding to histone H4-K20me2 and UBD binding to histone H2A-K15ub (Wilson et al. 2016).

In addition, many of the ub ligation enzymes themselves also contain UBDs. For example, RNF168, which is the E3 ligase responsible for H2A-K13/15ub generation in response to DNA damage, has three UBDs (a UMI and two MIU domains) in addition to a LR motif (LRM) nucleosome binding domain adjacent to the first UBD (Panier et al. 2012). This tandem LRM-UBD motif also recognises the nucleosomal H2A-K13/15ub mark. An equivalent LRM-UBD motif is found in the related RNF169 protein where it competes with 53BP1 binding to H2A-K13/15ub and acts as a negative regulator of RNF168 signalling (Chen et al. 2012a, Panier et al. 2012, Poulsen et al. 2012). However, it is less clear how this LRM-UBD motif serves the function of RNF168. It may serve to help the spread of RNF168 signalling by facilitating a feedback loop. Alternatively, it has been proposed that RNF168 generates rare K27-ub chains at damage sites (Gatti et al. 2015) and that RNF168's UBDs may serve to orient the acceptor ub in such a way as to direct formation of this unusual linkage.

SUMO interaction modules

There are three known SUMO binding domains: the SUMO interacting motifs (SIMs) (Hecker et al. 2006) and two more recently identified domains, myeloproliferative and mental retardation-type (MYM) (Guzzo et al. 2014) and ZZ zinc fingers (Danielsen et al. 2012, Diehl et al. 2016). The best characterized of these are the SIMs which form short alpha helices from stretches of hydrophobic amino acids with (V/I/L)x (V/I/L) (V/I/L) sequence homology (Hecker et al. 2006). These hydrophobic alpha helices bind a groove on the surface of SUMO. Additional contacts from acidic residues proximal to the SIM can provide some degree of binding specificity between SUMO isoforms (Hecker et al. 2006) and in some cases promote isoform-specific modification (Meulmeester et al. 2008). Further, phosphorylation of SIM-proximal residues can similarly promote specific SUMO isoform interactions and provides an interesting level of crosstalk between modification pathways (Stehmeier and Muller 2009). Finally, in the same way as UBDs in the ub system, SIMs found in SUMO enzymes can both aid SUMO conjugation and promote isoform-specific modification (Tatham et al. 2005). For example, both SUMO2/3 preference and chain formation by the SUMO E3 ligase ZNF451 requires the enzyme's N-terminal SIMs (Cappadocia et al. 2015, Eisenhardt et al. 2015). Many DSB repair proteins contain putative SIMs, but few have been extensively characterized. One exception is the SIM found in the homology directed recombination factor RAD51, which is essential for RAD51 filament formation, in the process of homology directed repair (HDR) of DNA double-strand breaks (Shima et al. 2013).

Like the SIM domain, the MYM zinc finger occupies the same binding surface on SUMO (Guzzo et al. 2014). In contrast, the ZZ zinc finger domain binds a distinct patch on the opposite surface of SUMO (Diehl et al. 2016). This opens up the potential for a single SUMO molecule to be bound simultaneously by two different SUMO binding domains. In addition, it has been shown that the ZZ domain preferentially binds SUMO1 over SUMO2/3 (Danielsen et al. 2012). The relatively recent discovery of the SUMO-binding capacity of MYM and ZZ domains means that these remain largely understudied. However, one MYM-containing protein, ZMYM3, has been reported to have a role in the DSB response by regulating BRCA1 recruitment to damage sites (Leung et al. 2017). It is not yet known if SUMO binding via the MYM domain is important for this function, although it is intriguing given that the SUMO modification pathway is known to be required for BRCA1 recruitment to DSBs (Galanty et al. 2009, Morris et al. 2009).

Some proteins are multi-mono SUMOylated, but the function this serves is not well understood. As SUMO-SIM interactions are relatively weak, it may be that multiple mono-SUMO conjugates on the same substrate improves protein–protein interactions, or promotes multiple co-ordinated interactions within a complex of SIM containing proteins. This would seem likely given the tendency of the SUMO enzyme cascade to modify protein groups and complexes (Psakhye and Jentsch 2012). Using a multi-mono SUMO2 mimic as bait, several proteins have been identified as multi-mono SUMO2 interactors, including several proteins that have roles in DSB repair, such as BLM and 53BP1 (Aguilar-Martinez et al. 2015).

Ub-like and SUMO-like domains

Some proteins contain domains that mimic the structure of ub or SUMO. Ub-like domains (UbLs) are frequently found in ubiquitin specific proteases (USPs) and are also present in a family of non-proteasomal ub-binding proteins, such as HR23A/B and Plic-1. These non-proteasomal ub-binding proteins interact with ubiquitinated substrates through their UBA domains while using their UbL domains to interact with the proteasome (reviewed in Su and Lau 2009). In USP enzymes, the UbL can link deubiquitinating activity to the proteasome (e.g. in USP14), but they also have more varied roles. For example, in USP4 the internal UbL contacts the catalytic domain and competes with ub itself, while in USP7 two of its five UbL domains promote a conformational change that promotes ub binding and catalysis (reviewed in Kim and Sixma 2017).

SUMO-like domains (SLDs) in mammalian cells are relatively understudied. One example of an SLD-containing protein is the regulatory factor UAF1 that heterodimerises with a number of DUBs, including USP1. Of UAF1's two SLDs, the second mimics SUMO to enable interaction with SIMs in the Fanconi anaemia protein, FANCI and the replication DNA clamp PCNA (Yang et al. 2011). The SLDs in UAF1 also mediate interaction with the DNA repair factor RAD51AP1 and are essential for formation of a UAF1-RAD51AP1-RAD51 ternary complex that is required for homology directed recombination (Liang et al. 2016).

Small modifier deconjugation

The modification of proteins by ub or SUMO is a dynamic process and the ability to edit, modify or remove these modifications is catalysed by a diverse array of specific ub or SUMO proteases. These proteases are required to process small modifier maturation from inactive precursors, to replenish the pools of free small modifiers and to edit or remove the modifications on substrates.

Deubiquitinating enzymes

In humans there are six families of deubiquitinating enzymes (DUBs) totalling ~100 DUBs across the genome (reviewed in Komander et al. 2009, Mevissen and Komander 2017). The vast majority of these are cysteine proteases which utilise a catalytic dyad or triad of crucial amino acids to promote nucleophilic attack from the catalytic cysteine towards the ub-isopeptide bond (Komander and Barford 2008). The five families of cysteine proteases are: ub-specific peptidases (USPs), ub C-terminal hydrolases (UCHs), ovarian tumour proteases (OTUs), Machado-Joseph disease protein domain proteases (MJDs) and the MIU-containing novel DUB family (MINDY). The final family of DUBs is the zinc metalloprotease family of Jab1/MPN domain-associated metalloproteases (JAMMs). This family of enzymes coordinate two zinc ions and catalyse ub bond breakage by zinc activation of a water molecule to promote nucleophilic attack of the ub-linkage (Sato et al. 2008).

Many of the DUBs contain additional binding or interaction domains which help to promote linkage or substrate specificity. The identification of ub-linkage-specific DUBs has enabled mechanistic insight into ub chain complexity. In general, the JAMM proteases demonstrate a preference for K63-ub (Cooper et al. 2009) and MINDY DUBs specificity for K48-chains (Abdul Rehman et al. 2016), while the USP family, with a few notable exceptions, largely shows no linkage preference. However, substrate specificity may also be achieved through large multi-subunit complexes that contain DUBs, perhaps best characterised by the H2B-K120ub specific DUB activity of the multi-subunit SAGA complex. Two other significant multi-subunit complexes in the DNA damage response include the proteasome and associated DUBs (Butler et al. 2012), and the BRCA1-A complex containing DUB BRCC36 (Sobhian et al. 2007).

Removal of ub from substrates can prevent proteasomal recognition and degradation, resulting in protein stabilisation. In addition, the regulation of non-degradative ub signalling by DUBs can alter ub-chain length or ub-chain linkage composition, leading to changes in protein-complex formation, protein structure, localisation, or even changes in enzyme activity. Indeed, DUBs themselves are regulated by ub and SUMO. For example, the DUB catalytic activity of ATXN3 is enhanced by ubiquitination (Todi et al. 2009), whereas SUMOylation of USP28 is inhibitory to DUB activity (Zhen et al. 2014).

The proteasome

Most regulated protein degradation is performed by the ub-proteasome system. The proteasome is a large, multi-protein complex composed of the 20S cylindrical core particle of four stacked rings around a pore, responsible for the proteolysis reactions, and the 19S regulatory particle of 19 proteins. Together these make up the 26S proteasome particle (Budenholzer et al. 2017). The 19S binds ubiquitinated substrates, or recognizes UbL domains though ub-binding proteins in its lid structure. Two DUBs within the lid, including Rpn11/POH1/PDSM14, then deubiquitinate the substrates. The AAA (ATPase associated with various cellular activities) ATPase base subunits of the 19S perform substrate unfolding and translocate substrates to the core particle for degradation.

DeSUMOylases

Far fewer SUMO proteases are found in eukaryotic genomes than ub proteases. At present, there are nine SUMO proteases identified in humans: six members of the C48 cysteine protease family (SENP1/2/3/5/6/7); one DUB–USPL1; and two other proteases, DeSi1/2 (Schulz et al. 2012, Shin et al. 2012). DeSUMOylases have three activities: first, SUMO maturation in which the C-terminal extensions are removed from pro-SUMO to expose the di-glycine termini needed for conjugation; second, deconjugation of SUMO monomers; and third, deconjugation of SUMO-polymers, including roles in chain editing. SENP1 is the dominant protease for SUMO maturation, but can also deconjugate monomers and to some extent polymers (Mendes et al. 2016). SENP2/3/5 deconjugates monomers (Gong and Yeh 2006), while SENP6/7 has a preference for SUMO2/3 polymers (Shen et al. 2009). SENP enzymes also have some isoform specificity with SENP1 being the dominant SUMO1 protease, while all SENPs can deconjugate SUMO2/3 (Mendes et al. 2016). Isoform specificity may also be enforced at the level of localisation as SUMO deconjugating enzymes distribute to distinct sub-cellular locations (reviewed in Mukhopadhyay and Dasso 2007). For example, SENP2 localises to nuclear pores, SENP3/5 to nucleoli, SENP7 to chromatin and USPL1 to Cajal bodies.

Mixed chain deconjugation

Ub-SUMO mixed linkages provide a third class of modification that would also require editing. Relatively little is known about the cellular regulation of these linkages and if, or how, they impact DUB and SENP recognition and enzymatic activity. Three DUB enzymes are capable of disassembling the ub component from SUMO polymers in vitro (USP11, USP7 and UCHL3), but whether this occurs in cells is less clear, and if they have any specificity for SUMO-ub hybrid linkages is unknown (Bett et al. 2015, Hendriks et al. 2015a, Lecona et al. 2016). Moreover it is unclear whether ub incorporation into SUMO2/3 chains affects SENP activity.

VCP/p97-ATPase and ubiquitin-mediated disassembly

The VCP/p97 protein is a member of the AAA family. While this protein is not a DUB protease, it contains two hexameric ATPase rings and utilises ATP hydrolysis to extract Ub-modified proteins from membranes, chromatin or large multi-subunit complexes (reviewed in Dantuma et al. 2014). Once extracted, proteins are either recycled or degraded by the proteasome. Interestingly, VCP/p97 binds co-factors that contain both UBD and SIM domains and is therefore able to act at the interface and crossover between ub and SUMO pathways (reviewed in Nie and Boddy 2016). In concert with the actions of SUMO-targeted ub-ligases, such as RNF4, which generate hybrid SUMO-ub chains at sites of damage, VCP/p97 is a powerful regulator that can control the release of protein complexes from chromatin.

CHROMATIN PRIMING FOR DNA REPAIR

Foundations of chromatin structure

DNA DSBs can cause cell death or cellular transformation and thus organisms have evolved mechanisms to repair these lesions. DNA DSBs are repaired by two main mechanisms. Throughout the cell cycle, the process of non-homologous end joining predominates. Here, DNA ends are ligated together, often after minimal processing of the DNA ends. The second mechanism of homology directed recombination occurs exclusively in late S-phase and G2. In this case, DNA on either side of the break is digested back, or resected, the ssDNA is bound by RPA and then by RAD51, which is used by the cell to identify the sister chromatid which is used for accurate repair. These processes do not occur on naked DNA but must negotiate the chromatin in which DNA is wound round an octamer of histone proteins termed a nucleosome. Nucleosomes are the dynamic hubs for enzymatic modifiers, chromatin remodellers and the landing platforms for protein complexes which regulate DNA function, access, packing and ultimately DNA repair. Chromatin context is crucial for DNA repair outcome, and the first challenge the cellular machinery meets is to deal with the underlying chromatin structure in which a DSB occurs.

The basic nucleosome consists of two copies of each of the core histones H2A, H2B, H3 and H4. The linker histone H1 binds at the DNA entry and exit points and stabilizes nucleosomes and thereby can promote higher-order chromatin architecture. Within the genome there are multiple variants for each of these histones and together they provide a complex array of subtle and not-so-subtle variations in nucleosome structure that occurs at the level of histone composition. In addition, all histones have long flexible N-terminal tails that extend away from the nucleosome body and which are highly modified by an array of PTMs, including phosphorylation, methylation, acetylation, PARylation and modification with small proteins such as ub and SUMO. This gives rise to a highly coordinated and intricate epigenetic code.

The differences between repair in open, active euchromatin compared to closed, repressive heterochromatin have been reviewed elsewhere (Murray et al. 2012, Watts 2016). Modifications by both SUMO and ub are associated with directing changes between chromatin states (Nickel and Davie 1989, Wang et al. 2004, Uchimura et al. 2006), and ub and SUMO modification sites have been identified in histones by mass spectrometry screens. Only a handful have been functionally examined, with the best studied being several ub sites on H2A and H2B. Histone ubiquitination in a nucleosomal context is remarkably site specific both for the E3 ligases responsible for laying down the ub mark and for the readers of these marks (reviewed in Uckelmann and Sixma 2017). In the core nucleosome, H2A has long tails at both N- and C-termini, which can be specifically modified by conjugation of ub at three major sites: K13/K15 by RNF168 (Mattiroli et al. 2012), K118/K119 by the polycomb repressive complex-1 (PRC1) (Nickel and Davie 1989), and K125/K127/K129 by BRCA1-BARD1 (Kalb et al. 2014a). The majority of H2A ubiquitination in the cell is at the K119 site (Nickel and Davie 1989), which is ubiquitinated by the RING1A/B E3 ligase in a heterodimeric complex with BMI1 or MEL18 as part of the PRC1 and is associated with transcriptional gene repression and heterochromatin (Wang et al. 2004). This may be partly due to the ability of H2A-K119ub to enhance binding of the linker histone H1 to nucleosomes (Jason et al. 2005) and support higher-order chromatin compaction.

Whilst H2A-K119ub is associated with repressive chromatin, the corresponding ubiquitination site in H2B, K120Ub, is found to be enriched at actively transcribed genes (Nickel and Davie 1989) and is mediated by the heterodimeric E3 ligase RNF20-RNF40 (Zhu et al. 2005). H2B-K120Ub supports increased access to DNA by promoting both local and higher-order chromatin decompaction (Fierz et al. 2011) through specific interactions between a glutamine patch in ub (Q16 and Q18) and basic residues in the surrounding nucleosome environment (Debelouchina et al. 2017). In this model, the ub moiety is proposed to form a 'wedge' between nucleosomes, thereby promoting chromatin relaxation (Debelouchina et al. 2017). This wedge effect cannot be duplicated by replacement of ub with SUMO (Chandrasekharan et al. 2009), suggesting that this is due to specific interactions and is more complex than a simple steric hindrance model. In contrast, in yeast the N-terminus of H2B has been reported to be SUMOylated at K6/7 and this is associated with transcriptional repression in yeast (Nathan et al. 2006).

The SUMO pathway also contributes to the determination of chromatin state, with the best-characterised examples indicating that SUMO contributes through influencing the function of chromatin-associated proteins. Indeed, in the SUMO proteome, proteins that have roles in chromatin architecture are among the most highly SUMOylated. SUMO modifications are more generally associated with a repressive chromatin phenotype, playing a central role in heterochromatin maintenance (Uchimura et al. 2006) and in transcriptional repression in various contexts (Neyret-Kahn et al. 2013). In addition, multiple repressor complexes show evidence of SUMOylation, and these complexes are enriched with SIM-bearing proteins (Garcia-Dominguez and Reyes 2009). For example, the transcriptional repressor and heterochromatin nucleator protein KRAB-associated protein 1 (KAP1, also referred to as TRIM28) is one of the most abundantly SUMOylated proteins in human cells (Hendriks and Vertegaal 2016). This is perhaps due in part to KAP1's auto-SUMO ligase activity found in its PHD domain. KAP1 is most heavily SUMOylated in its C-terminal bromodomain (Ivanov et al. 2007), which is in turn recognised by SIMs in the nucleosome remodelling deacetylase (NuRD) chromatin remodeller subunit chromodomain helicase DNA binding protein 3 (CHD3), and in the methyltransferase SET domain bifurcated 1 (SETDB1) (Ivanov et al. 2007). Recruitment of the NuRD complex promotes histone deacetylation via histone deacetylases (HDACs) and chromatin remodelling via CHD helicases, and recruitment of SETDB1 promotes H3K9me3. Together these changes promote gene silencing and chromatin compaction. In addition, a major interactor of KAP1 is heterochromatin protein 1α (HP1α), which binds to H3K9me3 (Ryan et al. 1999). This can therefore set up a positive feedback loop so that spreading of H3K9me3 along chromatin promotes further recruitment of HP1α proteins which in turn recruit additional KAP1 molecules, allowing propagation of a repressive chromatin state (Groner et al. 2010).

Preparing chromatin for repair

Immediately following DNA damage, chromatin undergoes a rapid, transient compaction in the environment local to the DSB before relaxing to allow access for DNA repair factors. This initial repressive state may prevent unwanted movement of the DSB in order to maintain the relationship between DNA ends, contribute to local transcriptional silencing, and recruit required remodelling complexes with essential activities; all of which may serve to prepare the histone code for repair (reviewed in Gursoy-Yuzugullu et al. 2016). A number of mechanisms have been proposed to drive this transient compaction event and the subsequent remodelling to an open permissive environment, including roles for chromatin remodellers and modifiers such as the NuRD complex, HDACs, lysine demethylases, the E3 ub-ligase PRC1 and the auto-SUMO ligase KAP1. Together these promote ubiquitination of H2A-K118/K119, histone deacetylation and increased H3K9me2/3 resulting in tighter nucleosome packing.

PRC1 in early chromatin compaction following DNA damage

PRC1 is rapidly recruited to sites of DNA damage following ionising radiation (IR) (Ismail et al. 2010) (**Figure 9.3**). In response to IR, this recruitment is partially dependent on SUMOylation of PRC1 by another polycomb protein, CBX4, which is a SUMO E3 ligase (Ismail et al. 2012). The SUMO site is on the BMI1 component of PRC1 at K88, and mutation of this to K88R abrogates BMI1 recruitment. Once at damage sites, the functional heterodimeric E3 ub-ligase of the PRC1 complex, BMI1 and RING1A/B, ubiquitinates H2A at K119 (Wang et al. 2004, Ismail et al. 2010, Ginjala et al. 2011). In response to damage, this PRC1-mediated H2A-K119ub mark contributes both to local transcriptional repression (Chou et al. 2010) and damage-responsive ub signalling (Ismail et al. 2010, Densham et al. 2016). In addition, the PRC1 component, BMI1, indirectly promotes transcriptional silencing by recruitment of the HECT E3 ligase UBR5. In turn, UBR5 interacts with, and ubiquitinates, components of the facilitates chromatin transcription (FACT) complex to block FACT enrichment and transcriptional elongation at damage loci (Sanchez et al. 2016). Moreover, recent findings indicate that PRC1 modification of H2A-K119ub can recruit the related polycomb repressor complex 2 (PRC2), which could

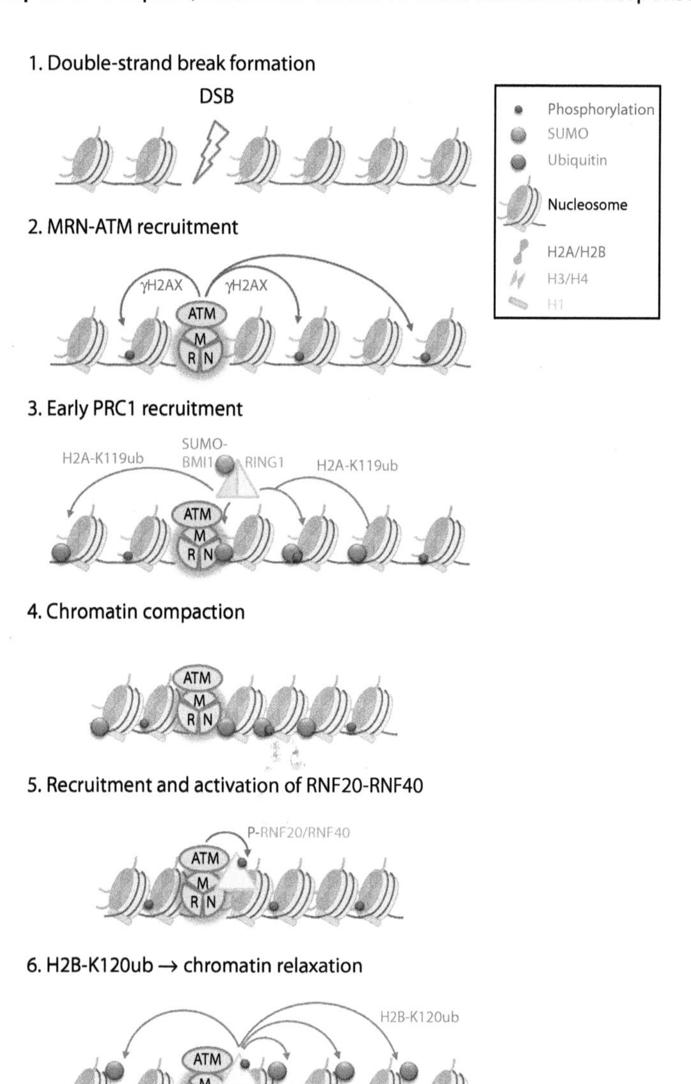

Figure 9.3 Ub signalling in early chromatin changes following DNA damage. DNA DSBs are bound by the MRE11-RAD50-NBS1 (MRN) complex. Subsequent ATM signalling lays down the DNA damage phosphorylation mark γH2AX. Early SUMO recruitment of the PRC1 E3 ub ligase complex (RING1-BMI1) promotes local H2A-K119ub modifications. This is associated with local chromatin compaction and transcriptional repression adjacent to the break. The MRN complex also recruits E3 ub ligases RNF20-RNF40, which is activated by ATM phosphorylation. RNF20-RNF40 promotes H2B-K120ub resulting in chromatin relaxation. It is unclear whether H2B-K120ub overrides the H2A-K119ub mark or if H2A-K119ub is removed by DUBs at this stage.

enhance spreading of PRC1/2 complexes adjacent to DSBs and propagation of the H2A-K119ub mark (Blackledge et al. 2014).

The PRC1 RING1B-BMI1 complex also has a wider signalling role in DNA repair (Ismail et al. 2010, Ginjala et al. 2011). Some reports suggest that the early recruitment of BMI1, perhaps that promoted by both PARP activity and SUMO (Ismail et al. 2012), is required for the initial recruitment of ATM (Pan et al. 2011, Wu et al. 2011). However, ATM signalling is required to sustain BMI1 localization at DSBs (Ginjala et al. 2011), and this sustained localization is required for HDR. Indeed, loss or inhibition of BMI1 does not block γH2AX formation but does prevent the formation of ub foci and the subsequent recruitment of late-arriving proteins such as BRCA1 and 53BP1 to sites of DNA damage (Ismail et al. 2010, 2013). Since ub signalling, but not RNF8 recruitment, is impaired (Ismail et al. 2013), it may be that mono-ubiquitination by PRC1 acts as a priming event for further polyubiquitination by RNF8 (Mattiroli et al. 2012) or alters chromatin to allow access for RNF8-mediated histone H1 ubiquitination (Thorslund et al. 2015). Finally, loss of BMI1 is associated with checkpoint activation following DNA damage (Liu et al. 2009).

KAP1 in chromatin repression following damage

Following damage, repressive complexes consisting of KAP1, HP1, H3K9 methyltransferases and macroH2A variants, are rapidly recruited by signals including PARP-mediated PARylation events (**Figure 9.4**). These complexes promote methyltransferase modification of H3K9me3, which sets up a feedback loop for further recruitment of these repressive complexes through the interaction with HP1's chromodomain (Ayrapetov et al. 2014). Recruitment of the highly SUMOylated KAP1 complex to damage sites (White et al. 2006) can initially spread for tens

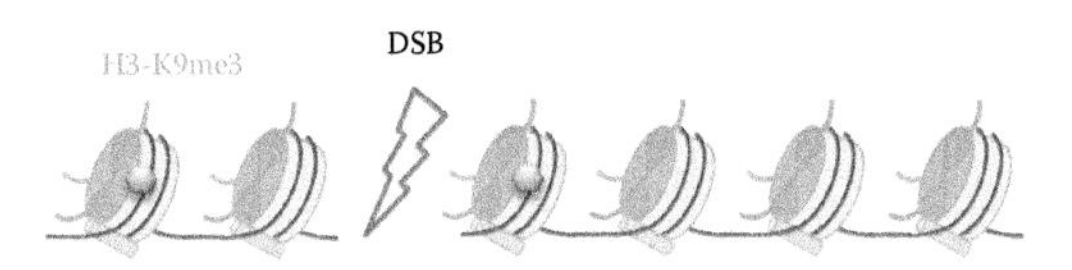

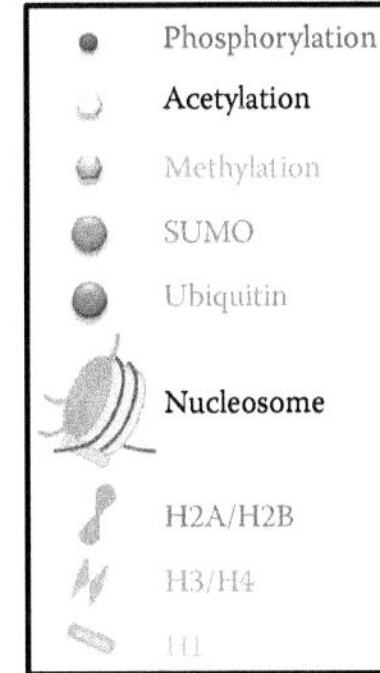

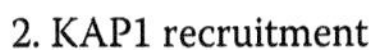

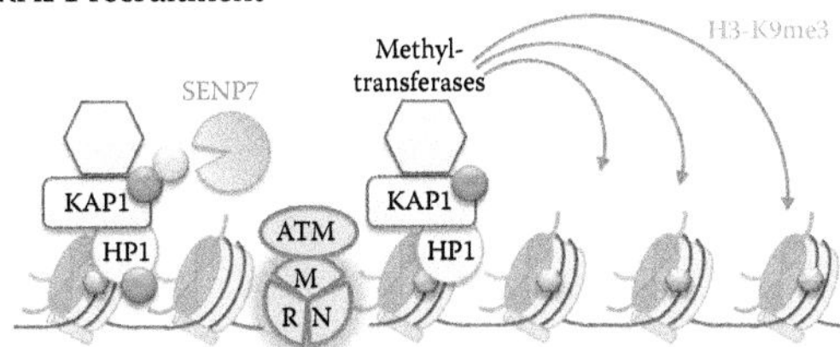

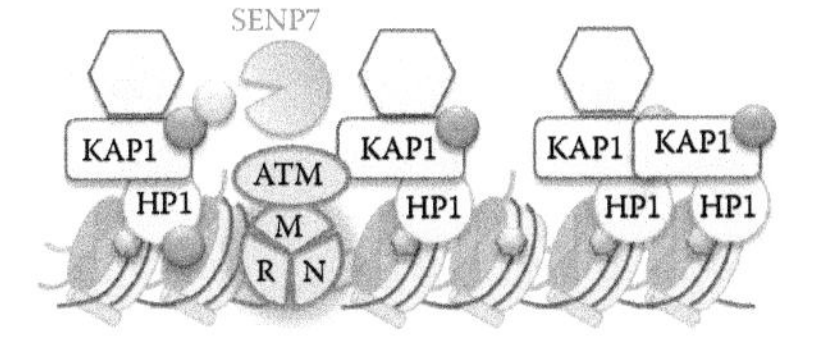

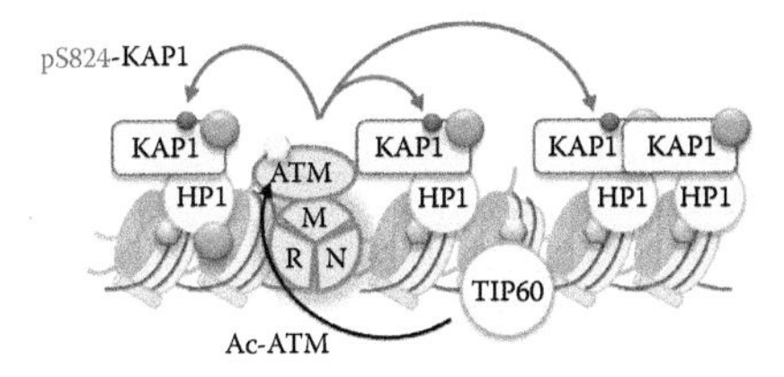

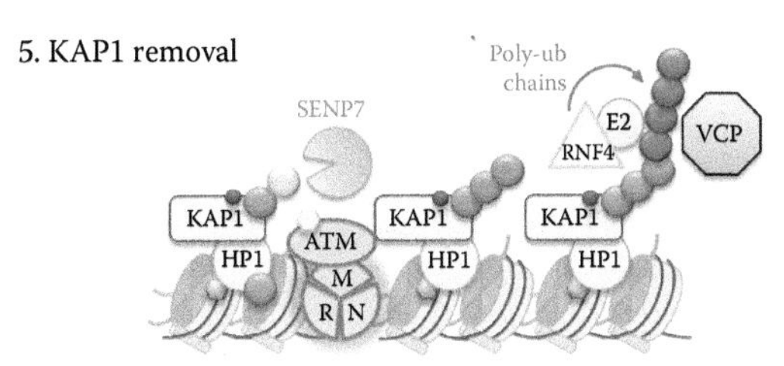

Figure 9.4 **SUMO signalling in early chromatin changes following DNA damage.** The repressive HP1-KAP1-methyl transferase complex is rapidly recruited to DSB sites, where it binds the H3K9me3 mark. Recruitment is also dependent on PARylation events (not shown). KAP1 is an auto SUMO E3 ligase and both KAP1 and HP1 are SUMOylated. The SUMO protease SENP7 acts to trim SUMO chains on KAP1. SUMO-KAP1 recruits methyltransferases which promote the spreading of H3K9me3 and further association of the HP1-KAP1 complex leading to chromatin compaction. The acetyl-transferase complex TIP60 also binds H3K9me3 and acetylates and activates ATM, which in turn phosphorylates and releases KAP1 from chromatin. In addition, RNF4 recognises and ubiquitinates extended SUMO chains on KAP1. These hybrid chains are a substrate for the AAA-ATPase VCP, which promotes KAP1 extraction from chromatin. Removal of repressive complexes leads to chromatin relaxation.

of kilobases adjacent to the lesion due to this feedback loop. In order for this damage-induced repression to be transient, the cell must then attenuate KAP1's repressive influence to shift to an open chromatin state permissive for repair. First, the H3K9me3 domains generated adjacent to the break bind and activate the TIP60 acetyltransferase complex (Sun et al. 2009, Ayrapetov et al. 2014). In turn, TIP60 acetylates and activates the master regulator of the DNA damage response, ATM (Sun et al. 2009). Once activated, ATM phosphorylates and releases KAP1 from chromatin, leading to dissociation of the complex, including the H3K9 methyltransferase, Suv39h1, responsible for H3K9me2/3 (Ayrapetov et al. 2014). Further, ATM-mediated phosphorylation of KAP1 at S824 adjacent to its bromodomain (Goodarzi et al. 2011) disrupts the KAP1-SUMO~SIM interaction with CHD3 of the NuRD complex. Removal of CHD3 promotes chromatin relaxation and allows DNA repair (Goodarzi et al. 2011), similar to that seen on KAP1 depletion (Ziv et al. 2006). The slower repair seen in heterochromatin correlates with the increased density of repressive complexes (reviewed in

Murray et al. 2012) and therefore ATM-mediated KAP1 phosphorylation has a greater impact on repair kinetics in these more condensed regions (Goodarzi et al. 2011).

An important node of chromatin condensation regulation comes through the control of deSUMOylation by SUMO proteases. The chromatin-associated SUMO protease SENP7 is enriched in heterochromatin (Maison et al. 2012) and can deSUMOylate both KAP1 and HP1α (Garvin et al. 2013, Maison et al. 2012). SENP7 is proposed to restrain the KAP1-SUMO~CHD3 interaction by removing KAP1-SUMO conjugates. Cells depleted for SENP7 show increased chromatin association of the NuRD component CHD3, a failure to relax chromatin following DNA damage and decreased HDR (Garvin et al. 2013). Conversely, cells depleted for CHD3 no longer require SENP7 to promote chromatin relaxation, HDR and resistance to IR (Garvin et al. 2013). While loss of SENP7 reduces accessibility in active open euchromatin, the most severe effects are seen in heterochromatin where DNA damage markers persist (Garvin et al. 2013), presumably due to the inability to promote chromatin decompaction and subsequent repair. Finally, the repressive KAP1~SUMO mark is also counteracted by the SUMO-targeted ub-ligase (STUbL) RNF4-VCP/p97 pathway. Both RNF4 and VCP/p97 interact with pS824-KAP1~SUMO, and RNF4 promotes the ub-mediated degradation of SUMOylated KAP1 (Kuo et al. 2014), providing a further mechanism for chromatin decompaction.

H2B-Ub in chromatin decompaction

In addition to the canonical RNF20-RNF40 modification found in open chromatin, H2B is ubiquitinated at K120Ub by RNF20-RNF40 following DNA damage (Moyal et al. 2011, Nakamura et al. 2011). In contrast to the role of PRC1-UBR5 in removing the histone chaperone FACT from damage sites (Sanchez et al. 2016), one of the components of the FACT complex, SUP16H, binds directly to RNF20 and is required to recruit the E3 ub ligase to the DSB (Oliveira et al. 2014). RNF20-RNF40 is activated by ATM phosphorylation and ubiquitinates H2B-K120. This H2B-K120ub mark is required for recruitment of subsequent DNA repair factors (Moyal et al. 2011), such as BRCA1 and RAD51, and for resection and HDR (Nakamura et al. 2011), however, this is likely to be an indirect effect of H2B-K120ub in promoting chromatin relaxation. Indeed, relaxation relieves the requirement for RNF20 in HDR (Nakamura et al. 2011) in keeping with a role for H2B-K120ub in opening chromatin structure.

In summary, the direct modification of histones by ub E3 ligases promotes chromatin compaction (H2A-K119ub) and decompaction (H2B-K120ub) events (**Figure 9.3**). Together with the indirect action of SUMO modifications on the rapid and transient recruitment of repressive complexes to chromatin around DSBs, small modifiers play a critical role in regulating the initial chromatin changes following a DSB. Indeed, chromatin compaction is critical for DSB-associated transcriptional repression. It may also serve to strip the chromatin of other factors and prime the modification landscape ready for a new palette of modifications by DSB response factors. However, chromatin compaction must necessarily be overcome in order to permit repair. Intriguingly, many proteins involved in chromatin organization contain SUMO conjugation sites and many of these appear to be deSUMOylated following methyl methanesulfonate treatment (Hendriks 2015a, 2015b) supporting a likely role for SUMO modifications in regulating chromatin changes following DNA damage.

Damage-responsive small modifier cascades

Initiating the DNA damage response

The signalling and repair of a DNA DSB involve the modification of the chromatin that surrounds it via a cascade of modifications and protein recruitments. Simply put, these events lead to the eventual association of 53BP1 and its effector proteins, and BRCA1. Cell cycle determinants then dictate whether resection, mediated by CtIP-MRE11 and BRCA1, lead to the initiation of HDR or whether effectors of the 53BP1-complex promote NHEJ and inhibit resection. Layers of recruitment and regulation are governed by phosphorylation and small modifier conjugation and deconjugation, and these ultimately promote these outcomes. Furthermore, there are separable activities of different BRCA1 complexes. The steps responsible for the recruitment of BRCA1 that are then responsible for resection promotion (often called BRCA1-C) are not yet clear, while one of the complexes, known as the BRCA1-A complex, is associated with repression of resection and HDR.

One of the earliest sensors of a DNA DSB is the MRE11-RAD50-NBS1 (MRN) complex that binds the broken DNA ends with high avidity (**Figure 9.5**). MRN in complex with DNA ends promotes ATM dimerization and activation at break sites (Lee and Paull 2005). ATM is the master kinase in the DSB response that phosphorylates many components of the DSB repair

1. ATM phosphorylation signalling initiates the DSB damage response

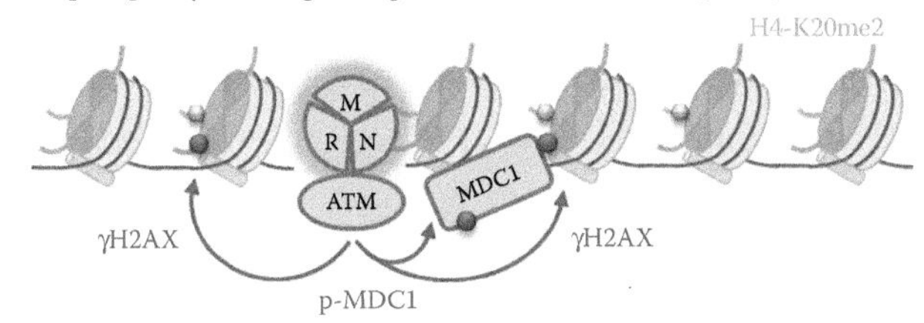

2. Phospho-MDCl recruitment of RNF8 initiates ubiquitin signalling on Hl

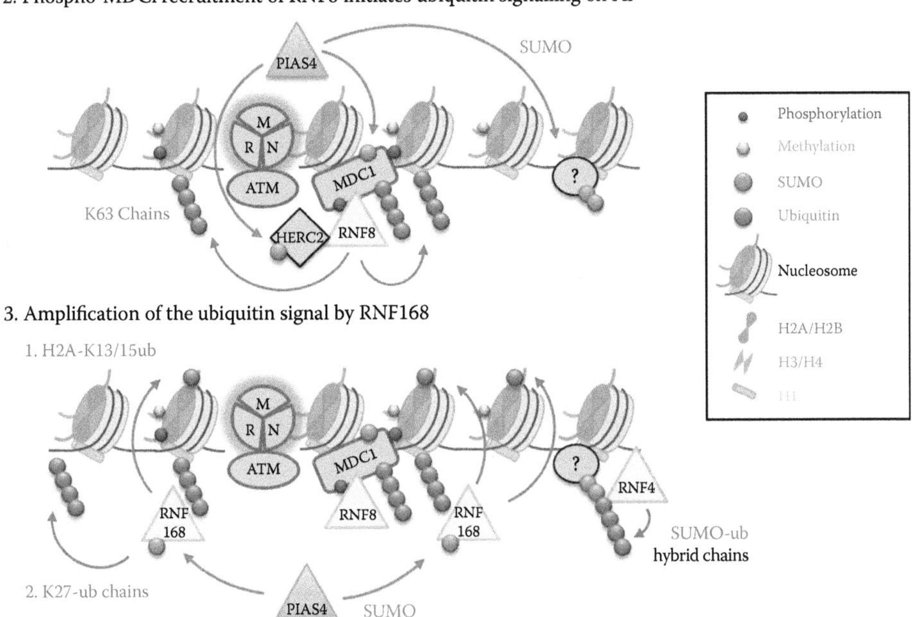

4. Recruitment of repair decision factors (53BP1 and BRCAl) by specific ubiquitin signals

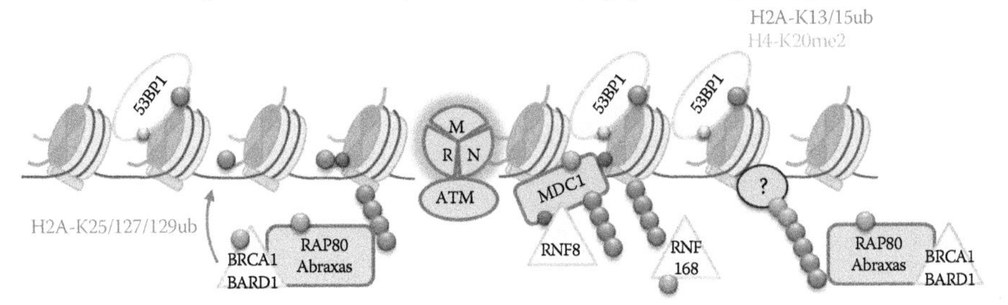

Figure 9.5 **Ub and SUMO in the DNA double-strand break response.** DNA DSBs are bound by the MRE11-RAD50-NBS1 (MRN) complex. Subsequent ATM signalling lays down the DNA damage phosphorylation mark γH2AX, which is in turn recognised by the scaffold protein MDC1. ATM phosphorylates MDC1, which recruits the E3 ub ligase RNF8. RNF8 lays down K63-ub chains on histone H1. PIAS4 SUMOylates a number of DSB factors, including HERC2, which enhances RNF8 ligase activity. RNF8-mediated K63 chains recruit RNF168, which generates K27-ub chains and lays down the H2A-K13/15ub mark recognised by 53BP1. Tandem binding of 53BP1 to H2A-K13/15ub and H4K20me2 allows specific recognition of damage-modified nucleosomes. In contrast, the BRCA1-BARD1 E3 ub ligase is recruited in complex with Abraxas-RAP80 to K63-chains and hybrid SUMO-ub chains via SIM and UBD domains in RAP80.

pathway. One of its earliest substrates is the histone variant H2AX phosphorylated at S139 (Burma et al. 2001); the phospho-form is often referred to as γH2AX. Large γH2AX domains are rapidly established around the break site, which serve to mark the lesion, amplify the response and provide a docking site for response factors. The mediator of damage checkpoint protein 1 (MDC1) binds γH2AX through its BRCT domain and is recruited to chromatin where it then undergoes phosphorylation by ATM (Stewart et al. 2003). MDC1 is a key-landing scaffold that coordinates the formation of damage factors into foci and initiates the ub signalling cascade to recruit proteins that regulate the choice of repair pathway. This cascade involves the E3 ub ligases RNF8, RNF168, RNF4 and HERC2, and culminates in the recruitment of the BRCA1-A complex and 53BP1.

A number of these early proteins (MDC1, HERC2 and RNF168) are modified by the activity of the SUMO E3 ligase PIAS4 following DNA damage (Danielsen et al. 2012, Luo et al. 2012, Yin et al. 2012). In addition, RNF8, the ub E2 UBE2N (also known as Ubc13), and the demethylase JMJD1C, which binds to RNF8, are also SUMOylated at multiple sites by an as yet unknown ligase (s) (Danielsen et al. 2012, Watanabe et al. 2013, Hendriks et al. 2017). All these proteins are co-located in time and space at damaged foci and therefore may act as an example of the 'SUMO spray model', where a whole protein complex undergoes indiscriminate SUMOylation (Psakhye and Jentsch 2012). However, SUMOylation of HERC2 mediates a specific structural change in which the ZZ-type zinc-finger in HERC2 undergoes a conformational change driven by an intramolecular interaction with the SUMO-modified form of HERC2. This conformational change regulates its ability to interact with and enhance RNF8 activity (Danielsen et al. 2012).

RNF8 is recruited to damage sites by interaction of its forkhead-associated domain with ATM-phosphoylated MDC1. Once at DSBs, RNF8, in complex with the E2 ub-conjugating enzyme UBE2N, modifies the linker histone H1 with K63-ub chains (Thorslund et al. 2015). It was originally proposed that RNF8 also modifies H2A, but in vitro studies showed that RNF8 is unable to recognize nucleosomal H2A (Mattiroli et al. 2012) and that in the absence of H1, K63 chains are largely absent from damage foci in spite of proficient RNF8 recruitment (Thorslund et al. 2015). Instead, RNF8 signalling promotes the recruitment of another E3 ub-ligase, RNF168, which binds K63-ub chains via its UBDs (Doil et al. 2009, Stewart et al. 2009, Panier et al. 2012), and also the nucleosome acidic patch where it catalyses mono-ub of H2A at K13/K15 to promote recruitment of the DSB factor 53BP1 (Doil et al. 2009, Fradet-Turcotte et al. 2013). The node of K13/15 ubiquitination is critical for regulation of 53BP1 accumulation. TIP60 acetylation of H2AK15 blocks 53BP1 binding, as this modification is mutually exclusive with H2AK15ub (Jacquet et al. 2016). RNF168 recognizes its own H2A-K13/15ub mark and thereby auto-propagates this signal along chromatin (Chen et al. 2012a, Panier et al. 2012, Poulsen et al. 2012). Furthermore, RNF168 appears particularly prone to auto-degradation and at least two DUBs, USP7 and USP34, are required to maintain RNF168 levels for efficient DSB response (Sy et al. 2013, Zhu et al. 2015). Both RNF168 transcription and protein stability is also supported by the SUMO E3 ligase PIAS4, including direct SUMOylation of RNF168 following IR (Danielsen et al. 2012).

The DUB OTUB1 is recruited to laser line DNA damage sites (Nakada et al. 2010), but its effects on the damage response are not due to its enzymatic activity. OTUB1 forms an inhibitory complex with the E2 enzyme UBE2N, which prevents efficient generation of K63-ub polymers (Nakada et al. 2010). It is not clear whether it limits RNF8 or RNF168 products, as RNF168 recruitment appears normal, yet K63-ub chains are disrupted (Nakada et al. 2010). In the presence of an ATM inhibitor, OTUB1 depletion increases the expansion of 53BP1 foci and improves HDR reporter activity (Nakada et al. 2010), whereas OTUB1 overexpression reduces 53BP1 foci formation (Mosbech et al. 2013).

RNF168 activity to generate mono-ub H2A-K15ub is required for recruitment of the double-strand break factor, 53BP1. Ub conjugated to H2AK15 is dynamic through a large spatial range but becomes constrained upon 53BP1 binding (Wilson et al. 2016). A 28-residue ub-dependent recruitment (UDR) motif (aa1604-1631) in 53BP1 threads in between the ub moiety on H2A-K15 and a H4-H2B cleft in the nucleosome to provide specific recognition and binding of H2A-K15ub (Wilson et al. 2016). Additional contacts from the adjacent tudor domain in 53BP1 with the H4K20me2 modification stabilises the 53BP1-nucleosome interaction and together provide exquisite specificity for 53BP1 recruitment to the damage-modified nucleosome.

Like RNF8, RNF168 has also been shown to work with UBE2N to generate K63-ub chains (Doil et al. 2009, Stewart et al. 2009). However, the relevance of this in cells is debated since overexpression of RNF168 mainly produces K27-ub chains of unknown function (Gatti et al. 2015), and mono-ub of H2A at K15 is sufficient for 53BP1 recruitment. Indeed, in UBE2N knockout cells, which cannot make K63-ub chains and do not form 53BP1 foci following damage, over-expression of RNF168, but not RNF8, is sufficient to restore 53BP1 foci (Thorslund et al. 2015). While none of this excludes a role for RNF168 generation of K63-ub chains, it is clear that these are not the major products of the E3 ub-ligase.

The RNF8 ub cascade not only promotes RNF168-53BP1 recruitment to DSBs but also stimulates recruitment of the BRCA1-A complex. The RAP80 component of the BRCA1-A complex is responsible for complex recruitment through both ub and SUMO-binding motifs (Kim et al. 2007, Sobhian et al. 2007, Wang et al. 2007). RAP80 is itself extensively SUMOylated, but what role this plays in DSB repair is currently unknown (Yan et al. 2007). RAP80 not only binds K63-ub chains via a dual UIM (Sobhian et al. 2007, Sims and Cohen 2009) but also utilises a SIM, which together with the UIMs, bind ub-SUMO hybrid chains with high affinity (Sims and Cohen 2009). It has been demonstrated that both the SIM and UIMs in RAP80 are important for BRCA1-A complex recruitment (Guzzo et al. 2012, Hu et al. 2012), and the whole BRCA1-A complex binds hybrid SUMO-K63ub chains in cell lysates (Guzzo et al. 2012). These hybrid chains are largely uncharacterized but could be generated by RNF4, which has been shown to produce hybrid chains in vitro (Branigan et al. 2015). RNF168 has also been detected to bind hybrid SUMO-K63ub chains, although the significance of this in the DNA damage response is yet to be determined (Shire et al. 2016).

Further work is needed to establish whether there is a specific substrate(s) on which hybrid SUMO-K63ub chains are formed and how this signal integrates into the damage-signalling network.

Chromatin surrounding DSBs is enriched for ub and SUMO and there are emerging notions that critical factors may need to be defended from these modifications to promote their function. For example, following ATM-mediated phosphorylation, the DUB USP13 locates to sites of damage where it targets RAP80 and acts to defend it from ub modification (Li et al. 2017). This in turn allows RAP80 to interact with K63-ub chains (Li et al. 2017). It remains to be seen whether SUMO protease(s) also perform the same task to prevent hyper-SUMOylation.

The BRCA1-A complex itself contains both a heterodimeric RING E3 ub-ligase, BRCA1-BARD1, and a JAMM family DUB in BRCA1/2 containing complex 36 (BRCC36). BRCC36 was the first DUB to be characterized as having a role in the DSB repair response (Dong et al. 2003) and is an integral member of the BRCA1-A complex consisting of BRCA1, RAP80, Abraxas, BRE, MERIT40 and BRCC45. Loss of BRCC36 disrupts formation of the BRCA1-A complex and results in reduced BRCA1 recruitment to IR-induced foci, defects in G2/M checkpoint and mild radiosensitivity (Dong et al. 2003, Sobhian et al. 2007, Ng et al. 2016). The DUB function of BRCC36 is dependent on the interaction with Abraxas and BRCC45 subunits of the BRCA1-A complex. A similar phenomenon occurs with another JAMM domain protease POH1 (also known as PSMD14), which is only active in the context of the proteasome lid (Patterson-Fortin et al. 2010). The exact substrate(s) on which BRCC36 acts are not fully known, although BRCC36 has a preference for K63-ub chains similar to other JAMM family members (Cooper et al. 2009, Shao et al. 2009). Knockdown of BRCC36 results in spreading of K63-ub at sites of damage, suggesting its role is to limit the spread of this modification and of subsequent 53BP1 accumulation. H2AX ubiquitination is increased on depletion of BRCC36, as is the ubiquitination of RAP80 (Shao et al. 2009, Patterson-Fortin et al. 2010).

BRCC36 acts as a HDR antagonist, with BRCC36 depletion increasing HDR while decreasing NHEJ repair of SceI-induced breaks (Coleman and Greenberg 2011, Ng et al. 2016). Consistent with this role, loss, BRCC36 also increases chromosome breakage and sister chromatid exchanges (Coleman and Greenberg 2011). The increase in HDR has been attributed to MRE11-CtIP–dependent hyper-resection that occurs in the absence of BRCC36, as these cells show excessive recruitment of RPA to sites of damage, high levels of phosphorylated RPA and an increase in single-stranded DNA marked by BrdU foci (Coleman and Greenberg 2011, Ng et al. 2016).

Independently from BRCC36, the DUBs USP26 and USP37 are recruited to sites of damage and counter RNF168-mediated ubiquitination at chromatin, processing H2A-ub (Typas et al. 2015). Surprisingly, the activity of these DUBs does not influence 53BP1 accumulation, but instead prevents the excessive spreading of RAP80-BRCA1, suggesting an influence on the BRCA1-A complex through H2A-ub that is independent of at least a subset of K63-ub chains (which appear unaffected by USP26 and USP37). These effects are functional in that increased sequestration of BRCA1 into BRCA1-A in USP26/37-depleted cells has a negative influence on HDR that can be overcome by disrupting the BRAC1-A complex by RAP80 depletion (Typas et al. 2015).

Small modifier regulation of damage factor residency at damage sites

The importance of the ordered and timely removal of damage proteins at key stages of the DSB response is critical for normal progression and final repair. Proteins that persist at damage sites beyond their usual 'lifetime' have a detrimental effect on downstream events. Both SUMO and ub modification pathways have important roles in providing temporal control of protein stability and release of factors from DSBs. Removal utilizes both the ub-proteasome pathway and the ub-VCP/p97 AAA-ATPase pathway, which extracts proteins from complexes. In addition, while SUMO has historically been thought to act as a signalling mark, rather than a degradative mark, crosstalk with the ub pathway challenges this view. Indeed, cells in which the proteasome has been inhibited by treatment with MG132 rapidly accumulate high molecular weight adducts for both ub and SUMO modifications (Schimmel et al. 2008). The discovery of specialized ub E3 enzymes which recognize and ubiquitinate SUMO2/3 polymers has brought much-needed insight into the link between SUMOylation and proteasomal degradation (Sun et al. 2007, Tatham et al. 2008).

RNF4-MDC1 removal

MDC1 is a critical early scaffold in the DNA damage response and is required to initiate the ub signalling cascade. However, prolonged association of MDC1 at damage sites leads to a reduction in HDR, thought to be via increased 53BP1 association (Luo et al. 2012), and removal of MDC1 is required for subsequent repair steps to proceed (Galanty et al. 2012, Luo et al. 2012, Yin et al. 2012). Therefore, temporal control of MDC1 association and clearance from damage sites must be finely balanced, and SUMO modification of MDC1 is an important factor in this temporal control (**Figure 9.6**).

Following IR, MDC1 is SUMOylated at a single major site, K1840, by PIAS4. Mutation of this site to K1840R delays MDC1 clearance (Galanty et al. 2012, Luo et al. 2012) and fails to rescue the radiosensitivity seen in MDC1-deficient cells. This SUMOylation on MDC1 promotes interaction with the STUbL RNF4 (Luo et al. 2012, Yin et al. 2012). Four tandem SIMs in RNF4 recognize SUMO2/3 polymers with a preference for chains with four or more SUMO moieties. Alternatively the SIMs may also recognize adjacent multi-mono-SUMO modifications (Sun et al. 2007, Tatham et al. 2008, Kung et al. 2014). These SIMs are required for SUMO-dependent recruitment of RNF4 to DSBs (Galanty et al. 2012, Yin et al. 2012, Vyas et al. 2013). RNF4 E3 ub ligase activity requires homo-dimerisation to occur, and this makes RNF4 an exquisite sensor of polySUMO chains (Liew et al. 2010). Once recruited to MDC1, RNF4 ubiquitinates SUMOylated MDC1 marking it for proteasomal turnover (Galanty et al. 2012).

Two DUBs have been reported to antagonise RNF4's activity: Ataxin 3 (ATXN3) (Pfeiffer et al. 2017) and USP11 (Hendriks et al. 2015a). USP11 can efficiently process hybrid SUMO-ub chains (Hendriks et al. 2015a), while ATXN3 has been shown to antagonise RNF4-dependent ubiquitination of MDC1. Depletion of ATXN3 slows MDC1 exchange kinetics at laser lines and results in defective downstream ub signalling that ultimately results in failure in both HDR and NHEJ (Pfeiffer et al. 2017). ATXN3 is rapidly recruited to sites of laser-induced damage and DSB-associated Lac arrays in a SIM-, but not UIM-, dependent fashion (Nishi et al. 2014, Pfeiffer et al. 2017). It is intriguing to note that the DUB activity of ATXN3 is stimulated by free SUMO1 in vitro raising the notion that non-covalent SUMO interactions may regulate catalysis (Pfeiffer et al. 2017). ATXN3 is also a substrate adapter for the AAA ATPase complex

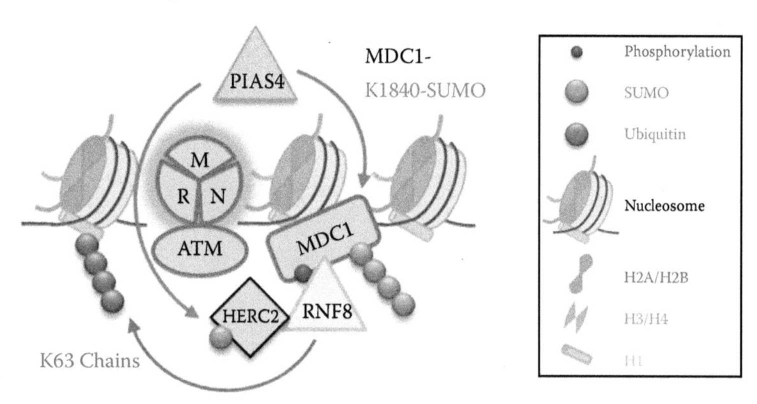

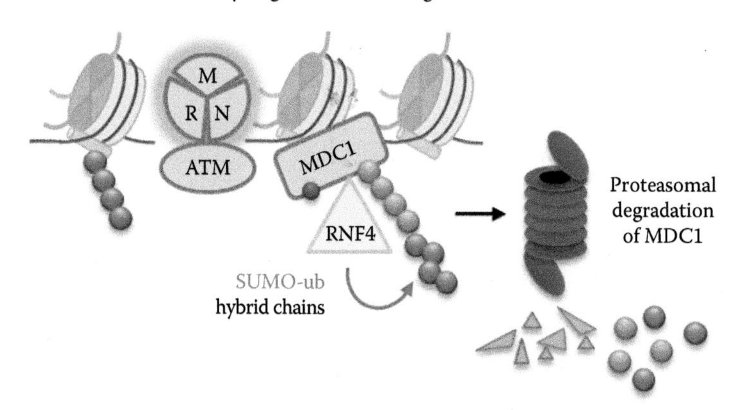

Figure 9.6 RNF4-mediated MDC1 removal. MDC1 is a critical early scaffold protein in the DNA DSB response. It is required for RNF8 recruitment which initiates the subsequent ub-signalling cascade. In addition, MDC1 is SUMOylated by PIAS4. SUMO~MDC1 is recognised and ubiquitinated by the SUMO-targeted E3 ub ligase (STUbL) RNF4, leading to chromatin extraction and proteasomal mediated degradation of MDC1. The failure to clear MDC1 in a timely manner is inhibitory to late stages of DNA repair.

VCP/p97 (Doss-Pepe et al. 2003), but this interaction is not required for ATXN3 recruitment to DSBs (Pfeiffer et al. 2017). ATXN3 is mutated in spinocerebellar ataxia-type 3 (SCA3) through CAG/polyQ repeat expansion (Kawaguchi et al. 1994), enabling interactions with proteins in vivo that do not occur with the wild-type protein. Mutant ATXN3 inappropriately binds and deactivates polynucleotide kinase 3′-phosphatase (PNKP), a protein essential for single-strand break repair, resulting in chronic ATM signalling and cellular apoptosis, thus providing a basis for neurodegeneration in SCA3 (Chatterjee et al. 2015, Gao et al. 2015). Whether mutant ATXN3 remains capable of regulating MDC1 residence time and influencing DNA DSB repair has yet to be investigated.

In the DSB response, RNF4 activity is not restricted to MDC1, and has also been implicated in the ub modification and proteasomal clearance of BRCA1, 53BP1, BLM, RPA, EXO1 and KAP1 (Galanty et al. 2012, Vyas et al. 2013, Kuo et al. 2014, Bohm et al. 2015, Bologna et al. 2015). All of these proteins are SUMOylated, demonstrating how widespread SUMO-ub signalling crosstalk may be in regulating the repair pathway.

VCP/p97-mediated removal

Some ubiquitinated proteins are physically removed from their surroundings by the AAA ATPase complex VCP/p97, often prior to their proteasomal degradation. VCP/p97 binds K48-ub conjugated proteins where it can either act as a scaffold for ub ligases or DUBs to modify the ub signal (Richly et al. 2005), or it can utilise its ATP segregase activity to remodel protein complexes (Ramadan et al. 2007). Importantly, it can act on nuclear proteins in the context of chromatin and with the replisome machinery (Dantuma et al. 2014). VCP/p97 and its adapters are recruited to DSBs through interactions with ub (Acs et al. 2011, Meerang et al. 2011). In yeast dual SUMO/ub, recognition by a VCP/p97 adapter is essential for DSB repair (Nie et al. 2012, Bergink et al. 2013) and this may have a similar activity in mammalian cells (Bergink et al. 2013).

Depletion of VCP/p97 sensitises cells to IR (Meerang et al. 2011). VCP/p97 recruitment to laser-induced DNA damage was abrogated following RNF8 depletion (Acs et al. 2011, Meerang et al. 2011). Consistent with a failure in signalling downstream of RNF8, depletion of VCP/p97 partially impaired BRCA1 and 53BP1 recruitment and almost completely abrogated the formation of RAD51 foci (Meerang et al. 2011). RNF8-dependent ubiquitination of L3MBTL1 promotes VCP/p97-mediated removal of L3MBTL1, thereby unmasking the 53BP1 H4K20me2 binding site (Acs et al. 2011). This is a highly regulated process, as the DUB, OTUB2 antagonises RNF8-dependent ubiquitination of L3MBTL1 (Kato et al. 2014). As a result, depletion of OTUB2 causes excessive clearance of L3MBTL1 and spreading of 53BP1, which reduces DNA resection, RAD51 foci formation and HDR. Consistent with this, OTUB2-depleted cells show increased sensitivity to camptothecin (Kato et al. 2014).

The 26S proteasome

The proteasome plays several roles in the DSB response (Krogan et al. 2004, Stone and Morris 2013). Proteasomal-mediated degradation is often the fate of proteins that encounter ligases such as RNF4 or are processed by VCP/p97. Moreover without the continuous ub-conjugate processing of the proteasome, cells are starved of the raw material, free ub, with which to signal DSBs at chromatin. This effect is demonstrated by the failure of the damage response in the presence of proteasome inhibitors (Murakawa et al. 2007), and the restoration of signalling in the presence of these inhibitors by the ectopic expression of ub (Butler et al. 2012).

In addition to this critical indirect role, components of the proteasomal degradative 20S proteolytic core and the 19S regulatory particle, which recognises, unfolds, and deubiquitylates substrates, locate to DNA damaged laser lines (see Stone and Morris 2013 for further review). The 26S recruitment depends on the activity of DSB-repair ub ligases (Butler et al. 2012, Galanty et al. 2012) and some substrates may be degraded in situ. The 19S carries the DUB POH1, a JAMM-type protease responsible for the cleavage *en masse* of ub chains from substrates bound for the 20S core (Verma et al. 2002). In addition, it has specificity for K63-ub chains. POH1 DUB activity opposes RNF8/RNF168 ubiquitination events that promote 53BP1 accumulation and in turn NHEJ. Its activity is additive, rather than redundant, with the K63-linkage-specific DUB BRCC36. Remarkably, acute POH1 depletion can restore 53BP1 accumulation and NHEJ repair to cells with low levels of RNF8, RNF168 or 53BP1 proteins, without impacting their protein levels (Butler et al. 2012) indicating its role in dampening ub signalling in the response. Indeed POH1, in turn, acts to restrict 53BP1 spread (Butler et al. 2012). In this capacity, it is also critical to allowing the formation of the 53BP1-devoid core at the centre of DSB foci in which resection occurs (Kakarougkas et al. 2013). Intriguingly, increased 53BP1 spreading also occurs on depletion of the E3 ligases TRIP12 and UBR5, which act to regulate the levels of RNF168 (Gudjonsson et al. 2012). Thus, both the levels of E3 ligase components and their products determine the amplitude of the 53BP1 response.

The proteasome also contains the small nucleotide-like component DSS1, which is critical to BRCA2-mediated loading of RAD51 (Liu et al. 2010), and proteasome recruitment may contribute to an increased concentration of this factor to promote RAD51 loading.

Small modifiers in non-homologous end joining

The majority of simple DSBs are repaired throughout the cell cycle by the NHEJ pathway without the need for a homologous template to guide repair. This pathway has been reviewed elsewhere (Yang et al. 2016, Chang et al. 2017) (and in Chapter 7 in this book). Briefly, following a DSB, the abundant nuclear Ku heterodimer, like the MRN complex, binds directly to the broken ends of DNA with high affinity. The heterodimer consists of Ku70 and Ku80 (also called Ku86) proteins, which form a ring-like structure that encircles the DNA (Walker et al. 2001) and recruits the DNA-protein kinase catalytic subunit (DNA-PKcs). Among other roles, DNA-PKcs facilitates end processing to generate clean blunt DNA ends by enzymes such as Artemis or PNKP, and promotes recruitment of the XRCC4-ligase IV complex to catalyse DNA repair. The Ku heterodimer serves to prevent extensive resection during end processing. Ku proteins are removed post-NHEJ repair to prevent interference with other DNA processes such as transcription or replication but can also be removed pre-NHEJ repair in order to mediate a switch to HDR and relieve the Ku barrier to allow long-range resection (reviewed in Postow 2011). Several E3 ub ligases are implicated in processing Ku proteins from DNA. RNF138 promotes Ku ubiquitination and subsequent degradation to drive Ku removal to facilitate the switch to resection and the promotion of HDR (Ismail et al. 2015). Interestingly, RNF138 is specifically recruited to sites of damage through its zinc fingers that preferentially interact with 5′ or 3′ single-stranded DNA overhangs, suggesting it may remove Ku from ends at which active MRE11 has initiated resection (Ismail et al. 2015).

Several ligases may be responsible for the clearance of Ku proteins after repair is complete. Ku80 is ubiquitinated with K48 chains when bound to DNA ends (Postow et al. 2008). RNF8 can ubiquitinate Ku80 and depletion of RNF8 leads to sustained persistence of Ku80 at laser-induced damage (Feng and Chen 2012). Other data suggest a cullin-ub ligase(s) may be responsible. Treatment of cells with MLN4924, a drug that specifically inhibits conjugation of the ub-like protein NEDD8 to target proteins, including preventing the NEDD8-mediated activation of cullin-E3 ligases, prevents the release of Ku from DNA damage sites (Brown et al. 2015), and the cullin ub E3 ligase SCF-Fbxl12 complex is required for Ku80 modification in *Xenopus* extracts (Postow and Funabiki 2013). While Ku80 is ultimately turned over by the proteasome, the removal of Ku80 from damage sites requires K48-ub chains but not the proteasome (Postow et al. 2008). Recently, a role for VCP/p97 and its ub-binding receptors in removing K48-ub–modified Ku80 from sites of damage has been revealed, in which loss of VCP/p97 impairs NHEJ and the early steps of HDR correlating with retention of Ku on DNA (van den Boom et al. 2016).

The Ku proteins are also modified by SUMO. In yeast, SUMOylation enhances Ku70 association with DNA (Hang et al. 2014). It is less clear if this effect of SUMO on Ku70-DNA-association is conserved in mammals, but Ku70 is stabilised following overexpression of SUMO conjugation components (Yurchenko et al. 2008) and SUMOylated-Ku70 interacts with a SIM-containing peptide (Li et al. 2010), although endogenous SIM interactors have yet to be identified. However, SUMO interactions are likely to play an important role in NHEJ, as exogenous overexpression of SIM-containing peptides inhibits NHEJ and increases cellular sensitivity to IR (Li et al. 2010). The finding that loss of the tandem SIM-containing E3 ub ligase, RNF4, reduces NHEJ is consistent with these observations (Galanty et al. 2012). Indeed, SUMO-targeted ubiquitination of Ku70 by RNF4 and the subsequent extraction of Ku70 is an attractive model not yet tested.

Finally, SUMOylation of another NHEJ factor, XRCC4, is linked to its nuclear localisation (Yurchenko et al. 2006). Removal of the K210 SUMOylation site in XRCC4 by mutation to arginine results in XRCC4 cytoplasmic retention coupled with increased cellular radiosensitivity and incomplete V(D)J recombination. These phenotypes can be overcome by fusing SUMO to the C-terminus of XRCC4 (Yurchenko et al. 2006).

Remodelling chromatin for resection

Arguably, the most critical decision for DSB repair pathway choice is in the control of resection lengths, as this determines which repair pathway can be utilised by the cell. NHEJ repair favours minimal resection, while microhomology-mediated end joining (MMEJ or alt-NHEJ) requires short limited resection to expose 5–25 bp regions of microhomology to promote end joining. In contrast, HDR requires long stretches of resection to promote RPA-coated ssDNA 3′ overhangs. In HDR by gene conversion, which is favoured as the most error-free repair option by utilising a sister-chromatid homology template, RPA is exchanged for RAD51 to form a RAD51-DNA filament that is required for homology searching and strand invasion. In contrast, extended or hyper-resection is associated with HDR by single-strand annealing (SSA), where homologous regions on the same strand are exposed and annealed, and is characterised by the replacement of RPA with RAD52. However, SSA is highly mutagenic, as the genomic information between homologous regions is lost during repair. Numerous factors have been implicated in controlling resection lengths including a role for small modifier proteins.

Small modifier regulation of resection initiation

Resection is itself a multistep process (reviewed in Symington 2016). Processing to trim or clean DNA ends is sometimes required for non-HDR pathways, NHEJ and MMEJ, and may utilise short-range resection enzymes, such as MRE11 and CtIP in some circumstances. MRE11 possesses both endo-nuclease and 3′-5′ exo-nuclease activity. In contrast, while the yeast homologue of CtIP, Sae2, demonstrates 5′-3′ exo-nuclease activity, human CtIP only demonstrates specialist 5′ flap endonuclease activity (Makharashvili et al. 2014). While this specialist activity is important for cleaving branched DNA structures and is proposed to be relevant for end processing (Makharashvili et al. 2014), CtIP is more usually associated with a role in enhancing the nuclease activity of MRE11 (Sartori et al. 2007). Initial resection by MRE11-CtIP commits repair of the lesion to HDR pathways, but a further step of long-range resection, by enzymes such as EXO1 or BLM-DNA2, is required for this repair to be efficient. The 5′ resection of DSBs produces single-stranded 3′ overhangs which are bound and protected by the single-stranded DNA binding protein RPA. RPA is then later exchanged for RAD51 through the activity of BRCA2 to form the RAD51 filament required for homology template searching. Several components of the resection machinery are regulated by ub and SUMO (for review, see Himmels and Sartori 2016).

Following damage, CtIP undergoes phosphorylation at S327 which promotes an interaction with the BRCT domains of BRCA1, often referred to as the BRCA1-C complex (Yun and Hiom 2009). Cells expressing a mutant form of CtIP that can no longer be phosphorylated at S327 and no longer interacts with BRCA1 have decreased HDR and reduced ssDNA formation post-damage in S phase, all indicative of a resection defect (Yu and Chen 2004, Yun and Hiom 2009). The phospho-dependent interaction of BRCA1 with CtIP has been proposed to also promote BRCA1-mediated ubiquitination of CtIP. In this case, the ub modification promotes the association of CtIP with chromatin and mediates a G2/M checkpoint response (Yu et al. 2006). Recently, a further ligase has been proposed to induce CtIP ubiquitination responsible for promoting its accrual at sites of DNA damage. In a large ub E2 and E3 screen, RNF138 and UBE2D family members were implicated in promoting DNA repair (Schmidt et al. 2015). The identified ubiquitination sites modified by RNF138 on CtIP were mutated and shown to be required for adequate CtIP accrual and the promotion of HDR (Schmidt et al. 2015). Thus, two independent studies place the E3 ub ligase RNF138 early in the damage response, where it appears to clear Ku proteins from DNA ends (Ismail et al. 2015) and promote CtIP accrual (Schmidt et al. 2015).

In addition, CtIP ubiquitination promotes its degradation. Signals for degradation come from E3 ligases SIAH-1 (Germani et al. 2003), APC/C^{Cdh1} (Lafranchi et al. 2014), and cullin3-KLHL15 (Ferretti et al. 2016). APC/C^{Cdh1} has major roles in controlling cell-cycle dependent changes in CtIP stability but is also linked to delayed clearance of CtIP from damaged foci (Lafranchi et al. 2014). A conserved KEN box in CtIP mediates the interaction with the Cdh1 subunit of the APC complex and ubiquitination drives proteasomal degradation (Lafranchi et al. 2014). The cullin-KLHL15 E3 ub ligase complex interaction with CtIP is mediated through interaction with a conserved tripeptide motif (FRY) in CtIP (Ferretti et al. 2016). Here, proteasomal degradation regulates CtIP turnover to fine-tune resection lengths and acts as another mechanism for preventing excessive resection and thereby restricting the chances of mutagenic repair.

CtIP is also regulated by the DUB USP4, although not through an intuitive mechanism. USP4 is unusual in that it is itself highly ubiquitinated on lysines and on non-conventional cysteine residues (Wijnhoven et al. 2015). USP4 is recruited to laser lines (Nishi et al. 2014, Liu et al. 2015, Wijnhoven et al. 2015) and interacts with both CtIP and the MRN complex (Wijnhoven et al. 2015). However, when it is ubiquitinated, the ub moieties prevent interaction with MRN and CtIP. Thus, auto-deubiquitination promotes the USP4 interaction with MRN and CtIP, and promotes CtIP accumulation to damage sites, and in turn, end resection (Liu et al. 2015, Wijnhoven et al. 2015).

UCHL5 is required for the DNA end resection promoted by BLM and EXO1. UCHL5 deubiquitinates the INO80 component NFRKB and may prevent its degradation. It is not currently understood how the INO80 complex regulates DNA end resection, but in agreement with the role for NFRKB, depletion of several INO80 components reduce long-range resection and HDR (Nishi et al. 2014). Structurally related DUB enzymes UCHL1/UCHL2 are also recruited to laser lines and their depletion similarly reduces DNA repair but how they function in the pathway is currently unclear (Nishi et al. 2014).

Many of the proteins which have a direct role in resection are modified by SUMO. For example, CtIP has been found to be SUMOylated in reconstituted systems, and the yeast homologue of CtIP, Sae2, is SUMOylated at a conserved K97 residue following DNA damage (Sarangi et al. 2015). Mutation of this Sae2-K97 site impaired DSB processing and repair in yeast (Sarangi et al. 2015) but the role of CtIP SUMOylation in mammals has yet to be fully characterised. Another example is MRE11, which is SUMOylated during the DSB response triggered by infection with adenovirus 5 (Ad5) (Sohn and Hearing 2012). SUMO site-mapping screens have identified multiple SUMO sites on MRN components. In addition, BLM is modified by SUMO following IR or replication stress (Eladad et al. 2005, Vyas et al. 2013) and SUMO-modified BLM promotes RAD51 foci formation after replication fork collapse (Ouyang et al. 2009). Another long-range resection enzyme, EXO1, is SUMOylated by PIAS4, which leads to a reduction in protein stability. Indeed, EXO1 stability can be improved by removal of the SUMOylation sites (Bologna et al. 2015).

Together, the SUMO-rich modifications found in resection enzymes raise the possibility of specific 'readers' of these modifications. Intriguingly, like RAD51, the structure-specific endonuclease scaffold protein SLX4 bears SIMs critical to its function (Guervilly et al. 2015, Ouyang et al. 2015). These SIMs are required for laser-induced damage localisation of SLX4 and for cellular resistance to camptothecin. While the SLX4-SIMs bind SUMOylated MRN and RPA, these interactions are not sufficient for SLX4 damage recruitment, which is thought to be mediated by other, as yet unknown, SUMO-modified proteins (Ouyang et al. 2015).

Likewise, the richness of SUMO modifications in resection and recombination fuels the potential for STUbL-mediated regulation of these pathways. Indeed, several studies have shown that depletion of RNF4 is associated with a reduction in RAD51 foci and decreased HDR (Galanty et al. 2012, Luo et al. 2012, Yin et al. 2012). While some suggest that the repair defect on RNF4 loss is due to increased persistence of MDC1 and 53BP1 at damage sites and therefore, reduced resection (Luo et al. 2012, Yin et al. 2012), others propose a model where RPA residency at resected DNA is regulated by RNF4-mediated turnover of SUMO-RPA1, resulting in reduced RAD51 loading (Galanty et al. 2012).

H2A-Ub modification and the 53BP1-BRCA1 relationship

One of the key relationships governing repair pathway choice is the relative balance of contributions from two central gatekeepers of resection, 53BP1 and BRCA1. The 53BP1 protein is a major scaffold for numerous effector proteins, is associated with a role in NHEJ and has an inhibitory effect on resection (Bakr et al. 2016). In contrast, BRCA1 favours HDR pathways by promoting resection through direct interactions with CtIP and by regulating chromatin remodelling at damage foci. The balance between 53BP1 and BRCA1 changes during the cell cycle (**Figure 9.7**). In G1, the balance is in favour of 53BP1, and BRCA1 is not retained at damage foci. In S and G2, the balance shifts toward BRCA1, which promotes remodelling of 53BP1 at damage foci to create a permissive chromatin environment for resection to occur. Much of this interplay is regulated by changes in the underlying chromatin environment and specifically by H2A-ub events. Following DNA damage and formation of a DNA double-strand break, H2A is mono-ub at multiple sites on both N- and C-terminal tails by RNF168 (H2A-K13/15ub) (Mattiroli et al. 2012), PRC1 (Wang et al. 2004, Ismail et al. 2010, Ginjala et al. 2011) and by BRCA1-BARD1 (H2A-K125/127/129ub) E3 ub ligases (Kalb et al. 2014a). As previously described, 53BP1 binds specifically to H2A-K15ub following DNA damage. Recruitment of 53BP1 and its effector proteins is inhibitory to resection (Bunting et al. 2010, Zimmermann et al. 2013), thereby promoting repair by NHEJ, and correlating with 53BP1's role in immune development, class switching and V(D)J recombination. However, in order to promote HDR mechanisms, the cell must find ways to relieve the 53BP1-mediated block on resection. Indeed, removal of 53BP1 by siRNA increases resection lengths and HDR by gene conversion, and a

In G1, 53BP1 complexes inhibit BRCA1 signalling and recruitment of resection enzymes

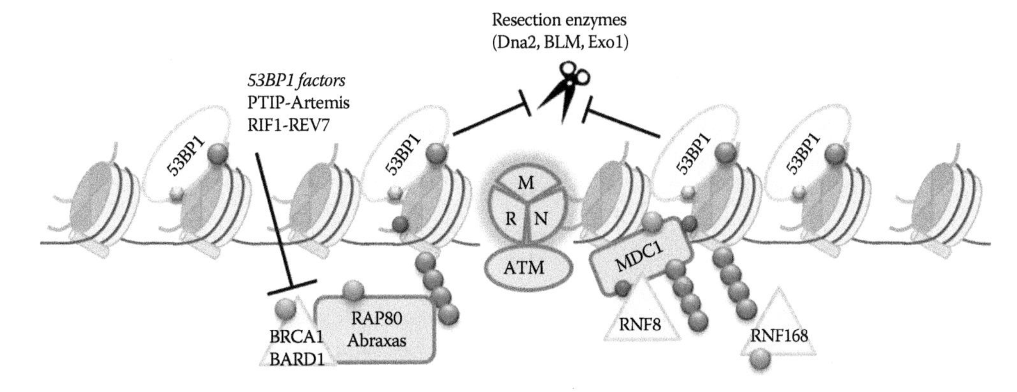

In S/G2, BRCA1 promotes SMARCAD1-dependent remodelling to remove the 53BP1 block on long-range resection

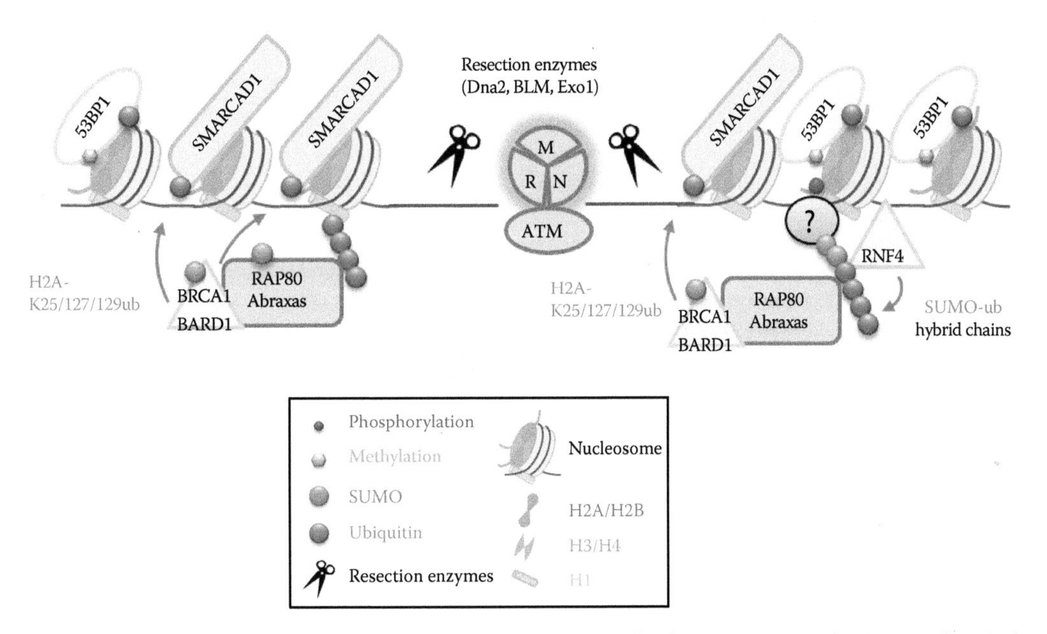

Figure 9.7 BRCA1 ligase-mediated chromatin changes promote resection. The relationship between 53BP1 and BRCA1 is cell-cycle dependent. In G1, the 53BP1 complexes inhibit BRCA1 retention at damage sites and block resection, favouring repair by NHEJ. In S/G2, BRCA1 recruitment promotes remodelling of 53BP1 complexes to the periphery of damage foci. This remodelling is dependent on BRCA1 E3 ub ligase activity to promote H2A-K125/127/129ub modification and ub-associated recruitment of the chromatin remodeller SMARCAD1. This remodelling relieves the 53BP1 block on resection and allows recruitment of long-range resection enzymes, Dna2, BLM, Exo1.

recent report suggests that high levels of damage results in 53BP1 exhaustion in the cell and a corresponding increase in RAD52 foci formation and use of SSA (Ochs et al. 2016), presumably due to hyper-resection.

BRCA1-BARD1 mediates ubiquitination of H2A-K125/K127/K129 (Kalb et al. 2014a) in response to DNA damage. It is thought that, like PRC1, BRCA1-BARD1 binds to the nucleosome acidic patch via an arginine anchor to promote ubiquitination of H2A (Buchwald et al. 2006), but unlike PRC1, specifically ubiquitinates the extreme C-terminus of H2A (Kalb et al. 2014a). In 2014, the first structure demonstrating the interaction of an ub E2-E3 ligase complex with the nucleosome substrate was solved to provide insight into how site specificity of the PRC1 RING1-BMI1 heterodimer for H2A-K119 is achieved (McGinty et al. 2014). Here, an arginine anchor in the RING E3 binds directly to the nucleosome acidic patch, while the E2 forms additional contacts with the nucleosome to position the E2 active site directly over H2A-K119. The arginine anchor is found in many diverse proteins that bind the nucleosome acidic patch, including E3 ligases RNF20-RNF40, RNF168 and BRCA1-BARD1 (Leung et al. 2014, Mattiroli et al. 2014). What determines the strict specificity of each of these ligases for discrete nucleosomal lysines remains to be resolved, but seems likely to be due to additional contributions from residues proximal to the arginine anchor in the RING domains, and from residues in the E2 partner. This is typified by data modelling the binding of the yeast homologue of RNF20-RNF40 (Bre1) with Rad6 in complex with the nucleosome (Gallego et al. 2016). Here, Bre1-Rad6 binds the nucleosome in an orientation that is an 180° rotation from the BMI1-RING1B-UbcH5a orientation even though both E3 dimers utilise an arginine anchor to bind the nucleosome acidic patch (McGinty et al. 2014, Gallego et al. 2016). It will be intriguing to solve the BRCA1-BARD1-E2 nucleosome structure given that it specifically ubiquitinates the extreme C-terminus of the highly flexible H2A tail, making it difficult to model the interaction of the substrate lysines with the E2 active site in the context of the nucleosome.

Recent evidence suggests that BRCA1-mediated extreme C-terminal ubiquitination of H2A is partially responsible for the recruitment of the chromatin remodeller, SMARCAD1, via its N-terminal ub-binding CUE domains, to sites of laser-induced damage (Densham et al. 2016). Once present, SMARCAD1 promotes the movement of 53BP1 to the periphery of damage foci in a manner that requires intact CUE domains and ATPase activity. It remains to be determined whether this movement is due to SMARCAD1's ability to promote nucleosome eviction, sliding or exchange (Awad et al. 2010). In any case, SMARCAD1's activity relieves the 53BP1 block sufficiently to allow long-range resection to occur. These data are consistent with experiments in yeast, which demonstrated that the yeast homologue of SMARCAD1, Fun30, was implicated in promoting long-range resection by facilitating the activity of Exo1 (Chen et al. 2012b, Costelloe et al. 2012), and also showed that the requirement for Fun30 is lessened if rad9, the yeast orthologue of 53BP1, is removed (Chen et al. 2012a, 2012b, Adkins et al. 2013). There is also evidence from yeast demonstrating that nucleosomes remain on single-stranded DNA post-resection and that Fun30 binds these nucleosomes with high affinity (Adkins et al. 2017). This may prove to be a key step towards understanding how remodelling events at damage foci lead to 53BP1 eviction from the foci core and the relationship of these events with resection and HDR. SMARCAD1 co-purifies with various other chromatin remodelling factors (Rowbotham et al. 2011), some of which also promote 53BP1 re-positioning (Alagoz et al. 2015) and additionally, a number of other chromatin remodellers (e.g. INO80C, SWI/SNF, RSC, NuA4 and SWR1C) show enhanced recruitment following resection at DSBs and are likely to play roles in this repositioning process (van Attikum et al. 2004, Shim et al. 2007, Bennett et al. 2013).

In recent years, advances in super-resolution microscopy have brought much-needed insight into the 3D structures of damage-associated foci (Chapman et al. 2012, Kakarougkas et al. 2013). 53BP1 forms a focus around the damage site that is remodelled in S phase to reveal a central core that contains BRCA1 and also the single-strand DNA binding protein RPA, which is indicative of resection. This movement of 53BP1 requires the proteasomal DUB POH1, BRCA1 BRCT domains (Kakarougkas et al. 2013) and BRCA1 ligase activity (Densham et al. 2016). The spread of damage-induced signalling and resection might be thought of as a linear event, extending either side from sites of damage along the DNA and spreading γH2AX and amplification of the ub signal by formation of K63-ub chains along the DNA in *cis*. However, since DNA is topologically constrained in three dimensions in cells, it may be more efficient to spread DNA damage signalling through direct three-dimensional contacts than via a strict linear pathway along DNA. Understanding the role of remodelling factors and the three-dimensional chromatin structures of damage foci will provide valuable insight and inform future directions in this field.

The role of BRCA1 ligase activity in cancer predisposition remains controversial (Shakya et al. 2011, Zhu et al. 2011, Li et al. 2016). However, this may be, in part, explained by a differential requirement for BRCA1-BARD1 ligase activity depending on the type and chromatin context of DNA damage, which differs with different damaging agents (Densham et al. 2016). Evidence suggests BRCA1-BARD1 ligase activity is not required for resection or HDR per se, but for the ability to promote remodelling of the chromatin environment surrounding the damage to allow long-range resection, and/or access of recombination proteins. Since different DNA damages cause different lesions in different chromatin environments, it may be that the requirement of ligase activity to promote a chromatin environment permissive for HDR is less important in the context of a stalled replication fork or interstrand cross-link repair, where other mechanisms exist to regulate chromatin structure.

The SUMO pathway also influences the recruitment of both BRCA1 and 53BP1 to sites of damage (Galanty et al. 2009, Morris et al. 2009), and SUMO modification of both proteins occurs in response to IR and a range of DNA damaging agents (Galanty et al. 2009, Morris et al. 2009, Yin et al. 2012, Hendriks et al. 2015b). BRCA1 is SUMOylated at an N-terminal consensus site at K119, and modification by SUMO enhances BRCA1-BARD1 ub ligase activity in vitro (Morris et al. 2009). It is mechanistically unclear how SUMO modification enhances ligase activity, but it could be related to a similar potentiation of ligase activity seen following auto-ubiquitination (Mallery et al. 2002); alternatively, it could enhance E2 processivity via E2-SIM interactions. It is also unclear whether the SUMO-mediated potentiation of BRCA1 ligase activity plays any role in regulating BRCA1's function in SMARCAD1 recruitment and 53BP1 repositioning (Densham et al. 2016). Finally, it is important to note that SUMOylated-BRCA1 is a substrate for RNF4 (Vyas et al. 2013) and therefore may be important for BRCA1 clearance from damage sites.

H2A deubiquitinases

Growing evidence suggests that 53BP1 acts not simply to block resection but also to fine-tune resection lengths (Ochs et al. 2016). Therefore, the ability to control the spread and localisation of 53BP1 is critical for maintaining appropriate resection lengths, since both too little (resulting in error-prone NHEJ and MMEJ) and too much resection (resulting in SSA) could promote genomic instability. In the context of the underlying ub signals on H2A, this would argue for tight regulation of the E3 ligases responsible for laying down the mark, and also the requirement for DUB erasers of the ub marks (reviewed in Vissers et al. 2008). To this end, DUBs USP51 (Wang et al., 2016b) and USP3 (Sharma et al. 2014) have been shown to remove the H2A-K13/15ub mark, and BAP1-ASXL1 removes H2A-K118/119ub (Sahtoe et al. 2016).

Cells in which USP51 is depleted show an increase in spontaneous DNA damage and in the formation of BRCA1 and 53BP1 foci (Wang et al. 2016b). Associated with this are cell cycle defects showing a reduction in the proportion of cells in S phase and a corresponding increase in G1 cells. Likewise, overexpression of USP51 reduces ub conjugates identified by the FK2 antibody, and BRCA1, 53BP1 and RNF169 foci formation but strikingly does not affect recruitment of RNF168 to IRIF. This is consistent with a role for USP51 in antagonising RNF168 signalling. Indeed, USP51-depleted cells show increased H2A-ub in cells expressing H2A-K118/119R but not those expressing H2A-K13/15R, suggesting it has specific activity against H2A-K13/15ub. Using a H2A-K15ub–specific antibody, Wang et al. showed that H2A-K15ub and 53BP1 persist at damage sites longer in cells in which USP51 was depleted. One might predict that persistence of 53BP1 at damage sites would lead to reduced resection and HDR, but, surprisingly, USP51-deficient cells are radiosensitive and yet show higher levels of HDR and NHEJ in reporter assays (Wang et al. 2016b).

Moreover, USP3 has also been shown to antagonise RNF168-mediated ub of H2A/H2AX at K13/K15 (Sharma et al. 2014). However, since USP3 also has activity against the H2A-K118/K119ub site of γH2AX, and overexpression of USP3 prevented the formation of RNF168 and downstream factors into damage foci, it is unclear whether the activity of USP3 to antagonise RNF168 signalling is direct or indirect (Mosbech et al. 2013, Wang et al. 2016a, 2016b).

The BRCA1-associated protein 1 (BAP1) is a DUB component of the Polycomb repressor-DUB complex that specifically cleaves H2A-K118/K119ub (Scheuermann et al. 2012) and not the H2A-K13/15ub mark deposited by RNF168 (Sahtoe et al. 2016). BAP1 is recruited to laser lines and has been detected by ChIP on chromatin adjacent to DSBs (Ismail et al. 2014, Nishi et al. 2014, Yu et al. 2014). The recruitment of BAP1 to DSBs is rapid and dependent on RNF8-RNF168 signalling, and BAP1 is required for HDR but not NHEJ repair (Ismail et al. 2014, Yu et al. 2014). BAP1 was originally identified as an interactor of the BRCA1 RING domain (Jensen et al. 1998) where it has been reported to have non-catalytic functions in disrupting BRCA1-BARD1 interactions and consequently inhibits BRCA1 ligase activity (Nishikawa et al. 2009). It is not known how much this activity or the H2A DUB function contributes to BAP1's role in HDR, but loss of BAP1 results in hypersensitivity to PARP inhibitors in human cell lines and DT40 knockouts (Nishikawa et al. 2009, Ismail et al. 2014, Yu et al. 2014). Consistent with this, BAP1 is mutated in a number of cancers, including mesothelioma, melanoma and renal cancers (Luchini et al. 2016). Indeed, cancer-associated mutants of BAP1 do not recruit to sites of damage, and mesothelioma cell lines expressing mutant BAP1 fail to clear γH2AX efficiently, suggesting defective DNA repair resolution (Ismail et al. 2014).

A number of other DUBs have also been implicated in regulating H2A deubiquitination but are not as well characterised as those described earlier. These include USP16, MYSM1, USP44 and USP17L2/DUB3. USP16 is a H2A DUB that is able to de-ubiquitinate both K118/K119 and K13/K15 in vitro. Some groups have reported a role for USP16 in ub signalling during cell cycle progression (Joo et al. 2007) and others a role in DSB repair (Zhang et al. 2014), while some suggest that the role in DSB signalling is limited (Mosbech et al. 2013, Nishi et al. 2014, Wang et al. 2016b). This may suggest a specialised role for USP16 in DSB repair that is not yet well understood. MYSM1 is a JAMM family DUB that has H2A de-ubiquitinase activity and is involved in transcriptional regulation (Zhu et al. 2007). MYSM1 is recruited to sites of damage (Nishi et al. 2014), but what role MYSM1 plays in DSB repair remains to be defined. Both USP44 and USP29 antagonise RNF168 recruitment to damage sites, but only USP44 has been shown to deubiquitinate DSB-induced H2A-ub (Mosbech et al. 2013). USP44 is itself recruited to laser-induced DNA damage lines (Mosbech et al. 2013). Finally, USP17L2/DUB3 overexpression reduces the mono-ub of H2AX (Delgado-Diaz et al. 2014). In cells, this results in a loss of signalling downstream of RNF168 (53BP1, BRCA1 and RAD51) without affecting γH2AX, MDC1 or RNF8 accrual (Delgado-Diaz et al. 2014). Finally, USP21 and USP22 deubiquitinate H2A in transcriptional regulation, but any link to the DSB response are unclear (Nakagawa et al. 2008, Zhang et al. 2008). USP22 is more commonly associated with the larger SAGA deubiquitinase complex, which specifically removes H2B-K120ub (Morgan et al. 2016, Ramachandran et al. 2016). There are emerging reports of the role of the SAGA complex in DNA repair linked to early changes in chromatin state (Ramachandran et al. 2016). In addition, USP22 overexpression in colon carcinoma is associated with decreased H2B mono-ub (Melo-Cardenas et al. 2016), and overexpression is seen in a number of other cancers, making it an attractive target for drug development.

Ub and SUMO-mediated control of recombination

Resection for HDR is promoted in the appropriate cell cycle phase, S and G2, by the activity of CDK2, which activates CtIP (Escribano-Diaz et al. 2013). In addition, the cell cycle uses ub signalling to control the recombination phase of HDR. Interaction of BRCA1 with PALB2 contributes to PALB2's recruitment of BRCA2 to sites of DNA damage critical for repair (Sy et al. 2009, Zhang et al. 2009). Intriguingly, in G1, the BRCA1 interaction with PALB2-BRCA2 is inhibited to prevent inappropriate BRCA2 function. This is regulated via an E3 ub ligase composed of KEAP1-cullin3 (CUL3)-RBX1, which ubiquitinates the BRCA1 interaction site in PALB2, suppressing PALB2s interaction with BRCA1. The interaction would be inhibited in all cell cycle stages but for the activity of the DUB enzyme USP11, which antagonises the CRL3-KEAP1–mediated ubiquitination of PALB2. USP11 is degraded in G1 phase of the cell cycle, and its degradation is increased further on DNA damage, providing a mechanism for the reduced BRCA1-PALB2 interaction in this phase (Orthwein et al. 2015).

Ub also has a positive and direct impact on PALB2. In a further intriguing link to H2A ubiquitination, PALB2 indirectly recognizes histone ubiquitination by physically associating with ub-bound RNF168 though a PALB2 interaction domain in RNF168 (Luijsterburg et al. 2017). These findings are initially surprising in linking PALB2-BRCA2 to the chromatin-ub signal in DSB repair, suggesting that the presence of nucleosomes is not mutually exclusive with resection and homology searching. Indeed the idea that nucleosomes may remain associated with resected DNA has recently been supported by in vitro studies (Adkins et al. 2017).

A number of the UCHL family of DUBs have been linked to recombination. UCHL3 has been reported to interact with and de-ubiquitinate RAD51 (Luo et al. 2016). Indeed, depletion of UCHL3 is associated with an increase in polyub-RAD51, which does not affect RAD51 stability, suggesting that this is not a degradative poly-ub signal. Instead de-ubiquitination by UCHL3 is required for RAD51 interaction with BRCA2 to promote RPA exchange and DNA loading. The DNA damage-dependent activity of UCHL3 is regulated by phosphorylation of a single site within UCHL3's catalytic domain by ATM. UCHL3 activity does not affect the recruitment of repair factors upstream of RAD51, but loss of UCHL3 abrogates RAD51 foci formation resulting in sensitivity to PARPi and reduced HDR efficiency. Conversely, increased UCHL3 expression in MCF7 cells increases resistance to IR and PARPi, and in breast cancer, UCHL3 is overexpressed, correlating with poor survival (Luo et al. 2016).

SUMO also plays a critical role at this stage of HDR. The ssDNA binding protein subunit RPA70/RPA1 has a critical SUMO-modification site which is required for subsequent RAD51 accumulation (Dou et al. 2010). This is particularly intriguing given that yeast-2-hybrid data established a direct interaction between RAD51 and SUMO (Shen et al. 1996). Later, work identified a conserved SIM (VAVV$_{261\text{-}264}$) in the C-terminus of RAD51, which upon mutation abrogated RAD51 accumulation at laser lines and reduced HDR (Shima et al. 2013). While a RPA-SUMO~RAD51 model is appealing (Dou et al. 2010), it remains to be seen whether this is a genuine interaction module. Indeed, given the SUMO-rich modifications of the resection machinery, a number of other candidates could equally well support SUMO-mediated RAD51 accumulation. Moreover, the SUMO protease SENP6 binds RPA and EXO1 to promote their hypoSUMOylation (Dou et al. 2010, Bologna et al. 2015) and SENP6 dissociation from RPA enables its modification to promote RAD51 exchange (Dou et al. 2010).

These findings demonstrate that ub and SUMO play critical, and often precise, roles in promoting the RAD51 loading phase of recombination which is so critical to accurate DSB repair.

DISEASE ASSOCIATIONS OF SMALL MODIFIER BIOLOGY OF DNA DSB REPAIR

Small modifiers and cancer

Both ub and SUMO pathways have been linked to cancer phenotypes, either through the influence of direct modifications on oncogenes or tumour suppressors, or through epigenetic or genetic alterations of ub and SUMO pathway components. For example: a lower level of ub-modified H2B is observed in some colon cancers compared to normal colon tissue (Wang et al. 2015); in acute myeloid leukaemia (AML) overexpression, the PRC1 E3 ub ligase component BMI1 is associated with more aggressive disease and poor outcomes (Chowdhury et al. 2007, Saudy et al. 2014), while overexpression of the ub E3 RNF168 in tumours results in excessive recruitment of 53BP1 and its effector proteins RIF1 and REV7, resulting in a shift in the DSB repair balance from HDR toward NHEJ and an increased resistance to IR (Chroma et al. 2017).

Components of the SUMO conjugation machinery, in particular the E1 and E2 enzymes, are upregulated in a number of cancers, including breast, brain, colorectal, lung and ovarian (Zhu et al. 2010, Dong et al. 2013, Bellail et al. 2014). Paradoxically, the deSUMOylating SENP enzymes are also upregulated in a number of cancers (Cashman et al. 2014, Ma et al. 2014, Cheng et al. 2017). While this may seem counterintuitive, some SENP enzymes are required for SUMO maturation and others promote the recycling SUMO and as such may be required to provide the free SUMO needed for rapid SUMOylation that occurs when cells encounter stress. SUMOylation can disable the activity of a number of oncogenic transcription factors, and high SENP levels may enforce a hypoSUMOylated state and transcriptional function (Bawa-Khalfe and Yeh 2010).

Since Ub and SUMO pathways are involved in multiple ways in cellular stress responses, a speculative conclusion is that cancer cells, which experience high levels of endogenous stress, exhibit upregulation of these modification networks to improve their survival.

BRCA1

Genetic alterations in the E3 ub ligase BRCA1 are associated with an increase in breast and ovarian cancer susceptibility (Castilla et al. 1994), although the role of the ligase function per se remains controversial (Shakya et al. 2011, Zhu et al. 2011). A mouse model bearing a mutation in the zinc-ligating residue of the RING (C61G) exhibits tumour susceptibility when the remaining wild-type allele is lost in the adult mammary gland (Drost et al. 2011). The tumours exhibit a level of genome instability comparable to animals, in which the entire BRCA1 gene is lost (Drost et al. 2011). The substitution at C61G reduces interaction with its N-terminal binding partner, BARD1, and impacts the stability of both proteins. BRCA1 zinc-ligating substitution variants show reduced HDR (Ruffner et al. 2001, Anantha et al. 2017). Surprisingly, unlike full gene BRCA1 knockouts, tumours carrying C61G-BRCA1 rapidly become resistant to HDR-directed therapies. Animals lacking BRCA1 exon 2 also express a protein disrupted within the RING region that does not interact with or support BARD1, but is stable itself (Li et al. 2016). Embryonic lethality of exon 2 deleted ('RING-less') mice are rescued by 53BP1 loss and do not develop tumours, supporting the conclusion that the ligase acts to suppress 53BP1 inhibitory effects. Cells from a conditional model of exon 2 loss show proficient RAD51 foci and HDR (Li et al. 2016), as do cells from del185AG animals, which express a RINGless-BRCA1 protein (expressed from an internal ATG) (Drost et al. 2016), and human cells expressing a RINGless BRCA1 expressed in therapy resistant cells, also from a downstream ATG (Wang et al. 2016a). These proteins reconstitute the majority of BRCA1 function, presumably the most important being the ability to localise to sites of damage and to redistribute PALB2-BRCA2-RAD51, but they lack its ligase function. Thus, at least in HDR-resistant tumours, BRCA1 E3 ub ligase function is dispensable. It is possible that the ligase role is entirely dispensable in tumour development, as animals bearing the E2-disruptive variant I26A-BRCA1 are not tumour prone (Shakya et al. 2011); however, animals lacking BRCA1 exhibit proliferation defects and apoptosis in a manner that can be rescued by H2A-ub (Zhu et al. 2011), the target of BRCA1-BARD1 required to counter 53BP1 (Densham et al. 2016). The most generous conclusion in favour of the BRCA1 E3 ub ligase activity having a role in cancer is that it may suppress tumour development, but that even low-level activity is able to do so, whereas it is not needed, or easily overcome, in therapy resistance.

The BRCA1 BRCT mediates interaction with several phosphorylated proteins, in separable complexes (Abraxas in BRCA1-A, BRIP/FANCJ in BRCA1-B and CtIP in BRCA1-C). Variants in the BRCTs that interfere with BRCA1 complex formation show reduced HDR in the majority of studies (Ruffner et al. 2001, Dever et al. 2011, Anantha et al. 2017). Similarly, alternative splicing of BRCA1 that generates a variant lacking part of the BRCT domains (Δ17-19) results in an inability to interact with both Abraxas and CtIP (and therefore both the A and C complex). This variant is deficient in both HDR and NHEJ repair (Sevcik et al. 2013). As BRCA1 substitutions within the BRCT domain have the potential to disrupt interaction with CtIP (BRCA1-C), Abraxas (BRCA1-A) and BRIP1/FANCJ (BRCA1-B) and therefore disrupt BRCA1 in all three complexes, it is difficult to infer a specific complex function from these variants.

The BRCA1-A complex re-distributes BRCA1-BARD1 via RAP80-dependent recognition of damage-induced ub/SUMO conjugates, and through sequestering BRCA1-BARD1 is thought to restrain its resection activities (Typas et al. 2015). Loss of BRCA1-A subunits results in hyper-resection and genomic instability (Coleman and Greenberg 2011, Hu et al. 2011). Clinically, the BRCA1-A complex is disrupted by at least three mechanisms: (1) mutations/deletions in BRCA1 that render it unable to interact with Abraxas, (2) mutations/deletions in RAP80 that prevent the ub-dependent interaction/recruitment of the A complex and (3) mutations of Abraxas that disrupt BRCA1 interaction. *RAP80* is genetically disrupted in reports that have examined tumour material, and the ovarian cancer cell line TOV21G is deficient in RAP80 expression due to a truncating mutation and promoter hypermethylation. BRCA1 and other BRCA1-A complex subunits do not localise to DSBs in these cells which are hypersensitive to IR, but these phenotypes can be restored by the expression of WT RAP80 (Bian et al. 2012). An in-frame deletion of three amino acids within the UIM domain of RAP80 has been identified in a cohort of Finnish breast cancer families. This mutation disrupts RAP80-ub interaction and recruitment of BRCA1 to sites of damage and promotes genomic instability (Nikkila et al. 2009). The R361Q mutation in Abraxas that was also identified in a Finnish cohort of breast cancer families disrupts the nuclear localisation of Abraxas and therefore interaction with BRCA1 (Solyom et al. 2012). Loss of function variants in Abraxas have also been identified in ovarian cancer (Pennington et al. 2014).

It is notable that these clinical interpretations of deficiencies in the BRCA1-A complex in cancer are not in entire agreement with the cell biology, which has found the BRCA1-A complex is associated with HDR suppression and only mild IR sensitivity (Coleman and Greenberg 2011, Hu et al. 2011, Typas et al. 2015). Few clinically oriented reports go into sufficient detail to conclude 'over-resection', and the majority find a defect in DNA repair associated with BRCA1-A complex disruption. Further work exploring the dependence of these tumours on repair that uses extended resection, such as single-stand annealing, and explores their potential sensitivity to RAD52 inhibitors would be informative.

Ub components of the polycomb complex in cancer

BMI1, a core component of the epigenetic repressor PRC1 E3 ub ligase, is considered an oncogene, particularly in hematological malignancies (reviewed in Sahasrabuddhe 2016), and its overexpression in AML is associated with more aggressive disease and poor outcomes (Chowdhury et al. 2007).

The ub mark laid down by PRC1, namely H2A-K119ub, is removed by the DUB, BAP1 (Harbour et al. 2010, Sahtoe et al. 2016). Bi-allelic mutations in BAP1 are associated with various malignancies, including mesothelioma, uveal melanoma intrahepatic cholangiocarcinoma and clear cell renal cancers (Harbour et al. 2010, Testa et al. 2011, Pena-Llopis et al. 2012, Rai et al. 2016), establishing BAP1 as tumour suppressor. A summary analysis of various studies which investigated BAP1-positive compared to BAP1-negative tumours found significantly increased all-cause mortality, cancer-specific mortality and risk of recurrence in all the BAP1 negative tumour types. One exception to this was mesothelioma, in which the presence of BAP1 mutations correlated with a better prognosis (Luchini et al. 2016).

In addition to the epigenetic transcriptional influence likely to be changed through BAP1 loss, the protein has many other non-epigenetic roles that may promote transformation (Carbone et al. 2013). Nevertheless, from a DNA repair perspective, it is interesting to note that cancer-associated mutants of BAP1 prevent its recruitment to sites of damage, and mesothelioma cell lines expressing mutant BAP1 fail to clear γH2AX efficiently, suggesting defective DNA repair resolution (Ismail et al. 2014) and therefore poor DNA repair may contribute to its role as a suppressor.

Small modifier links to neurodegenerative diseases

The ub protease system is associated with a number of neurodegenerative diseases. A hemizygous missense mutation (c.1670A>T; p.Glu557Val) in UBA1, one of the two critical ub-activating E1 enzymes, has been associated with spinal muscular atrophy X-linked 2 (SMAX2), a rare lethal disorder classed as a motor sensory neuronopathy (Dlamini et al. 2013). Likewise, mutations have been identified in various DNA damage-associated E3 ligases that have been linked to neurodegenerative disease. For example, the E3 ligase HERC2 is associated with the activation of RNF8 in the damage response (Bekker-Jensen et al. 2010) and with the regulation of CHK1 (Yuan et al. 2014, Zhu et al. 2014). Identification of a homozygous missense mutation in HERC2 (c.1781C>T, p.Pro594Leu) is associated with autosomal recessive mental retardation-38 (MRT38) and developmental delay (Puffenberger et al. 2012). The mutation results in a loss of protein stability and a consequent reduction in HERC2 protein levels.

The E3 ligase RNF168 responsible for signalling 53BP1 deposition at damage sites is, like 53BP1 itself, associated with immunodeficiency. Patients with RNF168 bialleleic mutations exhibit the immunodeficiency syndrome RIDDLE (radiosensitivity, immunodeficiency, dysmorphic features, and learning difficulties). RIDDLE was found to be associated with mutations in RNF168 and shares clinical features with ataxia telangiectasia (Stewart et al. 2007). Cells lacking RNF168 showed defective DSB repair corresponding with failure of 53BP1 recruitment and reduced BRCA1 recruitment (Stewart et al. 2007, 2009, Doil et al. 2009). In contrast, mutations in RNF8 have not been linked to heredity disease, although RNF8 overexpression has been shown to predict breast cancer (Lee et al. 2016).

Finally, loss of a chromosome region that includes the gene for the JAMM-type DUB, BRCC36, that forms part of the BRCA1-A complex (Dong et al. 2003), is associated with Moyamoya angiopathy, often found with neurofibromatosis, and Down syndrome, and shares some clinical features that are reminiscent of chromosome breakage syndromes (Miskinyte et al. 2011).

Inhibitors of small modifier pathways

The development of small molecule inhibitors to modulate small modifier pathways for clinical use has been an exciting area of research in recent years. There is a diverse range of cellular processes regulated by small modifiers and dysregulation of these small modifier pathways is associated with clinical disease. In addition the abundance of enzymes associated with small modifier cycles provides a large number of potential targets for small molecule inhibitor design.

Proteasome inhibitors

The best and most successful small molecule modulation of the ub pathway to date has been the development of reversible proteasome inhibitors, such as Bortezomib (VELCADE) (reviewed in Arkwright et al. 2017) and Carlfilzomib (Ruschak et al. 2011). Cancer cells are more reliant on proteasome function than normal healthy cells and therefore there is a narrow but effective therapeutic window for proteasome inhibition. Both Bortezomib and Carlfilzomib specifically bind and inhibit the proteasomal chymotrypsin b5 catalytic subunit. However, Bortezomib has pleiotropic effects resulting in dose-limiting toxicities, and both inherent and acquired drug resistance remain significant clinical obstacles. Experimental data show resistance mechanisms demonstrating upregulation of the chymotrypsin b5 subunit expression in Bortezomib-resistant cell lines (Oerlemans et al. 2008) or the occurrence of mutations in the chymotrypsin b5 subunit that prevent Bortezomib binding (Oerlemans et al. 2008, Franke et al. 2012). To some extent, combination therapy with drugs such as gemcitabine, dexamethasone or doxorubicin has been explored to increase the therapeutic window, and Bortezomib can improve the sensitivity of cancer cells to irradiation or conventional therapies. There has been a flurry of proteasome inhibitor interest in recent years, including development of oral inhibitors such as Ixazomib. Results from phase III trials for Ixazomib in relapsed, refractory melanoma were promising (Moreau et al. 2016), and Ixazomib is currently in phase I trials for solid tumours [NCT02942095] and phase II trials for treatment in multiple myeloma [NCT02924272].

Deubiquitinating enzyme inhibitors

To date, while there has been growing interest in the development of specific DUB inhibitors for clinical use, very few have been in clinical trials. Inhibitors against the DUB, USP7, have been most widely studied due to links with regulating p53 stability, and there are a number of preclinical inhibitors in the market. USP7 also modulates stability of E3 ub ligases RAD18 and RNF168 (Zhu et al. 2015), and is upregulated in p53-defective chronic lymphocytic leukaemia cells. Inhibiting USP7 in these cells impairs HDR and sensitises these chemoresistant CLL cells to DNA cross-linking agents (Agathanggelou et al. 2017).

The USP13-specific inhibitor Spautin-1 leads to an increase in RAP80 K63 ubiquitination. This had no effect on RAP80 stability but impairs DNA damage-induced RAP80 and BRCA1 foci formation. Consequently, Spautin-1 treatment results in impaired HDR and renders ovarian cancer cells more sensitive to Cisplatin and PARP inhibition (Li et al. 2017). Spautin-1 is currently in pre-clinical development but could be useful as a chemosensitiser for treatment of HDR-defective tumours.

A recent experimental paper identified DUBs as new molecular targets for phenethyl isothiocyanate (PEITC), a naturally occurring isothiocyanate that is found in some cruciferous vegetables and thought of as a potential therapeutic for chemoprevention of cancers. Intriguingly, of the 10 DUBs that were identified to be sensitive to PEITC, 7 have been described as having roles in DNA repair and/or chromatin remodelling (Lawson et al. 2017). As PEITC occurs in vegetables such as broccoli and kale, and are known to have chemoprotective roles, this opens up an interesting area for further drug development and study.

SENP inhibitors

Given the essential role SUMO plays in the repair of DSBs, compounds that inhibit the activity of SENPs may play an important role in improving response to chemotherapy or preventing the onset of chemoresistance. Inhibition of SENPs may deplete the free pool of SUMO, prevent maturation of pro-SUMO and promote hyperSUMOylation of certain substrates. The ensuing disruption in SUMO homeostasis could prevent cancer cells from mounting a normal DSB response either in response to endogenous stresses or DNA damaging chemotherapies. Development of small molecules that inhibit SENP enzymes has lagged behind DUB inhibitors. Currently, only a handful of compounds have been developed that target deSUMOylases, the majority of which are targeted at SENP1. Small molecule SENP inhibitors increase global SUMO conjugates in cells and have anti-proliferative action in some cell lines (Qiao et al. 2011, Wu et al. 2016). The majority of these compounds have only been tested in a small subset of SENPs and not against cysteine protease DUB enzymes. Most inhibit more than one SUMO protease and might be expected to have some non-specific activity on a proportion of DUBs. Those that are in experimental use have reactive chemical groups and are not suitable as clinical compounds (Kumar and Zhang 2015), and most compounds have also not been tested for their inhibitory action towards DSB repair thus far.

Ub conjugation inhibitors (E1, E2 and E3 inhibitors)

The experimental ub E1 inhibitor of UBA1, PYR-41, irreversibly binds the ub E1 active site. Inhibiting the first step of the ub enzyme cascade has a number of impacts on cells, including an increase in the levels of p53 (Yang et al. 2009). In relation to the DNA damage response, it is intriguing to note that PYR-41 treatment, like depletion of UBA1, but not UBA6, impairs 53BP1 foci formation following exposure to IR. Early ub-independent steps of the DSB response, such as MDC1 recruitment and γH2AX formation, are not affected by UBA1 depletion or inhibition. Indeed, PYR-41 treatment leads to persistent γH2AX foci, which is indicative of delayed or impaired repair (Moudry et al. 2012). The HDAC inhibitor Largozole and its derivatives also inhibit UBA1 (Ungermannova et al. 2012), but although HDAC inhibitors are in clinical use, this agent is not.

The precise ub E2–E3 pairings for many of the DNA damage E3 ligases remain to be determined. However, UBE2N (Ubc13) is the E2 responsible for generating specific K63 chains with the relevant E3. Since K63 chain formation is a critical component of the DNA damage response, it is interesting to note that the specific UBE2N inhibitor NSC697923 has a marked effect on the DNA damage response (Hodge et al. 2015) and promotes potent apoptosis in neuroblastoma cell lines (Cheng et al. 2014). Targeting UBE2N has been proposed as a therapeutic approach in myelodysplastic syndromes (MDSs) and AML in which the E3 ligase TRAF6 is overexpressed. TRAF6 is a key enzyme in innate immune signalling that catalyses K63-ub chains on TRAF6 and its substrates. Upon treatment with NSC697923, TRAF6 switched from making K63 chains to generating K48 chains by utilising an alternate E2 following UBE2N inhibition, leading to proteasome-mediated degradation of TRAF6 targets. This resulted in changing the ub mark from a signalling entity to a degradative one (Barreyro et al. 2016). It remains to be seen if this innovative approach can similarly switch K63 chain formation to factor degradation following DNA damage.

Drug targets against E3 ub-conjugating enzymes are harder to design and rationalise, in particular to the RING class of E3 ligases that make up the majority of those involved in the

DNA damage response, which rely on protein–protein interactions with the E2 and ub. The best example of a successful RING E3 inhibitor to date is of the Nutlin family of inhibitors that target the p53 E3 RING-type E3 ligase MDM2, and do so by preferentially binding to the p53-binding pocket of MDM2 (Vassilev et al. 2004). In terms of E3 ligases involved in the DNA damage response, the development of BMI1 inhibitors has shown the most promise. There is a large body of preclinical data supporting the use of the BMI1 inhibitor PTC-209 in colorectal (Kreso et al. 2014), breast and ovarian (Dimri et al. 2016) and prostate cancers, amongst others. However, due to limited potency and poor pharmacokinetic properties, this compound has not entered clinical trials. In contrast, the second-generation BMI1 inhibitor, PTC596, is currently in phase I clinical trial (NCT02404480).

In recent years, attention has been focused on synthetic lethality approaches for BRCA1-null tumours, in particular with the development and use of PARP inhibitors (reviewed in Helleday 2011 and also discussed further in Chapter 6). Given the recently revealed role of the ub ligase in countering 53BP1 and in promoting HDR (Densham et al. 2016), PARP inhibition, at least in part, targets the defect conferred by the absence of the Ub ligase function of BRCA1.

Finally, the NEDD8 E1 inhibitor, MLN4924 (Pevonedistat), has undergone a number of phase I clinical trials against relapsed/refractory multiple myeloma, melanoma and solid tumours with promising potential (Bhatia et al. 2016, Sarantopoulos et al. 2016, Shah et al. 2016). NEDD8 is critical for activating cullin ub-ligases, which have been implicated in the removal of Ku proteins from DNA ends as NHEJ concludes (Postow and Funabiki 2013, Brown et al. 2015).

SUMMARY

Work of recent years has seen a rapid elucidation of the roles that ub and SUMO signalling perform in promoting mammalian cell responses to, and repair of, DNA double-strand breaks. What is now emerging is a story of multiple, intricate and crosstalking regulation that communicates between signalling pathways, and is integrated with cell cycle stage and chromatin state to promote the correct repair outcome. Future insights will come from studying these proteins as co-operative modifiers and as part of the DDR response network. The complexity of the ub code, mediated by different chain topologies and by post-translational modifications of ub itself, can trigger specific cellular responses, and it seems likely that SUMO isoforms will convey as wide a range of signals. Given the importance of DSB repair in pathways relevant to health and disease, together with the potential for aspects of these pathways to offer opportunities for drug design, further examination of small modifier biology is a rich seam for advancing future therapeutic approaches to improving human health.

REFERENCES

Abdul Rehman, S. A., Kristariyanto, Y. A., Choi, S. Y., Nkosi, P. J., Weidlich, S., Labib, K., Hofmann, K., Kulathu, Y. 2016. MINDY-1 is a member of an evolutionarily conserved and structurally distinct new family of deubiquitinating enzymes. *Mol Cell* 63: 146–155.

Acs, K., Luijsterburg, M. S., Ackermann, L., Salomons, F. A., Hoppe, T., Dantuma, N. P. 2011. The AAA-ATPase VCP/p97 promotes 53BP1 recruitment by removing L3MBTL1 from DNA double-strand breaks. *Nat Struct Mol Biol* 18: 1345–1350.

Adkins, N. L., Niu, H., Sung, P., Peterson, C. L. 2013. Nucleosome dynamics regulates DNA processing. *Nat Struct Mol Biol* 20: 836–842.

Adkins, N. L., Swygert, S. G., Kaur, P., Niu, H., Grigoryev, S. A., Sung, P., Wang, H., Peterson, C. L. 2017. Nucleosome-like, single-stranded DNA (ssDNA)-histone octamer complexes and the implication for DNA double strand break repair. *J Biol Chem* 292: 5271–5281.

Agathanggelou, A., Smith, E., Davies, N. J., Kwok, M., Zlatanou, A., Oldreive, C. E., Mao, J., Da Costa, D. et al. 2017. USP7 inhibition alters homologous recombination repair and targets CLL cells independent of ATM/p53 functional status. *Blood* 130: 156–166.

Aguilar-Martinez, E., Chen, X., Webber, A., Mould, A. P., Seifert, A., Hay, R. T., Sharrocks, A. D. 2015. Screen for multi-SUMO-binding proteins reveals a multi-SIM-binding mechanism for recruitment of the transcriptional regulator ZMYM2 to chromatin. *Proc Natl Acad Sci USA* 112: E4854–E4863.

Alagoz, M., Katsuki, Y., Ogiwara, H., Ogi, T., Shibata, A., Kakarougkas, A., Jeggo, P. 2015. SETDB1, HP1 and SUV39 promote repositioning of 53BP1 to extend resection during homologous recombination in G2 cells. *Nucleic Acids Res* 43: 7931–7944.

Anantha, R. W., Simhadri, S., Foo, T. K., Miao, S., Liu, J., Shen, Z., Ganesan, S., Xia, B. 2017. Functional and mutational landscapes of BRCA1 for homology-directed repair and therapy resistance. *Elife* 6: e21350.

Arkwright, R., Pham, T. M., Zonder, J. A., Dou, Q. P. 2017. The preclinical discovery and development of bortezomib for the treatment of mantle cell lymphoma. *Expert Opin Drug Discov* 12: 225–235.

Awad, S., Ryan, D., Prochasson, P., Owen-Hughes, T., Hassan, A. H. 2010. The Snf2 homolog Fun30 acts as a homodimeric ATP-dependent chromatin-remodeling enzyme. *J Biol Chem* 285: 9477–9484.

Ayrapetov, M. K., Gursoy-Yuzugullu, O., Xu, C., Xu, Y., Price, B. D. 2014. DNA double-strand breaks promote methylation of histone H3 on lysine 9 and transient formation of repressive chromatin. *Proc Natl Acad Sci USA* 111: 9169–9174.

Bakr, A., Kocher, S., Volquardsen, J., Petersen, C., Borgmann, K., Dikomey, E., Rothkamm, K., Mansour, W. Y. 2016. Impaired 53BP1/RIF1 DSB mediated end-protection stimulates CtIP-dependent end resection and switches the repair to PARP1-dependent end joining in G1. *Oncotarget* 7: 57679–57693.

Barreyro, L., Sampson, A. M., Bolanos, L., Niederkorn, M., Pujato, M., Smith, M. A., Tomoya, M. et al. 2016. Inhibition of UBE2N as a therapeutic approach in myelodysplastic syndromes (MDS) and acute myeloid leukemia (AML). *Blood* 128: 579.

Bawa-Khalfe, T., Yeh, E. T. 2010. SUMO losing balance: SUMO proteases disrupt SUMO homeostasis to facilitate cancer development and progression. *Genes Cancer* 1: 748–752.

Bayer, P., Arndt, A., Metzger, S., Mahajan, R., Melchior, F., Jaenicke, R., Becker, J. 1998. Structure determination of the small ubiquitin-related modifier SUMO-1. *J Mol Biol* 280: 275–286.

Bekker-Jensen, S., Danielsen, J. R., Fugger, K., Gromova, I., Nerstedt, A., Bartek, J., Lukas, J., Mailand, N. 2010. HERC2 coordinates ubiquitin-dependent assembly of DNA repair factors on damaged chromosomes. *Nat Cell Biol* 12: 80–86.

Bellail, A. C., Olson, J. J., Hao, C. 2014. SUMO1 modification stabilizes CDK6 protein and drives the cell cycle and glioblastoma progression. *Nature Commun* 5: 4234.

Bennett, G., Papamichos-Chronakis, M., Peterson, C. L. 2013. DNA repair choice defines a common pathway for recruitment of chromatin regulators. *Nature Commun* 4: 2084.

Bergink, S., Ammon, T., Kern, M., Schermelleh, L., Leonhardt, H., Jentsch, S. 2013. Role of Cdc48/p97 as a SUMO-targeted segregase curbing Rad51–Rad52 interaction. *Nat Cell Biol* 15: 526–532.

Bett, J. S., Ritorto, M. S., Ewan, R., Jaffray, E. G., Virdee, S., Chin, J. W., Knebel, A. et al. 2015. Ubiquitin C-terminal hydrolases cleave isopeptide- and peptide-linked ubiquitin from structured proteins but do not edit ubiquitin homopolymers. *Biochem J* 466: 489–498.

Bhatia, S., Pavlick, A. C., Boasberg, P., Thompson, J. A., Mulligan, G., Pickard, M. D., Faessel, H., Dezube, B. J., Hamid, O. 2016. A phase I study of the investigational NEDD8-activating enzyme inhibitor pevonedistat (TAK-924/MLN4924) in patients with metastatic melanoma. *Invest New Drugs* 34: 439–449.

Bian, C., Wu, R., Cho, K., Yu, X. 2012. Loss of BRCA1-A complex function in RAP80 null tumor cells. *PLOS ONE* 7: e40406.

Blackledge, N. P., Farcas, A. M., Kondo, T., King, H. W., McGouran, J. F., Hanssen, L. L., Ito, S. et al. 2014. Variant PRC1 complex-dependent H2A ubiquitylation drives PRC2 recruitment and polycomb domain formation. *Cell* 157: 1445–1459.

Bohm, S., Mihalevic, M. J., Casal, M. A., Bernstein, K. A. 2015. Disruption of SUMO-targeted ubiquitin ligases Slx5-Slx8/RNF4 alters RecQ-like helicase Sgs1/BLM localization in yeast and human cells. *DNA Repair (Amst)* 26: 1–14.

Bologna, S., Altmannova, V., Valtorta, E., Koenig, C., Liberali, P., Gentili, C., Anrather, D. et al. 2015. Sumoylation regulates EXO1 stability and processing of DNA damage. *Cell Cycle* 14: 2439–2450.

Branigan, E., Plechanovova, A., Jaffray, E. G., Naismith, J. H., Hay, R. T. 2015. Structural basis for the RING-catalyzed synthesis of K63-linked ubiquitin chains. *Nat Struct Mol Biol* 22: 597–602.

Brown, J. S., Lukashchuk, N., Sczaniecka-Clift, M., Britton, S., le Sage, C., Calsou, P., Beli, P., Galanty, Y., Jackson, S. P. 2015. Neddylation promotes ubiquitylation and release of Ku from DNA-damage sites. *Cell Rep* 11: 704–714.

Brzovic, P. S., Rajagopal, P., Hoyt, D. W., King, M. C., Klevit, R. E. 2001. Structure of a BRCA1–BARD1 heterodimeric RING-RING complex. *Nat Struct Bioly* 8: 833–837.

Buchwald, G., van der Stoop, P., Weichenrieder, O., Perrakis, A., van Lohuizen, M., Sixma, T. K. 2006. Structure and E3-ligase activity of the Ring-Ring complex of polycomb proteins Bmi1 and Ring1b. *EMBO J* 25: 2465–2474.

Budenholzer, L., Cheng, C. L., Li, Y., Hochstrasser, M. 2017. Proteasome structure and assembly. *J Mol Biol* 429: 3500–3524.

Bunting, S. F., Callen, E., Wong, N., Chen, H. T., Polato, F., Gunn, A., Bothmer, A. et al. 2010. 53BP1 inhibits homologous recombination in Brca1-deficient cells by blocking resection of DNA breaks. *Cell* 141: 243–254.

Burma, S., Chen, B. P., Murphy, M., Kurimasa, A., Chen, D. J. 2001. ATM phosphorylates histone H2AX in response to DNA double-strand breaks. *J Biol Chem* 276: 42462–42467.

Butler, L. R., Densham, R. M., Jia, J., Garvin, A. J., Stone, H. R., Shah, V., Weekes, D., Festy, F., Beesley, J., Morris, J. R. 2012. The proteasomal de-ubiquitinating enzyme POH1 promotes the double-strand DNA break response. *EMBO J* 31: 3918–3934.

Campbell, S. J., Edwards, R. A., Leung, C. C., Neculai, D., Hodge, C. D., Dhe-Paganon, S., Glover, J. N. 2012. Molecular insights into the function of RING finger (RNF)-containing proteins hRNF8 and hRNF168 in Ubc13/Mms2-dependent ubiquitylation. *J Biol Chem* 287: 23900–23910.

Cappadocia, L., Pichler, A., Lima, C. D. 2015. Structural basis for catalytic activation by the human ZNF451 SUMO E3 ligase. *Nat Struct Mol Biol* 22: 968–975.

Carbone, M., Yang, H., Pass, H. I., Krausz, T., Testa, J. R., Gaudino, G. 2013. BAP1 and cancer. *Nat Rev Cancer* 13: 153–159.

Cashman, R., Cohen, H., Ben-Hamo, R., Zilberberg, A., Efroni, S. 2014. SENP5 mediates breast cancer invasion via a TGFbetaRI SUMOylation cascade. *Oncotarget* 5: 1071–1082.

Castilla, L. H., Couch, F. J., Erdos, M. R., Hoskins, K. F., Calzone, K., Garber, J. E., Boyd J. et al. 1994. Mutations in the BRCA1 gene in families with early-onset breast and ovarian cancer. *Nat Genet* 8: 387–391.

Chandrasekharan, M. B., Huang, F., Sun, Z. W. 2009. Ubiquitination of histone H2B regulates chromatin dynamics by enhancing nucleosome stability. *Proc Natl Acad Sci USA* 106: 16686–16691.

Chang, H. H. Y., Pannunzio, N. R., Adachi, N., Lieber, M. R. 2017. Non-homologous DNA end joining and alternative pathways to double-strand break repair. *Nat Rev Mol Cell Biol* 18: 495–506.

Chapman, J. R., Sossick, A. J., Boulton, S. J., Jackson, S. P. 2012. BRCA1-associated exclusion of 53BP1 from DNA damage sites underlies temporal control of DNA repair. *J Cell Sci* 125: 3529–3534.

Chatterjee, A., Saha, S., Chakraborty, A., Silva-Fernandes, A., Mandal, S. M., Neves-Carvalho, A., Liu, Y. et al. 2015. The role of the mammalian DNA end-processing enzyme polynucleotide kinase 3′-phosphatase in spinocerebellar ataxia type 3 pathogenesis. *PLoS Genet* 11: e1004749.

Chen, J., Feng, W., Jiang, J., Deng, Y., Huen, M. S. 2012a. Ring finger protein RNF169 antagonizes the ubiquitin dependent signaling cascade at sites of DNA damage. *J Biol Chem* 287: 27715–27722.

Chen, X., Cui, D., Papusha, A., Zhang, X., Chu, C. D., Tang, J., Chen, K., Pan, X., Ira, G. 2012b. The Fun30 nucleosome remodeller promotes resection of DNA double-strand break ends. *Nature* 489: 576–580.

Chen, Z., Sui, J., Zhang, F., Zhang, C. 2015. Cullin family proteins and tumorigenesis: Genetic association and molecular mechanisms. *J Cancer* 6: 233–242.

Cheng, J., Fan, Y. H., Xu, X., Zhang, H., Dou, J., Tang, Y., Zhong, X. et al. 2014. A small-molecule inhibitor of UBE2N induces neuroblastoma cell death via activation of p53 and JNK pathways. *Cell Death Dis* 5: e1079.

Cheng, J., Su, M., Jin, Y., Xi, Q., Deng, Y., Chen, J., Wang, W. et al. 2017. Upregulation of SENP3/SMT3IP1 promotes epithelial ovarian cancer progression and forecasts poor prognosis. *Tumour Biol* 39: 1010428317694543.

Chou, D. M., Adamson, B., Dephoure, N. E., Tan, X., Nottke, A. C., Hurov, K. E., Gygi, S. P., Colaiacovo, M. P., Elledge, S. J. 2010. A chromatin localization screen reveals poly (ADP ribose)-regulated recruitment of the repressive polycomb and NuRD complexes to sites of DNA damage. *Proc National Academy Sciences USA* 107: 18475–18480.

Chowdhury, M., Mihara, K., Yasunaga, S., Ohtaki, M., Takihara, Y., Kimura, A. 2007. Expression of Polycomb-group (PcG) protein BMI-1 predicts prognosis in patients with acute myeloid leukemia. *Leukemia* 21: 1116–22.

Chroma, K., Mistrik, M., Moudry, P., Gursky, J., Liptay, M., Strauss, R., Skrott, Z. et al. 2017. Tumors overexpressing RNF168 show altered DNA repair and responses to genotoxic treatments, genomic instability and resistance to proteotoxic stress. *Oncogene* 36: 2405–2422.

Coleman, K. A., Greenberg, R. A. 2011. The BRCA1-RAP80 complex regulates DNA repair mechanism utilization by restricting end resection. *J Biol Chem* 286: 13669–13680.

Cooper, E. M., Cutcliffe, C., Kristiansen, T. Z., Pandey, A., Pickart, C. M., Cohen, R. E. 2009. K63-specific deubiquitination by two JAMM/MPN+ complexes: BRISC-associated Brcc36 and proteasomal Poh1. *EMBO J* 28: 621–631.

Costelloe, T., Louge, R., Tomimatsu, N., Mukherjee, B., Martini, E., Khadaroo, B., Dubois, K. et al. 2012. The yeast Fun30 and human

SMARCAD1 chromatin remodellers promote DNA end resection. *Nature* 489: 581–584.
Danielsen, J. R., Povlsen, L. K., Villumsen, B. H., Streicher, W., Nilsson, J., Wikstrom, M., Bekker-Jensen, S., Mailand, N. 2012. DNA damage-inducible SUMOylation of HERC2 promotes RNF8 binding via a novel SUMO-binding Zinc finger. *J Cell Biol* 197: 179–187.
Dantuma, N. P., Acs, K., Luijsterburg, M. S. 2014. Should I stay or should I go: VCP/p97-mediated chromatin extraction in the DNA damage response. *Exper Cell Res* 329: 9–17.
Debelouchina, G. T., Gerecht, K., Muir, T. W. 2017. Ubiquitin utilizes an acidic surface patch to alter chromatin structure. *Nat Chem Biol* 13: 105–110.
Delgado-Diaz, M. R., Martin, Y., Berg, A., Freire, R., Smits, V. A. 2014. Dub3 controls DNA damage signalling by direct deubiquitination of H2AX. *Mol Oncol* 8: 884–893.
Densham, R. M., Garvin, A. J., Stone, H. R., Strachan, J., Baldock, R. A., Daza-Martin, M., Fletcher, A. et al. 2016. Human BRCA1-BARD1 ubiquitin ligase activity counteracts chromatin barriers to DNA resection. *Nat Struct Mol Biol* 23: 647–655.
Desterro, J. M., Thomson, J., Hay, R. T. 1997. Ubch9 conjugates SUMO but not ubiquitin. *FEBS Lett* 417: 297–300.
Dever, S. M., Golding, S. E., Rosenberg, E., Adams, B. R., Idowu, M. O., Quillin, J. M., Valerie, N., Xu, B., Povirk, L. F., Valerie, K. 2011. Mutations in the BRCT binding site of BRCA1 result in hyper-recombination. *Aging* 3: 515–532.
Diehl, C., Akke, M., Bekker-Jensen, S., Mailand, N., Streicher, W., Wikstrom, M. 2016. Structural analysis of a complex between small ubiquitin-like modifier 1 (SUMO1) and the ZZ domain of CREB-binding protein (CBP/p300) reveals a new interaction surface on SUMO. *J Biol Chem* 291: 12658–12672.
Dikic, I., Wakatsuki, S., Walters, K. J. 2009. Ubiquitin-binding domains–from structures to functions. *Nat Rev Mol Cell Biol* 10: 659–671.
Dimri, M., Kang, M., Dimri, G. P. 2016. A miR-200c/141-BMI1 autoregulatory loop regulates oncogenic activity of BMI1 in cancer cells. *Oncotarget* 7: 36220–36234.
Dlamini, N., Josifova, D. J., Paine, S. M., Wraige, E., Pitt, M., Murphy, A. J., King, A. et al. 2013. Clinical and neuropathological features of X-linked spinal muscular atrophy (SMAX2) associated with a novel mutation in the UBA1 gene. *Neuromuscul Disord: NMD* 23: 391–398.
Doil, C., Mailand, N., Bekker-Jensen, S., Menard, P., Larsen, D. H., Pepperkok, R., Ellenberg, J. et al. 2009. RNF168 binds and amplifies ubiquitin conjugates on damaged chromosomes to allow accumulation of repair proteins. *Cell* 136: 435–446.
Dong, M., Pang, X., Xu, Y., Wen, F., Zhang, Y. 2013. Ubiquitin-conjugating enzyme 9 promotes epithelial ovarian cancer cell proliferation *in vitro*. *Int J Mol Sci* 14: 11061–11071.
Dong, Y., Hakimi, M. A., Chen, X., Kumaraswamy, E., Cooch, N. S., Godwin, A. K., Shiekhattar, R. 2003. Regulation of BRCC, a holoenzyme complex containing BRCA1 and BRCA2, by a signalosome-like subunit and its role in DNA repair. *Mol Cell* 12: 1087–1099.
Doss-Pepe, E. W., Stenroos, E. S., Johnson, W. G., Madura, K. 2003. Ataxin-3 interactions with rad23 and valosin-containing protein and its associations with ubiquitin chains and the proteasome are consistent with a role in ubiquitin-mediated proteolysis. *Mol Cell Biol* 23: 6469–6483.
Dou, H., Huang, C., Singh, M., Carpenter, P. B., Yeh, E. T. 2010. Regulation of DNA repair through deSUMOylation and SUMOylation of replication protein A complex. *Mol Cell* 39: 333–345.
Drost, R., Bouwman, P., Rottenberg, S., Boon, U., Schut, E., Klarenbeek, S., Klijn, C. et al. 2011. BRCA1 RING function is essential for tumor suppression but dispensable for therapy resistance. *Cancer Cell* 20: 797–809.
Drost, R., Dhillon, K. K., van der Gulden, H., van der Heijden, I., Brandsma, I., Cruz, C., Chondronasiou, D. et al. 2016. BRCA1185delAG tumors may acquire therapy resistance through expression of RING-less BRCA1. *J Clin Invest* 126: 2903–2918.
Eisenhaber, B., Chumak, N., Eisenhaber, F., Hauser, M.-T. 2007. The ring between ring fingers (RBR) protein family. *Genome Biol* 8: 209.
Eisenhardt, N., Chaugule, V. K., Koidl, S., Droescher, M., Dogan, E., Rettich, J., Sutinen, P. et al. 2015. A new vertebrate SUMO enzyme family reveals insights into SUMO-chain assembly. *Nat Struct Mol Biol* 22: 959–967.
Eladad, S., Ye, T. Z., Hu, P., Leversha, M., Beresten, S., Matunis, M. J., Ellis, N. A. 2005. Intra-nuclear trafficking of the BLM helicase to DNA damage-induced foci is regulated by SUMO modification. *Hum Mol Genet* 14: 1351–1365.
Elia, A. E., Boardman, A. P., Wang, D. C., Huttlin, E. L., Everley, R. A., Dephoure, N., Zhou, C., Koren, I., Gygi, S. P., Elledge, S. J. 2015. Quantitative proteomic atlas of ubiquitination and acetylation in the DNA damage response. *Mol Cell* 59: 867–881.
Escribano-Diaz, C., Orthwein, A., Fradet-Turcotte, A., Xing, M., Young, J. T., Tkac, J., Cook, M. A. et al. 2013. A cell cycle-dependent regulatory circuit composed of 53BP1-RIF1 and BRCA1-CtIP controls DNA repair pathway choice. *Mol Cell* 49: 872–883.
Feng, L., Chen, J. 2012. The E3 ligase RNF8 regulates KU80 removal and NHEJ repair. *Nat Struct Mol Biol* 19: 201–206.
Ferretti, L. P., Himmels, S. F., Trenner, A., Walker, C., von Aesch, C., Eggenschwiler, A., Murina, O. et al. 2016. Cullin3-KLHL15 ubiquitin ligase mediates CtIP protein turnover to fine-tune DNA-end resection. *Nat Commun* 7: 12628.
Fierz, B., Chatterjee, C., McGinty, R. K., Bar-Dagan, M., Raleigh, D. P., Muir, T. W. 2011. Histone H2B ubiquitylation disrupts local and higher-order chromatin compaction. *Nat Chem Biol* 7: 113–119.
Fradet-Turcotte, A., Canny, M. D., Escribano-Diaz, C., Orthwein, A., Leung, C. C. Y., Huang, H., Landry, M. C. et al. 2013. 53BP1 is a reader of the DNA-damage-induced H2A Lys 15 ubiquitin mark. *Nature* 499: 50–54.
Franke, N. E., Niewerth, D., Assaraf, Y. G., van Meerloo, J., Vojtekova, K., van Zantwijk, C. H., Zweegman, S. et al. 2012. Impaired bortezomib binding to mutant beta 5 subunit of the proteasome is the underlying basis for bortezomib resistance in leukemia cells. *Leukemia* 26: 757–768.
Galanty, Y., Belotserkovskaya, R., Coates, J., Jackson, S. P. 2012. RNF4, a SUMO-targeted ubiquitin E3 ligase, promotes DNA double-strand break repair. *Genes Dev* 26: 1179–1195.
Galanty, Y., Belotserkovskaya, R., Coates, J., Polo, S., Miller, K. M., Jackson, S. P. 2009. Mammalian SUMO E3-ligases PIAS1 and PIAS4 promote responses to DNA double-strand breaks. *Nature* 462: 935–939.
Gallego, L. D., Ghodgaonkar Steger, M., Polyansky, A. A., Schubert, T., Zagrovic, B., Zheng, N., Clausen, T., Herzog, F., Kohler, A. 2016. Structural mechanism for the recognition and ubiquitination of a single nucleosome residue by Rad6-Bre1. *Proc Natl Acad Sci USA* 113: 10553–10558.
Gao, R., Liu, Y., Silva-Fernandes, A., Fang, X., Paulucci-Holthauzen, A., Chatterjee, A., Zhang, H. L. et al. 2015. Inactivation of PNKP by mutant ATXN3 triggers apoptosis by activating the DNA damage-response pathway in SCA3. *PLoS Genet* 11: e1004834.
Garcia-Dominguez, M., Reyes, J. C. 2009. SUMO association with repressor complexes, emerging routes for transcriptional control. *Biochim Biophys Acta* 1789: 451–459.
Garvin, A. J., Densham, R., Blair-Reid, S. A., Pratt, K. M., Stone, H. R., Weekes, D., Lawrence, K. J., Morris, J. R. 2013. The deSUMOylase SENP7 promotes chromatin relaxation for homologous recombination DNA repair. *Embo Reports* 14: 975–983.
Gatti, M., Pinato, S., Maiolica, A., Rocchio, F., Prato, M. G., Aebersold, R., Penengo, L. 2015. RNF168 promotes noncanonical K27 ubiquitination to signal DNA damage. *Cell Rep* 10: 226–238.
Germani, A., Prabel, A., Mourah, S., Podgorniak, M. P., Di Carlo, A., Ehrlich, R., Gisselbrecht, S., Varin-Blank, N., Calvo, F., Bruzzoni-Giovanelli, H. 2003. SIAH-1 interacts with CtIP and promotes its degradation by the proteasome pathway. *Oncogene* 22: 8845–8851.
Ginjala, V., Nacerddine, K., Kulkarni, A., Oza, J., Hill, S. J., Yao, M., Citterio, E., van Lohuizen, M., Ganesan, S. 2011. BMI1 is recruited to DNA breaks and contributes to DNA damage induced H2A ubiquitination and repair. *Mol Cell Biol* 31: 1972–1982.
Goldstein, G., Scheid, M., Hammerling, U., Boyse, E. A., Schlesinger, D. H., Niall, H. D. 1975. Isolation of a polypeptide that has

lymphocyte-differentiating properties and is probably represented universally in living cells. *Proc Natl Acad Sci USA* 72: 11–15.

Gong, L., Yeh, E. T. H. 2006. Characterization of a family of nucleolar SUMO-specific proteases with preference for SUMO-2 or SUMO-3. *J Biol Chem* 281: 15869–15877.

Goodarzi, A. A., Kurka, T., Jeggo, P. A. 2011. KAP-1 phosphorylation regulates CHD3 nucleosome remodeling during the DNA double-strand break response. *Nat Struct Mol Biol* 18: 831–839.

Groner, A. C., Meylan, S., Ciuffi, A., Zangger, N., Ambrosini, G., Denervaud, N., Bucher, P., Trono, D. 2010. KRAB-zinc finger proteins and KAP1 can mediate long-range transcriptional repression through heterochromatin spreading. *Plos Genetics* 6: e1000869.

Gudjonsson, T., Altmeyer, M., Savic, V., Toledo, L., Dinant, C., Grofte, M., Bartkova, J. et al. 2012. TRIP12 and UBR5 suppress spreading of chromatin ubiquitylation at damaged chromosomes. *Cell* 150: 697–709.

Guervilly, J. H., Takedachi, A., Naim, V., Scaglione, S., Chawhan, C., Lovera, Y., Despras, E. et al. 2015. The SLX4 complex is a SUMO E3 ligase that impacts on replication stress outcome and genome stability. *Mol Cell* 57: 123–137.

Gursoy-Yuzugullu, O., House, N., Price, B. D. 2016. Patching broken DNA: Nucleosome dynamics and the repair of DNA breaks. *J Mol Biol* 428: 1846–1860.

Guzzo, C. M., Berndsen, C. E., Zhu, J. M., Gupta, V., Datta, A., Greenberg, R. A., Wolberger, C., Matunis, M. J. 2012. RNF4-dependent hybrid SUMO-ubiquitin chains are signals for RAP80 and thereby mediate the recruitment of BRCA1 to sites of DNA damage. *Sci Signal* 5: ra88.

Guzzo, C. M., Ringel, A., Cox, E., Uzoma, I., Zhu, H., Blackshaw, S., Wolberger, C., Matunis, M. J. 2014. Characterization of the SUMO-binding activity of the myeloproliferative and mental retardation (MYM)-type zinc fingers in ZNF261 and ZNF198. *PLOS ONE* 9: e105271.

Hang, L. E., Lopez, C. R., Liu, X., Williams, J. M., Chung, I., Wei, L., Bertuch, A. A., Zhao, X. 2014. Regulation of Ku-DNA association by Yku70 C-terminal tail and SUMO modification. *J Biol Chem* 289: 10308–10317.

Harbour, J. W., Onken, M. D., Roberson, E. D., Duan, S., Cao, L., Worley, L. A., Council, M. L., Matatall, K. A., Helms, C., Bowcock, A. M. 2010. Frequent mutation of BAP1 in metastasizing uveal melanomas. *Science* 330: 1410–1413.

Hecker, C. M., Rabiller, M., Haglund, K., Bayer, P., Dikic, I. 2006. Specification of SUMO1- and SUMO2-interacting motifs. *J Biol Chem* 281: 16117–16127.

Helleday, T. 2011. The underlying mechanism for the PARP and BRCA synthetic lethality: Clearing up the misunderstandings. *Mol Oncol* 5: 387–393.

Hendriks, I. A., Lyon, D., Young, C., Jensen, L. J., Vertegaal, A. C., Nielsen, M. L. 2017. Site-specific mapping of the human SUMO proteome reveals co-modification with phosphorylation. *Nat Struct Mol Biol* 24: 325–336.

Hendriks, I. A., Schimmel, J., Eifler, K., Olsen, J. V., Vertegaal, A. C. 2015a. Ubiquitin-specific protease 11 (USP11) deubiquitinates hybrid small ubiquitin-like modifier (SUMO)-ubiquitin chains to counteract RING finger protein 4 (RNF4). *J Biol Chem* 290: 15526–15537.

Hendriks, I. A., Treffers, L. W., Verlaan-de Vries, M., Olsen, J. V., Vertegaal, A. C. 2015b. SUMO-2 orchestrates chromatin modifiers in response to DNA damage. *Cell Rep* S2211-1247(15)00179-5 10.1016/j.celrep.2015.02.033.

Hendriks, I. A., Vertegaal, A. C. 2016. A comprehensive compilation of SUMO proteomics. *Nat Rev Mol Cell Biol* 17: 581–595.

Hershko, A., Heller, H., Elias, S., Ciechanover, A. 1983. Components of ubiquitin-protein ligase system. Resolution, affinity purification, and role in protein breakdown. *J Biol Chem* 258: 8206–8014.

Himmels, S. F., Sartori, A. A. 2016. Controlling DNA-end resection: An emerging task for ubiquitin and SUMO. *Front Genet* 7: 152.

Hodge, C. D., Edwards, R. A., Markin, C. J., McDonald, D., Pulvino, M., Huen, M. S., Zhao, J., Spyracopoulos, L., Hendzel, M. J., Glover, J. N. 2015. Covalent inhibition of Ubc13 affects ubiquitin signaling and reveals active site elements important for targeting. *ACS Chem Biol* 10: 1718–1728.

Hu, X., Paul, A., Wang, B. 2012. Rap80 protein recruitment to DNA double-strand breaks requires binding to both small ubiquitin-like modifier (SUMO) and ubiquitin conjugates. *J Biol Chem* 287: 25510–25519.

Hu, Y., Scully, R., Sobhian, B., Xie, A., Shestakova, E., Livingston, D. M. 2011. RAP80-directed tuning of BRCA1 homologous recombination function at ionizing radiation-induced nuclear foci. *Genes Develt* 25: 685–700.

Husnjak, K., Dikic, I. 2012. Ubiquitin-binding proteins: Decoders of ubiquitin-mediated cellular functions. *Annu Rev Biochem* 81: 291–322.

Ismail, I. H., Andrin, C., McDonald, D., Hendzel, M. J. 2010. BMI1-mediated histone ubiquitylation promotes DNA double-strand break repair. *J Cell Biol* 191: 45–60.

Ismail, I. H., Davidson, R., Gagne, J. P., Xu, Z. Z., Poirier, G. G., Hendzel, M. J. 2014. Germline mutations in BAP1 impair its function in DNA double-strand break repair. *Cancer Res* 74: 4282–4294.

Ismail, I. H., Gagne, J. P., Caron, M. C., McDonald, D., Xu, Z., Masson, J. Y., Poirier, G. G., Hendzel, M. J. 2012. CBX4-mediated SUMO modification regulates BMI1 recruitment at sites of DNA damage. *Nucleic Acids Res* 40: 5497–5510.

Ismail, I. H., Gagne, J. P., Genois, M. M., Strickfaden, H., McDonald, D., Xu, Z., Poirier, G. G., Masson, J. Y., Hendzel, M. J. 2015. The RNF138 E3 ligase displaces Ku to promote DNA end resection and regulate DNA repair pathway choice. *Nat Cell Biol* 17: 1446–1457.

Ismail, I. H., McDonald, D., Strickfaden, H., Xu, Z., Hendzel, M. J. 2013. A small molecule inhibitor of polycomb repressive complex 1 inhibits ubiquitin signaling at DNA double-strand breaks. *J Biol Chem* 288: 26944–26954.

Ivanov, A. V., Peng, H., Yurchenko, V., Yap, K. L., Negorev, D. G., Schultz, D. C., Psulkowski, E. et al. 2007. PHD domain-mediated E3 ligase activity directs intramolecular sumoylation of an adjacent bromodomain required for gene silencing. *Mol Cell* 28: 823–837.

Jacquet, K., Fradet-Turcotte, A., Avvakumov, N., Lambert, J. P., Roques, C., Pandita, R. K., Paquet, E. et al. 2016. The TIP60 complex regulates bivalent chromatin recognition by 53BP1 through direct H4K20me binding and H2AK15 acetylation. *Mol Cell* 62: 409–421.

Jason, L. J., Finn, R. M., Lindsey, G., Ausio, J. 2005. Histone H2A ubiquitination does not preclude histone H1 binding, but it facilitates its association with the nucleosome. *J Biol Chem* 280: 4975–4982.

Jensen, D. E., Proctor, M., Marquis, S. T., Gardner, H. P., Ha, S. I., Chodosh, L. A., Ishov, A. M. et al. 1998. BAP1: A novel ubiquitin hydrolase which binds to the BRCA1 RING finger and enhances BRCA1-mediated cell growth suppression. *Oncogene* 16: 1097–1112.

Joo, H. Y., Zhai, L., Yang, C., Nie, S., Erdjument-Bromage, H., Tempst, P., Chang, C., Wang, H. 2007. Regulation of cell cycle progression and gene expression by H2A deubiquitination. *Nature* 449: 1068–1072.

Kagey, M. H., Melhuish, T. A., Wotton, D. 2003. The polycomb protein Pc2 is a SUMO E3. *Cell* 113: 127–137.

Kakarougkas, A., Ismail, A., Katsuki, Y., Freire, R., Shibata, A., Jeggo, P. A. 2013. Co-operation of BRCA1 and POH1 relieves the barriers posed by 53BP1 and RAP80 to resection. *Nucleic Acids Res* 41: 10298–10311.

Kalb, R., Latwiel, S., Irem Baymaz, H., Jansen, P. W. T. C., Mueller, C. W., Vermeulen, M., Mueller, J. 2014b. Histone H2A monoubiquitination promotes histone H3 methylation in Polycomb repression. *Nat Struct Mol Biol* 21: 569–571.

Kalb, R., Mallery, D. L., Larkin, C., Huang, J. T., Hiom, K. 2014a. BRCA1 is a histone-H2A-specific ubiquitin ligase. *Cell Rep* 8: 999–1005.

Kamitani, T., Kito, K., Nguyen, H. P., Fukuda-Kamitani, T., Yeh, E. T. H. 1998. Characterization of a second member of the sentrin family of ubiquitin-like proteins. *J Biol Chem* 273: 11349–11353.

Karvonen, U., Jaaskelainen, T., Rytinki, M., Kaikkonen, S., Palvimo, J. J. 2008. ZNF451 is a novel PML body- and SUMO-associated transcriptional coregulator. *J Mol Biol* 382: 585–600.

Kato, K., Nakajima, K., Ui, A., Muto-Terao, Y., Ogiwara, H., Nakada, S. 2014. Fine-tuning of DNA damage-dependent ubiquitination by OTUB2 supports the DNA repair pathway choice. *Molecular Cell* 53: 617–630.

Kawaguchi, Y., Okamoto, T., Taniwaki, M., Aizawa, M., Inoue, M., Katayama, S., Kawakami, H. et al. 1994. CAG expansions in a novel gene for Machado-Joseph disease at chromosome 14q32.1. *Nat Genet* 8: 221–228.

Kim, H., Chen, J., Yu, X. 2007. Ubiquitin-binding protein RAP80 mediates BRCA1-dependent DNA damage response. *Science* 316: 1202–1205.

Kim, R. Q., Sixma, T. K. 2017. Regulation of USP7: A high incidence of E3 complexes. *J Mol Biol* 429: 3395–3408.

Kim, W., Bennett, E. J., Huttlin, E. L., Guo, A., Li, J., Possemato, A., Sowa, M. E. et al. 2011. Systematic and quantitative assessment of the ubiquitin-modified proteome. *Mol Cell* 44: 325–340.

Komander, D., Barford, D. 2008. Structure of the A20 OTU domain and mechanistic insights into deubiquitination. *Biochem J* 409: 77–85.

Komander, D., Clague, M. J., Urbe, S. 2009. Breaking the chains: Structure and function of the deubiquitinases. *Nat Rev Mol Cell Biol* 10: 550–563.

Kreso, A., van Galen, P., Pedley, N. M., Lima-Fernandes, E., Frelin, C., Davis, T., Cao, L. et al. 2014. Self-renewal as a therapeutic target in human colorectal cancer. *Nat Med* 20: 29–36.

Krogan, N. J., Lam, M. H., Fillingham, J., Keogh, M. C., Gebbia, M., Li, J., Datta, N. et al. 2004. Proteasome involvement in the repair of DNA double-strand breaks. *Mol Cell* 16: 1027–1034.

Kumar, A., Zhang, K. Y. 2015. Advances in the development of SUMO specific protease (SENP) inhibitors. *Comput Struct Biotechnol J* 13: 204–211.

Kung, C. C., Naik, M. T., Wang, S. H., Shih, H. M., Chang, C. C., Lin, L. Y., Chen, C. L., Ma, C., Chang, C. F., Huang, T. H. 2014. Structural analysis of poly-SUMO chain recognition by the RNF4-SIMs domain. *Biochem J* 462: 53–65.

Kuo, C. Y., Li, X., Kong, X. Q., Luo, C., Chang, C. C., Chung, Y. Y., Shih, H. M., Li, K. K., Ann, D. K. 2014. An arginine-rich motif of ring finger protein 4 (RNF4) oversees the recruitment and degradation of the phosphorylated and SUMOylated Kruppel-associated box domain-associated protein 1 (KAP1)/TRIM28 protein during genotoxic stress. *J Biol Chem* 289: 20757–20772.

Lafranchi, L., de Boer, H. R., de Vries, E. G., Ong, S. E., Sartori, A. A., van Vugt, M. A. 2014. APC/C (Cdh1) controls CtIP stability during the cell cycle and in response to DNA damage. *Embo J* 33: 2860–2879.

Lawson, A. P., Bak, D. W., Shannon, D. A., Long, M. J. C., Vijaykumar, T., Yu, R., El Oualid, F., Weerapana, E., Hedstrom, L. 2017. Identification of deubiquitinase targets of isothiocyanates using SILAC-assisted quantitative mass spectrometry. *Oncotarget* 8: 51296–51316.

Lecona, E., Rodriguez-Acebes, S., Specks, J., Lopez-Contreras, A. J., Ruppen, I., Murga, M., Munoz, J., Mendez, J., Fernandez-Capetillo, O. 2016. USP7 is a SUMO deubiquitinase essential for DNA replication. *Nat Struct Mol Biol* 23: 270–277.

Lee, H. J., Li, C. F., Ruan, D., Powers, S., Thompson, P. A., Frohman, M. A., Chan, C. H. 2016. The DNA damage transducer RNF8 facilitates cancer chemoresistance and progression through twist activation. *Mol Cell* 63: 1021–1033.

Lee, J. H., Paull, T. T. 2005. ATM activation by DNA double-strand breaks through the Mre11-Rad50-Nbs1 complex. *Science* 308: 551–554.

Leung, J. W., Agarwal, P., Canny, M. D., Gong, F., Robison, A. D., Finkelstein, I. J., Durocher, D., Miller, K. M. 2014. Nucleosome acidic patch promotes RNF168- and RING1B/BMI1-dependent H2AX and H2A ubiquitination and DNA damage signaling. *PLoS Genet* 10: e1004178.

Leung, J. W., Makharashvili, N., Agarwal, P., Chiu, L. Y., Pourpre, R., Cammarata, M. B., Cannon, J. R. et al. 2017. ZMYM3 regulates BRCA1 localization at damaged chromatin to promote DNA repair. *Genes Dev* 31: 260–274.

Li, M., Cole, F., Patel, D. S., Misenko, S. M., Her, J., Malhowski, A., Alhamza, A. et al. 2016. 53BP1 ablation rescues genomic instability in mice expressing 'RING-less' BRCA1. *EMBO Rep* 17: 1532–1541.

Li, Y., Luo, K., Yin, Y., Wu, C., Deng, M., Li, L., Chen, Y., Nowsheen, S., Lou, Z., Yuan, J. 2017. USP13 regulates the RAP80-BRCA1 complex dependent DNA damage response. *Nat Commun* 8: 15752.

Li, Y. J., Stark, J. M., Chen, D. J., Ann, D. K., Chen, Y. 2010. Role of SUMO:SIM-mediated protein-protein interaction in non-homologous end joining. *Oncogene* 29: 3509–3518.

Liang, F. S., Longerich, S., Miller, A. S., Tang, C., Buzovetsky, O., Xiong, Y., Maranon, D. G., Wiese, C., Kupfer, G. M., Sung, P. 2016. Promotion of RAD51-mediated homologous DNA pairing by the RAD51AP1-UAF1 complex. *Cell Rep* 15: 2118–2126.

Liew, C. W., Sun, H., Hunter, T., Day, C. L. 2010. RING domain dimerization is essential for RNF4 function. *Biochem J* 431: 23–29.

Liu, H., Zhang, H., Wang, X., Tian, Q., Hu, Z., Peng, C., Jiang, P. et al. 2015. The deubiquitylating enzyme USP4 cooperates with CtIP in DNA double-strand break end resection. *Cell Rep* 13: 93–107.

Liu, J., Cao, L., Chen, J., Song, S., Lee, I. H., Quijano, C., Liu, H. et al. 2009. Bmi1 regulates mitochondrial function and the DNA damage response pathway. *Nature* 459: 387–92.

Liu, J., Doty, T., Gibson, B., Heyer, W. D. 2010. Human BRCA2 protein promotes RAD51 filament formation on RPA-covered single-stranded DNA. *Nat Struct Mol Biol* 17: 1260–1262.

Luchini, C., Veronese, N., Yachida, S., Cheng, L., Nottegar, A., Stubbs, B., Solmi, M. et al. 2016. Different prognostic roles of tumor suppressor gene BAP1 in cancer: A systematic review with meta-analysis. *Genes Chromosomes Cancer* 55: 741–749.

Luijsterburg, M. S., Typas, D., Caron, M. C., Wiegant, W. W., van den Heuvel, D., Boonen, R. A., Couturier, A. M., Mullenders, L. H., Masson, J. Y., van Attikum, H. 2017. A PALB2-interacting domain in RNF168 couples homologous recombination to DNA break-induced chromatin ubiquitylation. *Elife* 6: e20922 .

Luo, K., Zhang, H., Wang, L., Yuan, J., Lou, Z. 2012. Sumoylation of MDC1 is important for proper DNA damage response. *EMBO J* 31: 3008–3019.

Luo, K. T., Li, L., Li, Y. H., Wu, C. M., Yin, Y. J., Chen, Y. P., Deng, M., Nowsheen, S., Yuan, J., Lou, Z. K. 2016. A phosphorylation-deubiquitination cascade regulates the BRCA2-RAD51 axis in homologous recombination. *Genes Dev* 30: 2581–2595.

Ma, C., Wu, B., Huang, X., Yuan, Z., Nong, K., Dong, B., Bai, Y., Zhu, H., Wang, W., Ai, K. 2014. SUMO-specific protease 1 regulates pancreatic cancer cell proliferation and invasion by targeting MMP-9. *Tumour Biol* 35: 12729–12735.

Maison, C., Romeo, K., Bailly, D., Dubarry, M., Quivy, J. P., Almouzni, G. 2012. The SUMO protease SENP7 is a critical component to ensure HP1 enrichment at pericentric heterochromatin. *Nat Struct Mol Biol* 19: 458–460.

Makharashvili, N., Tubbs, A. T., Yang, S. H., Wang, H., Barton, O., Zhou, Y., Deshpande, R. A. et al. 2014. Catalytic and noncatalytic roles of the CtIP endonuclease in double-strand break end resection. *Mol Cell* 54: 1022–1033.

Mallery, D. L., Vandenberg, C. J., Hiom, K. 2002. Activation of the E3 ligase function of the BRCA1/BARD1 complex by polyubiquitin chains. *Embo J* 21: 6755–6762.

Mattiroli, F., Uckelmann, M., Sahtoe, D. D., van Dijk, W. J., Sixma, T. K. 2014. The nucleosome acidic patch plays a critical role in RNF168-dependent ubiquitination of histone H2A. *Nat Commun* 5: 3291.

Mattiroli, F., Vissers, J. H. A., van Dijk, W. J., Ikpa, P., Citterio, E., Vermeulen, W., Marteijn, J. A., Sixma, T. K. 2012. RNF168 ubiquitinates K13-15 on H2A/H2AX to drive DNA damage signaling. *Cell* 150: 1182–1195.

Matunis, M. J., Coutavas, E., Blobel, G. 1996. A novel ubiquitin-like modification modulates the partitioning of the Ran-GTPase-activating protein RanGAP1 between the cytosol and the nuclear pore complex. *J Cell Biol* 135: 1457–1470.

McGinty, R. K., Henrici, R. C., Tan, S. 2014. Crystal structure of the PRC1 ubiquitylation module bound to the nucleosome. *Nature* 514: 591–596.

Meerang, M., Ritz, D., Paliwal, S., Garajova, Z., Bosshard, M., Mailand, N., Janscak, P., Hubscher, U., Meyer, H., Ramadan, K. 2011. The ubiquitin-selective segregase VCP/p97 orchestrates the response to DNA double-strand breaks. *Nature Cell Biol* 13: 1376–1382.

Melo-Cardenas, J., Zhang, Y., Zhang, D. D., Fang, D. 2016. Ubiquitin-specific peptidase 22 functions and its involvement in disease. *Oncotarget* 7: 44848–44856.

Mendes, A. V., Grou, C. P., Azevedo, J. E., Pinto, M. P. 2016. Evaluation of the activity and substrate specificity of the human SENP family of SUMO proteases. *Biochim Biophys Acta* 1863: 139–1347.

Meulmeester, E., Kunze, M., Hsiao, H. H., Urlaub, H., Melchior, F. 2008. Mechanism and consequences for paralog-specific sumoylation of ubiquitin-specific protease 25. *Mol Cell* 30: 610–619.

Mevissen, T. E. T., Komander, D. 2017. Mechanisms of deubiquitinase specificity and regulation. *Annu Rev Biochem* 86: 159–192.

Miskinyte, S., Butler, M. G., Herve, D., Sarret, C., Nicolino, M., Petralia, J. D., Bergametti, F. et al. 2011. Loss of BRCC3 deubiquitinating enzyme leads to abnormal angiogenesis and is associated with syndromic moyamoya. *Am J Hum Genet* 88: 718–728.

Moreau, P., Masszi, T., Grzasko, N., Bahlis, N. J., Hansson, M., Pour, L., Sandhu, I. et al. 2016. Oral ixazomib, lenalidomide, and dexamethasone for multiple myeloma. *N Engl J Med* 374: 1621–1634.

Morgan, M. T., Haj-Yahya, M., Ringel, A. E., Bandi, P., Brik, A., Wolberger, C. 2016. Structural basis for histone H2B deubiquitination by the SAGA DUB module. *Science* 351: 725–728.

Morris, J. R., Boutell, C., Keppler, M., Densham, R., Weekes, D., Alamshah, A., Butler, L. et al. 2009. The SUMO modification pathway is involved in the BRCA1 response to genotoxic stress. *Nature* 462: 886–890.

Mosbech, A., Lukas, C., Bekker-Jensen, S., Mailand, N. 2013. The deubiquitylating enzyme USP44 counteracts the DNA double-strand break response mediated by the RNF8 and RNF168 ubiquitin ligases. *J Biol Chem* 288: 16579–16587.

Moudry, P., Lukas, C., Macurek, L., Hanzlikova, H., Hodny, Z., Lukas, J., Bartek, J. 2012. Ubiquitin-activating enzyme UBA1 is required for cellular response to DNA damage. *Cell Cycle* 11: 1573–1582.

Moyal, L., Lerenthal, Y., Gana-Weisz, M., Mass, G., So, S., Wang, S. Y., Eppink, B. et al. 2011. Requirement of ATM-dependent monoubiquitylation of histone H2B for timely repair of DNA double-strand breaks. *Mol Cell* 41: 529–542.

Mukhopadhyay, D., Dasso, M. 2007. Modification in reverse: The SUMO proteases. *Trends Biochem Sci* 32: 286–295.

Murakawa, Y., Sonoda, E., Barber, L. J., Zeng, W., Yokomori, K., Kimura, H., Niimi, A. et al. 2007. Inhibitors of the proteasome suppress homologous DNA recombination in mammalian cells. *Cancer Res* 67: 8536–8543.

Murray, J. M., Stiff, T., Jeggo, P. A. 2012. DNA double-strand break repair within heterochromatic regions. *Biochem Soc Trans* 40: 173–178.

Nakada, S., Tai, I., Panier, S., Al-Hakim, A., Iemura, S., Juang, Y. C., O'Donnell, L. et al. 2010. Non-canonical inhibition of DNA damage-dependent ubiquitination by OTUB1. *Nature* 466: 941–946.

Nakagawa, T., Kajitani, T., Togo, S., Masuko, N., Ohdan, H., Hishikawa, Y., Koji, T. et al. 2008. Deubiquitylation of histone H2A activates transcriptional initiation via trans-histone cross-talk with H3K4 di- and trimethylation. *Genes Dev* 22: 37–49.

Nakamura, K., Kato, A., Kobayashi, J., Yanagihara, H., Sakamoto, S., Oliveira, D. V. N. P., Shimada, M. et al. 2011. Regulation of homologous recombination by RNF20-dependent H2B ubiquitination. *Mol Cell* 41: 515–528.

Nathan, D., Ingvarsdottir, K., Sterner, D. E., Bylebyl, G. R., Dokmanovic, M., Dorsey, J. A., Whelan, K. A. et al. 2006. Histone sumoylation is a negative regulator in *Saccharomyces cerevisiae* and shows dynamic interplay with positive-acting histone modifications. *Genes Dev* 20: 966–976.

Neyret-Kahn, H., Benhamed, M., Ye, T., Le Gras, S., Cossec, J. C., Lapaquette, P., Bischof, O. et al. 2013. Sumoylation at chromatin governs coordinated repression of a transcriptional program essential for cell growth and proliferation. *Genome Res* 23: 1563–1579.

Ng, H. M., Wei, L. Z., Lan, L., Huen, M. S. Y. 2016. The Lys (63)-deubiquitylating enzyme BRCC36 limits DNA break processing and repair. *J Biol Chem* 291: 16197–16207.

Nickel, B. E., Davie, J. R. 1989. Structure of polyubiquitinated histone H2A. *Biochemistry* 28: 964–968.

Nie, M. H., Aslanian, A., Prudden, J., Heideker, J., Vashisht, A. A., Wohlschlegel, J. A., Yates, J. R., Boddy, M. N. 2012. Dual recruitment of Cdc48 (p97)-Ufd1-Npl4 ubiquitin-selective segregase by small ubiquitin-like modifier protein (SUMO) and ubiquitin in SUMO-targeted ubiquitin ligase-mediated genome stability functions. *J Biol Chem* 287: 29610–29619.

Nie, M., Boddy, M. N. 2016. Cooperativity of the SUMO and ubiquitin pathways in genome stability. *Biomolecules* 6: 14.

Nikkila, J., Coleman, K. A., Morrissey, D., Pylkas, K., Erkko, H., Messick, T. E., Karppinen, S. M., Amelina, A., Winqvist, R., Greenberg, R. A. 2009. Familial breast cancer screening reveals an alteration in the RAP80 UIM domain that impairs DNA damage response function. *Oncogene* 28: 1843–1852.

Nishi, R., Wijnhoven, P., le Sage, C., Tjeertes, J., Galanty, Y., Forment, J. V., Clague, M. J., Urbe, S., Jackson, S. P. 2014. Systematic characterization of deubiquitylating enzymes for roles in maintaining genome integrity. *Nat Cell Biol* 16: 1016–1026. 1–8.

Nishikawa, H., Wu, W., Koike, A., Kojima, R., Gomi, H., Fukuda, M., Ohta, T. 2009. BRCA1-associated protein 1 interferes with BRCA1/BARD1 RING heterodimer activity. *Cancer Res* 69: 111–119.

Ochs, F., Somyajit, K., Altmeyer, M., Rask, M. B., Lukas, J., Lukas, C. 2016. 53BP1 fosters fidelity of homology-directed DNA repair. *Nat Struct Mol Biol* 23: 714–721.

Oerlemans, R., Franke, N. E., Assaraf, Y. G., Cloos, J., van Zantwijk, I., Berkers, C. R., Scheffer, G. L. et al. 2008. Molecular basis of bortezomib resistance: Proteasome subunit beta 5 (PSMB5) gene mutation and overexpression of PSMB5 protein. *Blood* 112: 2489–2499.

Ohtake, F., Saeki, Y., Sakamoto, K., Ohtake, K., Nishikawa, H., Tsuchiya, H., Ohta, T., Tanaka, K., Kanno, J. 2015. Ubiquitin acetylation inhibits polyubiquitin chain elongation. *EMBO Rep* 16: 192–201.

Oliveira, D. V., Kato, A., Nakamura, K., Ikura, T., Okada, M., Kobayashi, J., Yanagihara, H., Saito, Y., Tauchi, H., Komatsu, K. 2014. Histone chaperone FACT regulates homologous recombination by chromatin remodeling through interaction with RNF20. *J Cell Sci* 127: 763–772.

Orthwein, A., Noordermeer, S. M., Wilson, M. D., Landry, S., Enchev, R. I., Sherker, A., Munro, M. et al. 2015. A mechanism for the suppression of homologous recombination in G1 cells. *Nature* 528: 422–426.

Ouyang, J., Garner, E., Hallet, A., Nguyen, H. D., Rickman, K. A., Gill, G., Smogorzewska, A., Zou, L. 2015. Noncovalent interactions with SUMO and ubiquitin orchestrate distinct functions of the SLX4 complex in genome maintenance. *Mol Cell* 57: 108–122.

Ouyang, K. J., Woo, L. L., Zhu, J., Huo, D., Matunis, M. J., Ellis, N. A. 2009. SUMO modification regulates BLM and RAD51 interaction at damaged replication forks. *PLoS Biol* 7: e1000252.

Owerbach, D., McKay, E. M., Yeh, E. T., Gabbay, K. H., Bohren, K. M. 2005. A proline-90 residue unique to SUMO-4 prevents maturation and sumoylation. *Biochem Biophys Res Commun* 337: 517–520.

Pan, M. R., Peng, G., Hung, W. C., Lin, S. Y. 2011. Monoubiquitination of H2AX protein regulates DNA damage response signaling. *J Biol Chem* 286: 28599–28607.

Panier, S., Ichijima, Y., Fradet-Turcotte, A., Leung, C. C., Kaustov, L., Arrowsmith, C. H., Durocher, D. 2012. Tandem protein interaction modules organize the ubiquitin-dependent response to DNA double-strand breaks. *Mol Cell* 47: 383–395.

Patterson-Fortin, J., Shao, G., Bretscher, H., Messick, T. E., Greenberg, R. A. 2010. Differential regulation of JAMM domain deubiquitinating enzyme activity within the RAP80 complex. *J Biol Chem* 285: 30971–30981.

Paul, A., Wang, B. 2017. RNF8- and Ube2S-dependent ubiquitin lysine 11-linkage modification in response to DNA damage. *Mol Cell* 66: 458–472. e5.

Pena-Llopis, S., Vega-Rubin-de-Celis, S., Liao, A., Leng, N., Pavia-Jimenez, A., Wang, S., Yamasaki, T. et al. 2012. BAP1 loss defines a new class of renal cell carcinoma. *Nat Genet* 44: 751–759.

Pennington, K. P., Walsh, T., Harrell, M. I., Lee, M. K., Pennil, C. C., Rendi, M. H., Thornton, A. et al. 2014. Germline and somatic mutations in homologous recombination genes predict platinum response and survival in ovarian, fallopian tube, and peritoneal carcinomas. *Clin Cancer Res* 20: 764–7675.

Pfeiffer, A., Luijsterburg, M. S., Acs, K., Wiegant, W. W., Helfricht, A., Herzog, L. K., Minoia, M. et al. 2017. Ataxin-3 consolidates the MDC1-dependent DNA double-strand break response by counteracting the SUMO-targeted ubiquitin ligase RNF4. *Embo J* 36: 1066–1083.

Pichler, A., Fatouros, C., Lee, H., Eisenhardt, N. 2017. SUMO conjugation - A mechanistic view. *Biomol Concepts* 8: 13–36.
Plechanovova, A., Jaffray, E. G., Tatham, M. H., Naismith, J. H., Hay, R. T. 2012. Structure of a RING E3 ligase and ubiquitin-loaded E2 primed for catalysis. *Nature* 489: 115–U35.
Postow, L. 2011. Destroying the ring: Freeing DNA from Ku with ubiquitin. *FEBS Lett* 585: 2876–2882.
Postow, L., Funabiki, H. 2013. An SCF complex containing Fbxl12 mediates DNA damage-induced Ku80 ubiquitylation. *Cell Cycle* 12: 587–595.
Postow, L., Ghenoiu, C., Woo, E. M., Krutchinsky, A. N., Chait, B. T., Funabiki, H. 2008. Ku80 removal from DNA through double strand break-induced ubiquitylation. *J Cell Biol* 182: 467–479.
Potts, P. R., Yu, H. 2005. Human MMS21/NSE2 is a SUMO ligase required for DNA repair. *Mol Cell Biol* 25: 7021–7032.
Poulsen, M., Lukas, C., Lukas, J., Bekker-Jensen, S., Mailand, N. 2012. Human RNF169 is a negative regulator of the ubiquitin-dependent response to DNA double-strand breaks. *J Cell Biol* 197: 189–199.
Pruneda, J. N., Littlefield, P. J., Soss, S. E., Nordquist, K. A., Chazin, W. J., Brzovic, P. S., Klevit, R. E. 2012. Structure of an E3:E2~Ub complex reveals an allosteric mechanism shared among RING/U-box ligases. *Mol Cell* 47: 933–942.
Psakhye, I., Jentsch, S. 2012. Protein group modification and synergy in the SUMO pathway as exemplified in DNA repair. *Cell* 151: 807–820.
Puffenberger, E. G., Jinks, R. N., Wang, H., Xin, B., Fiorentini, C., Sherman, E. A., Degrazio, D. et al. 2012. A homozygous missense mutation in HERC2 associated with global developmental delay and autism spectrum disorder. *Hum Mutat* 33: 1639–1646.
Qiao, Z., Wang, W., Wang, L., Wen, D., Zhao, Y., Wang, Q., Meng, Q., Chen, G., Wu, Y., Zhou, H. 2011. Design, synthesis, and biological evaluation of benzodiazepine-based SUMO-specific protease 1 inhibitors. *Bioorg Med Chem Lett* 21: 6389–6392.
Rabellino, A., Andreani, C., Scaglioni, P. P. 2017. The role of PIAS SUMO E3-ligases in cancer. *Cancer Res* 77: 1542–1547.
Rai, K., Pilarski, R., Cebulla, C. M., Abdel-Rahman, M. H. 2016. Comprehensive review of BAP1 tumor predisposition syndrome with report of two new cases. *Clin Genet* 89: 285–294.
Ramachandran, S., Haddad, D., Li, C., Le, M. X., Ling, A. K., So, C. C., Nepal, R. M. et al. 2016. The SAGA deubiquitination module promotes DNA repair and class switch recombination through ATM and DNAPK-mediated gammaH2AX formation. *Cell Rep* 15: 1554–1565.
Ramadan, K., Bruderer, R., Spiga, F. M., Popp, O., Baur, T., Gotta, M., Meyer, H. H. 2007. Cdc48/p97 promotes reformation of the nucleus by extracting the kinase Aurora B from chromatin. *Nature* 450: 1258–1262.
Richly, H., Rape, M., Braun, S., Rumpf, S., Hoege, C., Jentsch, S. 2005. A series of ubiquitin binding factors connects CDC48/p97 to substrate multiubiquitylation and proteasomal targeting. *Cell* 120: 73–84.
Rowbotham, S. P., Barki, L., Neves-Costa, A., Santos, F., Dean, W., Hawkes, N., Choudhary, P. et al. 2011. Maintenance of silent chromatin through replication requires SWI/SNF-like chromatin remodeler SMARCAD1. *Mol Cell* 42: 285–296.
Ruffner, H., Joazeiro, C. A., Hemmati, D., Hunter, T., Verma, I. M. 2001. Cancer-predisposing mutations within the RING domain of BRCA1:loss of ubiquitin protein ligase activity and protection from radiation hypersensitivity. *Proc Natl Acad Sci USA* 98: 5134–5139.
Ruschak, A. M., Slassi, M., Kay, L. E., Schimmer, A. D. 2011. Novel proteasome inhibitors to overcome bortezomib resistance. *J Natl Cancer Inst* 103: 1007–1017.
Ryan, R. F., Schultz, D. C., Ayyanathan, K., Singh, P. B., Friedman, J. R., Fredericks, W. J., Rauscher 3rd, F. J. 1999. KAP-1 corepressor protein interacts and colocalizes with heterochromatic and euchromatic HP1 proteins: A potential role for Kruppel-associated box-zinc finger proteins in heterochromatin-mediated gene silencing. *Mol Cell Biol* 19: 4366–4378.
Sahasrabuddhe, A. A. 2016. BMI1: A biomarker of hematologic malignancies. *Biomark Cancer* 8: 65–75.
Sahtoe, D. D., van Dijk, W. J., Ekkebus, R., Ovaa, H., Sixma, T. K. 2016. BAP1/ASXL1 recruitment and activation for H2A deubiquitination. *Nat Commun* 7: 10292.
Sampson, D. A., Wang, M., Matunis, M. J. 2001. The small ubiquitin-like modifier-1 (SUMO-1) consensus sequence mediates Ubc9 binding and is essential for SUMO-1 modification. *J Biol Chem* 276: 21664–21669.
Sanchez, A., De Vivo, A., Uprety, N., Kim, J., Stevens, S. M. Jr., Kee, Y. 2016. BMI1-UBR5 axis regulates transcriptional repression at damaged chromatin. *Proc Natl Acad Sci USA* 113: 11243–11248.
Sarangi, P., Steinacher, R., Altmannova, V., Fu, Q., Paull, T. T., Krejci, L., Whitby, M. C., Zhao, X. 2015. Sumoylation influences DNA break repair partly by increasing the solubility of a conserved end resection protein. *PLoS Genet* 11: e1004899.
Sarantopoulos, J., Shapiro, G. I., Cohen, R. B., Clark, J. W., Kauh, J. S., Weiss, G. J., Cleary, J. M. et al. 2016. Phase I study of the investigational NEDD8-activating enzyme inhibitor pevonedistat (TAK-924/MLN4924) in patients with advanced solid tumors. *Clin Cancer Res* 22: 847–857.
Sartori, A. A., Lukas, C., Coates, J., Mistrik, M., Fu, S., Bartek, J., Baer, R., Lukas, J., Jackson, S. P. 2007. Human CtIP promotes DNA end resection. *Nature* 450: 509–514.
Sato, Y., Yoshikawa, A., Mimura, H., Yamashita, M., Yamagata, A., Fukai, S. 2009. Structural basis for specific recognition of Lys 63-linked polyubiquitin chains by tandem UIMs of RAP80. *EMBO J* 28: 2461–2468.
Sato, Y., Yoshikawa, A., Yamagata, A., Mimura, H., Yamashita, M., Ookata, K., Nureki, O., Iwai, K., Komada, M., Fukai, S. 2008. Structural basis for specific cleavage of Lys 63-linked polyubiquitin chains. *Nature* 455: 358–362.
Saudy, N. S., Fawzy, M. M., Azmy, E., Coda, E. F., Eneen, A., Salam, E. M. A. 2014. BMI1 gene expression in myeloid leukemias and its impact on prognosis. *Blood Cells Mol Dis* 53: 194–198.
Scheuermann, J. C., Gutierrez, L., Muller, J. 2012. Histone H2A monoubiquitination and Polycomb repression: The missing pieces of the puzzle. *Fly (Austin)* 6: 162–168.
Schimmel, J., Larsen, K. M., Matic, I., van Hagen, M., Cox, J., Mann, M., Andersen, J. S., Vertegaal, A. C. 2008. The ubiquitin-proteasome system is a key component of the SUMO-2/3 cycle. *Mol Cell Proteomics* 7: 2107–2122.
Schmidt, C. K., Galanty, Y., Sczaniecka-Clift, M., Coates, J., Jhujh, S., Demir, M., Cornwell, M., Beli, P., Jackson, S. P. 2015. Systematic E2 screening reveals a UBE2D-RNF138-CtIP axis promoting DNA repair. *Nat Cell Biol* 17: 1458–1470.
Schulman, B. A., Harper, J. W. 2009. Ubiquitin-like protein activation by E1 enzymes: The apex for downstream signalling pathways. *Nat Rev Mol Cell Biol* 10: 319–331.
Schulz, S., Chachami, G., Kozaczkiewicz, L., Winter, U., Stankovic-Valentin, N., Haas, P., Hofmann, K. et al. 2012. Ubiquitin-specific protease-like 1 (USPL1) is a SUMO isopeptidase with essential, non-catalytic functions. *Embo Reports* 13: 930–938.
Sevcik, J., Falk, M., Macurek, L., Kleiblova, P., Lhota, F., Hojny, J., Stefancikova, L. et al. 2013. Expression of human BRCA1Delta17-19 alternative splicing variant with a truncated BRCT domain in MCF-7 cells results in impaired assembly of DNA repair complexes and aberrant DNA damage response. *Cell Signal* 25: 1186–1193.
Shah, J. J., Jakubowiak, A. J., O'Connor, O. A., Orlowski, R. Z., Harvey, R. D., Smith, M. R., Lebovic, D. et al. 2016. Phase I study of the novel investigational NEDD8-activating enzyme inhibitor pevonedistat (MLN4924) in patients with relapsed/refractory multiple myeloma or lymphoma. *Clin Cancer Res* 22: 34–43.
Shakya, R., Reid, L. J., Reczek, C. R., Cole, F., Egli, D., Lin, C. S., deRooij, D. G. et al. 2011. BRCA1 tumor suppression depends on BRCT phosphoprotein binding, but not its E3 ligase activity. *Science* 334: 525–528.
Shao, G., Lilli, D. R., Patterson-Fortin, J., Coleman, K. A., Morrissey, D. E., Greenberg, R. A. 2009. The Rap80-BRCC36 de-ubiquitinating enzyme complex antagonizes RNF8-Ubc13-dependent ubiquitination events at DNA double strand breaks. *Proc Natl Acad Sci USA* 106: 3166–31671.
Sharma, N., Zhu, Q., Wani, G., He, J., Wang, Q.-E., Wani, A. A. 2014. USP3 counteracts RNF168 via deubiquitinating H2A and gamma H2AX at lysine 13 and 15. *Cell Cycle* 13: 106–114.
Shen, L. N., Geoffroy, M. C., Jaffray, E. G., Hay, R. T. 2009. Characterization of SENP7, a SUMO-2/3-specific isopeptidase. *Biochem J* 421: 223–230.

Shen, Z., Pardington-Purtymun, P. E., Comeaux, J. C., Moyzis, R. K., Chen, D. J. 1996. UBL1, a human ubiquitin-like protein associating with human RAD51/RAD52 proteins. *Genomics* 36: 271–279.

Shim, E. Y., Hong, S. J., Oum, J. H., Yanez, Y., Zhang, Y., Lee, S. E. 2007. RSC mobilizes nucleosomes to improve accessibility of repair machinery to the damaged chromatin. *Mol Cell Biol* 27: 1602–1613.

Shima, H., Suzuki, H., Sun, J. Y., Kono, K., Shi, L., Kinomura, A., Horikoshi, Y. et al. 2013. Activation of the SUMO modification system is required for the accumulation of RAD51 at sites of DNA damage. *J Cell Sci* 126: 5284–5292.

Shin, E. J., Shin, H. M., Nam, E., Kim, W. S., Kim, J. H., Oh, B. H., Yun, Y. 2012. DeSUMOylating isopeptidase: A second class of SUMO protease. *Embo Rep* 13: 339–346.

Shire, K., Wong, A. I., Tatham, M. H., Anderson, O. F., Ripsman, D., Gulstene, S., Moffat, J., Hay, R. T., Frappier, L. 2016. Identification of RNF168 as a PML nuclear body regulator. *J Cell Sci* 129: 580–591.

Sims, J. J., Cohen, R. E. 2009. Linkage-specific avidity defines the lysine 63-linked polyubiquitin-binding preference of rap80. *Mol Cell* 33: 775–783.

Sobhian, B., Shao, G., Lilli, D. R., Culhane, A. C., Moreau, L. A., Xia, B., Livingston, D. M., Greenberg, R. A. 2007. RAP80 targets BRCA1 to specific ubiquitin structures at DNA damage sites. *Science* 316: 1198–1202.

Sohn, S. Y., Hearing, P. 2012. Adenovirus regulates sumoylation of Mre11-Rad50-Nbs1 components through a paralog-specific mechanism. *J Virol* 86: 9656–9665.

Solyom, S., Aressy, B., Pylkas, K., Patterson-Fortin, J., Hartikainen, J. M., Kallioniemi, A., Kauppila, S. et al. 2012. Breast cancer-associated Abraxas mutation disrupts nuclear localization and DNA damage response functions. *Sci Transl Med* 4: 122ra23.

Stehmeier, P., Muller, S. 2009. Phospho-regulated SUMO interaction modules connect the SUMO system to CK2 signaling. *Mol Cell* 33: 400–409.

Stewart, G. S., Panier, S., Townsend, K., Al-Hakim, A. K., Kolas, N. K., Miller, E. S., Nakada, S. et al. 2009. The RIDDLE syndrome protein mediates a ubiquitin-dependent signaling cascade at sites of DNA damage. *Cell* 136: 420–434.

Stewart, G. S., Stankovic, T., Byrd, P. J., Wechsler, T., Miller, E. S., Huissoon, A., Drayson, M. T., West, S. C., Elledge, S. J., Taylor, A. M. 2007. RIDDLE immunodeficiency syndrome is linked to defects in 53BP1-mediated DNA damage signaling. *Proc Natl Acad Sci USA* 104: 16910–16915.

Stewart, G. S., Wang, B., Bignell, C. R., Taylor, A. M., Elledge, S. J. 2003. MDC1 is a mediator of the mammalian DNA damage checkpoint. *Nature* 421: 961–966.

Stone, H. R., Morris, J. R. 2013. DNA damage emergency: Cellular garbage disposal to the rescue? *Oncogene* 33(7): 805–813.

Streich, F. C. Jr., Lima, C. D. 2016. Capturing a substrate in an activated RING E3/E2-SUMO complex. *Nature* 536: 304–308.

Su, V., Lau, A. F. 2009. Ubiquitin-like and ubiquitin-associated domain proteins: Significance in proteasomal degradation. *Cell Mol Life Sci CMLS* 66: 2819–2833.

Sun, H., Leverson, J. D., Hunter, T. 2007. Conserved function of RNF4 family proteins in eukaryotes: Targeting a ubiquitin ligase to SUMOylated proteins. *Embo J* 26: 4102–4012.

Sun, Y., Jiang, X., Xu, Y., Ayrapetov, M. K., Moreau, L. A., Whetstine, J. R., Price, B. D. 2009. Histone H3 methylation links DNA damage detection to activation of the tumour suppressor Tip60. *Nat Cell Biol* 11: 1376–1382.

Swaney, D. L., Beltrao, P., Starita, L., Guo, A., Rush, J., Fields, S., Krogan, N. J., Villen, J. 2013. Global analysis of phosphorylation and ubiquitylation cross-talk in protein degradation. *Nat Methods* 10: 676–682.

Sy, S. M., Huen, M. S., Chen, J. 2009. PALB2 is an integral component of the BRCA complex required for homologous recombination repair. *Proc Natl Acad Sci USA* 106: 7155–7160.

Sy, S. M., Jiang, J., Deng, W. S. O. Y., Huen, M. S. 2013. The ubiquitin specific protease USP34 promotes ubiquitin signaling at DNA double-strand breaks. *Nucleic Acids Res* 41: 8572–8580.

Symington, L. S. 2016. Mechanism and regulation of DNA end resection in eukaryotes. *Crit Rev Biochem Mol Biol* 51: 195–212.

Tammsalu, T., Matic, I., Jaffray, E. G., Ibrahim, A. F., Tatham, M. H., Hay, R. T. 2014. Proteome-wide identification of SUMO2 modification sites. *Science Signaling* 7: rs2.

Tatham, M. H., Geoffroy, M. C., Shen, L., Plechanovova, A., Hattersley, N., Jaffray, E. G., Palvimo, J. J., Hay, R. T. 2008. RNF4 is a poly-SUMO-specific E3 ubiquitin ligase required for arsenic-induced PML degradation. *Nat Cell Biol* 10: 538–546.

Tatham, M. H., Jaffray, E., Vaughan, O. A., Desterro, J. M., Botting, C. H., Naismith, J. H., Hay, R. T. 2001. Polymeric chains of SUMO-2 and SUMO-3 are conjugated to protein substrates by SAE1/SAE2 and Ubc9. *J Biol Chem* 276: 35368–35374.

Tatham, M. H., Kim, S., Jaffray, E., Song, J., Chen, Y., Hay, R. T. 2005. Unique binding interactions among Ubc9, SUMO and RanBP2 reveal a mechanism for SUMO paralog selection. *Nat Struct Mol Biol* 12: 67–74.

Testa, J. R., Cheung, M., Pei, J., Below, J. E., Tan, Y., Sementino, E., Cox, N. J. et al. 2011. Germline BAP1 mutations predispose to malignant mesothelioma. *Nat Genet* 43: 1022–1025.

Thorslund, T., Ripplinger, A., Hoffmann, S., Wild, T., Uckelmann, M., Villumsen, B., Narita, T. et al. 2015. Histone H1 couples initiation and amplification of ubiquitin signalling after DNA damage. *Nature* 527: 389–393.

Todi, S. V., Winborn, B. J., Scaglione, K. M., Blount, J. R., Travis, S. M., Paulson, H. L. 2009. Ubiquitination directly enhances activity of the deubiquitinating enzyme ataxin-3. *Embo J* 28: 372–382.

Typas, D., Luijsterburg, M. S., Wiegant, W. W., Diakatou, M., Helfricht, A., Thijssen, P. E., van den Broek, B., Mullenders, L. H., van Attikum, H. 2015. The de-ubiquitylating enzymes USP26 and USP37 regulate homologous recombination by counteracting RAP80. *Nucleic Acids Res* 43: 6919–6933.

Uchimura, Y., Ichimura, T., Uwada, J., Tachibana, T., Sugahara, S., Nakao, M., Saitoh, H. 2006. Involvement of SUMO modification in MBD1- and MCAF1-mediated heterochromatin formation. *J Biol Chem* 281: 23180–23190.

Uckelmann, M., Sixma, T. K. 2017. Histone ubiquitination in the DNA damage response. *DNA Repair (Amst)* 56: 92–101.

Ungermannova, D., Parker, S. J., Nasveschuk, C. G., Wang, W., Quade, B., Zhang, G., Kuchta, R. D., Phillips, A. J., Liu, X. 2012. Largazole and its derivatives selectively inhibit ubiquitin activating enzyme (e1). *PLOS ONE* 7: e29208.

van Attikum, H., Fritsch, O., Hohn, B., Gasser, S. M. 2004. Recruitment of the INO80 complex by H2A phosphorylation links ATP-dependent chromatin remodeling with DNA double-strand break repair. *Cell* 119: 777–788.

van den Boom, J., Wolf, M., Weimann, L., Schulze, N., Li, F., Kaschani, F., Riemer, A. et al. 2016. VCP/p97 extracts sterically trapped Ku70/80 Rings from DNA in double-strand break repair. *Mol Cell* 64:189–198.

van der Veen, A. G., Ploegh, H. L. 2012. Ubiquitin-like proteins. *Annu Rev Biochem* 81: 323–357.

Vassilev, L. T., Vu, B. T., Graves, B., Carvajal, D., Podlaski, F., Filipovic, Z., Kong, N. et al. 2004. *In vivo* activation of the p53 pathway by small-molecule antagonists of MDM2. *Science* 303: 844–848.

Verma, R., L. Aravind, R. Oania, W. H. McDonald, J. R. Yates, 3rd, E. V. Koonin, R. J. Deshaies. 2002. Role of Rpn11 metalloprotease in deubiquitination and degradation by the 26S proteasome. *Science* 298:611–615.

Vissers, J. H., Nicassio, F., van Lohuizen, M., Di Fiore, P. P., Citterio, E. 2008. The many faces of ubiquitinated histone H2A: Insights from the DUBs. *Cell Div* 3: 8.

Vyas, R., Kumar, R., Clermont, F., Helfricht, A., Kalev, P., Sotiropoulou, P., Hendriks, I. A. et al. 2013. RNF4 is required for DNA double-strand break repair *in vivo*. *Cell Death Differ* 20: 490–502.

Walker, J. R., Corpina, R. A., Goldberg, J. 2001. Structure of the Ku heterodimer bound to DNA and its implications for double-strand break repair. *Nature* 412: 607–614.

Wang, B., Elledge, S. J. 2007. Ubc13/Rnf8 ubiquitin ligases control foci formation of the Rap80/Abraxas/Brca1/Brcc36 complex in response to DNA damage. *Proc Natl Inst Sci USA* 104: 20759–20763.

Wang, B., Matsuoka, S., Ballif, B. A., Zhang, D., Smogorzewska, A., Gygi, S. P., Elledge, S. J. 2007. Abraxas and RAP80 form a BRCA1 protein complex required for the DNA damage response. *Science* 316: 1194–1198.

Wang, H., Wang, L., Erdjument-Bromage, H., Vidal, M., Tempst, P., Jones, R. S., Zhang, Y. 2004. Role of histone H2A ubiquitination in Polycomb silencing. *Nature* 431: 873–878.

Wang, Y., Krais, J. J., Bernhardy, A. J., Nicolas, E., Cai, K. Q., Harrell, M. I., Kim, H. H. et al. 2016a. RING domain-deficient BRCA1 promotes PARP inhibitor and platinum resistance. *J Clin Invest* 126: 3145–3157.

Wang, Z. J., Zhu, L. L., Guo, T. J., Wang, Y. P., Yang, J. L. 2015. Decreased H2B monoubiquitination and overexpression of ubiquitin-specific protease enzyme 22 in malignant colon carcinoma. *Human Pathol* 46: 1006–1014.

Wang, Z., Zhang, H., Liu, J., Cheruiyot, A., Lee, J. H., Ordog, T., Lou, Z., You, Z., Zhang, Z. 2016b. USP51 deubiquitylates H2AK13,15ub and regulates DNA damage response. *Genes Dev* 30: 946–959.

Watanabe, S., Watanabe, K., Akimov, V., Bartkova, J., Blagoev, B., Lukas, J., Bartek, J. 2013. JMJD1C demethylates MDC1 to regulate the RNF8 and BRCA1-mediated chromatin response to DNA breaks. *Nature Struct Mol Biol* 20: 1425–1433.

Watts, F. Z. 2016. Repair of DNA double-strand breaks in heterochromatin. *Biomolecules* 6: 47.

Weissman, A. M. 2001. Themes and variations on ubiquitylation. *Nat Rev Mol Cell Biol* 2: 169–178.

Werner, A., Flotho, A., Melchior, F. 2012. The RanBP2/RanGAP1* SUMO1/Ubc9 complex is a multisubunit SUMO E3 ligase. *Mol Cell* 46: 287–298.

White, D. E., Negorev, D., Peng, H., Ivanov, A. V., Maul, G. G., Rauscher 3rd, F. J. 2006. KAP1, a novel substrate for PIKK family members, colocalizes with numerous damage response factors at DNA lesions. *Cancer Research* 66: 11594–11599.

Wijnhoven, P., Konietzny, R., Blackford, A. N., Travers, J., Kessler, B. M., Nishi, R., Jackson, S. P. 2015. USP4 auto-deubiquitylation promotes homologous recombination. *Mol Cell* 60: 362–373.

Wilson, M. D., Benlekbir, S., Fradet-Turcotte, A., Sherker, A., Julien, J. P., McEwan, A., Noordermeer, S. M., Sicheri, F., Rubinstein, J. L., Durocher, D. 2016. The structural basis of modified nucleosome recognition by 53BP1. *Nature* 536: 100–103.

Wu, C. Y., Kang, H. Y., Yang, W. L., Wu, J., Jeong, Y. S., Wang, J., Chan, C. H. et al. 2011. Critical role of monoubiquitination of histone H2AX protein in histone H2AX phosphorylation and DNA damage response. *J Biol Chem* 286: 30806–30815.

Wu, J., Lei, H., Zhang, J., Chen, X., Tang, C., Wang, W., Xu, H., Xiao, W., Gu, W., Wu, Y. 2016. Momordin Ic, a new natural SENP1 inhibitor, inhibits prostate cancer cell proliferation. *Oncotarget* 7: 58995–9005.

Yan, J., Yang, X. P., Kim, Y. S., Joo, J. H., Jetten, A. M. 2007. RAP80 interacts with the SUMO-conjugating enzyme UBC9 and is a novel target for sumoylation. *Biochem Biophys Res Commun* 362: 132–138.

Yang, K., Guo, R., Xu, D. 2016. Non-homologous end joining: Advances and frontiers. *Acta Biochim Biophys Sin (Shanghai)* 48: 632–640.

Yang, K. L., Moldovan, G. L., Vinciguerra, P., Murai, J., Takeda, S., D'Andrea, A. D. 2011. Regulation of the Fanconi anemia pathway by a SUMO-like delivery network. *Genes Dev* 25: 1847–1858.

Yang, Y., Kitagaki, J., Wang, H., Hou, D. X., Perantoni, A. O. 2009. Targeting the ubiquitin-proteasome system for cancer therapy. *Cancer Sci* 100: 24–28.

Yau, R., Rape, M. 2016. The increasing complexity of the ubiquitin code. *Nat Cell Biol* 18: 579–586.

Yin, Y. L., Seifert, A., Chua, J. S., Maure, J. F., Golebiowski, F., Hay, R. T. 2012. SUMO-targeted ubiquitin E3 ligase RNF4 is required for the response of human cells to DNA damage. *Genes Dev* 26: 1196–1208.

Yu, H., Pak, H., Hammond-Martel, I., Ghram, M., Rodrigue, A., Daou, S., Barbour, H. et al. 2014. Tumor suppressor and deubiquitinase BAP1 promotes DNA double-strand break repair. *Proc Natl Inst Sci USA* 111: 285–290.

Yu, X., Chen, J. 2004. DNA damage-induced cell cycle checkpoint control requires CtIP, a phosphorylation-dependent binding partner of BRCA1 C-terminal domains. *Mol Cell Biol* 24: 9478–9486.

Yu, X., Fu, S., Lai, M., Baer, R., Chen, J. 2006. BRCA1 ubiquitinates its phosphorylation-dependent binding partner CtIP. *Genes Dev* 20: 1721–1726.

Yuan, J., Luo, K., Deng, M., Li, Y., Yin, P., Gao, B., Fang, Y., Wu, P., Liu, T., Lou, Z. 2014. HERC2-USP20 axis regulates DNA damage checkpoint through Claspin. *Nucleic Acids Res* 42: 13110–13121.

Yun, M. H., Hiom, K. 2009. CtIP-BRCA1 modulates the choice of DNA double-strand-break repair pathway throughout the cell cycle. *Nature* 459: 460–463.

Yurchenko, V., Xue, Z., Gama, V., Matsuyama, S., Sadofsky, M. J. 2008. Ku70 is stabilized by increased cellular SUMO. *Biochem Biophys Res Commun* 366: 263–268.

Yurchenko, V., Xue, Z., Sadofsky, M. J. 2006. SUMO modification of human XRCC4 regulates its localization and function in DNA double-strand break repair. *Mol Cell Biol* 26: 1786–1794.

Zhang, F., Ma, J., Wu, J., Ye, L., Cai, H., Xia, B., Yu, X. 2009. PALB2 links BRCA1 and BRCA2 in the DNA-damage response. *Curr Biol* 19: 524–529.

Zhang, X. Y., Pfeiffer, H. K., Thorne, A. W., McMahon, S. B. 2008. USP22, an hSAGA subunit and potential cancer stem cell marker, reverses the polycomb-catalyzed ubiquitylation of histone H2A. *Cell Cycle* 7: 1522–1524.

Zhang, Z., Yang, H., Wang, H. 2014. The histone H2A deubiquitinase USP16 interacts with HERC2 and fine-tunes cellular response to DNA damage. *J Biol Chem* 289: 32883–32894.

Zhen, Y., Knobel, P. A., Stracker, T. H., Reverter, D. 2014. Regulation of USP28 deubiquitinating activity by SUMO conjugation. *J Biol Chem* 289: 34838–34850.

Zhu, B., Zheng, Y., Pham, A. D., Mandal, S. S., Erdjument-Bromage, H., Tempst, P., Reinberg, D. 2005. Monoubiquitination of human histone H2B: The factors involved and their roles in HOX gene regulation. *Mol Cell* 20: 601–611.

Zhu, M., Zhao, H., Liao, J., Xu, X. 2014. HERC2/USP20 coordinates CHK1 activation by modulating CLASPIN stability. *Nucleic Acids Res* 42: 13074–13081.

Zhu, P., Zhou, W., Wang, J., Puc, J., Ohgi, K. A., Erdjument-Bromage, H., Tempst, P., Glass, C. K., Rosenfeld, M. G. 2007. A histone H2A deubiquitinase complex coordinating histone acetylation and H1 dissociation in transcriptional regulation. *Mol Cell* 27: 609–621.

Zhu, Q., Pao, G. M., Huynh, A. M., Suh, H., Tonnu, N., Nederlof, P. M., Gage, F. H., Verma, I. M. 2011. BRCA1 tumour suppression occurs via heterochromatin-mediated silencing. *Nature* 477: 179–184.

Zhu, Q., Sharma, N., He, J., Wani, G., Wani, A. A. 2015. USP7 deubiquitinase promotes ubiquitin-dependent DNA damage signaling by stabilizing RNF168. *Cell Cycle* 14: 1413–1425.

Zhu, S., Sachdeva, M., Wu, F., Lu, Z., Mo, Y. Y. 2010. Ubc9 promotes breast cell invasion and metastasis in a sumoylation-independent manner. *Oncogene* 29: 1763–1772.

Zimmermann, M., F. Lottersberger, S. B. Buonomo, A. Sfeir, T. de Lange. 2013. 53BP1 regulates DSB repair using Rif1 to control 5′ end resection. *Science* 339(6120): 700–704.

Ziv, Y., Bielopolski, D., Galanty, Y., Lukas, C., Taya, Y., Schultz, D. C., Lukas, J., Bekker-Jensen, S., Bartek, J., Shiloh, Y. 2006. Chromatin relaxation in response to DNA double-strand breaks is modulated by a novel ATM- and KAP-1 dependent pathway. *Nat Cell Biol* 8: 870–876.

Transcription in the Context of Genome Stability Maintenance

10

Marco Saponaro

Transcription is the process through which cells express the information encoded in their genomes, producing molecules of ribonucleic acid (RNA). Cells actually produce multiple types of RNA molecules, thanks to the activity of a series of specialised RNA polymerase machineries, each specifically responsible for producing a subset of all the transcripts:

- RNA polymerase I (RNAPI) transcribes the rRNA (5.8S, 18S and 28S in mammals).
- RNA polymerase II (RNAPII) transcribes mRNA, lncRNA, microRNA, piRNA and most of the snRNA and snoRNA.
- RNA polymerase III (RNAPIII) transcribes tRNA, 5S rRNA and the remaining snRNA and snoRNA.
- RNA polymerase IV and V (RNAPIV and RNAPV), found only in plant cells, orchestrate non-coding RNA-mediated gene silencing and appear to have evolved as specialised forms of RNAPII (Tucker et al. 2010).

These RNA molecules can be required either directly for the synthesis of proteins (mRNA, rRNA and tRNA), or be non-coding RNA (ncRNA), indicating that they are not translated into a protein nor required for general protein synthesis:

rRNAs (ribosomal RNA) have structural and functional roles in ribosomal protein organisation and ribosome activity (Ban et al. 2000). The large ribosome subunit contains three named rRNAs in mammals (5S, 5.8S and 28S), while the small ribosome subunit contains one molecule named 18S. The rRNAs 5.8S, 18S and 28S are transcribed as a unique pre-rRNA that is then processed to produce the different molecules; moreover, they are arranged in rDNA clusters, with 35–175 copies of each unit per cluster, mapped in humans on the short arms of the acrocentric chromosomes (chromosomes 13, 14, 15, 21 and 22) (Stults et al. 2008). The rRNA 5S is instead present in a single location on chromosome 1, arranged in a cluster of, on average, approximately 100 repeats (Stults et al. 2008).

mRNA (messenger RNA) is produced from protein-coding genes and serve as a template for translation. Recent estimates indicate that the human genome contains approximately 20,000 protein-coding genes (Ezkurdia et al. 2014), although the actual number of transcripts produced is exponentially increased by variations in transcription start sites, transcription termination sites and the specific composition of the mature mRNA thanks to alternative splicing events (Djebali et al. 2012, Nellore et al. 2016). The average length of protein-coding genes is 53.6 kb, with a wide range from a few hundred bases (several histone genes) to 2.4 megabases (DMD gene).

tRNAs (transfer RNA, 79–90 nucleotides long) are responsible for bringing amino acids to the ribosome complex, with a single amino acid specifically bound to each tRNA 3′ end. As a tRNA pairs its three-nucleotide anti-codon to the three-nucleotide codon present in the mRNA, it is able to align a precise amino acid to the specific codon in the mRNA, translating the nucleotide sequence of the mRNA into the amino acid sequence of the protein.

ncRNAs are divided in multiple subtypes based on functions and length:

- miRNAs (micro RNA, 21–24 nucleotides long) are involved in RNA silencing and post-transcriptional regulation of gene expression (He and Hannon 2004).
- piRNAs (Piwi-interacting RNA, 26–31 nucleotides long) form RNA-protein complexes with Piwi proteins, important for epigenetic and post-transcriptional gene silencing of retrotransposons and other genetic elements (Czech and Hannon 2016).

- snRNAs (small nuclear RNA, up to 150 nucleotides long) are involved in the processing of pre-mRNA and regulation of transcription factors and of the RNAPII (Valadkhan and Gunawardane 2013).
- snoRNAs (small nucleolar RNA, 80 to 1000 nucleotides) are important to guide precise nucleotide modifications in rRNAs, tRNAs and snRNAs (Dupuis-Sandoval et al. 2015).
- lncRNAs (long non-coding RNA, >200 nucleotides up to approximately 1 Mb) have been implicated in the regulation of gene expression, splicing, translation, imprinting, X-chromosome inactivation and telomere stability (Quinn and Chang 2016).

Apart from these transcripts, it has been shown that regulatory elements, like enhancers, are also transcribed (eRNA, enhancer RNA), although it is not really clear whether all eRNA have specific direct roles in regulating the transcription of their target genes (Li et al. 2016), in addition antisense transcription is generally produced on the transcription start sites and transcription termination sites when transcripts are produced (Core et al. 2008). As there is such a broad range of transcripts that any higher eukaryotic cell can produce, it becomes difficult to accurately define how much of each species' genome is capable of actually being transcribed. This is because the transcription profiles, in particular for higher eukaryotic cells, are dictated both by cell identity and by the multiplicity of internal and external stimuli that each cell receives. Nevertheless, by combining large-scale RNA-Seq studies in multiple cell lines, it was shown that at least up to 75% of the human genome is capable of being transcribed (Djebali et al. 2012). We do not know whether all these transcripts have functional roles in cells, or whether part of these transcripts might be produced due to mistakes in recognising appropriate transcription regulatory elements, leading to the erroneous recruitment of the RNA polymerase complexes on non-transcripts codifying regions. Importantly, however, the majority of these transcripts appear to be processed (spliced and/or poly-adenylated), features typically associated with RNAPII transcripts, suggesting that the RNAPII might be transcribing genomes more globally beyond the 3% of the human genome that, for example, encodes for protein-coding genes (Djebali et al. 2012).

All these data clearly give only a limited idea of the complexity of the transcription process, with hundreds of the factors involved working in opposing directions, to activate or repress it, throughout the majority of the genome. There are considerable variations in these effects with different cell types and throughout the lifetime of the cell. Therefore, it comes as no surprise that transcription can directly impact other processes that may concomitantly occur in our genome, such as chromatin remodelling, DNA damage repair and DNA replication. In this chapter we will focus on how transcription is directly involved or can influence if not antagonise the latter two processes.

TRANSCRIPTION AND DNA DAMAGE REPAIR

Transcriptional response to DNA damage

The cellular response to DNA damage involves a broad change in the transcription programs of cells, with many genes and transcripts up- or down-regulated to support the cellular response to the insult, but also with changes in the stability of specific transcripts (Venkata Narayanan et al. 2017). This has been known for a considerable time as, for example, many transcription factors are targeted by post-translational modifications such as phosphorylation by the DNA damage checkpoint kinases ATM and ATR (Matsuoka et al. 2007, Elia et al. 2015). However, more recently it has become evident that transcription is also finely tuned in response to DNA damage. For example, it has been shown that in response to multiple DNA damaging treatments, BRCA1 interacts with BCLAF1 in an ATM/ATR phosphorylation–dependent manner (Savage et al. 2014). BCLAF1 is a transcription factor shown to be involved in Bcl2 signalling and mRNA splicing that can interact with γH2AX (Merz et al. 2007, Sarras et al. 2010, Lee et al. 2012). In response to DNA damage, BRCA1 and BCLAF1 are recruited together to a series of DNA damage response genes, regulating the splicing and hence the expression of the transcribed mRNA, in order to maintain adequate protein expression levels in DNA damaging conditions (Savage et al. 2014). However, BRCA1 has also a more global impact on transcription and genome stability, as BRCA1 is a RNAPII transcription associated factor able to associate with actively transcribing RNAPII and regulating NF-kB responsive genes expression (Gardini et al. 2014). Moreover, it is involved in the cellular response to transcription-associated DNA damage, through the interaction with a series of transcription factors like SETX, FACT and TCEA2; BRCA1 is also required for the correct recruitment of SETX to termination sites to disassemble R-loops for the correct termination of the transcribing RNAPII (Hill et al. 2014,

Hatchi et al. 2015). An R-loop is a three-stranded nucleic acid structure, whereby the paired molecules are a strand of DNA and its complementary RNA, with the non-template single-stranded DNA displaced. Their impact on genome stability maintenance will be described in further details later in the chapter.

Transcription regulation in the context of DNA damage

Transcription regulation in response to double-strand breaks

DNA damage also has a direct impact on RNA polymerase progression, affecting transcription in some cases only in *cis* in the close proximity of the DNA damage site, and in other cases more globally, depending on the type of lesion present but also on the specific RNA polymerase involved. In the case of double-strand breaks, for example, we observe a global shutdown only of RNAPI transcription due to an ATM-dependent chromatin reorganisation (Kruhlak et al. 2007, Harding et al. 2015). With regard to RNAPII, the shutdown involves specifically only the transcripts that contain the lesion (Iacovoni et al. 2010). Also in this case, because of an ATM-dependent chromatin reorganisation through the phosphorylation of PBAF and the requirement of the E3 ubiquitin ligases RNF8 and RNF168, this leads to the ubiquitylation of the histone H2A, a modification commonly involved in transcription repression and chromatin condensation (Shanbhag et al. 2010, Kakarougkas et al. 2014). (Further details of other chromatin modifications occurring at sites of double-strand breaks are described in Chapter 9 of this book.) More recently, it has been shown that the presence of double-strand breaks leads to a downregulation of pre-existing transcription through reduced transcription initiation, in a process that depends, in this case, on the DNA damage checkpoint kinase ATM (Iannelli et al. 2017). Finally, transcription is also important for regulating the choice of the double-strand break repair pathway, homologous recombination or non-homologous end joining, as recent findings indicate that transcribed regions are preferentially repaired via error-free homologous recombination, while non-transcribed ones are preferentially repaired by the more mutagenic non-homologous end joining. Potentially, this has great importance as to how cells manage to preserve the genetic information encoded in their genomes (Aymard et al. 2014).

Transcription regulation in response to ultraviolet damage

In the case of ultraviolet (UV)-induced DNA damage the lesions generated impair RNA polymerase progression globally, with a characteristic shutdown of transcript production that recovers over time, although a considerable number of stress responsive transcripts are actually induced (Christians and Hanawalt 1993, Proietti-De-Santis et al. 2006, Williamson et al. 2017). On one hand, when the RNAPII encounters such lesions, it directly stimulates the recruitment of nucleotide excision repair factors at the site of damage (described later in more detail). On the other hand, the presence of these lesions leads to a great global slowdown of its progression (Munoz et al. 2009, Williamson et al. 2017). This slowdown of RNAPII elongation rates affects the splicing efficiency of the transcribed mRNA: it changes the rates with which alternative exons are included or excluded from the mRNA, whether introns are retained in the mRNA, and whether the transcript includes alternative last exons, changing in this way the actual termination site used for the mRNA (Munoz et al. 2009, Dujardin et al. 2014, Andrade-Lima et al. 2015, Williamson et al. 2017). Alternative splicing occurs as a direct consequence of RNAPII's capability to progress through the damaged chromatin and therefore is affected by the cell's ability to repair the DNA lesions, controlled by the DNA damage checkpoint kinase ATR (Andrade-Lima et al. 2015, Munoz et al. 2017, Williamson et al. 2017). Importantly, one noteworthy thing is that the inclusion or exclusion of alternative spliced exons can change the sequence of a specific protein, with potential dramatic effects on its activity and roles, as in the case of the pro-apoptotic factor Bcl-X (Taylor et al. 1999). In parallel, the inclusion of alternative last exons can generate shorter or longer transcripts that may have gained or lost important functional domains (Williamson et al. 2017). Intriguingly, the inclusion of an alternative last exon in the case of the *ASCC3* gene generates a shorter transcript that no longer functions as a protein-coding mRNA, but instead as an lncRNA, with *ASCC3* the first example of a gene that can encode for both mRNA and lncRNA (Williamson et al. 2017). This shorter isoform of *ASCC3* is important for the cellular response to UV damage, and in particular for the recovery of RNAPII transcription after damage (Williamson et al. 2017). These recent findings indicate, therefore, that the slowdown of the RNAPII in the presence of UV damage–induced DNA lesions, and consequently with that all the changes that occur in mRNA composition with alternative splicing events and the preferential expression of shorter isoforms instead of full-length ones, are part of a well-established and functional process that supports the cellular response to DNA damage. The complexity of this transcriptional response, and its relevance for cells' ability to recover from DNA damage, demonstrates how cells can potentially use the transcription machinery to sense and properly react to DNA lesions. How much this is a cellular response specific for UV damage (or UV-like induced DNA lesions, perhaps), or whether this can be observed also in other DNA damaging conditions, represents a future research challenge that needs to be undertaken.

Transcription in support of double-strand break repair

As mentioned, the cellular response to double-strand breaks includes a global shutdown of rDNA transcription and a local shutdown in *cis* for RNAPII transcribed genes. However, in recent years, it has become more evident that there is a direct contribution of transcription to double-strand break repair. The first indication of such a role came from the finding that 53BP1 foci are sensitive to RNase treatment (Pryde et al. 2005). Specifically, treatment with RNAseA but not RNAseH (hence degradation of RNA and not dismantling of R-loops) before fixation of the cells for immunofluorescence reduces 53BP1 signal at sites of double-strand breaks (Pryde et al. 2005). This result indicates a structural role for RNA at sites of DNA breaks, required for the correct recruitment and establishment of 53BP1 foci. In addition to this, it was shown that RNA is actually synthesised at the site of double-strand breaks and that it is this site-specific RNA that is required for the correct formation of 53BP1 foci, with the involvement of the RNA interference machinery in this process (Francia et al. 2012, Francia et al. 2016). More recently, a role for RNA produced directly at the site of double-strand break was also presented in the budding yeast *S. cerevisiae*. In this case, it was shown how, following DNA resection but before the loading of RPA, RNAPII is loaded on the double-strand break sites to produce RNA that then forms an R-loop structure with the single-stranded overhanging DNA. This intermediate step in the homologous recombination process appears to be required in yeast to preserve repeat regions before homologous recombination may take place (Ohle et al. 2016). We do not currently know whether, in higher eukaryotes, an R-loop intermediate stage is required before homologous recombination, although it has also been shown that in higher eukaryotes RNA is produced at the site of double-strand break, and this is crucial for the DNA damage response (Pryde et al. 2005, Francia et al. 2012).

Thus, a picture emerges of how transcription is regulated, but also of the fact that it is directly involved in double-strand break repair. It is, therefore, clearly a very complex process, our understanding of which is rapidly evolving. On one side, evidence clearly indicates that transcription undergoes shutdown both at the global (RNAPI) and local level (RNAPII) through the involvement of the DNA damage kinase ATM (Kruhlak et al. 2007, Iacovoni et al. 2010, Shanbhag et al. 2010, Kakarougkas et al. 2014, Harding et al. 2015), whilst other data indicate that transcripts are actually produced at double-strand break sites and such transcripts are required for the correct establishment of the DNA damage response (Pryde et al. 2005, Francia et al. 2012). Moreover, transcription helps to determine specific repair pathway choices, again in the case of double-strand breaks (Aymard et al. 2014). To add to this complex scenario, it was also shown that the activity of DNA-PK, rather than the presence of the double-strand break per se, inhibits RNAPII progression over lesions, suggesting that the RNAPII could actually overcome double-strand breaks (Pankotai et al. 2012). Further studies will be required to understand exactly what the destiny of the pre-existing transcribing RNAPII that encounters a lesion and that of the potentially newly loaded enzyme are, as well as the contribution of DNA damage response factors in the whole process.

RNA as a substrate for homologous recombination

An additional and completely independent role for transcription in DNA damage repair, more specifically in double-strand break repair, has been described in the model organism *S. cerevisiae*. In 2007, the Resnick group presented evidence to show that RNA molecules complementary to the broken ends of the DNA could actually be used directly as a template for double-strand break repair (Storici et al. 2007). Following this first report, the authors further detailed that this is a homologous recombination process that requires Rad52 and, in particular, its inverse strand activity (Keskin et al. 2014, Mazina et al. 2017). As previously mentioned, these results have been identified only in yeast and we currently do not know whether higher eukaryote RNA molecules can also be used as a template for homologous recombination-directed double-strand break repair. However, these data reveal an exciting, novel use for RNA molecules. As they actually contain all the genetic information encoded in a genomic locus, they could act as 'last-resort option' templates for the repair of DNA lesions. When we consider that 75% of the human genome has been found to be a substrate for transcription

(Djebali et al. 2012), it is immediately obvious that such a repair option could provide cells with valuable substrates to repair virtually any double-strand break in their genome. Further studies will be required to determine whether such a possibility is indeed available for human cells.

Transcription's role in UV-induced DNA damage repair

UV light produces a range of DNA lesions, including cyclobutane pyrimidine dimers and pyrimidine 6–4 pyrimidone photoproducts (6–4PPs) (Marteijn et al. 2014). These can stall RNA polymerases during transcription, but can also arrest DNA replication fork progression, with the risk of inducing double-strand breaks following replication fork collapse (Rastogi et al. 2010). These lesions are mainly repaired by the nucleotide excision repair (NER) pathway, which may be divided into two subtypes, based on how the damage is recognised, but share the mechanism by which the actual lesion is then processed and repaired. These two pathways are global genome nucleotide excision repair (GG-NER) and transcription-coupled nucleotide excision repair (TC-NER). (This is discussed in detail in Chapter 5.) In GG-NER, the entire genome is scanned for distortions of the double helix induced by the presence of DNA lesions by a complex formed by Rad23, XPC and CETN2 (Marteijn et al. 2014). Following initial binding by this complex, there is recruitment of the DDB1-DDB2 complex to stabilise XPC binding to the damage site, and further to recruit the remaining NER components. Once the lesion is recognised, the other components of the cascade will (1) unwind the DNA helix around the lesion site; (2) create incisions on the DNA strand that contains the lesion thanks to the endonuclease complex formed by XPF and ERCC1, and excise the damaged fragment while protecting the undamaged strand by loading of RPA; and (3) fill in the gap through the action of DNA polymerases and ligases (Marteijn et al. 2014). In TC-NER, the lesion is directly recognised by an RNAPII complex, becoming stalled at sites of DNA damage (Marteijn et al. 2014). This has a great advantage for the repair kinetics of UV-induced lesions, as it is faster on the transcribed strand, where they are recognised by the TC-NER, than on the non-transcribed strand, where they will be randomly recognised and repaired via the GG-NER (Bohr et al. 1985, Taschner et al. 2010). When RNAPII encounters a UV-induced damage site, it will directly recruit CSB and the complex formed by UVSSA and USP7, followed by the recruitment of CSA (Fousteri et al. 2006, Schwertman et al. 2012, Marteijn et al. 2014, Boeing et al. 2016). This whole complex will backtrack the RNAPII in order to expose the DNA lesion and make it accessible for repair by the core components of the NER pathway. Once the DNA lesion has been repaired, the RNAPII will be allowed to progress and continue transcribing (Marteijn et al. 2014). If the RNAPII, however, blocks the repair, it will undergo proteasome-dependent degradation after ubiquitylation by the E3 ubiquitin ligase NEDD4 (Anindya et al. 2007). As previously described in the case of double-strand breaks, UV-induced lesions also impair transcription progression. In addition, there is direct involvement of RNAPII and of transcription in the recruitment of DNA damage repair factors, as well as supporting and coordinating the cellular response to the lesions. Historically, the contribution of the RNAPII in NER has been known for longer than its contribution to double-strand break repair, with many factors identified to be specifically involved in this process. Intriguingly, many genetic syndromes are linked to mutations in components of the NER pathway, and they will be described later in this chapter.

TRANSCRIPTION-INDUCED GENOME INSTABILITY

Connections between RNAPII transcription and DNA replication

When a cell decides to duplicate, generating two daughter cells, it has to replicate its DNA in a process that is achieved exclusively during the S phase in a precise moment during cell cycle progression. RNA transcription, however, does not have this same restriction, as it can happen in all cell cycle stages from G1 to G2. There are, of course, changes in the transcripts produced throughout the cell cycle, and this applies also to S phase. Actually, many genes required for the DNA replication process, like components of the replication machinery and the histones required to pack the newly replicated DNA into chromatin, are specifically upregulated or expressed during the S phase of the cell cycle (van der Meijden et al. 2002). Indeed, the only cell cycle phase in which transcription is restricted is during M phase, when RNAPII is allowed to complete elongation while inhibiting new initiation events (Liang et al. 2015). This allows removal of the RNAPII complexes from the chromatin to reduce delay in progression through the cycle (Liang et al. 2015). Therefore, when a cell undergoes the duplication of its genome, there needs to be some level of coordination between DNA replication and RNA transcription to avoid reciprocal interference, as there will be parts of the genome that are being transcribed, even at high levels, when DNA replication is occurring. Some published data actually support the idea that the two processes are mutually exclusive, occurring in different parts of the nucleus or at different times throughout S phase (Wei et al. 1998, Meryet-Figuiere et al. 2014). However, there are also other data that show different results about this coordination, whereby the two processes can mutually coexist (Hassan et al. 1994). While there is not a definitive answer as to how the two processes occur over different genomic regions at different times throughout S phase, and how this might be regulated, it is definitely clear that replication and transcription are very closely intertwined. Much evidence, for example, highlights how transcribed regions are generally replicated early during S phase, and it can be observed that DNA replication origins are proximal to active transcription start sites and co-localise with active histone marks (Schüebeler et al. 2002, Karnani et al. 2010, Dellino et al. 2013, Pourkarimi et al. 2016). There are a number of explanations for this preferential enrichment of replication initiation next to active transcription sites:

- Actively transcribed regions are in an open chromatin conformation, making these genomic regions easier to access for the replication machineries.
- Some transcription-associated factors also play a role in DNA replication; therefore, there could be a direct exchange of proteins between the two processes.
- By first replicating actively transcribed regions, cells ensure all the genetic information required for their proper functioning is passed to the daughter cells.

Nevertheless, whether or not there is a higher-order organisation that coordinates transcription and replication to avoid reciprocal interference, there are many indications that transcription and replication can interfere with each other, and this induces DNA damage. Studies in model organisms have provided important information on this subject. For example, in *E. coli*, the most highly expressed transcripts during rapid growth are encoded on the leading strand of the genome, suggesting a specific genomic organisation to avoid direct head-to-head collisions and conflicts between transcription and replication (Brewer 1988). However, as DNA replication in bacteria is much faster than transcription, there is still the risk of head-to-tail collisions between the two processes. In 2008, Pomerantz and O'Donnell showed, using an in vitro purified system, how in the case of head-to-tail collisions, the replication machinery was able to bypass a stalled RNA polymerase utilising the transcribed RNA as a new primer to complete DNA replication, but leaving a small gap behind (Pomerantz and O'Donnell 2008). Following these initial findings, the same authors also characterised in vitro head-to-head collisions between transcription and replication, showing that the replication machinery remains stable on the DNA at the collision site, and will resume replication once the RNA polymerase has fallen off or has been removed by other factors (Pomerantz and O'Donnell 2010). These results suggest that head-to-head collisions represent the greatest impediment for the replication machinery progression, while head-to-tail ones can be easily overcome. In vivo, however, double-strand breaks arise in a mutant context unable to restart backtracked RNA polymerases. This is because there is an accumulation of gaps in the DNA due to the replication restarting downstream of the permanently stalled RNA polymerases (Dutta et al. 2011). These gaps would then hybridise with the RNA forming an R-loop. Intriguingly, no double strand breaks appear upon head-to-head collisions in vivo (Dutta et al. 2011). From work in yeast, we know, for example, that transcription units represent preferential pause sites for DNA replication (Azvolinsky et al. 2009). In higher eukaryotes, however, the scenario is not as well defined as for the model organisms, because of current technical limitations in performing these kinds of experiments with human cells, for example. We see, however, much more evidence for transcription interfering with the DNA replication with increased genome instability linked to transcriptional activity. Initial strong evidence comes in particular from the analysis of genomic fragile sites, which are chromosomal regions susceptible to breaks or gaps in response to low levels of replication stress. There are two main types of fragile sites, based on their frequency: rare fragile sites (caused generally by nucleotide repeat expansions) and common fragile sites (CFS). The first description of common fragile sites was in 1984, when Glover et al. (1984) demonstrated how some genomic regions were more prone than others to breaks when cells were treated with low doses of aphidicolin (a DNA polymerase alpha inhibitor). Since this first observation, many studies have supported the general idea that CFS are the regions in the genome that might be more difficult to replicate, explaining their increased sensitivity to low doses of aphidicolin; high doses of aphidicolin would impair all replication forks, whereas low doses are able to affect only replication in some specifically difficult-to-replicate regions (Debacker and Kooy 2007, Debatisse et al. 2012). Over the years, these CFS have been characterised, with great attention paid to identifying the factors involved in protecting and preventing genome instability at CFS, but also to identifying the mechanisms underlying this fragility (Casper et al. 2002, Pirzio et al. 2008, Bergoglio et al. 2013, Lu et al. 2013, Fu et al. 2015, Somyajit et al. 2015). The fragility of CFS is also clearly relevant for human health, as the sites appear to be hotspots of breakage sites observed in cancers, in this way disrupting the expression of oncogenes or onco-suppressors present in these regions. Significantly, these are among the most commonly rearranged regions in human cancer cell lines (Debacker and Kooy 2007, Bignell et al. 2010, Hazan et al. 2016).

As mentioned, although many mechanisms have been proposed to explain the propensity of CFS to undergo breakage, none was able to provide a unique explanation for their instability. A potential explanation came from the study of the coordination of transcription and replication over extremely long genes. As transcripts can also be several hundreds of kilobases long, and over one megabase in a few extreme cases, when we consider that transcription elongates at an average rate of 2 to 3 kilobases per minute depending on the cell lines (Singh and Padgett 2009, Saponaro et al. 2014), it becomes obvious that it will take several hours for a single RNAPII to transcribe through a gene of that length. Indeed, the Tora group showed that over some of these extremely long genes in CFS, transcription is still occurring while they are replicated, suggesting that breakages specifically occur over these genes because of collisions between transcription and replication (Helmrich et al. 2011). Moreover, this group showed that treatment with aphidicolin exacerbates the risk of these collisions and leads to an accumulation of R-loops (Helmrich et al. 2011). In 2013, Le Tallec et al. mapped CFS in multiple cell lines and presented evidence that a general common feature for CFS was actually the enrichment for large genes, although these genes were not specifically highly expressed ones. All this suggests that if a general mechanism is responsible for the fragility of CFS, such a mechanism is somehow transcription dependent. This would also explain why in different cell lines we observe distinct genomic regions sensitive to aphidicolin, as this depends on which parts of the genome are transcribed (Le Tallec et al. 2013).

However, CFS are not the only fragile sites linked to transcription. In 2013, Barlow et al. identified a new class of fragile sites, called early fragile sites (ERFS). Different from the CFS, ERFS appear to be regions prone to breakage when cells are treated with high doses of hydroxyurea but not with aphidicolin (Barlow et al. 2013). Moreover, their fragility increases upon ATR inhibition and Myc deregulation, and, similarly to CFS, they also overlap with genomic sites commonly lost in cancers; ERFS also overlap with transcribed genes, but in this case in particular, with highly expressed genes (Barlow et al. 2013). These results again provide a link between transcription and genome instability, indicating that although vital and essential, transcription needs to be carefully regulated in order to avoid interfering with genome stability.

Correlation between RNAPI and RNAPIII transcription and DNA replication interference

Although, so far, we have highlighted the impact of RNAPII transcription on the maintenance of genome stability, it is actually not the only RNA polymerase complex linked to increased DNA damage. rDNAs are arranged in clusters of up to >100 repeats over five genomic sites, and this makes them easy hotspots for genomic recombination events (Stults et al. 2009). Moreover, much evidence has shown that throughout evolution, rDNA clusters represent difficult-to-replicate regions and, indeed, each repeat of the cluster contains, towards the 3′ end of the pre-rRNA coding region, a so-called replication fork barrier (RFB) (Brewer and Fangman 1988, Akamatsu and Kobayashi 2015). The RFB generally inhibits DNA replication fork progression only for forks coming in the direction opposite to which the rRNA is transcribed, in order to avoid head-to-head collisions between the two machineries and to coordinate the two processes (Brewer and Fangman 1988, Akamatsu and Kobayashi 2015). Intriguingly, the loss of the RecQ helicase BLM, mutated in the genetic syndrome Bloom, has been shown to destabilise the rDNA cluster, probably not only due to its role in regulating homologous recombination but also because of a direct interaction with RNAPI and a role in regulating rRNA transcription (Killen et al. 2009, Grierson et al. 2012, Grierson et al. 2013). However, BLM is not the only DNA damage response factor that has been so far associated with rRNA transcription; for example, SHPRH and BRCA1 have also shown roles in this process (Johnston et al. 2016, Lee et al. 2017).

The evidence presented thus far shows how specifically RNAPI and RNAPII transcription can induce genome instability. However, although not yet fully proven in higher eukaryotes, at least in the model organism *S. cerevisiae*, several lines of evidence highlight the fact that RNAPIII transcription is also associated with perturbation of DNA replication progression, genome instability and DNA damage (Desphande and Newlon 1996, Admire et al. 2006, Fachinetti et al. 2010, Sabouri et al. 2012). An important consideration in the context of RNAPIII transcription is its crosstalk with RNAPII transcription, clearly proven in higher eukaryotes. Apart from being responsible for the transcription of ncRNA important for the regulation of RNAPII transcription, multiple groups have presented evidence of RNAPII and RNAPII-associated transcription factors, recruited adjacent to RNAPIII on its target transcripts (Oler et al. 2010, Raha et al. 2010). Although we currently do not know what exact role RNAPII and cofactors have in regulating RNAPIII transcription, it might still be possible to hypothesise that alterations in RNAPII transcription might also impact RNAPIII transcription.

In conclusion, as DNA replication and RNA transcription both use the same substrate, it doesn't come as a surprise that these processes can interfere with each other, especially in defective contexts. From now on we will detail some of the mechanistic details on how transcription can lead to increased genome instability.

Transcription-associated mutagenesis

The first indication that the actual process of transcribing a gene comes at the cost of increased genome instability derives from two studies in bacteria. In 1971, a first study in *E. coli* showed how the activation of transcription on the β-galactosidase gene increased the rate of mutations when cells were treated with alkylating agents (Brock 1971). Similar results were presented in the same year in another study showing how activating transcription increases the rates of reversion to Lac+ induced by the acridine mutagen ICR-191, with the authors suggesting that 'some aspect of transcription or translation, or both, in the neighbourhood of the ICR-191-induced mutation stimulated reversion by ICR-191' (Herman and Dworkin 1971). This was the first suggestion that in the process of transcribing a gene, DNA damage can be induced. However, it was more than 20 years later that a similar phenotype was described in eukaryotic cells. Datta and Jinks-Robertson (1995) showed how high levels of transcription are associated with an increase in the rate of spontaneous mutations. Over the course of the years, using the model organism *S. cerevisiae*, this group and others have shown how this increase of mutagenesis, dependent on transcription (called transcription-associated mutagenesis or TAM), is increased in the context of defective homologous recombination and nucleotide excision repair. Furthermore, highly transcribed genes are more prone to the formation of apurinic and apyrimidinic sites due to the misincorporation of dUTP instead of dTTP, GC transversions are increased in transcribed genes, and transcription increases topoisomerase I mutagenic processing of misincorporated ribonucleotides (Morey et al. 2000, Kim and Jinks-Robertson 2009, Kim et al. 2011, Alexander et al. 2013, Park et al. 2012, Cho et al. 2013).

Transcription-associated mutagenesis has been less well characterised in human cells than in model organisms. We know, however, that defective NER is also involved in increased mutagenesis in human cells, as cytosine deamination is more frequent over transcribed strands in NER-defective contexts (Hendriks et al. 2010). However, we also know that the accumulation of R-loops increases activation-induced cytidine deaminase (AID) mutagenesis, with important functional and pathological consequences (R-loops AID-induced mutagenesis will be described in more detail later in this chapter) (Gomez-Gonzalez and Aguilera 2007, Robbiani et al. 2008, Ruiz et al. 2011). Although the exact mechanisms of transcription-associated mutagenesis are not yet fully understood, genetic characterisation allows the definition of processes which exacerbate the problem. The connections with defective NER and the increased rates of mutagenesis on the transcribed strands indicate that lesions encountered by the RNAPII while transcribing are repaired through a more error-prone mechanism than GG-NER, although it is not fully clear what the mechanism is. The general implications of R-loop accumulation will be described in the next section, but it is important to highlight that R-loops induce genome instability also by increasing the rate of mutagenesis over transcribed regions. Finally, increased mutagenesis can also occur following transcription–replication collisions, as highlighted by mutations in MLL2/KMT2D, described in greater detail later in this chapter (Kantidakis et al. 2016).

Transcription-associated recombination and transcription–replication collisions

R-loop–dependent genome instability

As in the case of transcription-associated mutagenesis, the first evidence that active transcription can induce gross chromosomal rearrangements came from studies in bacteria. In 1977, Ikeda and Kobayashi showed that RNA polymerase was involved in genetic recombination in a *recA* (bacterial Rad51 homologue) independent pathway, proving that the actual process of transcription together with the RNA polymerase was directly involved in these recombination events. This study was followed a couple of years later by a further one showing how transcription was actually able to promote recombination independently of *recA*, and proposing that it was transcription elongation that was important for these recombination events (Ikeda and Matsumoto 1979). In their model, Ikeda and Matsumoto envisaged that the RNA produced during transcription was forming a helix with the transcribed strand, leaving the non-transcribed strand available to engage with a free end from another molecule. In support of this model, the authors showed later that the DNA substrate of the recombination events contained RNA, suggesting that R-loops generated during the transcription process might be directly involved in the recombination event (Matsumoto and Ikeda 1983). Although the transcription model proposed by these authors turned out not to be fully correct, their model of how transcription can induce increased recombination was actually close to later findings.

Following these initial publications, similar data were also shown for eukaryotic cells, proving that transcription is indeed associated with an increase in recombination events (Keil and Roeder 1984, Nickoloff and Reynolds 1990, Nickoloff 1992). It was, however, only a few years later, with the characterisation of the *Hpr1* gene in *S. cerevisiae*, that transcription-associated recombination gathered much more attention and interest, and the mechanisms behind this phenomenon were identified. Andres Aguilera and Hannah Klein (1988) initially identified *Hpr1* (hyper-recombination 1) as one out of eight mutations with increased mitotic intrachromosomal recombination compared to a wild-type strain. *Hpr1* was initially hypothesised to be involved in controlling and inhibiting intrachromosomal crossovers (Aguilera and Klein 1989). However, after further characterisation, *Hpr1* appeared to be a general transcription factor, and, in agreement with a role in transcription mutations in the RNAPII machinery, was able to rescue the hyper-recombination phenotype (Zhu et al. 1995, Fan et al. 1996). These findings somehow linked the *Hpr1* genome instability defect to RNAPII transcription, definitely confirmed after proving that *Hpr1* was indeed a novel transcription elongation factor involved in the nuclear export of mRNA (Chavez and Aguilera 1997, Schneiter et al. 1999). While many groups described in more detail the role of *Hpr1* in transcription, being part of a larger complex (the THO-TREX complex) involved in binding newly synthesised RNA and guiding its export outside the nucleus, it was still not clear how exactly this was related to its role in genome instability (Chavez et al. 2000, Jimeno et al. 2002, Strasser et al. 2002). It was finally in 2003 that this got resolved with a seminal publication from the Aguilera lab (Huertas and Aguilera 2003). With a series of elegant experiments, the authors showed how, in the absence of *Hpr1*, cells accumulated co-transcriptionally formed R-loops (Huertas and Aguilera 2003). While on one side R-loops inhibit RNAPII progression because the RNAPII becomes 'hooked' to the chromatin behind, explaining why there was a transcriptional defect, on the other these R-loops are the reason for the increased genome instability, as the overexpression of RNaseH (able to specifically dismantle R-loops) can rescue the hyper-recombination phenotype (Huertas and Aguilera 2003). This study represented the first one in which a clear mechanism on how defective transcription was able to induce genome instability was presented. R-loops were also found accumulating in other mutants belonging to the same pathway, and they were responsible as well for their hyper-recombination phenotype (Luna et al. 2005). More importantly, R-loops do represent a direct barrier for the progression of the DNA replication machinery, inducing replication stress (Gomez-Gonzalez et al. 2011). However, there has also been evidence that the accumulation of R-loops can lead to unscheduled DNA origin activations, at least in the rDNA locus in yeast, and these could contribute to the increased R-loop–dependent genome instability (Stuckey et al. 2015). Importantly, it is noteworthy to mention that not all the transcription-defective contexts that can lead to increased genome instability are actually dependent on an increase of R-loops, indicating that there are also other ways in which defective transcription can affect genome instability (Felipe-Abrio et al. 2015).

However, R-loop accumulation is a problem not only restricted to yeast cells, as also in human cells, mutants in the THO-TREX pathway, as well as in the splicing machinery, show accumulation of R-loops and R-loop–dependent genome instability (Li and Manley 2005, Dominguez-Sanchez et al. 2011, Gan et al. 2011). The relevance of an efficient transcription for genome stability maintenance became evident after a siRNA screening performed by Paulsen et al. (2009) aimed to identify novel contexts with increased genome instability in human cells. In this work, the authors identified hundreds of genes, that when depleted, showed an increase in γH2AX foci number. Unsurprisingly, analysing the function of these genes, revealed a clear enrichment for proteins involved in the metabolism of the nucleic acids. However, interestingly, the largest group represented factors involved in RNA metabolism, with gene ontology enrichment for splicing and RNA processing (Paulsen et al. 2009). Furthermore, in some of the gene knock-downs tested, the overexpression of RNaseH suppressed DNA damage levels completely or at least partially, indicating that genome instability was dependent on an excessive accumulation of R-loops (Paulsen et al. 2009). All these clearly indicated that deregulation of transcription, at many different stages, can have a direct global impact on genome stability; and also that R-loop accumulation, with its increase in DNA damage levels, is a general feature of many transcription-defective contexts. We will focus later in this section on some of these contexts and on how they are important for genome stability, highlighting their direct impact also on human health.

R-loops, however, can drive genome instability not only by directly impairing DNA replication fork progression. R-loops can also increase the levels of transcription-associated mutagenesis, as the displaced strand becomes more accessible to the action of AID (Gomez-Gonzalez and Aguilera 2007). This is, however, also a functionally relevant process, as R-loops arise at the immunoglobulin locus and are required for class switch recombination stimulating AID activity (Yu et al. 2003). On the other hand, an increase in AID-induced double-strand breaks in this locus can also lead to genomic translocations in the case of inappropriate recombination events, such as the c-myc/IgH translocation (Robbiani et al. 2008, Ruiz et al. 2011). It was recently shown that R-loops can be processed by components of the TC-NER pathway, generating double-strand breaks through the action of the endonucleases XPF and XPG, expanding the mechanisms through which an accumulation of R-loops increases genome instability (Sollier et al. 2014). Finally, R-loops have been shown to drive trinucleotide expansion, like in the cases of Friedreich ataxia and fragile X syndrome, promoting gene silencing through enrichment of the negative histone modification H3K9me2 (Groh et al. 2014). All these clearly highlight the relevance of controlling R-loop persistence in chromatin in order to preserve genome stability.

Among the contexts with increased genome instability, mutations in the yeast gene *Sen1* gathered a lot of attention over the years, as this is the homologue of the human gene SETX, which is mutated in the neurological genetic syndromes ataxia-ocular apraxia 2 (AOA2) and juvenile amyotrophic lateral sclerosis 4 (ALS4) (Chen et al. 2004, Moreira et al. 2004). Sen1 was first described to be part of the NRD complex and involved in the transcription termination of snRNAs and snoRNAs (Steinmetz et al. 2001). However, mutations of the *Sen1* gene present a hyper-recombination phenotype due to an accumulation of R-loops, and mutants require DNA damage repair factors for survival (Mischo et al. 2011). However, this role of *Sen1* in genome stability maintenance is, independent from its role in the NRD complex, as *Sen1* localizes and progresses with the replication forks and is required to remove R-loops that could stall or arrest DNA replication fork progression (Alzu et al. 2012). Also, the higher eukaryote homolog SETX is involved in transcription termination and genome stability maintenance. SETX defective cells are hypersensitive to a range of DNA-damaging agents and present constitutive oxidative DNA damage (Suraweera et al. 2007). SETX mutants accumulate R-loops, and SETX has a role in reducing R-loops levels specifically in the region just upstream of the termination pausing site, to favour XRN2 transcription termination (Skourti-Stathaki et al. 2011). Intriguingly, although R-loop accumulation is clearly visible in several tissues from a mouse model of autosomal recessive cerebellar ataxia knockout for SETX, these are not clearly visible in brain tissues, opening the question of why mutations in SETX are associated in particular with neurological syndromes (Yeo et al. 2014).

Last, the processing of the RNA polymerase through the DNA produces positive and negative supercoiling in front of and behind the RNA polymerase, and these supercoils are counteracted by the action of the topoisomerases. Indeed, it has been shown, both in yeast and in humans, that these topoisomerases are essential to counteract transcription-induced genome instability. In humans, TOP1 is required to counteract R-loops accumulations that can perturb DNA replication fork progression and increase DNA damage by also regulating the splicing factor ASF (Tuduri et al. 2009). In yeast, TOP2 binds in proximity of actively transcribed regions and preserves them from accumulating DNA damage during DNA replication (Bermejo et al. 2009).

R-loops in telomere stability and transcription regulation

In eukaryotes, the telomeric regions contain RNA molecules called TERRA that protect chromosome ends from telomere loss (Azzalin et al. 2007). Components of the THO complex are recruited to the telomeres to control the level of R-loop formation of TERRA, and preserve telomeres by processing of EXO1 (Pfeiffer et al. 2013). However, this protective mechanism appears to be disrupted when telomeres become short in yeast, contributing to the activation of a DNA damage response and the maintenance of these short telomeres through a homologous recombination–mediated mechanism (Graf et al. 2017).

So far we have mentioned how the accumulation and deregulation of R-loops can impact genome instability. However, over the last five years, it has become clear that R-loops play important functional roles in transcription biology. When in 2013 the Chedin group published the first genome-wide mapping of R-loops in higher eukaryotes, it became immediately obvious that R-loops could play a more general role in transcription, as they were enriched in the initiation and termination regions of transcripts where they form and disappear following transcription activity (Ginno et al. 2013, Sanz et al. 2016). Indeed, it was soon after shown that R-loops in the 3′ end of genes are required for the correct establishment of the Ago2-dependent H3K9me2 histone modification, important to promote efficient transcription termination (Skourti-Stathaki et al. 2014). More recently, it was shown that R-loops in the 5′ end of genes recruit the histone acetyltransferase TIP60, in this way antagonising the recruitment of the polycomb complex PRC2, which is important to keep the chromatin in an open conformation status and allowing gene expression (Chen et al. 2015).

While it is clear that an accumulation of unscheduled R-loops drives genome instability by many mechanisms there is also increasing evidence of their functional roles in many processes from class switch recombination to the regulation of transcription and of chromatin. This makes it virtually impossible to consider R-loops only as a deleterious by-product of transcription, but more likely as another transcription-related step that needs to be timely and spatially controlled to avoid it interfering with the other processes taking place in the genome.

Transcription stress–induced genome instability

RNAPII progression through transcripts is not even and constant, but changes, for example, in correspondence of the intron-exon junctions where the RNAPII slows down and pauses to support alternative exons inclusion (Kwak et al. 2013). However, while a physiological level of slowdown and pausing is required to support the processing of nascent transcripts, an excessive accumulation of slow or paused RNAPII reveals a defective transcription progression, and this is indicative of increased transcription stress (Saponaro et al. 2014). The first evidence that transcription stress can also impact genome instability comes from studies of the Svejstrup group, characterising the role of RECQL5 in RNAPII transcription. RECQL5 is a RecQ DNA helicase, involved, as with the other four members of the family, in genome stability maintenance and cancer suppression, and was identified as a new RNAPII interacting factor in chromatin in a mass spectrometry screening by the Svejstrup group (Hu et al. 2007, Aygün et al. 2008, Chu and Hickson 2009). Intriguingly, RECQL5 not only interacts with, but can also inhibit, RNAPII transcription in in vitro reconstituted reactions (Aygün et al. 2009). RECQL5, indeed, appears to inhibit transcription progression in vivo, as upon its knockdown the RNAPII elongates globally faster, being the first mutant in higher eukaryotes in which the RNAPII presents such a phenotype (Saponaro et al. 2014). However, although elongating faster, RNAPII transcription is not fully efficient and presents an increase of pausing and stalling polymerases (Saponaro et al. 2014). These high levels of transcription stress are, however, extremely deleterious for human cells, as the genome instability observed in the absence of RECQL5 localises in proximity of the transcription stress sites, enriched in particular in long genes in CFS. This was the first proof that an excessive accumulation of stalled/paused RNAPII can also be a driver of genome instability (Saponaro et al. 2014). RECQL5 represents the first factor presenting such a phenotype but is not the only one. Soon after, the same group showed how transcription stress occurs also upon the knock-out of MLL2/KMT2D, a histone H3K4 methyl-transferase mutated in the genetic syndrome Kabuki and cancer driver in many different tumour types (Ng et al. 2010, Kantidakis et al. 2016). In the case of the deletion of MLL2/KMT2D, DNA damage occurs in short highly transcribed genes overlapping with ERFS that present an uncoupling between the RNAPII levels and the amount of transcripts produced, indicating the presence of unproductive RNAPII, hence slow/stalled/paused RNAPII (Kantidakis et al. 2016). This increased DNA damage also generates a wave of mutagenesis in cells, as genes identified as having the highest levels of γH2AX in the deletion also specifically accumulated point mutations, providing a mechanism explaining the cancer driver role of MLL2/KMT2D in multiple tumour types (Kantidakis et al. 2016). The mechanisms behind the transcription stress appear, however, to be different between RECQL5 and MLL2/KMT2D: RECQL5 controls the RNAPII elongation rate, protecting it by an excessive risk of stalling and pausing, dangerous for cells in particular in long genes in CFS, whereas MLL2/KMT2D is responsible for restoring the correct levels of H3K4 methylation of the histones next to the RNAPII, which seems to be required for a smooth progression of the RNAPII through the chromatin, in particular over highly transcribed genes (Saponaro et al. 2014, Kantidakis et al. 2016). Although the only mutants with increased transcription stress so far presented in higher eukaryotes, it is unlikely that RECQL5 and MLL2/KMT2D are the only transcription factors required to reduce transcription stress levels and preserve genome stability, and this could actually be a more common problem for many transcription factors specifically required for the transcription elongation stage.

Oncogenes and transcription-associated genome instability

The connection between carcinogenesis and transcription-associated genome instability is, however, not only restricted to mutations in the cancer driver MLL2/KMT2D. In recent years, deregulated transcription activity has been associated with oncogene-induced replication stress and genome instability. Researchers have shown, for example, that the oncogene-induced replication stress observed with upregulation of cyclin E, or when a mutated HRAS-Val12 is expressed, are due to increased levels of transcription and R-loops (Jones et al. 2013, Kotstantis et al. 2016). Similarly, Stork et al. (2016) showed that R-loop–induced genome instability arises when cells are stimulated with oestrogen. These data indicate that transcription could be directly involved in generating a wave of DNA damage observed in pre-cancerous lesions and that drives tumourigenesis (Halazonetis et al. 2008). In support of this hypothesis, there are also other well-known oncogenes that in recent years have been linked to transcription-induced genome instability. BRCA1 for example, as aforementioned, is associated with actively transcribing RNAPII and has proven to be fundamental for the recruitment of SETX on termination sites to inhibit R-loop accumulation and breakages in the transcription termination region (Hatchi et al. 2015). BRCA2 has been shown to interact with the RNA export factor TREX2 and is required to counteract the accumulation of R-loops and genome instability (Bhatia et al. 2014). The Fanconi pathway is involved in reducing the impact of transcription replication collisions and the accumulation of R loops, and is able to dismantle R-loops in vitro thanks to the branch migration activity of FANCM (Schwab et al. 2015). These last publications support the growing evidence that DNA damage response and repair factors can directly regulate RNAPII transcription progression in their role of preserving genome instability. However, whether more of them are also transcription regulatory factors needs to be proved.

RNAPII transcription-dependent induced double-strand breaks

Finally, in our description of how transcription can induce genome instability, it is important to note that following estradiol treatment, topoisomerase IIb can generate a double-strand breaks to activate gene transcription (Ju et al. 2006). This finding has been further extended to other signalling pathways identifying that such a similar activation of topoisomerase IIb in response to androgen treatment can, for example, generate double-strand breaks on multiple target genes. Importantly, when this happens on the TMPRSS2 and ERG genes, an illegitimate recombination event can lead to a translocation event with the formation of the TMPRSS2-ERG fusion gene identified in prostate cancers (Haffner et al. 2010). These translocation events are not isolated cases, as transcription-associated double-strand breaks have been identified in many different cell types, representing another general mechanism through which RNAPII transcription can induce DNA damage and increase genome instability (Schwer et al. 2016).

Implications for human health for transcription factors involved in genome stability maintenance

In the course of this chapter, we have mentioned many transcription factors whose deregulation has a direct impact on human health. In the list of the factors whose mutations have such an impact, we will not mention the countless number of proteins altered in cancers, as we will focus only on genes that when mutated give rise to a specific genetic syndrome. However, this is intended to be an exhaustive catalogue of factors that are critical to allow transcription to function without interfering with the other processes taking place in the chromatin and causing genome instability. We will start focusing in this sense on factors involved in the TC-NER pathway. Their names indeed refer to the genetic syndromes in which they have been found mutated: Cockayne syndrome (CSA/B) and xeroderma pigmentosum (XPA-G, V) (also discussed in Chapter 5). Xeroderma pigmentosum is a rare autosomal recessive disorder in which individuals are extremely sensitive to sunlight because of their inability to repair UV-induced DNA damage; this predisposes patients to develop multiple skin cancers during

Table 10.1 **Dual activities on transcription and of transcription in genome stability maintenance.**

	Inhibitory	Activating
Double-strand break repair	• Global inhibition of RNAPI transcription • Inhibition *in cis* of RNAPII transcription only on genes with a double-strand break	• Requirement of RNA transcription on the site of damage for the correct establishment of the DNA damage response
UV-induced DNA damage	• Global transcription shutdown, and in the case of RNAPII transcription, also a global slowdown	• Recruitment and establishment of TC-NER • Changes in mRNA composition due to alternative splicing and alternative last exons events
DNA replication	• Replication origins are enriched in proximity of actively transcribed regions	• Transcribed regions are enriched in CFS and ERFS
R-loops	• Induce hyper-recombination and DNA damage • Inhibit DNA replication fork progression • Induce increased mutagenesis • Induce unscheduled replication origins activation • Induce telomere instability	• Are required for RNAPII transcription termination • Are required for preserving the activation of gene expression

their lives, and consequently individuals are forced to live avoiding any direct exposure to sunlight. Although individuals diagnosed with Cockayne syndrome also present increased sensitivity to sunlight, they are characterised by a much more severe neurodevelopmental and neurodegenerative phenotype. Indeed, recent data support a role for CSB in the expression of neuronal-specific transcripts and in neuronal differentiation, suggesting that CSB has also a more general role in transcription regulation beyond its role in TC-NER, explaining in this way the neurodegenerative disorder (Wang et al. 2014). Other disorders occurring because of mutations in some of the factors previously mentioned are AOA2 and ALS4 due to mutations of SETX, and Aicardi-Goutieres syndrome (AGS). AGS is characterised by a severe auto-inflammatory response due to a defect in the removal of ribonucleotides from the DNA that leads to a constitutive upregulation of interferon-responsive genes via an innate immune response driven by cGAS/STING (Günther et al. 2015, Mackenzie et al. 2016). Consequently, many mutations in genes involved in the processing of misincoporated ribonucleotides in the DNA, but also in the antiviral innate immune response, have been identified in AGS patients, namely TREX1, RNASEH2A, RNASEH2B, RNASEH2C, SAMHD1, ADAR1 and IFIH1 (Livingston and Crow 2016). Last, we have already mentioned that mutations in MLL2/KMT2D have been found associated with Kabuki syndrome, a rare genetic disorder characterised by a wide range of developmental defects and mental retardation (Ng et al. 2010). But mutations in MLL2/KMT2D are also found in many cancer types, and indeed MLL2/KMT2D has been identified as one of the most frequently mutated genes in cancers and the driver for many tumour types (Gonzalez-Perez et al. 2013, Froimchuk et al. 2017). BRCA1 and BRCA2 were initially identified as major breast and ovarian cancer susceptibility genes, but have since then been associated with predisposition to many different tumour types (Hall et al. 1990, Wooster et al. 1994). We have also shown how BRCA1 and BRCA2 are directly involved in regulating RNAPII transcription progression to reduce its negative impact on genome instability. Tightly related to the functional roles of BRCA1 and BRCA2 in DNA damage repair, and also involved in counteracting transcription associated genome instability, is the Fanconi anaemia complex. As indicated by the name, individuals were initially identified with a severe bone marrow failure with cancer predisposition, (particularly leukaemia). Mutations in more than 20 genes when mutated can give rise to Fanconi anaemia (Palovcak et al. 2017) (Chapter 12).

CONCLUSION

In this chapter, we intended to show some of the complexity of the crosstalk between transcription and genome stability maintenance, highlighting how transcription can support DNA replication and DNA damage repair pathways, but can also potentially impair DNA replication, inducing genome instability (Table 10.1). We have described the types of transcriptions occurring in each cell, both in terms of which RNA polymerases transcribe specific regions but also the large variety of transcripts that can be produced. We have shown how in response to DNA damage the RNA polymerases become carefully regulated, and how transcription helps and supports the DNA damage repair pathways. We have highlighted how transcription and replication are somehow connected, with replication initiating preferentially in proximity of actively transcribed regions, and described the large range of circumstances in which transcription can interfere with DNA replication inducing DNA damage. We have mentioned many factors known for their roles in DNA damage repair, which recently have been found to act as general transcription-associated factors too, with potential great implications for diseases associated with their mutations.

In conclusion, it is clear that transcription plays a fundamental role in genome stability maintenance, functioning most of the time to preserving it, but becoming a danger when not executed properly. In the future, we will have to understand more about the correlations, and potentially the crosstalk, between transcription, DNA damage repair pathways and DNA replication to understand how they can avoid interfering with each other and, maybe, also support each other.

REFERENCES

Admire, A., Shanks, L., Danzl, N., Wang, M., Weier, U., Stevens, W., Hunt, E. & Weinert, T. 2006. Cycles of chromosome instability are associated with a fragile site and are increased by defects in DNA replication and checkpoint controls in yeast. *Genes Dev*, 20, 159–73.

Aguilera, A. & Klein, H. L. 1988. Genetic control of intrachromosomal recombination in Saccharomyces cerevisiae. I. Isolation and genetic characterization of hyper-recombination mutations. *Genetics*, 119, 779–90.

Aguilera, A. & Klein, H. L. 1989. Genetic and molecular analysis of recombination events in Saccharomyces cerevisiae occurring in the presence of the hyper-recombination mutation hpr1. *Genetics*, 122, 503–17.

Akamatsu, Y. & Kobayashi, T. 2015. The human RNA polymerase I transcription terminator complex acts as a replication Fork Barrier that coordinates the progress of replication with rRNA transcription activity. *Mol Cell Biol*, 35, 1871–81.

Alexander, M. P., Begins, K. J., Crall, W. C., Holmes, M. P. & Lippert, M. J. 2013. High levels of transcription stimulate transversions at GC base pairs in yeast. *Environ Mol Mutagen*, 54, 44–53.

Alzu, A., Bermejo, R., Begnis, M., Lucca, C., Piccini, D., Carotenuto, W., Saponaro, M. et al. 2012. Senataxin associates with replication forks to protect fork integrity across RNA-polymerase-II-transcribed genes. *Cell*, 151, 835–46.

Andrade-Lima, L. C., Veloso, A., Paulsen, M. T., Menck, C. F. & Ljungman, M. 2015. DNA repair and recovery of RNA synthesis following exposure to ultraviolet light are delayed in long genes. *Nucleic Acids Res*, 43, 2744–56.

Anindya, R., Aygun, O. & Svejstrup, J. Q. 2007. Damage-induced ubiquitylation of human RNA polymerase II by the ubiquitin ligase Nedd4, but not Cockayne syndrome proteins or BRCA1. *Mol Cell*, 28, 386–97.

Aygun, O., Svejstrup, J. & Liu, Y. 2008. A RECQ5-RNA polymerase II association identified by targeted proteomic analysis of human chromatin. *Proc Natl Acad Sci U S A*, 105, 8580–4.

Aygun, O., Xu, X., Liu, Y., Takahashi, H., Kong, S. E., Conaway, R. C., Conaway, J. W. & Svejstrup, J. Q. 2009. Direct inhibition of RNA polymerase II transcription by RECQL5. *J Biol Chem*, 284, 23197–203.

Aymard, F., Bugler, B., Schmidt, C. K., Guillou, E., Caron, P., Briois, S., Iacovoni, J. S. et al. 2014. Transcriptionally active chromatin recruits homologous recombination at DNA double-strand breaks. *Nat Struct Mol Biol*, 21, 366–74.

Azvolinsky, A., Giresi, P. G., Lieb, J. D. & Zakian, V. A. 2009. Highly transcribed RNA polymerase II genes are impediments to replication fork progression in Saccharomyces cerevisiae. *Mol Cell*, 34, 722–34.

Azzalin, C. M., Reichenbach, P., Khoriauli, L., Giulotto, E. & Lingner, J. 2007. Telomeric repeat containing RNA and RNA surveillance factors at mammalian chromosome ends. *Science*, 318, 798–801.

Ban, N., Nissen, P., Hansen, J., Moore, P. B. & Steitz, T. A. 2000. The complete atomic structure of the large ribosomal subunit at 2.4 A resolution. *Science*, 289, 905–20.

Barlow, J. H., Faryabi, R. B., Callen, E., Wong, N., Malhowski, A., Chen, H. T., Gutierrez-Cruz, G. et al. 2013. Identification of early replicating fragile sites that contribute to genome instability. *Cell*, 152, 620–32.

Bergoglio, V., Boyer, A. S., Walsh, E., Naim, V., Legube, G., Lee, M. Y., Rey, L. et al. 2013. DNA synthesis by Pol eta promotes fragile site stability by preventing under-replicated DNA in mitosis. *J Cell Biol*, 201, 395–408.

Bermejo, R., Capra, T., Gonzalez-Huici, V., Fachinetti, D., Cocito, A., Natoli, G., Katou, Y. et al. 2009. Genome-organizing factors Top2 and Hmo1 prevent chromosome fragility at sites of S phase transcription. *Cell*, 138, 870–84.

Bhatia, V., Barroso, S. I., Garcia-Rubio, M. L., Tumini, E., Herrera-Moyano, E. & Aguilera, A. 2014. BRCA2 prevents R-loop accumulation and associates with TREX 2 mRNA export factor PCID2. *Nature*, 511, 362–5.

Bignell, G. R., Greenman, C. D., Davies, H., Butler, A. P., Edkins, S., Andrews, J. M., Buck, G. et al. 2010. Signatures of mutation and selection in the cancer genome. *Nature*, 463, 893–8.

Boeing, S., Williamson, L., Encheva, V., Gori, I., Saunders, R. E., Instrell, R., Aygun, O. et al. 2016. Multiomic analysis of the UV-induced DNA damage response. *Cell Rep*, 15, 1597–1610.

Bohr, V. A., Smith, C. A., Okumoto, D. S. & Hanawalt, P. C. 1985. DNA repair in an active gene: Removal of pyrimidine dimers from the DHFR gene of CHO cells is much more efficient than in the genome overall. *Cell*, 40, 359–69.

Brewer, B. J. 1988. When polymerases collide: Replication and the transcriptional organization of the E. *coli chromosome. Cell*, 53, 679–86.

Brewer, B. J. & Fangman, W. L. 1988. A replication fork barrier at the 3′ end of yeast ribosomal RNA genes. *Cell*, 55, 637–43.

Brock, R. D. 1971. Differential mutation of the beta-galactosidase gene of Escherichia coli. *Mutat Res*, 11, 181–6.

Casper, A. M., Nghiem, P., Arlt, M. F. & Glover, T. W. 2002. ATR regulates fragile site stability. *Cell*, 111, 779–89.

Chavez, S. & Aguilera, A. 1997. The yeast HPR1 gene has a functional role in transcriptional elongation that uncovers a novel source of genome instability. *Genes Dev*, 11, 3459–70.

Chavez, S., Beilharz, T., Rondon, A. G., Erdjument-Bromage, H., Tempst, P., Svejstrup, J. Q., Lithgow, T. & Aguilera, A. 2000. A protein complex containing Tho2, Hpr1, Mft1 and a novel protein, Thp2, connects transcription elongation with mitotic recombination in Saccharomyces cerevisiae. *EMBO J*, 19, 5824–34.

Chen, P. B., Chen, H. V., Acharya, D., Rando, O. J. & Fazzio, T. G. 2015. R loops regulate promoter-proximal chromatin architecture and cellular differentiation. *Nat Struct Mol Biol*, 22, 999–1007.

Chen, Y. Z., Bennett, C. L., Huynh, H. M., Blair, I. P., Puls, I., Irobi, J., Dierick, I. et al. 2004. DNA/RNA helicase gene mutations in a form of juvenile amyotrophic lateral sclerosis (ALS4). *Am J Hum Genet*, 74, 1128–35.

Cho, J. E., Kim, N., Li, Y. C. & Jinks-Robertson, S. 2013. Two distinct mechanisms of Topoisomerase 1-dependent mutagenesis in yeast. *DNA Repair (Amst)*, 12, 205–11.

Christians, F. C. & Hanawalt, P. C. 1993. Lack of transcription-coupled repair in mammalian ribosomal RNA genes. *Biochemistry*, 32, 10512–8.

Chu, W. K. & Hickson, I. D. 2009. RecQ helicases: Multifunctional genome caretakers. *Nat Rev Cancer*, 9, 644–54.

Core, L. J., Waterfall, J. J. & Lis, J. T. 2008. Nascent RNA sequencing reveals widespread pausing and divergent initiation at human promoters. *Science*, 322, 1845–8.

Czech, B. & Hannon, G. J. 2016. One loop to rule them all: The Ping-Pong cycle and piRNA-guided silencing. *Trends Biochem Sci*, 41, 324–337.

Datta, A. & Jinks-Robertson, S. 1995. Association of increased spontaneous mutation rates with high levels of transcription in yeast. *Science*, 268, 1616–9.

Debacker, K. & Kooy, R. F. 2007. Fragile sites and human disease. *Hum Mol Genet*, 16 Spec No. 2, R150–8.

Debatisse, M., Le Tallec, B., Letessier, A., Dutrillaux, B. & Brison, O. 2012. Common fragile sites: Mechanisms of instability revisited. *Trends Genet*, 28, 22–32.

Dellino, G. I., Cittaro, D., Piccioni, R., Luzi, L., Banfi, S., Segalla, S., Cesaroni, M., Mendoza-Maldonado, R., Giacca, M. & Pelicci, P. G. 2013. Genome-wide mapping of human DNA-replication origins: Levels of transcription at ORC1 sites regulate origin selection and replication timing. *Genome Res*, 23, 1–11.

Deshpande, A. M. & Newlon, C. S. 1996. DNA replication fork pause sites dependent on transcription. *Science*, 272, 1030–3.

Djebali, S., Davis, C. A., Merkel, A., Dobin, A., Lassmann, T., Mortazavi, A., Tanzer, A. et al. 2012. Landscape of transcription in human cells. *Nature*, 489, 101–8.

Dominguez-Sanchez, M. S., Barroso, S., Gomez-Gonzalez, B., Luna, R. & Aguilera, A. 2011. Genome instability and transcription elongation impairment in human cells depleted of THO/TREX. *PLoS Genet*, 7, e1002386.

Dujardin, G., Lafaille, C., De La Mata, M., Marasco, L. E., Munoz, M. J., Le Jossic-Corcos, C., Corcos, L. & Kornblihtt, A. R. 2014. How slow RNA polymerase II elongation favors alternative exon skipping. *Mol Cell*, 54, 683–90.

Dupuis-Sandoval, F., Poirier, M. & Scott, M. S. 2015. The emerging landscape of small nucleolar RNAs in cell biology. *Wiley Interdiscip Rev RNA*, 6, 381–97.

Dutta, D., Shatalin, K., Epshtein, V., Gottesman, M. E. & Nudler, E. 2011. Linking RNA polymerase backtracking to genome instability in E. *coli*. *Cell*, 146, 533–43.

Elia, A. E., Boardman, A. P., Wang, D. C., Huttlin, E. L., Everley, R. A., Dephoure, N., Zhou, C., Koren, I., Gygi, S. P. & Elledge, S. J. 2015. Quantitative Proteomic Atlas of Ubiquitination and Acetylation in the DNA Damage Response. *Mol Cell*, 59, 867–81.

Ezkurdia, I., Juan, D., Rodriguez, J. M., Frankish, A., Diekhans, M., Harrow, J., Vazquez, J., Valencia, A. & Tress, M. L. 2014. Multiple evidence strands suggest that there may be as few as 19,000 human protein-coding genes. *Hum Mol Genet*, 23, 5866–78.

Fachinetti, D., Bermejo, R., Cocito, A., Minardi, S., Katou, Y., Kanoh, Y., Shirahige, K., Azvolinsky, A., Zakian, V. A. & Foiani, M. 2010. Replication termination at eukaryotic chromosomes is mediated by Top2 and occurs at genomic loci containing pausing elements. *Mol Cell*, 39, 595–605.

Fan, H. Y., Cheng, K. K. & Klein, H. L. 1996. Mutations in the RNA polymerase II transcription machinery suppress the hyperrecombination mutant hpr1 delta of Saccharomyces cerevisiae. *Genetics*, 142, 749–59.

Felipe-Abrio, I., Lafuente-Barquero, J., Garcia-Rubio, M. L. & Aguilera, A. 2015. RNA polymerase II contributes to preventing transcription-mediated replication fork stalls. *EMBO J*, 34, 236–50.

Fousteri, M., Vermeulen, W., Van Zeeland, A. A. & Mullenders, L. H. 2006. Cockayne syndrome A and B proteins differentially regulate recruitment of chromatin remodeling and repair factors to stalled RNA polymerase II in vivo. *Mol Cell*, 23, 471–82.

Francia, S., Cabrini, M., Matti, V., Oldani, A. & D'adda Di Fagagna, F. 2016. DICER, DROSHA and DNA damage response RNAs are necessary for the secondary recruitment of DNA damage response factors. *J Cell Sci*, 129, 1468–76.

Francia, S., Michelini, F., Saxena, A., Tang, D., De Hoon, M., Anelli, V., Mione, M., Carninci, P. & D'adda Di Fagagna, F. 2012. Site-specific DICER and DROSHA RNA products control the DNA-damage response. *Nature*, 488, 231–5.

Froimchuk, E., Jang, Y. & Ge, K. 2017. Histone H3 lysine 4 methyltransferase KMT2D. *Gene*, 627, 337–342.

Fu, H., Martin, M. M., Regairaz, M., Huang, L., You, Y., Lin, C. M., Ryan, M. et al. 2015. The DNA repair endonuclease Mus81 facilitates fast DNA replication in the absence of exogenous damage. *Nat Commun*, 6, 6746.

Gan, W., Guan, Z., Liu, J., Gui, T., Shen, K., Manley, J. L. & Li, X. 2011. R-loop-mediated genomic instability is caused by impairment of replication fork progression. *Genes Dev*, 25, 2041–56.

Gardini, A., Baillat, D., Cesaroni, M. & Shiekhattar, R. 2014. Genome-wide analysis reveals a role for BRCA1 and PALB2 in transcriptional co-activation. *EMBO J*, 33, 890–905.

Ginno, P. A., Lim, Y. W., Lott, P. L., Korf, I. & Chedin, F. 2013. GC skew at the 5′ and 3′ ends of human genes links R-loop formation to epigenetic regulation and transcription termination. *Genome Res*, 23, 1590–600.

Glover, T. W., Berger, C., Coyle, J. & Echo, B. 1984. DNA polymerase alpha inhibition by aphidicolin induces gaps and breaks at common fragile sites in human chromosomes. *Hum Genet*, 67, 136–42.

Gomez-Gonzalez, B. & Aguilera, A. 2007. Activation-induced cytidine deaminase action is strongly stimulated by mutations of the THO complex. *Proc Natl Acad Sci U S A*, 104, 8409–14.

Gomez-Gonzalez, B., Garcia-Rubio, M., Bermejo, R., Gaillard, H., Shirahige, K., Marin, A., Foiani, M. & Aguilera, A. 2011. Genome-wide function of THO/TREX in active genes prevents R-loop-dependent replication obstacles. *EMBO J*, 30, 3106–19.

Gonzalez-Perez, A., Perez-Llamas, C., Deu-Pons, J., Tamborero, D., Schroeder, M. P., Jene-Sanz, A., Santos, A. & Lopez-Bigas, N. 2013. IntOGen-mutations identifies cancer drivers across tumor types. *Nat Methods*, 10, 1081–2.

Graf, M., Bonetti, D., Lockhart, A., Serhal, K., Kellner, V., Maicher, A., Jolivet, P., Teixeira, M. T. & Luke, B. 2017. Telomere Length Determines TERRA and R-Loop Regulation through the Cell Cycle. *Cell*, 170, 72–85 e14.

Grierson, P. M., Acharya, S. & Groden, J. 2013. Collaborating functions of BLM and DNA topoisomerase I in regulating human rDNA transcription. *Mutat Res*, 743-744, 89–96.

Grierson, P. M., Lillard, K., Behbehani, G. K., Combs, K. A., Bhattacharyya, S., Acharya, S. & Groden, J. 2012. BLM helicase facilitates RNA polymerase I-mediated ribosomal RNA transcription. *Hum Mol Genet*, 21, 1172–83.

Groh, M., Lufino, M. M., Wade-Martins, R. & Gromak, N. 2014. R-loops associated with triplet repeat expansions promote gene silencing in Friedreich ataxia and fragile X syndrome. *PLoS Genet*, 10, e1004318.

Gunther, C., Kind, B., Reijns, M. A., Berndt, N., Martinez-Bueno, M., Wolf, C., Tungler, V. et al. 2015. Defective removal of ribonucleotides from DNA promotes systemic autoimmunity. *J Clin Invest*, 125, 413–24.

Haffner, M. C., Aryee, M. J., Toubaji, A., Esopi, D. M., Albadine, R., Gurel, B., Isaacs, W. B. et al. 2010. Androgen-induced TOP2B-mediated double-strand breaks and prostate cancer gene rearrangements. *Nat Genet*, 42, 668–75.

Halazonetis, T. D., Gorgoulis, V. G. & Bartek, J. 2008. An oncogene-induced DNA damage model for cancer development. *Science*, 319, 1352–5.

Hall, J. M., Lee, M. K., Newman, B., Morrow, J. E., Anderson, L. A., Huey, B. & King, M. C. 1990. Linkage of early-onset familial breast cancer to chromosome 17q21. *Science*, 250, 1684–9.

Harding, S. M., Boiarsky, J. A. & Greenberg, R. A. 2015. ATM Dependent Silencing Links Nucleolar Chromatin Reorganization to DNA Damage Recognition. *Cell Rep*, 13, 251–9.

Hassan, A. B., Errington, R. J., White, N. S., Jackson, D. A. & Cook, P. R. 1994. Replication and transcription sites are colocalized in human cells. *J Cell Sci*, 107(Pt 2), 425–34.

Hatchi, E., Skourti-Stathaki, K., Ventz, S., Pinello, L., Yen, A., Kamieniarz-Gdula, K., Dimitrov, S. et al. 2015. BRCA1 recruitment to transcriptional pause sites is required for R-loop-driven DNA damage repair. *Mol Cell*, 57, 636–647.

Hazan, I., Hofmann, T. G. & Aqeilan, R. I. 2016. Tumor Suppressor Genes within Common Fragile Sites Are Active Players in the DNA Damage Response. *PLoS Genet*, 12, e1006436.

He, L. & Hannon, G. J. 2004. MicroRNAs: Small RNAs with a big role in gene regulation. *Nat Rev Genet*, 5, 522–31.

Helmrich, A., Ballarino, M. & Tora, L. 2011. Collisions between replication and transcription complexes cause common fragile site instability at the longest human genes. *Mol Cell*, 44, 966–77.

Hendriks, G., Calleja, F., Besaratinia, A., Vrieling, H., Pfeifer, G. P., Mullenders, L. H., Jansen, J. G. & De Wind, N. 2010. Transcription-dependent cytosine deamination is a novel mechanism in ultraviolet light-induced mutagenesis. *Curr Biol*, 20, 170–5.

Herman, R. K. & Dworkin, N. B. 1971. Effect of gene induction on the rate of mutagenesis by ICR-191 in Escherichia coli. *J Bacteriol*, 106, 543–50.

Hill, S. J., Rolland, T., Adelmant, G., Xia, X., Owen, M. S., Dricot, A., Zack, T. I. et al. 2014. Systematic screening reveals a role for BRCA1 in the response to transcription-associated DNA damage. *Genes Dev*, 28, 1957–75.

Hu, Y., Raynard, S., Sehorn, M. G., Lu, X., Bussen, W., Zheng, L., Stark, J. M. et al. 2007. RECQL5/Recql5 helicase regulates homologous recombination and suppresses tumor formation via disruption of Rad51 presynaptic filaments. *Genes Dev*, 21, 3073–84.

Huertas, P. & Aguilera, A. 2003. Cotranscriptionally formed DNA:RNA hybrids mediate transcription elongation impairment and transcription-associated recombination. *Mol Cell*, 12, 711–21.

Iacovoni, J. S., Caron, P., Lassadi, I., Nicolas, E., Massip, L., Trouche, D. & Legube, G. 2010. High-resolution profiling of gammaH2AX around DNA double strand breaks in the mammalian genome. *EMBO J*, 29, 1446–57.

Iannelli, F., Galbiati, A., Capozzo, I., Nguyen, Q., Magnuson, B., Michelini, F., D'alessandro, G. et al. 2017. A damaged genome's transcriptional landscape through multilayered expression profiling around in situ-mapped DNA double-strand breaks. *Nat Commun*, 8, 15656.

Ikeda, H. & Kobayashi, I. 1977. Involvement of DNA-dependent RNA polymerase in a recA-independent pathway of genetic recombination in Escheria coli. *Proc Natl Acad Sci U S A*, 74, 3932–6.

Ikeda, H. & Matsumoto, T. 1979. Transcription promotes recA-independent recombination mediated by DNA-dependent RNA polymerase in Escherichia coli. *Proc Natl Acad Sci U S A*, 76, 4571–5.

Jimeno, S., Rondon, A. G., Luna, R. & Aguilera, A. 2002. The yeast THO complex and mRNA export factors link RNA metabolism with transcription and genome instability. *EMBO J*, 21, 3526–35.

Johnston, R., D'costa, Z., Ray, S., Gorski, J., Harkin, D. P., Mullan, P. & Panov, K. I. 2016. The identification of a novel role for BRCA1 in regulating RNA polymerase I transcription. *Oncotarget*, 7, 68097–68110.

Jones, R. M., Mortusewicz, O., Afzal, I., Lorvellec, M., Garcia, P., Helleday, T. & Petermann, E. 2013. Increased replication initiation and conflicts with transcription underlie Cyclin E-induced replication stress. *Oncogene*, 32, 3744–53.

Ju, B. G., Lunyak, V. V., Perissi, V., Garcia-Bassets, I., Rose, D. W., Glass, C. K. & Rosenfeld, M. G. 2006. A topoisomerase IIbeta-mediated dsDNA break required for regulated transcription. *Science*, 312, 1798–802.

Kakarougkas, A., Ismail, A., Chambers, A. L., Riballo, E., Herbert, A. D., Kunzel, J., Lobrich, M., Jeggo, P. A. & Downs, J. A. 2014. Requirement for PBAF in transcriptional repression and repair at DNA breaks in actively transcribed regions of chromatin. *Mol Cell*, 55, 723–32.

Kantidakis, T., Saponaro, M., Mitter, R., Horswell, S., Kranz, A., Boeing, S., Aygun, O. et al. 2016. Mutation of cancer driver MLL2 results in transcription stress and genome instability. *Genes Dev*, 30, 408–20.

Karnani, N., Taylor, C. M., Malhotra, A. & Dutta, A. 2010. Genomic study of replication initiation in human chromosomes reveals the influence of transcription regulation and chromatin structure on origin selection. *Mol Biol Cell*, 21, 393–404.

Keil, R. L. & Roeder, G. S. 1984. Cis-acting, recombination-stimulating activity in a fragment of the ribosomal DNA of S. *cerevisiae*. *Cell*, 39, 377–86.

Keskin, H., Shen, Y., Huang, F., Patel, M., Yang, T., Ashley, K., Mazin, A. V. & Storici, F. 2014. Transcript-RNA-templated DNA recombination and repair. *Nature*, 515, 436–9.

Killen, M. W., Stults, D. M., Adachi, N., Hanakahi, L. & Pierce, A. J. 2009. Loss of Bloom syndrome protein destabilizes human gene cluster architecture. *Hum Mol Genet*, 18, 3417–28.

Kim, N., Huang, S. N., Williams, J. S., Li, Y. C., Clark, A. B., Cho, J. E., Kunkel, T. A., Pommier, Y. & Jinks-Robertson, S. 2011. Mutagenic processing of ribonucleotides in DNA by yeast topoisomerase I. *Science*, 332, 1561–4.

Kim, N. & Jinks-Robertson, S. 2009. dUTP incorporation into genomic DNA is linked to transcription in yeast. *Nature*, 459, 1150–3.

Kotsantis, P., Silva, L. M., Irmscher, S., Jones, R. M., Folkes, L., Gromak, N. & Petermann, E. 2016. Increased global transcription activity as a mechanism of replication stress in cancer. *Nat Commun*, 7, 13087.

Kruhlak, M., Crouch, E. E., Orlov, M., Montano, C., Gorski, S. A., Nussenzweig, A., Misteli, T., Phair, R. D. & Casellas, R. 2007. The ATM repair pathway inhibits RNA polymerase I transcription in response to chromosome breaks. *Nature*, 447, 730–4.

Kwak, H., Fuda, N. J., Core, L. J. & Lis, J. T. 2013. Precise maps of RNA polymerase reveal how promoters direct initiation and pausing. *Science*, 339, 950–3.

Le Tallec, B., Millot, G. A., Blin, M. E., Brison, O., Dutrillaux, B. & Debatisse, M. 2013. Common fragile site profiling in epithelial and erythroid cells reveals that most recurrent cancer deletions lie in fragile sites hosting large genes. *Cell Rep*, 4, 420–8.

Lee, D., An, J., Park, Y. U., Liaw, H., Woodgate, R., Park, J. H. & Myung, K. 2017. SHPRH regulates rRNA transcription by recognizing the histone code in an mTOR-dependent manner. *Proc Natl Acad Sci U S A*, 114, E3424–E3433.

Lee, Y. Y., Yu, Y. B., Gunawardena, H. P., Xie, L. & Chen, X. 2012. BCLAF1 is a radiation-induced H2AX-interacting partner involved in gamma-H2AX-mediated regulation of apoptosis and DNA repair. *Cell Death Dis*, 3, e359.

Li, W., Notani, D. & Rosenfeld, M. G. 2016. Enhancers as non-coding RNA transcription units: Recent insights and future perspectives. *Nat Rev Genet*, 17, 207–23.

Li, X. & Manley, J. L. 2005. Inactivation of the SR protein splicing factor ASF/SF2 results in genomic instability. *Cell*, 122, 365–78.

Liang, K., Woodfin, A. R., Slaughter, B. D., Unruh, J. R., Box, A. C., Rickels, R. A., Gao, X., Haug, J. S., Jaspersen, S. L. & Shilatifard, A. 2015. Mitotic Transcriptional Activation: Clearance of Actively Engaged Pol II via Transcriptional Elongation Control in Mitosis. *Mol Cell*, 60, 435–45.

Livingston, J. H. & Crow, Y. J. 2016. Neurologic Phenotypes Associated with Mutations in TREX1, RNASEH2A, RNASEH2B, RNASEH2C, SAMHD1, ADAR1, and IFIH1: Aicardi-Goutieres Syndrome and Beyond. *Neuropediatrics*, 47, 355–360.

Lu, X., Parvathaneni, S., Hara, T., Lal, A. & Sharma, S. 2013. Replication stress induces specific enrichment of RECQ1 at common fragile sites FRA3B and FRA16D. *Mol Cancer*, 12, 29.

Luna, R., Jimeno, S., Marin, M., Huertas, P., Garcia-Rubio, M. & Aguilera, A. 2005. Interdependence between transcription and mRNP processing and export, and its impact on genetic stability. *Mol Cell*, 18, 711–22.

Mackenzie, K. J., Carroll, P., Lettice, L., Tarnauskaite, Z., Reddy, K., Dix, F., Revuelta, A. et al. 2016. Ribonuclease H2 mutations induce a cGAS/STING-dependent innate immune response. *EMBO J*, 35, 831–44.

Marteijn, J. A., Lans, H., Vermeulen, W. & Hoeijmakers, J. H. 2014. Understanding nucleotide excision repair and its roles in cancer and ageing. *Nat Rev Mol Cell Biol*, 15, 465–81.

Matsumoto, T. & Ikeda, H. 1983. Role of R loops in recA-independent homologous recombination of bacteriophage lambda. *J Virol*, 45, 971–6.

Matsuoka, S., Ballif, B. A., Smogorzewska, A., Mcdonald, E. R. 3rd, Hurov, K. E., Luo, J., Bakalarski, C. E. et al. 2007. ATM and ATR substrate analysis reveals extensive protein networks responsive to DNA damage. *Science*, 316, 1160–6.

Mazina, O. M., Keskin, H., Hanamshet, K., Storici, F. & Mazin, A. V. 2017. Rad52 Inverse Strand Exchange Drives RNA-Templated DNA Double-Strand Break Repair. *Mol Cell*, 67, 19–29 e3.

Meryet-Figuiere, M., Alaei-Mahabadi, B., Ali, M. M., Mitra, S., Subhash, S., Pandey, G. K., Larsson, E. & Kanduri, C. 2014. Temporal separation of replication and transcription during S-phase progression. *Cell Cycle*, 13, 3241–8.

Merz, C., Urlaub, H., Will, C. L. & Luhrmann, R. 2007. Protein composition of human mRNPs spliced in vitro and differential requirements for mRNP protein recruitment. *RNA*, 13, 116–28.

Mischo, H. E., Gomez-Gonzalez, B., Grzechnik, P., Rondon, A. G., Wei, W., Steinmetz, L., Aguilera, A. & Proudfoot, N. J. 2011. Yeast Sen1 helicase protects the genome from transcription-associated instability. *Mol Cell*, 41, 21–32.

Moreira, M. C., Klur, S., Watanabe, M., Nemeth, A. H., Le Ber, I., Moniz, J. C., Tranchant, C. et al. 2004. Senataxin, the ortholog of a yeast RNA helicase, is mutant in ataxia-ocular apraxia 2. *Nat Genet*, 36, 225–7.

Morey, N. J., Greene, C. N. & Jinks-Robertson, S. 2000. Genetic analysis of transcription-associated mutation in Saccharomyces cerevisiae. *Genetics*, 154, 109–20.

Munoz, M. J., Nieto Moreno, N., Giono, L. E., Cambindo Botto, A. E., Dujardin, G., Bastianello, G., Lavore, S. et al. 2017. Major Roles for Pyrimidine Dimers, Nucleotide Excision Repair, and ATR in the Alternative Splicing Response to UV Irradiation. *Cell Rep*, 18, 2868–2879.

Munoz, M. J., Perez Santangelo, M. S., Paronetto, M. P., De La Mata, M., Pelisch, F., Boireau, S., Glover-Cutter, K. et al. 2009. DNA damage regulates alternative splicing through inhibition of RNA polymerase II elongation. *Cell*, 137, 708–20.

Nellore, A., Jaffe, A. E., Fortin, J. P., Alquicira-Hernandez, J., Collado-Torres, L., Wang, S., Phillips, R. A. et al. 2016. Human splicing diversity and the extent of unannotated splice junctions across human RNA-seq samples on the Sequence Read Archive. *Genome Biol*, 17, 266.

Ng, S. B., Bigham, A. W., Buckingham, K. J., Hannibal, M. C., Mcmillin, M. J., Gildersleeve, H. I., Beck, A. E. et al. 2010. Exome sequencing identifies MLL2 mutations as a cause of Kabuki syndrome. *Nat Genet*, 42, 790–3.

Nickoloff, J. A. 1992. Transcription enhances intrachromosomal homologous recombination in mammalian cells. *Mol Cell Biol*, 12, 5311–8.

Nickoloff, J. A. & Reynolds, R. J. 1990. Transcription stimulates homologous recombination in mammalian cells. *Mol Cell Biol*, 10, 4837–45.

Ohle, C., Tesorero, R., Schermann, G., Dobrev, N., Sinning, I. & Fischer, T. 2016. Transient RNA-DNA Hybrids Are Required for Efficient Double-Strand Break Repair. *Cell*, 167, 1001–1013 e7.

Oler, A. J., Alla, R. K., Roberts, D. N., Wong, A., Hollenhorst, P. C., Chandler, K. J., Cassiday, P. A. et al. 2010. Human RNA polymerase III transcriptomes and relationships to Pol II promoter chromatin and enhancer-binding factors. *Nat Struct Mol Biol*, 17, 620–8.

Palovcak, A., Liu, W., Yuan, F. & Zhang, Y. 2017. Maintenance of genome stability by Fanconi anemia proteins. *Cell Biosci*, 7, 8.

Pankotai, T., Bonhomme, C., Chen, D. & Soutoglou, E. 2012. DNAPKcs-dependent arrest of RNA polymerase II transcription in the presence of DNA breaks. *Nat Struct Mol Biol*, 19, 276–82.

Park, C., Qian, W. & Zhang, J. 2012. Genomic evidence for elevated mutation rates in highly expressed genes. *EMBO Rep*, 13, 1123–9.

Paulsen, R. D., Soni, D. V., Wollman, R., Hahn, A. T., Yee, M. C., Guan, A., Hesley, J. A. et al. 2009. A genome-wide siRNA screen reveals diverse cellular processes and pathways that mediate genome stability. *Mol Cell*, 35, 228–39.

Pfeiffer, V., Crittin, J., Grolimund, L. & Lingner, J. 2013. The THO complex component Thp2 counteracts telomeric R-loops and telomere shortening. *EMBO J*, 32, 2861–71.

Pirzio, L. M., Pichierri, P., Bignami, M. & Franchitto, A. 2008. Werner syndrome helicase activity is essential in maintaining fragile site stability. *J Cell Biol*, 180, 305–14.

Pomerantz, R. T. & O'donnell, M. 2008. The replisome uses mRNA as a primer after colliding with RNA polymerase. *Nature*, 456, 762–6.

Pomerantz, R. T. & O'donnell, M. 2010. Direct restart of a replication fork stalled by a head-on RNA polymerase. *Science*, 327, 590–2.

Pourkarimi, E., Bellush, J. M. & Whitehouse, I. 2016. Spatiotemporal coupling and decoupling of gene transcription with DNA replication origins during embryogenesis in C. *elegans*. *Elife*, 5.

Proietti-De-Santis, L., Drane, P. & Egly, J. M. 2006. Cockayne syndrome B protein regulates the transcriptional program after UV irradiation. *EMBO J*, 25, 1915–23.

Pryde, F., Khalili, S., Robertson, K., Selfridge, J., Ritchie, A. M., Melton, D. W., Jullien, D. & Adachi, Y. 2005. 53BP1 exchanges slowly at the sites of DNA damage and appears to require RNA for its association with chromatin. *J Cell Sci*, 118, 2043–55.

Quinn, J. J. & Chang, H. Y. 2016. Unique features of long non-coding RNA biogenesis and function. *Nat Rev Genet*, 17, 47–62.

Raha, D., Wang, Z., Moqtaderi, Z., Wu, L., Zhong, G., Gerstein, M., Struhl, K. & Snyder, M. 2010. Close association of RNA polymerase II and many transcription factors with Pol III genes. *Proc Natl Acad Sci U S A*, 107, 3639–44.

Rastogi, R. P., Richa, K. A., Tyagi, M. B. & Sinha, R. P. 2010. Molecular mechanisms of ultraviolet radiation-induced DNA damage and repair. *J Nucleic Acids*, 2010, 592980.

Robbiani, D. F., Bothmer, A., Callen, E., Reina-San-Martin, B., Dorsett, Y., Difilippantonio, S., Bolland, D. J. et al. 2008. AID is required for the chromosomal breaks in c-myc that lead to c-myc/IgH translocations. *Cell*, 135, 1028–38.

Ruiz, J. F., Gomez-Gonzalez, B. & Aguilera, A. 2011. AID induces double-strand breaks at immunoglobulin switch regions and c-MYC causing chromosomal translocations in yeast THO mutants. *PLoS Genet*, 7, e1002009.

Sabouri, N., Mcdonald, K. R., Webb, C. J., Cristea, I. M. & Zakian, V. A. 2012. DNA replication through hard-to-replicate sites, including both highly transcribed RNA Pol II and Pol III genes, requires the S. *pombe Pfh1 helicase*. *Genes Dev*, 26, 581–93.

Sanz, L. A., Hartono, S. R., Lim, Y. W., Steyaert, S., Rajpurkar, A., Ginno, P. A., Xu, X. & Chedin, F. 2016. Prevalent, Dynamic, and Conserved R-Loop Structures Associate with Specific Epigenomic Signatures in Mammals. *Mol Cell*, 63, 167–78.

Saponaro, M., Kantidakis, T., Mitter, R., Kelly, G. P., Heron, M., Williams, H., Soding, J., Stewart, A. & Svejstrup, J. Q. 2014. RECQL5 controls transcript elongation and suppresses genome instability associated with transcription stress. *Cell*, 157, 1037–49.

Sarras, H., Alizadeh Azami, S. & Mcpherson, J. P. 2010. In search of a function for BCLAF1. *Sci World J*, 10, 1450–61.

Savage, K. I., Gorski, J. J., Barros, E. M., Irwin, G. W., Manti, L., Powell, A. J., Pellagatti, A. et al. 2014. Identification of a BRCA1-mRNA splicing complex required for efficient DNA repair and maintenance of genomic stability. *Mol Cell*, 54, 445–59.

Schneiter, R., Guerra, C. E., Lampl, M., Gogg, G., Kohlwein, S. D. & Klein, H. L. 1999. The Saccharomyces cerevisiae hyperrecombination mutant hpr1Delta is synthetically lethal with two conditional alleles of the acetyl coenzyme A carboxylase gene and causes a defect in nuclear export of polyadenylated RNA. *Mol Cell Biol*, 19, 3415–22.

Schubeler, D., Scalzo, D., Kooperberg, C., Van Steensel, B., Delrow, J. & Groudine, M. 2002. Genome-wide DNA replication profile for Drosophila melanogaster: A link between transcription and replication timing. *Nat Genet*, 32, 438–42.

Schwab, R. A., Nieminuszczy, J., Shah, F., Langton, J., Lopez Martinez, D., Liang, C. C., Cohn, M. A., Gibbons, R. J., Deans, A. J. & Niedzwiedz, W. 2015. The Fanconi Anemia Pathway Maintains Genome Stability by Coordinating Replication and Transcription. *Mol Cell*, 60, 351–61.

Schwer, B., Wei, P. C., Chang, A. N., Kao, J., Du, Z., Meyers, R. M. & Alt, F. W. 2016. Transcription-associated processes cause DNA double-strand breaks and translocations in neural stem/progenitor cells. *Proc Natl Acad Sci U S A*, 113, 2258–63.

Schwertman, P., Lagarou, A., Dekkers, D. H., Raams, A., Van Der Hoek, A. C., Laffeber, C., Hoeijmakers, J. H. et al. 2012. UV-sensitive syndrome protein UVSSA recruits USP7 to regulate transcription-coupled repair. *Nat Genet*, 44, 598–602.

Shanbhag, N. M., Rafalska-Metcalf, I. U., Balane-Bolivar, C., Janicki, S. M. & Greenberg, R. A. 2010. ATM-dependent chromatin changes silence transcription in cis to DNA double-strand breaks. *Cell*, 141, 970–81.

Singh, J. & Padgett, R. A. 2009. Rates of in situ transcription and splicing in large human genes. *Nat Struct Mol Biol*, 16, 1128–33.

Skourti-Stathaki, K., Kamieniarz-Gdula, K. & Proudfoot, N. J. 2014. R-loops induce repressive chromatin marks over mammalian gene terminators. *Nature*, 516, 436–9.

Skourti-Stathaki, K., Proudfoot, N. J. & Gromak, N. 2011. Human senataxin resolves RNA/DNA hybrids formed at transcriptional pause sites to promote Xrn2-dependent termination. *Mol Cell*, 42, 794–805.

Sollier, J., Stork, C. T., Garcia-Rubio, M. L., Paulsen, R. D., Aguilera, A. & Cimprich, K. A. 2014. Transcription-coupled nucleotide excision repair factors promote R-loop-induced genome instability. *Mol Cell*, 56, 777–85.

Somyajit, K., Saxena, S., Babu, S., Mishra, A. & Nagaraju, G. 2015. Mammalian RAD51 paralogs protect nascent DNA at stalled forks and mediate replication restart. *Nucleic Acids Res*, 43, 9835–55.

Steinmetz, E. J., Conrad, N. K., Brow, D. A. & Corden, J. L. 2001. RNA-binding protein Nrd1 directs poly(A)-independent 3′-end formation of RNA polymerase II transcripts. *Nature*, 413, 327–31.

Storici, F., Bebenek, K., Kunkel, T. A., Gordenin, D. A. & Resnick, M. A. 2007. RNA-templated DNA repair. *Nature*, 447, 338–41.

Stork, C. T., Bocek, M., Crossley, M. P., Sollier, J., Sanz, L. A., Chedin, F., Swigut, T. & Cimprich, K. A. 2016. Co-transcriptional R-loops are the main cause of estrogen-induced DNA damage. *Elife*, 5.

Strasser, K., Masuda, S., Mason, P., Pfannstiel, J., Oppizzi, M., Rodriguez-Navarro, S., Rondon, A. G. et al. 2002. TREX is a conserved complex coupling transcription with messenger RNA export. *Nature*, 417, 304–8.

Stuckey, R., Garcia-Rodriguez, N., Aguilera, A. & Wellinger, R. E. 2015. Role for RNA:DNA hybrids in origin-independent replication priming in a eukaryotic system. *Proc Natl Acad Sci U S A*, 112, 5779–84.

Stults, D. M., Killen, M. W., Pierce, H. H. & Pierce, A. J. 2008. Genomic architecture and inheritance of human ribosomal RNA gene clusters. *Genome Res*, 18, 13–8.

Stults, D. M., Killen, M. W., Williamson, E. P., Hourigan, J. S., Vargas, H. D., Arnold, S. M., Moscow, J. A. & Pierce, A. J. 2009. Human rRNA gene clusters are recombinational hotspots in cancer. *Cancer Res*, 69, 9096–104.

Suraweera, A., Becherel, O. J., Chen, P., Rundle, N., Woods, R., Nakamura, J., Gatei, M. et al. 2007. Senataxin, defective in ataxia oculomotor apraxia type 2, is involved in the defense against oxidative DNA damage. *J Cell Biol*, 177, 969–79.

Taschner, M., Harreman, M., Teng, Y., Gill, H., Anindya, R., Maslen, S. L., Skehel, J. M., Waters, R. & Svejstrup, J. Q. 2010. A role for checkpoint kinase-dependent Rad26 phosphorylation in transcription-coupled DNA repair in Saccharomyces cerevisiae. *Mol Cell Biol*, 30, 436–46.

Taylor, J. K., Zhang, Q. Q., Wyatt, J. R. & Dean, N. M. 1999. Induction of endogenous Bcl-xS through the control of Bcl-x pre-mRNA splicing by antisense oligonucleotides. *Nat Biotechnol*, 17, 1097–100.

Tucker, S. L., Reece, J., Ream, T. S. & Pikaard, C. S. 2010. Evolutionary history of plant multisubunit RNA polymerases IV and V: Subunit origins via genome-wide and segmental gene duplications, retrotransposition, and lineage-specific subfunctionalization. *Cold Spring Harb Symp Quant Biol*, 75, 285–97.

Tuduri, S., Crabbe, L., Conti, C., Tourriere, H., Holtgreve-Grez, H., Jauch, A., Pantesco, V. et al. 2009. Topoisomerase I suppresses genomic instability by preventing interference between replication and transcription. *Nat Cell Biol*, 11, 1315–24.

Valadkhan, S. & Gunawardane, L. S. 2013. Role of small nuclear RNAs in eukaryotic gene expression. *Essays Biochem*, 54, 79–90.

Van Der Meijden, C. M., Lapointe, D. S., Luong, M. X., Peric-Hupkes, D., Cho, B., Stein, J. L., Van Wijnen, A. J. & Stein, G. S. 2002. Gene profiling of cell cycle progression through S-phase reveals sequential expression of genes required for DNA replication and nucleosome assembly. *Cancer Res*, 62, 3233–43.

Venkata Narayanan, I., Paulsen, M. T., Bedi, K., Berg, N., Ljungman, E. A., Francia, S., Veloso, A., Magnuson, B., Di Fagagna, F. D., Wilson, T. E. & Ljungman, M. 2017. Transcriptional and post-transcriptional regulation of the ionizing radiation response by ATM and p53. *Sci Rep*, 7, 43598.

Wang, Y., Chakravarty, P., Ranes, M., Kelly, G., Brooks, P. J., Neilan, E., Stewart, A., Schiavo, G. & Svejstrup, J. Q. 2014. Dysregulation of gene expression as a cause of Cockayne syndrome neurological disease. *Proc Natl Acad Sci U S A*, 111, 14454–9.

Wei, X., Samarabandu, J., Devdhar, R. S., Siegel, A. J., Acharya, R. & Berezney, R. 1998. Segregation of transcription and replication sites into higher order domains. *Science*, 281, 1502–6.

Williamson, L., Saponaro, M., Boeing, S., East, P., Mitter, R., Kantidakis, T., Kelly, G. P. et al. 2017. UV Irradiation Induces a Non-coding RNA that Functionally Opposes the Protein Encoded by the Same Gene. *Cell*, 168, 843–855 e13.

Wooster, R., Neuhausen, S. L., Mangion, J., Quirk, Y., Ford, D., Collins, N., Nguyen, K. et al. 1994. Localization of a breast cancer susceptibility gene, BRCA2, to chromosome 13q12-13. *Science*, 265, 2088–90.

Yeo, A. J., Becherel, O. J., Luff, J. E., Cullen, J. K., Wongsurawat, T., Jenjaroenpun, P., Kuznetsov, V. A., Mckinnon, P. J. & Lavin, M. F. 2014. R-loops in proliferating cells but not in the brain: Implications for AOA2 and other autosomal recessive ataxias. *PLOS ONE*, 9, e90219.

Yu, K., Chedin, F., Hsieh, C. L., Wilson, T. E. & Lieber, M. R. 2003. R-loops at immunoglobulin class switch regions in the chromosomes of stimulated B cells. *Nat Immunol*, 4, 442–51.

Zhu, Y., Peterson, C. L. & Christman, M. F. 1995. HPR1 encodes a global positive regulator of transcription in Saccharomyces cerevisiae. *Mol Cell Biol*, 15, 1698–708.

RNA Binding Proteins and the DNA Damage Response

11

Roger J. A. Grand

INTRODUCTION

RNA-binding proteins (RBPs), defined as proteins that bind to double- and single-stranded RNA, are a principal component of ribonucleoprotein complexes. RBPs contain one or more copies of various well-defined structural motifs, such as RNA recognition motifs (RRMs), quasi-RRMs which have some similarity to the structure of RRMs but bind RNA through a different conformation, RGG (arginine-glycine-glycine) boxes, dsRNA binding domains, K homology (KH) domains, DEAD/DEAH (aspartic acid–glutamic acid–alanine–aspartic acid/aspartic acid–glutamic acid–alanine–histidine) motifs and zinc fingers, all of which can bind RNA (**Figure 11.1**). Structural analysis of the RNA binding domains suggests that they are relatively flexible and bind to RNA in either a sequence-dependent or independent manner (Thickman et al. 2007). Furthermore, it has been suggested that RNA binding proteins have a high degree of structural disorder, particularly in regions between RNA binding motifs, and this property is significant in the interaction with RNA (Basu and Bahadur 2016).

RBPs are integral to multiple cellular pathways, such as those essential for gene and protein expression, and mRNA transport and localisation. In particular, they play major roles in the post-transcriptional processing of mRNAs, being central to splicing, capping, polyadenylation, mRNA stabilisation, localisation and translation; they can also act as destabilizers and modifiers. Early estimates had suggested that there were in excess of 500 genes encoding RBPs in mammalian cells, each with the ability to bind specific mRNAs and interact with other proteins. However, more recent studies have given figures appreciably in excess of 1,500 for the number of RBPs in the human genome, equivalent to approximately 8% of all protein-coding genes (Ashburner et al. 2000, Gerstberger et al. 2014). In addition, it was suggested that these RBPs comprise something in the order of 600 distinct structural classes (Gerstberger et al. 2014).

RBPs are present in both the cytoplasm and nucleus of eukaryotic cells, with the majority of those RBPs present in the nucleus being associated with pre-mRNAs and mRNAs in ribonucleoprotein (RNP) particles. RNPs are the functional forms in which pre-mRNAs and mRNAs are present in the cell, such that the RBP components determine the export, stability and translation of the mRNA. As well as binding to specific mRNAs, RNA binding proteins constitute much of the spliceosome, the exon junction complex, the cap binding complex and the TREX (transcription–export) complex. However, over the last few years, it has become apparent that RNA binding proteins have functions far beyond their roles in RNA metabolism. In this chapter, I have concentrated on the ways in which RBPs impinge on the cellular DNA damage response (DDR); this occurs at a number of levels, for example, through post-transcriptional gene regulation following DNA damage, through direct effects at DNA damage foci and through direct interaction with DDR proteins and long non-coding RNAs (lncRNAs). In addition, RBPs have been linked to a number of diseases, including cancers and multiple neurodegenerative diseases, although it is not clear to what extent this is related to their roles in the DDR. The literature relevant to RNA binding proteins is immense and even when limited to their role in the DNA damage response, it is appreciable. Therefore, here some areas of published research are covered in detail, whereas others are considered only in passing. This is a reflection of my own interest rather than an indication of their scientific or physiological importance. A detailed description of the relationship between transcription and DNA damage is presented in the previous chapter, and readers requiring an in-depth

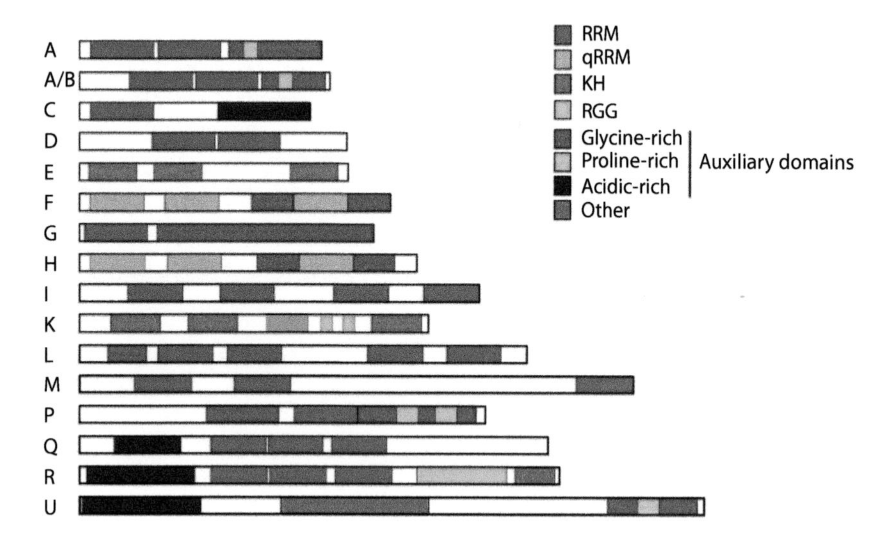

Figure 11.1 Comparison of the structures of hnRNPs. hnRNPs are listed on the left-hand side of the figure. Positions of the conserved domains are shown in different colours. RRM, RNA recognition motif; qRRM, quasi-RNA recognition motif; KH, hnRNPK homology domain; RGG, arginine-glycine-glycine–rich domain. RNA binding is generally through an RRM or quasi RRM (quasi RRMs, although broadly similar to classical RRMs, lack RNP consensus sequences and therefore have a different mode of RNA recognition). hnRNPU binds RNA through its RGG motif. (From Geuens, T. et al. 2016, *Hum Genet* 135(8): 851–867.)

discussion of the roles of RNA polymerases and transcription per se in the context of DNA damage and genome instability are referred there. A number of other recent reviews on the subject of RNA binding proteins and the DNA damage response can be recommended for readers in need of greater detail or a different viewpoint (Dutertre et al. 2014, D'Alessandro and d'Adda di Fagagna 2017, Nishida et al. 2017).

INDIRECT ACTIONS OF RNA BINDING PROTEINS ON THE DNA DAMAGE RESPONSE

RNA binding proteins, protein expression and DNA damage

In response to DNA damage, there is an immediate widespread but transient repression of gene transcription (Mayne and Lehmann 1982, Rockx et al. 2000, Muñoz et al. 2009, Reinhardt et al. 2011, Williamson et al. 2017). For example, Rieger and Chu (2004) have shown in a detailed analysis that the expression of 24% of all genes changes in response to ionising radiation (IR) and 32% change following ultraviolet (UV) treatment. Reduced transcriptional activity after UV radiation, for example, coincides with a marked reduction of RNA polymerase II hypophosphorylated in the C-terminal domain; this is the form of RNA polymerase II considered to be involved in initiation rather than elongation (Rockx et al. 2000). However, a more recent study indicates that there is also an initial marked reduction in transcript elongation, followed by a slower decrease in initiation, following UV irradiation (Williamson et al. 2017). Modified binding of TATA-binding protein (TBP) to DNA as well as alterations in the chromatin structure after UV irradiation have also been reported (Vichi et al. 1997, Adam et al. 2013). Such an overall reduction in protein expression after DNA damage is to be expected in order to reduce the production of mutated proteins. However, there is an obvious requirement, at the same time, for specific proteins to deal with the DNA damage and elicit repair or apoptosis if the damage is too great. In general, this is achieved by stabilisation of pre-mRNAs and mRNAs by association with specific RBPs. Maintenance of expression of DDR proteins following DNA damage may also be achieved by the selective translation of particular mRNAs (Braunstein et al. 2009, Powley et al. 2009). Following UV radiation, it has been demonstrated that upstream ORFs in the 5′ untranslated regions of these mRNAs are central to the selection of particular mRNAs for translation and this process can be regulated by DNA-PK activity (Powley et al. 2009). Following UV radiation, mRNAs encoding ERCC1, ERCC5, DDB1, XPA, XPD and OGG1 are preferentially translated, for example (Powley et al. 2009). In another study, it was shown that ionising radiation enhances the translation of a set of mRNA encoding proteins involved in DNA repair and cell survival, at early times after damage, followed by a reduction in translation at later times. The initial response, in this study, was through the activation of the MAP kinases ERK1 and ERK2, whereas the later inhibition of translation after IR requires activation of ATM (Braunstein et al. 2009).

It should be noted, however, that DNA damage (by UV radiation and hydroxyurea treatment, for example) leads to transient, widespread repression of 3′ mRNA processing (Kleiman and Manley 1999, 2001). This has been shown to be due to the formation of a complex between the mRNA polyadenylation factor cleavage stimulating factor-50 (CstF-50) and BARD1 and BRCA1 (Kleiman and Manley 2001, Mirkin et al. 2008). Further studies have demonstrated the presence of p53 in CstF/BARD1 complexes after DNA damage, linking 3′ mRNA processing to the p53 signalling pathway (Nazeer et al. 2011). Significantly, in these experiments, it was found that although there was an inverse correlation between p53 level and general 3′ mRNA cleavage, there was no effect on the polyadenylation of DDR genes (Nazeer et al. 2011). Although these appear to be major mechanisms determining protein levels after DNA damage, there is also considerable evidence to suggest that RBPs can play a significant role in the processes of downregulating and upregulating expression. Some of these pathways are discussed in the following sections.

Although there is a widespread inhibition of 3′ end formation, leading to repression of mRNA synthesis, in response to DNA damage, as mentioned earlier, this appears not to apply to p53 itself. Due to an interaction between hnRNP H/F and a G-quadruplex structure located downstream from the p53 cleavage site, 3′ end processing of p53 is successfully accomplished ensuring expression of p53 after DNA damage (Decorsière et al. 2011). As well as the inhibition of 3′ end processing affecting protein expression after DNA damage, it is also clear that the adoption of alternative splicing and changes in splicing efficiency are of considerable importance (see review by Dutertre et al. 2011). It has now been shown that U1snRNA, a component of U1snRNP which plays an important role in splicing and 3′ end processing, regulates intronic alternative cleavage and polyadenylation during the DDR, with downregulation of U1snRNA regulating alternative cleavage and polyadenylation (Devany et al. 2016). The activation of intronic alternative cleavage and polyadenylation results in widespread production of truncated transcripts, and this affects genes involved in the DDR in particular (Devany et al. 2016). Some of the alternatively spliced variants contain premature stop codons, which can be subject to nonsense-mediated decay (see Section 'Nonsense-mediated decay') (Dutertre et al. 2011). Importantly, RNA polymerase II subunit A is ubiquitylated in response to UV DNA damage and rapidly degraded (Bregman et al. 1996, Ratner et al. 1998), leading to a general reduction in transcript level. Ubiquitylation and subsequent degradation require phosphorylation of a site in the C-terminal domain of RNA polymerase II, and this is considered characteristic of the elongating pol II (Ratner et al. 1998, Mitsui and Sharp 1999).

RNA stability and the DNA damage response

There is a general reduction in the stability of many mRNAs after DNA damage, although this is accompanied by an increase in the stability of others. Using a global analysis technique, it was shown that, in response to UV radiation, for the majority of genes, both changes in transcription and RNA turnover are involved in their altered expression (Fan et al. 2002). In this study, mRNA stabilisation was a major cause of increased expression of a number of genes, including those involved in the DDR, whereas mRNA destabilisation was responsible for decreasing the expression of an appreciably larger group of genes (Fan et al. 2002). This was in addition to transcriptional regulation, which also affects a significant group of genes after UV radiation (Fan et al. 2002). In a study of the effects of IR, it was shown that some transcripts were stabilised and/or synthesised, whereas others were destabilised in a fashion comparable to the effect of UV. Increased RNA synthesis and stability was shown to be ATM and p53 dependent (Canman et al. 1998, Venkata Narayanan et al. 2017).

The marked decrease in mRNA stability and translation following DNA damage, leading to a reduction in protein expression, is determined by a number of factors. Certain RBPs bind mRNAs and promote their stabilisation; therefore, dissociation of this interaction leads to reduced mRNA levels. For example, the RNA binding protein HuR (human antigen R) associates with a large number of coding and non-coding transcripts and regulates their splicing, stabilisation and translation (reviewed, e.g., by Grammatikakis et al. 2017). HuR binds to U- and AU-rich regions of RNAs. HuR level has been shown to be reduced by miRNAs which target the 3′ UTR and coding region of the mRNA, but its properties are mainly determined by post-translational modification. Phosphorylation of HuR by Cdks regulates the sub-cellular distribution of the protein; for example, Cdk1 phosphorylation of S^{202} causes retention of HuR in the nucleus. UV activation of ATR results in phosphorylation of Cdk1, abolishing its binding to HuR, which then translocates into the cytoplasm where it associates with p21 mRNA and enhances its translation (Al-Khalaf and Aboussekhra 2014). HuR is also phosphorylated at multiple sites by Chk2 (Abdelmohsen et al. 2007, Masuda et al. 2011). After hydrogen peroxide treatment, HuR was seen to be phosphorylated, primarily at S^{100}, and to dissociate from *SIRT1* mRNA, leading to its decay. In a further study, in response to IR Chk2-dependent phosphorylation of

HuR caused its dissociation from multiple target mRNAs and resulted in increased cell survival. The very large number of mRNAs associated with HuR in *wt* cells and in irradiated cells has been listed in Masuda et al. (2011). These authors found 1263 mRNAs associated with HuR in untreated cells but only 246 after irradiation, indicative of reduced stabilisation (Masuda et al. 2011). Ubiquitylation is also known to regulate both the level and activity of HuR. Thus, ubiquitylation following DNA damage causes degradation of HuR and subsequent decreased level of *XIAP* mRNA (Lucchesi et al. 2016). Ubiquitylation with a K29 polyubiquitin chain at K^{313} and K^{326}, however, causes HuR to dissociate from p21, MKP1 and SIRT1 target mRNAs, reducing their stability (Zhou et al. 2013). The ATP-dependent release of HuR from the mRNAs is mediated by a complex of p97 and UBXD8, and it has been suggested that these proteins could also regulate stability of other RNAs by ubiquitylation of other RBPs (Zhou et al. 2013).

miRNAs play a significant role in determining the DDR response by regulating mRNA levels. The level of a number of miRNAs is enhanced by RBPs such as KSRP (KH-type splicing regulatory protein) (Zhang et al. 2011). Activation of ATM leads to phosphorylation of, and binding to, KSRP, which is then able to promote pri-miRNA processing and miRNA expression (Zhang et al. 2011). miRNAs are discussed in more detail in a later section dealing with non-coding RNAs.

As mentioned earlier, there is an obvious requirement for DDR proteins to be exempted from the general reduction in protein expression after DNA damage. A further way in which this is achieved is by the ability of some RBPs to promote mRNA translation. This is particularly applicable to p53, where it has been shown that a number of RNA binding proteins, such as HuR, ribosomal protein L26 and HDM2, increase the translation of p53 mRNA (Takagi et al. 2005, Candeias et al. 2008, Gajjar et al. 2012). Of particular interest is the complex relationship between HDM2, HDMX, p53 and mRNAs. HDM2 binds to p53 mRNA following DNA damage and phosphorylation of HDM2 at S^{395} by ATM (Gajjar et al. 2012). p53 mRNA is also required for localisation of HDM2 to the nucleoli and SUMOylation in DNA damaged cells. Disruption of the p53 mRNA/HDM2 interaction prevents p53 stabilisation and activation following DNA damage (Gajjar et al. 2012). Moreover, HDMX (also known as MDM4), which is generally considered to be a negative regulator of p53 transcriptional activity, is phosphorylated at S^{403} by ATM and binds p53 mRNA, producing a conformation which enhances its interaction with HDM2 and the induction of p53 synthesis (Malbert-Colas et al. 2014). It has also been shown that both HDMX and HDM2 bind to the same p53 internal ribosome entry sequences (IRES) and act as IRES trans-acting factors, resulting in an altered structure of p53 mRNA and a stimulation of translation (Malbert-Colas et al. 2014). In addition to HDM2 and HDMX acting as RNA binding proteins, p53 itself can bind RNA, and this has been shown to be of particular importance in the control of p21 mRNA translation in some circumstances (Riley and Maher 2007, Mlynarczyk and Fåhraeus 2014). p53 binds to the 5′ UTR of HDMX mRNA and suppresses its translation (Tournillon et al. 2017). It appears that this interaction is through the central DNA binding domain of p53, but there is also an essential requirement for an N-terminal suppression domain (aa37-aa43) to inhibit HDMX synthesis (Tournillon et al. 2017). Thus, it is clear that p53 is a member of that rather rare class of proteins which can bind both the DNA and RNA of a target gene, giving considerable subtlety of regulation. Under normal conditions, HDM2 translocates p53 to the cytoplasm, where it is ubiquitylated by HDM2 with the assistance of HDMX. After DNA damage, p53, HDM2 and HDMX are phosphorylated at multiple sites, promoting the dissociation of the complexes and upregulation of p53 expression. Phosphorylation of HDM2 by ATM also allows its binding to p53 mRNA, SUMOylation of HDM2 and accumulation of HDM2 in the nucleolus. HDM2 bound to p53 mRNA enhances its translation and causes the inhibition of HDM2 E3 ligase activity. Phosphorylation of HDMX by ATM promotes its association with the p53 mRNA-HDM2 complex through binding to the mRNA, favouring p53 protein expression. The interaction between p53 and HDMX mRNA is not affected by DNA damage and probably functions to keep HDMX levels low when p53 has been activated.

The spliceosome and DNA repair

Because the response to DNA damage is so widespread within the cell, it involves pathways at all levels of gene expression. RNA binding proteins are integral to many of these pathways. Pre-mRNA splicing, by which introns are removed from a pre-mRNA and the exons ligated, occurs in the spliceosome (**Figure 11.2**). This is a large nuclear multi-component complex which comprises five small nuclear ribonucleoproteins (snRNPs) – U1, U2, U4, U5 and U6 – together with a large number of auxiliary proteins, many of which are also RNA binding proteins

(Wahl et al. 2009). In the mammalian system, almost all RNA polymerase II–transcribed genes, with the exception of histones, contain introns and therefore have the potential for alternative splicing. This can give rise to distinct mRNAs which may encode variant proteins with quite distinct properties (Giono et al. 2016, Kelemen et al. 2013, Shkreta and Chabot 2015).

Alternative splicing following DNA damage

Transcription and splicing at the sites of DNA damage are generally appreciably reduced to facilitate access to repair proteins. However, it is now apparent that alternative splicing and expression of a number of DDR genes occurs after damage, giving rise to proteins with modified functions (e.g. Giono et al. 2016, Magnuson et al. 2016). Alternative splicing of other genes can give rise to mRNAs containing premature stop codons, which are unstable (Dutertre et al. 2011). These truncated products can be substrates for nonsense-mediated decay (see Section 11.1.6). The alternative splicing can be regulated in a number of ways; for example, by splice site selection by mRNA binding by hnRNPs, the turnover of splicing regulators, phosphorylation of splicing regulators, exon skipping regulated by SR proteins and an inhibition of transcriptional elongation caused by phosphorylation of RNA polymerase II C-terminal domain (Muñoz et al. 2009, Moore et al. 2010). It has recently been shown that, in response to UV, there is a slowdown in elongation of transcripts and subsequent shift from expression of longer mRNAs to shorter forms through the use of splice sites which give rise to shorter last exons (Williamson et al. 2017). In addition, it has been shown that there are changes to the spliceosome itself after damage, resulting in alternatively spliced proteins, and this has been linked to the activation of ATM (Tresini et al. 2015). A complex comprising BRCA1, BCLAF1, Prp8, U2AF65, U2AF35, SF3B1 and other components of the mRNA-splicing machinery is activated in response to DNA damage (Savage et al. 2014). The

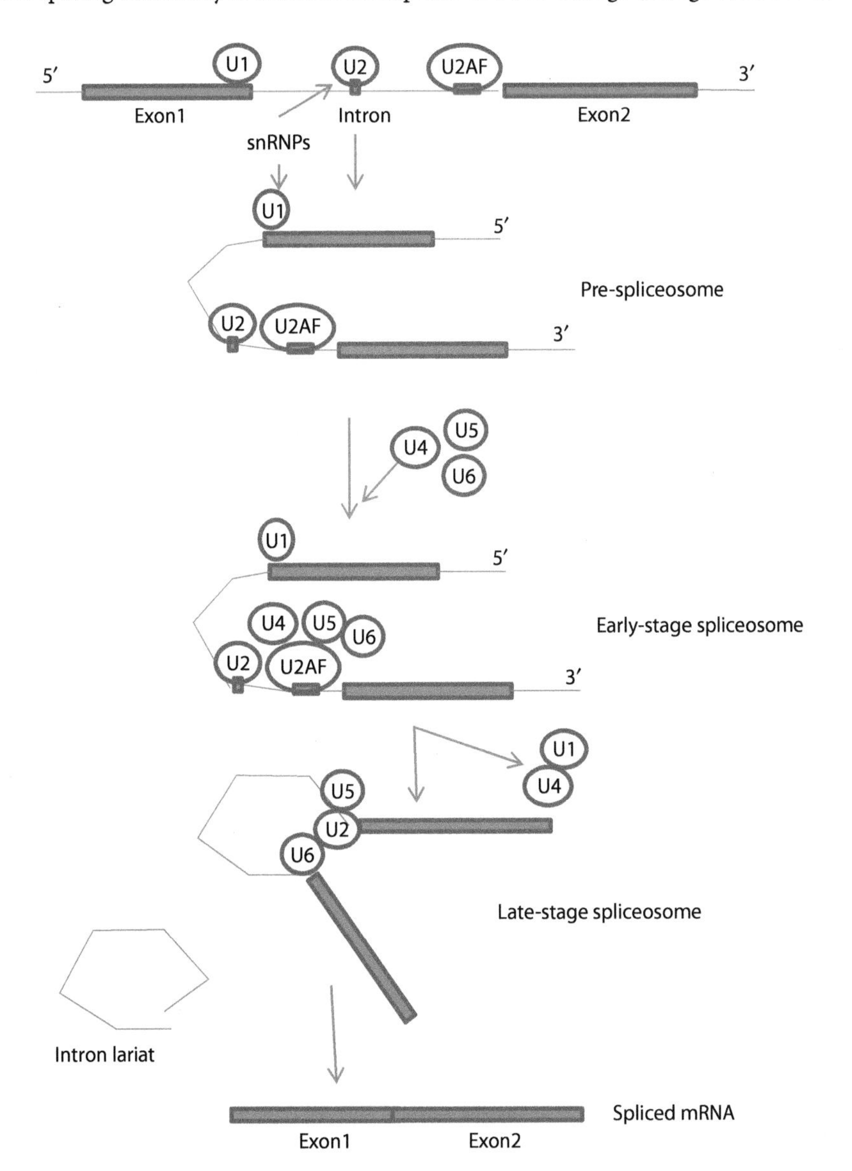

Figure 11.2 Spliceosome assembly. The major components of the spliceosome are assembled on the pre-mRNA in two stages. U1, U2, U4, U5 and U6 are the five major snRNPs that make up the spliceosome. A large number of accessory proteins are involved but not shown. U2AF is made up of U2AF2 (U2AF65) and U2AF1 (U2AF35).

complex regulates pre-mRNA splicing of a number of genes involved in the DDR, such as EXO1, BACH and ATRIP, by increasing the stability of their mRNAs (Savage et al. 2014). The importance of BCLAF1 in the regulation of splicing has been confirmed by the observation that it forms part of a complex with SNIP1 (SMAD nuclear interacting protein-1) and SKIP (Ski-interacting protein) to regulate splicing of the *Ccnd1* (encoding cyclin D1) gene (Bracken et al. 2008). The complex is recruited to the 3′ ends of the cyclin D1 gene and cyclin D1 RNA. SNIP1 and SKIP are required for the further recruitment of the 3′ splice site recognition factor U2AF65 (Bracken et al. 2008). A similar complex is essential for p21, but not PUMA, splicing and expression after DNA damage and in normally growing cells (Chen et al. 2011). In this study, SKIP was shown to interact with U2AF65, a subunit of the U2AF splice site recognition complex, and recruit it to the p21 gene and mRNA. Dutertre and colleagues (2010) have shown that, following DNA damage, alternative spliced transcripts of HDM2 are produced independently of p53. After camptothecin treatment, for example, there is a marked reduction in the interaction between EWSR1 and YB-1, a spliceosome-associated factor. This results in co-transcriptional exon skipping of the HDM2 gene and reduced protein expression but increased p53 level (Dutertre et al. 2010). In the same study, the authors listed a large number of exons that are skipped following camptothecin treatment and dissociation of the EWSR1/YB-1 complex. These include DDR proteins such as RIF1, BRCA2, RAD51C and BRIP1 as well as a number of RBPs (Dutertre et al. 2010). A further more extensive list linking genes which are alternatively spliced in response to particular DNA damaging agents in specific cell lines has been reported by Shkreta and Chabot (2015). These include a significant number of pro- and anti-apoptotic genes, such as caspases 2, 8 and 9; Bcl-x; APAF1 and DIABLO; and DDR and cell cycle regulatory genes such as GADD45, NBS1, RAD17, ERCC1, HDM2, HDMX and CDC25C (Shkreta and Chabot 2015). In many cases, however, the functional significance of the alternative splicing has not been evaluated.

Many of these changes in RNA splicing in response to DNA damage result from post-translational modifications of splicing factors (reviewed by Shkreta and Chabot 2015). Large-scale studies, which have examined proteins phosphorylated after DNA damage, have identified a number of splicing factors and other RNA binding proteins as substrates. For example, it was shown that Chk2 can interact with and phosphorylate CDK11, which, in turn, stimulates pre-mRNA splicing (Choi et al. 2014). CDK11 associates with a number of proteins involved in RNA splicing such as RNPS1, 9G8 and cyclin L. In cells depleted of CDK11 pre-mRNA splicing is reduced, whereas over-expression results in increased splicing (Choi et al. 2014). Phosphorylation of the serine/arginine-rich (SR) family of proteins is known to regulate their distribution and therefore their activity. These proteins are important regulators of splicing with roles in splice site selection, spliceosome assembly, and constitutive and alternative splicing. Tip60 regulates phosphorylation of SRSF2 by inhibiting the nuclear localisation of the SRPK1 and SRPK2 kinases (Edmond et al. 2011). In addition, it enhances proteasomal degradation of SRSF2 by acetylating K^{52} in the RNA recognition motif. Relevantly, it was shown that acetylation and phosphorylation of SRSF2, following cisplatin-induced DNA damage, regulates protein levels and the pre-mRNA splicing of the *caspase 8* gene (Edmond et al. 2011). Other post-translational modifications, such as ubiquitylation, SUMOylation and methylation, are also known to regulate splicing factors which impinge on the DDR (reviewed by Shkreta and Chabot 2015). SRSF1 binds to the SUMO E2 conjugating enzyme Ubc9, enhancing global SUMOylation (Pelisch et al. 2010). It also interacts, through its RNA recognition motif, with the E3 ligase PIAS1, regulating PIAS1-mediated SUMOylation (Pelisch et al. 2010). As mentioned at other points in this chapter, it is becoming clear that PARylation is of great significance as a post-translational modification, particularly in relation to the DDR. The splicing factors RBMX, hnRNPA1 and the non-Pou domain-containing octamer-binding protein (NONO, nuclear RNA-binding protein $p54^{nrb}$) all bind to PAR, and this may result in their recruitment to sites of damage. For example, NONO and SFPQ interact and have a number of roles in RNA metabolism as well as being implicated in the DDR. NONO is transiently recruited to sites of damage in a PARP-dependent manner through binding to PAR (Krietsch et al. 2012). Its depletion inhibits NHEJ and stimulates HR; in addition, using siRNA and PARP inhibitors, it has been shown that PARP and NONO function in the same NHEJ pathway (Krietsch et al. 2012). These latter splicing factors are discussed in greater detail in the section entitled 'Additional RNA binding proteins involved in the DNA damage response', and PARylation in the section 'RNA binding proteins at DNA damage foci'.

A further link between splicing and the DDR is the observation that depletion of splicing factors themselves can cause DNA damage (Paulsen et al. 2009). This is particularly relevant to proteins involved in the cellular response to the formation of R-loops. Depletion of splicing factors, such as SR proteins and hnRNPs, increases the cellular concentration of free RNA, which can then hybridise DNA to template strands, giving rise to R-loops, subsequently causing genomic instability. Similarly, inactivation or depletion of the serine/arginine-rich splicing factors SRSF1 and SRSF2 favours the formation of R-loops and subsequent DNA damage (Li and Manley 2006). R-loops are considered in more detail in the section 'RNA binding proteins and the prevention of R-loops' and Chapter 10.

Sub-cellular distribution of RNA binding proteins and DNA damage

Although hnRNPs and splicing factors are predominantly nuclear, most are able to shuttle between the nucleus and the cytoplasm; in addition, RBPs are commonly relocated within the nucleus following DNA damage. SR factors are normally in the nucleoplasm but also accumulate in 'splicing speckles,' which have been suggested to be reservoirs of splicing proteins (Biamonti and Caceres 2009). Following phosphorylation, SR proteins are translocated into the nucleoplasm, where they are involved in transcription and mRNA maturation. Biamonti and Vourc'h (2010) have suggested that SRSF1, SRSF9, hnRNPK and SAFB are recruited to nuclear stress bodies which are active transcription sites after DNA damage. In other studies, it has been demonstrated that treatment of cells with a topoisomerase II inhibitor results in translocation of a number of RBPs involved in alternative splicing, such as Sam68 and the SR proteins SC35 and ASF, from the nucleoplasm to nuclear granules which also contain phosphorylated RNA polymerase II and are sites of transcription (Busà et al. 2010).

There is also considerable evidence to indicate that certain RNA binding proteins are recruited to DNA damage foci. These include hnRNPC, which is recruited to damage sites and has been suggested to be required for the pre-mRNA splicing and expression of BRCA1, BRCA2 and RAD51 (Anantha et al. 2013), and RBMX, which is required for BRCA2 expression (Adamson et al. 2012). Other RNA binding proteins observed at damage sites include hnRNPUL1, FUS, PSF, TAF15 and EWSR1; this is discussed in more detail in the section entitled 'Direct interactions between RNA binding proteins and DNA damage response components'. A number of serine/arginine-rich RBPs are relocalised to specific areas of the nucleolus, and this relocalisation has been linked to modified splicing after DNA damage. Similarly, EWSR1 transiently localises to nucleoli after DNA damage, resulting in alternative splicing of DDR genes encoding Chk2, c-Abl and MAP4K2, and reduced protein expression (Paronetto et al. 2011). HDM2 also localises to the nucleolus after DNA damage and this requires the interaction of the protein with p53 mRNA (Gajjar et al. 2012). This promotes HDM2 SUMOylation and the stabilisation of p53.

Shuttling of RBPs between the nucleus and the cytoplasm in response to DNA damage can be seen as a mechanism for the control of mRNA maturation and translation. PTB translocation into the cytoplasm after DNA damage contributes to the co-ordinated translation of p53 isoforms (Grover et al. 2008). Similarly, UV-induced redistribution of hnRNPA1 to the cytoplasm, regulated by phosphorylation by MKK (3/6)-p38, affects alternative splicing (van der Houven van Oordt et al. 2000, Guil et al. 2006). HuR is also exported from the nucleus after DNA damage (Kim et al. 2010). Activation of Chk1 results in the phosphorylation and inactivation of Cdk1 and this allows export of HuR into the cytoplasm, where it can act as a regulator of alternative splicing as discussed above (Izquierdo 2008, Kim et al. 2010).

RNA binding proteins and the prevention of R-loops

As mentioned earlier, RBPs are essential for the prevention of R-loop formation (Aguilera and García-Muse 2012). Transcriptional R-loops are formed when an RNA transcript hybridises with the transcribed DNA template leaving the other DNA strand unpaired (Huertas and Aguilera 2003). R-loops are particularly susceptible to DNA damaging agents and can provide a block to incoming replication forks. R-loops tend to form in regions of G-rich DNA repeats and where negative supercoiling accumulates behind the transcriptional machinery. A number of mechanisms are present to reduce the occurrence of R-loops, and most involve RNA binding proteins. As soon as mRNA molecules are released from the transcriptional apparatus, they are bound by RNA binding proteins to form RNPs, reducing the potential for R-loop formation. DNA topoisomerase I is active to prevent the negative supercoiling induced by polymerases by relaxing and transiently cleaving the DNA. Many of the proteins associated with TOP1 are RBPs; SRSF1 and SRSF3, for example, co-operate with TOP1 to prevent R-loop formation (Li and Manley 2005, Tuduri et al. 2009). TOP1 has also been shown to associate with a large number of other RNA binding proteins, including hnRNPs and members of the spliceosome (Czubaty et al. 2005). The THO-TREX complex, containing a number of RBPs, also plays a central role in the prevention of R-loops, as it has been shown in a number of organisms that mutation of its components favours R-loop formation (Huertas and Aguilera 2003, Masuda et al. 2005). This complex is involved in coupling transcription and RNA processing to nuclear export of transcripts (Bermejo et al. 2011, Aguilera and García-Muse 2012). RECQL5 helicase interacts with DNA topoisomerase I and promotes its SUMOylation by the E3 ligase PIAS1 together with SRSF1 (Li et al. 2015). This facilitates the binding of TOP1 to RNA polymerase II and the recruitment of RNA splicing factors to chromatin which is being transcribed, reducing the possibility of R-loop formation (Li et al. 2015). The addition of splicing inhibitors or depletion of proteins which associate with splicing factors also leads to R-loop formation and DDR activation (e.g. Wan et al. 2015). R-loops are discussed in greater detail in Chapter 10.

Nonsense-mediated decay

Nonsense-mediated decay (NMD) has been considered a pathway that controls the degradation of RNAs that terminate prematurely, RNAs with nonsense mutations and mRNAs from rearranged genes, amongst others (Kervestin and Jacobson 2012). The major NMD pathway components are Upf proteins: Upf1 is an RNA helicase, and Upf2 and Upf3 are required for its ATP binding and activation. Although NMD has been seen as a process for removing unwanted and unstable mRNAs, it also appears to have specific roles in the regulation of the DDR. For example, it has been shown that *RAD55, RAD51, RAD54* and *RAD57* transcript and protein levels, in *Saccharomyces cerevisiae*, are regulated by NMD (Janke et al. 2016). Loss of NMD components results in increased homologous recombination and resistance to DNA damaging agents (Janke et al. 2016).

RNA binding proteins and telomeres

Telomeres are particularly important in the maintenance of long-term genetic integrity. A number of RNA binding proteins play significant roles in telomere maintenance. This occurs at several levels: RBPs bind directly to TERT (telomerase reverse transcriptase), RBPs regulate expression of proteins involved in telomere maintenance, bind to telomeric components and regulate the localisation of relevant lncRNAs (recently reviewed by Nishida et al. 2017). hnRNPA1 displaces RPA from single-stranded telomeric DNA after DNA replication (Flynn et al. 2011). It functions in concert with protection of telomeres 1 (POT1) and the lncRNA TERRA (telomeric repeat-containing RNA) to displace RPA from telomeric ssDNA after DNA replication and promote telomere capping to preserve genomic integrity (Flynn et al. 2011, Liu et al. 2017). Further study has shown the binding of hnRNPA1 to telomeric RNA G-quadruplex with the binding site for hnRNPA1 located in the G-quadruplex RNA loops (Liu et al. 2017). hnRNPA1, together with hnRNPA/B and hnRNPF, is also involved in the stabilisation of TERRA (López de Silanes et al. 2010). In addition, FUS interacts with the G-quadruplex consisting of telomere DNA and TERRA to regulate histone modifications in the telomere DNA (Takahama et al. 2015). hnRNPU and hnRNPC associate with TERT, contributing to the regulation of telomere shortening (Fu and Collins 2007). hnRNPA18 binds to telomerase to maintain its activity (Zhang et al. 2016), whilst hnRNPD binds to the promoter of TERT to activate its transcription (Pont et al. 2012). RNA binding proteins are also integral components of the shelterin telomere complex which protects chromosome 3′ ends from the DNA replication and repair machineries but also allows telomerase activity for chromosome 3′end elongation (Hockemeyer and Collins 2015). The loss of telomeric repeat sequences or deficiencies in telomeric proteins can result in chromosome fusion and lead to chromosome instability.

NON-CODING RNAs AND THE DNA DAMAGE RESPONSE

It is now apparent that a number of small non-coding RNA molecules have important roles in the regulation of many cellular pathways. The major classes of these have been suggested to be micro RNAs (miRNA), small interfering RNAs (siRNA), Piwi-interacting RNAs (piRNA) and long non-coding RNAs (lncRNA) (Wilson and Doudna 2013, Zhao and Lin 2015). RNAs from each of these groups have specific functions within the cell, without being translated into proteins or peptides. miRNA, siRNA and piRNA pathways have a number of features in common. All rely on the formation of a ribonucleoprotein complex containing an Argonaute protein and an ssRNA molecule of 19–30 nucleotides, which can silence genes through specific base pairing. miRNAs derive from the genome, whereas siRNAs can be produced endogenously or exogenously, for example during viral infection (Carthew and Sontheimer 2009).

The small non-coding RNAs act in the RNA interference (RNAi) pathways. RNAi precursors are processed by double-stranded RNA-specific endoribonucleases type III, such as Drosha and Dicer, which generate dsRNA molecules (Hannon 2002, Wilson and Doudna 2013). miRNAs probably contribute to the regulation of most cellular pathways through silencing of genes by mRNA degradation and/or inhibition of translation. A large set of miRNAs is induced following DNA damage, with various miRNAs able to regulate cell cycle checkpoints and/or apoptosis (Sharma and Misteli 2013, Zhang and Peng 2015). Depletion of miRNA processing proteins such as Dicer and the Argonaute protein, Ago2, reduces the ability of the cell to respond to DNA damage. Significantly, both ATM and BRCA1 are involved in miRNA production through phosphorylation of, or interaction with, the Drosha complex (Zhang et al. 2011, Kawai and Amano 2012). In particular, after DNA damage, there is a widespread increase in miRNAs such that the level of approximately a quarter of all miRNAs is induced by an ATM-dependent mechanism (Zhang et al. 2011). ATM binds to and phosphorylates KH-type splicing regulatory protein (KSRP), an RNA binding protein, at S^{274}, S^{670} and S^{134}. KSRP enhances pri-miRNA processing by increasing pri-miRNA–Drosha interaction (Trabucchi et al. 2009, Zhang et al. 2011). However, it has also been reported that BRCA1 is phosphorylated by ATM after DNA damage and interacts with Drosha, DEAD box RNA helicase p68 (DDX5) and pri-miRNAs to facilitate the processing of pri-miRNAs (Kawai and Amano 2012). In addition, BRCA1 appears to interact directly with a number of pri-miRNAs through its DNA binding domain (aa448–1048) (Kawai and Amano 2012). Other DDR proteins, such as p53, are also involved in the production and regulation of miRNAs (Suzuki and Miyazono 2011). The relationship between p53 and miRNAs is primarily at the transcriptional level and so is not really relevant to this chapter. However, just to give one example, after DNA damage, p53 activation causes expression of miR-34 which inhibits cell survival and cell cycle progression (Bommer et al. 2007). As well as being regulated by DDR proteins, miRNAs are able, in turn, to control expression of DDR components. For example, BRCA1, ATM, Rad51 and H2AX levels are all reduced in level in response to appropriate miRNAs (Sharma and Misteli 2013).

Following DNA damage, a further set of RNAs has been reported to play a significant role. These have been termed Drosha- and Dicer-dependent small RNAs (DDRNAs) (Francia et al. 2012, Sharma and Misteli 2013). They are transcribed from DNA sequences close to the site of damage and are involved in the regulation of DDR focus formation. DDR focus formation requires the activity of RNA polymerase II–dependent transcription, whilst RNase A treatment inhibits the production of foci (Francia et al. 2012). A variety of possible roles for the DDRNAs have been proposed, including acting as templates for DNA polymerases, as guides for the recruitment of repair proteins or as scaffolds for the repair foci themselves (Chowdhury et al. 2013).

It is now apparent that lncRNAs are expressed in response to DNA damage as well as being important regulators of multiple other cellular pathways (Vance and Ponting 2014). These highly heterogeneous molecules are greater than 200 nucleotides in length. Most lncRNAs are transcribed by RNA polymerase II and can regulate gene expression in a number of ways, such as by transcriptional and post-transcriptional regulation and chromatin modification. Although it has been estimated that the human genome potentially encodes many thousands of lncRNAs, one particular set has been associated with p53 and is therefore highly relevant to the DDR (Sharma and Misteli 2013). Thus, several lncRNAs act to regulate p53 activity and/or protein level, directly or indirectly. For example, MEG3 and LIRR1 expression can both lead to increased p53 activity, probably through action on HDM2 (Zhou et al. 2007). Alternatively, other lncRNAs act as p53 'effectors'. lincRNA-p21, for example, appears to be a major mediator of p53 gene repression, interacting with hnRNPK (Huarte et al. 2010). The PANDAR (also known as PANDA; p21 associated ncRNA DNA damage activated) lncRNA is induced in response to DNA damage (Hung et al. 2011). It is encoded upstream of the p21 promoter and is expressed as a result of p53 binding to a p53RE between the p21 and PANDAR promoters (Hung et al. 2011). It appears to exert its effect through interaction with nuclear transcription factor alpha (NFYA), which regulates expression of pro-apoptotic and pro-survival genes. DINO is also transcribed from the promoter region of the p21 gene (*CDKN1A*) (Schmitt et al. 2016). This lncRNA binds to the C-terminal region of p53, resulting in its stabilisation. It has been proposed that the p53/lncRNA complex binds to the promoters of normal p53-regulated genes, such as *p21* and *GADD45A*, and enhances their expression, such that in the absence of the lncRNA there is reduced induction of a number of DDR genes (Schmitt et al. 2016). A further lncRNA, NORAD, is induced in response to DNA damage in a p53-dependent manner (Lee et al. 2016). $NORAD^{-/-}$ cells show marked chromosomal instability with a tendency to gain or lose chromosomes. This is considered to be due to direct interaction with the RNA binding proteins Pumilio2 (PUM2) and Pumilio1 (PUM1) (Lee et al. 2016).

Other lncRNAs are directly involved in the DDR, independently of p53. PCAT-1-expressing cells show deficiencies in HR through repression of BRCA2 expression, leaving cells sensitive to PARP inhibitors (Prensner et al. 2014). It has recently been shown that the lncRNA, DNA damage-sensitive RNA1 (*DDSR1*), is induced in response to double-strand breaks in an ATM-NF-κB-dependent manner (Sharma et al. 2015). Furthermore, DNA damage signalling, in particular homologous recombination, is impaired in cells lacking DDSR1. It was shown that DDSR1 binds to hnRNPUL1 and probably inhibits DNA end resection (Sharma et al. 2015). Aberrant accumulation of BRCA1 and RAP80 were seen at sites of DSBs in the absence of DDSR1. A scheme has been proposed in which DDSR1 interacts with the BRCA1-RAP80 complex and regulates its binding to damaged chromatin directly after damage. At the same time PARP-mediated PARylation is required for assembly of DDR proteins, such as hnRNPUL1, at the sites of damage (Hong et al. 2013) (see the sections dealing with hnRNPU, RNA binding proteins and DNA damage foci). To what extent DDSR1 binding affects PARylation and vice versa is unclear. DDSR1 also negatively regulates p53-dependent gene expression (Sharma et al. 2015).

Recently, it has been shown that an RNA encoded by the first 3 exons of the *ASCC3* DNA helicase gene (together with a unique fourth exon) acts as a ncRNA and has the ability to counteract the function of the full-length ASCC3 protein following UV irradiation and other forms of DNA damage which produce bulky adducts (Williamson et al. 2017). ASSC3, probably together with other components of the ASCC (activating signal co-integrator 1 complex) complex, appears to act as a widespread transcriptional suppressor at later times, following DNA damage. Expression of the short ASCC3 ncRNA aids recovery from the inhibition of transcription, induced by full-length ASCC3, after UV treatment (Williamson et al. 2017).

DIRECT INTERACTIONS BETWEEN RNA BINDING PROTEINS AND DNA DAMAGE RESPONSE COMPONENTS

As well as playing significant roles in the regulation of the DNA damage response through their roles in RNA metabolism, RBPs can directly impact the DDR, largely through protein–protein interactions. A major target appears to be p53, although numerous other significant associations have been identified. In the following section, a number of hnRNPs and other RBPs will be considered with emphasis placed on their impact on the DDR.

Heterogeneous nuclear ribonucleoproteins

Heterogeneous nuclear ribonucleoproteins (hnRNPs) are a family of proteins which associate with pre-mRNAs and mRNAs in the nucleus and cytoplasm of the cell to form ribonucleoprotein complexes (RNPs) (Dreyfuss et al. 2002, Geuens et al. 2016) (**Figure 11.1**). The protein components of these RNPs are involved in pre-mRNA processing and regulate mRNA export, localisation, translation and stability. hnRNP complexes can also contain, as well as 'traditional' hnRNPs, snRNPs and splicing factors (Dreyfuss et al. 2002, Geuens et al. 2016). The protein complexes can be extremely large, with numerous different proteins in various combinations and concentrations. A unique set of hnRNP proteins is considered to form on each mRNA, and this is dependent on the mRNA sequence and on the hnRNP proteins present in the nucleus during transcription. hnRNPs generally contain RNA binding sequences (variously called RBDs [RNA binding domains] or RRMs [RNA recognition motifs]), RRG motifs (containing one or more Arg-Arg-Gly sequences) and protein–protein interaction sites; they are also routinely post-translationally modified by phosphorylation and methylation. Although some hnRNPs have confirmed roles in the DDR, the evidence for others is relatively preliminary. However, observations that at least 16 hnRNP (A1, A2/B1, A18, AB, C1/C2, D, E1, E2, H1, HY3, K, L, M, R and U) mRNAs are induced by ionising radiation points to significant roles in various aspects of the damage response (Haley et al. 2009). The upregulation of at least two of them (hnRNPC1/C2 and hnRNPK) has been shown to be p53 dependent.

hnRNPA/B

The hnRNPA/B family members make up the majority of hnRNPs present in the cell. There is considerable homology between the proteins, and both hnRNPA and hnRNPB exist in a number of splice variants (Dreyfuss et al. 2002). hnRNPA1 is phosphorylated by DNA-PK at S^{95} and S^{192}; however, this appears to be primarily relevant to telomere maintenance rather than the DDR per se (Ting et al. 2009). Thus, the association between hnRNPA1 and DNA-PK, and hnRNPA1 phosphorylation by DNA-PK is cell cycle dependent. Significantly, DNA-PK-dependent hnRNP-A1 phosphorylation promotes the RPA-to-POT1 switch in single-stranded telomeric DNA. It has, therefore, been suggested that DNA-PK–mediated hnRNPA1 phosphorylation plays an important role in the protection of newly replicated telomeres to prevent telomeric aberrations (Flynn et al. 2011). There appears to be little direct evidence for the involvement of hnRNPA1 in the DDR, although it has been shown to accumulate in the cytoplasm following UV damage and to increase in expression after IR (Jen and Cheung 2003). Like hnRNPA1, hnRNPA2/B1 is increased in expression and relocated after IR or UV exposure (van der Houven van Oordt et al. 2000). More recently it has been demonstrated that the protein, together with PARP1, localises at blunt-ended DSBs in human chromosomes formed after prolonged incubation with proteases (Tchurikov et al. 2013). Protected 50–250 kb regions, which may correspond to DNA structures, possess co-ordinately expressed genes, with PARP1 and hnRNPA2/B1 at the termini. Following UV radiation, A18 hnRNP is induced and translocated to the cytoplasm where it associates with a number of stress-induced transcripts, including that of RPA32 (Yang and Carrier 2001).

hnRNPC

In a proteomics screen, hnRNPC1 and C2 were identified as proteins recruited to chromatin after DNA damage, although they do not appear to bind at sites of damage (Lee et al. 2005). However, it has been reported that there is a marked decrease in hnRNPC1/C2 transcripts and protein degradation following ionising radiation (Waterhouse et al. 1996). hnRNPC also associates with Ku86 in an RNA-dependent manner and can be phosphorylated by DNA-PK (Zhang et al. 2004). Data from these sorts of co-immunoprecipitation studies involving hnRNPs should be treated with caution, however, as these proteins all form heterodimeric aggregates in which they interact with one another. Thus, in the experiments described by Zhang and colleagues, precipitation with an antibody against hnRNPC showed interacting Ku86 but also immunoprecipitated hnRNPs K, J, H and F. Similarly, immunoprecipitation with a Ku antibody showed the same co-precipitating components (Zhang et al. 2004). It has been suggested that hnRNPC plays a significant role in the regulation of DDR pathways following IR (Haley et al. 2009), but at present, convincing experimental confirmation is lacking.

hnRNPF and hnRNPH

The level of expression of hnRNPH increases after UV irradiation, and it has been suggested that changes in the concentration of hnRNPH may be involved in the UV-induced internal ribosome entry site-mediated translation of serine hydroxymethyltransferase 1 (SHMT1) (Fox et al. 2009). hnRNPH and the related hnRNPF play a significant role in the regulation of p53 expression following DNA damage. In most cases, DNA damage results in the inhibition of mRNA 3′-end processing, contributing to the repression of mRNA synthesis. However, p53 mRNA 3′-end formation is unimpeded, and this has been attributed to an interaction between hnRNPH and hnRNPF and a G-quadruplex RNA structure downstream of the p53 polyadenylation site (Decorsière et al. 2011). Thus, these hnRNPs are able to regulate p53's role in apoptosis induction after UV radiation.

hnRNPG

hnRNPG (also known as RBMX) has largely been seen as a splicing factor, regulating the alternative splicing of the *tau* and *SMN* genes, for example (Moursy et al. 2014). However, it has also been demonstrated to be involved in cohesion and to associate with Scc1, Smc1 and with the cohesion regulator, Wap l (Matsunaga et al. 2012). Depletion of RBMX causes the loss of cohesin from the centromeric regions before anaphase, resulting in premature chromatid separation. RBMX has also been shown to localise transiently to sites of laser-induced DNA damage in a PARP-dependent manner (Adamson et al. 2012). It has been suggested that RBMX promotes HR by facilitating BRCA2 expression, possibly by effects on splicing (Adamson et al. 2012). It has also been shown that hnRNPG can bind and protect DNA ends from degradation by nucleases (Shin et al. 2007). This association is dependent on *wt* p53 and contributes to error-free end joining (Shin et al. 2007).

hnRNPK

hnRNPK is an RNA binding protein with very little homology to other RNPs and was originally characterised on the basis of its ability to bind exceptionally strongly to poly(C) sequences in RNA. Its role in the DNA damage response is, primarily, through the regulation of p53 activity, and this appears to be determined by multiple post-translational modifications (reviewed by Lu and Gao 2016). Following DNA damage by ionising radiation, its level of expression increases in an ATM-dependent manner (Moumen et al. 2005, 2013). hnRNPK functions as a p53 cofactor, such that in its absence there is a marked reduction in the transcription of p53-regulated genes (Moumen et al. 2005). Indeed, both p53 and hnRNPK are found at the promoters of p53-dependent genes (Moumen et al. 2005). The level of hnRNPK itself is regulated by HDM2, and it is presumed that phosphorylation by ATM inhibits the association of hnRNPK with HDM2 and therefore leads to its stabilisation (Moumen et al. 2013). Additional post-translational modifications of hnRNPK also affect its interaction with p53. Phosphorylation of hnRNPK at S^{379} by Aurora-A reduces binding to p53. Thus, following DNA damage when Aurora-A activity is reduced, there is an elevated interaction between hnRNPK and p53 (Hsueh et al. 2011). hnRNPK is also SUMOylated and methylated. SUMOylation has been suggested to be mediated by the E3 ligases Pc2/CBX4 or PIAS3 and is increased after DNA damage (Lee et al. 2012, Pelisch et al. 2012). SUMOylation inhibits ubiquitylation by reducing the affinity of hnRNPK for HDM2 (Lee et al. 2012). This has the effect of increasing p53 transcriptional activation. At a later timepoints after UV damage, it was shown that SENP2 removed SUMO from hnRNPK, leading to its destabilisation and cell cycle progression due to a reduction in p53 activity (Lee et al. 2012). Methylation of hnRNPK on arginine residues markedly increases after UV irradiation, enhancing its affinity for p53. Reduction of methylation results in reduced p53 transcriptional activity, seen as decreased p21 expression (Chen et al. 2008).

FUS/TLS (hnRNPP2)

hnRNPP2, more commonly known as FUS/TLS (fused in sarcoma/translocated in liposarcoma), has a number of well-characterised links to the DDR and has also been strongly implicated as a causative agent in the neurodegenerative diseases amyotrophic lateral sclerosis (ALS, also known as Lou Gehrig's disease) and frontotemporal lobar degeneration (FTLD) (discussed in 'RNA binding proteins, the DNA damage response and disease'). FUS is a member of the FET group of proteins which includes FUS, EWSR1 (Ewing sarcoma) and TAF15 (TATA binding associated factor 15) (**Figure 11.3**). FET proteins are multi-functional, having roles at most stages of protein expression (reviewed in Schwartz et al. 2015). They interact with DNA and RNA and have a number of common structural motifs. These include an RGG-rich region, an RNA recognition motif (RRM), zinc-binding domain and a prion-like domain (PrLD), rich in glutamine, glycine, serine and tyrosine residues.

FUS binds directly to RNA and single- and double-stranded DNA (Liu et al. 2013). It localises to laser-induced sites of DNA damage. This has been shown to occur very rapidly, apparently before the re-localisation of Ku70, γH2AX, phospho-ATM or NBS1 (Mastrocola et al. 2013, Wang et al. 2013, Rulten et al. 2014). Cells depleted of FUS have reduced ability to repair DSBs, and this is due to deficiencies in both HR and NHEJ (Wang et al. 2013). Furthermore, there is a reduced recruitment of γH2AX and 53BP1 to damage foci in the absence of FUS (Wang et al. 2013). Based on these observations, it is not surprising that FUS double knockout mice (FUS–/–) display signs of genomic instability (Hicks et al. 2000, Kuroda et al. 2000). The mice also exhibit radiation sensitivity, growth retardation and immunodeficiency (Hicks et al. 2000, Kuroda et al. 2000). Significantly, heterozygous FUS+/– mice appeared normal in these studies, with no signs of neurodegenerative disease (Kuroda et al. 2000). This has obvious implications for ALS patients who have a mutation in a single FUS allele. Studies have also been carried out on DNA repair in cells expressing FUS mutations similar to those seen in ALS patients. The variants tested were recruited to damage foci normally, but when they were transfected into *wt* FUS-depleted cells, homologous recombination was deficient, although NHEJ was appreciably less affected (Wang et al. 2013).

There is also a close relationship between FUS and PARP1. Inhibition of PARP reduces recruitment of FUS to laser-induced double-strand breaks, as has been noted for other RBPs, whereas inhibition of ATM or DNA-PK has little effect (Mastrocola et al. 2013, Rulten et al. 2014). FUS binds to poly (ADP-ribose), or PAR, chains in vitro through an RGG

Figure 11.3 The domain structure of TDP-43 and FUS. Different protein domains are indicated in different colours. Both proteins contain prion-like and RRM domains. NLS, nuclear localisation signal; RRM, RNA recognition motif; NES, nuclear export signal; Gly-rich, glycine-rich region; SYGQ-rich, glutamine-glycine-serine-tyrosine–rich region; RGG, arginine-glycine–rich region; ZnF, zinc finger motif. Amino acid numbers are shown. (From Lee, S., Kim, H.-J., 2015, *Exp Neurobiol* 24(1): 1–7.)

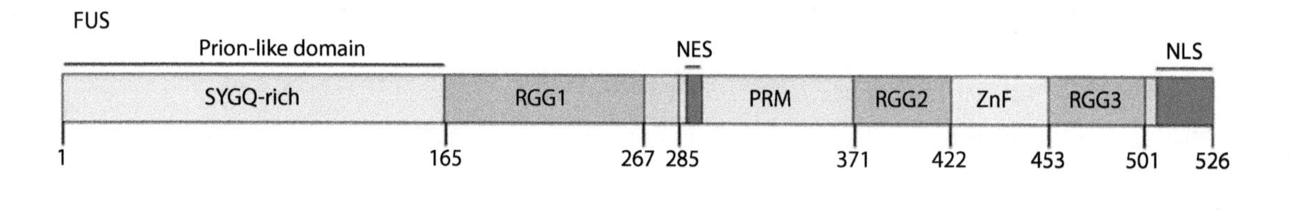

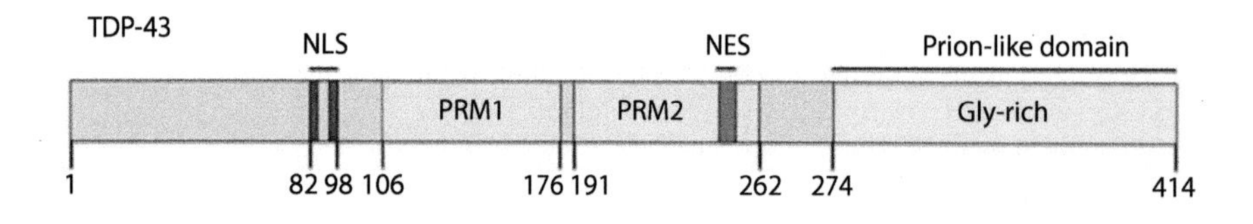

domain (Mastrocola et al. 2013, Rulten et al. 2014). Histone deacetylase 1 (HDAC1) has also been implicated in FUS recruitment to DSBs, such that in FUS-depleted cells, HDAC1 recruitment to sites of DSBs was appreciably reduced (Wang et al. 2013).

As well as playing a role at sites of DNA damage, FUS can also affect the damage response through its role in the regulation of transcription. Following IR, FUS is recruited to the cyclin D1 promoter where it inhibits protein expression, contributing to cell cycle arrest (Wang et al. 2008). This recruitment involves the interaction with a single-strand ncRNA induced following DNA damage, which is located at the 5′ regulatory region of the cyclin gene (*CCDN1*). FUS was also shown, in these studies, to bind to the histone acetyl transferases (HATs) CBP, p300 and Tip60, and inhibit their enzymatic activities (Wang et al. 2008). Two other FET proteins, EWSR1 and TAF15, have similar effects. A model has been proposed in which ncRNAs act as molecular binding partners for FUS (and other RNA binding proteins) causing its release from an inactive conformation. This facilitates the interaction of FUS with CBP resulting in an inhibition of HAT activity and repression of transcription (Wang et al. 2008).

FUS has other effects on transcription which will have repercussions for the DDR. It interacts with, and affects the activities of, RNA polymerases II and III, TFIID and TFIIIB (Tan and Manley 2009, Schwartz et al. 2012). In genome-wide screens, FUS has been shown to bind in excess of a thousand DNA promoter regions, suggesting a general role in transcriptional regulation (Tan et al. 2012). This will likely affect genes involved in the DDR.

hnRNPU-like proteins and hnRNPU

hnRNPUL1 and hnRNPUL2 both have appreciable homology to hnRNPU but appear to have significant additional roles specific to the DDR, as well as being involved in RNA transport (Bachi et al. 2000). hnRNPUL1 was originally identified as an interacting protein with the adenovirus 5 E1B55K oncoprotein (where it was known as E1B55K-associated protein 5 [E1B-AP5]) and was then shown to associate directly with p53 (Gabler et al. 1998, Barral et al. 2005). The interaction with p53 reduces the transcriptional activity of p53 such that p21 is not induced after DNA damage in the presence of transfected hnRNPUL1 (Barral et al. 2005). Further investigation has indicated that hnRNPUL1 and UL2 have direct roles in DNA end resection (Polo et al. 2012). Both proteins bind directly to NBS1 and can be identified in association with the MRN complex and CtIP (Polo et al. 2012). The interaction of hnRNPUL1 with NBS1 is directly dependent on methylation of arginine residues within the RGG domain by PRMT1 (Kzhyshkowska et al. 2001, Gurunathan et al. 2015). hnRNPUL1 and UL2 are recruited to sites of damage after ionising irradiation in an MRN- and methylation-dependent manner but are then very rapidly excluded. The localisation of hnRNPUL proteins in the nucleus of undamaged cells is dependent on the presence of RNA, but this is not necessary for recruitment to sites of damage. Rapid exclusion from sites of damage is dependent on active transcription but is unaffected by protein methylation (Polo et al. 2012, Gurunathan et al. 2015). Thus, recruitment of hnRNPUL1 predominates over exclusion upon increased damage when transcription is inhibited. Depletion of both hnRNPUL proteins impairs homologous recombination and ATR signalling. This appears to be due to the fact that they are required for recruitment of BLM to sites of damage. It was suggested that there may be two pools of hnRNPUL proteins: one involved in RNA metabolism and the other in DNA resection, although this has yet to be confirmed (Polo et al. 2012). hnRNPUL1 is also involved in PARP1 metabolism following DNA damage in that the two proteins associate and hnRNPUL1's recruitment to sites of damage requires its poly (ADP-ribosyl)ation. In addition, depletion of hnRNPUL1 enhances the concentration of PARP1 at the sites of DSBs (Hong et al. 2013). How these latter observations relate to hnRNPUL1's role in DNA end resection is not clear. In addition, hnRNPUL1 binds ATP directly through a Walker A motif at amino acids 428–436. This motif is also present in hnRNPUL2 and hnRNPU and is highly conserved, for example, in *Drosophila* hnRNPUL (Pratt and Grand, unpublished observations). hnRNPUL1 has also been linked to transcription, binding to BRD7 and being able to both repress and activate basic transcription (Kzhyshkowska et al. 2003).

Recently, it has been demonstrated that a novel long non-coding RNA, *DDSR1* (DNA damage-sensitive RNA1), which is induced in an ATM-dependent manner, associates with hnRNPUL1. Depletion of the lncRNA causes an increase in BRCA1 and RAP80 recruitment to laser-induced DSBs soon after damage, limiting homologous recombination (Sharma et al. 2015). Similarly, depletion of hnRNPUL1 causes an accumulation of BRCA1 and RAP80. It has been suggested, therefore, that hnRNPUL1's interaction with *DDSR1* plays a role in modulating HR by regulating BRCA1 and RAP80 access to DSBs (Sharma et al. 2015). Additionally, it has been shown that a second lncRNA, *CRNDE*, binds to hnRNPUL2 and directs its transport between the nucleus and cytoplasm. Cytoplasmic hnRNPUL2 that was present in the cytoplasm could interact with *CRNDE* and increase the stability of *CRNDE* RNA (Jiang et al. 2017).

The role of hnRNPU (also known as scaffold attachment factor A, SAF-A) in the DDR is much less well characterised. It is phosphorylated at S^{59} by DNA-PK in response to double-strand breaks, and by Plk1 during mitosis (Britton et al. 2009, Douglas et al. 2015). It has been suggested that hnRNPU contributes to the temporary inhibition of base excision repair (BER) of bi-stranded oxidised bases after ionising radiation so that NHEJ can take place unhindered (Hegde et al. 2016). Inhibition of DNA glycosylases by Ku is considered to occur by direct interaction. Phosphorylated hnRNPU is involved in the release of the DNA glycosylase NEIL1 from chromatin by direct interaction, inhibiting BER (Hegde et al. 2012, 2016). hnRNPU has also been shown to bind to most regulatory non-coding RNAs and to play an important role in the control of splicing through regulation of U2 snRNP maturation (Xiao et al. 2012). hnRNPU also forms a complex with the RNA polymerase II C-terminal domain which contributes to the regulation of the elongation activity of pol II (Kim and Nikodem 1999, Kukalev et al. 2005). hnRNPU also binds the *Oct4* promoter in ES cells where it regulates Oct4 expression (Vizlin-Hodzic et al. 2011); other interactions of hnRNPU with proteins involved in transcriptional regulation include binding to BRG1, WT1 and the Myc-Max complex (Spraggon et al. 2007). Whether these multiple associations are relevant to the DNA damage response is not apparent at present. It is also not clear to what extent interactions and activities, relevant to the DDR, observed for hnRNPU are also exhibited by hnRNPUL1 and hnRNPUL2 and vice versa.

Additional RNA binding proteins involved in the DNA damage response

As well as hnRNPs, other RBPs have direct roles in the cellular response to DNA damage. A number of these are discussed next, although this is far from an exhaustive list.

SRSF1, SRSF2 and SRSF3

Serine and arginine splicing factors (SRSF proteins) are essential regulators of splicing, being particularly important in splice selection for both constitutive and alternative splicing. In addition to their role in splicing, SR proteins are also involved in the maintenance of genome stability through the prevention of R-loops during transcription (Li and Manley 2005, Sollier et al. 2014). They comprise one or two RNA recognition motifs, followed by the RS domain containing numerous serine–arginine repeats. SRSF proteins are primarily nuclear but are able to shuttle between the nucleus and the cytoplasm. This distribution is determined largely by phosphorylation of amino acids in the RS domains by SRPK1 and SRPK2 kinases (Giannakouros et al. 2011). Phosphorylation is also responsible for the sub-nuclear distribution of SR splicing factors into nuclear speckles. SRSF1 binds to topoisomerase I and is phosphorylated by it (Rossi et al. 1996). It has been suggested that SRSF1 inhibits DNA nicking by TOP1 (Kowalska-Loth et al. 2002). SRSF1 is also hyperphosphorylated in response to chronic replication stress, and this has implications for splicing itself (Leva et al. 2012). The protein also regulates alternative splicing of HDM2 after DNA damage (Comiskey et al. 2015). SRSF2 (also known as SC35) is commonly mutated in chronic myelomonocytic leukaemia (CMML), and it has been shown that the protein with a mutation at P^{95} activates alternative splicing of CDC25C that is also activated in response to DNA damage (Skrdlant et al. 2016). This mutation affects the ability of SRSF2 to bind its canonical splicing enhancer sequences in RNA (Kim et al. 2015, Zhang et al. 2015). SRSF2 is upregulated by E2F1 in response to DNA damage with MMS and cyclophosphamide, and this leads to alternative splicing of a number of genes (such as *caspases 8* and *9* and *bcl-x*), favouring apoptosis (Merdzhanova et al. 2008). SRSF3 is also involved in alternative splicing, mRNA export and 3′ end processing. It binds to interphase, but not hyperphosphorylated, chromatin (Loomis et al. 2009). In cells depleted of SRSF3, large numbers of genes have altered expression and/or splicing and these include BRCA1, BRIP1 and RAD51 resulting in impaired homologous recombination (He and Zhang 2015). Depletion of SRSF3 results in reduced levels of cyclin D and E2F1 mRNA and protein, as well as exon skipping of the *HIPK2* gene (Kurokawa et al. 2014).

SFPQ and NONO

SFPQ (also known as splicing factor, proline and glutamine rich, polypyrimidine tract-binding protein-associated splicing factor, PTB-associated splicing factor and PSF) is a pre-mRNA splicing factor that also acts as a transcriptional co-repressor and is a component of the PERIOD complex involved in the generation of circadian rhythms. It binds to RNA and DNA and to RAD51D, and is involved in homologous recombination (Morozumi et al. 2009). SFPQ, in association with the closely related splicing factor NONO, is recruited to sites of DNA damage in association with the nuclear matrix protein Matrin 3 (Ha et al. 2011). Recruitment of NONO has been shown to be PARP-1 dependent and involves the binding of NONO, through an RNA recognition motif, to PAR chains (Krietsch et al. 2012). In this study, it was suggested that stimulation of NHEJ after PAR-dependent recruitment of NONO to DNA damage sites is accompanied by inhibition of homologous recombination (Krietsch et al. 2012). Recent evidence shows that SFPQ is also important for non-homologous end joining. The SFPQ-NONO complex can substitute for XLF in in vitro NHEJ assays, but all three proteins are required for efficient end joining in cell-based assays (Jaafar et al. 2017). Furthermore, it has been shown that NONO is necessary for protection against the effects of ionising radiation in mouse embryonic fibroblasts and in whole animal studies (Li et al. 2017).

Pre-cursor of mRNA processing factor (Prp19, also known as hPso4)

Prp19 is a multi-functional protein with roles in RNA splicing, DNA repair and cellular senescence (reviewed in Mahajan 2016). Notably, Prp19 has ubiquitin E3 ligase activity and it appears that this is central to its role in the spliceosome (Chan et al. 2003, Song et al. 2010). Prp19 ubiquitylates Prp3, which is a component of U4 snRNP and this increases its affinity for the U5 snRNP protein Prp8, leading to the stabilisation of the U4/U6/U5snRNP spliceosome complex (Song et al. 2010). Thus, Prp19, and other components of a core complex with Cdc5, PLRG1 and SPF27, are required for the first step in splicing by the spliceosome (Makarova et al. 2004). Moreover, a complex of Prp19 and the splicing factor U2AF65 associates with the large C-terminal domain of RNA polymerase II to link RNA splicing and transcription (David and Manley 2011). U2AF65 binds directly to the phosphorylated C-terminal domain, and this interaction results in increased recruitment of U2AF65 and Prp19 to the pre-mRNA. These observations have provided a mechanism by which RNA polymerase II enhances splicing and couples it to transcription (David and Manley 2011). Prp19 is also able to ubiquitylate itself, in particular after DNA damage, dissociating it from Cdc5 and PLRG1, and it has been suggested that this may be a mechanism by which the protein shifts from a role in splicing to a more direct one in the DDR (Mahajan 2016). Prp19 has been shown to be involved in various aspects of the DNA damage response and in the protection of cells from senescence and apoptosis (Dellago et al. 2012, Mahajan 2016). Prp19 is phosphorylated at S^{149} by ATM in response to various damaging agents, and this modification is required for a number of its biological roles (Dellago et al. 2012).

Of particular relevance to the theme of this book is the role of Prp19 in the DNA replication stress response (Abbas et al. 2014, Maréchal et al. 2014). Prp19 interacts with RPA at sites of DNA damage and is necessary for replication fork stability and restart (Abbas et al. 2014, Maréchal et al. 2014, Wan and Huang 2014). Prp19 activates ATR, but not ATM, signalling and is important for recovery of cells from arrest after replication stress and for fork progression on damaged DNA (Maréchal et al. 2014). In addition, Prp19, together with the E2 protein UbcH5c, ubiquitylates RPA32 and 70 with K63-linked chains following DNA damage. The K63-linked ubiquitin chains on RPA32 promote the localization of ATRIP at DNA damage sites, explaining how Prp19 promotes ATR activation as an ubiquitin ligase on RPA-ssDNA (Maréchal et al. 2014, Wan and Huang 2014). Significantly, it was demonstrated, using proteins with separation of function mutations, that the role of Prp19 in replication stress is quite distinct from that in the spliceosome (Maréchal et al. 2014).

As well as functioning in the response to replication stress Prp19 is also involved in the repair of DNA interstrand cross-links (Zhang et al. 2005). Cdc5, in association with Prp19, interacts with the WRN helicase, which is important in the early stages of processing of ICLs (Zhang et al. 2005). It has been suggested that the Prp19/Cdc5 complex could act as a docking site for other DDR proteins in the later stages of repair of ICLs after the initial processing stage (Mahajan 2016). In support of this, is the observation that Cdc5L binds to ATR and is necessary for downstream ATR signalling and correct S-phase checkpoint activation (Zhang et al. 2009).

Y-Box–binding protein 1 (YB-1)

YB-1 is a multi-functional RNA and DNA binding protein and is involved in pre-mRNA splicing, mRNA translation, DNA replication and transcription (reviewed in Eliseeva et al. 2011). YB-1 has been shown to associate with numerous DNA repair proteins, particularly those involved in base excision repair, nucleotide excision repair and mismatch repair (Gaudreault et al. 2004, Das et al. 2007). It binds directly to NEIL2 and forms a large complex with DNA ligase IIIα and DNA polymerase β (Das et al. 2007). Significantly, in response to UV-induced damage, YB-1 relocates from the cytoplasm to the nucleus and interacts with, and activates, NEIL2 (Das et al. 2007). YB-1 also associates with Nth1 and markedly increases its AP lyase activity in the early stages of BER (Marenstein et al. 2001). Amongst other interacting proteins, WRN, Ku80, MSH2, DNA polymerase δ and APE-1 have been identified (Gaudreault et al. 2004). YB-1 has been shown to be able to separate DNA duplexes, in particular those containing base mismatches and cisplatin-induced modifications (Gaudreault et al. 2004). Additionally, YB-1 has both exonuclease and endonuclease activity on double-stranded DNA (Izumi et al. 2001, Gaudreault et al. 2004). Increased expression of YB-1 also correlates with increased levels of PCNA, DNA ligase IIIα and DNA polymerase α as well as the cell cycle regulators cyclin A and B1, giving rise to increased cellular proliferation (Jurchott et al. 2003).

TAF15, TDP43 and EWSR1

TAF15 (TATA box-binding protein-associated factor 15), TDP43 (TAR DNA-binding protein) and EWSR1 (EWS [Ewing sarcoma] RNA-binding protein 1) are RNA binding proteins, commonly mutated in ALS (see Section 'RNA binding proteins, the DNA damage response and disease') (**Figure 11.3**). It has been suggested that loss of function of at least one of these proteins leads to an accumulation of transcription-associated DNA damage which contributes to clinical symptoms (Hill et al. 2016). Although the evidence for the roles of these three proteins in the DDR is not overwhelming, their importance in neuronal disease makes it worth summarising here. TAF15 localises to laser-induced sites of DNA damage in a PARP-dependent manner (Jungmichel et al. 2013, Izhar et al. 2015). The formation of TAF15 nucleolar cap structures is diminished following genotoxic stress with H_2O_2 or MMS in a PAR-dependent manner, possibly due to recruitment of the protein to sites of DNA damage (Jungmichel et al. 2013). Depletion of TDP43 (and FUS) leads to defective prevention or repair of transcription-associated DNA damage in a manner comparable to that seen for loss of BRCA1 (Hill et al. 2014, 2016). Similarly, TDP43 also localises to those sites of DNA damage with RNA polymerase II and it has been demonstrated that its role is in the prevention or repair of DNA damage resulting from R-loops (Hill et al. 2016). EWSR1, which like TAF15 is highly PARylated, regulates alternative splicing of the *ABL1*, *CHEK2* and *MAP4K2* genes (Paronetto et al. 2011, Jungmichel et al. 2013). Depletion of EWSR1 was shown to have the same effect in reducing the level of these proteins as DNA damage due to dissociation of EWSR1 from target mRNAs (Paronetto et al. 2011). Despite these observations, it is likely that aggregation of the mutated FET proteins is the primary cause of ALS rather than deficiencies in DNA repair pathways (see Section 'RNA binding proteins at DNA damage foci').

RNA BINDING PROTEINS AT DNA DAMAGE FOCI

It is now well established that the protein aggregates that form in neuronal cells in neurodegenerative diseases, such as ALS and FTLD, are largely composed of RNA binding proteins, the most common of which are the FET proteins (reviewed in Schwartz et al. 2015, March et al. 2016, Harrison and Shorter 2017); this is discussed in more detail in Section 'RNA binding proteins, the DNA damage response and disease'. These RNA binding proteins contain prion-like domains (PrLDs; also known as intrinsically disordered regions, IDRs). RNA binding proteins with PrLDs, which have been linked to neurodegenerative diseases, include the FET proteins, TDP-43, hnRNPA1 and hnRNPA2, ataxin 1 and ataxin 2 (March et al. 2016), and hnRNPUL1 (Pratt, Weishaupt and Grand, unpublished observations). These proteins are aggregation-prone and the presence of mutations in the PrLDs favours mis-folding and enhances the formation of aggregates and fibrils.

These aggregates appear to form by the process of liquid demixing (Altmeyer et al. 2015, Patel et al. 2015). Recent investigations have shown that membraneless organelles assemble by phase separation and behave as liquid droplets in an aqueous solution (reviewed by Aguzzi and Altmeyer 2016). The structures, formed by liquid demixing, are dynamic and can be transitory, although in the case of ALS, for example, they are irreversible. They can change in conformation and undergo multiple protein–protein interactions. In studies which have concentrated on FUS, it has been shown that the protein will form liquid compartments at the sites of DNA damage (Altmeyer et al. 2015, Patel et al. 2015). It has been known for some time that FUS localises to DNA damage foci, where it is phosphorylated by ATM (Gardiner et al. 2008). Its localisation is dependent on PARP1 activity and the presence of PAR chains (Mastrocola et al. 2013) (**Figure 11.4**). In in vitro studies, it has been demonstrated that the addition of PAR to solutions of FUS leads to the rapid formation of FUS droplets (Altmeyer et al. 2015, Patel et al. 2015). The association of FUS with PAR depends on an electrostatic interaction between the positively charged RGG domain of FUS and the negatively charged

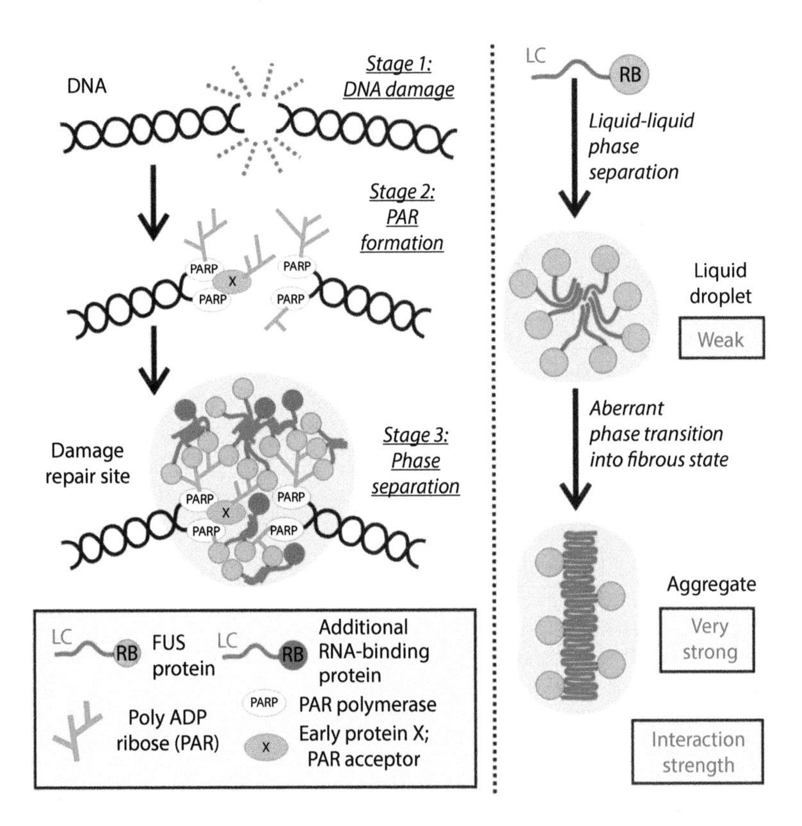

Figure 11.4 The possible molecular mechanisms responsible for the formation of DNA damage repair sites and disease-associated aggregates. Left-hand panel: FUS compartment formation after DNA damage is due to the action of PARP and localised PARylation. PARP1 and other proteins (X) are modified with PAR chains. PAR formation leads to FUS recruitment and initiates phase separation and compartment formation through low sequence complexity domains (LC domain) interactions. Other RNA binding proteins, such as EWS, hnRNPUL1, hnRNPUL2, hnRNPG and TAF15 are probably recruited, forming a compartment (focus) for DNA damage repair. Right-hand panel: FUS compartments form through phase separation from a concentrated solution of FUS, a reaction probably driven by weak interactions between prion-like LC domains. Liquid FUS droplets convert with time into an aggregated state, which presumably is associated with ALS-like disease. It is considered that certain mutations favour aggregation. LC domains are indicated in red; RB, RNA- and PAR-binding domains are indicated in blue. (From Patel, A. et al. 2015, *Cell* 162(5): 1066–1077.)

PAR (Altmeyer et al. 2015). Thus, a scenario has been envisaged in which PARP1 is recruited to sites of DNA damage within seconds and this then PARylates RNA binding proteins such as FUS. This facilitates the formation of a liquid droplet, allowing the accumulation of other DDR proteins. The PAR chains themselves can also act as a seed for the formation of a liquid droplet and protein recruitment. It has been suggested that DNA ends, single-strand DNA and/or RNA could also provide the seed for phase transition, the formation of liquid droplets and presumably the recruitment of DDR proteins (Aguzzi and Altmeyer 2016).

It is now clear that a number of RNA binding proteins, as well as FUS, localise to DNA damage foci. Thus, hnRNPUL1 and UL2 and RBMX are rapidly recruited to laser-induced damage stripes in a PARP1-dependent manner and are then quickly excluded (Adamson et al. 2012, Polo et al. 2012). It is tempting to speculate that these RNA binding proteins behave in a similar fashion to FUS and are integral to the very early stages in the formation of DNA damage foci (**Figure 11.4**). It is also apparent that while RNA plays a role in the recruitment of hnRNPUL proteins to the nucleus in the absence of DNA damage, it is not required for recruitment to damage foci (Polo et al. 2012). Furthermore, recruitment also occurs in the absence of RNA polymerase II activity (Polo et al. 2012). However, in spite of these interesting observations and suggestions, DNA damage foci form quite adequately in the absence of PARP1 and PARP2. There are, however, many more enzymes capable of PARylation than PARP1 and PARP2. This is a controversial area of research but appears to offer a physico-chemical explanation for how DNA damage foci are formed and confirmation of the importance of RNA binding proteins in the process.

RNA BINDING PROTEINS, THE DNA DAMAGE RESPONSE AND DISEASE

RNA binding proteins have been linked to numerous diseases, most notably to neurodegenerative diseases such as amyotrophic lateral sclerosis (ALS) and frontotemporal lobar degeneration with ubiquitin-positive inclusions (FTLD-U) (March et al. 2016, Harrison and Shorter 2017). Loss or gain of function of RBPs have also been implicated in muscular atrophies such as spinal muscular atrophy (SMA) and cancers, such as those attributable to translocations of FET family members. To what extent any of these diseases are due to aberrations in DDR functions of the RNA binding proteins is less clear. Fragile X syndrome (FXS) is the most common hereditary cause of mental retardation in humans (Grigsby 2016). FXS is attributable to the expansion, in excess of 200-fold, of a CGG triplet located in the 5′-UTR of the *FMR1* gene. Normally, there are between approximately 5 and 50 repeats. Hypermethylation of CpG islands gives rise to gene silencing and lack of expression of FMRP, which is an RNA binding protein, involved in mRNA metabolism. Similarly, SMA is caused by mutation or deletion of the telomeric copy of the survival motor neuron gene (*SMN1*). *SMN1* is required for correct RNP assembly.

Prion-like domains comprise amino acid sequences enriched in uncharged polar amino acids such as asparagine, glutamine, tyrosine and glycine, and they enable proteins to form self-templating fibrils (March et al. 2016, Harrison and Shorter 2017). PrLDs have been found to be particularly prevalent in classes of molecules under the general headings of 'RNA binding proteins' and 'DNA binding proteins'. Indeed, in a genome-wide search, approximately 1.2% of human proteins contained prion-like domains and of these, 30% (72/240) were RBPs (March et al. 2016). A number of RNA binding proteins containing PrLDs have been strongly associated with neurodegenerative diseases, such as ALS. These include the FET proteins (FUS/TLS, EWSR1 and TAF15), TDP-43, hnRNPA1 and A2 and hnRNPUL1, many of which appear to be intrinsically prone to mutation. Mutations in the PrLDs have been associated with deleterious protein mis-folding and aggregation (**Figure 11.4**).

There have been a limited number of reports which have implicated proteins associated with ALS in the DNA damage response. Recently it has been demonstrated that TDP43 and FUS are necessary for repair or prevention of transcription-associated DNA damage (Hill et al. 2016). These proteins localise at DNA damage foci, caused by UV irradiation, together with RNA polymerase II; they are also involved in the prevention or repair of R-loop–associated damage (Hill et al. 2016). It had previously been shown that FUS is recruited to UVA-induced DNA damage sites in a PARP-dependent fashion, and FUS and PARP bind to each other (Rulten et al. 2014). Furthermore, FUS carrying an R > G mutation at amino acid 521 (a mutation associated with ALS) has a reduced ability to localise to DNA damage foci (Rulten et al. 2014). In another report it was shown that FUS was necessary for homologous recombination and non-homologous end joining in tumour cell lines and primary neurons, and was recruited at very early times to sites of DNA damage (Wang et al. 2013). In these studies there was decreased recruitment of γH2AX, 53BP1 and phospho-Chk2 to DNA damage foci in the absence of FUS, as well as increased DNA damage as measured by comet assay (Wang et al. 2013). Interestingly, it was shown that FUS binds to HDAC1 and this interaction is enhanced following DNA damage. It was also noted that depletion of HDAC1 impaired DNA repair. Cells expressing FUS with mutations seen in ALS patients ($244^{R>C}$, $514^{R>S}$, $517^{H>Q}$ and $521^{R>C}$) were all deficient in HR, although the effects on NHEJ were only notable for the $514^{R>S}$ and $521^{R>C}$ mutants (and to a lesser extent $244^{R>C}$ mutant). Although recruitment of the FUS mutant proteins to damage foci was seen to be normal, it was found that in cells expressing $244^{R>C}$, $514^{R>S}$ and $521^{R>C}$, but not $517^{H>Q}$, FUS recruitment of phospho-ATM and NBS1 to DNA damage sites was impaired (Wang et al. 2013). When brain samples from patients with FUS mutations were analysed, increased γH2AX staining was observed compared to controls. It should be noted, however, that this could be indicative of increased apoptosis (Wang et al. 2013). In a transgenic mouse model, in which the animals expressed the $521^{R>C}$ mutant FUS protein, considerable DNA damage was identified in cortical and spinal motor neurons (Qiu et al. 2014). It was also observed that the interaction between mutant FUS and HDAC1 was reduced. Significantly, $521^{R>C}$ FUS bound to the *wt* FUS and inhibited its ability to interact with HDAC1. Extensive loss of motor neurons and dendritic area was seen in $521^{R>C}$ FUS mice of age 1–3 months (Qiu et al. 2014). In a screen of genes known to be involved in dendritic growth and synapse formation, in the same study, it was observed that the brain-derived neurotrophic factor (*Bdnf*) gene showed consistent DNA damage in the brains and spinal cords of $521^{R>C}$ FUS animals. Furthermore, $521^{R>C}$ FUS was observed to bind *Bdnf* mRNA more strongly than the *wt* protein, contributing to splicing defects and having implications for TrkB signalling (Qiu et al. 2014). In other animal studies, it has been shown that expression of truncated FUS or different FUS mutations also resulted in loss of spinal motor neurons (Huang et al. 2011, Sephton et al. 2014, Sharma et al. 2016). It is interesting to note, however, that $521^{R>C}$ FUS protein is largely localised in the nuclei of motor neurons in transgenic animals with very few cytoplasmic aggregates, whereas in ALS patients the mutant protein forms large cytoplasmic aggregates (Kwiatkowski et al. 2009, Vance et al. 2009, Huang et al. 2010). Reasons for this discrepancy are not immediately apparent, but it may be indicative of the age of the animals compared to the human patients, although it may also indicate the involvement of as-yet-unknown additional factors.

Mutations in hnRNPs have also been associated with ALS. hnRNPA1 and hnRNPA2B1 mutations have been identified in families suffering from inherited deterioration of brain, motor neuron and muscle function, and in one case of ALS (Kim et al. 2013). hnRNPA2 and

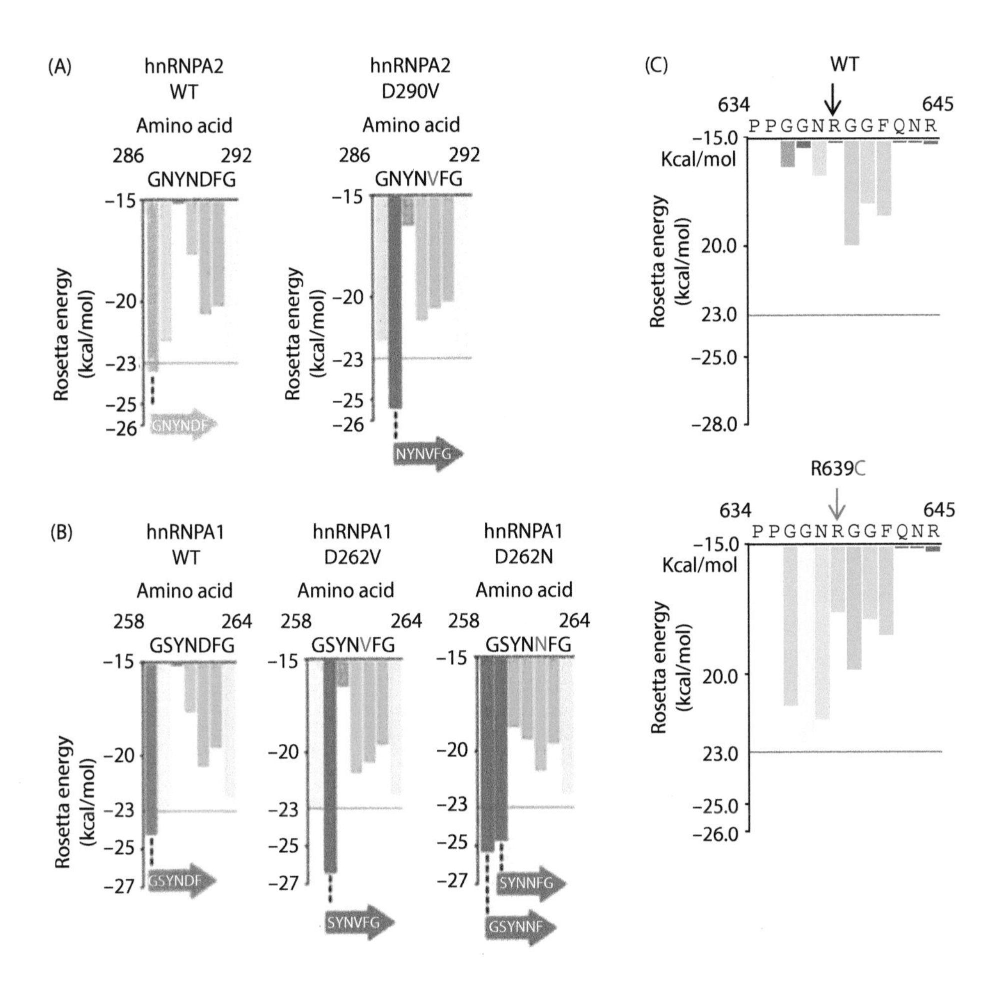

Figure 11.5 Potential impact of ALS-causing mutations in hnRNPA2/B1, hnRNPA1 and hnRNPUL1 on the amyloid fibril-forming ability of PrLDs. ZipperDB detected 6-amino-acid stretches within the core PrLDs for which the ALS disease mutations increased the predicted amyloid fibril-forming potential beyond (in the case of hnRNPA1 and hnRNPA2) or almost to (in the case of hnRNPUL1) the Rosetta threshold (taken as 23 kcal/mol). Amino acid and Rosetta energies are shown for the wt and mutated sequences in (A) hnRNPA2, (B) hnRNPA1, and (C) hnRNPUL1. (Data from Kim, H. J. et al. 2013, *Nature* 495(7442): 467–473; unpublished data from Pratt, Weishaupt and Grand.)

hnRNPA1 have a strong tendency to assemble into self-seeding fibrils, which is increased by the mutations observed in the patients. These mutations enhance a 'steric zipper' motif in the PrLD of the proteins which favour aggregate formation, and this is accompanied by the recruitment of *wt* protein into the aggregates (Kim et al. 2013) (**Figure 11.5**). Again, it is not clear whether the mutations impinge on the possible roles of these hnRNPs in telomere maintenance and the DDR, but if the proteins are localised in insoluble cytoplasmic aggregates, they are unlikely to be able to function in their normal pathways. My colleagues and I recently identified three ALS patients with mutations in hnRNPUL1: two siblings with an aa639$^{R>C}$ substitution and another patient from a second family with an aa468$^{R>C}$ substitution (K. Pratt, J. Weishaupt and R. Grand, unpublished observations). In a preliminary screen these mutations appeared to have no effect on the DDR in epithelial cells engineered to express only the mutated protein. However, the mutated proteins had a distinct impairment in their ability to bind single-strand DNA and RNA in in vitro assays (Pratt, Weishaupt and Grand, unpublished observations). Interestingly, the aa639$^{R>C}$ substitution, present in the predicted PrLD of hnRNPUL1, causes a significant increase in the potential of the protein to form a steric zipper (**Figure 11.5**). hnRNPU deletions have been identified in patients suffering from neurological conditions characterised by intellectual impairment, epilepsy, microcephaly, craniofacial abnormalities, seizures and limb abnormalities (Thierry et al. 2012, Bramswig et al. 2017). From the point of view of this review, the major question is to what extent deficiencies in the DDR caused by mutations in RBPs contribute to the clinical symptoms seen in patients with neurodegenerative diseases compared to the problems caused by aggregate formation in the cytoplasm of neuronal cells. At present, it appears that the gross effects of aggregation are likely to outweigh those relating to the DDR, but this view may change as we learn more about the direct role of the FET and other RNA binding proteins in the DDR.

CONCLUSION

In this chapter, I have tried to give a relatively broad overview of our current understanding of the contribution of RNA binding proteins to the DNA damage response. These are many and various, ranging from the regulation of expression of DDR proteins at transcription, to direct actions at DNA damage foci. As has been noted, for example, in the recent review by Giono et al. (2016), cells rely heavily on a transcriptional response to DNA damage; with increased transcription of proteins being required for an efficient damage response, and the inhibition of transcription of the majority other genes, as well as alternative splicing of an appreciable number of DDR mRNAs. RNA binding proteins play a central role in all these processes. However, here I have concentrated more on the direct roles of RNA binding proteins in the DDR through interactions with DDR proteins and recruitment to sites of damage. This recruitment seems to be largely dependent on the activity of PARP and PARylation. It is also interesting to note that a number of the proteins discussed here have been identified as significantly PARylated after different forms of genotoxic shock in a wide-ranging screen (Jungmichel et al. 2013). These include PARP1 itself, FUS, RBMX, TAF15, EWSR1, hnRNPA2B1, NONO, hnRNPU, hnRNPA1, SAFB and hnRNPUL1 (Jungmichel et al. 2013). Recruitment of the RBPs, for which information is available, tends to be transient with very rapid localisation at the damage site and then rapid dissociation (Adamson et al. 2012, Polo et al. 2012, Altmeyer et al. 2015). The roles of the proteins at the sites of damage are generally not clear, although it has been shown that hnRNPUL1 is required for DNA end resection in the repair of double-strand breaks (Polo et al. 2012). It has been suggested that some of these proteins may act as scaffolds for the formation of DNA damage foci by liquid demixing. A number of the RNA binding proteins also have PrLDs and have been shown to be involved in the generation of protein aggregates seen in patients suffering from ALS. One outstanding question, from a clinical point of view, is whether there is a link between the roles of RBPs in the formation of DNA damage foci and in the formation of ALS-linked aggregates. This is in addition to understanding whether the RBPs have specific functions at the sites of damage.

REFERENCES

Abbas, M., Shanmugam, I., Bsaili, M., Hromas, R., Shaheen, M. 2014. The role of the human psoralen 4 (hPso4) protein complex in replication stress and homologous recombination. *J Biol Chem* 289(20): 14009–14019.

Abdelmohsen, K., Pullmann, R. Jr., Lal, A., Kim, H. H., Galban, S., Yang, X., Blethrow, J. D. et al. 2007. Phosphorylation of HuR by Chk2 regulates SIRT1 expression. *Mol Cell* 25(4): 543–557.

Adam, S., Polo, S. E., Almouzni, G. 2013. Transcription recovery after DNA damage requires chromatin priming by the H3.3 histone chaperone HIRA. *Cell* 155(1): 94–106.

Adamson, B., Smogorzewska, A., Sigoillot, F. D., King, R. W., Elledge, S. J. 2012. A genome-wide homologous recombination screen identifies the RNA-binding protein RBMX as a component of the DNA-damage response. *Nat Cell Biol* 14(3): 318–328.

Aguilera, A., García-Muse, T. 2012. R loops: From transcription byproducts to threats to genome stability. *Mol Cell* 46(2): 115–124.

Aguzzi, A., Altmeyer, M. 2016. Phase separation: Linking cellular compartmentalization to disease. *Trends Cell Biol* 26(7): 547–558.

Al-Khalaf, H. H., Aboussekhra, A. 2014. ATR controls the UV-related upregulation of the CDKN1A mRNA in a Cdk1/HuR-dependent manner. *Mol Carcinog* 53(12): 979–987.

Altmeyer, M., Neelsen, K. J., Teloni, F., Pozdnyakova, I., Pellegrino, S., Grøfte, M., Rask, M. B. et al. 2015. Liquid demixing of intrinsically disordered proteins is seeded by poly(ADP-ribose). *Nat Commun* 6: 8088. doi: 10.1038/ncomms9088.

Anantha, R. W., Alcivar, A. L., Ma, J., Cai, H., Simhadri, S., Ule, J., König, J., Xia, B. 2013. Requirement of heterogeneous nuclear ribonucleoprotein C for BRCA gene expression and homologous recombination. *PLOS ONE* 8(4): e61368.

Ashburner, M., Ball, C. A., Blake, J. A., Botstein, D., Butler, H., Cherry, J. M., Davis, A. P. et al. 2000. Gene ontology: Tool for the unification of biology. *Nat Genet* 25(1): 25–8089.

Bachi, A., Braun, I. C., Rodrigues, J. P., Panté, N., Ribbeck, K., von Kobbe, C., Kutay, U. et al. 2000. The C-terminal domain of TAP interacts with the nuclear pore complex and promotes export of specific CTE-bearing RNA substrates. *RNA* 6(1): 136–158.

Barral, P. M., Rusch, A., Turnell, A. S., Gallimore, P. H., Byrd, P. J., Dobner, T., Grand, R. J. 2005. The interaction of the hnRNP family member E1B-AP5 with p53. *FEBS Lett* 579(13): 2752–2758.

Basu, S., Bahadur, R. P. 2016. A structural perspective of RNA recognition by intrinsically disordered proteins. *Cell Mol Life Sci* 73(21): 4075–4084.

Bermejo, R., Capra, T., Jossen, R., Colosio, A., Frattini, C., Carotenuto, W., Cocito, A. et al. 2011. The replication checkpoint protects fork stability by releasing transcribed genes from nuclear pores. *Cell* 146(2): 233–246.

Biamonti, G., Caceres, J. F. 2009. Cellular stress and RNA splicing. *Trends Biochem Sci* 34(3): 146–153.

Biamonti, G., Vourc'h, C. 2010. Nuclear stress bodies. *Cold Spring Harb Perspect Biol* 2(6): a000695.

Bommer, G. T., Gerin, I., Feng, Y., Kaczorowski, A. J., Kuick, R., Love, R. E., Zhai, Y. et al. 2007. p53-mediated activation of miRNA34 candidate tumor-suppressor genes. *Curr Biol* 17(15): 1298–1307.

Bracken, C. P., Wall, S. J., Barré, B., Panov, K. I., Ajuh, P. M., Perkins, N. D. 2008. Regulation of cyclin D1 RNA stability by SNIP1. *Cancer Res* 68(18): 7621–7628.

Bramswig, N. C., Lüdecke, H. J., Hamdan, F. F., Altmüller, J., Beleggia, F., Elcioglu, N. H., Freyer, C. et al. 2017. Heterozygous HNRNPU variants cause early onset epilepsy and severe intellectual disability. *Hum Genet* 136(7): 821–834.

Braunstein, S., Badura, M. L., Xi, Q., Formenti, S. C., Schneider, R. J. 2009. Regulation of protein synthesis by ionizing radiation. *Mol Cell Biol* 29(21): 5645–5656.

Bregman, D. B., Halaban, R., van Gool, A. J., Henning, K. A., Friedberg, E. C., Warren, S. L. 1996. UV-induced ubiquitination of RNA polymerase II: A novel modification deficient in Cockayne syndrome cells. *Proc Natl Acad Sci USA* 93(21): 11586–11590.

Britton, S., Froment, C., Frit, P., Monsarrat, B., Salles, B., Calsou, P. 2009. Cell nonhomologous end joining capacity controls SAF-A phosphorylation by DNA-PK in response to DNA double-strand breaks inducers. *Cell Cycle* 8(22): 3717–3722.

Busà, R., Geremia, R., Sette, C. 2010. Genotoxic stress causes the accumulation of the splicing regulator Sam68 in nuclear foci of transcriptionally active chromatin. *Nucleic Acids Res* 38(9): 3005–3018.

Candeias, M. M., Malbert-Colas, L., Powell, D. J., Daskalogianni, C., Maslon, M. M., Naski, N., Bourougaa, K., Calvo, F., Fåhraeus, R. 2008. P53 mRNA controls p53 activity by managing Mdm2 functions. *Nat Cell Biol* 10(9): 1098–1105.

Canman, C. E., Lim, D. S., Cimprich, K. A., Taya, Y., Tamai, K., Sakaguchi, K., Appella, E., Kastan, M. B., Siliciano, J. D. 1998. Activation of the ATM kinase by ionizing radiation and phosphorylation of p53. *Science* 281(5383): 1677–1679.

Carthew, R. W., Sontheimer, E. J. 2009. Origins and mechanisms of miRNAs and siRNAs. *Cell* 136(4): 642–655.

Chan, S. P., Kao, D. I., Tsai, W. Y., Cheng, S. C. 2003. The Prp19p-associated complex in spliceosome activation. *Science* 302(5643): 279–282.

Chen, Y., Zhang, L., Jones, K. A. 2011. SKIP counteracts p53-mediated apoptosis via selective regulation of p21Cip1 mRNA splicing. *Genes Dev* 25(7): 701–716.

Chen, Y., Zhou, X., Liu, N., Wang, C., Zhang, L., Mo, W., Hu, G. 2008. Arginine methylation of hnRNP K enhances p53 transcriptional activity. *FEBS Lett* 582(12): 1761–1765.

Choi, H. H., Choi, H. K., Jung, S. Y., Hyle, J., Kim, B. J., Yoon, K., Cho, E. J. et al. 2014. CHK2 kinase promotes pre-mRNA splicing via phosphorylating CDK11(p110). *Oncogene* 33(1): 108–115.

Chowdhury, D., Choi, Y. E., Brault, M. E. 2013. Charity begins at home: Non-coding RNA functions in DNA repair. *Nat Rev Mol Cell Biol* 14(3): 181–189.

Comiskey, D. F. Jr., C. D., Jacob, A. G., Singh, R. K., Tapia-Santos, A. S., Chandler, D. S. 2015. Splicing factor SRSF1 negatively regulates alternative splicing of MDM2 under damage. *Nucleic Acids Res* 43(8): 4202–4218.

Czubaty, A., Girstun, A., Kowalska-Loth, B., Trzcińska, A. M., Purta, E., Winczura, A., Grajkowski, W., Staroń, K. 2005. Proteomic analysis of complexes formed by human topoisomerase I. *Biochim Biophys Acta* 1749(1): 133–141.

D'Alessandro, G., d'Adda di Fagagna, F. 2017. Transcription and DNA damage: Holding hands or crossing swords? *J Mol Biol* pii: S0022-2836(16)30471-5

Das, S., Chattopadhyay, R., Bhakat, K. K., Boldogh, I., Kohno, K., Prasad, R., Wilson, S. H., Hazra, T. K. 2007. Stimulation of NEIL2-mediated oxidized base excision repair via YB-1 interaction during oxidative stress. *J Biol Chem* 282(39): 28474–28484.

David, C. J., Manley, J. L. 2011. The RNA polymerase C-terminal domain: A new role in spliceosome assembly. *Transcription* 2(5): 221–225.

Decorsière, A., Cayrel, A., Vagner, S., Millevoi, S. 2011. Essential role for the interaction between hnRNP H/F and a G quadruplex in maintaining p53 pre-mRNA 3′-end processing and function during DNA damage. *Genes Dev* 25(3): 220–225.

Dellago, H., Khan, A., Nussbacher, M., Gstraunthaler, A., Lämmermann, I., Schosserer, M., Mück, C. et al. 2012. ATM-dependent phosphorylation of SNEVhPrp19/hPso4 is involved in extending cellular life span and suppression of apoptosis. *Aging (Albany NY)* 4(4): 290–304.

Devany, E., Park, J. Y., Murphy, M. R., Zakusilo, G., Baquero, J., Zhang, X., Hoque, M., Tian, B., Kleiman, F. E. 2016. Intronic cleavage and polyadenylation regulates gene expression during DNA damage response through U1 snRNA. *Cell Discov* 2: 16013.

Douglas, P., Ye, R., Morrice, N., Britton, S., Trinkle-Mulcahy, L., Lees-Miller, S. P. 2015. Phosphorylation of SAF-A/hnRNP-U serine 59 by polo-like kinase 1 is required for mitosis. *Mol Cell Biol* 35(15): 2699–16713.

Dreyfuss, G., Kim, V. N., Kataoka, N. 2002. Messenger-RNA-binding proteins and the messages they carry. *Nat Rev Mol Cell Biol* 3(3): 195–205.

Dutertre, M., Lambert, S., Carreira, A., Amor-Guéret, M., Vagner, S. 2014. DNA damage: RNA-binding proteins protect from near and far. *Trends Biochem Sci* 39(3): 141–149.

Dutertre, M., Sanchez, G., De Cian, M. C., Barbier, J., Dardenne, E., Gratadou, L., Dujardin, G., Le Jossic-Corcos, C., Corcos, L., Auboeuf, D. 2010. Cotranscriptional exon skipping in the genotoxic stress response. *Nat Struct Mol Biol* 17(11): 1358–1366.

Dutertre, M., Sanchez, G., Barbier, J., Corcos, L., Auboeuf, D. 2011. The emerging role of pre-messenger RNA splicing in stress responses: Sending alternative messages and silent messengers. *RNA Biol* 8(5): 740–747.

Edmond, V., Moysan, E., Khochbin, S., Matthias, P., Brambilla, C., Brambilla, E., Gazzeri, S., Eymin, B. 2011. Acetylation and phosphorylation of SRSF2 control cell fate decision in response to cisplatin. *EMBO J* 30(3): 510–523.

Eliseeva, I. A., Kim, E. R., Guryanov, S. G., Ovchinnikov, L. P., Lyabin, D. N. 2011. Y-box-binding protein 1 (YB-1) and its functions. *Biochemistry (Mosc)* 76(13): 1402–1433.

Fan, J., Yang, X., Wang, W., Wood, W. H. 3rd, Becker, K. G., Gorospe, M. 2002. Global analysis of stress-regulated mRNA turnover by using cDNA arrays. *Proc Natl Acad Sci USA* 99(16): 10611–10616.

Flynn, R. L., Centore, R. C., O'Sullivan, R. J., Rai, R., Tse, A., Songyang, Z., Chang, S., Karlseder, J., Zou, L. 2011. TERRA and hnRNPA1 orchestrate an RPA-to-POT1 switch on telomeric single-stranded DNA. *Nature* 471(7339): 532–536.

Fox, J. T., Shin, W. K., Caudill, M. A., Stover, P. J. 2009. A UV-responsive internal ribosome entry site enhances serine hydroxymethyltransferase 1 expression for DNA damage repair. *J Biol Chem* 284(45): 31097–31108.

Francia, S., Michelini, F., Saxena, A., Tang, D., de Hoon, M., Anelli, V., Mione, M., Carninci, P., d'Adda di Fagagna, F. 2012. Site-specific DICER and DROSHA RNA products control the DNA-damage response. *Nature* 488(7410): 231–235.

Fu, D., Collins, K. 2007. Purification of human telomerase complexes identifies factors involved in telomerase biogenesis and telomere length regulation. *Mol Cell* 28(5): 773–785.

Gabler, S., Schütt, H., Groitl, P., Wolf, H., Shenk, T., Dobner, T. 1998. E1B 55-kilodalton-associated protein: A cellular protein with RNA-binding activity implicated in nucleocytoplasmic transport of adenovirus and cellular mRNAs. *J Virol* 72(10): 7960–7971.

Gajjar, M., Candeias, M. M., Malbert-Colas, L., Mazars, A., Fujita, J., Olivares-Illana, V., Fåhraeus, R. 2012. The p53 mRNA-Mdm2 interaction controls Mdm2 nuclear trafficking and is required for p53 activation following DNA damage. *Cancer Cell* 21(1): 25–35.

Gardiner, M., Toth, R., Vandermoere, F., Morrice, N. A., Rouse, J. 2008. Identification and characterization of FUS/TLS as a new target of ATM. *Biochem J* 415(2): 297–307.

Gaudreault, I., Guay, D., Lebel, M. 2004. YB-1 promotes strand separation *in vitro* of duplex DNA containing either mispaired bases or cisplatin modifications, exhibits endonucleolytic activities and binds several DNA repair proteins. *Nucleic Acids Res* 32(1): 316–327.

Gerstberger, S., Hafner, M., Tuschl, T. 2014. A census of human RNA-binding proteins. *Nat Rev Genet* 15(12): 829–845.

Geuens, T., Bouhy, D., Timmerman, V. 2016. The hnRNP family: Insights into their role in health and disease. *Hum Genet* 135(8): 851–867.

Giannakouros, T., Nikolakaki, E., Mylonis, I., Georgatsou, E. 2011. Serine-arginine protein kinases: A small protein kinase family with a large cellular presence. *FEBS J* 278(4): 570–586.

Giono, L. E., Nieto Moreno, N., Cambindo Botto, A. E., Dujardin, G., Muñoz, M. J., Kornblihtt, A. R. 2016. The RNA response to DNA damage. *J Mol Biol* 428(12): 2636–2651.

Grammatikakis, I., Abdelmohsen, K., Gorospe, M. 2017. Posttranslational control of HuR function. *Wiley Interdiscip Rev RNA* 8(1).

Grigsby, J. 2016. The fragile X mental retardation 1 gene (FMR1): Historical perspective, phenotypes, mechanism, pathology, and epidemiology. *Clin Neuropsychol* 30(6): 815–833.

Grover, R., Ray, P. S., Das, S. 2008. Polypyrimidine tract binding protein regulates IRES-mediated translation of p53 isoforms. *Cell Cycle* 7(14): 2189–2198.

Guil, S., Long, J. C., Cáceres, J. F. 2006. hnRNP A1 relocalization to the stress granules reflects a role in the stress response. *Mol Cell Biol* 26(15): 5744–5758.

Gurunathan, G., Yu, Z., Coulombe, Y., Masson, J. Y., Richard, S. 2015. Arginine methylation of hnRNPUL1 regulates interaction with NBS1 and recruitment to sites of DNA damage. *Sci Rep* 5: 10475.

Ha, K., Takeda, Y., Dynan, W. S. 2011. Sequences in PSF/SFPQ mediate radioresistance and recruitment of PSF/SFPQ-containing complexes to DNA damage sites in human cells. *DNA Repair (Amst)* 10(3): 252–9.

Haley, B., Paunesku, T., Protić, M., Woloschak, G. E. 2009. Response of heterogeneous ribonuclear proteins (hnRNP) to ionising radiation and their involvement in DNA damage repair. *Int J Radiat Biol* 85(8): 643–655.

Hannon, G. J. 2002. RNA interference. *Nature* 418(6894): 244–251.

Harrison, A. F., Shorter, J. 2017. RNA-binding proteins with prion-like domains in health and disease. *Biochem J* 474(8): 1417–1438.

He, X., Zhang, P. 2015. Serine/arginine-rich splicing factor 3 (SRSF3) regulates homologous recombination-mediated DNA repair. *Mol Cancer* 14: 158.

Hegde, M. L., Banerjee, S., Hegde, P. M., Bellot, L. J., Hazra, T. K., Boldogh, I., Mitra, S. 2012. Enhancement of NEIL1 protein-initiated oxidized DNA base excision repair by heterogeneous nuclear ribonucleoprotein U (hnRNP-U) via direct interaction. *J Biol Chem* 287(41): 34202–111.

Hegde, M. L., Dutta, A., Yang, C., Mantha, A. K., Hegde, P. M., Pandey, A., Sengupta, S. et al. 2016. Scaffold attachment factor A (SAF-A) and Ku temporally regulate repair of radiation-induced clustered genome lesions. *Oncotarget* 7(34): 54430–54444.

Hicks, G. G., Singh, N., Nashabi, A., Mai, S., Bozek, G., Klewes, L., Arapovic, D. et al. 2000. Fus deficiency in mice results in defective B-lymphocyte development and activation, high levels of chromosomal instability and perinatal death. *Nat Genet* 24(2): 175–179.

Hill, S. J., Mordes, D. A., Cameron, L. A., Neuberg, D. S., Landini, S., Eggan, K., Livingston, D. M. 2016. Two familial ALS proteins function in prevention/repair of transcription-associated DNA damage. *Proc Natl Acad Sci USA* 113(48): E7701–E7709.

Hill, S. J., Rolland, T., Adelmant, G., Xia, X., Owen, M. S., Dricot, A., Zack, T. I. et al. 2014. Systematic screening reveals a role for BRCA1 in the response to transcription-associated DNA damage. *Genes Dev* 28(17): 1957–1975.

Hockemeyer, D., Collins, K. 2015. Control of telomerase action at human telomeres. *Nat Struct Mol Biol* 22(11): 848–852.

Hong, Z., Jiang, J., Ma, J., Dai, S., Xu, T., Li, H., Yasui, A. 2013. The role of hnRPUL1 involved in DNA damage response is related to PARP1. *PLOS ONE* 8(4): e60208.

Hsueh, K. W., Fu, S. L., Huang, C. Y., Lin, C. H. 2011. Aurora-A phosphorylates hnRNPK and disrupts its interaction with p53. *FEBS Lett* 585(17): 2671–2675.

Huang, C., Zhou, H., Tong, J., Chen, H., Liu, Y. J., Wang, D., Wei, X., Xia, X. G. 2011. FUS transgenic rats develop the phenotypes of amyotrophic lateral sclerosis and frontotemporal lobar degeneration. *PLoS Genet* 7(3): e1002011.

Huang, E. J., Zhang, J., Geser, F., Trojanowski, J. Q., Strober, J. B., Dickson, D. W., Brown, R. H. Jr., Shapiro, B. E., Lomen-Hoerth, C. 2010. Extensive FUS-immunoreactive pathology in juvenile amyotrophic lateral sclerosis with basophilic inclusions. *Brain Pathol* 20(6): 1069–76.

Huarte, M., Guttman, M., Feldser, D., Garber, M., Koziol, M. J., Kenzelmann-Broz, D., Khalil, A. M. et al. 2010. A large intergenic noncoding RNA induced by p53 mediates global gene repression in the p53 response. *Cell* 142(3): 409–419.

Huertas, P., Aguilera, A. 2003. Cotranscriptionally formed DNA: RNA hybrids mediate transcription elongation impairment and transcription-associated recombination. *Mol Cell* 12(3): 711–721.

Hung, T., Wang, Y., Lin, M. F., Koegel, A. K., Kotake, Y., Grant, G. D., Horlings, H. M. et al. 2011. Extensive and coordinated transcription of noncoding RNAs within cell-cycle promoters. *Nat Genet* 43(7): 621–629.

Izhar, L., Adamson, B., Ciccia, A., Lewis, J., Pontano-Vaites, L., Leng, Y., Liang, A. C., Westbrook, T. F., Harper, J. W., Elledge, S. J. 2015. A systematic analysis of factors localized to damaged chromatin reveals PARP-dependent recruitment of transcription factors. *Cell Rep* 11(9): 1486–1500.

Izquierdo, J. M. 2008. Hu antigen R (HuR) functions as an alternative pre-mRNA splicing regulator of Fas apoptosis-promoting receptor on exon definition. *J Biol Chem* 283(27): 19077–19084.

Izumi, H., Imamura, T., Nagatani, G., Ise, T., Murakami, T., Uramoto, H., Torigoe, T. et al. 2001. Y box-binding protein-1 binds preferentially to single-stranded nucleic acids and exhibits 3′-->5′ exonuclease activity. *Nucleic Acids Res* 29(5): 1200–1207.

Jaafar, L., Li, Z., Li, S., Dynan, W. S. 2017. SFPQ•NONO and XLF function separately and together to promote DNA double-strand break repair via canonical nonhomologous end joining. *Nucleic Acids Res* 45(4): 1848–1859.

Janke, R., Kong, J., Braberg, H., Cantin, G., Yates, J. R. 3rd, Krogan, N. J., Heyer, W. D. 2016. Nonsense-mediated decay regulates key components of homologous recombination. *Nucleic Acids Res* 44(11): 5218–5230.

Jen, K. Y., Cheung, V. G. 2003. Transcriptional response of lymphoblastoid cells to ionizing radiation. *Genome Res* 13(9): 2092–2100.

Jiang, H., Wang, Y., Ai, M., Wang, H., Duan, Z., Wang, H., Zhao, L., Yu, J., Ding, Y., Wang, S. 2017. Long noncoding RNA CRNDE stabilized by hnRNPUL2 accelerates cell proliferation and migration in colorectal carcinoma via activating Ras/MAPK signaling pathways. *Cell Death Dis* 8(6): e2862.

Jungmichel, S., Rosenthal, F., Altmeyer, M., Lukas, J., Hottiger, M. O., Nielsen, M. L. 2013. Proteome-wide identification of poly(ADP-Ribosyl)ation targets in different genotoxic stress responses. *Mol Cell* 52(2): 272–285.

Jurchott, K., Bergmann, S., Stein, U., Walther, W., Janz, M., Manni, I., Piaggio, G., Fietze, E., Dietel, M., Royer, H. D. 2003. YB-1 as a cell cycle-regulated transcription factor facilitating cyclin A and cyclin B1 gene expression. *J Biol Chem* 278(30): 27988–27996.

Kawai, S., Amano, A. 2012. BRCA1 regulates microRNA biogenesis via the DROSHA microprocessor complex. *J Cell Biol* 197(2): 201–208.

Kelemen, O., Convertini, P., Zhang, Z., Wen, Y., Shen, M., Falaleeva, M., Stamm, S. 2013. Function of alternative splicing. *Gene* 514(1): 1–30.

Kervestin, S., Jacobson, A. 2012. NMD: A multifaceted response to premature translational termination. *Nat Rev Mol Cell Biol* 13(11): 700–712.

Kim, E., Ilagan, J. O., Liang, Y., Daubner, G. M., Lee, S. C., Ramakrishnan, A., Li, Y. et al. 2015. SRSF2 mutations contribute to myelodysplasia by mutant-specific effects on exon recognition. *Cancer Cell* 27(5): 617–630.

Kim, H. H., Abdelmohsen, K., Gorospe, M. 2010. Regulation of HuR by DNA damage response kinases. *J Nucleic Acids* 2010. pii: 981487. doi: 10.4061/2010/981487.

Kim, H. J., Kim, N. C., Wang, Y. D., Scarborough, E. A., Moore, J., Diaz, Z., MacLea, K. S. et al. 2013. Mutations in prion-like domains in hnRNPA2B1 and hnRNPA1 cause multisystem proteinopathy and ALS. *Nature* 495(7442): 467–473.

Kim, M. K., Nikodem, V. M. 1999. hnRNP U inhibits carboxy-terminal domain phosphorylation by TFIIH and represses RNA polymerase II elongation. *Mol Cell Biol* 19(10): 6833–6844.

Kleiman, F. E., Manley, J. L. 1999. Functional interaction of BRCA1-associated BARD1 with polyadenylation factor CstF-50. *Science* 285(5433): 1576–1579.

Kleiman, F. E., Manley, J. L. 2001. The BARD1-CstF-50 interaction links mRNA 3′ end formation to DNA damage and tumor suppression. *Cell* 104(5): 743–753.

Kowalska-Loth, B., Girstun, A., Piekiełko, A., Staroń, K. 2002. SF2/ASF protein inhibits camptothecin-induced DNA cleavage by human topoisomerase I. *Eur J Biochem* 269(14): 3504–3510.

Krietsch, J., Caron, M. C., Gagné, J. P., Ethier, C., Vignard, J., Vincent, M., Rouleau, M., Hendzel, M. J., Poirier, G. G., Masson, J. Y. 2012. PARP activation regulates the RNA-binding protein NONO in the DNA damage response to DNA double-strand breaks. *Nucleic Acids Res* 40(20): 10287–10301.

Kukalev, A., Nord, Y., Palmberg, C., Bergman, T., Percipalle, P. 2005. Actin and hnRNP U cooperate for productive transcription by RNA polymerase II. *Nat Struct Mol Biol* 12(3): 238–244.

Kuroda, M., Sok, J., Webb, L., Baechtold, H., Urano, F., Yin, Y., Chung, P. et al. 2000. Male sterility and enhanced radiation sensitivity in TLS(−/−) mice. *EMBO J* 19(3): 453–462.

Kurokawa, K., Akaike, Y., Masuda, K., Kuwano, Y., Nishida, K., Yamagishi, N., Kajita, K., Tanahashi, T., Rokutan, K. 2014. Downregulation of serine/arginine-rich splicing factor 3 induces G1 cell cycle arrest and apoptosis in colon cancer cells. *Oncogene* 33(11): 1407–1417.

Kwiatkowski, T. J. Jr., K. T., Bosco, D. A., Leclerc, A. L., Tamrazian, E., Vanderburg, C. R., Russ, C., Davis, A. et al. 2009. Mutations in the FUS/TLS gene on chromosome 16 cause familial amyotrophic lateral sclerosis. *Science* 323(5918): 1205–1208.

Kzhyshkowska, J., Rusch, A., Wolf, H., Dobner, T. 2003. Regulation of transcription by the heterogeneous nuclear ribonucleoprotein E1B-AP5 is mediated by complex formation with the novel bromodomain-containing protein BRD7. *Biochem J* 371(Pt 2): 385–393.

Kzhyshkowska, J., Schütt, H., Liss, M., Kremmer, E., Stauber, R., Wolf, H., Dobner, T. 2001. Heterogeneous nuclear ribonucleoprotein E1B-AP5 is methylated in its Arg-Gly-Gly (RGG) box and interacts with human arginine methyltransferase HRMT1L1. *Biochem J* 358(Pt 2): 305–314.

Lee, S., Kim, H.-J. 2015. Prion-like mechanism in amyotrophic lateral sclerosis: Are protein aggregates the key? *Exp Neurobiol* 24(1): 1–7.

Lee, S., Kopp, F., Chang, T. C., Sataluri, A., Chen, B., Sivakumar, S., Yu, H., Xie, Y., Mendell, J. T. 2016. Noncoding RNA NORAD regulates genomic stability by sequestering PUMILIO proteins. *Cell* 164(1–2): 69–80.

Lee, S. W., Lee, M. H., Park, J. H., Kang, S. H., Yoo, H. M., Ka, S. H., Oh, Y. M., Jeon, Y. J., Chung, C. H. 2012. SUMOylation of hnRNP-K is required for p53-mediated cell-cycle arrest in response to DNA damage. *EMBO J* 31(23): 4441–4452.

Lee, S. Y., Park, J. H., Kim, S., Park, E. J., Yun, Y., Kwon, J. 2005. A proteomics approach for the identification of nucleophosmin and heterogeneous nuclear ribonucleoprotein C1/C2 as chromatin-binding proteins in response to DNA double-strand breaks. *Biochem J* 388(Pt 1): 7–15.

Leva, V., Giuliano, S., Bardoni, A., Camerini, S., Crescenzi, M., Lisa, A., Biamonti, G., Montecucco, A. 2012. Phosphorylation of SRSF1 is modulated by replicational stress. *Nucleic Acids Res* 40(3): 1106–1117.

Li, M., Pokharel, S., Wang, J. T., Xu, X., Liu, Y. 2015. RECQ5-dependent SUMOylation of DNA topoisomerase I prevents transcription-associated genome instability. *Nat Commun* 6: 6720.

Li, S., Shu, F. J., Li, Z., Jaafar, L., Zhao, S., Dynan, W. S. 2017. Cell-type specific role of the RNA-binding protein, NONO, in the DNA double-strand break response in the mouse testes. *DNA Repair (Amst)* 51: 70–6778.

Li, X., Manley, J. L. 2005. Inactivation of the SR protein splicing factor ASF/SF2 results in genomic instability. *Cell* 122(3): 365–378.

Li, X., Manley, J. L. 2006. Cotranscriptional processes and their influence on genome stability. *Genes Dev* 20(14): 1838–1847.

Liu, X., Ishizuka, T., Bao, H. L., Wada, K., Takeda, Y., Iida, K., Nagasawa, K., Yang, D., Xu, Y. 2017. Structure-dependent binding of hnRNPA1 to Telomere RNA. *J Am Chem Soc* 139(22): 7533–7539.

Liu, X., Niu, C., Ren, J., Zhang, J., Xie, X., Zhu, H., Feng, W., Gong, W. 2013. The RRM domain of human fused in sarcoma protein reveals a non-canonical nucleic acid binding site. *Biochim Biophys Acta* 1832(2): 375–385.

Loomis, R. J., Naoe, Y., Parker, J. B., Savic, V., Bozovsky, M. R., Macfarlan, T., Manley, J. L., Chakravarti, D. 2009. Chromatin binding of SRp20 and ASF/SF2 and dissociation from mitotic chromosomes is modulated by histone H3 serine 10 phosphorylation. *Mol Cell* 33(4): 450–461.

López de Silanes, I., Stagno d'Alcontres, M., Blasco, M. A. 2010. TERRA transcripts are bound by a complex array of RNA-binding proteins. *Nat Commun* 1: 33.

Lu, J., Gao, F. H. 2016. Role and molecular mechanism of heterogeneous nuclear ribonucleoprotein K in tumor development and progression. *Biomed Rep* 4(6): 657–663.

Lucchesi, C., Sheikh, M. S., Huang, Y. 2016. Negative regulation of RNA-binding protein HuR by tumor-suppressor ECRG2. *Oncogene* 35(20): 2565–2573.

Magnuson, B., Bedi, K., Ljungman, M. 2016. Genome stability versus transcript diversity. *DNA Repair (Amst)* 44: 81–86.

Mahajan, K. 2016. hPso4/hPrp19: A critical component of DNA repair and DNA damage checkpoint complexes. *Oncogene* 35(18): 2279–2286.

Makarova, O. V., Makarov, E. M., Urlaub, H., Will, C. L., Gentzel, M., Wilm, M., Lührmann, R. 2004. A subset of human 35S U5 proteins, including Prp19, function prior to catalytic step 1 of splicing. *EMBO J* 23(12): 2381–2391.

Malbert-Colas, L., Ponnuswamy, A., Olivares-Illana, V., Tournillon, A. S., Naski, N., Fåhraeus, R. 2014. HDMX folds the nascent p53 mRNA following activation by the ATM kinase. *Mol Cell* 54(3): 500–511.

March, Z. M., King, O. D., Shorter, J. 2016. Prion-like domains as epigenetic regulators, scaffolds for subcellular organization, and drivers of neurodegenerative disease. *Brain Res* 1647: 9–18.

Maréchal, A., Li, J. M., Ji, X. Y., Wu, C. S., Yazinski, S. A., Nguyen, H. D., Liu, S., Jiménez, A. E., Jin, J., Zou, L. 2014. PRP19 transforms into a sensor of RPA-ssDNA after DNA damage and drives ATR activation via a ubiquitin-mediated circuitry. *Mol Cell* 53(2): 235–246.

Marenstein, D. R., Ocampo, M. T., Chan, M. K., Altamirano, A., Basu, A. K., Boorstein, R. J., Cunningham, R. P., Teebor, G. W. 2001. Stimulation of human endonuclease III by Y box-binding protein 1 (DNA-binding protein B). Interaction between a base excision repair enzyme and a transcription factor. *J Biol Chem* 276(24): 21242–9.

Mastrocola, A. S., Kim, S. H., Trinh, A. T., Rodenkirch, L. A., Tibbetts, R. S. 2013. The RNA-binding protein fused in sarcoma (FUS) functions downstream of poly(ADP-ribose) polymerase (PARP) in response to DNA damage. *J Biol Chem* 288(34): 24731–24741.

Masuda, K., Abdelmohsen, K., Kim, M. M., Srikantan, S., Lee, E. K., Tominaga, K., Selimyan, R. et al. 2011. Global dissociation of HuR-mRNA complexes promotes cell survival after ionizing radiation. *EMBO J* 30(6): 1040–1053.

Masuda, S., Das, R., Cheng, H., Hurt, E., Dorman, N., Reed, R. 2005. Recruitment of the human TREX complex to mRNA during splicing. *Genes Dev* 19(13): 1512–1517.

Matsunaga, S., Takata, H., Morimoto, A., Hayashihara, K., Higashi, T., Akatsuchi, K., Mizusawa, E. et al. 2012. RBMX: A regulator for maintenance and centromeric protection of sister chromatid cohesion. *Cell Rep* 1(4): 299–308.

Mayne, L. V., Lehmann, A. R. 1982. Failure of RNA synthesis to recover after UV irradiation: An early defect in cells from individuals with Cockayne's syndrome and xeroderma pigmentosum. *Cancer Res* 42(4): 1473–1478.

Merdzhanova, G., Edmond, V., De Seranno, S., Van den Broeck, A., Corcos, L., Brambilla, C., Brambilla, E., Gazzeri, S., Eymin, B. 2008. E2F1 controls alternative splicing pattern of genes involved in apoptosis through upregulation of the splicing factor SC35. *Cell Death Differ* 15(12): 1815–1823.

Mirkin, N., Fonseca, D., Mohammed, S., Cevher, M. A., Manley, J. L., Kleiman, F. E. 2008. The 3′ processing factor CstF functions in the DNA repair response. *Nucleic Acids Res* 36(6): 1792–1804.

Mitsui, A., Sharp P, A. 1999. Ubiquitination of RNA polymerase II large subunit signaled by phosphorylation of carboxyl-terminal domain. *Proc Natl Acad Sci USA* 96(11): 6054–6059.

Mlynarczyk, C., Fåhraeus, R. 2014. Endoplasmic reticulum stress sensitizes cells to DNA damage-induced apoptosis through p53-dependent suppression of p21 (CDKN1A). *Nat Commun* 8(5): 5067.

Moore, M. J., Wang, Q., Kennedy, C. J., Silver, P. A. 2010. An alternative splicing network links cell-cycle control to apoptosis. *Cell* 142(4): 625–5036.

Morozumi, Y., Takizawa, Y., Takaku, M., Kurumizaka, H. 2009. Human PSF binds to RAD51 and modulates its homologous-pairing and strand-exchange activities. *Nucleic Acids Res* 37(13): 4296–4307.

Moumen, A., Magill, C., Dry, K. L., Jackson, S. P. 2013. ATM-dependent phosphorylation of heterogeneous nuclear ribonucleoprotein K promotes p53 transcriptional activation in response to DNA damage. *Cell Cycle* 12(4): 698–704.

Moumen, A., Masterson, P., O'Connor, M. J., Jackson, S. P. 2005. hnRNP K: An HDM2 target and transcriptional coactivator of p53 in response to DNA damage. *Cell* 123(6): 1065–1078.

Moursy, A., Allain, F. H., Cléry, A. 2014. Characterization of the RNA recognition mode of hnRNP G extends its role in SMN2 splicing regulation. *Nucleic Acids Res* 42(10): 6659–6672.

Muñoz, M. J., Pérez Santangelo, M. S., Paronetto, M. P., de la Mata, M., Pelisch, F., Boireau, S., Glover-Cutter, K. et al. 2009. DNA damage regulates alternative splicing through inhibition of RNA polymerase II elongation. *Cell* 137(4): 708–720.

Nazeer, F. I., Devany, E., Mohammed, S., Fonseca, D., Akukwe, B., Taveras, C., Kleiman, F. E. 2011. p53 inhibits mRNA 3′ processing through its interaction with the CstF/BARD1 complex. *Oncogene* 30(27): 3073–3083.

Nishida, K., Kuwano, Y., Nishikawa, T., Masuda, K., Rokutan, K. 2017. RNA binding proteins and genome integrity. *Int J Mol Sci* 18(7). pii: E1341.

Paronetto, M. P., Miñana, B., Valcárcel, J. 2011. The Ewing sarcoma protein regulates DNA damage-induced alternative splicing. *Mol Cell* 43(3): 353–368.

Patel, A., Lee, H. O., Jawerth, L., Maharana, S., Jahnel, M., Hein, M. Y., Stoynov, S. et al. 2015. A liquid-to-solid phase transition of the ALS protein FUS accelerated by disease mutation. *Cell* 162(5): 1066–1077.

Paulsen, R. D., Soni, D. V., Wollman, R., Hahn, A. T., Yee, M. C., Guan, A. et al. 2009. A genome-wide siRNA screen reveals diverse cellular processes and pathways that mediate genome stability. *Mol Cell* 35: 228–229.

Pelisch, F., Gerez, J., Druker, J., Schor, I. E., Muñoz, M. J., Risso, G., Petrillo, E. et al. 2010. The serine/arginine-rich protein SF2/ASF regulates protein sumoylation. *Proc Natl Acad Sci USA* 107(37): 16119–16124.

Pelisch, F., Pozzi, B., Risso, G., Muñoz, M. J., Srebrow, A. 2012. DNA damage-induced heterogeneous nuclear ribonucleoprotein K sumoylation regulates p53 transcriptional activation. *J Biol Chem* 287(36): 30789–30799.

Polo, S. E., Blackford, A. N., Chapman, J. R., Baskcomb, L., Gravel, S., Rusch, A., Thomas, A. et al. 2012. Regulation of DNA-end resection by hnRNPU-like proteins promotes DNA double-strand break signaling and repair. *Mol Cell* 45(4): 505–516.

Pont, A. R., Sadri, N., Hsiao, S. J., Smith, S., Schneider, R. J. 2012. mRNA decay factor AUF1 maintains normal aging, telomere maintenance, and suppression of senescence by activation of telomerase transcription. *Mol Cell* 47(1): 5–15.

Powley, I. R., Kondrashov, A., Young, L. A., Dobbyn, H. C., Hill, K., Cannell, I. G., Stoneley, M. et al. 2009. Translational reprogramming following UVB irradiation is mediated by DNA-PKcs and allows selective recruitment to the polysomes of mRNAs encoding DNA repair enzymes. *Genes Dev* 23(10): 1207–1220.

Prensner, J. R., Chen, W., Iyer, M. K., Cao, Q., Ma, T., Han, S., Sahu, A. et al. 2014. PCAT-1, a long noncoding RNA, regulates BRCA2 and controls homologous recombination in cancer. *Cancer Res* 74(6): 1651–1660.

Qiu, H., Lee, S., Shang, Y., Wang, W. Y., Au, K. F., Kamiya, S., Barmada, S. J. et al. 2014. ALS-associated mutation FUS-R521C causes DNA damage and RNA splicing defects. *J Clin Invest* 124(3): 981–999.

Ratner, J. N., Balasubramanian, B., Corden, J., Warren, S. L., Bregman, D. B. 1998. Ultraviolet radiation-induced ubiquitination and proteasomal degradation of the large subunit of RNA polymerase II. Implications for transcription-coupled DNA repair. *J Biol Chem* 273(9): 5184–5189.

Reinhardt, H. C., Cannell, I. G., Morandell, S., Yaffe, M. B. 2011. Is post-transcriptional stabilization, splicing and translation of selective mRNAs a key to the DNA damage response? *Cell Cycle* 10(1): 23–27.

Rieger, K. E., Chu, G. 2004. Portrait of transcriptional responses to ultraviolet and ionizing radiation in human cells. *Nucleic Acids Res* 32(16): 4786–4803.

Riley, K. J., Maher, L. J. 3rd. 2007. p53 RNA interactions: New clues in an old mystery. *RNA* 13(11): 1825–1833.

Rockx, D. A., Mason, R., van Hoffen, A., Barton, M. C., Citterio, E., Bregman, D. B., van Zeeland, A. A., Vrieling, H., Mullenders, L. H. 2000. UV-induced inhibition of transcription involves repression of transcription initiation and phosphorylation of RNA polymerase II. *Proc Natl Acad Sci USA* 97(19): 10503–10508.

Rossi, F., Labourier, E., Forné, T., Divita, G., Derancourt, J., Riou, J. F., Antoine, E., Cathala, G., Brunel, C., Tazi, J. 1996. Specific phosphorylation of SR proteins by mammalian DNA topoisomerase I. *Nature* 381(6577): 80–82.

Rulten, S. L., Rotheray, A., Green, R. L., Grundy, G. J., Moore, D. A., Gómez-Herreros, F., Hafezparast, M., Caldecott, K. W. 2014. PARP-1 dependent recruitment of the amyotrophic lateral sclerosis-associated protein FUS/TLS to sites of oxidative DNA damage. *Nucleic Acids Res* 42(1): 307–314.

Savage, K. I., Gorski, J. J., Barros, E. M., Irwin, G. W., Manti, L., Powell, A. J., Pellagatti, A. et al. 2014. Identification of a BRCA1-mRNA splicing complex required for efficient DNA repair and maintenance of genomic stability. *Mol Cell* 54(3): 445–459.

Schmitt, A. M., Garcia, J. T., Hung, T., Flynn, R. A., Shen, Y., Qu, K., Payumo, A. Y. et al. 2016. An inducible long noncoding RNA amplifies DNA damage signaling. *Nat Genet* 48(11): 1370–1376.

Schwartz, J. C., Cech, T. R., Parker, R. R. 2015. Biochemical properties and biological functions of FET. *Annu Rev Biochem* 84: 355–379.

Schwartz, J. C., Ebmeier, C. C., Podell, E. R., Heimiller, J., Taatjes, D. J., Cech, T. R. 2012. FUS binds the CTD of RNA polymerase II and regulates its phosphorylation at Ser2. *Genes Dev* 26(24): 2690–2695.

Sephton, C. F., Tang, A. A., Kulkarni, A., West, J., Brooks, M., Stubblefield, J. J., Liu, Y. et al. 2014. Activity-dependent FUS dysregulation disrupts synaptic homeostasis. *Proc Natl Acad Sci USA* 111(44): E4769–E4778.

Sharma, A., Lyashchenko, A. K., Lu, L., Nasrabady, S.E., Elmaleh, M., Mendelsohn, M., Nemes, A., Tapia, J.C., Mentis, G.Z. 2016. Shneider: NAALS-associated mutant FUS induces selective motor neuron degeneration through toxic gain of function. *Nat Commun* 7:10465.

Sharma, V., Khurana, S., Kubben, N., Abdelmohsen, K., Oberdoerffer, P., Gorospe, M., Misteli, T. 2015. A BRCA1-interacting lncRNA regulates homologous recombination. *EMBO Rep* 16(11): 1520–10434.

Sharma, V., Misteli, T. 2013. Non-coding RNAs in DNA damage and repair. *FEBS Lett* 587(13): 1832–1839.

Shin, K. H., Kim, R. H., Kang, M. K., Kim, R. H., Kim, S. G., Lim, P. K., Yochim, J. M., Baluda, M. A., Park, N. H. 2007. p53 promotes the fidelity of DNA end-joining activity by, in part, enhancing the expression of heterogeneous nuclear ribonucleoprotein G. *DNA Repair (Amst)* 6(6): 830–840.

Shkreta, L., Chabot, B. 2015. The RNA splicing response to DNA damage. *Biomolecules* 5(4): 2935–2937.

Skrdlant, L., Stark, J. M., Lin, R. J. 2016. Myelodysplasia-associated mutations in serine/arginine-rich splicing factor SRSF2 lead to alternative splicing of CDC25C. *BMC Mol Biol* 17(1): 18.

Sollier, J., Stork, C. T., García-Rubio, M. L., Paulsen, R. D., Aguilera, A., Cimprich, K. A. 2014. Transcription-coupled nucleotide excision repair factors promote R-loop-induced genome instability. *Mol Cell* 56(6): 777–85.

Song, E. J., Werner, S. L., Neubauer, J., Stegmeier, F., Aspden, J., Rio, D., Harper, J. W., Elledge, S. J., Kirschner, M. W., Rape, M. 2010. The Prp19 complex and the Usp4Sart3 deubiquitinating enzyme control reversible ubiquitination at the spliceosome. *Genes Dev* 24(13): 1434–1447.

Spraggon, L., Dudnakova, T., Slight, J., Lustig-Yariv, O., Cotterell, J., Hastie, N., Miles, C. 2007. hnRNP-U directly interacts with WT1 and modulates WT1 transcriptional activation. *Oncogene* 26(10): 1484–1491.

Suzuki, H. I., Miyazono, K. 2011. Emerging complexity of microRNA generation cascades. *J Biochem* 149(1): 15–25.

Takagi, M., Absalon, M. J., McLure, K. G., Kastan, M. B. 2005. Regulation of p53 translation and induction after DNA damage by ribosomal protein L26 and nucleolin. *Cell* 123(1): 49–63.

Takahama, K., Miyawaki, A., Shitara, T., Mitsuya, K., Morikawa, M., Hagihara, M., Kino, K., Yamamoto, A., Oyoshi, T. 2015. G-quadruplex DNA- and RNA-specific-binding proteins engineered from the RGG domain of TLS/FUS. *ACS Chem Biol* 10(11): 2564–2569.

Tan, A. Y., Manley, J. L. 2009. The TET family of proteins: Functions and roles in disease. *J Mol Cell Biol* 1(2): 82–92.

Tan, A. Y., Riley, T. R., Coady, T., Bussemaker, H. J., Manley, J. L. 2012. TLS/FUS (translocated in liposarcoma/fused in sarcoma) regulates target gene transcription via single-stranded DNA response elements. *Proc Natl Acad Sci USA* 109(16): 6030–6035.

Tchurikov, N. A., Kretova, O. V., Fedoseeva, D. M., Sosin, D. V., Grachev, S. A., Serebraykova, M. V., Romanenko, S. A., Vorobieva, N. V., Kravatsky, Y. V. 2013. DNA double-strand breaks coupled with PARP1 and HNRNPA2B1 binding sites flank coordinately expressed domains in human chromosomes. *PLoS Genet* 9(4): e1003429.

Thickman, KR1, Sickmier, E. A., Kielkopf, C. L. 2007. Alternative conformations at the RNA-binding surface of the N-terminal U2AF(65) RNA recognition motif. *J Mol Biol* 366(3): 703–10.

Thierry, G., Bénéteau, C., Pichon, O., Flori, E., Isidor, B., Popelard, F., Delrue, M. A. et al. 2012. Molecular characterization of 1q44 microdeletion in 11 patients reveals three candidate genes for intellectual disability and seizures. *Am J Med Genet A* 158A(7): 1633–1640.

Ting, N. S., Pohorelic, B., Yu, Y., Lees-Miller, S. P., Beattie, T. L. 2009. The human telomerase RNA component, hTR, activates the DNA-dependent protein kinase to phosphorylate heterogeneous nuclear ribonucleoprotein A1. *Nucleic Acids Res* 37(18): 6105–6115.

Tournillon, A. S., López, I., Malbert-Colas, L., Findakly, S., Naski, N., Olivares-Illana, V., Karakostis, K., Vojtesek, B., Nylander, K., Fåhraeus, R. 2017. p53 binds the mdmx mRNA and controls its translation. *Oncogene* 36(5): 723–730.

Trabucchi, M., Briata, P., Garcia-Mayoral, M., Haase, A. D., Filipowicz, W., Ramos, A., Gherzi, R., Rosenfeld, M. G. 2009. The RNA-binding protein KSRP promotes the biogenesis of a subset of microRNAs. *Nature* 459(7249): 1010–1014.

Tresini, M., Warmerdam, D. O., Kolovos, P., Snijder, L., Vrouwe, M. G., Demmers, J. A., van IJcken, W. F. et al. 2015. The core spliceosome as target and effector of non-canonical ATM signalling. *Nature* 523(7558): 53–58.

Tuduri, S., Crabbé, L., Conti, C., Tourrière, H., Holtgreve-Grez, H., Jauch, A. et al. 2009. Topoisomerase I suppresses genomic instability by preventing interference between replication and transcription. *Nat Cell Biol* 11:1315–1324.

Vance, C., Rogelj, B., Hortobágyi, T., De Vos, K. J., Nishimura, A. L., Sreedharan, J., Hu, X. et al. 2009. Mutations in FUS, an RNA processing protein, cause familial amyotrophic lateral sclerosis type 6. *Science* 323(5918): 1208–1211.

Vance, K. W., Ponting, C. P. 2014. Transcriptional regulatory functions of nuclear long noncoding RNAs. *Trends Genet* 30(8): 348–355.

van der Houven van Oordt, W., Diaz-Meco, M. T., Lozano, J., Krainer, A. R., Moscat, J., Cáceres, J. F. 2000. The MKK(3/6)-p38-signaling cascade alters the subcellular distribution of hnRNP A1 and modulates alternative splicing regulation. *J Cell Biol* 149(2): 307–316.

Venkata Narayanan, I., Paulsen, M. T., Bedi, K., Berg, N., Ljungman, E. A., Francia, S. et al. 2017. Transcriptional and post-transcriptional regulation of the ionizing radiation response by ATM and p53. *Sci Rep* 7:43598.

Vichi, P., Coin, F., Renaud, J. P., Vermeulen, W., Hoeijmakers, J. H., Moras, D., Egly, J. M. 1997. Cisplatin- and UV-damaged DNA lure the basal transcription factor TFIID/TBP. *EMBO J* 16(24): 7444–43556.

Vizlin-Hodzic, D., Runnberg, R., Ryme, J., Simonsson, S., Simonsson, T. 2011. SAF-A forms a complex with BRG1 and both components are required for RNA polymerase II mediated transcription. *PLOS ONE* 6(12): e28049.

Wahl, M. C., Will, C. L., Lührmann, R. 2009. The spliceosome: Design principles of a dynamic RNP machine. *Cell* 136(4): 701–718.

Wan, L., Huang, J. 2014. The PSO4 protein complex associates with replication protein A (RPA) and modulates the activation of ataxia telangiectasia-mutated and Rad3-related (ATR). *J Biol Chem* 289(10): 6619–6626.

Wan, Y., Zheng, X., Chen, H., Guo, Y., Jiang, H., He, X., Zhu, X., Zheng, Y. 2015. Splicing function of mitotic regulators links R-loop-mediated DNA damage to tumor cell killing. *J Cell Biol* 209(2): 235–246.

Wang, W. Y., Pan, L., Su, S. C., Quinn, E. J., Sasaki, M., Jimenez, J. C., Mackenzie, I. R., Huang, E. J., Tsai, L. H. 2013. Interaction of FUS and HDAC1 regulates DNA damage response and repair in neurons. *Nat Neurosci* 16(10): 1383–1391.

Wang, X., Arai, S., Song, X., Reichart, D., Du, K., Pascual, G., Tempst, P., Rosenfeld, M. G., Glass, C. K., Kurokawa, R. 2008. Induced ncRNAs allosterically modify RNA-binding proteins in cis to inhibit transcription. *Nature* 454(7200): 126–130.

Waterhouse, N., Kumar, S., Song, Q., Strike, P., Sparrow, L., Dreyfuss, G., Alnemri, E. S., Litwack, G., Lavin, M., Watters, D. 1996. Heteronuclear ribonucleoproteins C1 and C2, components of the spliceosome, are specific targets of interleukin 1beta-converting enzyme-like proteases in apoptosis. *J Biol Chem* 271(46): 29335–29341.

Williamson, L., Saponaro, M., Boeing, S., East, P., Mitter, R., Kantidakis, T., Kelly, G. P. et al. 2017. UV irradiation induces a non-coding RNA that functionally opposes the protein encoded by the same gene. *Cell* 168(5): 843–855.

Wilson, R. C., Doudna, J. A. 2013. Molecular mechanisms of RNA interference. *Annu Rev Biophys* 42: 217–239.

Xiao, R., Tang, P., Yang, B., Huang, J., Zhou, Y., Shao, C., Li, H., Sun, H., Zhang, Y., Fu, X. D. 2012. Nuclear matrix factor hnRNP U/SAF-A exerts a global control of alternative splicing by regulating U2 snRNP maturation. *Mol Cell* 45(5): 656–668.

Yang, C., Carrier, F. 2001. The UV-inducible RNA-binding protein A18 (A18 hnRNP) plays a protective role in the genotoxic stress response. *J Biol Chem* 276(50): 47277–47284.

Zhang, C., Peng, G. 2015. Non-coding RNAs: An emerging player in DNA damage response. *Mutat Res Rev Mutat Res* 763: 202–211.

Zhang, J., Lieu, Y. K., Ali, A. M., Penson, A., Reggio, K. S., Rabadan, R., Raza, A., Mukherjee, S., Manley, J. L. 2015. Disease-associated mutation in SRSF2 misregulates splicing by altering RNA-binding affinities. *Proc Natl Acad Sci USA* 112(34): E4726–E4734.

Zhang, N., Kaur, R., Akhter, S., Legerski, R. J. 2009. Cdc5L interacts with ATR and is required for the S-phase cell-cycle checkpoint. *EMBO Rep* 10(9): 1029–1035.

Zhang, N., Kaur, R., Lu, X., Shen, X., Li, L., Legerski, R. J. 2005. The Pso4 mRNA splicing and DNA repair complex interacts with WRN for processing of DNA interstrand cross-links. *J Biol Chem* 280(49): 40559–40567.

Zhang, S., Schlott, B., Görlach, M., Grosse, F. 2004. DNA-dependent protein kinase (DNA-PK) phosphorylates nuclear DNA helicase II/RNA helicase A and hnRNP proteins in an RNA-dependent manner. *Nucleic Acids Res* 32(1): 1–10.

Zhang, X., Wan, G., Berger, F. G., He, X., Lu, X. 2011. The ATM kinase induces microRNA biogenesis in the DNA damage response. *Mol Cell* 41(4): 371–383.

Zhang, Y., Wu, Y., Mao, P., Li, F., Han, X., Zhang, Y., Jiang, S. et al. 2016. Cold-inducible RNA-binding protein CIRP/hnRNP A18 regulates telomerase activity in a temperature-dependent manner. *Nucleic Acids Res* 44(2): 761–775.

Zhao, X. Y., Lin, J. D. 2015. Long noncoding RNAs: A new regulatory code in metabolic control. *Trends Biochem Sci* 40(10): 586–596.

Zhou, H. L., Geng, C., Luo, G., Lou, H. 2013. The p97-UBXD8 complex destabilizes mRNA by promoting release of ubiquitinated HuR from mRNP. *Genes Dev* 27(9): 1046–1058.

Zhou, Y., Zhong, Y., Wang, Y., Zhang, X., Batista, D. L., Gejman, R., Ansell, P. J., Zhao, J., Weng, C., Klibanski, A. 2007. Activation of p53 by MEG3 non-coding RNA. *J Biol Chem* 282(34): 24731–24742.

DNA Replication and Inherited Human Disease

12

John J. Reynolds and Grant S. Stewart

INTRODUCTION

DNA replication is a fundamental cellular process that involves the accurate duplication of the genome during each passage through the cell cycle. A major challenge for the DNA replication machinery is to ensure that this occurs in an efficient and error-free manner, despite the numerous obstacles and difficulties that replication forks can encounter. Furthermore, this must occur in parallel to other fundamental cellular processes.

Obstacles to DNA replication cause a state called 'replication stress', which is loosely defined as the slowing or stalling of replication fork progression (Zeman and Cimprich 2014). Replication stress can arise from numerous sources, including, but not limited to, the collisions between the replication fork and DNA lesions (endogenous and exogenous), DNA synthesis through difficult-to-replicate genomic sequences and the misincorporation of ribonucleotides into DNA. To combat this replication stress, and to avoid genetic instability and/or cell death, cells have evolved an efficient replication stress response that stabilises and restarts stalled forks, signals and repairs DNA damage, and activates cell cycle checkpoints (Zeman and Cimprich 2014, Munoz and Mendez 2017). In addition, the cell employs dedicated DNA repair pathways during S phase, such as the Fanconi anaemia pathway, to resolve specific types of DNA damage that the replication fork can encounter (Ceccaldi et al. 2016).

During cellular proliferation, the cell also faces other types of problems that can compromise replication fidelity, such as DNA re-replication and/or insufficient numbers of replication origins which can lead to under-replicated DNA, aneuploidy, chromosomal instability and mitotic catastrophe. To prevent DNA re-replication, the cell tightly controls replication licensing and initiation to ensure that DNA origins (regions of the genome from which DNA replication initiates) are licenced solely during late mitosis and G1 phase of the cell cycle, and that origins can only fire once during S phase (Fragkos et al. 2015). Furthermore, the cell licences origins in approximately 3-, to 10-fold excess and prevents origins from all firing simultaneously, which allows the flexibility needed to safeguard complete and efficient replication of the genome (Fragkos et al. 2015). Another challenge the cell faces during DNA replication is the misincorporation of bases by DNA polymerases, which can cause mutations in subsequent rounds of the cell cycle. Therefore, the replication machinery has evolved several strategies to repair DNA mismatches as they occur (Kunkel and Erie 2015, Ganai and Johansson 2016). Finally, DNA replication is inextricably linked with other cellular processes that occur in parallel, and as such the cell must coordinate them so that they do not interfere with each other. For example, as DNA and RNA polymerases share the same DNA template, collisions between the replication and transcription machinery are inevitable, and several processes are employed to resolve these conflicts either before or after they occur (Hamperl and Cimprich 2016).

The importance of these cellular processes is highlighted by the fact that mutations in many components of these DNA replication and replication stress response pathways are the cause of inherited human diseases, such as Fanconi anaemia, Seckel syndrome and Meier-Gorlin syndrome (Alcantara and O'Driscoll 2014). Although these diseases display phenotypic variability, they broadly present with similar/overlapping symptoms, in particular neurological dysfunction, developmental defects and cancer predisposition, indicating common underlying pathological processes/defects.

Neurological dysfunction

Primary microcephaly is clinically defined as a significant decrease in the occipitofrontal circumference (OFC; head circumference), of more than 3 standard deviations below the age- and sex-matched means, detectable before 36 weeks of gestation (Woods and Parker 2013). As the growth of the skull is driven by the outward pressure of the developing brain, primary microcephaly is considered to be a disorder of impaired neurological development (Woods and Parker 2013, Alcantara and O'Driscoll 2014). Primary microcephaly can also be associated with other abnormalities of the structures of the brain, although the presence of these seem to be dependent on the underlying cause. The development of the human central nervous system is a complex process involving highly coordinated periods of rapid stem cell expansion, and neuronal proliferation, migration and differentiation, making it very sensitive to defects that impact upon genome stability, DNA replication and/or cell division (Alcantara and O'Driscoll 2014).

The adult brain faces different challenges than the developing brain as it is predominantly post-mitotic, and it possesses only a limited capacity to rejuvenate (Brazel and Rao 2004). Therefore, it is critical to maintain the functional integrity of post-mitotic neurons during the lifetime of the organism, and any defects that impact on protein stability and/or genome integrity can have a large impact on neuronal function. One type of neurological dysfunction in the adult brain common to genome instability syndromes is progressive neurodegeneration. Neurodegeneration can occur when the progressive loss of neuronal structure and function leads to a decline in the number of neurons due to increased cell death. There is a group of hereditary neurodegenerative diseases characterised by the progressive degeneration/atrophy of the cerebellum that predominantly manifests in children and young adults. The most striking and debilitating symptom of these disorders is ataxia, which is defined as the loss of coordination of muscle movement affecting movement, speech and balance, and these disorders are collectively referred to as autosomal recessive cerebellar ataxias (Anheim et al. 2012). Amyotrophic lateral sclerosis (ALS) is a second type of neurological dysfunction that afflicts the mature nervous system and is characterised by degeneration of motor neurons in the brain and spinal cord, leading to limb weakness, muscle wastage and atrophy (Coppede and Migliore 2015). Mutations in several RNA binding proteins involved in the DNA damage response have been associated with ALS (discussed in detail in Chapter 11).

Developmental defects

Growth dysfunction is a common manifestation of human disease associated with DNA replication abnormalities, and can be evident either before birth (intrauterine growth deficiency) or can manifest after birth as an individual develops (Klingseisen and Jackson 2011). In mammals, the size of an organism seems to be largely determined by total cell number. Therefore, any genetic defect that reduces the total number of cells produced during development, which can be caused by either a reduction in the number of cells generated or an increase in cell death, can inevitably lead to a smaller individual. Primordial dwarfism is a group of single-gene disorders in which growth is profoundly impaired prenatally and continues to be impaired postnatally (Klingseisen and Jackson 2011). As a whole, primordial dwarfism genes seem to be mostly involved in cell cycle progression, DNA synthesis and/or the maintenance of genome stability (Klingseisen and Jackson 2011).

In addition to growth retardation, other developmental abnormalities can also be associated with defective DNA replication and genome instability. The most common congenital abnormalities, other than short stature, are microcephaly, craniofacial anomalies and skeletal defects, although other organs such as the heart and kidneys can also be affected in these disorders. It is not clear why different diseases have differential impacts on different tissue types, however it is likely that it is, at least in part, dependent on the underlying cause of the disease. In addition, in the case of multi-systemic diseases, it is possible that the underlying defect is severe enough, or affects such a fundamental process, that it has a much broader impact on the maintenance and/or proliferation of multiple stem cells populations and/or tissue types.

Cancer predisposition

In contrast to neurodysfunction and growth deficiency, which can be considered to be diseases associated with a reduction in cell growth, an increase in cell dysfunction and/or an increase in cell death, cancer is a disease of uncontrolled cell proliferation and survival. There are many changes to fundamental cellular processes that need to happen before a cell completes its journey to malignancy, such as achieving replicative immortality and resistance to cell death. One of the driving forces behind this tumour evolution is genome instability, a hallmark of all cancers, which can lead to the accumulation of large numbers of genetic alterations (Hoeijmakers 2001, Hanahan and Weinberg 2011). DNA replication is a cellular process that is particularly vulnerable to genome instability, and replication stress is also a potent inducer of genome instability. It is therefore unsurprising that replication stress is a feature of both pre-cancerous and cancerous cells, and that many disorders associated with defective responses to replication stress, and/or defective DNA repair pathways, give rise to cancer predisposition (Gaillard et al. 2015). This relationship between cancer and replication has been exploited in cancer treatment and a large proportion of chemotherapeutic agents induce replication stress.

In this chapter, we describe the key cellular processes that regulate and impact upon DNA replication and the replication stress response, and will discuss how defects in these processes lead to human disease.

REGULATION OF DNA REPLICATION

To prevent genome instability, the processes of genome duplication and cell division are very tightly controlled. This is partly achieved by division of the cell cycle into functionally distinct phases, namely G1, S, G2 and M. This ensures that the cellular processes involved in the preparation, initiation, progression and completion of DNA replication, and also the condensation and separation of chromosomes, are temporally separated and proceed in an ordered and unidirectional fashion (DNA replication is also discussed in detail in Chapters 2, 3 and 6).

DNA origin licencing and activation

Preparation for DNA replication occurs exclusively in G1 phase of the cell cycle. During this time DNA replication origins are licenced in a process that involves the recruitment of the pre-replication complex (pre-RC) to chromatin (reviewed in Fragkos et al. 2015). The pre-RC consists of the origin recognition complex (ORC1-6), CDC6, CDT1 and the inactive MCM helicase complex (a hexamer consisting of the six subunits MCM2-7). Once origins have been licenced, the cells employ several mechanisms to ensure that origins are not relicensed or re-activated in S-phase, which would lead to DNA re-replication and genome instability (Arias and Walter 2007). For example, the replication inhibitor geminin (GMNN), which is expressed during S and G2 phases of the cell cycle, directly binds to CDT1 and restrains its ability to recruit the MCM complex to origins (Wohlschlegel et al. 2000). In addition, the replication stress response kinases ATR and CHK1 also function to suppress DNA origin re-licencing (Liu et al. 2007).

It has long been recognised that a far greater number of origins are licenced in G1 than will be activated during S phase. These 'dormant origins' mostly remain unused under normal conditions, but are used as a mechanism to prevent DNA under-replication by firing dormant origins close to stalled replication forks (Kawabata et al. 2011). This is evidenced by the fact that a reduction in the number of dormant origins can cause genome instability in cells even in the absence of exogenous replication stress, due to the breakage of regions of under-replicated DNA during mitosis (Kawabata et al. 2011).

Following formation of the pre-RC, origin activation is achieved by the formation of a second complex, called the pre-initiation complex (pre-IC) at the G1/S phase transition, and subsequent assembly and activation of the CMG (GINS-MCM-CDC45) helicase complex during S phase. This involves the conversion of the inactive MCM complex into two active single MCM hexamers, and the phosphorylation-dependent loading of multiple replication factors (including TOPBP1, MCM10, CDC45, RECQL4, GINS, Treslin, MTBP and DNA polymerase ε) onto origins (Heller et al. 2011, Fragkos et al. 2015).

Replisome assembly and DNA synthesis

The last stage in the formation of a functional replisome, before DNA synthesis is initiated, is the recruitment of other replication factors, such as RFC, PCNA, RPA, TIPIN-TIMELESS-CLASPIN, and the DNA polymerases (polymerase α, ε and δ). This converts the pre-IC into two functional replisomes that move bi-directionally from the activated origin (Masai et al. 2010). DNA replication begins with the synthesis of a short RNA primer by the primase subunit of polymerase α (Pol α), which is then elongated by polymerase ε and δ (Pol ε and Pol δ). Due to the 5′ to 3′ polarity of DNA synthesis, only one strand can be synthesised continuously in the same direction as the CMG helicase (3′-5′ direction), called the leading strand. The other strand, called the lagging strand, is synthesised in the opposite direction to the helicase in a discontinuous manner as a series of short nascent DNA strands called Okazaki fragments. Okazaki fragments are resolved by the lagging strand polymerase displacing the RNA primer associated with the preceding Okazaki fragment, which is then processed by one of several nucleases, such as DNA2 or FEN1, before being ligated by DNA ligase I (LIG1) (Balakrishnan and Bambara 2013). Another factor critical for DNA replication is PCNA (proliferating cell nuclear antigen), a homotrimeric ring-shaped clamp that encircles DNA. It interacts with, and stimulates the processivity and activity of, DNA polymerases ε and δ, DNA ligase I and FEN1 (Moldovan et al. 2007). Finally, once DNA replication is complete, the process of DNA termination, involving disassembly of the replisome, begins (discussed in detail in Chapter 3).

Human disorders of the DNA origin licencing and DNA replication machinery

Within the last decade mutations in multiple components of the pre-RC and pre-IC machinery, and the CMG helicase, have been found to be associated with human disease (Table 12.1). The majority of these mutations give rise to a primordial dwarfism disorder called Meier-Gorlin syndrome, although there are differing clinical presentations associated with some of the mutations.

Meier-gorlin syndrome

Meier-Gorlin syndrome (MGS) is a microcephalic primordial dwarfism disorder that typically presents with a combination of three clinical phenotypes: (1) reduced growth before and after birth (intrauterine and postnatal growth retardation); (2) small/underdeveloped ears (microtia); and (3) absent or underdeveloped knees (absent or hypoplastic patellae) (de Munnik et al. 2015). Beyond these three clinical phenotypes MGS is clinically variable and can present with a variety of different abnormalities. In particular, mild to severe microcephaly can be associated with MGS, although a significant proportion of cases do not exhibit this clinical phenotype (de Munnik et al. 2015). Other additional clinical symptoms that can be present with MGS include genitourinary abnormalities, congenital pulmonary emphysema, multiple facial abnormalities (including eyes, mouth, teeth), skeletal abnormalities (including rib cage, hands, limbs and feet) and craniosynostosis (premature fusion of different sections of the skull during growth). Currently MGS is known to be caused by mutations in the *ORC1*, *ORC4*, *ORC6*, *CDT1*, *CDC6*, *GMNN*, *CDC45* and *MCM5* genes. Mutations in the majority of these genes are inherited in an autosomal recessive fashion, although heterozygous mutations in *GMNN* have been shown to be autosomal dominant.

Consistent with the fact that ORC1, ORC4, ORC6, CDT1 and CDC6 are all essential components of the pre-RC, cell lines derived from patients with mutations in these proteins exhibit defects in DNA origin licensing, and defective entry into, and progression through, S phase (Bicknell et al. 2011a, Stiff et al. 2013). Furthermore, it was shown that cells from patients with mutations in *MCM5* (an essential component of the CMG helicase) also display delayed cell cycle progression (Vetro et al. 2017). Interestingly, the de novo mutations within *GMNN* that are associated with MGS lie within a critical destruction box motif that is targeted by the anaphase-promoting complex (APC) and is essential for its proteasome-dependent degradation at the end of mitosis (Burrage et al. 2015). As a consequence, these GMNN mutants are unable to be degraded efficiently, which reduces replication origin licencing via the sustained inhibition of CDT1.

Additional human disorders associated with mutations in components of the replisome

To date, mutations in several core DNA replication proteins associated with human diseases other than MGS have been reported. Mutations in *MCM4* have been shown to cause microcephaly and intrauterine and postnatal growth retardation, combined with immunodeficiency and adrenal insufficiency (Gineau et al. 2012, Hughes et al. 2012). The immunodeficiency manifested as a reduction in the number of natural killer (NK) cells (circulating cytotoxic lymphocytes). Cell lines derived from patients with *MCM4* mutations display defective cell cycle progression and elevated levels of genome instability following replication stress (Gineau et al. 2012). More recently, mutations in *GINS1* have also been associated with intrauterine and

Table 12.1 **Human diseases associated with defective regulation of DNA replication.**

Gene	Gene function	Disease	MOI	Clinical presentation	OMIM #	References
ORC1	Subunit of the origin recognition complex (ORC) required for DNA replication licencing	Meier-Gorlin syndrome 1 (MGS1)	AR	Intrauterine and postnatal growth retardation; microtia; absent or hypoplastic patellae; microcephaly; skeletal abnormalities; dysmorphic facial features	224690	Bicknell et al. (2011b)
ORC4	Subunit of the origin recognition complex (ORC) required for DNA replication licencing	Meier-Gorlin syndrome 2 (MGS2)	AR	Intrauterine and postnatal growth retardation; microtia; absent or hypoplastic patellae; microcephaly; skeletal abnormalities; dysmorphic facial features	613800	Bicknell et al. (2011a), Guernsey et al. (2011)
ORC6	Subunit of the origin recognition complex (ORC) required for DNA replication licencing	Meier-Gorlin syndrome 3 (MGS3)	AR	Intrauterine and postnatal growth retardation; microtia; absent or hypoplastic patellae; microcephaly; skeletal abnormalities; dysmorphic facial features	613803	Bicknell et al. (2011a)
CDT1	Replication licencing factor required for the assembly of pre-replication complexes (pre-RC)	Meier-Gorlin syndrome 4 (MGS4)	AR	Intrauterine and postnatal growth retardation; microtia; absent or hypoplastic patellae; microcephaly; skeletal abnormalities; dysmorphic facial features	613804	Bicknell et al. (2011a)
CDC6	Regulates DNA replication initiation and cell cycle checkpoints	Meier-Gorlin syndrome 5 (MGS5)	AR	Intrauterine and postnatal growth retardation; microtia; absent or hypoplastic patellae; microcephaly; skeletal abnormalities; dysmorphic facial features	613805	Bicknell et al. (2011a)
GMNN	DNA replication and the cell cycle regulator that also inhibits unscheduled DNA synthesis	Meier-Gorlin syndrome 6 (MGS6)	AD	Intrauterine and postnatal growth retardation; microtia; absent or hypoplastic patellae; microcephaly; skeletal abnormalities; dysmorphic facial features	616835	Burrage et al. (2015)
CDC45	Essential component of the CDC45-MCM-GINS (CMG) helicase complex	Meier-Gorlin syndrome 7 (MGS7)	AR	Intrauterine and postnatal growth retardation; microtia; absent or hypoplastic patellae; craniosynostosis; microcephaly; skeletal abnormalities; dysmorphic facial features	617063	Fenwick et al. (2016)
MCM5	Essential component of the CDC45-MCM-GINS (CMG) helicase complex	Meier-Gorlin syndrome 8 (MGS8)	?[b]	Intrauterine and postnatal growth retardation; microtia; absent or hypoplastic patellae; dysmorphic facial features; kidney abnormalities	617564	Vetro et al. (2017)
GINS1	Essential subunit of the GINS complex within the CDC45-MCM-GINS (CMG) helicase complex	Growth retardation with neutropenia and natural killer cell deficiency	AR	Intrauterine and postnatal growth retardation; immunodeficiency; mild facial dysmorphism	617827	Cottineau et al. (2017)
MCM2	Essential component of the CDC45-MCM-GINS (CMG) helicase complex	Deafness, autosomal dominant 70	AD	Progressive hearing loss leading to mild to profound deafness	616968	Gao et al. (2015)
MCM4	Essential component of the CDC45-MCM-GINS (CMG) helicase complex	Natural killer cell and glucocorticoid deficiency with DNA repair defect	AR	Intrauterine and postnatal growth retardation; microcephaly; immunodeficiency; adrenal insufficiency	609981	Gineau et al. (2012), Hughes et al. (2012)
POLE1	Catalytic subunit of the replicative polymerase ε, which has essential roles in DNA replication and DNA repair	Facial dysmorphism, immunodeficiency, livedo, and short stature	AR	Short stature; skin abnormalities; immunodeficiency; facial dysmorphia	615139	Pachlopnik Schmid et al. (2012)
POLE2	Component of the replicative polymerase ε, which has essential roles in DNA replication and DNA repair	Combined immunodeficiency	?[b]	Immunodeficiency; dysmorphic facial features; autoimmunity	NA[c]	Frugoni et al. (2016)

Table 12.1 (*Continued*) **Human diseases associated with defective regulation of DNA replication.**

Gene	Gene function	Disease	MOI	Clinical presentation	OMIM #	References
POLD1	Catalytic subunit of the replicative polymerase δ, which has essential roles in DNA replication and DNA repair	Mandibular hypoplasia, deafness, progeroid features, and lipodystrophy syndrome	AD	Lack of subcutaneous fat; progeroid appearance; multiple facial abnormalities; abnormalities of the skeleton and genitals; deafness; insulin resistance and diabetes	615381	Weedon et al. (2013)
RECQL4[a]	A 3′-5′ DNA-helicase that is required for efficient replication initiation and fork progression	Rothmund–Thomson syndrome	AR	Poikiloderma; premature ageing; skeletal abnormalities; short stature; cancer predisposition (in particular osteosarcoma and skin carcinoma); facial dysmorphia	268400	Kitao et al. (1999)
		RAPADILINO syndrome	AR	Radial ray defects; absent or hypoplastic patellae; cleft or high arched palate; diarrhoea; skeletal abnormalities; short stature; cancer predisposition (in particular lymphoma and osteosarcoma)	266280	Siitonen et al. (2003)
		Baller-Gerold syndrome	AR	Short stature; intellectual disability; facial abnormalities; skeletal abnormalities; craniosynostosis	218600	Van Maldergem (1993)
DNA ligase I/ LIG1	DNA ligase with critical roles in DNA repair and DNA replication	DNA ligase I deficiency	?[b]	Immunodeficiency; growth failure; UV sensitivity; telangiectasia	126391	Barnes et al. (1992)

Abbreviations: AD, autosomal dominant; AR, autosomal recessive; MOI, mode of inheritance; NA, not applicable.
[a]Mutations in these genes cause multiple distinct clinical presentations.
[b]Insufficient data to determine mode of inheritance.
[c]OMIM designation has not yet been given.

postnatal growth retardation and immunodeficiency, again manifesting as a loss of NK cells (Cottineau et al. 2017). Cells from these patients also exhibited impaired DNA replication, increased levels of DNA damage and a defective response to replication stress.

Interestingly, individuals with mutations in the POLE1 and POLE2 subunits of polymerase ε also suffer from diseases associated with immunodeficiency and, in the case of POLE1, growth deficiency (Pachlopnik Schmid et al. 2012, Frugoni et al. 2016). In contrast, a specific autosomal dominant mutation in POLD1, the catalytic subunit of POLδ, is associated with individuals with progeroid features (premature aging), deafness, mandibular hypoplasia (small jaw), lipodystrophy (inability to produce fat) and diabetes (Weedon et al. 2013), and mutations in *MCM2* have been associated with progressive hearing loss/deafness (Gao et al. 2015). It is also noteworthy that autosomal dominant mutations in the exonuclease domains of *POLE1* and *POLD1* also give rise to a high risk of developing colorectal cancer, which is due to loss of the proofreading activities of these replicative polymerases (discussed in a later section together with DNA mismatch repair).

Mutations within the active site of DNA ligase I have also been found in a single patient exhibiting growth failure, immunodeficiency, telangiectasia and UV sensitivity (Barnes et al. 1992). Cells from this patient were found to exhibit defective joining of Okazaki fragments and were also sensitive to a wide range of DNA damaging agents including alkylating agents, DNA cross-linking agents, ionising radiation, PARP inhibitors and UV light, reflecting the critical roles that DNA ligase I plays in both DNA replication and DNA repair (Teo et al. 1983, Levin et al. 2000). Furthermore, mutations in the core replisome factor *PCNA* have been associated with growth deficiency, microcephaly, neurodegeneration, spinocerebellar ataxia, developmental delay, telangiectasia, photosensitivity and a predisposition for developing malignancy (Baple et al. 2014). However, in contrast to ligase I-derived patient cells, PCNA patient-derived cells did not display defective DNA replication, but rather showed an impaired ability to repair certain types of DNA damage (Baple et al. 2014), most likely due to the roles PCNA plays in various DNA repair pathways (discussed further in Chapter 5).

Mutations in *RECQL4*, a member of the RECQ family of helicases, are associated with three distinct, but related, autosomal recessive disorders: Rothmund-Thomson syndrome,

RAPADILINO syndrome and Baller–Gerold syndrome (Croteau et al. 2014). The common clinical presentations amongst these disorders are growth deficiency and skeletal abnormalities, but there are noteworthy differences between the three diseases. As two other RECQ helicases, namely BLM and WRN, also give rise to human disease, these will all be discussed in a later section (RECQ helicases and human disease).

Therefore, mutations in various components of the replication machinery and replication-associated factors can be linked to a wide range of clinical presentations affecting multiple organ systems. However, a striking commonality amongst these diseases is the presence of growth deficiency, and often microcephaly, which is observed in patients with mutations in *ORC1*, *ORC4*, *ORC6*, *CDT1*, *CDC6*, *GMNN*, *CDC45*, *MCM5*, *GMNN*, *MCM4*, *RECQL4*, *POLE1*, *DNA ligase I*, and *PCNA*. The simplest explanation seems to be that the general growth failure and severe microcephaly seen in these individuals is caused by slowed/delayed cell cycle progression as a result of an inability to complete DNA replication, due to a combination of defective DNA origin licencing, impaired G1-S phase transition, a loss of dormant and active origins, and/or perturbed replication fork progression. It is therefore likely that the impaired growth and other developmental abnormalities exhibited by patients with MGS arise from a reduction in the total number of cells produced during development, due to an overall reduction in the cellular proliferative capacity, and/or elevated cell death resulting from genome instability.

However, the underlying causes of the other clinical abnormalities are not so readily apparent. Indeed, whilst the underdevelopment of the ears and knees is a defining clinical presentation of MGS and is observed in other human disorders associated with defective DNA replication, it is not as straightforward to link these clinical presentations to replication/cell cycle abnormalities. It is also striking that immunodeficiency has been associated with mutations in multiple DNA replication factors, namely *MCM4*, *GINS1*, *POLE1*, *POLE2* and *LIG1*, suggesting that the development and maintenance of the immune system is highly sensitive to inefficiencies of DNA replication, but again the reasons behind this are not clear. Despite this, it is likely that these clinical phenotypes are caused by a combination of the replication defects described earlier. For example, patients with mutations in *CDC45* and *RECQL4* can present with craniosynostosis (Van Maldergem 1993, Fenwick et al. 2016). The growth of the skull is tightly coordinated with brain development, and the maintenance of cranial sutures is controlled by a fine balance between osteogenic proliferation and differentiation (Twigg and Wilkie 2015). It has therefore been suggested that impaired cell cycle progression due to defective DNA replication may disrupt this proliferation–differentiation balance, leading to increased differentiation and premature fusion of cranial plates within the skull (Twigg and Wilkie 2015, Fenwick et al. 2016).

CELLULAR RESPONSE TO DNA REPLICATION STRESS

Sources of replication stress

Replication stress poses a significant threat to genome stability, and is loosely defined as anything that blocks, slows or stalls DNA replication fork progression (discussed in detail in Chapters 2 and 6) (Zeman and Cimprich 2014, Munoz and Mendez 2017). A major source of replication stress is the presence of DNA lesions, from either endogenous or exogenous sources. There is a wide variety of DNA lesions that can cause problems for the DNA replication machinery, including DNA interstrand cross-links (ICLs), DNA–protein cross-links (DPCs), bulky DNA adducts, oxidative base damage and DNA single-strand breaks (SSBs). Some of these lesions, such as DNA ICLs and DPCs, act as potent blocks for both the CMG helicase and replicative polymerases, whilst other lesions, such as bulky DNA adducts and oxidative base damage, can be bypassed by the CMG helicase, but impede progression of the DNA polymerases. Finally, collision of the replication machinery with an SSB will lead to replication fork collapse and the generation of DSB. Each of these DNA lesions are repaired in their own dedicated DNA repair pathways, limiting their impact on replication forks in normal cells. However, defective DNA repair will lead to replication stress.

Common fragile sites (CFSs) are specific loci within chromosomes that are prone to breakage upon low levels of replication stress (Durkin and Glover 2007). CFSs have clinical significance, as regions of the genome that are particularly sensitive to replication stress are frequently rearranged in tumour cells. In addition, research into CFS breakage has increased our understanding of what genomic features pose difficulties for the DNA replication machinery. For example, DNA secondary structures prone to form at specific genomic regions or DNA sequences can impede DNA replication (Zlotorynski et al. 2003). Furthermore, breakage at some CFSs has also been attributed to regions of low density of replication origins, or transcription–replication conflicts within large genes (Letessier et al. 2011, Hamperl and Cimprich 2016).

DNA replication is also very sensitive to imbalances in the levels of nucleotides available and, therefore, nucleotide depletion is a potent source of replication stress. An endogenous source of nucleotide depletion with clinical significance is the excessive firing of dormant DNA origins, which is a pathological mechanism commonly associated with oncogene-induced dysregulation of DNA replication initiation (Sorensen and Syljuasen 2012). Nucleotide depletion can also be induced exogenously by exposure to genotoxic agents such as the ribonucleotide reductase inhibitor hydroxyurea (HU) (Petermann et al. 2010).

Replication-stress response

To deal with replication stress, and to prevent genome instability, the cell employs several mechanisms within a coordinated cellular response (Munoz and Mendez 2017). One of the central factors within this replication-stress response is the protein kinase ATR (ATM- and Rad3-related) (Saldivar et al. 2017). ATR is a member of the PIKK (phosphoinositide-3-kinase-like kinase) family of kinases that phosphorylate S/TQ sites within many protein substrates. ATR is activated in response to a large variety of DNA lesions, and together with its downstream effector kinase, CHK1, initiates a wide-ranging signalling cascade to promote numerous cellular processes, such as activation of cell cycle checkpoints, stimulation of DNA repair, and regulation of origin firing and replication fork stability. As a consequence, ATR is essential for mammalian cell viability, with the ATR knockout mouse exhibiting embryonic lethality (Brown and Baltimore 2000) (also detailed in Chapter 2).

Activation of the ATR-CHK1 pathway

ATR activation is a multi-step process that is initiated by stretches of RPA-coated ssDNA, a type of DNA structure that is generated at a large proportion of stalled/collapsed replication forks by the uncoupling of the helicase and polymerase (Saldivar et al. 2017). ATR is first recruited to sites of RPA-coated ssDNA via its binding partner, ATRIP (ATR-interacting protein) (Cortez et al. 2001). Next, TOPBP1 and ETAA1 localise to stalled replication forks and function in parallel to directly stimulate ATR kinase activity (Niedzwiedz 2016). TOPBP1 localisation requires the RAD17–RFC clamp loader, as well as other factors such as the MRN complex and BLM helicase (Delacroix et al. 2007, Duursma et al. 2013, Blackford et al. 2015). ATR then activates CHK1 via direct phosphorylation on serine residues 317 and 345; post-translational modifications that are commonly used in laboratories as a read-out for ATR function (Liu et al. 2000). The efficient activation of CHK1 also requires other accessory factors, such as Claspin, RAD17, Timeless and TIPIN (Kumagai and Dunphy 2000, Bao et al. 2001, Unsal-Kacmaz et al. 2007). DONSON, a replication-associated factor, has also been shown to promote efficient activation of the ATR-CHK1 pathway (Reynolds et al. 2017). Together these demonstrate the complexity of the ATR-CHK1 pathway, with multiple levels of regulation.

Cell cycle checkpoint activation and inhibition of late origins

A major function of the ATR-CHK1 pathway is to activate cell cycle checkpoints, such as the intra-S phase and G2/M checkpoints, to allow cells time to repair/resolve any problems and to prevent the transmission of DNA damage through the cell cycle. Cell cycle arrest at the G2/M boundary is achieved by the phosphorylation of the CDC25 phosphatases (CDC25A, B, C) by CHK1, which either leads to their rapid degradation or translocation out of the nucleus (Furnari et al. 1997). CDC25 phosphatases function to regulate cell cycle progression by removing inhibitory phosphorylation present on CDKs (cyclin-dependent kinases) (Furnari et al. 1997). Degradation of CDC25A upon activation of the ATR-CHK1 pathway ensures the persistence of inhibitory phosphorylation on S-phase specific CDKs, preventing them from promoting cell cycle transition in the presence of unrepaired DNA damage.

Another important function of the ATR-CHK1 pathway within the intra-S phase checkpoint response is to inhibit new origin firing in the presence of replication stress. This gives the cell time to resolve the replication stress and/or DNA damage, and prevents newly fired origins from encountering unrepaired DNA lesions (Yekezare et al. 2013). The exact mechanism behind the ATR-dependent suppression of new origin firing in mammalian cells is still not completely understood. Nevertheless, it has been suggested to involve the CHK1-dependent inhibition of the interaction between Treslin and TOPBP1, thus preventing assembly of the CMG replicative helicase (Boos et al. 2011). However, complete suppression of all origins firing in a cell experiencing replication stress can be deleterious itself, as irreversibly stalled replications can only be rescued by the firing of a local dormant origin in some situations. Indeed, a reduction in the numbers of available dormant origins leads to increased cellular hypersensitivity to replication stress-inducing agents (Yekezare et al. 2013). Therefore, upon replication stress, the cell globally inhibits dormant origins, but activates origins local to stalled forks (Ge and Blow 2010, Yekezare et al. 2013).

A failure to either globally inhibit new origin firing and/or promote firing of local dormant origins upon replication stress risks a failure to complete DNA replication in a timely manner. This results in regions of under-replicated DNA being transmitted through to mitosis, which can lead to genetic instability and mitotic abnormalities. Therefore, cells have evolved specific pathways that can resolve under-replicated DNA, including the ability to continue performing DNA synthesis in late G2 and early mitosis (Fragkos and Naim 2017).

Regulation of replication fork stability and fork restart

In every cell cycle, some replication forks will stall due to endogenous replication stress, and in most cases, this is only a temporary pausing of the replication machinery. However, a more 'persistent' stalling can occur under certain circumstances. In this case, the intra-S phase checkpoint functions to stabilise stalled forks to ensure they remain functional and can restart after the cause of the stalling has been resolved/removed (Cortez 2015). However, if the checkpoint is defective, or the replication block has not been resolved after many hours, then replication fork collapse will occur (loosely defined as the inactivation of the replication fork) (Petermann et al. 2010). It is hypothesised that the ATR-CHK1 pathway functions to directly regulate different nucleases and DNA repair enzymes to prevent fork collapse and to promote fork restart, and loss of ATR or CHK1 leads to replication fork–associated DSBs, predominantly generated by the structure-specific nuclease MUS81-EME1 (Trenz et al. 2006, Forment et al. 2011).

A key step in replication fork restart is remodelling of the fork to form a Holliday junction–like structure called a 'chicken foot' (Neelsen and Lopes 2015). This process is known as replication fork reversal and can be observed in cells exposed to a wide range of replication stress-inducing agents (Zellweger et al. 2015). There are many aspects to the cellular mechanisms that promote replication fork stability, reversal and restart, and numerous enzymes have been implicated in these processes (Cortez 2015). Factors that are critical for promoting homologous recombination (HR), such as RAD51, BRCA2/PALB2, BRCA1 and the RAD51 paralogue complexes, have been shown to be essential for maintaining replication fork stability and restart (Petermann et al. 2010, Adelman et al. 2013, Zellweger et al. 2015). Three members of the SWI/SNF-related family of helicases (SMARCAL1, ZRANB3 and RAD54) have also been shown to promote fork reversal (Bugreev et al. 2011, Betous et al. 2012, Vujanovic et al. 2017). In particular, SMARCAL1 can catalyse fork reversal, and is recruited to, and stimulated by, RPA-coated ssDNA at stalled replication forks (Bansbach et al. 2009, Betous et al. 2012).

Several RecQ helicases are also important for replication fork stability and fork restart. Both BLM (3′-5′ helicase) and WRN (3′-5′ helicase and exonuclease) are recruited to replication forks upon replication stress in an ATR-dependent manner, and loss of WRN or BLM compromises replication fork restart (Davies et al. 2007, Murfuni et al. 2013). Furthermore, DNA2 functions with WRN to process reversed forks and promote replication fork restart (Thangavel et al. 2015). Additional pathways for the stabilisation of stalled forks and promotion of replication fork reversal seem to be mediated by RECQL1 (another RecQ helicase), which is regulated by the activity of PARP1, and the DNA helicase DDX11 in combination with Timeless (Berti et al. 2013, Cali et al. 2016).

The Fanconi anaemia (FA) pathway (discussed in more detail in a dedicated section later in this chapter) also has important roles in promoting replication fork restart following fork stalling (Schlacher et al. 2012). In particular, two FA-associated DNA helicases, FANCM and FANCJ, and a core FA factor, FANCD2 (in cooperation with CtIP), function to promote efficient replication fork recovery following replication stress (Schwab et al. 2010, Yeo et al. 2014, Raghunandan et al. 2015). The MRN complex also has a critical role in promoting replication fork restart, and it has been suggested that the essential function of the MRN complex is to protect cells against endogenous replication stress generated during development by resolving replication fork intermediates (Bruhn et al. 2014).

There are other factors required for replication fork stability, and it is expected that more will be discovered. Indeed, it has been shown that loss of DONSON leads to severe replication-associated genome instability due to the cleavage of stalled replication forks by MUS81 and XPF (Reynolds et al. 2017). Furthermore, two E3 ubiquitin ligases, TRAIP and RFWD3, have recently been demonstrated to promote replication fork recovery upon replication stress (Elia et al. 2015, Harley et al. 2016, Hoffmann et al. 2016).

Like most cellular DNA repair processes, uncontrolled and excessive fork remodelling and fork reversal seems to be as deleterious as the absence of fork remodelling, or inappropriate fork reversal, which can all lead to DNA replication errors (Couch and Cortez 2014). Therefore, the factors that promote replication fork reversal exist in equilibrium with factors that suppress fork reversal. Indeed, ATR-dependent phosphorylation of SMARCL1 suppresses its activity, and both depletion and overexpression of SMARCAL1 causes replication-associated DNA damage (Bansbach et al. 2009, Couch et al. 2013). Furthermore, BLM also acts to suppress inappropriate recombination, and loss of BLM leads to a large increase in sister chromatid exchange events (Chaganti et al. 1974). Moreover, a novel RPA-like ssDNA binding factor, RADX, was found to play an important role in ensuring the correct level of fork remodelling at collapsed replication forks by regulating RAD51 activity (Dungrawala et al. 2017). Loss of RADX results in increased replication fork collapse due to the excessive accumulation of RAD51 at replication forks. Together this emphasises the importance of ensuring a correct balance between different/opposing repair mechanisms/activities, and demonstrates that there can be too much of a good thing in terms of recombination at a replication fork.

Protection of replication forks against nuclease-mediated degradation

In addition to preventing fork collapse and promoting fork restart, numerous factors also function to protect stalled/collapsed replication forks from excessive nucleolytic degradation. Uncontrolled replication fork degradation, catalysed by MRE11, is seen in the absence of functional HR and Fanconi anaemia pathways (Schlacher et al. 2012). The loading of RAD51 onto reversed replication forks seems to be required to block access of MRN and other nucleases (Schlacher et al. 2012). The RECQ helicase RECQL5 (Kim et al. 2015) and the TLS polymerase REV1 (Yang et al. 2015) have also been shown to suppress MRE11 dependent fork resection, and WRN prevents EXO1-mediated fork degradation (Iannascoli et al. 2015). Furthermore, a replication associated factor, BOD1L, also was shown to prevent DNA2-mediated degradation of replication forks upon replication stress, potentially by suppressing the action of anti-recombinases, such as BLM and FBH1 (Higgs et al. 2015). Interestingly, interactions between many of the fork protection factors exist, raising the possibility of the existence of a single 'fork protection complex', with subunits that function to protect against different nucleases (Higgs et al. 2015).

Resolving Replication–Transcription Conflicts and R-Loops

Another significant source of endogenous replication stress and genome instability is collisions between replication forks and transcription bubbles (further discussed in Chapter 10). Although replication and transcription processes occur at different times, and are spatially separated as much as possible, collisions occur as RNA and DNA polymerases occupy the same DNA molecule (Hamperl and Cimprich 2016). Replication–transcription conflicts can arise in one of several different ways. First, the replication and transcription machinery can either collide in a 'head-on' or 'co-directional' manner (Srivatsan et al. 2010). Second, replication–transcription conflicts can occur due to the formation/persistence of secondary structures containing RNA:DNA hybrids, called R-loops (Gan et al. 2011). R-loops can form during normal transcription, and it is believed that they have physiological roles in multiple cellular processes but can also arise from replication–transcription collisions (R-loops are discussed in detail in Chapters 10 and 11). Furthermore, it is known that the presence of high levels of R-loops impede replication fork progression and can cause replication-associated DNA damage and genome instability, although the underlying mechanisms behind this are not currently clear (Hamperl and Cimprich 2016).

The cell employs several strategies to prevent replication–transcription conflicts. Transcription-coupled nucleotide excision repair (TC-NER; detailed in Chapter 5) is activated when RNA POL II stalls due to the presence of transcription blocking lesions in actively transcribed genes, and functions to remove the RNA polymerase complex and repair the DNA lesion. Loss of TC-NER factors results in prolonged transcription arrest and increased replication stress (Pani and Nudler 2017). Another factor that reduces replication–transcription conflicts is RECQL5, which functions to suppress stalling, pausing and backtracking of RNA polymerases during normal transcription, thereby reduced the chances of collision between replication and transcription machinery (Saponaro et al. 2014).

The cell also has several mechanisms to prevent/resolve the formation of R-loops. RNaseH enzymes can directly cleave DNA:RNA hybrids (Wahba et al. 2011) and Senataxin (SETX) is a DNA/RNA helicase that associates with replication forks and can resolve R-loops (Yuce and West 2013). SETX also functionally interacts with BRCA1, and loss of SETX, BRCA1 or BRCA2 results in an accumulation of R-loops (Skourti-Stathaki et al. 2011, Bhatia et al. 2014). Furthermore, another cellular pathway that resolves R-loops is the FA pathway, which is activated by R-loop–induced replication–transcription conflicts, and promotes FANCM-dependent resolution of replication blocking DNA:RNA hybrids (Schwab et al. 2015).

Defects in the replication stress response pathway and human disease

Mutations in many of the factors involved in the response to replication stress give rise to human diseases (Table 12.2). Although these diseases are distinct from each other and present with a range of clinical phenotypes, there are commonalities between them, such as microcephaly, short stature and cancer predisposition as discussed below.

Seckel syndrome and microcephalic dwarfism

Seckel syndrome (SCKL) is an autosomal recessive disorder characterised by intrauterine and postnatal growth delay resulting in short stature, severe microcephaly (with moderate to severe mental retardation), skeletal abnormalities and a characteristic 'bird-like' facial appearance. SCKL can also present with other abnormalities such as craniosynostosis. Mutations in several proteins involved in the cellular response to replication stress are associated with SCKL.

Hypomorphic mutations in *ATR* was the first identified genetic cause of SCKL (O'Driscoll et al. 2003). Since this initial discovery, several other SCKL patients with *ATR* mutations and a single SCKL patient with mutations in the ATR interacting protein (*ATRIP*) have been identified (Ogi et al. 2012). Cell lines derived from patients with *ATR* mutations (SCKL1) have markedly impaired replication fork progression and increased levels of replication fork instability, dormant origin firing and DNA damage (Mokrani-Benhelli et al. 2013). Furthermore, cells from both ATR and ATRIP SCKL patients exhibit defective activation of the ATR-dependent DNA damage response (Ogi et al. 2012).

Additionally, in a study in which the original patient-associated hypomorphic *ATR* mutation was modelled in mice, it was found that the ATR-SCKL mice phenocopied the clinical presentation of SCKL1 patients (Murga et al. 2009). In particular, the ATR-SCKL mice exhibited severe growth delay and microcephaly, and they accumulated increased levels of replication stress and DNA damage during embryogenesis, resulting in a significant increase in apoptosis in proliferating cells within the developing brain (Murga et al. 2009). This supports the idea that the developing brain is exquisitely sensitive to increased replication stress during the period of rapid neuronal cell expansion. Interestingly, the increase in endogenous replication stress and DNA damage was only observed during embryogenesis, and was not detected in the tissues of adult mice. As the adult ATR-SCKL mice exhibited progressive symptoms indicative of a premature ageing phenotype, this suggests that problems experienced during embryogenesis can have a negative impact upon the health of an adult individual, potentially by causing loss of stem cell populations (Murga et al. 2009). Indeed, this is also supported by the observation that whilst the presence of an ATR deficiency during embryogenesis leads to a significantly shortened life expectancy, loss of ATR in one-month-old mice does not lead to a reduced lifespan (Murga et al. 2009).

The discovery of a microcephalic primordial dwarfism (MPD; intrauterine and postnatal growth delay and microcephaly) disorder caused by mutations in *DONSON*, expands the spectrum of diseases associated with defective/inefficient ATR activity (Reynolds et al. 2017). *DONSON* mutations cause severe microcephaly, mild to moderate intrauterine and postnatal growth delay and, in some patients, skeletal abnormalities. A very severe form of MPD (microcephaly-micromelia syndrome), characterised by profound microcephaly with abnormal brain structures, severe interuterine growth retardation, limb and lung abnormalities, and neonatal lethality, was also found to be caused by mutation of *DONSON* (Evrony et al. 2017, Reynolds et al. 2017). DONSON-patient derived cell lines exhibited defective activation of the intra-S phase checkpoint, decreased replication fork stability, increased DNA damage and elevated genome instability. Although the disease caused by mutation of *DONSON* is distinct from Seckel syndrome, there is a significant overlap of clinical phenotypes with patients with mutations in ATR/ATRIP, particularly microcephaly and growth deficiency.

Table 12.2 **Human diseases associated with a defective cellular response to replication stress.**

Gene	Gene function	Disease	MOI	Clinical presentation	OMIM #	References
ATR	Protein kinase that regulates the cellular response to replication stress, including cell cycle checkpoints and replisome stability	Seckel syndrome 1 (SCKL1)	AR	Severe microcephaly; intellectual disability; intrauterine growth deficiency; dwarfism; skeletal defects; characteristic 'bird-headed' facial appearance	210600	O'Driscoll et al. (2003)
ATRIP	ATR-interacting protein required for efficient activation of the ATR-CHK1 pathway	Seckel syndrome	AR	Severe microcephaly; dwarfism; short stature; skeletal abnormalities; characteristic 'bird-headed' facial appearance	NA[c]	Ogi et al. (2012)
CtIP[a]	Plays a critical role in the cellular response to replication stress and DNA repair. CtIP is essential for resection of DSBs and also promotes replication forks restart	Seckel syndrome 2 (SCKL2)	AR	Microcephaly; short stature; digital anomalies; characteristic 'bird-headed' facial appearance	606744	Qvist et al. (2011)
		Jawad Syndrome	AR	Microcephaly; intellectual disability; digital anomalies; characteristic 'bird-headed' facial appearance	251255	Qvist et al. (2011)
DNA2[a]	5′-3′ helicase/endonuclease with critical roles in DNA replication, the cellular response to replication stress and DNA repair	Seckel syndrome 8 (SCKL8)	AR	Intrauterine and postnatal growth retardation; microcephaly; intellectual disability; characteristic 'bird-headed' facial appearance	615807	Shaheen et al. (2014)
		Progressive external ophthalmoplegia 6 (PEOA6)	AD	Muscle weakness of the lower limbs; exercise intolerance; ophthalmoplegia; loss of mitochondrial DNA within muscles	615156	Ronchi et al. (2013)
TRAIP	An E3 RING ubiquitin ligase that promotes checkpoint signalling and replication fork progression upon replication stress	Seckel syndrome 9 (SCKL9)	AR	Microcephaly; intrauterine growth retardation; developmental delay; dysmorphic facial features	616777	Harley et al. (2016)
MMS21/ NSMCE2	E3 SUMO-protein ligase subunit of the SMC5-SMC6 complex with roles in the cellular responses to DNA damage and replication stress	Seckel syndrome 10 (SCKL10)	AR	Severe growth deficiency; microcephaly; learning difficulties; dysmorphic facial features; primary gonadal failure; insulin resistant diabetes	617253	Payne et al. (2014)
NSMCE3	A subunit of the SMC5-SMC6 complex with roles in the cellular responses to DNA damage and replication stress	Lung disease, immunodeficiency, and chromosome breakage syndrome	AR	Progressive lung disease; immunodeficiency; mild facial dysmorphia; feeding difficulties and failure to thrive	608243	van der Crabben et al. (2016)
DONSON[a]	Replication fork protein required for replication fork stability and efficient activation of the ATR-CHK1	Microcephaly, short stature and limb abnormalities	AR	Disproportionate microcephaly; short stature; minor skeletal abnormalities	617604	Reynolds et al. (2017)
		Microcephaly-Micromelia Syndrome	AR	Severe microcephaly; intrauterine growth retardation; craniofacial anomalies; skeletal abnormalities; neonatal lethality	251230	Evrony et al. (2017)
NBS1/NBN	Subunit of the MRE11-RAD50-NBS1 complex with roles in DNA repair, cell cycle checkpoint activation and the response to replication stress	Nijmegen breakage syndrome (NBS)	AR	Microcephaly; intellectual disability; intrauterine growth retardation; immunodeficiency; facial dysmorphism; predisposition to malignancies (in particular lymphoma)	251260	Varon et al. (2001)

Table 12.2 (*Continued*) Human diseases associated with a defective cellular response to replication stress.

Gene	Gene function	Disease	MOI	Clinical presentation	OMIM #	References
RAD50	Subunit of the MRE11-RAD50-NBS1 complex with roles in DNA repair, cell cycle checkpoint activation and the response to replication stress	Nijmegen breakage syndrome- like disorder (NBSLD)	?[b]	Microcephaly; growth retardation; mild non-progressive ataxia	613078	Waltes et al. (2009)
MRE11	Subunit of the MRE11-RAD50-NBS1 complex with roles in DNA repair, cell cycle checkpoint activation and the response to replication stress	Ataxia-telangiectasia-like disorder 1 (ATLD1)	AR	Progressive cerebellar degeneration; spinocerebellar ataxia; oculomotor apraxia	604391	Stewart et al. (1999)
RNASEH2	Enzyme that specifically cleave ribonucleotides in RNA-DNA duplexes	Aicardi-Goutieres syndrome (AGS)	AR	Symptoms similar to congenital viral infection, including: microcephaly; intrauterine growth retardation; spasticity; cerebral and cerebellar atrophy; high levels of interferon α in the cerebrospinal fluid	610181 (AGS2); 610329 (AGS3); 610333 (AGS4)	Crow et al. (2006)
PCNA	A critical replisome factor that has multiple roles in DNA replication and DNA repair	Ataxia-telangiectasia-like disorder 2 (ATLD2)	AR	Neurodegeneration; spinocerebellar ataxia; developmental delay; postnatal growth retardation; premature ageing; telangiectasia; photosensitivity; predisposition to UV-induced malignancies	615919	Baple et al. (2014)
SETX[a]	RNA-DNA helicase with roles in RNA metabolism and genome stability	Ataxia-oculomotor apraxia 2 (AOA2)	AR	Cerebellar atrophy; spinocerebellar ataxia; oculomotor apraxia; axonal neuropathy	606002	Moreira et al. (2004)
		Amyotrophic lateral sclerosis 4, juvenile (ALS4)	AD	Chronic motor neuron disease; weakness of upper and lower limb muscles; difficulty walking; axonal degeneration; loss of spinal cord anterior horn cells	602433	Chen et al. (2004)
BLM/ RECQL3	A 3′-5′ DNA-helicase with pro- and anti-recombinase activities during DSB repair and the response to replication stress	Bloom syndrome (BS)	AR	Intrauterine and postnatal growth retardation; sun sensitivity; facial telangiectasia; hypo- and hyperpigmented skin; immunodeficiency; predisposition to many types of cancer; increased chromosomal instability and sister chromatid exchanges in patient cells; diabetes	210900	Ellis et al. (1995)
WRN/ RECQL2	A 3′-5′ DNA-helicase exonuclease that promotes DNA resection DSB repair and replication fork restart upon replication stress	Werner syndrome (WS)	AR	Growth deficiency presenting during adolescence; prematurely aged facial appearance; predisposition to malignancy (in particular soft-tissue sarcomas and osteosarcomas); skin abnormalities; osteoporosis; diabetes	277700	Yu et al. (1996)
RMI2	Part of the BLM-RMI1-RMI2-TOP3α complex with roles in the processing of homologous recombination intermediates	Bloom-Like Syndrome	?[b]	Growth deficiency; skin abnormalities; increased sister chromatid exchanges in patient cells	NA[c]	Hudson et al. (2016)
SMARCAL1	A SWI/SNF protein with critical roles in promoting fork restart	Schimke immuno-osseous dysplasia (SIOD)	AR	Short stature; spondylopiphyseal dysplasia; cerebral ischemia; facial dysmorphia; weakened immune system; abnormalities of the skeleton and kidneys; bone marrow failure	242900	Boerkoel et al. (2002)

Table 12.2 **(*Continued*) Human diseases associated with a defective cellular response to replication stress.**

Gene	Gene function	Disease	MOI	Clinical presentation	OMIM #	References
ERCC6L2/ Hebo	Member of the SNF2 family of helicase-like proteins with potential roles in DNA repair and mitochondrial function	Bone marrow failure syndrome 2	AR	Microcephaly; learning difficulties; bone marrow failure	615715	Tummala et al. (2014)
SPRTN	DNA-dependent protease required for the repair of DNA-protein cross-links	Ruijs-Aalfs syndrome	AR	Short stature; premature ageing; facial dysmorphia; multiple skeletal abnormalities; muscular atrophy; early-onset hepatocellular carcinoma	616200	Lessel et al. (2014)

Abbreviations: AD, autosomal dominant; AR, autosomal recessive; MOI, mode of inheritance; NA, not applicable.
[a]Mutations in these genes cause multiple distinct clinical presentations.
[b]Insufficient data to determine mode of inheritance.
[c]OMIM designation has not yet been given.

Seckel syndrome is also caused by bi-allelic mutations in the endonuclease *CtIP* (Qvist et al. 2011) and the helicase/exonuclease *DNA2* (Shaheen et al. 2014). Both CtIP and DNA2 have multiple roles in DNA replication and DNA repair, including promotion of DNA resection during HR and replication fork restart following replication stress. Like ATR, complete loss of both CtIP and DNA2 in mice is embryonically lethal (Chen et al. 2005, Lin et al. 2013). Interestingly, hypomorphic mutations in both *CtIP* and *DNA2* can cause distinct pathologies other than Seckel syndrome. Mutations in *CtIP* can cause a Seckel-like disorder called Jawad syndrome that is characterised by microcephaly, mental retardation and digital abnormalities, but differs from Seckel syndrome as these patients do not suffer from growth deficiency (Qvist et al. 2011). In addition, mutations in *DNA2* are also associated with progressive muscle weakness, external ophthalmoplegia (weakness of the eye muscles) and loss of mitochondrial DNA within muscles (Ronchi et al. 2013). As DNA2 is also required for efficient replication of mitochondrial DNA, it is possible that the differing clinical presentations associated with DNA2 mutations reflects the relative impact of the mutations on the nuclear and/or mitochondrial functions of DNA2 (Duxin et al. 2009).

Last, Seckel syndrome is also known to be caused by mutations in the E3 SUMO ligase *MMS21/NSMCE2* (Payne et al. 2014) and the E3 ubiquitin ligase *TRAIP* (Harley et al. 2016). MMS21/NSMCE2 is a critical subunit of the SMC5-SMC6 complex. Like all members of the SMC family, SMC5-SMC6 forms a characteristic ring structure that allows it to encircle DNA. Hypomorphic mutations in *MMS21/NSMCE2* were found in patients presenting with severe growth retardation, microcephaly, primary gonadal failure and diabetes with extreme insulin resistance (Payne et al. 2014). The SMC5-SMC6 complex has multiple roles in the DNA damage response and the cellular response to replication stress, including promoting HR and replication fork stability. NSMCE2 patient-derived cell lines display impaired S phase progression and increased replication fork stalling, genome stability and DNA damage following replication stress (Payne et al. 2014, Menolfi et al. 2015). Interestingly, mutations in *NSMCE3*, another subunit of the SMC5-SMC6 complex, also result in a disorder associated with severe progressive lung disease and immunodeficiency (van der Crabben et al. 2016). NSMCE3 patient-derived cell lines possess a markedly destabilised SMC5-SMC6 complex, and exhibit increased genome instability, hypersensivity to replication stress inducing agents and defective HR. It is not clear why there is such a difference in clinical presentations between mutations in *MMS21/NSMCE2* and *NSMCE3*, although one explanation could be the degree of destabilisation of the SMC5-SMC6 complex. The *MMS21/NSMCE2* mutations result in a mild destabilisation of the SMC5-SMC6 complex, whilst the *NSMCE3* mutations lead to an almost complete loss of the complex (Payne et al. 2014, van der Crabben et al. 2016).

Mutations in TRAIP were found in three patients presenting with intrauterine and postnatal growth delay and microcephaly (Harley et al. 2016). TRAIP is an E3 ubiquitin ligase that is rapidly recruited to stressed replication forks via an interaction with PCNA and is a critical component of the response to replication stress (Hoffmann et al. 2016). Cell lines derived from patients with TRAIP mutations exhibit impaired growth and S-phase progression, and experience increased replication fork asymmetry and stalling following UV irradiation, but not exposure to HU or MMC (Harley et al. 2016). This suggests that the patient-associated mutations in TRAIP result in defective repair/bypass of specific replication blocking lesions, rather than resulting in a general defect in replication fork stability.

It is striking that although ATR-ATRIP, DONSON, DNA2, CTIP, MMS21/NSMCE2 and TRAIP all have multiple distinct roles in the cellular processes that maintain genome stability, they all result in a similar clinical presentation when mutated, including short stature, severe microcephaly and skeletal abnormalities. It is also noteworthy that complete loss of ATR, ATRIP, DNA2, CTIP or SMC5-SMC6 are all embryonically lethal in mice, emphasising the essential nature of their functions (Brown and Baltimore 2000, Chen et al. 2005, Ju et al. 2013, Lin et al. 2013, Dickinson et al. 2016). One strong theme that has been identified is the impact of loss/mutation of these factors on DNA replication. Indeed, it is believed that the essential function of the ATR-CHK1 pathway is to prevent replication fork collapse, and it has been shown that ATR mutants that retain the ability to activate the G2/M checkpoint, but have lost the ability to promote replication fork restart following replication stress, are unable to support cell viability (Nam et al. 2011). Therefore, it is hypothesised that the severe growth deficiency, microcephaly and other developmental abnormalities exhibited by SCKL patients are caused by increased levels of replication stress and DNA damage during embryogenesis, leading to elevated cell death by apoptosis and/or reduced cellular proliferative capacity, and ultimately resulting in reduced total numbers of cells produced during development (Murga et al. 2009).

However, in addition to the growth deficiency and neurodevelopmental defects seen in SCKL patients, they can also present with a wide range of additional clinical phenotypes. For example, depending on the genetic cause, SCKL patients can exhibit progressive muscle weakness (*DNA2*), diabetes with extreme insulin resistance (*MMS21/NSMCE2*), and immunodeficiency (*NSMCE3*). It is possible that these reflect the loss of specific functions of the mutated genes. For example, impaired replication of mitochondrial DNA likely underlies the progressive muscle weakness and ophthalmoplegia seen in some individuals with *DNA2* mutations (Duxin et al. 2009). However, due to the small numbers of patients associated with some diseases, it is not always possible to make meaningful judgements on what constitutes a 'typical' clinical phenotype. It is therefore possible that the discovery of additional patients may reveal some genotype–phenotype correlations that have not yet been discerned.

MRN complex and human disease

The MRN (MRE11-RAD50-NBN) complex is an essential component of the cellular processes that maintain genome stability, and plays roles in the response to replication stress, the DNA damage response and HR. Mutations in all three components of the MRN complex have been associated with human disease (discussed in detail in Chapter 13). Nijmegen breakage syndrome (NBS) is caused by mutations in the NBN subunit of the MRN complex, and is characterised by moderate to severe microcephaly, growth retardation, immunodeficiency and an increased cancer predisposition (in particular lymphoma) (Varon et al. 2001). Mutations in *RAD50* have also been identified in a single patient with an NBS-like disease (NBSLD), characterised by moderate microcephaly, growth retardation and non-progressive ataxia (Waltes et al. 2009). A third disorder resulting from disruption of the MRN complex is ataxia-telangiectasia–like disorder (ATLD), which is caused by mutations in *MRE11* and is predominantly characterised by progressive cerebellar ataxia and oculomotor apraxia (Stewart et al. 1999). As its name suggests, the clinical presentation of ATLD is more similar to ataxia-telangiectasia (A-T), which is caused by mutations in the critical DDR and DSB repair factor, ATM. However, there is a certain degree of phenotypic variability that can be associated with mutations in the MRN complex. For example, ATLD patients with microcephaly alongside progressive cerebellar ataxia have been identified (Fernet et al. 2005), and individuals with *NBN* mutations have been found that lack all the features typically associated with NBS, except for infertility, despite having a cellular phenotype very similar to a typical NBS patient (Warcoin et al. 2009).

Due to the critical role that the MRN complex plays in promoting ATM-dependent DSB signalling and repair, and the overlapping symptoms of the human diseases, it was originally hypothesised that defective DSB repair, and an accumulation of unrepaired DSBs, was the underlying cause of the clinical symptoms in A-T, NBS, NBSLD and ATLD (Reynolds and Stewart 2013). In addition, complete loss of MRE11, RAD50 or NBN is embryonically lethal, and it was originally thought that unrepaired DSBs were the pathological lesion. However, it is becoming more evident that the MRN complex also has critical roles in DNA replication and the resolution of replication stress. Indeed, a study involving the generation of an inducible Nbn-deleted mouse found that the essential function of Nbn is associated with its role in promoting the resolution of replication intermediates, rather than the repair of DSBs (Bruhn et al. 2014). Based on this, it is likely that the conversion of these unresolved replication intermediates into DSBs, leading to genome instability and mitotic errors, underlies the loss of cell viability in the absence of functional NBN. Therefore, it is probable that genome instability caused by the defective resolution of endogenous replication stress is a major contributor to the microcephaly and intrauterine growth delay seen in NBS individuals, and this may also contribute to the clinical presentations seen in ATLD and NBSLD individuals.

RECQ helicases and human disease

The RECQ family of DNA helicases consists of five members (RECQL1, WRN/RECQL2, BLM/RECQL3, RECQL4 and RECQL5) that have multiple important roles in the preservation of genome stability (Croteau et al. 2014). So far mutations in three of the RECQ helicases, *RECQL4, BLM* and *WRN*, have been associated with autosomal recessive human disease.

Bloom's syndrome (BS) is caused by mutations in the BLM helicase and is primarily characterised by intrauterine and postnatal growth retardation, microcephaly, immunodeficiency and a high predisposition for developing cancer (Ellis et al. 1995). Patients with BS also suffer from sensitivity to sunlight and are at risk of developing abnormalities of multiple organ systems. The most striking and consistent feature of Bloom syndrome is growth deficiency, although the increased risk of cancer is the most debilitating symptom as patients develop multiple primary tumours, which are the most frequent cause of death (de Renty and Ellis 2017). Indeed, analysis of Bloom's syndrome patients have indicated that they are nearly 100-fold more likely to develop any type of cancer relative to the general population (de Renty and Ellis 2017). In addition, mutations in RMI2, a subunit of the BLM-RMI1-RMI2-TOP3α complex that processes HR intermediates, gives rise to growth deficiency (Hudson et al. 2016). WRN DNA helicase is mutated in Werner syndrome (WS), which has a markedly different clinical presentation to that of Bloom's syndrome. Werner syndrome is predominantly a premature ageing disorder, and the hallmarks of the disease are bilateral ocular cataracts, premature greying and thinning of scalp hair, abnormalities of the skin, osteoporosis and growth deficiency during adolescence (Yu et al. 1996, de Renty and Ellis 2017). Patients suffer from increased cancer predisposition, although in contrast to Bloom's syndrome, this is confined to specific cancer types, such as soft-tissue sarcomas and osteosarcomas (de Renty and Ellis 2017).

Mutations in RECQL4 cause three distinct, but related, autosomal recessive disorders: Rothmund-Thomson syndrome (RTS), RAPADILINO syndrome and Baller-Gerold syndrome (BGS). RTS is a clinically heterogeneous disease, and patients can present with a range of symptoms, including growth deficiency, skeletal abnormalities, radial ray defects (abnormalities of bones in the arms and hands), premature ageing and a predisposition to developing osteosarcoma and skin carcinoma (Kitao et al. 1999, Larizza et al. 2010). One clinical symptom consistently present in the majority of RTS individuals is poikiloderma, a facial skin rash characterised by hypopigmentation, hyperpigmentation, telangiectasia and atrophy. The second disorder, RAPADILINO, is associated with short stature, radial ray defects, dislocated joints, underdeveloped or absent knees, abnormalities of the palate, diarrhoea in infancy and an increased risk of developing lymphoma and osteosarcoma (Siitonen et al. 2003). In contrast, BGS is not associated with cancer predisposition, and instead primarily presents with radial ray defects and craniosynostosis, although patients can present with growth deficiency, poikiloderma, and abnormalities of the knees (Van Maldergem 1993). Therefore, the most consistent features of RTS, RAPADILINO and BGS are short stature and skeletal abnormalities. Despite a few unique characteristics, the three disorders share many clinical features and are considered to be part of the same spectrum of disorders. It is not clear why mutations in RECQL4 can result in three different diseases, as there doesn't seem to be any genotype–phenotype correlation between mutations within different domains of *RECQL4* and clinical presentation, and interestingly, several mutations are associated with all three diseases (Larizza et al. 2010).

The RECQ helicases all share highly conserved helicase domains, but each possesses unique domains and motifs that mediate protein–protein interactions and cellular sub-localisation. Therefore, whilst BLM, WRN and RECQL4 have overlapping roles in DNA replication and DNA repair, they also have unique functions, and loss of each individual factor cannot be compensated by the presence of the other RECQ helicases (reviewed in Croteau et al. 2014). For example, BLM has both pro- and anti-recombinogenic functions during the HR-mediated repair of DNA DSBs. Cells derived from Bloom's syndrome patients display increased spontaneous replication stress and genome instability, and exhibit a hyper-recombination phenotype (Chaganti et al. 1974, Ellis et al. 1995, Rassool et al. 2003). Like BLM, WRN functions to regulate HR, but unlike BLM, loss of WRN activity leads to a HR deficiency (Saintigny et al.

2002). Further, BLM, WRN and RECQL4 have all been implicated in the maintenance of telomeres, but RECQL4 is the only RECQ helicase that functions to promote mitochondrial DNA replication and repair (Croteau et al. 2012). Therefore, RECQL4-deficient cells exhibit loss of mitochondrial DNA in addition to telomere dysfunction (Chi et al. 2012, Croteau et al. 2014). Moreover, all three helicases have been implicated in promoting replication fork progression and fork stability in unstressed and stressed conditions, but only RECQL4 has a role in the initiation of DNA replication (Sangrithi et al. 2005).

Therefore, the growth deficiency, genome instability and increased cancer predisposition common to all three of the RECQ helicase syndromes probably reflects the impact of *BLM*, *WRN* and *RECQL4* mutations on common cellular processes, such as DNA replication. As discussed in this chapter, prenatal growth deficiency and microcephaly are frequently associated with diseases linked with impaired DNA replication. However, this is unlikely to be the whole story, as the RECQ helicases play many different roles in the cell. Indeed, defective DNA repair also underlies many developmental and cancer predisposition diseases (Alcantara and O'Driscoll 2014, Broustas and Lieberman 2014), and telomere dysfunction has been shown to lead to premature ageing and cancer (Chang et al. 2004, Calado and Young 2012). The unique functions of each RECQ helicase, and the differences in the underlying cellular defects associated with their loss, may also explain the different features of the diseases. For example, *BLM* mutations are associated with an increased risk of most types of cancer, whereas *WRN* and *REQL4* are prone to developing specific types of cancer, such as soft-tissue sarcoma and osteosarcoma. It is suggested that hyper-recombination could underlie the increased the risk of tumourigenesis in all tissue types observed in BS (Luo et al. 2000), whilst only specific tissue types may be at risk of developing cancer due to reduced HR. Furthermore, patient-derived cell lines with *WRN* and *RECQL4* mutations, but not *BLM* mutations, reach replicative senescence faster than normal, which may partly explain why features of premature ageing are more prominent in WS and RTS than BS (de Renty and Ellis 2017, Lu et al. 2017). Due to space limitations, we are unable to discuss the RECQ helicase disorders in more detail, but the similarities and differences of BS, WS and RTS are described in depth in reviews by de Renty and Ellis (2017) and Lu et al. (2017).

SNF2-family proteins and human disease

The SNF2 family of ATPases predominantly constitute a group of chromatin remodelling proteins with diverse roles in various cellular processes such as DNA replication and repair, HR and transcription (Flaus et al. 2006). SMARCAL1, ERCC6L2 and CSB are three members of the SNF2 family associated with human disease. CSB is a critical component of the TC-NER and mutations in *CSB* are associated with Cockayne's syndrome type B (CSB), which is characterised by growth deficiency, premature ageing, intellectual disability, UV light sensitivity and progressive cerebellar atrophy, and is discussed in more detail in Chapter 5.

Mutation of SMARCAL1 is associated with a disease called Schimke immuno-osseous dysplasia (SIOD) and is associated with growth retardation, heart and kidney defects, skeletal abnormalities, a weakened immune system, cerebral ischemia (loss of blood flow to the brain), and sometimes bone marrow failure (Boerkoel et al. 2002). SMARCAL1 is an 'annealing' DNA helicase and has the ability to reanneal RPA-coated ssDNA bubbles into dsDNA (Bansbach et al. 2009). SMARCAL1 is regulated by ATR and has been shown to prevent replication fork collapse, and promote fork regression and fork restart at stalled replication forks (Bansbach et al. 2009, Betous et al. 2012). Consistent with this, patient-derived cell lines exhibit increased levels of replication stress and impaired cell cycle progression. Therefore, impaired DNA replication is likely to contribute to the pathology of the disease. However, as SMARCAL1 also functions to maintain telomeres and regulate gene expression, telomere dysfunction and/or impaired/altered gene expression could also contribute to the development of SIOD (Baradaran-Heravi et al. 2012, Morimoto et al. 2016).

The third disease associated with the SNF2 family is bone marrow failure syndrome 2, which is caused by mutations in ERCC6L2 (Tummala et al. 2014). Patients with mutations in ERCC6L2 present with bone marrow failure in combination with microcephaly and learning difficulties. The function of ERCC6L2 has yet to be determined, although the cellular phenotypes of ERCC6L2-depleted cells suggests a role in the repair of mitochondrial and nuclear DNA.

Defective resolution of transcription–replication conflicts and human disease

There are numerous links between the defective resolution of transcription–replication conflicts and human disease. SETX is a helicase that removes R-loops, and has functions in promoting transcription termination and resolving transcription–replication conflicts

(Skourti-Stathaki et al. 2011, Yuce and West 2013). SETX is associated with two distinct disease pathologies. The first is ataxia-oculomotor apraxia 2 (AOA2), an autosomal recessive disorder characterised by progressive cerebellar atrophy, spinocerebellar ataxia, oculomotor apraxia and axonal neuropathy (Moreira et al. 2004), and the second is amyotrophic lateral sclerosis 4 (ALS4), an autosomal dominant chronic motor neuron disease that is associated with muscle weakness, difficulty walking, axonal degradation and loss of spinal cord cells (Chen et al. 2004). Additionally, RNASEH2 can directly resolve DNA:RNA hybrids and is mutated in the neuroinflammatory disorder Aicardi-Goutieres syndrome 4 (AGS4), a disease characterised by microcephaly, growth retardation, seizures, and cerebral and cerebellar atrophy (Crow et al. 2006). Furthermore, BRCA1, BRCA2 and components of the FA pathway have also been implicated in the resolution of R-loops and transcription–replication conflicts (Bhatia et al. 2014, Schwab et al. 2015).

Therefore, defective resolution of transcription–replication conflicts gives rise to a diverse range of symptoms, including neurodegeneration, neuroinflamation, chronic motor neurone disease, increased cancer predisposition, bone marrow failure, microcephaly and growth retardation. In addition, transcription–replication conflicts can potentially lead to several deleterious consequences, such as impaired DNA replication and cell cycle progression, and/or defective transcription and altered gene expression. Since most of the factors involved in the resolution of these conflicts have multiple roles in DNA replication and DNA repair, it will be difficult to delineate exactly which aspects of transcription-induced replication stress contribute to human disease.

Summary

As discussed in the current and previous sections, mutations in factors with critical roles within the DNA replication and replication stress pathways are associated with a large variety of human diseases. Interestingly, the majority of these diseases exhibit developmental deficiency. The ultimate cause of primary microcephaly and intrauterine growth deficiency is the failure to produce enough cells to form a 'normal'-sized individual. Therefore, the simplest unifying hypothesis is that impaired DNA replication leads to delayed cell cycle progression, genome instability and/or increased cell death, resulting in a reduction in the total numbers of cells during development. Indeed, study of an ATR-SCKL mouse model, demonstrated that ATR deficiency leads to increased levels of replication stress and apoptotic cell death specifically during embryogenesis, leading to severe growth delay and microcephaly (Murga et al. 2009). Furthermore, another study in mice concluded that the essential function of Nbn is to promote the resolution of replication intermediates, rather than the repair of DSBs (Bruhn et al. 2014).

However, proteins that function within the replication stress response typically have roles in multiple cellular processes, including transcription, mitochondrial DNA replication and repair, telomere maintenance, and nuclear DNA repair pathways; therefore, we cannot ascribe the developmental deficiency phenotypes solely to impaired DNA replication. It is also highly likely that deficiencies in these cellular processes also contribute to the wide range of other symptoms associated with mutations in replication stress response genes, such as premature ageing (*WRN, BLM, RECQL4, SMARCAL1, CSB*), neurodegeneration (*MRE11, SETX*), motor neuron disease (*SETX, RNASH2*), bone marrow failure (*SMARCAL1, ERCC6L2*), immunodeficiency (*NSMCE3, BLM, SMARCAL1, NBN*), progressive muscle weakness (*DNA2, SETX*), diabetes (*MMS21/NSMCE2*), lung abnormalities (*DONSON, NSMCE3*) and cancer predisposition (*BLM, WRN, RECQL4, NBN*).

An interesting observation is that cancer predisposition doesn't seem to be associated with mutations in 'core' DNA replication factors, or with SCKL, but is caused by mutations in proteins involved in multiple DNA replication and DNA repair pathways. A potential contributing factor to this could be whether the underlying molecular defect impacts on apoptotic pathways. For example, in the presence of functional apoptotic pathways, elevated levels of replication stress would lead to increased levels of apoptosis-induced cell death, preventing cells with genome instability from proliferating and protecting against tumorigenesis. However, if too much cell death occurs, this could lead to other pathologies, such as neurodegeneration, premature ageing or microcephaly. In contrast, loss of apoptosis would allow cells experiencing genome instability to survive and persist, increasing the risk of generating mutations that drive cancer development. Therefore, impaired DNA replication itself may not be sufficient to initiate tumorigenesis, but perhaps it is the misrepair of replication fork intermediates, or the absence of signals that promote cell cycle arrest and/or cell death, which drive tumour development. However, the mechanisms behind this are poorly understood and, indeed, in contrast to patients, mice with mutations in the MCM subunits do exhibit increased cancer predisposition as well as growth defects (Shima et al. 2007). Therefore, further research needs to be performed to increase our understanding of the factors that give rise to increased rates of cancer development.

DNA REPLICATION AND DNA INTERSTRAND CROSS-LINK REPAIR

DNA cross-links are lesions involving a covalent linkage between two strands of the DNA double helix. There are two types of DNA cross-links that can arise in a cell: intrastrand cross-links, in which two nucleotide residues within the same DNA strand are covalently linked, and interstrand cross-links (ICLs) in which two nucleotide residues from opposite DNA strands are covalently linked. ICLs are considered to be particularly cytotoxic as they inhibit essential processes such as DNA replication and transcription and, if left unrepaired, can lead to mutations, chromosomal aberrations, and mitotic catastrophe. DNA intrastrand cross-links are considered to be less genotoxic but can be mutagenic. DNA ICLs are primarily repaired by the Fanconi anaemia DNA repair pathway, whereas DNA intrastrand cross-links are typically repaired by the nucleotide excision repair (NER) pathway, which is discussed in detail in Chapter 5.

Sources of DNA interstrand cross-links (ICLs)

A major source of endogenous ICLs is thought to be reactive aldehydes, which can be generated as metabolic by-products of normal cellular processes, such as histone demethylation, or from the metabolism of ethanol (Shi et al. 2004, Lopez-Martinez et al. 2016). ICLs can also arise from unsaturated aldehydes found in cigarette smoke and vehicle exhaust fumes (Pang and Andreassen 2009). Indeed, mice deficient in the repair of ICLs are sensitive to acetaldehyde treatment (Langevin et al. 2011). Due to the cytotoxic nature of DNA ICLs, DNA cross-linking agents, such as platinum compounds (e.g. cisplatin), mitomycin C (MMC), diepoxybutane and psoralen, have been successfully used in the clinic in the treatment of cancers (Tomasz 1995, Kelland 2007, Lopez-Martinez et al. 2016).

Fanconi anaemia (FA) ICL repair pathway

Our understanding of how DNA ICLs are repaired has been primarily informed by research into the autosomal recessive disorder Fanconi anaemia (FA) and is a good example of how the study of human disease can result in advances in understanding the mechanisms of DNA repair pathways. FA is a rare genetic disorder characterised by bone marrow failure, multiple congenital anomalies and cancer predisposition, and is caused by mutations in one of 21 different genes that cooperate within overlapping DNA repair pathways (Tables 12.3 and 12.4) (Mamrak et al. 2017). A major function of the FA pathway is to coordinate the recognition and repair of DNA ICLs during DNA replication via a highly coordinated process that also involves the activity of other DNA repair pathways, such as HR, translesion synthesis (TLS) and NER. In addition to DNA ICL repair, the FA pathway also performs other important roles during DNA replication, including functioning to protect stalled replication forks from uncontrolled resection by nucleases, promoting replication fork recovery following replication stress, and resolving replication–transcription conflicts (Ceccaldi et al. 2016).

There are several models of how an ICL is repaired. The canonical FA pathway is active during S phase and functions to repair ICLs following collision of converging replication forks with an ICL (Zhang et al. 2015). The FA repair pathway then proceeds through the following steps: ICL recognition, ICL unhooking, lesion bypass by TLS, then HR (Figure 12.1). In addition, the FA pathway can be broadly characterised into three different groups. The first group consists of the FA core complex (FANCA, FANCB, FANCC, FANCE, FANCF, FANCG, FANCL, FANCM, FANCT, MHF1, MHF2, FAAP20, FAAP24, and FAAP100), which has essential roles in the initial stages of ICL repair (Ceccaldi et al. 2016). The second group is the FANCD2-FANCI complex, which is monoubiquitylated by the core complex upon ICL induction (Garcia-Higuera et al. 2001, Smogorzewska et al. 2007). FANCD2-FANCI is a central factor required for the recruitment of the third group of proteins, namely the downstream effector proteins, such as structure-specific nucleases and HR factors (Ceccaldi et al. 2016). Alternatively, it has recently been shown that ICLs can be bypassed by a replication fork through a traverse mechanism that is dependent on the FANCM-MHF complex, but is independent of FANCD2-FANCI

Table 12.3 Human diseases associated with defective homologous recombination.

Gene	Gene function	Disease	MOI	Clinical presentation	OMIM #	References
BRCA2/ FANCD1	Required for the assembly of RAD51-ssDNA nucleofilaments and has essential roles in homologous recombination, ICL repair and replication fork stability/protection	Fanconi anaemia subtype D1 (FANCD1)	AR	Severe intrauterine and postnatal growth retardation; microcephaly; predisposition to early onset leukaemia and solid tumours, such as brain tumours; VATER group of congenital defects	605724	Howlett et al. (2002)
PALB2/ FANCN	Binding partner of BRCA2 with essential roles in homologous recombination, ICL repair and replication fork protection	Fanconi anaemia subtype N (FANCN)	AR	Severe intrauterine and postnatal growth retardation; microcephaly; anaemia; predisposition to early onset cancer (including leukaemia and brain tumours); abnormalities of the kidneys and heart	610832	Reid et al. (2007), Xia et al. (2007)
RAD51C/ FANCO	RAD51 paralogue required for efficient homologous recombination and replication fork restart	Fanconi anaemia subtype O (FANCO)	AR	Short stature; congenital abnormalities of the kidneys and heart	613390	Vaz et al. (2010)
RAD51/ FANCR	An essential recombination factor with multiple roles in replication fork protection, fork restart and ICL repair	Fanconi anaemia subtype R (FANCR)	AD	Growth retardation; microcephaly, hydrocephalus; skeletal abnormalities; intellectual disability	617244	Ameziane et al. (2015)
XRCC2/ FANCU	RAD51 paralogue involved in homologous recombination	Fanconi anaemia subtype U (FANCU)	AR	Severe growth deficiency; microcephaly; abnormalities of the kidneys, skin and skeleton	617247	Shamseldin et al. (2012)
RFWD3/ FANCW	E3 ubiquitin ligase involved in homologous recombination and ICL repair	Fanconi anaemia subtype W (FANCW)	AR	Intrauterine and postnatal growth retardation; microcephaly; bone marrow failure; abnormalities of the kidneys and skeleton	617784	Knies et al. (2017)
SPIDR	Nuclear scaffolding protein with roles in the regulation of homologous recombination	Primary ovarian insufficiency (POI)	AR	Absent/delayed puberty; ovarian abnormalities	NA[b]	Smirin-Yosef et al. (2017)
MCM8	Part of the MCM8-MCM9 complex involved in homologous recombination, ICL repair and replication fork restart	Premature ovarian failure 10 (POF10)	AR	Premature ovarian and testicular failure; genital abnormalities.	612885	Alasiri et al. (2015)
MCM9	Part of the MCM8-MCM9 complex involved in homologous recombination, ICL repair and replication fork restart	Ovarian dysgenesis 4 (ODG4)	AR	Short stature; ovarian abnormalities; absent/delayed puberty	616185	Wood-Trageser et al. (2014)
FANCM	DNA helicase that functions in the recognition of DNA ICLs, and promotion of replication fork progression, stability and restart	Familial breast cancer	?[a]	Cancer predisposition (particularly breast cancer); toxicity to chemotherapy; early menopause	NA[b]	Catucci et al. (2017), Bogliolo et al. (2017)

Abbreviations: AD, autosomal dominant; AR, autosomal recessive; MOI, mode of inheritance; NA, not applicable.
[a]Insufficient data to determine mode of inheritance.
[b]OMIM designation has not yet been given.

and the FA core complex (Huang et al. 2013). This would allow DNA replication to continue unimpeded and is thought to leave behind an 'X-structure' with the ICL at the centre, which is then repaired by the canonical FA pathway (**Figure 12.1**).

ICL recognition

There seems to be multiple aspects to the recognition of DNA ICLs. An early event is the recognition of DNA ICLs by the DNA damage sensor protein UHRF1, which recruits FANCD2 to the site of damage (Liang and Cohn 2016). In parallel, FANCM, ATR, BRCA1/FANCS and FANCI (but not FANCD2) promote the recruitment of the FA core complex to ICLs (Castella et al. 2015, Xue et al. 2015). Efficient activation of the FA pathway also requires BLM

Table 12.4 Human diseases associated with defective repair of DNA-ICLs.

Gene	Gene function	Disease	MOI	Clinical presentation	OMIM #	References
FANCA	Subunit of the FA core complex with a central role in ICL repair	Fanconi anaemia subtype A (FANCA)	AR	Growth retardation; microcephaly; bone marrow failure; predisposition to cancer; abnormalities of the kidneys, heart, skin and skeleton	227650	Fanconi Anaemia/ Breast Cancer Consortium (1996)
FANCB	Subunit of the FA core complex with a central role in ICL repair	Fanconi anaemia subtype B (FANCB)	XLR	Growth retardation; microcephaly; bone marrow failure; abnormalities of the kidneys, heart, skin and skeleton; VACTERL-H group of congenital defects	300514	Meetei et al. (2004), McCauley et al. (2011)
FANCC	Subunit of the FA core complex with a central role in ICL repair	Fanconi anaemia subtype C (FANCC)	AR	Growth retardation; microcephaly; bone marrow failure; predisposition to cancer; abnormalities of the kidneys, heart, skin and skeleton	227645	Strathdee et al. (1992), Gavish et al. (1993)
FANCD2	Part of the FANCD2-FANCI complex with a role in ICL repair and the cellular response to replication stress	Fanconi anaemia subtype D2 (FANCD2)	AR	Growth retardation; microcephaly; bone marrow failure; predisposition to cancer; abnormalities of the kidneys, heart, skin and skeleton	227646	Timmers et al. (2001)
FANCE	Subunit of the FA core complex with a central role in ICL repair	Fanconi anaemia subtype E (FANCE)	AR	Growth retardation; microcephaly; bone marrow failure; predisposition to cancer; abnormalities of the kidneys, heart, skin and skeleton	600901	de Winter et al. (2000a)
FANCF	Subunit of the FA core complex with a central role in ICL repair	Fanconi anaemia subtype F (FANCF)	AR	Growth retardation; microcephaly; bone marrow failure; predisposition to cancer; abnormalities of the kidneys, heart, skin and skeleton	603467	de Winter et al. (2000b)
FANCG/ XRCC9	Subunit of the FA core complex with a central role in ICL repair	Fanconi anaemia subtype G (FANCG)	AR	Growth retardation; microcephaly; bone marrow failure; predisposition to cancer; abnormalities of the kidneys, heart, skin and skeleton	614082	de Winter et al. (1998)
FANCI	Part of the FANCD2-FANCI complex with a role in ICL repair and the cellular response to replication stress	Fanconi anaemia subtype I (FANCI)	AR	Growth retardation; microcephaly; bone marrow failure; predisposition to cancer; abnormalities of the kidneys, heart, skin and skeleton	609053	Dorsman et al. (2007)
FANCJ/ BRIP	5′-3′ DNA helicase with roles in ICL repair, homologous recombination and the cellular response to replication stress	Fanconi anaemia subtype J (FANCJ)	AR	Growth retardation; bone marrow failure; predisposition to cancer; abnormalities of the kidneys, heart, skin and skeleton	609054	Levitus et al. (2005)
FANCL	The E3 ubiquitin ligase subunit within the FA core complex that is required for the mono-ubiquitylation of FANCD2-FANCI upon the induction of ICLs	Fanconi anaemia subtype L (FANCL)	AR	Growth retardation; microcephaly; bone marrow failure; abnormalities of the kidneys, heart, skin and skeleton	614083	Meetei et al. (2003)
SLX4/ FANCP	Scaffolding protein for structure-specific nucleases (MUS81-EME1, SLX1 and XPF-ERCC1) with roles in DNA repair and the cellular response to replication stress	Fanconi anaemia subtype P (FANCP)	AR	Growth retardation; microcephaly; bone marrow failure; abnormalities of the kidneys, skin and skeleton	613951	Stoepker et al. (2011)
XPF/ ERCC4/ FANCQ[a]	Structure-specific nuclease with roles in DNA repair, homologous recombination and ICL repair	Fanconi anaemia subtype Q (FANCQ)	AR	Growth retardation; microcephaly; bone marrow failure; abnormalities of the heart and skin	615272	Bogliolo et al. (2013), Kashiyama et al. (2013)
		Xeroderma pigmentosum type F (XPF)	AR	Sensitivity to sunlight; increased risk of skin cancer	278760	Sijbers et al. (1996)
		Xeroderma pigmentosum type F/Cockayne syndrome (XPF/CS)	AR	High sensitivity to sunlight; increased risk of skin cancer; intellectual disability; cerebral atrophy; cerebellar ataxia; microcephaly; short stature	278760	Kashiyama et al. (2013)

Table 12.4 **(*Continued*) Human diseases associated with defective repair of DNA-ICLs.**

Gene	Gene function	Disease	MOI	Clinical presentation	OMIM #	References
ERCC1	Binding partner of the structure-specific nuclease XPF with roles in DNA repair	Cerebro-oculo-facio-skeletal syndrome 4 (COFS4)	AR	Intrauterine growth retardation; dysmorphic facial features; multiple skeletal abnormalities; microcephaly	610758	Jaspers et al. (2007), Kashiyama et al. (2013)
BRCA1/ FANCS	E3 ubiquitin ligase with critical roles in the regulation of DNA repair, DSB resection and cell cycle checkpoints	Fanconi anaemia subtype S (FANCS)	AR	Growth retardation; microcephaly; developmental delay intellectual disability; predisposition to cancer	617883	Sawyer et al. (2015)
UBE2T/ FANCT	E2 ubiquitin ligase required for the monoubiquitylation of FANCD2 upon the induction of replication stress	Fanconi anaemia subtype T (FANCT)	AR	Growth retardation; bone marrow failure; abnormalities of the skeleton	616435	Hira et al. (2015)
REV7/ FANCV	Subunit of the DNA polymerase zeta complex that functions in translesion synthesis	Fanconi anaemia subtype V (FANCV)	AR	Growth retardation; microcephaly; bone marrow failure; facial abnormalities; abnormalities of the kidneys	617243	Bluteau et al. (2016)
FAN1	Nuclease required for efficient ICL repair	Karyomegalic interstitial nephritis	AR	Kidney failure; abnormal liver function	614817	Zhou et al. (2012)

Abbreviations: AR, autosomal recessive; MOI, mode of inheritance; NA, not applicable; XLR, X-linked recessive.

[a]Mutations in these genes cause multiple distinct clinical presentations.

[b]OMIM designation has not yet been given.

(Hemphill et al. 2009). Finally, recent studies have also demonstrated that ICL repair requires the convergence of two replication forks (Zhang et al. 2015) and the BRCA1/FANCS dependent unloading of the CMG helicase at sites of ICL stalled replication forks (Long et al. 2014). Taken together, these data demonstrate the complexity of the process of ICL recognition.

Following recognition of the DNA ICL, the FANCD2-FANCI complex is monoubiquitylated by the E3 ubiquitin ligase FANCL (a subunit of the FA core complex) in combination with the E2 conjugating enzyme UBE2T/FANCT, and phosphorylated by ATR (Andreassen et al. 2004, Rickman et al. 2015). This monoubiquitylation of FANCD2-FANCI is considered to be a central event within the FA pathway and is critical for ICL repair, as mutation of lysine 561, which is the site of monoubiquitylation, sensitises cells to ICL-inducing agents (Garcia-Higuera et al. 2001). It has also been shown that deubiquitylation of FANCD2 by USP1-UAF is also required for efficient ICL repair (Oestergaard et al. 2007). However, despite the importance of the monoubiquitylation of FANCD2-FANCI, it is still not clear what exact role this modification plays in regulating ICL repair.

ICL unhooking

Following its monoubiquitylation and recruitment, FANCD2-FANCI then coordinates the 'unhooking' of the ICL by stimulating nucleolytic incisions on the same DNA strand on either side of the lesion (Zhang and Walter 2014). First, monoubiquitylated FANCD2 recruits SLX4/FANCP to the ICL (Yamamoto et al. 2011). SLX4/FANCP is a scaffold protein that interacts with multiple structure-specific nucleases that catalyse the unhooking of the ICL, including XPF/FANCQ-ERCC1, MUS81-EME1 and SLX1 (Klein Douwel et al. 2014, Zhang and Walter 2014). The XPF/FANCQ-ERCC1 complex has been shown to have a critical role in cleaving the DNA 3′ to the ICL (Klein Douwel et al. 2014), whilst SLX1 has been proposed to be the main candidate for the 5′ incision (Castor et al. 2013). MUS81-EME1 is proposed to have a role in resolving a small subset of repair intermediates (Klein Douwel et al. 2014). Furthermore, other structure-specific nucleases have been shown to have roles in ICL repair, including FAN1, SNM1A and SNM1B (Zhang and Walter 2014).

Despite our knowledge of the proteins involved in ICL unhooking and repair, the exact mechanisms underlying this process are still unclear. Interestingly, it is becoming apparent that alternative mechanisms exist that can resolve a subset of ICLs independently of the canonical FA DNA repair pathway. For example, it has been reported that ICLs induced by psoralen can be unhooked via N-glycosyl bond cleavage by the DNA glycosylase NEIL3, rather than nucleolytic incision (Semlow et al. 2016). Additionally, it was shown this NEIL3-dependent repair of ICLs

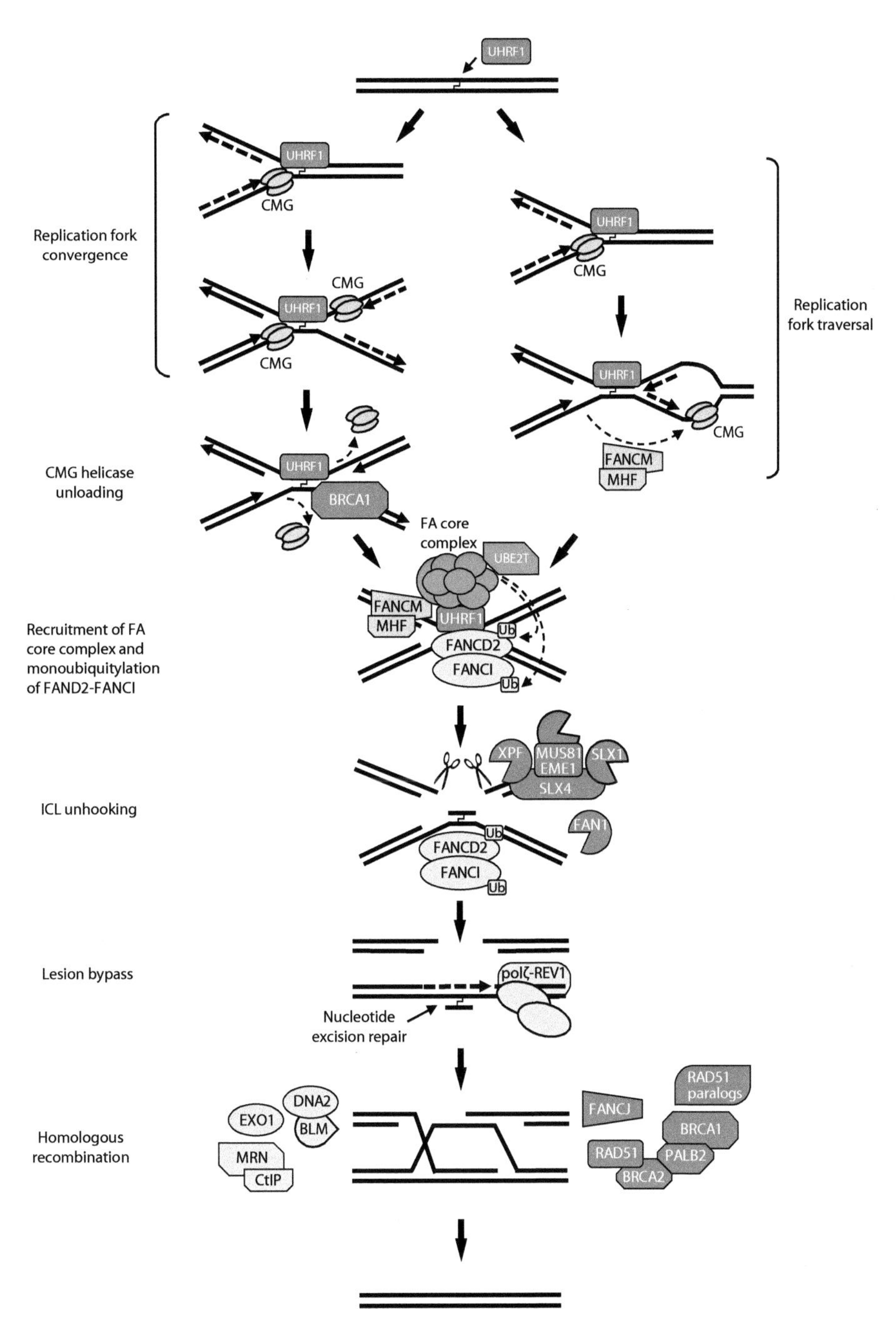

Figure 12.1 **The Fanconi anaemia ICL repair pathway.** The Fanconi anaemia (FA) pathway functions to repair DNA interstrand cross-links (ICLs) during DNA replication. **Detection of DNA ICLs:** There are several aspects to the recognition of DNA ICLs. UHRF1 is a DNA damage sensor that rapidly recognises and binds to DNA ICLs, and promotes the recruitment of downstream factors. Another important event for ICL recognition is the stalling of the replication machinery upon collision with the DNA lesion. At this stage, two mechanisms can operate to progress the repair pathway. First, the replication machinery can wait for a converging replication fork to stall on the opposite side of the ICL, at which point the CMG helicases are removed in a BRCA1-dependent manner. Alternatively, the FANCM-MHF complex can promote an ICL traversal mechanism to allow the CMG helicase to proceed with DNA replication past the damage. Either mechanism will result in an X-structure with the ICL at the centre. **Recruitment of FA core complex and monoubiquitylation of FAND2-FANCI:** The next stage of repair involves the recruitment of the FA core complex and FANCD2-FANCI complex. A central event of the FA pathway is the monoubiquitylation of FANCD2-FANCI by the E3 ubiquitin ligase FANCL (a subunit of the FA core complex) and the E2 conjugating enzyme UBE2T. Monoubiquitylated FANCD2-FANCI then coordinates downstream repair events within the pathway. **ICL unhooking:** Monoubiquitylated FANCD2-FANCI recruits the SLX4 complex (containing the SLX1, XPF-ERCC1, MUS81-EME structure specific nucleases) to the ICL. The SLX4 complex then 'unhooks' the DNA ICL by incising the DNA on either side of the lesion to generate a two-ended DSB. Other structure specific nucleases, such as FAN1, may also have important roles during ICL unhooking. **Lesion bypass:** Once the ICL has been unhooked, translesion synthesis (TLS) polymerases (such as polζ-REV1) then perform DNA synthesis across the ICL lesion, and the remaining DNA damage is removed by the nucleotide excision repair pathway. **Homologous recombination:** Finally, the resulting two-ended DSB is repaired by the error-free homologous recombination (HR) pathway. Briefly, HR first involves the generation of 3′ single stranded DNA by resection by multiple helicases and nucleases (e.g. MRN-CtIP, BLM-DNA2, EXO1). The next stage is formation of RAD51 nucleofilaments which is promoted by pro-recombinogenic factors such as BRCA1, BRCA2-PALB2, FANCJ and the RAD51 paralogues. The RAD51 nucleofilaments promote strand invasion of the undamaged DNA template, and repair is completed following DNA repair synthesis, resolution and DNA ligation. (Adapted from Lopez-Martinez, D. et al., 2016. *Cell Mol Life Sci* 73:3097–3114.)

does not require the FANCD2-FANCI complex, or involve either CMG unloading or DSB formation, leading to the hypothesis that this may be a favoured method of ICL unhooking, whenever possible.

Lesion bypass

Once the ICL has been unhooked from one of the strands, DNA is synthesised across the ICL lesion in preparation for repair of the broken DNA strand by HR. This process is called translesion synthesis (TLS) and involves specialised low-fidelity DNA polymerases that can accommodate the presence of bulky DNA adducts within their active sites, which are larger than those of replicative polymerases (discussed in detail in Chapter 4) (Yang and Woodgate 2007). Multiple polymerases have been implicated in ICL repair, although the precise contributions of each has been difficult to determine. Nevertheless, the polζ(Rev3-Rev7-POLD2-POLD3)-REV1 complex has an important role in ICL repair, and loss of this complex renders cells hypersensitive to ICL-inducing agents (Shen et al. 2006). Due to their error-prone nature (compared to replicative polymerases), TLS polymerases risk introducing mutations and therefore they are highly regulated to avoid inappropriate use. TLS polymerases are recruited to ICLs in a manner dependent on monoubiquitylated PCNA and the FA core complex, but not FANCD2-FANCI (Shen et al. 2006, Budzowska et al. 2015). Furthermore, it has also been shown that the E3 ubiquitin ligase RAD18, FANCJ and the DNA-dependent protease SPRTN also have various roles in the regulation of TLS (Song et al. 2010, Xie et al. 2010, Kim et al. 2013). Once the TLS polymerase has bypassed the lesion, the remaining ICL is then believed to be removed by NER (Ceccaldi et al. 2016).

Homologous recombination (HR)

Unhooking of the ICL generates a replication-associated DSB which can then be repaired by the error-free DSB repair (DSBR) pathway utilising HR (discussed in further detail in Chapter 6) (Heyer et al. 2010). The first stage of HR involves DNA end resection to generate RPA-coated 3′ ssDNA overhangs. This is initiated by CtIP together with the MRN complex, a process which is coordinated by both FANCD2 and BRCA1/FANCS (Murina et al. 2014). Long-range resection of the DSB is then carried out by the helicase-nuclease BLM-DNA2 complex, WRN helicase and the exonuclease EXO1 (Karanja et al. 2012, Sturzenegger et al. 2014). In addition, to avoid deleterious DNA resection, which would lead to genome instability, several factors, such as 53BP1 and RIF1, function to suppress excessive resection (Shibata 2017). The interplay between anti-resection and pro-resection factors, therefore, ensures that DSBs and replication forks are only resected when required.

Following formation of the RPA-coated ssDNA, the BRCA1/FANCS-BRCA2/FANCD1-PALB2/FANCN complex promotes displacement of RPA, and the formation of RAD51-coated ssDNA nucleofilaments, which catalyse strand invasion of the unbroken homologous template (Prakash et al. 2015). The formation of the RAD51 nucleofilament is essential for HR, and loss of RAD51 renders cells sensitive to ICL-inducing agents (Hanlon Newell et al. 2008). The process of RAD51 nucleofilament formation is another tightly regulated process, as uncontrolled HR can be deleterious (Heyer et al. 2010). As such, there are opposing factors that function to either stabilise or destabilise RAD51 nucleofilaments. The DNA helicases BLM, RECQL5 and FBH1 can disrupt RAD51 and function to suppress HR (Hu et al. 2007, Fugger et al. 2009). In opposition to these, BOD1L and the RAD51 paralogues, RAD51C/FANCO and XRCC2, function to stabilise RAD51 nucleofilaments, and loss of these factors results in cellular hypersensitivity to ICL inducing agents (Somyajit et al. 2012, Higgs et al. 2015, Park et al. 2016). Finally, following strand invasion and DNA repair synthesis, resolution and ligation occurs, repairing the DSB and completing FA pathway–dependent ICL repair.

Interestingly, in addition to promoting HR, the FA pathway seems to suppress the non-homologous end joining (NHEJ) DSBR pathway (Bunting and Nussenzweig 2010). NHEJ functions to repair DSBs by re-ligating two DNA ends together in a manner that does not require homology to be present (discussed in Chapter 7), and can therefore lead to the inappropriate ligation of DNA ends on different chromosomes, leading to chromosome translocations (Shibata 2017, Nussenzweig and Nussenzweig 2010). Indeed, NHEJ factors are inappropriately recruited to damaged/stalled replication forks in the absence of FANCD2, and the sensitivity of FA deficient cells to ICL inducing agents is supressed by inhibition of NHEJ factors (Bunting and Nussenzweig 2010). This is an important observation as it suggests that it is the formation of toxic chromosome fusions that kills cells in the absence of a functional FA pathway, rather than the lack of FA-dependent ICL repair.

Fanconi anaemia pathway and human disease

FA is a rare multisystem disease characterised by bone marrow failure, pre- and postnatal growth retardation, microcephaly, multiple congenital abnormalities and cancer predisposition. FA has an estimated incidence of 1 in 360,000 live births (Mamrak et al. 2017). FA was first described in the 1920s, and whilst it became clear in the 1980s that multiple FA complementation groups existed, it was not until 1992 that the first FA gene, FANC, was identified (Strathdee et al. 1992). Initially the rate of discovery for new FA genes was relatively slow, but as genetic and biochemical technologies improved, the rate of identification rapidly increased, and there are now 21 confirmed FA genes (Tables 12.3 and 12.4) (Gueiderikh et al. 2017). As known, FA genes account for approximately 95% of FA patients, it is likely that more complementation groups will be discovered in the near future (Lopez-Martinez et al. 2016). Due to the role of the FA pathway in repairing DNA ICLs, the classic diagnostic test for FA is the chromosome breakage assay to assess whether patient cells are hypersensitivity to ICL-inducing agents, such as diepoxybutane and MMC.

The most commonly presenting symptom of FA is the presence of haematological abnormalities. Over 75% of FA patients develop bone marrow failure in early life, and most patients develop other diseases of the blood, such as aplastic anaemia, myelodysplastic syndrome (MDS) and acute myeloid leukaemia (AML) (Kee and D'Andrea 2012). In addition, approximately half of FA patients are born with skeletal abnormalities, most commonly within the thumb and forearm, and also suffer from short stature caused by decreased levels of growth hormone production (Kee and D'Andrea 2012). These features constitute the classical FA phenotype, and together with hypersensitivity to ICL-inducing agents, are required for a true diagnosis of FA (Bogliolo and Surralles 2015, Gueiderikh et al. 2017). However, mutations in several genes cause a disease very similar to FA in most aspects, but lack bone marrow failure and leukaemia predisposition. These are, therefore, considered to be 'Fanconi anaemia–like' genes, rather than bona fide FA genes (Bogliolo and Surralles 2015). Furthermore, the question of whether *FANCM* is truly an FA gene has been debated for some time. Although one patient was initially reported to possess biallelic mutations in *FANCM*, this patient also possessed biallelic *FANCA* mutations (Singh et al. 2009). However, the recent discovery of biallelic FANCM mutations in numerous individuals who developed early onset breast cancer, but lacked any of the clinical presentations typical of FA individuals, suggests that FANCM is a cancer-predisposing gene but not a genuine FA gene (Bogliolo et al. 2017, Catucci et al. 2017).

FA has been primarily considered a childhood disease, and FA once had a very poor prognosis, as most patients did not survive into adulthood. However, due to advances in management and treatment of the disease, particularly in the treatment of bone marrow failure, there is an increasing number of surviving adult FA patients. Whilst this is a desirable situation, it has also brought different challenges, such as an increased risk of cancer (Kee and D'Andrea 2012). Indeed, 29% of FA patients develop a solid tumour, such as head and neck squamous cell cancers and cervical/gynaecological cancers, before the age of 48 (Rosenberg et al. 2003).

It is also interesting that mutations in the structure-specific nuclease FAN1 do not give rise to FA, but instead cause karyomegalic interstitial nephritis (Zhou et al. 2012), a kidney disease. It is not currently clear why this is, although it may reflect roles for FAN1 outside of FA-mediated ICL repair (Klein Douwel et al. 2014).

FA disease and hematopoietic stem cells

As discussed earlier, bone marrow failure is one of the hallmarks of FA. FA patients can be successfully treated by bone marrow transplantation, suggesting that there is a defect in the haematopoietic stem cells and progenitor cells, rather than in the maintenance of differentiated cells (Kee and D'Andrea 2012, Brosh et al. 2017). Indeed, analysis of bone marrow samples of FA patients revealed deficiencies in hematopoietic stem and progenitor cells (HSPCs) manifesting even before the onset of bone marrow failure (Ceccaldi et al. 2012). It has been proposed that the loss of HSPCs is due to constitutive activation of the p53/p21 DNA damage response triggered by unresolved replication stress and endogenous DNA damage, leading to cell cycle arrest and cell death, and ultimately resulting in severe aplastic anaemia (Ceccaldi et al. 2012, Walter et al. 2015).

There are several possible explanations of what the physiological source of DNA damage in FA cells is. It has been shown that FA cells exhibit mitochondrial dysfunction, and have elevated levels of reactive oxygen species and a concurrent increase in oxidative DNA damage (Kumari et al. 2014). Another potential source of DNA damage that drives hematopoietic stem cell attrition in FA patients is endogenous formaldehyde (Garaycoechea et al. 2012). Indeed, dysfunction of ALDH2, an enzyme involved in the detoxification of acetaldehyde, causes accelerated progression of bone marrow failure in FA patients (Hira et al. 2013), and leads to increased endogenous DNA damage, loss of hematopoietic stem cells, and progressive bone marrow failure in FA deficient mice. This strongly suggests that endogenous formaldehyde is a significant source of DNA damage in FA patients. In addition, as it has been shown that formaldehyde exposure can induce the formation of R-loops and that the FA pathway can resolve R-loop associated replication–transcription conflicts (Schwab et al. 2015), this raises the possibility that the accumulation of R-loops in FA patients also contributes to the loss of hematopoietic stem cells and bone marrow failure.

Cancer predisposition and the FA and HR pathways

Aside from bone marrow failure, a predisposition to cancer, in particular AML and head and neck squamous cell cancers, is another key feature of FA (Kee and D'Andrea 2012). Furthermore, one of the major roles that HR plays in human health is the prevention of tumorigenesis, and mutations in HR factors strongly predispose to cancer. Due to the substantial overlap between the FA and HR pathways, it is useful to discuss their roles in the prevention of cancer together.

Inherited monoallelic mutations in *BRCA1/FANCS, BRCA2/FANCD1, PALB2/FANCN, FANCJ, FANCM*, and the RAD51 paralogues (*RAD51B, RAD51C/FANCO, RAD51D, XRCC2/FANCU* and *XRCC3*) are associated with a high risk of developing familial breast and/or ovarian cancer (Metcalfe et al. 2010, Katsuki and Takata 2016, Bogliolo et al. 2017). Indeed, monoallelic *BRCA1* mutations confer a 60% lifetime risk of developing breast cancer and a 40% risk of developing ovarian cancer (Metcalfe et al. 2010). Loss of heterozygosity is believed to be an important factor in these HR defective cancers, as a significant proportion of tumours from individuals with monoallelic mutations in *BRCA1* or *BRCA2* have been reported to have lost the WT allele (King et al. 2007). Furthermore, inherited biallelic mutations in *BRCA1/FANCS, BRCA2*/FANCD1, *PALB2*/FANCN, *XRCC2/FANCU, FANCJ, XPF/FANCQ, SLX4/FANCP, RAD51/FANCR, RAD51C/FANCO* and *RFWD3/FANCW* (an E3 ubiquitin ligase required for efficient HR) give rise to FA or an FA-like disease (Howlett et al. 2002, Levitus et al. 2005, Reid et al. 2007, Shamseldin et al. 2012, Knies et al. 2017). In particular, biallelic mutations in *BRCA2/FANCD1* and *PALB2/FANCN* lead to a very severe disease presentation with a high risk of developing tumours in the early years of life. Another addition to the varied phenotypes associated with defective HR is ovarian abnormalities. Mutations in *MCM8* and *MCM9*, which promote HR in a variety of situations, have been identified in individuals with premature ovarian failure and short stature (Alasiri et al. 2015, Wood-Trageser et al. 2014). Additionally, a nuclear scaffolding protein that promotes HR, *SPIDR*, is also mutated in individuals with ovarian abnormalities (Smirin-Yosef et al. 2017).

One question that this raises is why are FA and HR pathway defects so strongly associated with cancer disposition. It is clear that HR is vitally important to prevent the process of tumorigenesis, and how HR deficiencies contribute to the development of cancer is under investigation (Katsuki and Takata 2016). To discuss this, it is important to understand that HR is an essential error-free DNA repair pathway with multiple roles in the maintenance of genome stability. It has critical roles in the repair of DNA DSBs (Jeggo et al. 2011) and DNA interstrand cross-links (Ceccaldi et al. 2016) as well as being required for replication fork restart, the resolution of replication intermediates that arise during DNA replication (Petermann et al. 2010) (Chapter 6), and the repair of programmed SPO11-generated DSBs formed during crossover in meiosis (Lam and Keeney 2014). It has long been appreciated that the accumulation of mutations, caused by persistent genome instability, is an important driver of cancer (Hoeijmakers 2001). Therefore, loss of the FA and HR pathways would result in the cell employing more error-prone repair pathways to deal with endogenous DNA damage, leading to increased genome instability and increased risk of generating mutations/translocations that drive tumorigenesis. For example, it has been shown that acquired loss of p53 in transformed FA cells promotes tumourigenesis (Ceccaldi et al. 2011). Additionally, NHEJ factors generate toxic chromosome fusions in the absence of a functional FA pathway, leading to increased cell death (Bunting and Nussenzweig 2010). Another possible contributing factor to the increased cancer predisposition associated with defective HR could be the increased breakage at CFSs arising from replication fork instability. Studies have shown that increased instability of CFSs due to replication stress leads to alterations in gene copy number within CFS-associated genes (Glover et al. 2017).

Crucially, the types of copy number variations that occurred in the presence of replication stress is reminiscent of those seen in cancer cells. Furthermore, chromosomal translocations involving CFS-associated genes could also contribute to the development of cancer (Roukos et al. 2013).

A second question that arises is why do deficiencies in the FA or HR pathways give rise to cancer in a tissue-specific manner. For example, breast and ovarian tissues are much more susceptible to developing cancers upon loss of HR. The ovarian failure seen in individuals with mutations in *MCM8*, *MCM9* and *SPIDR*, which is phenocopied in MCM8-MCM9–deficient mice (Lutzmann et al. 2012), could be explained by the critical role of HR in repairing programmed DSBs generated during meiosis. In the absence of efficient HR, these DSBs are unable to be correctly repaired, which would lead to oocyte cell death and therefore ovarian failure (Alasiri et al. 2015). This also raises the possibility that defective repair of meiotic DNA DSBs contributes to the development of ovarian cancer. In addition, it has been found that BRCA1 is required to prevent genome instability caused by the accumulation of DSBs induced by oestrogen and oestrogen metabolites (Savage et al. 2014). As breast and ovarian tissues are exposed to high levels of oestrogen, these tissues could be particularly sensitive to loss of BRCA1 function. Furthermore, as the stability of CFSs is dependent on cell type (Glover et al. 2017), it is possible that the instability of specific CFSs which only occur in certain tissue types could also contribute to the tissue-specific tumorigenesis. However, further research is required to answer this question.

Additionally, the different cancers which FA patients develop could also result from several factors. For example, the mechanisms underlying predisposition to AML may be similar to those that underlie bone marrow failure in FA patients. HSPCs will cycle between dormancy and proliferation at times of stress to replenish mature cells that are lost. The loss of the FA and the HR pathway would lead to increased DNA damage in the cycling HSPCs (Walter et al. 2015). This is proposed to be the reason for the loss of HSPCs, but it could also result in malignancy due to increased mutagenesis arising from error-prone repair of the persisting DNA damage. Furthermore, infection with high-risk human papilloma virus (HPV) is associated with a significant proportion of sporadic head and neck squamous cell cancers, particularly amongst those who have never smoked (Leemans et al. 2011). These HPV-associated head and neck squamous cell cancers have been noted to be similar to those manifesting in FA patients, and it has been suggested that defects in the FA pathway result in an increased predisposition to HPV-associated cancers (Park et al. 2010). However, this link is controversial, as different studies have reported either an increased or decreased prevalence of HPV in head and neck squamous cell cancers in FA patients in comparison to the general population (Kee and D'Andrea 2012).

DNA REPLICATION AND DNA–PROTEIN CROSS-LINK (DPC) PROTEOLYSIS REPAIR

Sources and types of DPCs

DNA–protein cross-links (DPCs) are a type of DNA lesion that is generated when proteins become covalently and irreversibly bound to DNA (Stingele and Jentsch 2015). There are two main types of DPCs. The first is non-enzymatic DPCs, which are caused by the non-specific covalent linkage of proteins typically found in close proximity to DNA (e.g. histones, transcription factors, DNA repair and replication factors) following exposure to cross-linking agents (Ide et al. 2015). Reactive aldehydes and reactive oxygen species are significant sources of endogenous non-enzymatic DPCs, whilst exogenous sources include ionising radiation, UV light and DNA cross-linking chemotherapeutic agents (Ide et al. 2015).

The second type of DPCs are enzymatic and are generated by certain enzymes that form transient covalent bonds with DNA during their normal reaction mechanism, such as DNA topoisomerases, DNA polymerases and DNA methyltransferases. These enzyme–DNA intermediates usually only exist for a very short time and are rapidly reversed as the enzyme completes its reaction cycle. However, under certain circumstances these reaction intermediates can become irreversible, and a stable DPC is produced. Enzymatic DPCs, such as topoisomerase 1 (TOP1) and topoisomerase 2 (TOP2), are also typically associated with DNA breaks (Hande 2003, Pommier 2009).

DPCs are thought to be a commonly occurring type of DNA lesion, with estimates of approximately 6,000 DPCs being generated spontaneously per cell each day (Oleinick et al. 1987). As DPCs present strong physical blocks for the progression of DNA replication and transcription, they pose a severe threat to genome stability.

Repair of DPCs

As any protein in contact with DNA could potentially form a DPC, there is a huge diversity in the chemistries and structures of DPCs, posing a challenge for the cell to recognise and repair them. Therefore, several mechanisms to repair DPCs have evolved.

Direct enzymatic removal of DPCs

Several DNA repair enzymes have evolved to specifically remove certain enzymatic DPCs. For example, tyrosyl-DNA phosphodiesterase 1 (TDP1) can hydrolyse the phosphotyrosyl linkage between the tyrosine residue within the TOP1 active site and the 3′ termini of the DNA SSB within TOP1-DPCs (Interthal et al. 2001). Additionally, TDP2 can cleave the phosphotyrosyl linkage between TOP2 and the 5′ termini of the DNA DSB within TOP2-DPCs (Cortes Ledesma et al. 2009). However, neither TDP1 nor TDP2 can remove full-length TOP1 or TOP2 enzymes from DNA, and repair of these DNA lesions first requires proteosomal degradation of the full-length TOP1 and TOP2 enzymes into small peptides (Lin et al. 2008, Gao et al. 2014). Removal of the TOP1-DPC by TDP1 then leaves a DNA nick that requires further processing and repair via the DNA SSBR pathway (discussed in Chapter 5), and removal of the TOP2-DPCs by TDP2 leaves a 'clean' DNA DSB that is predominantly repaired by non-homologous end joining (also discussed in Chapter 7).

Indirect repair of DPCs via the action of nucleases

Due to the diversity of structures and chemistries of DPCs, it is not possible for the cell to employ specific enzymes to repair each type of lesion that can arise. Therefore, the cell utilises repair mechanisms that can repair DPCs, irrespective of their structure and origin. NER is a versatile DNA repair pathway that can repair a wide range of structurally unrelated DNA lesions by unwinding the DNA surrounding the damage, and excising the DNA lesion via the action of structure-specific nucleases (Spivak 2015) (discussed further in Chapter 5). It is hypothesised that the NER pathway can repair DPC lesions as NER defective cell lines are sensitive to DPC inducing agents. Indeed, in vitro and in vivo analysis has shown that the NER pathway is able to remove DPCs that are less than 16 kDa in size (Nakano et al. 2007).

DPCs present at a DSB, such as TOP2-DPCs, can also be removed via the action of the MRN complex, in cooperation with CtIP and BRCA1, and involves multiple processing events (Aparicio et al. 2016). First, MRE11 incises the DNA downstream of the TOP2-DPC (Hoa et al. 2016). The incised DNA strand is then resected in a 3′ to 5′ direction, and a second incision is made on the DNA strand opposite to the original nick, excising the damaged DNA (Deshpande et al. 2016). After removal of the TOP2-DPC, the resulting DSB can then be repaired (Hoa et al. 2016). MRN is also able to repair TOP1-DPCs, and it seems to be able to repair DPC-associated DSBs irrespective of the type of protein linked to the DNA (Aparicio et al. 2016).

Therefore, NER and MRN-dependent DSBR represent two pathways that can repair a subset of DPCs, specifically those under 16 kDa, or those present at a DSB. However, loss of MRE11 or XPC (a critical NER factor) does not result in increased global levels of DPCs, suggesting that the DPCs repaired by these pathways only make up a small fraction of the total amount that accumulate in human cells (Vaz et al. 2016).

DNA replication-dependent DPC proteolytic repair

Recently, a DNA-dependent protease, SPRTN, has been identified as the central player in a novel DNA replication-associated repair pathway that repairs DPCs via proteolysis (Lopez-Mosqueda et al. 2016, Stingele et al. 2016, Vaz et al. 2016). SPRTN is a member of the superfamily of metallopeptidases that interacts with PCNA, ubiquitin and the ATPase p97/VCP (a key factor within the ubiquitin–proteasome system) (Vaz et al. 2016). Loss of SPRTN leads to a global accumulation of DPCs and renders cells sensitive to cross-linking agents as well as TOP1 and TOP2 inhibitors (Vaz et al. 2016, Maskey et al. 2017). Furthermore, replication fork progression is impeded in the absence of SPRTN, both spontaneously and in the presence of exogenous DPC-inducing agents (Vaz et al. 2016).

SPRTN-dependent DPC proteolytic repair functions within S phase and is initiated when SPRTN is recruited to a replication forks that have stalled upon collision with a DPC (Stingele et al. 2016). The protein within the DPC is then proteolytically degraded to a small peptide by SPRTN, which allows the CMG helicase to proceed (Duxin et al. 2014). TLS polymerases are proposed to perform DNA synthesis across the remaining lesion (Duxin et al. 2014). There are several regulatory mechanisms to prevent SPRTN from degrading any proteins stably bound to DNA, and ensure it only becomes activated once a DPC causes problems at a replication fork (Stingele et al. 2016). First, SPRTN exists in a monoubiquitylated form and needs to be de-ubiquitylated to be recruited to damaged forks. Second, SPRTN is only active as a protease in the presence of DNA, preventing inappropriate degradation of proteins not bound to DNA. Last, binding of SPRTN to ssDNA, which would be present at stalled forks, stimulates SPRTN activity, whilst binding to dsDNA induces SPRTN to autocleave itself, allowing it to 'switch off' following completion of repair (Stingele et al. 2016).

The significance of the SPRTN-dependent DPC proteolytic repair pathway lies in its versatility, as it can theoretically repair any DPC, irrespective of the identity of the cross-linked protein. Therefore, the combination of TDP1, TDP2, NER, MRN and SPRTN-dependent DPC repair provides multiple distinct mechanisms to repair any type of DPCs that can arise in the cell.

DPCs and human disease

The importance of repairing DPCs is highlighted by the existence of numerous diseases caused by mutations within components of each of the main DPC repair pathways/mechanisms (Table 12.2). Spinocerebellar ataxia with axonal neuropathy 1 (SCAN1) and autosomal recessive spinocerebellar ataxia 23 (SCAR23) are caused by mutations in TDP1 and TDP2, respectively. SCAN1 is characterized by progressive cerebellar ataxia and progressive atrophy of the muscles of the fingers and feet (Takashima et al. 2002) (discussed further in Chapter 5). Consistent with the role of TDP1 in the repair of TOP1-DPCs, SCAN1 cells have been shown to accumulate more DNA breaks in the presence of the TOP1 inhibitor camptothecin (El-Khamisy et al. 2005). Further analysis has also revealed that the repair of both transcription- and replication-dependent TOP1-DPCs are delayed in SCAN1 cells (El-Khamisy and Caldecott 2006). SCAR23 individuals present with intellectual disability, early onset seizures and progressive ataxia (Gomez-Herreros et al. 2014). SCAR23-derived cell lines exhibited cellular hypersensitivity and increased levels of DNA damage when incubated with etoposide, a TOP2-specific inhibitor (Gomez-Herreros et al. 2014). Crucially, TDP2-deficient cells are also defective for the transcription of TOP2-dependent genes (Gomez-Herreros et al. 2014). It has, therefore, been proposed that the neurodegeneration seen in SCAN1 and SCA23 individuals is caused by accumulating levels of transcription arrest due to unrepaired TOP1-DPCs and TOP2-DPCs (Sordet et al. 2009).

As discussed earlier in this chapter, and in detail in Chapter 13, mutations in all three components of the MRN complex have been associated with human disease: microcephaly, growth retardation, immunodeficiency and an increased cancer predisposition (in particular lymphoma) is associated with mutations in *NBN* (NBS); microcephaly, growth retardation and non-progressive ataxia results from mutations in *RAD50* (NBSLD); and mutations in *MRE11* gives rise to progressive cerebellar ataxia and oculomotor apraxia (ATLD).

Mutations in multiple components of the NER pathway give rise to several human diseases, including xeroderma pigmentosum (XP) and Cockayne syndrome (CS), depending on which NER factors are mutated (discussed in more detail in Chapter 5). XP is characterised by a 10,000-fold increased risk of developing sunlight induced skin cancer, and a 10-fold increased risk of developing internal tumours (DiGiovanna and Kraemer 2012). CS, however, is not associated with cancer predisposition, but is instead characterised by microcephaly, neurodegeneration and progressive ataxia, as well as premature ageing (Laugel 2013). The increased cancer predisposition in XP individuals is likely the result of error-prone TLS polymerases synthesising across unrepaired DNA adducts, thus increasing mutagenesis and driving tumourigenesis (Yang and Woodgate 2007). In contrast, CS is associated with loss of a specific NER sub-pathway dedicated to repairing DNA adducts that stall RNA POL II (TC-NER), and the neurological defects and premature ageing seen in CS individuals is thought to arise from the persistence of transcription-blocking DNA lesions (Ljungman and Lane 2004).

Ruijs-Aalfs syndrome (RJALS) which is characterised by premature aging and early onset hepatocellular carcinoma, has recently been discovered to be caused by mutations in *SPRTN* (Lessel et al. 2014, Maskey et al. 2014). Interestingly, a similar phenotype occurs in *Sprtn* knockout mice (Maskey et al. 2017). Due to defective DPC repair, patient-derived cell lines globally accumulate DPCs in their genome, leading to the persistence of DSBs in S phase. The presence of hepatocellular carcinomas is a striking aspect to RJALS, and it is likely that this reflects the large amounts of reactive aldehydes that are produced as by-products of various metabolic processes in the liver.

Therefore, mutations in factors involved in repair of DPCs can give rise to a range of clinical phenotypes, including cancer predisposition (XP, NBS, RJALS), premature ageing (RJALS, CS), neurodegeneration (SCAN1, SCAR23, CS, ATLD) and microcephaly (NBS, NBSLD, CS). These varied clinical presentations probably reflect the different mechanisms through which DPCs can arise, and what cellular processes the persisting unrepaired DPCs impact. For example, transcription abnormalities may underlie the neurodegeneration associated with these disorders. Due to their long-lived nature and limited regenerative capacity, post-mitotic neurons are believed to be particularly sensitive to transcription abnormalities, and the neurodegeneration seen in SCAN1, SCAR23 and CS is proposed to result from increasing levels of neuronal cell death triggered by accumulating levels of transcription-blocking lesions (El-Khamisy and Caldecott 2006, Weidenheim et al. 2009, Gomez-Herreros et al. 2014). It is also noteworthy that mutations in SETX, which functions to promote efficient transcription termination and resolving transcription–replication conflicts, are also associated with progressive neurodegeneration (Chen et al. 2004, Moreira et al. 2004). Additionally, transcription defects are believed to contribute to the progeroid features observed in CS patients, and it is tempting to speculate that similar transcription abnormalities also underlie the premature ageing of RJALS (Weidenheim et al. 2009). In turn, genome instability and mutagenesis are potent drivers of tumourigenesis, and it is likely that error-prone repair of DPCs in the absence of NER, and MRN- and SPRTN-dependent repair, leads to an accumulation of cancer-driving mutations. Furthermore, as MRN- and SPRTN- dependent repair mechanisms operate in S phase, they would be unable to compensate for loss of TDP1 or TDP2 in post-mitotic neurons. However, the presence of these alternative repair pathways would prevent TOP1- and TOP2-DPCs from accumulating in cycling cells, and could explain the lack of cancer predisposition in SCAN1 and SCAR23.

DNA REPLICATION AND DNA MISMATCH REPAIR

To maintain genome stability, it is essential that the genome is copied faithfully and accurately before it is transferred to the next generation, otherwise changes to the genetic material could result in deleterious mutations. Therefore, it is imperative that the replicative DNA polymerases are as accurate as possible to ensure faithful genome duplication (Ganai and Johansson 2016). The replicative polymerases POLα, POLδ, and POLε have very low rates of error, and it has been estimated that only 1 mutation per 10^{10} nucleotides occurs per cell cycle in eukaryotic cells (Kunkel and Erie 2015). Several factors contribute to this level of accuracy. First, the active sites of the replicative polymerases are structured such that the formation of the phosphodiester bond is dependent on the presence of the correct interactions between critical active site residues, the incoming nucleotide, and the 3′OH termini of the template DNA. An incorrect nucleotide present within the active site will not form these critical interactions and will therefore be ejected from the polymerase before phosphodiester bond formation (Ganai and Johansson 2016). Second, the replicative polymerases possess 3′–5′exonuclease activity that confers an intrinsic proofreading ability, allowing the polymerase to remove an incorrectly inserted nucleotide. Finally, if these mechanisms fail to prevent incorporation of an incorrect base, the cell employs the DNA mismatch repair (MMR) pathway that is dedicated to recognising and removing numerous types of replicative errors.

DNA mismatch repair

Mammalian MMR operates during S-phase and consists of four steps: mismatch recognition; nascent strand discrimination; incision and degradation of the nascent strand containing the mismatch; and gap filling and DNA ligation reviewed in Kunkel and Erie (2015). In the first stage, the most common DNA mismatches, namely single base mismatches and insertion/deletion mutations, are recognised by the MSH2-MSH6 complex (also called MUTSα). MSH2-MSH6 binds to the mismatch and undergoes a conformational change to encircle the DNA as a clamp, allowing it to slide along the DNA and recruit the PMS2-MLH1 complex (also called MUTLα) (Groothuizen and Sixma 2016). The next stage of repair is to identify the newly replicated strand that contains the mismatch. It is still not clear how strand discrimination occurs in eukaryotic cells, but it has been suggested that some replication intermediates may act as the signal used to identify the nascent strand (Kunkel and Erie 2015).

Following recognition of the mismatch and strand discrimination, PMS2-MLH1 then incises the nascent strand in a reaction that is dependent on the presence of PCNA and RFC (Kadyrov et al. 2006). This PMS2-MLH1 induced nick allows the nicked nascent strand to be degraded in one of several mechanisms depending on the position of the nick in relation to the mismatch. The nascent strand can be either be degraded by EXO1 in a 5′ to 3′ direction, or it can be excised by the 3′ to 5′ exonuclease activity of the replicative polymerases (Goellner et al. 2015). Alternatively, the mismatch can be displaced and removed following elongation by POLδ or POLε. Following removal of the mismatch, DNA synthesis fills the remaining gap and DNA ligation completes the repair. Loss of function of any of the MMR factors leads to a mutator phenotype and genome instability.

Other complexes can also contribute to MMR. Indeed, the MSH2-MSH3 complex (also called MUTSβ) can also repair large and small insertion/deletion mutations, although it is approximately 10 times less abundant than MSH2-MSH6 (Strand et al. 1995). Furthermore, MLH1-MLH2 (MUTLβ) and MLH1-MLH3 (MUTLγ) constitute two additional MUTL complexes that can also contribute to the repair of insertion/deletion mutations (Jiricny 2013). However, loss of the subunits unique to these complexes (i.e. MSH3, MLH2 and MLH3) leads to a milder mutator phenotype than loss of MSH2, MSH6, PMS2 or MLH1 (Kunkel and Erie 2015), suggesting that MSH2-MSH6 and PMS2-MLH1 are responsible for the repair of the majority of DNA mismatches.

Defects in DNA mismatch repair and cancer predisposition

The presence of unrepaired DNA mismatches in the genome is a serious threat to genome stability as mismatches can be converted to permanent mutations during subsequent rounds of DNA replication. The importance of the MMR pathway is highlighted by the fact that loss/mutation of MMR factors results in higher rates of spontaneous mutation, in turn leading to genome instability and cancer predisposition (Sijmons and Hofstra 2016), and several different inherited cancer predisposition syndromes are caused by mutations in MMR factors (Table 12.5).

Hereditary non-polyposis colorectal cancer (HNPCC), also called Lynch syndrome, is an autosomal dominant hereditary cancer syndrome caused by mutations in MSH2, MLH1, PMS2, MSH6 and MLH3. HNPCC is the most common form of hereditary colorectal cancer and is characterised by a very high risk of developing colorectal cancer and/or endometrial cancer (Poulogiannis et al. 2010, Lynch et al. 2015). Mismatch repair cancer syndrome (MMRCS) is the second cancer predisposition syndrome caused by loss of MMR, and is a rare autosomal recessive disorder caused by mutations in MSH2, MLH1, PMS2 and MSH6 (Vasen et al. 2014). MMRCS is an early onset cancer syndrome with a high mortality rate, and patients will usually present with a tumour in childhood or adolescence, and will likely develop several different types of cancer in their lifetimes (Vasen et al. 2014). MMRCS individuals are especially prone to haematological malignancies and cancers of the colon and/or brain, although they can develop other types of cancer as well (Sijmons and Hofstra 2016).

One of the hallmarks of cancers associated with MMR defects is the presence of microsatellite instability, which refers to changes in the number of repeats contained within microsatellites (Umar et al. 2004). Repetitive sequences within the microsatellites can cause strand slippage during DNA replication, resulting in insertion/deletion mutations. These are typically repaired by the MMR pathway, and therefore loss of MMR activity leads to significant alterations in the repeats within microsatellites. Microsatellite instability has proved to be a useful diagnostic marker for MMR-defective tumours. Unrepaired DNA mismatches that persist within coding genes can generate missense mutations and/or frameshift mutations during subsequent rounds of DNA replication, potentially leading to alterations in the functionality of the mutated protein (Mori et al. 2001, Boland and Goel 2010). It is believed that this mechanism underlies the highly elevated risk of tumorigenesis in MMR-defective patients, as an elevated rate of mutation would increase the chances of gaining a mutation in a tumour suppressor or oncogene.

Mutations in DNA polymerases and human disease

Specific mutations in the catalytic subunits of the replicative polymerases, *POLD1* and *POLE1* have also been associated with an autosomal dominant hereditary cancer syndrome that predisposes individuals to developing colorectal and endometrial cancers (Palles et al. 2013). These mutations were found in the exonuclease domains of *POLD1* and *POLE1*, therefore compromising the proofreading ability of the replicative polymerases. Consistent with this, *POLD1* and *POLE1* mutations result in a mutator phenotype associated with an accumulation

Table 12.5 Human diseases associated with defective repair of DNA mismatches.

Gene	Gene function	Disease	MOI	Clinical presentation	OMIM #	References
MSH2[b]	Part of the MUTSα (MSH2-MSH6) complex which is required for the recognition of base mismatches during MMR	Hereditary non-polyposis colorectal cancer type 1 (HNPCC1)	AD	Increased risk of colorectal and endometrial cancer	120435	Leach et al. (1993)
		Mismatch repair cancer syndrome	AR	Predisposition to developing several types of cancer early in life, including colorectal cancers, hematologic malignancies, and tumours of the brain and central nervous system	276300	Whiteside et al. (2002)
		Muir-Torre syndrome	AD	Sebaceous skin cancer; predisposition to other tumour types	158320	Kruse et al. (1996)
MLH1[b]	Part of the PMS2-MLH1 (MUTLα) complex with roles in DNA MMR	Hereditary non-polyposis colorectal cancer type 2 (HNPCC2)	?[a]	Increased risk of colorectal and endometrial cancer	609310	Papadopoulos et al. (1994)
		Mismatch repair cancer syndrome	AR	Predisposition to developing several types of cancer early in life, including colorectal cancers, hematologic malignancies, and tumours of the brain and central nervous system	276300	Poley et al. (2007)
		Muir-Torre syndrome	AD	Predisposition to sebaceous skin cancer and other tumour types	158320	Bapat et al. (1996)
PMS2[b]	Part of the PMS2-MLH1 (MUTLα) complex with roles in DNA MMR	Hereditary non-polyposis colorectal cancer type 4 (HNPCC4)	?[a]	Increased risk of colorectal and endometrial cancer	614337	Nicolaides et al. (1994)
		Mismatch repair cancer syndrome	AR	Predisposition to developing several types of cancer early in life, including colorectal cancers, hematologic malignancies, and tumours of the brain and central nervous system	276300	Auclair et al. (2007)
MSH6[b]	Part of the MUTSα (MSH2-MSH6) complex which is required for the recognition of base mismatches during MMR	Hereditary non-polyposis colorectal cancer type 5 (HNPCC5)	AD	Increased risk of colorectal and endometrial cancer	614350	Papadopoulos et al. (1995)
		Mismatch repair cancer syndrome	AR	Predisposition to developing several types of cancer early in life, including colorectal cancers, hematologic malignancies, tumours of the brain, and central nervous system.	276300	Auclair et al. (2007)
MLH3	Part of the MLH1-MLH3 complex with roles in the repair of insertion/deletion mutations during MMR	Hereditary non-polyposis colorectal cancer type 7 (HNPCC7)	?[a]	Increased risk of colorectal and endometrial cancer	614385	Liu et al. (2003)
POLD1	Catalytic subunit of the replicative polymerase δ, which has essential roles in DNA replication and DNA repair	Colorectal cancer susceptibility type 10	AD	Increased risk of colorectal and endometrial cancer	612591	Palles et al. (2013)
POLE1	Catalytic subunit of the replicative polymerase ε, which has essential roles in DNA replication and DNA repair	Colorectal cancer susceptibility type 12	AD	Increased risk of colorectal and endometrial cancer	615083	Palles et al. (2013)

Abbreviations: AD, autosomal dominant; AR, autosomal recessive; MOI, mode of inheritance.

[a]Insufficient data to determine mode of inheritance.

[b]Mutations in these genes cause multiple distinct clinical presentations.

of missense mutations, but not increased microsatellite instability, differentiating them from the MMR-defective cancer predisposition syndromes (Palles et al. 2013).

CONCLUSIONS

The last two decades have been witness to astonishing progress in our understanding of both the molecular mechanisms underpinning the cellular processes ensuring efficient and accurate DNA replication, and the human diseases caused by dysregulation of these pathways. A significant part of this has been driven by the discovery and study of numerous human diseases associated with mutations in DNA replication and associated factors. Here in this chapter we have aimed to briefly discuss our current understanding of these diseases. However, despite substantial progress, there are still many unanswered questions associated with these diseases.

As discussed throughout this chapter, a common pathology associated with impaired DNA replication is intrauterine growth deficiency and/or primary microcephaly. Due to the requirement to produce vast numbers of cells during embryogenesis, it is easy to understand why developmental processes are sensitive to defects that impact upon DNA replication and cell cycle progression. However, many DNA replication factors have multiple roles in the cell, and loss/mutation of these genes can impact on DNA replication in several ways, or can result in defects in multiple repair pathways, which are likely to contribute to the disease pathologies. Furthermore, it is important to consider other factors, such as growth hormone failure, when attempting to understand the molecular mechanisms that give rise to growth deficiency. Indeed, approximately half of FA individuals exhibit short stature resulting from growth hormone deficiency and hyperthyroidism, and thyroid treatment in FA children with short stature can lead to significant growth improvement (Kee and D'Andrea 2012). Growth hormone deficiency may also be relevant to some other diseases, but this is not universal as growth hormone treatment is unable to reverse the growth deficiency of MGS individuals in which the causative mutation has been identified (de Munnik et al. 2015).

It is also not clear why defects in DNA replication generally do not have a wider impact on multiple organ systems. Mutations in core DNA replication factors could conceivably have a global impact upon every tissue that is undergoing cellular proliferation. However, mutations in different replication complexes/pathways will often impact only on specific tissues. For example, why don't more DNA replication-defective disorders give rise to bone marrow failure or immunodeficiency? Also, why are different cancer predisposition diseases associated with the development of tissue-specific cancers? The reasons behind this phenotypic variability and the presence of specific tissue defects is poorly understood but is likely to arise from a combination factors.

Some tissues may be more sensitive to the loss of specific factors due to the types of damage they encounter. The presence of the large amounts of reactive aldehydes that are by-products of various metabolic processes in the liver, combined with its remarkable regenerative capacity, would mean that liver cells may be more prone to DPC-dependent fork stalling, and perhaps more dependent on SPRTN-dependent repair. Furthermore, depending on the nature of the underlying molecular defect different, redundant mechanisms/pathways may exist in different cell types.

The nature of the disease-causing mutation can also contribute to the disease pathology. As many DNA replication factors are essential, a large proportion of disease-causing mutations are hypomorphic, and conversely a small fraction seem to be dominant-negative. Therefore, individual mutations may have lost/gained different protein functions and may be associated with different disease presentations. However, this doesn't explain why the same mutation can give rise to very different clinical phenotypes in different patients. In this case, the presence of genetic modifiers and/or environmental factors may contribute to the clinical variability. For instance, the presence of dominant negative ALDH2 mutations correlates with more severe bone marrow failure in FA patients (Hira et al. 2013).

Another factor that may impact on whether diseases are associated with specific pathologies, such as cancer predisposition, is how quickly the disease progresses, and whether patients will live long enough to develop the full spectrum of potential clinical symptoms. For instance, whilst FA was traditionally considered to be a childhood disease as patients once rarely survived in adulthood, due to advances in disease management and treatment there is now an increasing number of surviving adult FA patients. However, it is now recognised that adult FA patients face different challenges, such as an increased risk of developing solid tumours (Kee and D'Andrea 2012). An important question about the pathologies that arise after birth in childhood, or adulthood, is whether they arise directly, or perhaps indirectly due to stresses occurring during embryogenesis, as suggested by the ATR-SCKL mouse model (Murga et al. 2009). The answer to this will have important implications for whether therapy options will be available to treat/manage certain diseases.

Many questions and challenges surrounding the treatment of patients also exist. How can we efficiently treat cancer in individuals that are hypersensitive to DNA damaging agents? Are there any strategies to halt or delay the progression of neurodegeneration? Can we use gene therapy to correct specific tissue defects associated with certain diseases? It is anticipated that further research, continued identification of novel human diseases, and improvements in the use of animal models will lead to a better understanding of the molecular defects underlying the disease pathologies, and ultimately allow novel/improved therapeutic strategies to be developed to enable treatment of human disease.

REFERENCES

Adelman, C. A., Lolo, R. L., Birkbak, N. J., Murina, O., Matsuzaki, K., Horejsi, Z., Parmar, K. et al. 2013. HELQ promotes RAD51 paralogue-dependent repair to avert germ cell loss and tumorigenesis. *Nature* 502: 381–4.

Alasiri, S., Basit, S., Wood-Trageser, M. A., Yatsenko, S. A., Jeffries, E. P., Surti, U., Ketterer, D. M. et al. 2015. Exome sequencing reveals MCM8 mutation underlies ovarian failure and chromosomal instability. *J Clin Invest* 125: 258–62.

Alcantara, D., O'Driscoll, M. 2014. Congenital microcephaly. *Am J Med Genet C Semin Med Genet* 166C: 124–39.

Ameziane, N., May, P., Haitjema, A., Van De Vrugt, H. J., Van Rossum-Fikkert, S. E., Ristic, D., Williams, G. J. et al. 2015. A novel Fanconi anaemia subtype associated with a dominant-negative mutation in RAD51. *Nat Commun* 6: 8829.

Andreassen, P. R., D'andrea, A. D., Taniguchi, T. 2004. ATR couples FANCD2 monoubiquitination to the DNA-damage response. *Genes Dev* 18: 1958–63.

Anheim, M., Tranchant, C., Koenig, M. 2012. The autosomal recessive cerebellar ataxias. *N Engl J Med* 366: 636–46.

Aparicio, T., Baer, R., Gottesman, M., Gautier, J. 2016. MRN, CtIP, and BRCA1 mediate repair of topoisomerase II-DNA adducts. *J Cell Biol* 212: 399–408.

Arias, E. E., Walter, J. C. 2007. Strength in numbers: Preventing rereplication via multiple mechanisms in eukaryotic cells. *Genes Dev* 21: 497–518.

Auclair, J., Leroux, D., Desseigne, F., Lasset, C., Saurin, J. C., Joly, M. O., Pinson, S. et al. 2007. Novel biallelic mutations in MSH6 and PMS2 genes: Gene conversion as a likely cause of PMS2 gene inactivation. *Hum Mutat* 28: 1084–90.

Balakrishnan, L., Bambara, R. A. 2013. Okazaki fragment metabolism. *Cold Spring Harb Perspect Biol* 5: a010173 .

Bansbach, C. E., Betous, R., Lovejoy, C. A., Glick, G. G., Cortez, D. 2009. The annealing helicase SMARCAL1 maintains genome integrity at stalled replication forks. *Genes Dev* 23: 2405–14.

Bao, S., Tibbetts, R. S., Brumbaugh, K. M., Fang, Y., Richardson, D. A., Ali, A., Chen, S. M., Abraham, R. T., Wang, X. F. 2001. ATR/ATM-mediated phosphorylation of human Rad17 is required for genotoxic stress responses. *Nature* 411: 969–74.

Bapat, B., Xia, L., Madlensky, L., Mitri, A., Tonin, P., Narod, S. A., Gallinger, S. 1996. The genetic basis of Muir-Torre syndrome includes the hMLH1 locus. *Am J Hum Genet* 59: 736–9.

Baple, E. L., Chambers, H., Cross, H. E., Fawcett, H., Nakazawa, Y., Chioza, B. A., Harlalka, G. V. et al. 2014. Hypomorphic PCNA mutation underlies a human DNA repair disorder. *J Clin Invest* 124: 3137–46.

Baradaran-Heravi, A., Cho, K. S., Tolhuis, B., Sanyal, M., Morozova, O., Morimoto, M., Elizondo, L. I. et al. 2012. Penetrance of biallelic SMARCAL1 mutations is associated with environmental and genetic disturbances of gene expression. *Hum Mol Genet* 21: 2572–87.

Barnes, D. E., Tomkinson, A. E., Lehmann, A. R., Webster, A. D., Lindahl, T. 1992. Mutations in the DNA ligase I gene of an individual with immunodeficiencies and cellular hypersensitivity to DNA-damaging agents. *Cell* 69: 495–503.

Berti, M., Ray Chaudhuri, A., Thangavel, S., Gomathinayagam, S., Kenig, S., Vujanovic, M., Odreman, F. et al. 2013. Human RECQ1 promotes restart of replication forks reversed by DNA topoisomerase I inhibition. *Nat Struct Mol Biol* 20: 347–54.

Betous, R., Mason, A. C., Rambo, R. P., Bansbach, C. E., Badu-Nkansah, A., Sirbu, B. M., Eichman, B. F., Cortez, D. 2012. SMARCAL1 catalyzes fork regression and Holliday junction migration to maintain genome stability during DNA replication. *Genes Dev* 26: 151–62.

Bhatia, V., Barroso, S. I., Garcia-Rubio, M. L., Tumini, E., Herrera-Moyano, E., Aguilera, A. 2014. BRCA2 prevents R-loop accumulation and associates with TREX-2 mRNA export factor PCID2. *Nature* 511: 362–5.

Bicknell, L. S., Bongers, E. M., Leitch, A., Brown, S., Schoots, J., Harley, M. E., Aftimos, S. et al. 2011a. Mutations in the pre-replication complex cause Meier-Gorlin syndrome. *Nat Genet* 43: 356–9.

Bicknell, L. S., Walker, S., Klingseisen, A., Stiff, T., Leitch, A., Kerzendorfer, C., Martin, C. A. et al. 2011b. Mutations in ORC1, encoding the largest subunit of the origin recognition complex, cause microcephalic primordial dwarfism resembling Meier-Gorlin syndrome. *Nat Genet* 43: 350–5.

Blackford, A. N., Nieminuszczy, J., Schwab, R. A., Galanty, Y., Jackson, S. P., Niedzwiedz, W. 2015. TopBP1 interacts with BLM to maintain genome stability but is dispensable for preventing BLM degradation. *Mol Cell* 57: 1133–41.

Bluteau, D., Masliah-Planchon, J., Clairmont, C., Rousseau, A., Ceccaldi, R., Dubois D'enghien, C., Bluteau, O. et al. 2016. Biallelic inactivation of REV7 is associated with Fanconi anemia. *J Clin Invest* 126: 3580–4.

Boerkoel, C. F., Takashima, H., John, J., Yan, J., Stankiewicz, P., Rosenbarker, L., Andre, J. L. et al. 2002. Mutant chromatin remodeling protein SMARCAL1 causes Schimke immuno-osseous dysplasia. *Nat Genet* 30: 215–20.

Bogliolo, M., Bluteau, D., Lespinasse, J., Pujol, R., Vasquez, N., D'enghien, C. D., Stoppa-Lyonnet, D., Leblanc, T., Soulier, J. , Surralles, J. 2017. Biallelic truncating FANCM mutations cause early-onset cancer but not Fanconi anemia. *Genet Med* doi:10.1038/gim.2017.124.

Bogliolo, M., Schuster, B., Stoepker, C., Derkunt, B., Su, Y., Raams, A., Trujillo, J. P. et al. 2013. Mutations in ERCC4, encoding the DNA-repair endonuclease XPF, cause Fanconi anemia. *Am J Hum Genet* 92: 800–6.

Bogliolo, M., Surralles, J. 2015. Fanconi anemia: A model disease for studies on human genetics and advanced therapeutics. *Curr Opin Genet Dev* 33: 32–40.

Boland, C. R., Goel, A. 2010. Microsatellite instability in colorectal cancer. *Gastroenterology* 138: 2073–87. e3.

Boos, D., Sanchez-Pulido, L., Rappas, M., Pearl, L. H., Oliver, A. W., Ponting, C. P., Diffley, J. F. 2011. Regulation of DNA replication through Sld3-Dpb11 interaction is conserved from yeast to humans. *Curr Biol* 21: 1152–7.

Brazel, C. Y., Rao, M. S. 2004. Aging and neuronal replacement. *Ageing Res Rev* 3: 465–83.

Brosh, R. M., Jr., Bellani, M., Liu, Y., Seidman, M. M. 2017. Fanconi anemia: A DNA repair disorder characterized by accelerated decline of the hematopoietic stem cell compartment and other features of aging. *Ageing Res Rev* 33: 67–75.

Broustas, C. G., Lieberman, H. B. 2014. DNA damage response genes and the development of cancer metastasis. *Radiat Res* 181: 111–30.

Brown, E. J., Baltimore, D. 2000. ATR disruption leads to chromosomal fragmentation and early embryonic lethality. *Genes Dev* 14: 397–402.

Bruhn, C., Zhou, Z. W., Ai, H., Wang, Z. Q. 2014. The essential function of the MRN complex in the resolution of endogenous replication intermediates. *Cell Rep* 6: 182–95.

Budzowska, M., Graham, T. G., Sobeck, A., Waga, S., Walter, J. C. 2015. Regulation of the Rev1-pol zeta complex during bypass of a DNA interstrand cross-link. *EMBO J* 34: 1971–85.

Bugreev, D. V., Rossi, M. J., Mazin, A. V. 2011. Cooperation of RAD51 and RAD54 in regression of a model replication fork. *Nucleic Acids Res* 39: 2153–64.

Bunting, S. F., Nussenzweig, A. 2010. Dangerous liaisons: Fanconi anemia and toxic nonhomologous end joining in DNA crosslink repair. *Mol Cell* 39: 164–6.

Burrage, L. C., Charng, W. L., Eldomery, M. K., Willer, J. R., Davis, E. E., Lugtenberg, D., Zhu, W. et al. 2015. De Novo GMNN mutations cause autosomal-dominant primordial dwarfism associated with meier-gorlin syndrome. *Am J Hum Genet* 97: 904–13.

Calado, R., Young, N. 2012. Telomeres in disease. *F1000 Med Rep* 4: 8.

Cali, F., Bharti, S. K., Di Perna, R., Brosh, R. M., Jr., Pisani, F. M. 2016. Tim/Timeless, a member of the replication fork protection complex, operates with the Warsaw breakage syndrome DNA helicase DDX11 in the same fork recovery pathway. *Nucleic Acids Res* 44: 705–17.

Castella, M., Jacquemont, C., Thompson, E. L., Yeo, J. E., Cheung, R. S., Huang, J. W., Sobeck, A., Hendrickson, E. A., Taniguchi, T. 2015. FANCI regulates recruitment of the FA core complex at sites of DNA damage independently of FANCD2. *PLoS Genet* 11: e1005563.

Castor, D., Nair, N., Declais, A. C., Lachaud, C., Toth, R., Macartney, T. J., Lilley, D. M., Arthur, J. S., Rouse, J. 2013. Cooperative control of Holliday junction resolution and DNA repair by the SLX1 and MUS81-EME1 nucleases. *Mol Cell* 52: 221–33.

Catucci, I., Osorio, A., Arver, B., Neidhardt, G., Bogliolo, M., Zanardi, F., Riboni, M. et al. 2017. Individuals with FANCM biallelic mutations do not develop Fanconi anemia, but show risk for breast cancer, chemotherapy toxicity and may display chromosome fragility. *Genet Med* doi:10.1038/gim.2017.123.

Ceccaldi, R., Briot, D., Larghero, J., Vasquez, N., Dubois D'enghien, C., Chamousset, D., Noguera, M. E. et al. 2011. Spontaneous abrogation of the G(2)DNA damage checkpoint has clinical benefits but promotes leukemogenesis in Fanconi anemia patients. *J Clin Invest* 121: 184–94.

Ceccaldi, R., Parmar, K., Mouly, E., Delord, M., Kim, J. M., Regairaz, M., Pla, M. et al. 2012. Bone marrow failure in Fanconi anemia is triggered by an exacerbated p53/p21 DNA damage response that impairs hematopoietic stem and progenitor cells. *Cell Stem Cell* 11: 36–49.

Ceccaldi, R., Sarangi, P., D'Andrea, A. D. 2016. The Fanconi anaemia pathway: New players and new functions. *Nat Rev Mol Cell Biol* 17: 337–49.

Chaganti, R. S., Schonberg, S., German, J. 1974. A manyfold increase in sister chromatid exchanges in Bloom's syndrome lymphocytes. *Proc Natl Acad Sci USA* 71: 4508–12.

Chang, S., Multani, A. S., Cabrera, N. G., Naylor, M. L., Laud, P., Lombard, D., Pathak, S., Guarente, L., Depinho, R. A. 2004. Essential role of limiting telomeres in the pathogenesis of Werner syndrome. *Nat Genet* 36: 877–82.

Chen, P. L., Liu, F., Cai, S., Lin, X., Li, A., Chen, Y., Gu, B., Lee, E. Y., Lee, W. H. 2005. Inactivation of CtIP leads to early embryonic lethality mediated by G1 restraint and to tumorigenesis by haploid insufficiency. *Mol Cell Biol* 25: 3535–42.

Chen, Y. Z., Bennett, C. L., Huynh, H. M., Blair, I. P., Puls, I., Irobi, J., Dierick, I. et al. 2004. DNA/RNA helicase gene mutations in a form of juvenile amyotrophic lateral sclerosis (ALS4). *Am J Hum Genet* 74: 1128–35.

Chi, Z., Nie, L., Peng, Z., Yang, Q., Yang, K., Tao, J., Mi, Y., Fang, X., Balajee, A. S., Zhao, Y. 2012. RecQL4 cytoplasmic localization: Implications in mitochondrial DNA oxidative damage repair. *Int J Biochem Cell Biol* 44: 1942–51.

Consortium, F. A. B. C. 1996. Positional cloning of the Fanconi anaemia group A gene. *Nat Genet* 14: 324–8.

Coppede, F., Migliore, L. 2015. DNA damage in neurodegenerative diseases. *Mutat Res* 776: 84–97.

Cortes Ledesma, F., El Khamisy, S. F., Zuma, M. C., Osborn, K., Caldecott, K. W. 2009. A human 5′-tyrosyl DNA phosphodiesterase that repairs topoisomerase-mediated DNA damage. *Nature* 461: 674–8.

Cortez, D. 2015. Preventing replication fork collapse to maintain genome integrity. *DNA Repair (Amst)* 32: 149–57.

Cortez, D., Guntuku, S., Qin, J., Elledge, S. J. 2001. ATR and ATRIP: Partners in checkpoint signaling. *Science* 294: 1713–6.

Cottineau, J., Kottemann, M. C., Lach, F. P., Kang, Y. H., Vely, F., Deenick, E. K., Lazarov, T. et al. 2017. Inherited GINS1 deficiency underlies growth retardation along with neutropenia and NK cell deficiency. *J Clin Invest* 127: 1991–2006.

Couch, F. B., Bansbach, C. E., Driscoll, R., Luzwick, J. W., Glick, G. G., Betous, R., Carroll, C. M. et al. 2013. ATR phosphorylates SMARCAL1 to prevent replication fork collapse. *Genes Dev* 27: 1610–23.

Couch, F. B., Cortez, D. 2014. Fork reversal, too much of a good thing. *Cell Cycle* 13: 1049–50.

Croteau, D. L., Popuri, V., Opresko, P. L., Bohr, V. A. 2014. Human RecQ helicases in DNA repair, recombination, and replication. *Annu Rev Biochem* 83: 519–52.

Croteau, D. L., Rossi, M. L., Canugovi, C., Tian, J., Sykora, P., Ramamoorthy, M., Wang, Z. M. et al. 2012. RECQL4 localizes to mitochondria and preserves mitochondrial DNA integrity. *Aging Cell* 11: 456–66.

Crow, Y. J., Leitch, A., Hayward, B. E., Garner, A., Parmar, R., Griffith, E., Ali, M. et al. 2006. Mutations in genes encoding ribonuclease H2 subunits cause Aicardi-Goutieres syndrome and mimic congenital viral brain infection. *Nat Genet* 38: 910–6.

Davies, S. L., North, P. S., Hickson, I. D. 2007. Role for BLM in replication-fork restart and suppression of origin firing after replicative stress. *Nat Struct Mol Biol* 14: 677–9.

de Munnik, S. A., Hoefsloot, E. H., Roukema, J., Schoots, J., Knoers, N. V., Brunner, H. G., Jackson, A. P., Bongers, E. M. 2015. Meier-Gorlin syndrome. *Orphanet J Rare Dis* 10: 114.

de Renty, C., Ellis, N. A. 2017. Bloom's syndrome: Why not premature aging? A comparison of the BLM and WRN helicases. *Ageing Res Rev* 33: 36–51.

de Winter, J. P., Leveille, F., Van Berkel, C. G., Rooimans, M. A., Van der Weel, L., Steltenpool, J., Demuth, I. et al. 2000a. Isolation of a cDNA representing the Fanconi anemia complementation group E gene. *Am J Hum Genet* 67: 1306–8.

de Winter, J. P., Rooimans, M. A., Van Der Weel, L., Van Berkel, C. G., Alon, N., Bosnoyan-Collins, L., De Groot, J. et al. 2000b. The Fanconi anaemia gene FANCF encodes a novel protein with homology to ROM. *Nat Genet* 24: 15–6.

de Winter, J. P., Waisfisz, Q., Rooimans, M. A., Van Berkel, C. G., Bosnoyan-Collins, L., Alon, N., Carreau, M. et al. 1998. The Fanconi anaemia group G gene FANCG is identical with XRCC9. *Nat Genet* 20: 281–3.

Delacroix, S., Wagner, J. M., Kobayashi, M., Yamamoto, K., Karnitz, L. M. 2007. The Rad9-Hus1-Rad1 (9-1-1) clamp activates checkpoint signaling via TopBP1. *Genes Dev* 21: 1472–7.

Deshpande, R. A., Lee, J. H., Arora, S., Paull, T. T. 2016. Nbs1 converts the human Mre11/Rad50 nuclease complex into an endo/exonuclease machine specific for protein-DNA adducts. *Mol Cell* 64: 593–606.

Dickinson, M. E., Flenniken, A. M., Ji, X., Teboul, L., Wong, M. D., White, J. K., Meehan, T. F. et al. 2016. High-throughput discovery of novel developmental phenotypes. *Nature* 537: 508–14.

Digiovanna, J. J., Kraemer, K. H. 2012. Shining a light on xeroderma pigmentosum. *J Invest Dermatol* 132: 785–96.

Dorsman, J. C., Levitus, M., Rockx, D., Rooimans, M. A., Oostra, A. B., Haitjema, A., Bakker, S. T. et al. 2007. Identification of the Fanconi anemia complementation group I gene, FANCI. *Cell Oncol* 29: 211–8.

Dungrawala, H., Bhat, K. P., Le Meur, R., Chazin, W. J., Ding, X., Sharan, S. K., Wessel, S. R., Sathe, A. A., Zhao, R., Cortez, D. 2017. RADX promotes genome stability and modulates chemosensitivity by regulating RAD51 at replication forks. *Mol Cell* 67: 374–86, e5.

Durkin, S. G., Glover, T. W. 2007. Chromosome fragile sites. *Annu Rev Genet* 41: 169–92.

Duursma, A. M., Driscoll, R., Elias, J. E., Cimprich, K. A. 2013. A role for the MRN complex in ATR activation via TOPBP1 recruitment. *Mol Cell* 50: 116–22.

Duxin, J. P., Dao, B., Martinsson, P., Rajala, N., Guittat, L., Campbell, J. L., Spelbrink, J. N., Stewart, S. A. 2009. Human Dna2 is a nuclear and mitochondrial DNA maintenance protein. *Mol Cell Biol* 29: 4274–82.

Duxin, J. P., Dewar, J. M., Yardimci, H., Walter, J. C. 2014. Repair of a DNA-protein crosslink by replication-coupled proteolysis. *Cell* 159: 346–57.

El-Khamisy, S. F., Caldecott, K. W. 2006. TDP1-dependent DNA single-strand break repair and neurodegeneration. *Mutagenesis* 21: 219–24.

El-Khamisy, S. F., Saifi, G. M., Weinfeld, M., Johansson, F., Helleday, T., Lupski, J. R., Caldecott, K. W. 2005. Defective DNA single-strand break repair in spinocerebellar ataxia with axonal neuropathy-1. *Nature* 434: 108–13.

Elia, A. E., Wang, D. C., Willis, N. A., Boardman, A. P., Hajdu, I., Adeyemi, R. O., Lowry, E., Gygi, S. P., Scully, R., Elledge, S. J. 2015. RFWD3-dependent ubiquitination of RPA regulates repair at stalled replication forks. *Mol Cell* 60: 280–93.

Ellis, N. A., Groden, J., Ye, T. Z., Straughen, J., Lennon, D. J., Ciocci, S., Proytcheva, M., German, J. 1995. The Bloom's syndrome gene product is homologous to RecQ helicases. *Cell* 83: 655–66.

Evrony, G. D., Cordero, D. R., Shen, J., Partlow, J. N., Yu, T. W., Rodin, R. E., Hill, R. S. et al. 2017. Integrated genome and transcriptome sequencing identifies a noncoding mutation in the genome replication factor DONSON as the cause of microcephaly-micromelia syndrome. *Genome Res* 27: 1323–35.

Fenwick, A. L., Kliszczak, M., Cooper, F., Murray, J., Sanchez-Pulido, L., Twigg, S. R., Goriely, A. et al. 2016. Mutations in CDC45, encoding an essential component of the pre-initiation complex, cause meier-gorlin syndrome and craniosynostosis. *Am J Hum Genet* 99: 125–38.

Fernet, M., Gribaa, M., Salih, M. A., Seidahmed, M. Z., Hall, J., Koenig, M. 2005. Identification and functional consequences of a novel MRE11 mutation affecting 10 Saudi Arabian patients with the ataxia telangiectasia-like disorder. *Hum Mol Genet* 14: 307–18.

Flaus, A., Martin, D. M., Barton, G. J., Owen-Hughes, T. 2006. Identification of multiple distinct Snf2 subfamilies with conserved structural motifs. *Nucleic Acids Res* 34: 2887–905.

Forment, J. V., Blasius, M., Guerini, I., Jackson, S. P. 2011. Structure-specific DNA endonuclease Mus81/Eme1 generates DNA damage caused by Chk1 inactivation. *PLOS ONE* 6: e23517.

Fragkos, M., Ganier, O., Coulombe, P., Mechali, M. 2015. DNA replication origin activation in space and time. *Nat Rev Mol Cell Biol* 16: 360–74.

Fragkos, M., Naim, V. 2017. Rescue from replication stress during mitosis. *Cell Cycle* 16: 613–33.

Frugoni, F., Dobbs, K., Felgentreff, K., Aldhekri, H., Al Saud, B. K., Arnaout, R., Ali, A. A. et al. 2016. A novel mutation in the POLE2 gene causing combined immunodeficiency. *J Allergy Clin Immunol* 137: 635–8. e1.

Fugger, K., Mistrik, M., Danielsen, J. R., Dinant, C., Falck, J., Bartek, J., Lukas, J., Mailand, N. 2009. Human Fbh1 helicase contributes to genome maintenance via pro- and anti-recombinase activities. *J Cell Biol* 186: 655–63.

Furnari, B., Rhind, N., Russell, P. 1997. Cdc25 mitotic inducer targeted by chk1 DNA damage checkpoint kinase. *Science* 277: 1495–7.

Gaillard, H., Garcia-Muse, T., Aguilera, A. 2015. Replication stress and cancer. *Nat Rev Cancer* 15: 276–89.

Gan, W., Guan, Z., Liu, J., Gui, T., Shen, K., Manley, J. L., Li, X. 2011. R-loop-mediated genomic instability is caused by impairment of replication fork progression. *Genes Dev* 25: 2041–56.

Ganai, R. A., Johansson, E. 2016. DNA replication-A matter of fidelity. *Mol Cell* 62: 745–55.

Gao, J., Wang, Q., Dong, C., Chen, S., Qi, Y., Liu, Y. 2015. Whole exome sequencing identified MCM2 as a novel causative gene for autosomal dominant nonsyndromic deafness in a Chinese family. *PLOS ONE* 10: e0133522.

Gao, R., Schellenberg, M. J., Huang, S. Y., Abdelmalak, M., Marchand, C., Nitiss, K. C., Nitiss, J. L., Williams, R. S., Pommier, Y. 2014. Proteolytic degradation of topoisomerase II (Top2) enables the processing of Top2. DNA and Top2.RNA covalent complexes by tyrosyl-DNA-phosphodiesterase 2 (TDP2). *J Biol Chem* 289: 17960–9.

Garaycoechea, J. I., Crossan, G. P., Langevin, F., Daly, M., Arends, M. J., Patel, K. J. 2012. Genotoxic consequences of endogenous aldehydes on mouse haematopoietic stem cell function. *Nature* 489: 571–5.

Garcia-Higuera, I., Taniguchi, T., Ganesan, S., Meyn, M. S., Timmers, C., Hejna, J., Grompe, M., D'Andrea, A. D. 2001. Interaction of the Fanconi anemia proteins and BRCA1 in a common pathway. *Mol Cell* 7: 249–62.

Gavish, H., Dos Santos, C. C., Buchwald, M. 1993. A Leu554-to-Pro substitution completely abolishes the functional complementing activity of the Fanconi anemia (FACC) protein. *Hum Mol Genet* 2: 123–6.

Ge, X. Q., Blow, J. J. 2010. Chk1 inhibits replication factory activation but allows dormant origin firing in existing factories. *J Cell Biol* 191: 1285–97.

Gineau, L., Cognet, C., Kara, N., Lach, F. P., Dunne, J., Veturi, U., Picard, C. et al. 2012. Partial MCM4 deficiency in patients with growth retardation, adrenal insufficiency, and natural killer cell deficiency. *J Clin Invest* 122: 821–32.

Glover, T. W., Wilson, T. E., Arlt, M. F. 2017. Fragile sites in cancer: More than meets the eye. *Nat Rev Cancer* 17: 489–501.

Goellner, E. M., Putnam, C. D., Kolodner, R. D. 2015. Exonuclease 1-dependent and independent mismatch repair. *DNA Repair (Amst)* 32: 24–32.

Gomez-Herreros, F., Schuurs-Hoeijmakers, J. H., Mccormack, M., Greally, M. T., Rulten, S., Romero-Granados, R., Counihan, T. J. et al. 2014. TDP2 protects transcription from abortive topoisomerase activity and is required for normal neural function. *Nat Genet* 46: 516–21.

Groothuizen, F. S., Sixma, T. K. 2016. The conserved molecular machinery in DNA mismatch repair enzyme structures. *DNA Repair (Amst)* 38: 14–23.

Gueiderikh, A., Rosselli, F., Neto, J. B. C. 2017. A never-ending story: The steadily growing family of the FA and FA-like genes. *Genet Mol Biol* 40: 398–407.

Guernsey, D. L., Matsuoka, M., Jiang, H., Evans, S., Macgillivray, C., Nightingale, M., Perry, S. et al. 2011. Mutations in origin recognition complex gene ORC4 cause Meier-Gorlin syndrome. *Nat Genet* 43: 360–4.

Hamperl, S., Cimprich, K. A. 2016. Conflict resolution in the Genome: How transcription and replication make it work. *Cell* 167: 1455–67.

Hanahan, D., Weinberg, R. A. 2011. Hallmarks of cancer: The next generation. *Cell* 144: 646–74.

Hande, K. R. 2003. Topoisomerase II inhibitors. *Cancer Chemother Biol Response Modif* 21: 103–25.

Hanlon Newell, A. E., Hemphill, A., Akkari, Y. M., Hejna, J., Moses, R. E., Olson, S. B. 2008. Loss of homologous recombination or non-homologous end-joining leads to radial formation following DNA interstrand crosslink damage. *Cytogenet Genome Res* 121: 174–80.

Harley, M. E., Murina, O., Leitch, A., Higgs, M. R., Bicknell, L. S., Yigit, G., Blackford, A. N. et al. 2016. TRAIP promotes DNA damage response during genome replication and is mutated in primordial dwarfism. *Nat Genet* 48: 36–43.

Heller, R. C., Kang, S., Lam, W. M., Chen, S., Chan, C. S., Bell, S. P. 2011. Eukaryotic origin-dependent DNA replication *in vitro* reveals sequential action of DDK and S-CDK kinases. *Cell* 146: 80–91.

Hemphill, A. W., Akkari, Y., Newell, A. H., Schultz, R. A., Grompe, M., North, P. S., Hickson, I. D. et al. 2009. Topo IIIalpha and BLM act within the Fanconi anemia pathway in response to DNA-crosslinking agents. *Cytogenet Genome Res* 125: 165–75.

Heyer, W. D., Ehmsen, K. T., Liu, J. 2010. Regulation of homologous recombination in eukaryotes. *Annu Rev Genet* 44: 113–39.

Higgs, M. R., Reynolds, J. J., Winczura, A., Blackford, A. N., Borel, V., Miller, E. S., Zlatanou, A. et al. 2015. BOD1L is required to suppress deleterious resection of stressed replication forks. *Mol Cell* 59: 462–77.

Hira, A., Yabe, H., Yoshida, K., Okuno, Y., Shiraishi, Y., Chiba, K., Tanaka, H. et al. 2013. Variant ALDH2 is associated with accelerated progression of bone marrow failure in Japanese Fanconi anemia patients. *Blood* 122: 3206–9.

Hira, A., Yoshida, K., Sato, K., Okuno, Y., Shiraishi, Y., Chiba, K., Tanaka, H. et al. 2015. Mutations in the gene encoding the E2 conjugating enzyme UBE2T cause Fanconi anemia. *Am J Hum Genet* 96: 1001–7.

Hoa, N. N., Shimizu, T., Zhou, Z. W., Wang, Z. Q., Deshpande, R. A., Paull, T. T., Akter, S. et al. 2016. Mre11 is essential for the removal of lethal topoisomerase 2 covalent cleavage complexes. *Mol Cell* 64: 580–92.

Hoeijmakers, J. H. 2001. Genome maintenance mechanisms for preventing cancer. *Nature* 411: 366–74.

Hoffmann, S., Smedegaard, S., Nakamura, K., Mortuza, G. B., Raschle, M., Ibanez, D. E., Opakua, A. et al. 2016. TRAIP is a PCNA-binding ubiquitin ligase that protects genome stability after replication stress. *J Cell Biol* 212: 63–75.

Howlett, N. G., Taniguchi, T., Olson, S., Cox, B., Waisfisz, Q., De Die-Smulders, C., Persky, N. et al. 2002. Biallelic inactivation of BRCA2 in Fanconi anemia. *Science* 297: 606–9.

Hu, Y., Raynard, S., Sehorn, M. G., Lu, X., Bussen, W., Zheng, L., Stark, J. M. et al. 2007. RECQL5/Recql5 helicase regulates homologous recombination and suppresses tumor formation via disruption of Rad51 presynaptic filaments. *Genes Dev* 21: 3073–84.

Huang, J., Liu, S., Bellani, M. A., Thazhathveetil, A. K., Ling, C., De Winter, J. P., Wang, Y., Wang, W., Seidman, M. M. 2013. The DNA translocase FANCM/MHF promotes replication traverse of DNA interstrand crosslinks. *Mol Cell* 52: 434–46.

Hudson, D. F., Amor, D. J., Boys, A., Butler, K., Williams, L., Zhang, T., Kalitsis, P. 2016. Loss of RMI2 increases genome instability and causes a bloom-like syndrome. *PLoS Genet* 12: e1006483.

Hughes, C. R., Guasti, L., Meimaridou, E., Chuang, C. H., Schimenti, J. C., King, P. J., Costigan, C., Clark, A. J., Metherell, L. A. 2012. MCM4 mutation causes adrenal failure, short stature, and natural killer cell deficiency in humans. *J Clin Invest* 122: 814–20.

Iannascoli, C., Palermo, V., Murfuni, I., Franchitto, A., Pichierri, P. 2015. The WRN exonuclease domain protects nascent strands from pathological MRE11/EXO1-dependent degradation. *Nucleic Acids Res* 43: 9788–803.

Ide, H., Nakano, T., Shoulkamy, M. I., Salem, A. M. H. 2015. Formation, repair, and biological effects of DNA–protein cross-link damage. In: Chen, C. C., editor, *Advances in DNA Repair, chap. 2*. InTech, Rijeka.

Interthal, H., Pouliot, J. J., Champoux, J. J. 2001. The tyrosyl-DNA phosphodiesterase Tdp1 is a member of the phospholipase D superfamily. *Proc Natl Acad Sci USA* 98: 12009–14.

Jaspers, N. G., Raams, A., Silengo, M. C., Wijgers, N., Niedernhofer, L. J., Robinson, A. R., Giglia-Mari, G. et al. 2007. First reported patient with human ERCC1 deficiency has cerebro-oculo-facio-skeletal syndrome with a mild defect in nucleotide excision repair and severe developmental failure. *Am J Hum Genet* 80: 457–66.

Jeggo, P. A., Geuting, V., Lobrich, M. 2011. The role of homologous recombination in radiation-induced double-strand break repair. *Radiother Oncol* 101: 7–12.

Jiricny, J. 2013. Postreplicative mismatch repair. *Cold Spring Harb Perspect Biol* 5: a012633.

Ju, L., Wing, J., Taylor, E., Brandt, R., Slijepcevic, P., Horsch, M., Rathkolb, B. et al. 2013. SMC6 is an essential gene in mice, but a hypomorphic mutant in the ATPase domain has a mild phenotype with a range of subtle abnormalities. *DNA Repair (Amst)* 12: 356–66.

Kadyrov, F. A., Dzantiev, L., Constantin, N., Modrich, P. 2006. Endonucleolytic function of MutLalpha in human mismatch repair. *Cell* 126: 297–308.

Karanja, K. K., Cox, S. W., Duxin, J. P., Stewart, S. A., Campbell, J. L. 2012. DNA2 and EXO1 in replication-coupled, homology-directed repair and in the interplay between HDR and the FA/BRCA network. *Cell Cycle* 11: 3983–96.

Kashiyama, K., Nakazawa, Y., Pilz, D. T., Guo, C., Shimada, M., Sasaki, K., Fawcett, H. et al. 2013. Malfunction of nuclease ERCC1-XPF results in diverse clinical manifestations and causes Cockayne syndrome, xeroderma pigmentosum, and Fanconi anemia. *Am J Hum Genet* 92: 807–19.

Katsuki, Y., Takata, M. 2016. Defects in homologous recombination repair behind the human diseases: FA and HBOC. *Endocr Relat Cancer* 23: T19–37.

Kawabata, T., Luebben, S. W., Yamaguchi, S., Ilves, I., Matise, I., Buske, T., Botchan, M. R., Shima, N. 2011. Stalled fork rescue via dormant replication origins in unchallenged S phase promotes proper chromosome segregation and tumor suppression. *Mol Cell* 41: 543–53.

Kee, Y., D'Andrea, A. D. 2012. Molecular pathogenesis and clinical management of Fanconi anemia. *J Clin Invest* 122: 3799–806.

Kelland, L. 2007. The resurgence of platinum-based cancer chemotherapy. *Nat Rev Cancer* 7: 573–84.

Kim, M. S., Machida, Y., Vashisht, A. A., Wohlschlegel, J. A., Pang, Y. P., Machida, Y. J. 2013. Regulation of error-prone translesion synthesis by Spartan/C1orf124. *Nucleic Acids Res* 41: 1661–8.

Kim, T. M., Son, M. Y., Dodds, S., Hu, L., Luo, G., Hasty, P. 2015. RECQL5 and BLM exhibit divergent functions in cells defective for the Fanconi anemia pathway. *Nucleic Acids Res* 43: 893–903.

King, T. A., Li, W., Brogi, E., Yee, C. J., Gemignani, M. L., Olvera, N., Levine, D. A. et al. 2007. Heterogenic loss of the wild-type BRCA allele in human breast tumorigenesis. *Ann Surg Oncol* 14: 2510–8.

Kitao, S., Shimamoto, A., Goto, M., Miller, R. W., Smithson, W. A., Lindor, N. M., Furuichi, Y. 1999. Mutations in RECQL4 cause a subset of cases of Rothmund-Thomson syndrome. *Nat Genet* 22: 82–4.

Klein Douwel, D., Boonen, R. A., Long, D. T., Szypowska, A. A., Raschle, M., Walter, J. C., Knipscheer, P. 2014. XPF-ERCC1 acts in Unhooking DNA interstrand crosslinks in cooperation with FANCD2 and FANCP/SLX4. *Mol Cell* 54: 460–71.

Klingseisen, A., Jackson, A. P. 2011. Mechanisms and pathways of growth failure in primordial dwarfism. *Genes Dev* 25: 2011–24.

Knies, K., Inano, S., Ramirez, M. J., Ishiai, M., Surralles, J., Takata, M., Schindler, D. 2017. Biallelic mutations in the ubiquitin ligase RFWD3 cause Fanconi anemia. *J Clin Invest* 27: 3013–27.

Kruse, R., Lamberti, C., Wang, Y., Ruelfs, C., Bruns, A., Esche, C., Lehmann, P. et al. 1996. Is the mismatch repair deficient type of Muir-Torre syndrome confined to mutations in the hMSH2 gene? *Hum Genet* 98: 747–50.

Kumagai, A., Dunphy, W. G. 2000. Claspin, a novel protein required for the activation of Chk1 during a DNA replication checkpoint response in Xenopus egg extracts. *Mol Cell* 6: 839–49.

Kumari, U., Ya Jun, W., Huat Bay, B., Lyakhovich, A. 2014. Evidence of mitochondrial dysfunction and impaired ROS detoxifying machinery in Fanconi anemia cells. *Oncogene* 33: 165–72.

Kunkel, T. A., Erie, D. A. 2015. Eukaryotic mismatch repair in relation to DNA replication. *Annu Rev Genet* 49: 291–313.

Lam, I., Keeney, S. 2014. Mechanism and regulation of meiotic recombination initiation. *Cold Spring Harb Perspect Biol* 7: a016634.

Langevin, F., Crossan, G. P., Rosado, I. V., Arends, M. J., Patel, K. J. 2011. Fancd2 counteracts the toxic effects of naturally produced aldehydes in mice. *Nature* 475: 53–8.

Larizza, L., Roversi, G., Volpi, L. 2010. Rothmund-Thomson syndrome. *Orphanet J Rare Dis* 5: 2.

Laugel, V. 2013. Cockayne syndrome: The expanding clinical and mutational spectrum. *Mech Ageing Dev* 134: 161–70.

Leach, F. S., Nicolaides, N. C., Papadopoulos, N., Liu, B., Jen, J., Parsons, R., Peltomaki, P. et al. 1993. Mutations of a mutS homolog in hereditary nonpolyposis colorectal cancer. *Cell* 75: 1215–25.

Leemans, C. R., Braakhuis, B. J., Brakenhoff, R. H. 2011. The molecular biology of head and neck cancer. *Nat Rev Cancer* 11: 9–22.

Lessel, D., Vaz, B., Halder, S., Lockhart, P. J., Marinovic-Terzic, I., Lopez-Mosqueda, J., Philipp, M. et al. 2014. Mutations in SPRTN cause early onset hepatocellular carcinoma, genomic instability and progeroid features. *Nat Genet* 46: 1239–44.

Letessier, A., Millot, G. A., Koundrioukoff, S., Lachages, A. M., Vogt, N., Hansen, R. S., Malfoy, B., Brison, O., Debatisse, M. 2011. Cell-type-specific replication initiation programs set fragility of the FRA3B fragile site. *Nature* 470: 120–3.

Levin, D. S., Mckenna, A. E., Motycka, T. A., Matsumoto, Y., Tomkinson, A. E. 2000. Interaction between PCNA and DNA ligase I is critical for joining of Okazaki fragments and long-patch base-excision repair. *Curr Biol* 10: 919–22.

Levitus, M., Waisfisz, Q., Godthelp, B. C., De Vries, Y., Hussain, S., Wiegant, W. W., Elghalbzouri-Maghrani, E. et al. 2005. The DNA helicase BRIP1 is defective in Fanconi anemia complementation group J. *Nat Genet* 37: 934–5.

Liang, C. C., Cohn, M. A. 2016. UHRF1 is a sensor for DNA interstrand crosslinks. *Oncotarget* 7: 3–4.

Lin, C. P., Ban, Y., Lyu, Y. L., Desai, S. D., Liu, L. F. 2008. A ubiquitin-proteasome pathway for the repair of topoisomerase I-DNA covalent complexes. *J Biol Chem* 283: 21074–83.

Lin, W., Sampathi, S., Dai, H., Liu, C., Zhou, M., Hu, J., Huang, Q. et al. 2013. Mammalian DNA2 helicase/nuclease cleaves G-quadruplex DNA and is required for telomere integrity. *EMBO J* 32: 1425–39.

Liu, E., Lee, A. Y., Chiba, T., Olson, E., Sun, P., Wu, X. 2007. The ATR-mediated S phase checkpoint prevents rereplication in mammalian cells when licensing control is disrupted. *J Cell Biol* 179: 643–57.

Liu, H. X., Zhou, X. L., Liu, T., Werelius, B., Lindmark, G., Dahl, N., Lindblom, A. 2003. The role of hMLH3 in familial colorectal cancer. *Cancer Res* 63: 1894–9.

Liu, Q., Guntuku, S., Cui, X. S., Matsuoka, S., Cortez, D., Tamai, K., Luo, G. et al. 2000. Chk1 is an essential kinase that is regulated by Atr and required for the G(2)/M DNA damage checkpoint. *Genes Dev* 14: 1448–59.

Ljungman, M., Lane, D. P. 2004. Transcription: Guarding the genome by sensing DNA damage. *Nat Rev Cancer* 4: 727–37.

Long, D. T., Joukov, V., Budzowska, M., Walter, J. C. 2014. BRCA1 promotes unloading of the CMG helicase from a stalled DNA replication fork. *Mol Cell* 56: 174–85.

Lopez-Martinez, D., Liang, C. C., Cohn, M. A. 2016. Cellular response to DNA interstrand crosslinks: The Fanconi anemia pathway. *Cell Mol Life Sci* 73: 3097–114.

Lopez-Mosqueda, J., Maddi, K., Prgomet, S., Kalayil, S., Marinovic-Terzic, I., Terzic, J., Dikic, I. 2016. SPRTN is a mammalian DNA-binding metalloprotease that resolves DNA-protein crosslinks. *Elife* 5: e21491.

Lu, L., Jin, W., Wang, L. L. 2017. Aging in Rothmund-Thomson syndrome and related RECQL4 genetic disorders. *Ageing Res Rev* 33: 30–5.

Luo, G., Santoro, I. M., McDaniel, L. D., Nishijima, I., Mills, M., Youssoufian, H., Vogel, H., Schultz, R. A., Bradley, A. 2000. Cancer predisposition caused by elevated mitotic recombination in Bloom mice. *Nat Genet* 26: 424–9.

Lutzmann, M., Grey, C., Traver, S., Ganier, O., Maya-Mendoza, A., Ranisavljevic, N., Bernex, F. et al. 2012. MCM8- and MCM9-deficient mice reveal gametogenesis defects and genome instability due to impaired homologous recombination. *Mol Cell* 47: 523–34.

Lynch, H. T., Snyder, C. L., Shaw, T. G., Heinen, C. D., Hitchins, M. P. 2015. Milestones of Lynch syndrome: 1895–2015. *Nat Rev Cancer* 15: 181–94.

Mamrak, N. E., Shimamura, A., Howlett, N. G. 2017. Recent discoveries in the molecular pathogenesis of the inherited bone marrow failure syndrome Fanconi anemia. *Blood Rev* 31: 93–9.

Masai, H., Matsumoto, S., You, Z., Yoshizawa-Sugata, N., Oda, M. 2010. Eukaryotic chromosome DNA replication: Where, when, and how? *Annu Rev Biochem* 79: 89–130.

Maskey, R. S., Flatten, K. S., Sieben, C. J., Peterson, K. L., Baker, D. J., Nam, H. J., Kim, M. S. et al. 2017. Spartan deficiency causes accumulation of Topoisomerase 1 cleavage complexes and tumorigenesis. *Nucleic Acids Res* 45: 4564–76.

Maskey, R. S., Kim, M. S., Baker, D. J., Childs, B., Malureanu, L. A., Jeganathan, K. B., Machida, Y., Van Deursen, J. M., Machida, Y. J. 2014. Spartan deficiency causes genomic instability and progeroid phenotypes. *Nat Commun* 5: 5744.

Mccauley, J., Masand, N., Mcgowan, R., Rajagopalan, S., Hunter, A., Michaud, J. L., Gibson, K. et al. 2011. X-linked VACTERL with hydrocephalus syndrome: Further delineation of the phenotype caused by FANCB mutations. *Am J Med Genet A* 155A: 2370–80.

Meetei, A. R., De Winter, J. P., Medhurst, A. L., Wallisch, M., Waisfisz, Q., Van, D. E., Vrugt, H. J. et al. 2003. A novel ubiquitin ligase is deficient in Fanconi anemia. *Nat Genet* 35: 165–70.

Meetei, A. R., Levitus, M., Xue, Y., Medhurst, A. L., Zwaan, M., Ling, C., Rooimans, M. A. et al. 2004. X-linked inheritance of Fanconi anemia complementation group B. *Nat Genet* 36: 1219–24.

Menolfi, D., Delamarre, A., Lengronne, A., Pasero, P., Branzei, D. 2015. Essential roles of the Smc5/6 complex in replication through natural pausing sites and endogenous DNA damage tolerance. *Mol Cell* 60: 835–46.

Metcalfe, K., Lubinski, J., Lynch, H. T., Ghadirian, P., Foulkes, W. D., Kim-Sing, C., Neuhausen, S. et al. 2010. Family history of cancer and cancer risks in women with BRCA1 or BRCA2 mutations. *J Natl Cancer Inst* 102: 1874–8.

Mokrani-Benhelli, H., Gaillard, L., Biasutto, P., Le Guen, T., Touzot, F., Vasquez, N., Komatsu, J. et al. 2013. Primary microcephaly, impaired DNA replication, and genomic instability caused by compound heterozygous ATR mutations. *Hum Mutat* 34: 374–84.

Moldovan, G. L., Pfander, B., Jentsch, S. 2007. PCNA, the maestro of the replication fork. *Cell* 129: 665–79.

Moreira, M. C., Klur, S., Watanabe, M., Nemeth, A. H., Le Ber, I., Moniz, J. C., Tranchant, C. et al. 2004. Senataxin, the ortholog of a yeast RNA helicase, is mutant in ataxia-ocular apraxia 2. *Nat Genet* 36: 225–7.

Mori, Y., Yin, J., Rashid, A., Leggett, B. A., Young, J., Simms, L., Kuehl, P. M., Langenberg, P., Meltzer, S. J., Stine, O. C. 2001. Instabilotyping: Comprehensive identification of frameshift mutations caused by coding region microsatellite instability. *Cancer Res* 61: 6046–9.

Morimoto, M., Choi, K., Boerkoel, C. F., Cho, K. S. 2016. Chromatin changes in SMARCAL1 deficiency: A hypothesis for the gene expression alterations of Schimke immuno-osseous dysplasia. *Nucleus* 7: 560–71.

Munoz, S., Mendez, J. 2017. DNA replication stress: From molecular mechanisms to human disease. *Chromosoma* 126: 1–15.

Murfuni, I., Nicolai, S., Baldari, S., Crescenzi, M., Bignami, M., Franchitto, A., Pichierri, P. 2013. The WRN and MUS81 proteins limit cell death and genome instability following oncogene activation. *Oncogene* 32: 610–20.

Murga, M., Bunting, S., Montana, M. F., Soria, R., Mulero, F., Canamero, M., Lee, Y., Mckinnon, P. J., Nussenzweig, A., Fernandez-Capetillo, O. 2009. A mouse model of ATR-Seckel shows embryonic replicative stress and accelerated aging. *Nat Genet* 41: 891–8.

Murina, O., Von Aesch, C., Karakus, U., Ferretti, L. P., Bolck, H. A., Hanggi, K., Sartori, A. A. 2014. FANCD2 and CtIP cooperate to repair DNA interstrand crosslinks. *Cell Rep* 7: 1030–8.

Nakano, T., Morishita, S., Katafuchi, A., Matsubara, M., Horikawa, Y., Terato, H., Salem, A. M. et al. 2007. Nucleotide excision repair and homologous recombination systems commit differentially to the repair of DNA-protein crosslinks. *Mol Cell* 28: 147–58.

Nam, E. A., Zhao, R., Cortez, D. 2011. Analysis of mutations that dissociate G(2) and essential S phase functions of human ataxia telangiectasia-mutated and Rad3-related (ATR) protein kinase. *J Biol Chem* 286: 37320–7.

Neelsen, K. J., Lopes, M. 2015. Replication fork reversal in eukaryotes: From dead end to dynamic response. *Nat Rev Mol Cell Biol* 16: 207–20.

Nicolaides, N. C., Papadopoulos, N., Liu, B., Wei, Y. F., Carter, K. C., Ruben, S. M., Rosen, C. A. et al. 1994. Mutations of two PMS homologues in hereditary nonpolyposis colon cancer. *Nature* 371: 75–80.

Niedzwiedz, W. 2016. Activating ATR, the devil's in the dETAA1l. *Nat Cell Biol* 18: 1120–2.

Nussenzweig, A., Nussenzweig, M. C. 2010. Origin of chromosomal translocations in lymphoid cancer. *Cell* 141: 27–38.

O'Driscoll, M., Ruiz-Perez, V. L., Woods, C. G., Jeggo, P. A., Goodship, J. A. 2003. A splicing mutation affecting expression of ataxia-telangiectasia and Rad3-related protein (ATR) results in Seckel syndrome. *Nat Genet* 33: 497–501.

Oestergaard, V. H., Langevin, F., Kuiken, H. J., Pace, P., Niedzwiedz, W., Simpson, L. J., Ohzeki, M., Takata, M., Sale, J. E., Patel, K. J. 2007. Deubiquitination of FANCD2 is required for DNA crosslink repair. *Mol Cell* 28: 798–809.

Ogi, T., Walker, S., Stiff, T., Hobson, E., Limsirichaikul, S., Carpenter, G., Prescott, K. et al. 2012. Identification of the first ATRIP-deficient patient and novel mutations in ATR define a clinical spectrum for ATR-ATRIP Seckel Syndrome. *PLoS Genet* 8: e1002945.

Oleinick, N. L., Chiu, S. M., Ramakrishnan, N., Xue, L. Y. 1987. The formation, identification, and significance of DNA-protein cross-links in mammalian cells. *Br J Cancer Suppl* 8: 135–40.

Pachlopnik Schmid, J., Lemoine, R., Nehme, N., Cormier-Daire, V., Revy, P., Debeurme, F., Debre, M. 2012. Polymerase epsilon1 mutation in a human syndrome with facial dysmorphism, immunodeficiency, livedo, and short stature ('FILS syndrome'). *J Exp Med* 209: 2323–30.

Palles, C., Cazier, J. B., Howarth, K. M., Domingo, E., Jones, A. M., Broderick, P., Kemp, Z. et al. 2013. Germline mutations affecting the proofreading domains of POLE and POLD1 predispose to colorectal adenomas and carcinomas. *Nat Genet* 45: 136–44.

Pang, Q., Andreassen, P. R. 2009. Fanconi anemia proteins and endogenous stresses. *Mutat Res* 668: 42–53.

Pani, B., Nudler, E. 2017. Mechanistic insights into transcription coupled DNA repair. *DNA Repair (Amst)* 56: 42–50.

Papadopoulos, N., Nicolaides, N. C., Liu, B., Parsons, R., Lengauer, C., Palombo, F., D'Arrigo, A. et al. 1995. Mutations of GTBP in genetically unstable cells. *Science* 268: 1915–7.

Papadopoulos, N., Nicolaides, N. C., Wei, Y. F., Ruben, S. M., Carter, K. C., Rosen, C. A., Haseltine, W. A. et al. 1994. Mutation of a mutL homolog in hereditary colon cancer. *Science* 263: 1625–9.

Park, J. W., Pitot, H. C., Strati, K., Spardy, N., Duensing, S., Grompe, M., Lambert, P. F. 2010. Deficiencies in the Fanconi anemia DNA damage response pathway increase sensitivity to HPV-associated head and neck cancer. *Cancer Res* 70: 9959–68.

Park, J. Y., Virts, E. L., Jankowska, A., Wiek, C., Othman, M., Chakraborty, S. C., Vance, G. H., Alkuraya, F. S., Hanenberg, H., Andreassen, P. R. 2016. Complementation of hypersensitivity to DNA interstrand crosslinking agents demonstrates that XRCC2 is a Fanconi anaemia gene. *J Med Genet* 53: 672–80.

Payne, F., Colnaghi, R., Rocha, N., Seth, A., Harris, J., Carpenter, G., Bottomley, W. E. et al. 2014. Hypomorphism in human NSMCE2 linked to primordial dwarfism and insulin resistance. *J Clin Invest* 124: 4028–38.

Petermann, E., Orta, M. L., Issaeva, N., Schultz, N., Helleday, T. 2010. Hydroxyurea-stalled replication forks become progressively inactivated and require two different RAD51-mediated pathways for restart and repair. *Mol Cell* 37: 492–502.

Poley, J. W., Wagner, A., Hoogmans, M. M., Menko, F. H., Tops, C., Kros, J. M., Reddingius, R. E., Meijers-Heijboer, H., Kuipers, E. J., Dinjens, W. N. 2007. Biallelic germline mutations of mismatch-repair genes: A possible cause for multiple pediatric malignancies. *Cancer* 109: 2349–56.

Pommier, Y. 2009. DNA topoisomerase I inhibitors: Chemistry, biology, and interfacial inhibition. *Chem Rev* 109: 2894–902.

Poulogiannis, G., Frayling, I. M., Arends, M. J. 2010. DNA mismatch repair deficiency in sporadic colorectal cancer and Lynch syndrome. *Histopathology* 56: 167–79.

Prakash, R., Zhang, Y., Feng, W., Jasin, M. 2015. Homologous recombination and human health: The roles of BRCA1, BRCA2, and associated proteins. *Cold Spring Harb Perspect Biol* 7: a016600.

Qvist, P., Huertas, P., Jimeno, S., Nyegaard, M., Hassan, M. J., Jackson, S. P., Borglum, A. D. 2011. CtIP mutations cause Seckel and Jawad syndromes. *PLoS Genet* 7: e1002310.

Raghunandan, M., Chaudhury, I., Kelich, S. L., Hanenberg, H., Sobeck, A. 2015. FANCD2, FANCJ and BRCA2 cooperate to promote replication fork recovery independently of the Fanconi Anemia core complex. *Cell Cycle* 14: 342–53.

Rassool, F. V., North, P. S., Mufti, G. J., Hickson, I. D. 2003. Constitutive DNA damage is linked to DNA replication abnormalities in Bloom's syndrome cells. *Oncogene* 22: 8749–57.

Reid, S., Schindler, D., Hanenberg, H., Barker, K., Hanks, S., Kalb, R., Neveling, K. et al. 2007. Biallelic mutations in PALB2 cause Fanconi anemia subtype FA-N and predispose to childhood cancer. *Nat Genet* 39: 162–4.

Reynolds, J. J., Bicknell, L. S., Carroll, P., Higgs, M. R., Shaheen, R., Murray, J. E., Papadopoulos, D. K. et al. 2017. Mutations in DONSON disrupt replication fork stability and cause microcephalic dwarfism. *Nat Genet* 49: 537–49.

Reynolds, J. J., Stewart, G. S. 2013. A nervous predisposition to unrepaired DNA double strand breaks. *DNA Repair (Amst)* 12: 588–99.

Rickman, K. A., Lach, F. P., Abhyankar, A., Donovan, F. X., Sanborn, E. M., Kennedy, J. A., Sougnez, C. et al. 2015. Deficiency of UBE2T, the E2 ubiquitin ligase necessary for FANCD2 and FANCI ubiquitination, causes FA-T subtype of Fanconi anemia. *Cell Rep* 12: 35–41.

Ronchi, D., Di Fonzo, A., Lin, W., Bordoni, A., Liu, C., Fassone, E., Pagliarani, S. et al. 2013. Mutations in DNA2 link progressive myopathy to mitochondrial DNA instability. *Am J Hum Genet* 92: 293–300.

Rosenberg, P. S., Greene, M. H., Alter, B. P. 2003. Cancer incidence in persons with Fanconi anemia. *Blood* 101: 822–6.

Roukos, V., Burman, B., Misteli, T. 2013. The cellular etiology of chromosome translocations. *Curr Opin Cell Biol* 25: 357–64.

Saintigny, Y., Makienko, K., Swanson, C., Emond, M. J., Monnat, R. J., Jr. 2002. Homologous recombination resolution defect in werner syndrome. *Mol Cell Biol* 22: 6971–8.

Saldivar, J. C., Cortez, D., Cimprich, K. A. 2017. The essential kinase ATR: Ensuring faithful duplication of a challenging genome. *Nat Rev Mol Cell Biol* 18: 622–36.

Sangrithi, M. N., Bernal, J. A., Madine, M., Philpott, A., Lee, J., Dunphy, W. G., Venkitaraman, A. R. 2005. Initiation of DNA replication requires the RECQL4 protein mutated in Rothmund-Thomson syndrome. *Cell* 121: 887–98.

Saponaro, M., Kantidakis, T., Mitter, R., Kelly, G. P., Heron, M., Williams, H., Soding, J., Stewart, A., Svejstrup, J. Q. 2014. RECQL5 controls transcript elongation and suppresses genome instability associated with transcription stress. *Cell* 157: 1037–49.

Savage, K. I., Matchett, K. B., Barros, E. M., Cooper, K. M., Irwin, G. W., Gorski, J. J., Orr, K. S. et al. 2014. BRCA1 deficiency exacerbates estrogen-induced DNA damage and genomic instability. *Cancer Res* 74: 2773–84.

Sawyer, S. L., Tian, L., Kahkonen, M., Schwartzentruber, J., Kircher, M., Majewski, J., Dyment, D. A. et al. 2015. Biallelic mutations in BRCA1 cause a new Fanconi anemia subtype. *Cancer Discov* 5: 135–42.

Schlacher, K., Wu, H., Jasin, M. 2012. A distinct replication fork protection pathway connects Fanconi anemia tumor suppressors to RAD51-BRCA1/2. *Cancer Cell* 22: 106–16.

Schwab, R. A., Blackford, A. N., Niedzwiedz, W. 2010. ATR activation and replication fork restart are defective in FANCM-deficient cells. *EMBO J* 29: 806–18.

Schwab, R. A., Nieminuszczy, J., Shah, F., Langton, J., Lopez Martinez, D., Liang, C. C., Cohn, M. A., Gibbons, R. J., Deans, A. J., Niedzwiedz, W. 2015. The Fanconi anemia pathway maintains genome stability by coordinating replication and transcription. *Mol Cell* 60: 351–61.

Semlow, D. R., Zhang, J., Budzowska, M., Drohat, A. C., Walter, J. C. 2016. Replication-dependent unhooking of DNA interstrand cross-links by the NEIL3 glycosylase. *Cell* 167: 498–511, e14.

Shaheen, R., Faqeih, E., Ansari, S., Abdel-Salam, G., Al-Hassnan, Z. N., Al-Shidi, T., Alomar, R., Sogaty, S., Alkuraya, F. S. 2014. Genomic analysis of primordial dwarfism reveals novel disease genes. *Genome Res* 24: 291–9.

Shamseldin, H. E., Elfaki, M., Alkuraya, F. S. 2012. Exome sequencing reveals a novel Fanconi group defined by XRCC2 mutation. *J Med Genet* 49: 184–6.

Shen, X., Jun, S., O'neal, L. E., Sonoda, E., Bemark, M., Sale, J. E., Li, L. 2006. REV3 and REV1 play major roles in recombination-independent repair of DNA interstrand cross-links mediated by monoubiquitinated proliferating cell nuclear antigen (PCNA). *J Biol Chem* 281: 13869–72.

Shi, Y., Lan, F., Matson, C., Mulligan, P., Whetstine, J. R., Cole, P. A., Casero, R. A., Shi, Y. 2004. Histone demethylation mediated by the nuclear amine oxidase homolog LSD1. *Cell* 119: 941–53.

Shibata, A. 2017. Regulation of repair pathway choice at two-ended DNA double-strand breaks. *Mutat Res* 803–805: 51–5.

Shima, N., Alcaraz, A., Liachko, I., Buske, T. R., Andrews, C. A., Munroe, R. J., Hartford, S. A., Tye, B. K., Schimenti, J. C. 2007. A viable allele of

Mcm4 causes chromosome instability and mammary adenocarcinomas in mice. *Nat Genet* 39: 93–8.

Siitonen, H. A., Kopra, O., Kaariainen, H., Haravuori, H., Winter, R. M., Saamanen, A. M., Peltonen, L., Kestila, M. 2003. Molecular defect of RAPADILINO syndrome expands the phenotype spectrum of RECQL diseases. *Hum Mol Genet* 12: 2837–44.

Sijbers, A. M., De Laat, W. L., Ariza, R. R., Biggerstaff, M., Wei, Y. F., Moggs, J. G., Carter, K. C. et al. 1996. Xeroderma pigmentosum group F caused by a defect in a structure-specific DNA repair endonuclease. *Cell* 86: 811–22.

Sijmons, R. H., Hofstra, R. M. 2016. Review: Clinical aspects of hereditary DNA Mismatch repair gene mutations. *DNA Repair (Amst)* 38: 155–62.

Singh, T. R., Bakker, S. T., Agarwal, S., Jansen, M., Grassman, E., Godthelp, B. C., Ali, A. M. et al. 2009. Impaired FANCD2 monoubiquitination and hypersensitivity to camptothecin uniquely characterize Fanconi anemia complementation group M. *Blood* 114: 174–80.

Skourti-Stathaki, K., Proudfoot, N. J., Gromak, N. 2011. Human senataxin resolves RNA/DNA hybrids formed at transcriptional pause sites to promote Xrn2-dependent termination. *Mol Cell* 42: 794–805.

Smirin-Yosef, P., Zuckerman-Levin, N., Tzur, S., Granot, Y., Cohen, L., Sachsenweger, J., Borck, G. et al. 2017. A biallelic mutation in the homologous recombination repair gene SPIDR is associated with human gonadal dysgenesis. *J Clin Endocrinol Metab* 102: 681–8.

Smogorzewska, A., Matsuoka, S., Vinciguerra, P., Mcdonald, E. R., 3rd, Hurov, K. E., Luo, J., Ballif, B. A. et al. 2007. Identification of the FANCI protein, a monoubiquitinated FANCD2 paralog required for DNA repair. *Cell* 129: 289–301.

Somyajit, K., Subramanya, S., Nagaraju, G. 2012. Distinct roles of FANCO/RAD51C protein in DNA damage signaling and repair: Implications for Fanconi anemia and breast cancer susceptibility. *J Biol Chem* 287: 3366–80.

Song, I. Y., Palle, K., Gurkar, A., Tateishi, S., Kupfer, G. M., Vaziri, C. 2010. Rad18-mediated translesion synthesis of bulky DNA adducts is coupled to activation of the Fanconi anemia DNA repair pathway. *J Biol Chem* 285: 31525–36.

Sordet, O., Redon, C. E., Guirouilh-Barbat, J., Smith, S., Solier, S., Douarre, C., Conti, C. et al. 2009. Ataxia telangiectasia mutated activation by transcription- and topoisomerase I-induced DNA double-strand breaks. *EMBO Rep* 10: 887–93.

Sorensen, C. S., Syljuasen, R. G. 2012. Safeguarding genome integrity: The checkpoint kinases ATR, CHK1 and WEE1 restrain CDK activity during normal DNA replication. *Nucleic Acids Res* 40: 477–86.

Spivak, G. 2015. Nucleotide excision repair in humans. *DNA Repair (Amst)* 36: 13–8.

Srivatsan, A., Tehranchi, A., Macalpine, D. M., Wang, J. D. 2010. Co-orientation of replication and transcription preserves genome integrity. *PLoS Genet* 6: e1000810.

Stewart, G. S., Maser, R. S., Stankovic, T., Bressan, D. A., Kaplan, M. I., Jaspers, N. G., Raams, A., Byrd, P. J., Petrini, J. H., Taylor, A. M. 1999. The DNA double-strand break repair gene hMRE11 is mutated in individuals with an ataxia-telangiectasia-like disorder. *Cell* 99: 577–87.

Stiff, T., Alagoz, M., Alcantara, D., Outwin, E., Brunner, H. G., Bongers, E. M., O'Driscoll, M., Jeggo, P. A. 2013. Deficiency in origin licensing proteins impairs cilia formation: Implications for the aetiology of Meier-Gorlin syndrome. *PLoS Genet* 9: e1003360.

Stingele, J., Bellelli, R., Alte, F., Hewitt, G., Sarek, G., Maslen, S. L., Tsutakawa, S. E. et al. 2016. Mechanism and regulation of DNA-protein crosslink repair by the DNA-dependent metalloprotease SPRTN. *Mol Cell* 64: 688–703.

Stingele, J., Jentsch, S. 2015. DNA-protein crosslink repair. *Nat Rev Mol Cell Biol* 16: 455–60.

Stoepker, C., Hain, K., Schuster, B., Hilhorst-Hofstee, Y., Rooimans, M. A., Steltenpool, J., Oostra, A. B. et al. 2011. SLX4, a coordinator of structure-specific endonucleases, is mutated in a new Fanconi anemia subtype. *Nat Genet* 43: 138–41.

Strand, M., Earley, M. C., Crouse, G. F., Petes, T. D. 1995. Mutations in the MSH3 gene preferentially lead to deletions within tracts of simple repetitive DNA in *Saccharomyces cerevisiae*. *Proc Natl Acad Sci USA* 92: 10418–21.

Strathdee, C. A., Gavish, H., Shannon, W. R., Buchwald, M. 1992. Cloning of cDNAs for Fanconi's anaemia by functional complementation. *Nature* 358: 434.

Sturzenegger, A., Burdova, K., Kanagaraj, R., Levikova, M., Pinto, C., Cejka, P., Janscak, P. 2014. DNA2 cooperates with the WRN and BLM RecQ helicases to mediate long-range DNA end resection in human cells. *J Biol Chem* 289: 27314–26.

Takashima, H., Boerkoel, C. F., John, J., Saifi, G. M., Salih, M. A., Armstrong, D., Mao, Y. et al. 2002. Mutation of TDP1, encoding a topoisomerase I-dependent DNA damage repair enzyme, in spinocerebellar ataxia with axonal neuropathy. *Nat Genet* 32: 267–72.

Teo, I. A., Arlett, C. F., Harcourt, S. A., Priestley, A., Broughton, B. C. 1983. Multiple hypersensitivity to mutagens in a cell strain (46BR) derived from a patient with immuno-deficiencies. *Mutat Res* 107: 371–86.

Thangavel, S., Berti, M., Levikova, M., Pinto, C., Gomathinayagam, S., Vujanovic, M., Zellweger, R. et al. 2015. DNA2 drives processing and restart of reversed replication forks in human cells. *J Cell Biol* 208: 545–62.

Timmers, C., Taniguchi, T., Hejna, J., Reifsteck, C., Lucas, L., Bruun, D., Thayer, M. et al. 2001. Positional cloning of a novel Fanconi anemia gene, FANCD2. *Mol Cell* 7: 241–8.

Tomasz, M. 1995. Mitomycin C: Small, fast and deadly (but very selective). *Chem Biol* 2: 575–9.

Trenz, K., Smith, E., Smith, S., Costanzo, V. 2006. ATM and ATR promote Mre11 dependent restart of collapsed replication forks and prevent accumulation of DNA breaks. *EMBO J* 25: 1764–74.

Tummala, H., Kirwan, M., Walne, A. J., Hossain, U., Jackson, N., Pondarre, C., Plagnol, V., Vulliamy, T., Dokal, I. 2014. ERCC6L2 mutations link a distinct bone-marrow-failure syndrome to DNA repair and mitochondrial function. *Am J Hum Genet* 94: 246–56.

Twigg, S. R., Wilkie, A. O. 2015. A genetic-pathophysiological framework for craniosynostosis. *Am J Hum Genet* 97: 359–77.

Umar, A., Boland, C. R., Terdiman, J. P., Syngal, S., De, L. A., Chapelle, A., Ruschoff, J. et al. 2004. Revised Bethesda Guidelines for hereditary nonpolyposis colorectal cancer (Lynch syndrome) and microsatellite instability. *J Natl Cancer Inst* 96: 261–8.

Unsal-Kacmaz, K., Chastain, P. D., Qu, P. P., Minoo, P., Cordeiro-Stone, M., Sancar, A., Kaufmann, W. K. 2007. The human Tim/Tipin complex coordinates an Intra-S checkpoint response to UV that slows replication fork displacement. *Mol Cell Biol* 27: 3131–42.

Van Der Crabben, S. N., Hennus, M. P., Mcgregor, G. A., Ritter, D. I., Nagamani, S. C., Wells, O. S., Harakalova, M. et al. 2016. Destabilized SMC5/6 complex leads to chromosome breakage syndrome with severe lung disease. *J Clin Invest* 126: 2881–92.

Van Maldergem, L. 1993. Baller-Gerold syndrome. In: Pagon, R. A., Adam, M. P., Ardinger, H. H., Wallace, S. E., Amemiya, A., Bean, L. J. H., Bird, T. D. et al., editors, *GeneReviews*. Seattle, University of Washington. Available from https://www.ncbi.nlm.nih.gov/books/NBK1204/.

Varon, R., Reis, A., Henze, G., Von Einsiedel, H. G., Sperling, K., Seeger, K. 2001. Mutations in the Nijmegen breakage syndrome gene (NBS1) in childhood acute lymphoblastic leukemia (ALL). *Cancer Res* 61: 3570–2.

Vasen, H. F., Ghorbanoghli, Z., Bourdeaut, F., Cabaret, O., Caron, O., Duval, A., Entz-Werle, N. et al. 2014. Guidelines for surveillance of individuals with constitutional mismatch repair-deficiency proposed by the European Consortium 'Care for CMMR-D' (C4CMMR-D). *J Med Genet* 51: 283–93.

Vaz, B., Popovic, M., Newman, J. A., Fielden, J., Aitkenhead, H., Halder, S., Singh, A. N. et al. 2016. Metalloprotease SPRTN/DVC1 orchestrates replication-coupled DNA-protein crosslink repair. *Mol Cell* 64: 704–719.

Vaz, F., Hanenberg, H., Schuster, B., Barker, K., Wiek, C., Erven, V., Neveling, K. et al. 2010. Mutation of the RAD51C gene in a Fanconi anemia-like disorder. *Nat Genet* 42: 406–9.

Vetro, A., Savasta, S., Russo Raucci, A., Cerqua, C., Sartori, G., Limongelli, I., Forlino, A. et al. 2017. MCM5: A new actor in the link between DNA replication and Meier-Gorlin syndrome. *Eur J Hum Genet* 25: 646–50.

Vujanovic, M., Krietsch, J., Raso, M. C., Terraneo, N., Zellweger, R., Schmid, J. A., Taglialatela, A. et al. 2017. Replication fork slowing and reversal upon DNA damage require PCNA polyubiquitination and ZRANB3 DNA translocase activity. *Mol Cell* 67: 882–90, e5.

Wahba, L., Amon, J. D., Koshland, D., Vuica-Ross, M. 2011. RNase H and multiple RNA biogenesis factors cooperate to prevent RNA: DNA hybrids from generating genome instability. *Mol Cell* 44: 978–88.

Walter, D., Lier, A., Geiselhart, A., Thalheimer, F. B., Huntscha, S., Sobotta, M. C., Moehrle, B. et al. 2015. Exit from dormancy provokes DNA-damage-induced attrition in haematopoietic stem cells. *Nature* 520: 549–52.

Waltes, R., Kalb, R., Gatei, M., Kijas, A. W., Stumm, M., Sobeck, A., Wieland, B. et al. 2009. Human RAD50 deficiency in a Nijmegen breakage syndrome-like disorder. *Am J Hum Genet* 84: 605–16.

Warcoin, M., Lespinasse, J., Despouy, G., Dubois d'Enghien, C., Lauge, A., Portnoi, M. F., Christin-Maitre, S., Stoppa-Lyonnet, D., Stern, M. H. 2009. Fertility defects revealing germline biallelic nonsense NBN mutations. *Hum Mutat* 30: 424–30.

Weedon, M. N., Ellard, S., Prindle, M. J., Caswell, R., Lango Allen, H., Oram, R., Godbole, K. et al. 2013. An in-frame deletion at the polymerase active site of POLD1 causes a multisystem disorder with lipodystrophy. *Nat Genet* 45: 947–50.

Weidenheim, K. M., Dickson, D. W., Rapin, I. 2009. Neuropathology of Cockayne syndrome: Evidence for impaired development, premature aging, and neurodegeneration. *Mech Ageing Dev* 130: 619–36.

Whiteside, D., Mcleod, R., Graham, G., Steckley, J. L., Booth, K., Somerville, M. J., Andrew, S. E. 2002. A homozygous germ-line mutation in the human MSH2 gene predisposes to hematological malignancy and multiple cafe-au-lait spots. *Cancer Res* 62: 359–62.

Wohlschlegel, J. A., Dwyer, B. T., Dhar, S. K., Cvetic, C., Walter, J. C., Dutta, A. 2000. Inhibition of eukaryotic DNA replication by geminin binding to Cdt1. *Science* 290: 2309–12.

Wood-Trageser, M. A., Gurbuz, F., Yatsenko, S. A., Jeffries, E. P., Kotan, L. D., Surti, U., Ketterer, D. M. et al. 2014. MCM9 mutations are associated with ovarian failure, short stature, and chromosomal instability. *Am J Hum Genet* 95: 754–62.

Woods, C. G., Parker, A. 2013. Investigating microcephaly. *Arch Dis Child* 98: 707–13.

Xia, B., Dorsman, J. C., Ameziane, N., De Vries, Y., Rooimans, M. A., Sheng, Q., Pals, G. et al. 2007. Fanconi anemia is associated with a defect in the BRCA2 partner PALB2. *Nat Genet* 39: 159–61.

Xie, J., Litman, R., Wang, S., Peng, M., Guillemette, S., Rooney, T., Cantor, S. B. 2010. Targeting the FANCJ-BRCA1 interaction promotes a switch from recombination to poleta-dependent bypass. *Oncogene* 29: 2499–508.

Xue, X., Sung, P., Zhao, X. 2015. Functions and regulation of the multitasking FANCM family of DNA motor proteins. *Genes Dev* 29: 1777–88.

Yamamoto, K. N., Kobayashi, S., Tsuda, M., Kurumizaka, H., Takata, M., Kono, K., Jiricny, J., Takeda, S., Hirota, K. 2011. Involvement of SLX4 in interstrand cross-link repair is regulated by the Fanconi anemia pathway. *Proc Natl Acad Sci USA* 108: 6492–6.

Yang, W., Woodgate, R. 2007. What a difference a decade makes: Insights into translesion DNA synthesis. *Proc Natl Acad Sci USA* 104: 15591–8.

Yang, Y., Liu, Z., Wang, F., Temviriyanukul, P., Ma, X., Tu, Y., Lv, L. et al. 2015. FANCD2 and REV1 cooperate in the protection of nascent DNA strands in response to replication stress. *Nucleic Acids Res* 43: 8325–39.

Yekezare, M., Gomez-Gonzalez, B., Diffley, J. F. 2013. Controlling DNA replication origins in response to DNA damage–inhibit globally, activate locally. *J Cell Sci* 126: 1297–306.

Yeo, J. E., Lee, E. H., Hendrickson, E. A., Sobeck, A. 2014. CtIP mediates replication fork recovery in a FANCD2-regulated manner. *Hum Mol Genet* 23: 3695–705.

Yu, C. E., Oshima, J., Fu, Y. H., Wijsman, E. M., Hisama, F., Alisch, R., Matthews, S. et al. 1996. Positional cloning of the Werner's syndrome gene. *Science* 272: 258–62.

Yuce, O., West, S. C. 2013. Senataxin, defective in the neurodegenerative disorder ataxia with oculomotor apraxia 2, lies at the interface of transcription and the DNA damage response. *Mol Cell Biol* 33: 406–17.

Zellweger, R., Dalcher, D., Mutreja, K., Berti, M., Schmid, J. A., Herrador, R., Vindigni, A., Lopes, M. 2015. Rad51-mediated replication fork reversal is a global response to genotoxic treatments in human cells. *J Cell Biol* 208: 563–79.

Zeman, M. K., Cimprich, K. A. 2014. Causes and consequences of replication stress. *Nat Cell Biol* 16: 2–9.

Zhang, J., Dewar, J. M., Budzowska, M., Motnenko, A., Cohn, M. A., Walter, J. C. 2015. DNA interstrand cross-link repair requires replication-fork convergence. *Nat Struct Mol Biol* 22: 242–7.

Zhang, J., Walter, J. C. 2014. Mechanism and regulation of incisions during DNA interstrand cross-link repair. *DNA Repair (Amst)* 19: 135–42.

Zhou, W., Otto, E. A., Cluckey, A., Airik, R., Hurd, T. W., Chaki, M., Diaz, K. et al. 2012. FAN1 mutations cause karyomegalic interstitial nephritis, linking chronic kidney failure to defective DNA damage repair. *Nat Genet* 44: 910–5.

Zlotorynski, E., Rahat, A., Skaug, J., Ben-Porat, N., Ozeri, E., Hershberg, R., Levi, A., Scherer, S. W., Margalit, H., Kerem, B. 2003. Molecular basis for expression of common and rare fragile sites. *Mol Cell Biol* 23: 7143–51.

Ataxia Telangiectasia and Ataxia Telangiectasia–Like Disorders

13

A. Malcolm R. Taylor

Understanding the genetic basis and the cellular characteristics of ataxia telangiectasia and ataxia telangiectasia–like disorders has contributed greatly to our knowledge of cellular responses to DNA damage, particularly DNA double-strand breaks. Interestingly, the relationship of these genes and cellular defects to the clinical features of these disorders has been more difficult to elucidate.

ATAXIA TELANGIECTASIA

Ataxia telangiectasia (A-T) is a rare disorder with a prevalence of about 3 per million in the population (Woods et al. 1990), and in the United Kingdom with a population of ~65 million, there are estimated to be about 200 cases. We believe that the ascertainment of A-T in the United Kingdom is close to 100%; therefore, this estimate of the number of cases fits well with the known number of diagnosed cases. A-T is inherited in an autosomal recessive manner so that both parents are obligate carriers of a mutation in the *ATM* gene, although there are rare instances of a de novo germ line mutation in *ATM* contributing to the phenotype.

A-T can be diagnosed in infancy and commonly before the age of 2 years, when most children have started walking (Sedgwick and Boder 1991). Parents notice some unusual movements of the head or drooling, and further tests can confirm the diagnosis. A-T is a progressive neurological disease, meaning that as time passes, the symptoms worsen significantly. It is also a disease principally of the cerebellum, where there is a degeneration of this part of the brain that co-ordinates movement and speech. At the age of 2 to 3 years, the affected child can walk and run with almost no noticeable difference from an unaffected individual, but gradually, the progressive loss of movement control leads to problems with both upper and lower limbs (ataxia). Holding an item such as a cup or knife and fork and completing the fine movements required to use them becomes very difficult. Similarly, the loss of movement control of the legs leads to difficulty in walking so that by the early teenage years, the A-T patient needs to use a wheelchair for mobility. Other neurological features of A-T include tremor, chorea, athetosis, dystonia and difficulty with swallowing (dysphagia) (Lefton-Greif et al. 2000). Also, by this age, and an inevitable feature of A-T, there is the progressive difficulty in moving the eyes (oculomotor apraxia). This is because the cerebellum also controls eye movements and the degeneration in A-T results in a very complex pattern of eye movement disorder. The effect on speech is also severe, resulting in much slower and distorted speech that can be difficult to follow for the listener. Despite these physical setbacks, A-T children appear to be intellectually normal. Further gross progression of the disease, as measured by neurological examination, probably stops in the late teenage years or early adulthood (Crawford et al. 2000).

Neuropathology has shown that the hallmark of A-T is a primary cerebellar cortical degeneration with severe loss of Purkinje cells and granular layer cells. With increasing age, these changes progress and other parts of the brain can be affected (Sedgwick and Boder 1991, Verhagen et al. 2012). However, A-T may be more than a degenerative disorder. Purkinje cells have been seen in unusual positions in the cerebellum in both A-T patients and in A-T mice, possibly indicating the presence of a developmental defect as well.

The telangiectasia part of the name derives from the presence in some patients of very obvious dilated blood vessels on the bulbar conjunctiva of the eye. This has nothing to do with the abnormal eye movements, and the cause of the telangiectasia, which can also affect the skin and other parts of the body including the brain, is unknown (Carrillo et al. 2009). Another consistent feature of A-T is the elevated level of serum alpha fetoprotein (AFP), the cause of which is unknown, but nevertheless is a useful tool in helping to confirm the diagnosis.

Interestingly, whereas diagnoses were once made exclusively in early childhood, nowadays new diagnoses are being made at any age up to 60-plus years, although this latter group is a small minority of patients. The reason for the later diagnosis in all such cases is the atypical clinical presentation of features giving a milder form of the disorder. A lot has been gained in understanding the role of the ATM protein in disease development by studying this range of presentations.

As well as the cerebellar degeneration and consequent progressive neurological disease, there are several other clinical features typical of A-T, but apparently unrelated to the central nervous system (CNS). The most important of these is an immunodeficiency that can be seen in all typical A-T patients. There can be deficiency of IgA and IgG2 levels, and a small proportion of patients require immunoglobulin replacement. Reduced B and T cell numbers and lymphopenia are the most frequently recorded features. Spectratyping of T cell repertoires has shown some restriction of TCRV Beta usage in some patients. A small proportion of patients also show decreased IgA and IgG levels together with hyper IgM, suggesting a more obvious defect in class switch recombination in such patients (Pan-Hammarstrom et al. 2003, 2006). This has also been observed in ATM-deficient mice (Lumsden et al. 2004, Reina-San-Martin 2004). It is not known what the basis is for this hyper IgM syndrome in a minority of patients, and it is not associated with particular *ATM* mutations. Possibly, there is an effect of a modifying gene(s) in such individuals. The immunodeficiency in A-T patients may be described as congenitally aged and is not progressive (Exley et al. 2011, Carney et al. 2012). Interestingly, A-T patients with no ATM kinase activity have a markedly more severe immunological phenotype than those expressing a low level of ATM activity (Staples et al. 2008).

Does this cellular immunodeficiency have a clinical effect on the person with A-T? Undoubtedly in some cases the answer is yes, in that it can result in an increase in upper respiratory tract infections (McGrath-Morrow 2010). This can give rise to respiratory complications such as acute and chronic respiratory tract infection and problems with aspiration/swallowing that can lead to bronchiectasis. The rare hyper IgM patients appear to show shorter survival (van Os et al. 2017). There are also some nasty disfiguring skin lesions and granulomas that the immunodeficiency also contributes to in A-T, possibly as a consequence of autoimmunity (Chiam et al. 2011). However, not all A-T patients with the typical, classical form of the disorder show susceptibility to serious infections, and this again may be the influence of modifying genes.

More will be said next about the clinical variation seen in ataxia telangiectasia.

CELLULAR FEATURES OF ATAXIA TELANGIECTASIA AND FUNCTIONS OF ATM

(See Table 13.1.) A-T patient cells show good evidence of spontaneously occurring damage to their genetic material as a result of loss of function of ATM. This includes elevated levels of very specific chromosome translocations in cultured T lymphocytes, involving breakage in the sites of different immune system genes in different chromosomes (chromosome 7, site of the TCR beta and gamma chain genes; chromosome 14, the site of the TCR alpha chain gene) and illegitimate recombination events between them to give the translocations. The consequence of at least some of the translocations is to provide the lymphocytes with a proliferative advantage; consequently, clones of translocation cells that grow in size over time can be observed in the patient's blood. At this stage of clonal growth, fusion of the telomeres of pairs of chromosomes can be seen, resulting in dicentric chromosome formation (Metcalfe et al. 1996). Part of this process, presumably, is the loss of telomeric sequences during clonal growth, which contributes to the dicentric formation as a result of telomeric fusions. These spontaneously occurring abnormalities are an indicator of a role for ATM in maintaining correct immune system gene rearrangement and also in telomere maintenance. Indeed ATM, together with RAG2, plays an important role in regulating recombination between immune system loci (Bredemeyer et al. 2006, Chaumeil et al. 2013).

The most striking cellular feature of ataxia telangiectasia patients is the increased sensitivity to ionising radiation (IR) and radiomimetic agents. Exposure of patients to radiotherapy has resulted in death or disfigurement (Byrd et al. 2012). Increased radio sensitivity can be readily

Table 13.1 **Comparison of clinical features in A-T, ATLD, NBS and NBSLD.**

Clinical feature	A-T (ATM-classical)	ATLD (MRE11)	NBS (NBN)	NBSLD (RAD50)
Progressive neurodegeneration	++	++	–	–
Microcephaly	–	– [a]	+	+
Immunodeficiency	+	±	+	?
CSR defect	+	+	+	?
Telangiectasia	+	–	–	–
Elevated serum AFP	+	±	–	–
Lymphoid tumours	++	–	++	?
Solid tumours	++	–	?	?
Breast cancer	++	?	?	?

Note: CSR = class switch recombination; ++ = effect/measure is large; + = effect/measure is less but clearly present; – = effect/measure is absent.

[a]Unusual patients with MRE11 mutations described with microcephaly.

measured by cell survival assays (Byrd et al. 2012, Worth et al. 2013) or in chromosome damage assays, particularly using the G2 assay, where cells are irradiated just a few hours before harvest at mitosis, when they are in the G2 phase of the cell cycle (Worth et al. 2013). The level of chromatid-type damage in lymphocytes, say, from a typical A-T patient, is about 10 times that seen in a normal control given the same dose of IR. This provides good confirmation of the diagnosis (taken with the clinical features) in typical cases of A-T.

Another important function of ATM, therefore, is to protect our cells from exogenously induced DNA damage of the type(s) caused by IR, principally DNA double-strand breaks. It is likely that most of this IR-like damage is actually endogenous and caused by, for example, reactive oxygen species.

Endogenous DNA double strand breaks occur as part of the normal process of immune system gene rearrangement, and this involves non-homologous end joining. NHEJ and homologous recombination repair (HR) (Hustedt and Durocher 2017) are also involved in the repair of induced DNA DSBs caused by IR. Cells from patients with deficiency in the NHEJ pathway proteins Ligase IV or XRCC4 are unusually radiosensitive. Similarly, loss of the HRR pathway proteins RAD51, BRCA1 or BRCA2 also causes unusual sensitivity to IR. Therefore, sensitivity of cells to IR is associated with loss of the ability to repair DNA DSBs, implying a defect in either NHEJ, HR or both. Indeed, a defect in both NHEJ and HR may underlie at least some of the features of ataxia telangiectasia. (HR and NHEJ are discussed in Chapters 6 and 7.)

In the G1 phase of the cell cycle, the repair of the great majority of DNA DSBs, even in A-T cells, takes place in just a few hours (fast kinetics) and is carried out by NHEJ, and a small proportion of probably 'hard-to-repair' DSBs persist and require longer for repair. In A-T cells, this hard-to-repair proportion is noticeably greater than in controls indicating an important role for ATM in the repair of this subgroup of breaks. Such breaks may be in close proximity to regions of heterochromatin, and require further modulation before repair can be achieved. Slower re-joining of these breaks can be seen following both G1 and G2 irradiation in the absence of ATM; it is likely that ATM plays a role in both of these slower components that involve NHEJ (in G1) and HR (in G2), respectively (Beucher et al. 2009).

The ATM protein and ATM signalling

A huge amount has been written about the ATM protein. The *ATM* coding sequence is 9168 bp long and codes for one of the cell's largest proteins of 3,056 amino acids (~370 kD). ATM protein is a PI-3–kinase like serine-threonine protein kinase, with a C- terminal kinase domain (**Figure 13.1**). ATM phosphorylates a very large number of target proteins in the cell, many as part of the cellular signalling response to damage. The cellular response to IR-induced DNA double-strand breaks normally triggers the MRN complex (MRE11-RAD50-Nibrin (Nibrin is also called NBN or NBS1); see detail later), which activates ATM kinase signalling, resulting

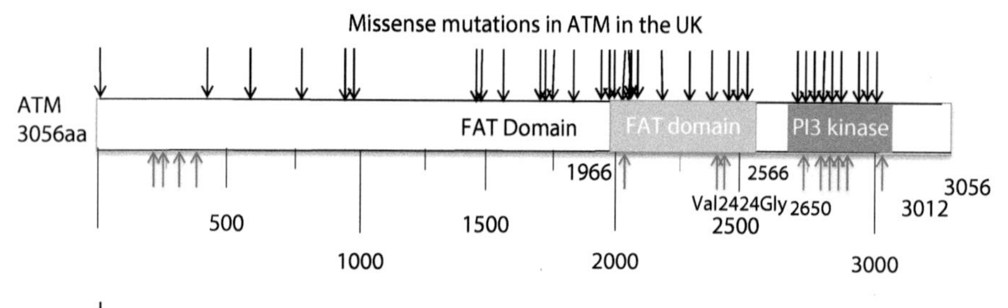

Missense mutations resulting in absence of ATM signaling activity

Missense mutations resulting in the presence of some ATM signaling activity

Figure 13.1 Depiction of ATM protein, showing the locations of missense mutations in A-T patients in the Untied Kingdom. This indicates that these mutations occur across the whole of the coding sequence. The more restricted positions of mutations that allow expression of some ATM kinases are indicated.

in autophosphorylation of ATM at several sites (Bakkenist and Kastan 2003). The signalling cascade results in repair or apoptosis. However, in mouse cells, at least, MRN is not essential for initial ATM activation, although its presence does optimise ATM-dependent responses (Hartlerode et al. 2015). As well as being activated by DNA DSBs, ATM can also be activated by oxidative stress to allow phosphorylation of targets such as p53 and Chk2. MRN is not essential for this, suggesting that there is more than one pathway to ATM activation (Guo et al. 2010).

Normal ATM signalling can be measured in any cell type by irradiating the cells and then examining the ability of the activated ATM to phosphorylate any number of cellular targets; phosphorylation can be detected using an appropriate phospho-specific antibody. In normal cells, ATM will phosphorylate the targets tested. In contrast, doing this with cells from typical A-T patients reveals the absence of ATM protein and therefore absence of phosphorylation of ATM targets. As well as absence of ATM in some A-T patients, in other A-T patients, the presence of ATM protein at either a normal or reduced level also occurs; the protein may be either kinase dead or, alternatively, retain the presence of some ATM signalling. Figure 13.2

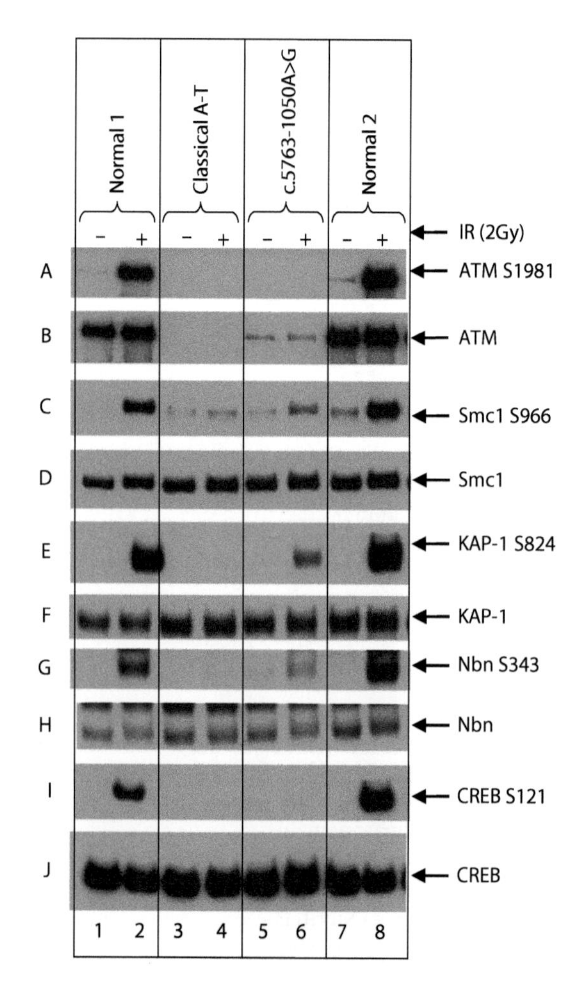

Figure 13.2 Lysate from a typical A-T patient without any ATM signalling and also lysate from a patient with a leaky splice site mutation c.5763-1050A>G (p.Pro1922fs), resulting in expression of a low level of normal ATM that gives a reduced level of signalling.

shows lysate from a typical A-T patient without any ATM signalling and also lysate from a patient with a leaky splice site mutation c.5763-1050A>G (p.Pro1922fs) resulting in expression of a low level of normal ATM that gives a reduced level of signalling (McConville et al. 1996, Stewart et al. 2001, Sutton et al. 2004).

ATM mutation and A-T patient heterogeneity

The cause of ataxia telangiectasia is the loss of function of the ATM protein. The typical (or classical) form of the disease is most severe and results from this total loss of function of the ATM protein. The basis of this can be the total absence of the ATM protein (perhaps as a consequence of the presence of biallelic truncating *ATM* mutations). On the other hand, biallelic mutations might include a missense mutation that allows expression of mutant ATM protein, but without activity. Either way, it is the total loss of ATM activity or signalling that correlates with the typical more severe form of the disorder rather than the absence of ATM protein per se.

There are, therefore, functional assays for the degree of radio sensitivity in A-T patient cells. These include the colony-forming ability following IR or chromosome damage following irradiation at G2 stage of the cell cycle. Appropriate positive (a cell line from a normal individual) and negative controls are essential in measuring this. The appropriate negative controls are cells from a patient with no ATM expression as a consequence of carrying two truncating mutations that result in instability and loss of ATM protein expression in the cell. The same controls would be used in a signalling assay to measure phosphorylation of ATM targets. In the A-T patients that carry the leaky splice site mutation described earlier, the low level of signalling results in a less severe form of A-T, a slightly later onset, longevity, an improved immune response and protection against the development of tumours (see later).

Similarly, the presence of one of a small number of missense mutations can result in expression of mutant ATM protein but with some activity (**Figure 13.1**) (Taylor et al. 2015); again, this can give a slightly milder form of the disease. There are many questions here that remain unanswered. Is it possible for a larger amount of mutant ATM protein to have a greater effect on reducing the severity of the clinical phenotype compared with a low level of normal ATM protein, say, from a leaky splice site mutation? Are some mutant ATM proteins better than others at phosphorylating the same targets? Would this be seen as a clinical difference? Do different ATM proteins phosphorylate different combinations of ATM targets? Would this result in an observable clinical difference?

We know that patients with the same ATM leaky splice site mutation show pretty well the same clinical presentation and also a similar progression and live longer; this is because there are quite a few of these individuals (McConville et al. 1996). Those with particular missense mutations are rarer, although there is a group of milder A-T patients in the United Kingdom with the same interesting mutation c.7271T>G; p.Val2424Gly (also see later). This mutant ATM protein has retained signalling/activity and A-T patients carrying this mutation have a milder form of the disorder (Stankovic et al. 1998, Stewart et al. 2001). Recently a small group of Canadian Mennonite A-T patients was reported who were homozygous for a different *ATM* missense mutation c.6200C>A; p.Ala2067Asp, and they also had a clinically milder form of the disorder although with a slightly different spectrum of clinical characteristics (Saunders-Pullman et al. 2012).

Broadly speaking, therefore, there is a good general genotype–phenotype correlation in ataxia telangiectasia and defining this is facilitated by the presence of both functional and signalling/activity assays. In the United Kingdom, about a third of A-T patients show some residual ATM activity and this correlates well with a clinically milder phenotype.

The later onset or slower progression of the neurological disease in the presence of either a low level of normal ATM, or the presence of mutant ATM with limited activity, suggests that ATM activity is neurologically important. However, there are many patients with mutant ATM that have no measurable activity and, in this circumstance, there is the typical neurological progression. Interestingly, patients without any activity also show the largest increase in radio sensitivity. So, does the loss of ATM result in unrepaired DNA breaks and is it these that cause the neurological degeneration, or is the neurodegeneration caused by some other mechanism via ATM loss?

The relationship between ATM activity/radio sensitivity and neurology in A-T is unclear

Absence of ATM protein is associated with the most severe neurological form of A-T. In the absence of ATM, radio sensitivity is also greatest at both the cellular and whole organism level.

In contrast, the greater the level of ATM activity/signalling in ATM mutant cells, the lower the level of radio sensitivity and the milder the neurological clinical presentation. Generally, in A-T, an increased radio sensitivity is associated with decreased ATM signalling. However, we do not know what the threshold level is for ATM with activity that will result in detectable signalling. It is quite possible that useful signalling activity might be present at levels below detection by current assays, and one important goal of current research is to improve the sensitivity of these signalling assays. As we have seen, this increase in radio sensitivity may result from both an ATM-related defect in homologous recombination repair and an ATM-related defect in end joining.

The reverse is not always the case: that the milder the neurological presentation of A-T, the greater the level of ATM activity/signalling. Indeed, there are rare A-T patients with the classical radio sensitive cellular phenotype with no ATM, but who have a milder clinical picture, therefore begging the question of the role of ATM in the neurological phenotype (Worth et al. 2013). Again, this is part of the amazing heterogeneity in A-T. Is this just an unusual exception? The important conclusion, in this particular case, is that the affected individual could get by without ATM and did not have the severe neurological effects seen in classical A-T. One may speculate that this ability to survive without ATM may be due to the loss of a second protein that nullifies the effect of ATM loss. Is there, therefore, some hope for treating classical A-T patients?

In answer to the question of whether unrepaired DNA DSBs are associated with cerebellar ataxia in A-T or ATLD (A-T–like disorder), we have to say that we do not know. It is unclear, therefore, what the role of ATM is in the neurological presentation in A-T. Reduction of ATM signalling is also associated with mutation of the MRE11-RAD50-NBN complex in ATLD, Nijmegen breakage syndrome (NBS) and Nijmegen breakage syndrome–like disorder (NBSLD). Similarities can be observed between A-T and ATLD (with respect to neurology) or A-T and NBS (with respect to tumour predisposition and immunodeficiency). These are discussed in more detail later. All these proteins are multifunctional, making it extremely difficult to relate one particular clinical feature to one particular function.

MRN DISORDERS

A-T–like disorder (MRE11 708aa)

MRN is the MRE11-RAD50-NBN protein complex, and there is another disorder much rarer than A-T that is important for our understanding of A-T and ATM. It is A-T–like disorder, so called because the neurology is indistinguishable from A-T (**Table 13.1**) and it is caused by mutation of the *MRE11* gene component of MRN (Hernandez et al. 1993, Klein et al. 1996, Stewart et al. 1999, Taylor et al. 2004). The age of onset is also in early childhood, with similar presenting features as A-T. The rate of progress is usually slower and, again, abnormal eye movements and dysarthria are present. There is, however, no telangiectasia. Affected individuals require wheelchair use by teenage years. Only 20 or so patients have been reported with ATLD (Delia et al. 2004, Fernet et al. 2005, Yoshida et al. 2014) but, interestingly, there is some striking variation in the clinical presentation. Two patients, compound heterozygous for mutations in the *MRE11* gene, have been reported with severe microcephaly (Matsumoto et al. 2011), a feature more commonly associated with Nijmegen breakage syndrome (see later). Two young siblings with ATLD in a different family both developed lung cancer, although it is not clear how the *MRE11* mutations were related to this tumour development (Uchisaka et al. 2009).

At the cellular level, radio sensitivity by colony forming assay is intermediate between A-T and normal, and similarly, there is an intermediate level of chromosomal radio sensitivity compared with A-T measured as G2 damage (**Table 13.2**). There is also an immunodeficiency different to that in A-T that can be demonstrated at the cellular level. Loss of MRE11 results in abnormal switch recombination junctions following class switch recombination with increased microhomology and additional sequences inserted at the junction points (Lahdesmaki et al. 2004). Interestingly, additional nucleotides are also seen at the telomere fusion points in ATLD cells (Tankimanova et al. 2012). This suggests a role for MRE11 in alternative end joining (A-EJ) processes, and that these processes may underlie the fusion of short dysfunctional telomeres. MRN is important for ATM activation at dysfunctional telomeres.

Why does mutation of *MRE11* result in the same neurological phenotype as mutation of *ATM*? What are the major functions of the *ATM* and *MRE11* genes that might give these

Table 13.2 Comparison of cellular features of A-T, ATLD, NBS and NBSLD.

Cellular feature	A-T (ATM-classical)	ATLD (MRE11)	NBS (NBN)	NBSLD (RAD50)
ATM signalling defect	+++	++	++	++
Chromosome 7 and 14 translocations	+++	+++	+++	+++
Increased chromosomal R/S	+++	++	++	++
Decreased survival following IR	+++	++	++	++
Radioresistant DNA synthesis	+++	+++	+++	+++
Defect in functional MRE11	–	++	++	++
Defect in functional MRN	–	++	++	++
Telomere defects	+++	++	++	?

Note: +++ = effect/measure is large; ++ = effect/measure is less but clearly present; – = effect/measure is absent.

similarities? Are there differences in the clinical presentations between A-T and ATLD? What other functions might MRE11 have that contribute to the clinical phenotype?

The MRN complex is crucial in the initiating steps of DNA double-strand break repair, homologous recombination repair and classical and alternative non-homologous end joining (see Chapters 6 and 7). Two MRE11 and two RAD50 molecules form a heterotetramer important for DNA binding, bridging DNA ends and interaction with NBN. MRE11 has both NBN and RAD50 binding domains, and NBN binds ATM (Figure 13.3), resulting in ATM activation and signalling following the production of DNA DSBs. The RAD50 component is the largest protein (1312 aa) of the complex and as a dimer allows tethering and bridging of two DNA ends. MRE11 (708 aa) also forms a dimer through an N-terminal domain that has nuclease function, with both single-stranded endonuclease and 3′-5′ ds DNA exonuclease function. MRE11 also controls 5′-3′ resection; an initial step of HR is the production of a 3′ single-stranded DNA overhang at the broken DNA end that involves MRN dependent nucleolytic activity.

Figure 13.3 **Depiction of the components of the MRN complex, MRE11, RAD50 and NBN indicating where the reported mutations are for ATLD, NBS and NBSLD.** (Adapted from Hopfner, K. P. et al. 2000 Jun 23, *Cell* 101(7): 789–800; Williams, R. S. et al. 2010, *Cell* 135(1): 97–109; Damiola, F. et al. 2014 Jun 3, *Breast Cancer Res* 16(3): R58; Lafrance-Vanasse, J. et al. 2015, *Prog Biophys Mol Biol* 117(2–3): 182–193; Kim, J. H. et al. 2017 Jan 10, *Cell Rep* 18(2): 496–507.)

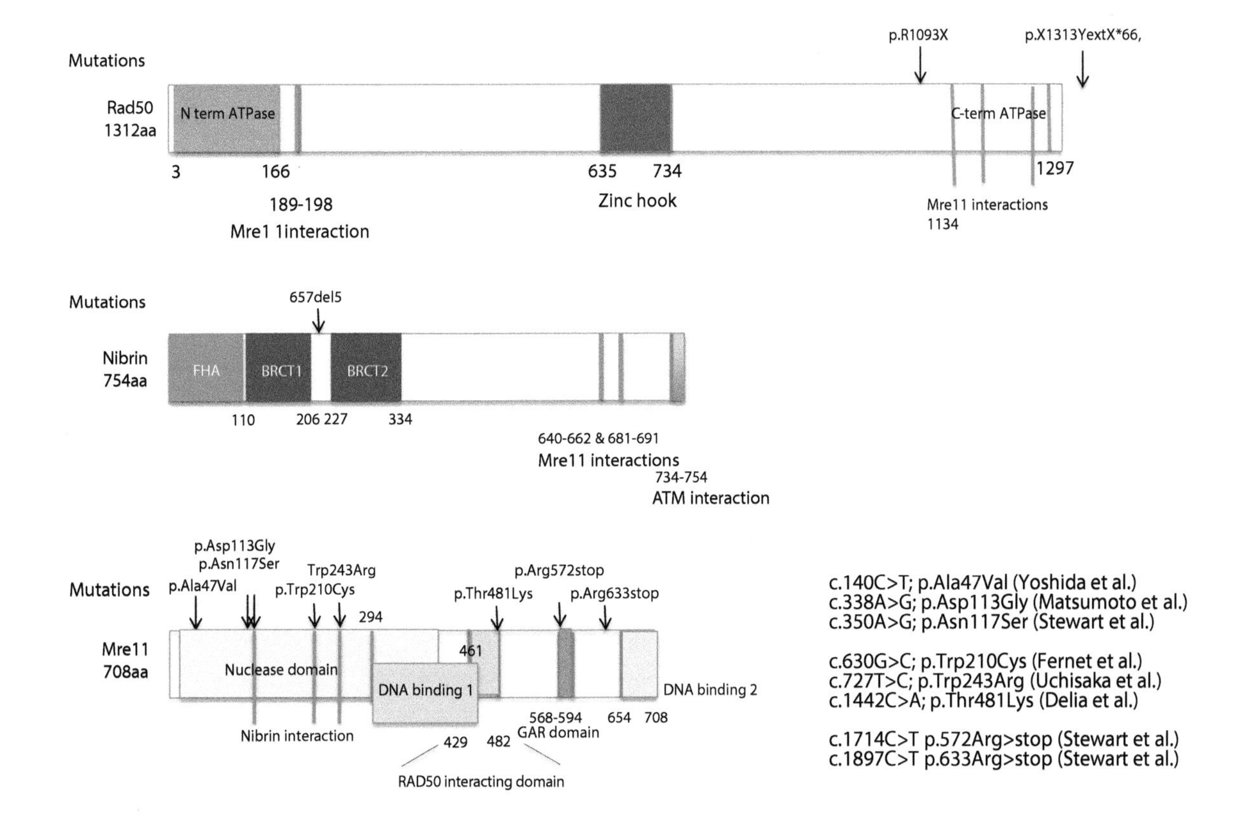

In an experimental system, different RAD50 mutations were shown to affect RAD50 ATP states (Hopfner et al. 2000, Deshpande et al. 2016). These can affect the rate at which the RAD50 ATPase switches the MRE11-RAD50 complex from ATP bound and closed to ATP free and open state. The closed state was associated with reduced ATPase and nuclease activity but increased DNA binding, while the open state was associated more with MRE11 nuclease and end resection activity (Deshpande et al. 2016). Altering the ATPase activity of RAD50 could affect preference for one repair pathway or another, thus homologous recombination repair rather than non-homologous end joining.

The dimerisation of MRE11 is essential for stable DNA binding. Therefore, an important function of the MRN complex is to act early in the first step of DNA DSB repair and to activate the ATM protein. Without MRE11, there is deficient ATM signalling. Indeed, in an experimental system, the most severe mutations in the MRE11 interaction domain of NBN prevented the MRE11–NBN interaction and resulted in loss of viability and indicating the requirement of this interaction for ATM activation (Kim et al. 2017) (and see later).

Experience shows that whereas classical A-T patients may be devoid of any ATM protein in their cells, all ATLD patients express mutant MRE11 protein at a reduced level, with a reduced level of the two other complex components, RAD50 and NBN. As a consequence of this reduction in the MRN complex, cells from all of these patients show reduced ATM signalling. In the case of ATLD, in order to investigate how different patient *MRE11* mutations impact the cellular phenotype, the mutations have been modelled in other systems. For example, both the N-terminal Asn117Ser and Trp210Cys mutant MRE11 proteins in ATLD patients impair the essential NBN binding described earlier (Williams et al. 2008), and this is probably true for p.Asp113Gly. The Trp243Arg mutant MRE11 probably affects both the nuclease activity of MRE11 and its ability to bind NBN (Schiller et al. 2012). Allowing expression of some of these mutant MRE11 proteins to normal levels has shown that they either have normal ATM signalling (e.g. the deleted form of MRE11 Arg633Ter) or have an intrinsic reduction in signalling (e.g. MRE11Trp243Arg). It has been concluded that there are different mechanisms by which ATM signalling is reduced in the presence of mutant MRE11. These are either as a consequence of a low-level expression of the complex (rather like the ATM splice site mutation) as in the case of Arg633Ter, or, alternatively, occur by a destabilisation of the M-R-N interactions caused by the intrinsic *MRE11* mutations such as the Trp243Arg (Regal et al. 2013). Human MRE11 mutations that impact the MRE11-NBN interaction also result in reduced MRE11 dimerisation. The MRE11 Leu72 position was identified as necessary for both MRE11 dimer stability and interaction with NBN, and this reduction in interaction with NBN leads to loss of ATM signalling. There is, therefore, a significant link between MRE11 dimerisation, MRE11-NBN interaction and ATM signalling in mammalian cells (Regal et al. 2013). In contrast, studying the same Trp243Arg mutation but in *S. pombe*, Limbo et al. (2012) showed that this mutated protein was not defective in inactivating Tel1, the ATM equivalent in *S. pombe*. The Trp243Arg MRE11 retained nuclease and DNA binding activity in vitro.

Common to both A-T (with loss of ATM function) and ATLD patients (with deficient MRE11) is deficient ATM signalling. Because MRE11 patients always have some ATM signalling by virtue of expressing some MRE11, the consequences of ATM signalling loss might be expected to be less in ATLD compared with classical A-T. Indeed, the neurological presentation may be later and progress more slowly. Therefore, although there is a correlation of neurology with the level of ATM signalling, there is no firm evidence that loss of signalling causes the neurological abnormality.

The reason for this retained MRE11 function is because cells without any of the separate components of the MRN complex (MRE11, RAD50 or NBN) cannot survive; they are essential genes, which ATM is not. Whereas the cell/organism can survive without *ATM*, it cannot survive when *MRE11* (or *NBN* or *RAD50*) is lost (Xiao and Weaver 1997, Luo et al. 1999, Zhu et al. 2001). Indeed, the fact that every ATLD patient carries one *MRE11* mutation that allows expression of MRE11 that contributes to the survival of the cell/organism means there must be a function(s) of MRE11 more fundamental for the cell than activation of ATM. That MRE11 (and also NBN and RAD50) are likely to have functions independent of ATM is also clear from the observation that in the absence of ATM the effects of hypomorphic mutations of *MRE11*, *RAD50* and *NBN* are lethal (Williams et al. 2002, Theunissen et al. 2003, Morales et al. 2005).

We can ask the question of what function of MRN (MRE11) is essential for viability. Homozygosity for the H129N MRE11 mutation in the nuclease domain in mice causes embryonic lethality, indicating an essential role for MRE11 nuclease activity. Interestingly, the H129N mutant protein is stable and able to interact in the MRN complex. However, the H129N/Del mouse cells shows a high level of chromosomal abnormalities, indicating a role for nuclease activity in preventing this genomic instability. The H129N mutant cells could carry out ATM signalling, indicating that the nuclease activity of MRE11 is not required for ATM activation. However, H129N cells are as radiosensitive as MRE11 Del/Del cells, indicating a role for nuclease function in the radiation response. Indeed, examining DNA DSB repair more directly using pulsed field gel electrophoresis showed H129N cells to be deficient in this repair, consistent with the idea that radio sensitivity most likely results from a deficient repair process involving MRE11 nuclease activity. Overall, nuclease activity of MRE11 is required for repair of IR-induced DNA DSBs, and this is separate from its role in ATM activation. Further, there is evidence that it may be a combined defect in DNA repair (both NHEJ and HRR for the repair of DNA DSB) in cells showing absence of ATM and presence of a hypomorphic MRN complex (NBN) mutation (Balestrini et al., 2016).

MRE11 is part of the MRN (MRE11-RAD50-NBN) complex, each component being an essential gene. For both Nijmegen breakage syndrome (caused by NBN mutation) and NBSLD (caused by RAD50 mutation), one of the mutant alleles allows expression of some NBN and RAD50, respectively. Interestingly, cells from both NBS and NBSLD patients show a reduction in ATM signalling, but they do not show the same neurological abnormality as A-T or MRE11 despite MRE11, RAD50 and NBN being part of the same complex. ATLD (MRE11), NBS (NBN) and NBSLD (RAD50) have quite different clinical phenotypes. The basis for this observation remains a mystery.

Nijmegen breakage syndrome (NBN 754 aa)

NBS patients are characterised by microcephaly, short stature, recurrent upper respiratory tract infections, immunodeficiency (**Table 13.1**) and a greatly increased risk of lymphoid tumours (see later).

The immunodeficiency is manifest as a severe hypogammaglobulinaemia found in a proportion of patients, and an IgA deficiency (Wolska-Kuśnierz et al. 2015). As in A-T, the most frequent finding is a reduction in absolute numbers of both B and T cells observed in the majority of patients. With respect to class switch recombination, like ATLD, there is a trend towards an increased usage of microhomology (≥4 bp) at the switch junctions in NBS patients (Pan et al. 2002, Lahdesmaki et al. 2004). In addition, class switch recombination is diminished in NBN-deficient murine B cells (Kracker et al. 2005, Reina-San-Martin et al. 2005). Chromosome translocations involving chromosomes 7 and 14 are observed in 10%–15% of stimulated T cells, as with A-T. NBS patient cells are unusually sensitive to IR by colony-forming assay and are also chromosomally radiosensitive (**Table 13.2**) (Taalman et al. 1983, Varon et al. 2017).

NBS has a very restricted geographical origin, with the vast majority of patients being of Slavic origin. It is caused by mutation of the *NBN* gene and all patients of Slavic origin are homozygous for the NBN mutation c.657_661del5, indicating the importance of this founder effect. A limited number, probably a dozen or so, of other NBS-associated NBN mutations have been identified worldwide. In the United Kingdom, there is homozygosity for c.698_701del4, p.Lys233SerfsTer4 NBN mutation, but this is extremely rare.

As stated earlier, NBN is an essential gene and so some mutant form of NBN is required for cellular viability; the 657del5 mutation results in a truncated NBN protein (p70 as opposed to p95). Indeed, the p70 truncated form of NBN is sufficient for survival. This is because the N-terminally deleted (loss of FHA and BRCT domains) p70 protein can still interact with MRE11 through a more C-terminal domain (aa433-754) (see **Figure 13.3**) to give some residual functional MRN complex (Maser et al. 2001). Unlike RAD50 and MRE11, NBN has no DNA binding capacity or enzyme activity; its role is to interact with other proteins, such as MDC1 and CtIP, through its N-terminal FHA and BRCT domains; CtIP then activates MRE11 endonuclease activity. NBN is also essential for the recruitment of MRE11, and recent work showed that mutation of the 108 aa MRE11 interacting domain of NBN alone was sufficient to result in loss of viability, indicating the essential nature of this interaction in the MRN complex (Kim et al. 2017).

NBS 657del5 cells show radio-resistant DNA synthesis, indicating that NBS1 p70 does not confer full function within the MRE11 complex. Clearly, the FHA and BRCT domains of NBN are required for the MRE11 complex to activate the S-phase checkpoint. Therefore, while 657del5 allows survival of NBS patients, the loss of NBN interactions through its N-terminus is likely to contribute to the pathology of the disease (Carney et al. 1998, Varon et al. 1998, Maser et al. 2001).

There are some very rare milder forms of NBS with different mutations. One such example is a patient homozygous for the NBN mutation c.741_742dupGG with a slightly milder phenotype where a transcript from in frame alternative splicing may have coded for a partially functional p70 NBN (Varon et al. 2006). Another example is a family with biallelic truncating *NBN* mutations (p.Tyr110Ter and pTrp375Ter) were reported where there were no features of NBS, although both affected individuals had fertility problems and, curiously, showed the cellular radio sensitivity phenotype typical of NBS (Warcoin et al. 2009).

The function of NBN is to interact with MRE11, which itself binds to RAD50. As indicated earlier, the MRE11Leu72 position was identified as necessary for both MRE11 dimer stability and interaction with NBN, and this reduction in interaction with NBN led to loss of ATM signalling. The most severe mutations in the MRE11 interaction domain of NBN prevented the MRE11–NBN interaction, resulting in loss of viability (Lafrance-Vanasse et al. 2015, Kim et al. 2017).

Nijmegen breakage syndrome–like disorder

NBSLD is caused by biallelic mutation of RAD50 (**Figure 13.3**), and only one case has so far been published, although others exist (Waltes et al. 2009). The case was originally published as a then 4-year-old girl of non-consanguineous descent with an unknown NBS-like disorder (Barbi et al. 1991). She showed microcephaly, learning difficulty, bird-like face and short stature, but had no problem with infections, immunodeficiency or cancer predisposition (**Table 13.1**). Spontaneous chromosomal instability was present that included characteristic translocations between chromosomes 7 and 14, cellular hypersensitivity to ionizing radiation and radio resistant DNA synthesis (**Table 13.2**). By age 15, there was evidence of impaired senso-motor coordination, which appeared as a subtle unsteadiness in straight-line walking, resembling a slight and non-progressive ataxia. Lymphocyte counts, subset distributions and lymphocyte response to mitogens were within normal limits, as was the level of serum alpha-fetoprotein. RAD50-deficient cells showed G1/S and intra S-phase checkpoint defects and accumulation in G2 after irradiation. RAD50-deficient cells from the patient showed deficient ATM signalling and increased radio sensitivity (by survival and chromosomally) very similar to ATLD.

Radio sensitivity, immunodeficiency, learning difficulty

In addition to mutation of ATM and the ATM-activating MRN complex, it is possible that other genes in the ATM pathway, when mutated, might cause an A-T related disorder.

Two patients with radio sensitivity, immunodeficiency, dysmorphic features and learning difficulty have so far been described with biallelic mutations in the *ATM* pathway gene, *RNF168*. In the first, an individual with absence of IgG, and requiring IgG replacement, also showed a growth deficiency and an unusual ataxic gait. The patient's cultured skin fibroblasts could be shown to be unusually radiosensitive by colony-forming assay. An increased level of unrepaired chromatid-type damage following G2 irradiation of his blood lymphocytes could be demonstrated. The RNF168-deficient cells showed a defect in the G2/M checkpoint and a mild intra-S phase checkpoint (RDS) defect following exposure to IR. Compared with ataxia telangiectasia, the increased radio sensitivity is mild, as measured by both colony-forming assay and G2 irradiation (Stewart et al. 2007, 2009). Another patient has been described who showed some clinical similarities to the first. A small increased radio sensitivity was demonstrated in lymphoblastoid cells in suspension from the patient (Devgan et al. 2011).

CANCER IN A-T AND ATLD

A-T Patients

In 296 consecutively analysed A-T patients, tumours were identified in 66 cases (~22%) (Reiman et al. 2011). The range of tumour types associated with A-T is quite remarkable and reminiscent of the spread of tumours in patients with Li-Fraumeni disorder and p53 mutation, although they show a different tumour spectrum. Tumour types in A-T that occur in either early childhood or adulthood are mainly lymphoid, although of different origins in childhood. Strikingly, T cell prolymphocytic leukaemia commonly occurs in young adult A-T patients. In

Table 13.3 Tumour types in A-T and A-T heterozygotes.

A-T Patients	A-T Heterozygotes
Lymphoid tumours	
AILD	
T-PLL	
T-ALL	
B cell lymphoma	
T cell lymphoma	
Burkitt-like lymphoma	
Hodgkin lymphoma	
Myeloma	
Non-lymphoid tumours	Non-lymphoid tumours
Myeloid leukaemia	
Brain tumours	
Astrocytoma	
Gangioglioma	
Medulloblastoma	
Carcinomas	Carcinomas
Breast	Breast
Hepatocellular carcinoma	Gastric
Pancreatic	Pancreatic
Thyroid	Prostate
Ectopic pituitary	Colorectal(?)

addition, brain tumours (in A-T children), endocrine tumours (thyroid, pancreatic) and, again strikingly, breast cancers in young adult A-T women (see Table 13.3) are present.

The variety of tumours in Li-Fraumeni disorder is associated with loss of the different tumour suppressor functions of p53, such as induction of apoptosis, cell cycle arrest, and cellular senescence. Similarly, it is likely that the different tumours in A-T have different causes related to loss of different tumour suppressor functions of ATM. There is now good evidence of a very substantially increased risk of breast cancer in both A-T patients at a young age and in *ATM* mutation carriers. A recent study showed the risk of an A-T patient at age 50 was 45%, comparable to the risk of being a BRCA1 carrier (Reiman et al. 2011).

It is possible that there is more than one mechanism for cancer predisposition in A-T. For example, it is likely that T-PLL (T-cell-prolymphocytic leukemia) arises as a result of an illegitimate recombination event in a lymphoid cell involving a break in an immune system gene and an oncogene to give a chromosome translocation that results in a growth advantage. The increased frequency of illegitimate translocations associated with A-T results in the increased predisposition to T-PLL and at an earlier age. In contrast, breast cancer could arise by a completely different mechanism either involving loss of heterozygosity and loss of function or, indeed, in a gain of function mutation. As a possible example of the latter, it is clear that the risk of breast cancer is not the same for all *ATM* mutations and is increased, for example, in the presence of the c.7271T>G (p.Val2424Gly) mutant ATM protein.

Interestingly, there is a reduced risk of lymphoid tumours in children whose cells express either a low level of normal ATM protein (e.g. from a leaky splice site mutation) or mutant protein (Reiman et al. 2011). Whether this is a tumour protective effect of ATM kinase activity or some other activity of ATM is not known.

Cancer risks in patients with ATLD

Apart from the unusual ATLD family showing two cases of lung cancer (Uchisaka et al. 2009), there is no evidence of an increased risk of cancer in ATLD patients, although the number of patients reported is small. If there is a tumour protective effect of ATM kinase activity in A-T, then the same protection might be expected in ATLD that also has residual ATM activity. Certainly there is no evidence for an increased risk of lymphoid tumours in this group of patients, although the numbers are small. Consequently, there may be lower frequency of tumours in ATLD patients compared with classical A-T. Consistent with this, a mouse model of the human MRE11 p.Arg633 stop mutation did not produce any lymphomas in these mice (Theunissen et al. 2003).

Cancer in NBS patients

As with A-T, NBS patients are particularly predisposed to lymphoid tumours, more or less equally split between B and T cell lymphomas. The risk of lymphoma is approximately 40% for NBS patients before age 20 (Varon et al. 2017), which is a higher rate than for A-T. Regarding the predisposition, it has to be borne in mind that most NBS patients are homozygous for the same mutation and do not show the allelic heterogeneity of A-T. Like A-T, other tumours in NBS include brain tumours (medulloblastoma and glioma) as well as rhabdomyosarcoma. Overall, there is probably a significantly greater cancer risk in NBS compared with A-T, although NBS patients do not show complete overlap with A-T in terms of tumour type. Whether there are cancer risks in NBSLD is not known because of the rarity of the disorder.

A-T Heterozygotes

It is now well established that female *ATM* mutation carriers have an increased risk of developing breast cancer (Geoffroy-Perez et al. 2001, Olsen et al. 2001, Thompson et al. 2005). The relative risk for breast cancer was 2.23 compared with the general population in the United Kingdom but was 4.94 in those under age 50 (Thompson et al. 2005). The increased risk of breast cancer is associated with both *ATM* truncating and missense mutations, but the spectrum of pathogenic variants includes a high proportion of missense mutations, as is the case for mutations in *CHEK2*, but different to BRCA1/2 where most of the susceptibility alleles are protein-truncating variants. Truncating variants in *ATM* were more strongly associated with risk of oestrogen receptor (ER)-positive than ER-negative disease (Decker et al. 2017). More recent data suggest that there is a further increased risk of breast cancer in carriers of particular types of *ATM* mutations, for example, the c.7271T>G, p.Gly2424Val *ATM* missense mutation (Goldgar et al. 2011, Southey et al. 2016). There was also some evidence of excess risks of colorectal cancer (RR = 2.54%, 95% CI = 1.06–6.09) and stomach cancer (RR = 3.39%, 95% CI = 0.86–13.4) in carriers (Thompson et al. 2005), and ATM has been identified as a clear pancreatic cancer susceptibility gene (Roberts et al. 2012, 2016; Bakker and de Winter 2012). More recent work has also confirmed the association of loss of function *ATM* mutations with gastric cancer, pancreatic and prostate cancer (Helgasson et al. 2015). Interestingly, there is no evidence of an increased risk of lymphoid tumours in carriers (Table 13.3).

MRN Heterozygotes

The evidence for increased risk of cancer associated with heterozygosity for an NBN mutation is not as strong as for ATM, although increased risk of breast, prostate, medulloblastoma and melanoma in NBN heterozygotes has been reported (Varon et al. 2017). More recently, however, MRE11, RAD50 and NBN were reported as intermediate risk breast cancer susceptibility alleles (Damiola et al. 2014).

A-T, ATLD, NBS AND NBSLD COMPARED

The closest clinical similarity may be between A-T and ATLD, although there is no evidence for tumour predisposition in ATLD patients, apart from an unusual family with lung cancer. The differences in clinical features of ATLD, NBS and NBSLD are perhaps unexpected considering that they each involve mutation of a protein that is part of a single complex. However, the rarity of both ATLD and NBSLD, on the one hand, and the genetic homogeneity of NBS, on the other, are confounding factors in understanding this.

In the commonest disorder, A-T, the ATM gene, which is more than 4 times larger than MRE11, has been shown to be mutated in hundreds of places across its entire coding sequence to give A-T, which is seen worldwide. Over 250 different ATM mutations have been seen in A-T patients in the United Kingdom alone. In contrast, MRE11, RAD50 and NBN are essential genes. Consequently, the combinations of mutations that can give rise to recessive disorders of these genes is restricted because each has to produce sufficient protein to allow viability of the cell and organism. ATLD (MRE11) and NBSLD (RAD50) are, indeed, very rare, and NBS (NBN) has arisen from a single founder mutation in over 90% of cases. There are only a dozen or so other known NBN mutations giving typical NBS, presumably all of them resulting in a C-terminal protein that can interact with MRE11. Therefore, the number and type of mutation that can give NBS is restricted by the fact that expression of NBN with an MRE11 interacting domain is required for viability of the organism. It is possible that the heterozygote frequency might be significantly higher than can be estimated from the frequency of affected individuals, although inspection of the ExAC database of exome sequencing of >60,000 individuals from different populations does not suggest this is the case. Interestingly, the ratio of the MRE11Arg572Ter, the most common loss of function MRE11 mutation in the ExAC database, to the MRE11 missense mutation Asn117Ser (with retained function) is 9:1.

Can we say what the cause of the cerebellar neurodegeneration is in A-T? In particular, is it a consequence of the DNA BSB repair defect? Interestingly, the A-T mouse does not show the neurological characteristic which is the hallmark of A-T (Barlow et al. 1996b). The reason for this is not understood. The simplest explanation is that the cellular defect associated with ATM loss, inability to repair DNA DSB, is not the cause of the neurological degeneration even in A-T. But this would need us to assume an equivalence between humans and mice that may not exist. It is not that mice cannot develop progressive cerebella ataxia; indeed, paradoxically, this is part of the phenotype of a murine model of a brain-specific NBN knock down that shows both microcephaly and also progressive cerebellar ataxia (Frappart et al. 2005). Therefore, we know that the mouse can develop neurological features similar to ataxia telangiectasia in response to loss of a protein closely associated with ATM in the DNA DSB response – just not ATM itself.

The complexities of the multifunctional nature of these individual proteins, the essential nature of some but not other genes and the rarity of different MRE11-RAD50-NBN mutations all combine to contribute to the difficulty of relating one particular clinical feature to one particular function. Nevertheless, little by little, progress will be made in achieving this, and hopefully this new understanding will bring promise of treatment especially for the most common of these disorders, ataxia telangiectasia.

SUMMARY

This chapter discussed the disorders associated with defects in the ATM pathway, principally ataxia telangiectasia but also the quite different disorders resulting from deficiency in each of the components of the MRN complex. An interesting added feature of these components, in contrast to ATM, is that each is an essential protein and therefore the mutation is present as a hypomorphic allele in the three groups of MRN patients. They are very rare and NBS is distinguished by being, largely, the consequence of a founder mutation. In ataxia telangiectasia, there are hundreds of ATM mutations. The group of patients is large enough, the types of mutation wide enough and the clinical features varied enough so that good phenotype-genotype correlations are apparent, and indeed there is evidence for the effects of modifying genes.

Careful analysis of mutations in these patients and the cellular features associated with each disorder has contributed enormously to our understanding of the DNA damage response pathways in humans. With respect to the disorders, less progress has been made in understanding how ATM loss, for example, results in cerebellar degeneration, and the mouse model A-T has not been very helpful in this. Similarly, it is not known why MRE11 deficiency leads to an A-T–like disorder clinically (although there are exceptions) and NBN deficiency to a different clinical endpoint. It is also likely that additional functions of these proteins will be revealed to explain these features.

ATM can be lost or reduced in the cell in ways not caused directly by ATM mutation and this may result in new associated disorders being recognised. Are there other ATM pathway genes awaiting discovery? We do not know the answer to this question, and bearing in mind the heterogeneity of clinical presentation described earlier for these disorders, we cannot guess what their phenotype might be, although some cellular features in common are expected.

REFERENCES

Bakkenist, C. J., Kastan, M. B. 2003. DNA damage activates ATM through intermolecular autophosphorylation and dimer association. *Nature* 421: 499–506.

Bakker, J. L., de Winter, J. P. 2012 Jan. A role for ATM in hereditary pancreatic cancer. *Cancer Discov* 2(1): 14–5.

Balestrini, A., Nicolas, L., Yang-lott, K., Guryanova, O. A., Levine, R. L., Bassing, C. H., Chaudhuri, J., Petrini, J. H. J. 2016 Feb. Defining ATM-independent functions of the MRE11 complex with a novel mouse model. *Mol Cancer Res* 14(2): 185–195.

Barbi, G., Scheres, J. M., Schindler, D., Taalman, R. D., Rodens, K., Mehnert, K., Muller, M., Seyschab, H. 1991. Chromosome instability and x-ray hypersensitivity in a microcephalic and growth-retarded child. *Am J Med Genet* 40: 44–50.

Barlow, C., Hirotsune, S., Paylor, R., Liyanage, M., Eckhaus, M., Collins, F., Shiloh, Y., et al. 1996 Jul 12. ATM-deficient mice: A paradigm of ataxia telangiectasia. *Cell* 86(1): 159–171.

Beucher, A., Birraux, J., Tchouandong, L., Barton, O., Shibata, A., Conrad, S., Goodarzi, A. A., Krempler, A., Jeggo, P. A., Löbrich, M. 2009 Nov 4. ATM and Artemis promote homologous recombination of radiation-induced DNA double-strand breaks in G2. *EMBO J* 28(21): 3413–3427.

Bredemeyer, A. L., Sharma, G. G., Huang, C. Y., Helmink, B. A., Walker, L. M., Khor, K. C., Nuskey, B., et al. 2006 Jul 27. ATM stabilizes DNA double-strand-break complexes during V(D)J recombination. *Nature* 442(7101): 466–470.

Byrd, P. J., Srinivasan, V., Last, J. I., Smith, A., Biggs, P., Carney, E. F., Exley, A., et al. 2012. Severe reaction to radiotherapy for breast cancer as the presenting feature of ataxia telangiectasia. *Br J Cancer* 106(2): 262–268.

Carney, E. F. S., Moss, P. A. V., Taylor, A. M. 2012. Classical ataxia telangiectasia patients have a congenitally aged immune system with high expression of CD95. *J Immunol* 189(1): 261–268.

Carney J. P., Maser, R. S., Olivares, H., Davis, E. M., Le Beau, M., Yates, J. R. 3rd, Hays, L., Morgan, W. F., Petrini, J. H. 1998. The hMRE11/hRAD50 protein complex and Nijmegen breakage syndrome: Linkage of double-strand break repair to the cellular DNA damage response. *Cell* 93(3): 477–486.

Carrillo, F., Schneider, S. A., Taylor, A. M., Srinivasan, V., Kapoor, R., Bhatia, K. P. 2009. Prominent oromandibular dystonia and pharyngeal telangiectasia in atypical ataxia telangiectasia. *Cerebellum* 8(1): 22–27.

Chaumeil, J., Micsinai, M., Ntziachristos, P., Roth, D. B., Aifantis, I., Kluger, Y., Deriano, L., Skok, J. A. 2013. The RAG2 C-terminus and ATM protect genome integrity by controlling antigen receptor gene cleavage. *Nat Commun* 4: 2231.

Chiam, L. Y., Verhagen, M. M., Haraldsson, A., Wulffraat, N., Driessen, G. J., Netea, M. G., Weemaes, C. M., Seyger, M. M., van Deuren, M. 2011. Cutaneous granulomas in ataxia telangiectasia and other primary immunodeficiencies: Reflection of inappropriate immune regulation? *Dermatology* 223(1): 13–19.

Crawford, T. O., Mandir, A. S., Lefton-Greif, M. A., Goodman, S. N., Sengul, H., Lederman, H. M. 2000. Quantitative neurologic assessment of ataxia telangiectasia. *Neurology* 54: 1505–1509.

Damiola, F., Pertesi, M., Oliver, J., Le Calvez-Kelm, F., Voegele, C., Young, E. L., Robinot, N., et al. 2014 Jun 3. Rare key functional domain missense substitutions in MRE11A, RAD50, and NBN contribute to breast cancer susceptibility: Results from a Breast Cancer Family Registry case-control mutation-screening study. *Breast Cancer Res* 16(3): R58.

Decker, B., Allen, J., Luccarini, C., Pooley, K. A., Shah, M., Bolla, M. K., Wang, Q. et al. 2017 Aug 4. Rare, protein-truncating variants in ATM, CHEK2 and PALB2, but not XRCC2, are associated with increased breast cancer risks. *J Med Genet.* doi: 10.1136/jmedgenet-2017-104588. [Epub]

Delia, D., Piane, M., Buscemi, G., et al. 2004. MRE11 mutations and impaired ATM-dependent responses in an Italian family with ataxia-telangiectasia-like disorder. *Hum Mol Genet* 13: 2155–2163.

Deshpande, R. A., Williams, G. J., Limbo, O., Williams, R. S., Kuhnlein, J., Lee, J. H., Classen, S., et al. 2016 Apr 1. ATP-driven RAD50 conformations regulate DNA tethering, end resection, and ATM checkpoint signaling. *EMBO J* 35(7): 791.

Devgan, S. S., Sanal, O., Doil, C., Nakamura, K., Nahas, S. A., Pettijohn, K., Bartek, J., Lukas, C., Lukas, J., Gatti, R. A. 2011. Homozygous deficiency of ubiquitin-ligase ring-finger protein RNF168 mimics the radiosensitivity syndrome of ataxia-telangiectasia. *Cell Death Differ* 18(9): 1500–1506.

Exley, A. R., Buckenham, S., Hodges, E., Hallam, R., Byrd, P., Last, J., Trinder, C., et al. 2011. Premature ageing of the immune system underlies immunodeficiency in ataxia telangiectasia. *Clin Immunol* 140(1): 26–36.

Fernet, M., Gribaa, M., Salih, M. A., Seidahmed, M. Z., Hall, J., Koenig, M. 2005. Identification and functional consequences of a novel MRE11 mutation affecting 10 Saudi Arabian patients with the ataxia telangiectasia-like disorder. *Hum Mol Genet* 14: 307–318.

Frappart, P. O., Tong, W. M., Demuth, I., Radovanovic, I., Herceg, Z., Aguzzi, A., Digweed, M., Wang, Z. Q. 2005 May. An essential function for NBS1 in the prevention of ataxia and cerebellar defects. *Nat Med* 11(5): 538–44.

Geoffroy-Perez, B., Janin, N., Ossian, K., Laug, A., Croquette, M. F., Griscelli, C., et al. 2001. Cancer risk in heterozygotes for ataxia-telangiectasia. *Int J Cancer* 93: 288–293.

Goldgar, D. E., Healey, S., Dowty, J. G., Da Silva, L., Chen, X., Spurdle, A. B., Terry, M. B. et al. 2011 Jul 25. Rare variants in the ATM gene and risk of breast cancer. *Breast Cancer Res* 13(4): R73.

Guo, Z., Kozlov, S., Lavin, M. F., Person, M. D., Paull, T.T. 2010. ATM activation by oxidative stress. *Science* 330(6003):517–521.

Hartlerode, A. J., Morgan, M. J., Wu, Y., Buis, J., Ferguson, D. O. 2015. Recruitment and activation of the ATM kinase in the absence of DNA-damage sensors. *Nat Struct Mol Biol* 22(9): 736–744.

Helgason, H., Rafnar, T., Olafsdottir, H. S., Jonasson, J. G., Sigurdsson, A., Stacey, S. N., Jonasdottir, A., et al. 2015 Aug. Loss-of-function variants in ATM confer risk of gastric cancer. *Nat Genet* 47(8): 906–10.

Hernandez, D., McConville, C. M., Stacey, M., Woods, C. G., Brown, M. M., Shutt, P., Rysiecki, G., Taylor, A. M. 1993. A family showing no evidence of linkage between the ataxia telangiectasia gene and chromosome 11q22-23. *J Med Genet* 30(2): 135–140.

Hopfner, K. P., Karcher, A., Shin, D. S., Craig, L., Arthur, L. M., Carney, J. P., Tainer, J. A. 2000 Jun 23. Structural biology of RAD50 ATPase: ATP-driven conformational control in DNA double-strand break repair and the ABC-ATPase superfamily. *Cell* 101(7): 789–800.

Hustedt, N., Durocher, D. 2017. The control of DNA repair by the cell cycle. *Nat Cell Biol* 19: 1–9.

Kim, J. H., Grosbart, M., Anand, R., Wyman, C., Cejka, P., Petrini, J. H. J. 2017 Jan 10. The MRE11-Nbs1 interface is essential for viability and tumor suppression. *Cell Rep* 18(2): 496–507.

Klein, C., Wenning, G. K., Quinn, N. P., Marsden, C. D. 1996. Ataxia without telangiectasia masquerading as benign hereditary chorea. *Mov Disord* 11(2): 217–220.

Kracker, S., Bergmann, Y., Demuth, I., Frappart, P. O., Hildebrand, G., Christine, R., Wang, Z. Q., Sperling, K., Digweed, M., Radbruch, A. 2005 Feb 1. Nibrin functions in Ig class-switch recombination. *Proc Natl Acad Sci USA* 102(5): 1584–1589.

Lafrance-Vanasse, J., Williams, G. J., Tainer, J. 2015 Mar. Envisioning the dynamics and flexibility of MRE11-RAD50-Nbs1 complex to decipher its roles in DNA replication and repair. *Prog Biophys Mol Biol* 117(2–3): 182–193.

Lahdesmaki, A., Taylor, A. M., Chrzanowska, K. H., Pan-Hammarstrom, Q. 2004. Delineation of the role of the MRE11 complex in class switch recombination. *J Biol Chem* 279: 16479–16487.

Lefton-Greif, M. A., Crawford, T. O., Winkelstein, J. A., Loughlin, G. M., Koerner, C. B., Zahurak, M., Lederman, H. M. 2000. Oropharyngeal dysphagia and aspiration in patients with ataxia-telangiectasia. *J Pediatr* 136(2): 225–231.

Limbo, O., Moiani, D., Kertokalio, A., Wyman, C., Tainer, J. A., Russell, P. 2012 Dec. MRE11 ATLD17/18 mutation retains Tel1/ATM activity but blocks DNA double-strand break repair. *Nucleic Acids Res* 40(22): 11435–49.

Luo, G., Yao, M. S., Bender, C. F., Mills, M., Bladl, A. R., Bradley, A. 1999 Jun 22. Petrini JH Disruption of mRAD50 causes embryonic stem cell lethality, abnormal embryonic development, and sensitivity to ionizing radiation. *Proc Natl Acad Sci USA* 96(13): 7376–7381.

Lumsden, J. M., McCarty, T., Petiniot, L. K., Shen, R., Barlow, C., Wynn, T. A., Morse, H. C. 3rd, et al. 2004 Nov 1. Immunoglobulin class switch recombination is impaired in ATM-deficient mice. *J Exp Med* 200(9):1111–1121.

Maser, R. S., Zinkel, R., Petrini, J. H. 2001 Apr. An alternative mode of translation permits production of a variant NBS1 protein from the common Nijmegen breakage syndrome allele. *Nat Genet* 27(4): 417–421.

Matsumoto, Y., Miyamoto, T., Sakamoto, H., Izumi, H., Nakazawa, Y., Ogi, T., Tahara, H., et al. 2011. Two unrelated patients with MRE11A mutations and Nijmegen breakage syndrome-like severe microcephaly. *DNA Repair (Amst)* 10(3): 314–321.

McConville, C. M., Stankovic, T., Byrd, P. J., McGuire, G. M., Yao, Q. Y., Lennox, G. G., Taylor, M. R. 1996. Mutations associated with variant phenotypes in ataxia-telangiectasia. *Am J Hum Genet* 59(2): 320–330.

McGrath-Morrow, S. A., Gower, W. A., Rothblum-Oviatt, C., Brody, A. S., Langston, C., Fan, L. L., Lefton-Greif, M. A., et al. 2010. Evaluation and management of pulmonary disease in ataxia-telangiectasia. *Pediatr Pulmonol* 45(9): 847–859.

Metcalfe, J. A., Parkhill, J., Campbell, L., Stacey, M., Biggs, P., Byrd, P. J., Taylor, A. M. R. 1996. Accelerated telomere shortening in ataxia telangiectasia. *Nature Genet* 13: 350–353.

Morales, M., Theunissen, J. W., Kim, C. F., Kitagawa, R., Kastan, M. B., Petrini, J. H. 2005. The RAD50S allele promotes ATM-dependent DNA damage responses and suppresses ATM deficiency: Implications for the MRE11 complex as a DNA damage sensor. *Genes Dev* 19: 3043–3054.

Olsen, J. H., Hahnemann, J. M., Borresen-Dale, A. L., Brondum-Nielsen, K., Hammarstrom, L., Kleinerman, R., Kleinerman, R., Kaariainen, H., et al. 2001. Cancer in patients with ataxia-telangiectasia and in their relatives in the nordic countries. *J Natl Cancer Inst* 93: 121–127.

Pan, Q., Petit-Frére, C., Lähdesmäki, A., Gregorek, H., Chrzanowska, K. H., Hammarström, L. 2002 May. Alternative end joining during switch recombination in patients with ataxia-telangiectasia. *Eur J Immunol* 32(5): 1300–1308.

Pan-Hammarström, Q., Dai, S., Zhao, Y., van Dijk-Härd, I. F., Gatti, R. A., Børresen-Dale, A. L., Hammarström, L. 2003. ATM is not required in somatic hypermutation of VH, but is involved in the introduction of mutations in the switch mu region. *J Immunol* 170(7): 3707–3716.

Pan-Hammarström, Q., Lähdesmäki, A., Zhao, Y., Du, L., Zhao, Z., Wen, S., Ruiz-Perez, V. L., Dunn-Walters, D. K., Goodship, J. A., Hammarström, L. 2006. Disparate roles of ATR and ATM in immunoglobulin class switch recombination and somatic hypermutation. *J Exp Med* 203(1): 99–110. Erratum in: J Exp Med. 2006; 203(1):251.

Regal, J. A., Festerling, T. A., Buis, J. M., Ferguson, D. O. 2013 Dec 20. Disease-associated MRE11 mutants impact ATM/ATR DNA damage signaling by distinct mechanisms. *Hum Mol Genet* 22(25): 5146–5159.

Reiman, A., Srinivasan, V., Barone, G., Last, J. I., Wootton, L. L., Davies, E. G., Verhagen, M. M., et al. 2011. Lymphoid tumours and breast cancer in ataxia telangiectasia;substantial protective effect of residual ATM kinase activity against childhood tumours. *Br J Cancer* 105(4): 586–591.

Reina-San-Martin, B., Chen, H. T., Nussenzweig, A., Nussenzweig, M. C. 2004 Nov 1. ATM is required for efficient recombination between immunoglobulin switch regions. *J Exp Med* 200(9): 1103–1110.

Reina-San-Martin, B., Nussenzweig, M. C., Nussenzweig, A., Difilippantonio, S. 2005 Feb 1. Genomic instability, endoreduplication, and diminished Ig class-switch recombination in B cells lacking Nbs1. *Proc Natl Acad Sci USA* 102(5): 1590–1595.

Roberts, N. J., Jiao, Y., Yu, J., Kopelovich, L., Petersen, G. M., Bondy, M. L., Gallinger, S., et al. 2012 Jan. ATM mutations in patients with hereditary pancreatic cancer. *Cancer Discov* 2(1): 41–46.

Roberts, N. J., Norris, A. L., Petersen, G. M., Bondy, M. L., Brand, R., Gallinger, S., Kurtz, R. C. et al. 2016 Feb. Whole genome sequencing defines the genetic heterogeneity of familial pancreatic cancer. *Cancer Discov* 6(2):166–175.

Saunders-Pullman, R., Raymond, D., Stoessl, A. J., Hobson, D., Nakamura, K., Pullman, S., Lefton, D., et al. 2012. Variant ataxia-telangiectasia presenting as primary-appearing dystonia in Canadian Mennonites. *Neurology* 78(9): 649–657.

Schiller, C. B., Lammens, K., Guerini, I., Coordes, B., Feldmann, H., Schlauderer, F., Möckel, C., et al. 2012 Jun 17. Structure of MRE11-Nbs1 complex yields insights into ataxia-telangiectasia-like disease mutations and DNA damage signaling. *Nat Struct Mol Biol* 19(7): 693–700.

Sedgwick, R. P., Boder, E. 1991. Ataxia telangiectasia. In: de Jong, J. M. B. V., editor, *Handbook of Clinical Neurology: Hereditary Neuropathies and Spinocerebellar Atrophies*, 347–342. Elsevier Science Publishers BV, Oxford.

Southey, M. C., Goldgar, D. E., Winqvist, R., Pylkäs, K., Couch, F., Tischkowitz, M., Foulkes, W. D., et al. 2016 Dec. PALB2, CHEK2 and ATM rare variants and cancer risk: Data from COGS. *J Med Genet* 53(12): 800–811.

Stankovic, T., Kidd, A. M., Sutcliffe, A., McGuire, G. M., Robinson, P., Weber, P., Bedenham, T., et al. 1998. ATM mutations and phenotypes in ataxia-telangiectasia families in the British Isles: Expression of mutant ATM and the risk of leukemia, lymphoma, and breast cancer. *Am J Hum Genet* 62(2): 334–345.

Staples, E. R., McDermott, E. M., Reiman, A., Byrd, P. J., Ritchie, S., Taylor, A. M. R., Davies, E. G. 2008. Immunodeficiency in ataxia telangiectasia is strongly correlated with the presence of two null mutations in the ATM gene. *Clin Exp Immunol* 153: 214–220.

Stewart, G. S., Last, J. I., Stankovic, T., Haites, N., Kidd, A. M., Byrd, P. J., Taylor, A. M. 2001. Residual ataxia telangiectasia mutated protein function in cells from ataxia telangiectasia patients, with 5762ins137 and 7271T-->G mutations, showing a less severe phenotype. *J Biol Chem* 276: 30133–30141.

Stewart, G. S., Maser, R. S., Stankovic, T., Bressan, D. A., Kaplan, M. I., Jaspers, N. G., Raams, A., Byrd, P. J., Petrini, J. H., Taylor, A. M. et al. 1999. The DNA double-strand break repair gene hMRE11 is mutated in individuals with an ataxia-telangiectasia-like disorder. *Cell* 99: 577–587.

Stewart, G. S., Panier, S., Townsend, K., Al-Hakim, A. K., Kolas, N. K., Miller, E. S., Nakada, S., et al. 2009. The RIDDLE syndrome protein mediates a ubiquitin-dependent signaling cascade at sites of DNA damage. *Cell* 136(3): 420–434.

Stewart, G. S., Stankovic, T., Byrd, P. J., Wechsler, T., Miller, E. S., Huissoon, A., Drayson, M. T., West, S. C., Elledge, S. J., Taylor, A. M. R. 2007. RIDDLE immunodeficiency syndrome is linked to defects in 53BP1-mediated DNA damage signaling. *Proc Natl Acad Sci USA* 104: 16910–16915.

Sutton, I. J., Last, J. I., Ritchie, S. J., Harrington, H. J., Byrd, P. J., Taylor, A. M. 2004. Adult-onset ataxia telangiectasia due to ATM 5762ins137 mutation homozygosity. *Ann Neurol.* 55(6): 891–895.

Taalman, R. D., Jaspers, N. G., Scheres, J. M., de Wit, J., Hustinx, T. W. 1983. Hypersensitivity to ionizing radiation, *in vitro*, in a new chromosomal breakage disorder, the Nijmegen breakage syndrome. *Mutat Res* 112(1): 23–32.

Tankimanova, M., Capper, R., Letsolo, B. T., Rowson, J., Jones, R. E., Britt-Compton, B., Taylor, A. M., Baird, D. M. 2012 Mar. MRE11 modulates the fidelity of fusion between short telomeres in human cells. *Nucleic Acids Res* 40(6): 2518–2526.

Taylor, A. M., Groom, A., Byrd, P. J. 2004. Ataxia-telangiectasia-like disorder (ATLD) – Its clinical presentation and molecular basis. *DNA Repair (Amst)* 3(8–9): 1219–1225.

Taylor, A. M. R., Lam, Z., Last, J. I., Byrd, P. J. 2015. Ataxia telangiectasia: More variation at clinical and cellular levels. *Clin Genet* 87: 199–208.

Theunissen, J. W., Kaplan, M. I., Hunt, P. A., Williams, B. R., Ferguson, D. O., Alt, F. W., Petrini, J. H. et al. 2003. Checkpoint failure and chromosomal instability without lymphomagenesis in MRE11(ATLD1/ATLD1) mice. *Mol Cell* 12: 1511–1523.

Thompson, D., Duedal, S., Kirner, J., McGuffog, L., Last, J., Reiman, A., Byrd, P., Taylor, M., Easton, D. F. 2005. Cancer risks and mortality in heterozygous ATM mutation carriers. *J Natl Cancer Inst* 97(11): 813–822.

Uchisaka, N., Takahashi, N., Sato, M., Kikuchi, A., Mochizuki, S., Imai, K., Nonoyama, S., et al. 2009. Two brothers with ataxia-telangiectasia-like disorder with lung adenocarcinoma. *J Pediatr* 155(3): 435–4438.

van Os, N. J. H., Jansen, A. F. M., van Deuren, M., Haraldsson, A., van Driel, N. T. M., Etzioni, A., van der Flier, M., et al. 2017 May. Ataxia-telangiectasia: Immunodeficiency and survival. *Clin Immunol* 178: 45–55.

Varon, R., Demuth, I., Chrzanowska, K. H. 1999 May 17 (updated 2017 Feb 2). Nijmegen breakage syndrome. In: Pagon, R. A., Adam, M. P., Ardinger, H. H., Wallace, S. E., Amemiya, A., Bean, L. J. H. et al. editors. *GeneReviews®[Internet]*. Seattle, University of Washington. https://www.ncbi.nlm.nih.gov/books/NBK1176/.

Varon, R., Dutrannoy, V., Weikert, G., Tanzarella, C., Antoccia, A., Stöckl, L., Spadoni, E., et al. 2006 Mar 1. Mild Nijmegen breakage syndrome phenotype due to alternative splicing. *Hum Mol Genet* 15(5): 679–689.

Varon, R., Vissinga, C., Platzer, M., Cerosaletti, K. M., Chrzanowska, K. H., Saar, K., Beckmann, G., et al. 1998. Nibrin, a novel DNA double-strand break repair protein, is mutated in Nijmegen breakage syndrome. *Cell* 93(3): 467–476.

Verhagen, M. M., Martin, J. J., van Deuren, M., Ceuterick-de Groote, C., Weemaes, C. M., Kremer, B. H., Taylor, M. A., Willemsen, M. A., Lammens, M. 2012. Neuropathology in classical and variant ataxia-telangiectasia. *Neuropathology* 32(3): 234–244.

Waltes, R., Kalb, R., Gatei, M., Kijas, A. W., Stumm, M., Sobeck, A., Wieland, B., et al. 2009 May. Human RAD50 deficiency in a Nijmegen breakage syndrome-like disorder. *Am J Hum Genet* 84(5): 605–616.

Warcoin, M. 1., Lespinasse, J., Despouy, G., Dubois d'Enghien, C., Laugé, A., Portnoï, M. F., Christin-Maitre, S., Stoppa-Lyonnet, D., Stern, M. H. 2009 Mar. Fertility defects revealing germline biallelic nonsense NBN mutations. *Hum Mutat* 30(3): 424–430.

Williams, B. R., Mirzoeva, O. K., Morgan, W. F., Lin, J., Dunnick, W., Petrini, J. H. 2002. A murine model of Nijmegen breakage syndrome. *Curr Biol* 12: 648–653.

Williams, R. S., Moncalian, G., Williams, J. S., Yamada, Y., Limbo, O., Shin, D. S., Groocock, L. M., et al. 2008 Oct 3. MRE11 dimers coordinate DNA end bridging and nuclease processing in double-strand-break repair. *Cell* 135(1): 97–109.

Wolska-Kuśnierz, B., Gregorek, H., Chrzanowska, K., Piątosa, B., Pietrucha, B., Heropolitańska-Pliszka, E., Pac, M., et al. 2015 Aug. Nijmegen breakage syndrome: Clinical and immunological features, long-term outcome and treatment options – A retrospective analysis. *J Clin Immunol* 35(6): 538–549.

Woods, C. G., Bundey, S. E., Taylor, A. M. 1990. Unusual features in the inheritance of ataxia telangiectasia. *Hum Genet* 84(6): 555–562.

Worth, P., Srinivasan, V., Smith, A., Last, J. I., Wootton, L. L., Biggs, P. M., Davies, N. P., Carney, E. F., Byrd, P. J., Taylor, A. M. R. 2013. Very mild presentation in an adult with the classical cellular phenotype of ataxia telangiectasia. *Mov Disord* 28(4): 524–528.

Xiao, Y., Weaver, D. T. 1997 Aug 1. Conditional gene targeted deletion by Cre recombinase demonstrates the requirement for the double-strand break repair MRE11 protein in murine embryonic stem cells. *Nucleic Acids Res* 25(15): 2985–2991.

Yoshida, T., Awaya, T., Shibata, M., Kato, T., Numabe, H., Kobayashi, J., Komatsu, K., Heike, T. 2014 July. Hypergonadotropic hypogonadism and hypersegmented neutrophils in a patient with ataxia-telangiectasia-like disorder: Potential diagnostic clues? *Am J Med Genet A* 164A(7): 1830–1834.

Zhu, J., Petersen, S., Tessarollo, L., Nussenzweig, A. 2001 Jan 23. Targeted disruption of the Nijmegen breakage syndrome gene NBS1 leads to early embryonic lethality in mice. *Curr Biol* 11(2): 105–109.

DNA Repair Mechanisms in Stem Cells and Implications during Ageing

14

Rachel Bayley and Paloma Garcia

INTRODUCTION

The origins of mature and specialised cells found within multicellular organisms can all be traced back to stem cells. Stem cells are defined by two main properties: (1) their ability to reproduce themselves over long time periods while maintaining an unspecialised phenotype and (2) their ability to generate daughter cells capable of becoming specialised and differentiated cells. In mammalian organisms, embryonic stem cells (ESCs) can be found within the inner cell mass of the embryo, where they proliferate and differentiate to generate all the tissues of the developing organism. Following embryogenesis, a population of adult stem cells (ASCs) reside in the fully formed tissues where they internally repair and replenish lost or damaged cells throughout the organism's lifespan. Aside from these naturally occurring types of stem cells, another type has been produced in vitro by reprogramming adult cells to an immature 'stem cell-like' state to give rise to induced pluripotent stem cells (iPSCs) (Takahashi and Yamanaka 2006). iPSCs have been generated with a view to studying stem cell differentiation and function and to be used as a source of stem cells for the treatment of human disease. Given that stem cells have the potential to produce all the mature and specialised cells required for the development of an organism, it is essential they maintain genomic stability through the use of efficient DNA repair pathways. If DNA repair in stem cells is compromised, this could have potentially harmful effects in the stem cell pool itself, or further downstream in the production of defective and genomically unstable mature cells leading to a range of diseases.

As stem cells are essential at all stages of life, long-term stem cell function must be preserved, and one of the ways this is achieved is through efficient DNA repair mechanisms. Prolonged stem cell function is important from a clinical perspective because disturbances in this can lead to disease development, for example, perturbed functioning of haematopoietic stem cells (HSCs) responsible for producing mature blood cells is associated with the development of blood disorders, including cancer and pre-cancerous conditions in humans (Risitano et al. 2007). Interestingly, many of these diseases have an increased incidence in the ageing population, highlighting that changes to stem cell function over time are an important area of study for the prevention and treatment of age-related diseases (Aul et al. 2001). It has been hypothesised that one of the reasons stem cell function declines with age is due to the accumulation of DNA damage over time (Behrens et al. 2014). This could be due to changes within the stem cell microenvironment (known as the niche) with ageing (Drummond-Barbosa 2008), but also changes in DNA damage sensing and repair mechanisms within these cells, leading to increased levels of DNA damage (Beerman et al. 2014). In order to prevent and treat diseases associated with a decline in stem cell function during ageing, it is essential to have knowledge of the DNA repair mechanisms in stem cells and how they are affected by ageing.

Similarly to other cell types, stem cells are constantly exposed to exogenous and endogenous DNA damaging agents and must be prepared to respond by activating DNA repair mechanisms to prevent genomic instability. For example, it has been estimated that each human cell can experience up to 10,000 endogenous oxidative DNA damage events alone each day (Bernstein et al. 2013), highlighting the requirement for repair pathways to combat these potentially harmful events. In many cases, endogenous DNA damaging events occur as part of normal cellular processes such as DNA replication and metabolism, which can generate reactive oxygen species (Chance et al. 1979), but nonetheless possess the potential

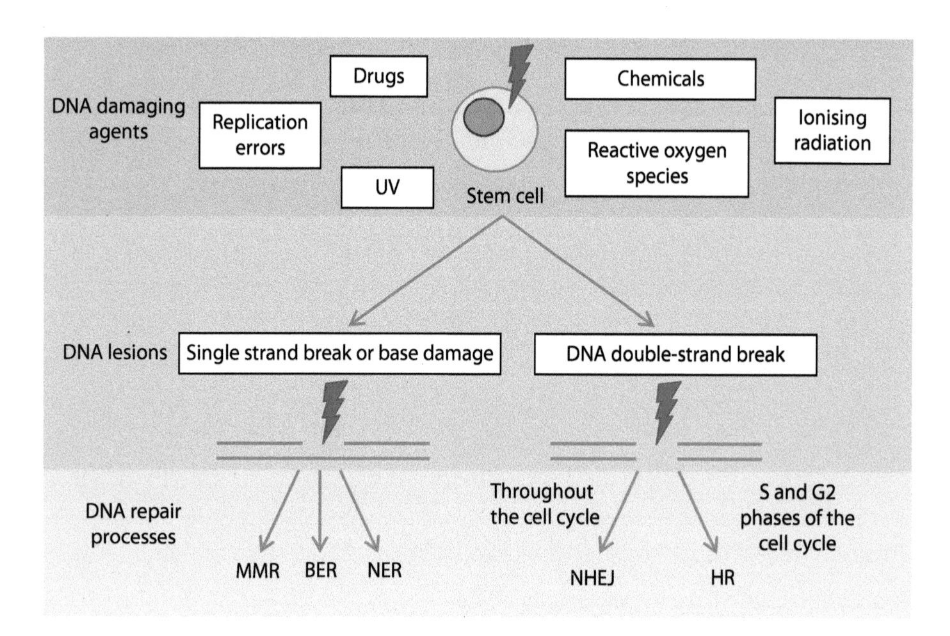

Figure 14.1 Summary of the DNA repair mechanisms used by stem cells to combat different types of DNA damage. Stem cells are constantly being exposed to endogenous and exogenous sources of DNA damage, including errors in DNA replication, reactive oxygen species, ionising radiation, UV, chemicals and drugs. To combat this DNA damage, stem cells can activate DNA repair pathways depending upon the type of lesion. Base mismatches, base adducts and single-strand breaks present in the DNA can be repaired by MMR, NER and BER, respectively, whereas double-strand breaks in the DNA can be repaired by NHEJ and HR. UV, ultraviolet; MMR, mismatch repair; BER, base excision repair; NER, nucleotide excision repair; NHEJ, non-homologous end joining; HR, homologous recombination.

to be mutagenic. In addition to this, there is also the potential for stem cells to be affected by exogenous sources of DNA damage, including exposure to chemicals, ultraviolet (UV) light and ionising radiation (IR).

Stem cells use a number of pathways to sense and repair damaged DNA, allowing maintenance of genomic stability and ultimately preserving their lifelong function (Figure 14.1). These processes are not dissimilar to those employed by other types of mammalian cells and are activated in response to the type of DNA lesion. They can be broadly divided into (1) single-stranded DNA repair mechanisms, including mismatch repair (MMR), nucleotide excision repair (NER) and base excision repair (BER) (Caldecott 2008); and (2) double-stranded DNA repair mechanisms, including non-homologous end joining (NHEJ) and homologous recombination (HR) (Shrivastav et al. 2008). These processes are important in the three main types of stem cells: (1) ESCs, (2) ASCs and (3) iPSCs. For example, unrepaired DNA in ESCs could lead to abnormalities within the developing embryo leading to birth defects, as observed in many inherited disorders of DNA repair (Knoch et al. 2012). If DNA repair is compromised later in life in ASCs, this could have detrimental effects by inhibiting the efficiency of tissue repair leading to a decline in tissue function. For example in satellite cells, stem cells of the adult skeletal muscle, a decreased regeneration capacity and increased markers of DNA damage can be observed during ageing (Sousa-Victor et al. 2014). Last, given the potential of iPSCs to be used for regenerative medicine purposes, it is important that these cells not accumulate damaged DNA during their generation and maintenance, as this could affect their long-term function and medical use.

Given the requirement for efficient DNA repair in stem cells, and the potential clinical implications for changes to this within the ageing human population, this chapter will discuss DNA repair mechanisms used by stem cells, how these processes change during ageing and what the consequences of this are for the pathogenesis and treatment of human diseases with a stem cell component.

DNA REPAIR MECHANISMS IN STEM CELLS

In general, stem cells display an elevated DNA repair capacity which decreases upon cell differentiation (Maynard et al. 2008, Saretzki et al. 2008). Aside from this observation, little is known about how DNA repair is regulated in stem cells compared to somatic cells, and also if this regulation is similar in different types of stem cells. However, it has been noted that the sensitivity of stem cells to DNA damage is highly dependent on the overall regenerative capacity of their resident tissue and also their proliferation rate (Mandal et al. 2011). Consequently, the slowing of the stem cell proliferation rate is thought to increase resistance to DNA damaging agents, believed to be important for maintaining tissue homeostasis. Under certain conditions in vivo, stem cells remain in a quiescent state (Li and Clevers 2010), until they are required to proliferate in situations which require tissue regeneration and repair. Their quiescent nature

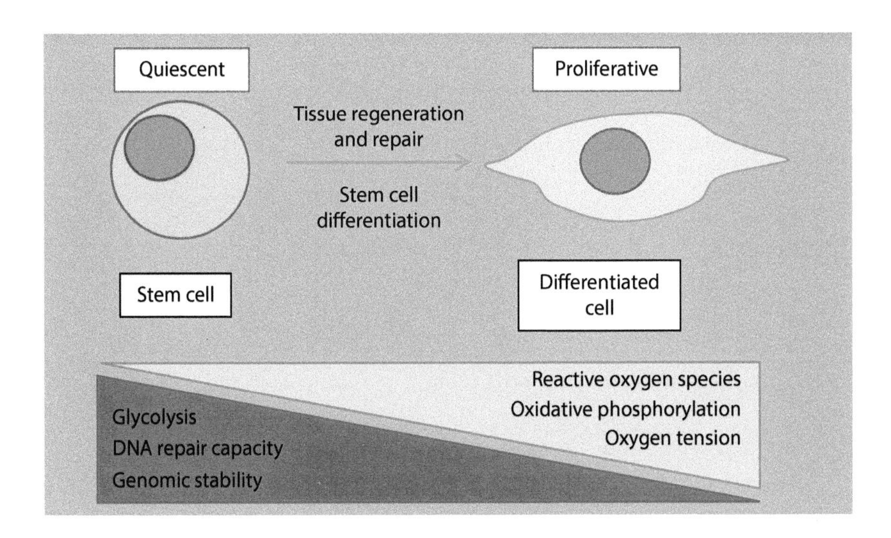

Figure 14.2 **Metabolic, proliferative and DNA repair properties of stem cells compared to differentiated cells.** Stem cells usually remain in a quiescent state in locations of low oxygen tension in which glycolytic metabolism predominates. Stem cells have an elevated DNA repair capacity which is thought to contribute to genomic stability within these cells. In situations such as tissue injury, stem cells are induced to cycle to produce differentiated cells required for tissue repair and regeneration. Differentiated cells show a decreased capacity for DNA repair and metabolise mainly through the mitochondria using oxidative phosphorylation. This generates reactive oxygen species which could contribute to DNA damage within these cells.

permits stem cells to rely on metabolism by anaerobic glycolysis, which in turn suppresses mitochondrial oxidation (Simsek et al. 2010). This is thought to be a protective mechanism allowing stem cells to limit their production of potentially DNA damaging ROS via oxidative phosphorylation and allowing maintenance of genomic stability. In addition, many stem cell populations reside in locations where the oxygen tension is very low, further limiting their exposure to oxidative molecules and decreasing their requirements for use of mitochondria. For example, HSCs are located in the bone marrow (BM) and are exposed to local oxygen tensions as low as 1.3% within their niche environment (Spencer et al. 2014). Overall, these adaptations of metabolism and cell cycle status equip stem cells with the tools to maintain minimal levels of DNA damage (Figure 14.2). However when DNA damage does occur, stem cells have a number of ways in which they can respond.

Mismatch repair

During normal DNA replication, DNA polymerases function to duplicate the entire genome of stem cells, but even with the proofreading activity within this system, the process is not without error (Kunkel and Erie 2005). To combat DNA polymerase errors during normal stem cell replication and also during unscheduled DNA synthesis, the DNA mismatch repair pathway is required (also discussed in Chapter 12). This pathway is also important for apoptosis and cell cycle checkpoint responses when stem cells are exposed to certain types of DNA damaging agents to prevent stem cells with damaged DNA from proceeding through the cell cycle (Stojic et al. 2004). The MMR pathway is conserved from *E. coli* to humans and begins with identification of the mismatched base and binding of MLH1, before the recognition of a newly synthesised DNA strand (Kunkel and Erie 2005). The mismatched strand is then removed by an exonuclease and DNA is re-synthesised and ligated to complete the repair process.

Stem cells have an elevated capacity for MMR and display enhanced expression of proteins important for this process (MSH2, MSH6, MLH1 and PMS2) when compared to somatic cells (Lin et al. 2014). Similar observations have been made when comparing mature CD34– cells with multipotent CD34+ cells found in human umbilical cord blood, in which the same gene set was found to be upregulated in the multipotent cells (Casorelli et al. 2007). Interestingly, when CD34+ cells are expanded in culture these differences are lost, illustrating how removing stem cells from their natural niche can alter their DNA repair properties. In addition, the efficiency of DNA repair by MMR has been shown to be increased when comparing human ESCs and iPSCs to parental un-reprogrammed fibroblasts (Lin et al. 2014), suggesting that elevated MMR capacity is associated with the re-programming process.

Base excision repair

Base excision repair is used by stem cells when small base modifications or single-strand breaks occur in their DNA (detailed in Chapter 5). There are a wide variety of small base modifications that require repair, for example, alkylation, oxidation and deamination. These modifications can occur naturally through spontaneous decay of DNA, but can also be more frequent when cells are exposed to environmental hazards such as radiation, chemicals or cytostatic drugs (Blaisdell et al. 2001). Initiation of BER occurs when 1 of the 11 known DNA glycosylases identifies and subsequently removes the damaged base, making way for processing of the remaining abasic site to continue the BER process (Kim and Wilson 2012).

Studies in mouse embryonic stem cells have revealed that genes involved in BER (Ape1, XRCC1, PCNA, Parp-1, UNG2 and DNA ligase III) are highly upregulated in these cells when compared to mouse embryonic fibroblasts, which also correlates with higher BER activity measured using an in vitro short-patch BER assay (Tichy et al. 2011). Previous work has also shown that during murine muscle development when myotubes are formed from differentiating myoblasts, expression of the BER genes DNA ligase I and II and XRCC1 are downregulated and this reduces the cells' ability to carry out BER in response to hydrogen peroxide exposure (Narciso et al. 2007). Similarly, in the murine brain, neural stem and progenitor cells are observed to express high levels of DNA glycosylases, allowing them to efficiently remove DNA lesions by BER, an ability which is lost following in vitro differentiation to mature neuronal cells using the generation of neurospheres (Hildrestrand et al. 2007).

Nucleotide excision repair

When stem cells are confronted with DNA damage which causes distortion of the DNA helix structure such as the formation of bulky DNA adducts following exposure to UV light, they use the nucleotide excision repair pathway to repair such lesions. This repair process is broadly divided into two pathways according to which areas of the DNA they act upon: (1) global genome NER (GG-NER) for repair of DNA across the entire genome and (2) transcription-coupled NER (TC-NER) for repair of specific regions on the transcribed strand of active genes (Fousteri and Mullenders 2008, Petruseva et al. 2014) (detailed in Chapter 5). In GG-NER, damage is sensed by the XPC protein which forms a multiprotein complex that moves along DNA to identify lesions. In TC-NER, damage recognition is indirect via the stalling of RNA polymerase II during transcript elongation of active genes.

Similarly to other DNA repair pathways, expression of genes involved in NER has been shown to decrease when stem cells differentiate. This has been observed in many types of human stem cells, including neuronal stem cell differentiation into neurons (Nouspikel and Hanawalt 2000) and monocytic lineage cell differentiation into mature macrophages (Hsu et al. 2007). More specifically, it was observed that NER kinetics are significantly slower in terminally differentiated human neurons when compared to neural precursor cells, with neurons taking 3 days to almost fully repair DNA damage and precursor cells achieving the same level of repair within a few hours (Nouspikel and Hanawalt 2000). A loss of NER activity during stem cell differentiation has been shown using a system of monocyte differentiation into macrophages to be due to insufficient ubiquitination of NER proteins, which is also accompanied by changes in the phosphorylation status of the ubiquitin-activating enzyme E1 (Nouspikel 2007). Although studies in mouse embryonic stem cells have shown that NER is active in these cells, they appear to display an increased sensitivity to DNA damage induced by exposure to high doses of UV damage relative to mouse embryonic fibroblasts (de Waard et al. 2008), indicating a potential saturation of this pathway during exposure to high levels of DNA damage.

Non-homologous end joining

DNA repair by non-homologous end joining is used by stem cells when DNA double-strand breaks are induced by a number of agents, two common ones being exposure to ROS and IR (Mahaney et al. 2009). NHEJ repairs DNA by binding and processing DNA ends at the site of a break, followed by ligation of the ends to reform the helix structure. Repair by NHEJ does not require a DNA template, leading this repair process to be considered an error-prone mechanism of repairing DNA DSBs (detailed in Chapter 7). A common feature of repair by NHEJ is a small number of nucleotides added or removed at the re-joining site; hence, repair is not always completely accurate. During DNA repair by NHEJ, a DNA DSB is recognised by the Ku protein, which acts as a scaffold to recruit other components of the core NHEJ machinery (Davis and Chen 2013). There is also Ku-independent recruitment of other key NHEJ proteins to the Ku-DNA complex such as XRCC4, XLF, DNA ligase IV and APLF which function to form a stable complex to allow completion of repair.

It has been observed in human embryonic stem cell (hESC) lines that the kinetics of repair by NHEJ is slower when compared to differentiated cells generated from the same line (Adams et al. 2010). Although the speed of repair was slower in hESCs, it was found that repair by NHEJ was much more accurate and this accuracy decreased upon cell differentiation. It was concluded that these differences could be explained by the fact that NHEJ in hESCs showed a high dependency on XRCC4 but was independent of proteins known to be important for NHEJ in differentiated cells including ATM, DNA-PK and PARP. In contrast, it has been shown that hESC lines exposed to IR require DNA-PK for repair by NHEJ in G2, before passing through the G2/M checkpoint, and that this type of NHEJ was extremely error-prone (Bogomazova et al. 2011), suggesting that under conditions of excessive DNA damage the fidelity of repair by NHEJ is reduced in hESCs.

It is believed that many types of adult stem cells in vivo mainly repair DNA DSBs using the NHEJ repair pathway, as they are mostly quiescent and unable to use the HR pathway (Cheung and Rando 2013). It has been hypothesised that this use of the more error-prone NHEJ pathway by stem cells could lead to inaccuracies in repair with potentially detrimental effects (Mohrin et al. 2010). For example, lung basal stem cells implicated in the pathogenesis of lung cancer have been shown to preferentially repair their DNA using NHEJ, which has been suggested to contribute to genomic instability in these cells, leading to disease development (Weeden et al. 2017). Further to this, it has been shown that quiescent HSCs of the blood system mainly repair radiation-induced DSBs using NHEJ, resulting in the acquisition of mutations and persistence of aberrant cell clones in vivo (Mohrin et al. 2010). This appearance of aberrant clones is a prominent feature of pre-malignant and malignant blood disorders, illustrating that DNA repair in stem cells by perturbed functioning of the NHEJ pathway can directly contribute to disease phenotypes.

Homologous recombination

Repair of DNA by homologous recombination is another mechanism used by stem cells to repair DNA DSBs and is considered a more accurate repair mechanism when compared to NHEJ. This is because HR can only be performed when sister chromatids can act as a DNA template during S and G2 phases of the cell cycle, allowing error-free repair (Branzei and Foiani 2008) (also described in Chapter 6). There are three main phases of repair by HR: (1) pre-synapsis in which DNA ends are recognised and processed by the MRE11-RAD50-NBS1 (MRN) complex, (2) synapsis in which RAD51 bound to single-stranded DNA invades the undamaged DNA strand and (3) post-synapsis (which has not been fully characterised) in which the final repair steps are completed through production of crossover or non-crossover DNA products (Renkawitz et al. 2014).

There is little known regarding use of the HR pathway for repair of DNA DSBs in vivo by stem cells; however, a number of in vitro studies have provided some insight into this repair pathway. Embryonic stem cells have a unique cell cycle profile, in that they have very short G1 and G2 phases, lack a G1 checkpoint and spend around 75% of their cycle time in S phase (Savatier et al. 2002). Based on this high proliferative activity, it may be anticipated that embryonic stem cells would use HR over NHEJ as they often have the sister chromatid template available. One study reported that indeed this is the case and showed that 81% of DSBs in ESCs were repaired by HR, with the remaining 19% being repaired by NHEJ (Francis and Richardson 2007). Further work using a different system in ESCs yielded similar results, with the authors reporting 75% of DSBs being accurately repaired by HR (Yang et al. 2004). Evidence from iPSC studies shows that HR is also crucial to the reprogramming process, as it is the major pathway for repairing reprogramming-induced DSBs allowing generation of iPSCs (Gonzalez et al. 2013).

EFFECTS OF AGEING ON STEM CELLS

Mammalian ageing is a consequence of complex and cell-specific changes over time which overall contribute to the functional decline of the organism. The process of ageing is known to be caused by interactions between a number of genetic and environmental factors and furthermore is the most common risk factor for a large number of human diseases (Niccoli and Partridge 2012). Although in recent years the human lifespan is increasing (Wilmoth et al. 2000, Vaupel 2010), the human healthspan is not changing, such that although people are living longer, many of their last years are spent in poor health (Hung et al. 2011). Consequently, understanding the ageing process and ways to slow it down are crucial areas of research, and one aspect of this is the effects of ageing on stem cells. In fact, stem cell exhaustion is a predominant feature of ageing and is now considered to be a molecular hallmark of the ageing process (Lopez-Otin et al. 2013). A hallmark of ageing has three main properties: (1) it must be present during normal aging, (2) experimental aggravation should accelerate the ageing process and (3) experimental amelioration should slow the normal ageing process and increase healthy lifespan. In the case of stem cells, not all functional changes that occur during ageing are associated with the onset of age-related diseases. However, they do alter the way stem cells respond to stimuli, which in some cases can lead to ineffective tissue repair and regeneration following an insult (Sousa-Victor et al. 2014).

Features of ageing

Many ageing phenotypes have been hypothesised to be caused by cells with excessive amounts of DNA damage which has accumulated over time. The free radical theory states that ageing is the result of ROS production which exceeds the capacity of the detoxification systems used by cells, ultimately leading to irreversible oxidative damage which impacts physiological function (Beckman and Ames 1998). An important feature of aged cells is the process of cellular senescence, in which proliferative capacity has been reached and cells go into a state of permanent cell cycle arrest (Hayflick and Moorhead 1961). Cells remain metabolically active and viable but can no longer replicate, which for stem cells would be a loss of their very important and defining self-renewal function. The DNA damage response itself is an important mediator of human cellular senescence, either through the shortening of telomeres through insufficient repair, or through its activation by external stimuli such as oxidative stress and IR through activation of p53 and Rb (Kulju and Lehman 1995, Narita et al. 2003). Having a senescent stem cell is not only detrimental in terms of lack of self-renewal, but senescent cells can also secrete factors which have been shown to induce local inflammation associated with ageing (Coppe et al. 2010), which may perpetuate the onset of inflammatory diseases. Cellular senescence in stem cells could also be an explanation for many of the physical phenotypes associated with ageing in humans. For example, sarcopenia, or loss of skeletal muscle mass and strength, is common during ageing and is associated with a loss of regenerative capacity of skeletal muscle stem cells (Sousa-Victor et al. 2014).

DNA repair and haematopoietic stem cell ageing

One system in which the effects of ageing on its stem cell populations has been well characterised is the haematopoietic system (Geiger et al. 2013), in which HSCs produce all of the functional blood cells required by an organism for its entire lifespan. There are many advantages to using HSC function as a model for ageing stem cells, including easy identification by cell surface markers, well-established assays to measure HSC function in vitro, and HSC transplantation models for the determination of true in vivo stem cell function. In adults, HSCs are predominately located in the BM within their own niche, providing them with an environment rich in soluble mediators and other neighbouring cell types which allow them to self-renew and differentiate as required (Morrison and Scadden 2014). HSCs are a rare cell population and are estimated to be only be 1 in 10,000 cells of the BM and 1 in 100,000 cells in the peripheral blood (Ng et al. 2009).

A number of differences in HSC function have been observed when comparing young HSC with old HSC in the absence of any disease phenotype (**Figure 14.3**). In terms of HSC number, it has been noted in both mouse and human studies that the actual number of HSCs increases with age (Sudo et al. 2000, Pang et al. 2011), which is thought to be due to an enhanced proliferative capacity exhibited by older HSCs. Despite this, when the ability of old HSCs to reconstitute the entire blood system through transplantation experiments is assessed, it can be seen that these cells individually have a decreased regenerative capacity under such conditions (Sudo et al. 2000). To summarise these results, it appears that although old HSCs are still able to expand and sustain blood cell production, they appear to have functional defects rendering them unable to be used for transplantation, stemming from a decrease in their ability to self-renew effectively. The mechanism resulting in increased HSC number with age is not known, but it has been suggested that this is a compensatory mechanism to combat the loss of function. Aside from these features, old HSCs exhibit differences in their cell potential in terms of lineage differentiation. Young HSCs appear to have a balanced differentiation potential for myeloid and lymphoid cells of the immune system. In contrast, old HSCs show a skewed differentiation potential towards cells of the myeloid lineage, which may account for the age-related decline in adaptive immune responses (Pang et al. 2011).

In vivo, HSCs are maintained in the G0 stage of the cell cycle, which allows them to remain quiescent and inactive for long periods of time. This quiescent status was believed to be a protective mechanism to allow these long-lived cells to maintain genome stability during their lifespan. However, more recently, it has been shown in terms of DNA repair that this

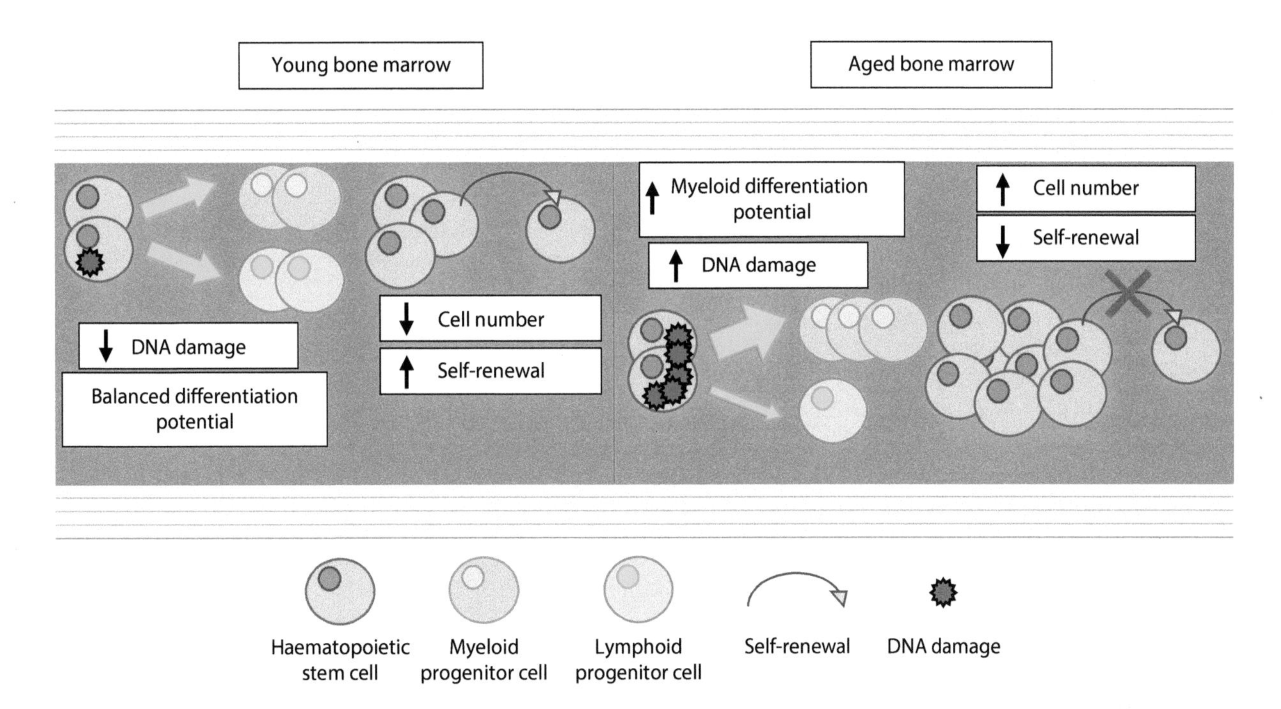

Figure 14.3 **Changes to the properties of haematopoietic stem cells in the bone marrow during ageing.** During the process of ageing, a number of changes occur in the haematopoietic stem cell (HSC) population. During ageing, the actual number of HSCs increases, but this is parallel with an overall decrease in their self-renewal capacity. Functionally, older HSCs display a skewed differentiation potential towards cells of the myeloid lineage when compared to young HSCs. Also with ageing, the levels of detectable DNA damage within the HSC population increase.

quiescent status can also be detrimental to HSCs. Work has shown that quiescent HSCs mainly repair DNA DSBs using the NHEJ pathway, which can be error-prone and induce significant genomic re-arrangements within these cells (Mohrin et al. 2010). Studies have shown that mutant HSCs generated by NHEJ-induced errors are able to survive in vivo and contribute to abnormalities within the haematopoietic system, illustrating that mis-repaired DNA in HSCs can affect the whole blood system. Under certain conditions of physiological stress, such as blood loss, proliferation and activation of HSCs is required in order to replenish lost mature blood cells. In both young and old HSCs, impaired activation of the G1-S cell cycle checkpoint in response to DNA damage can be observed during HSC proliferation (Moehrle et al. 2015), which could allow cells with damaged DNA to continue cycling and contribute to a pool of mutated blood cell progenitors. In old HSCs, this process of entering the cell cycle has been associated with increased levels of replication stress, characterised by chromosomal abnormalities and changes in replication fork dynamics (Flach et al. 2014). Furthermore, it has been observed that when old HSCs re-enter quiescence, residual DNA damage is still present, suggesting that old HSCs are less efficient at DNA repair, which could result in the persistence and clonal expansion of cells with genetic abnormalities. In fact, it has been observed during ageing that clonal haematopoiesis is more frequent, this being the process by which a single mutant haematopoietic stem or progenitor cell makes a significant contribution to the overall pool of mature blood cell lineages (Adams et al. 2015).

Studies in murine and human HSCs have indicated a role for the DNA damage response in the ageing phenotype of these cells. First, in aged murine HSCs an increased amount of DNA damage in the form of DSBs has been observed using the marker gamma-H2AX, when compared to young HSCs (Rossi et al. 2007). This study suggests that DNA damage accumulates over the lifespan of HSCs, and work from others has identified that this could be due to replication stress during HSC cycling (Flach et al. 2014). This work has been followed up by looking in aged human CD34+ HSCs and haematopoietic progenitors, where a similar result was found (Rube et al. 2011). Another study in murine cells used comet assays to directly show that there was a significant increase in the amount of DNA DSBs in old HSCs when compared to young HSCs (Beerman et al. 2014), providing further support for the accumulation of DNA damage in HSCs with ageing.

Using studies of mice deficient in certain DNA repair proteins, it has been observed that DNA repair in HSCs is not essential in order to maintain HSC numbers in the BM (Nijnik et al. 2007), which is contrary to proliferating progenitor cells. However, it was shown that DNA repair is essential to maintain HSC function over time (Nijnik et al. 2007, Rossi et al. 2007), as when HSCs deficient in DNA repair proteins were transplanted into host animals, a rapid HSC exhaustion was observed unless young donors were used. Repair of DNA DSBs is critical to maintaining HSC function, and one of the main mechanisms used by HSCs for this is

NHEJ (Mohrin et al. 2010). The requirement for NHEJ in HSCs is evident from studies using mice deficient in DNA ligase IV, a protein central to repair of DSBs by NHEJ (Nijnik et al. 2007). HSCs from these animals display increased DNA DSBs and evidence of replication stress in the form of elevated gamma-H2AX foci, which is associated with progressive loss of HSCs and impaired function in transplantation assays. Another mechanism used by HSCs to repair DNA DSBs is HR, which can only be used when HSCs enter the cell cycle. The importance of HR for HSC function is evident from murine studies using knockout mouse models for proteins involved in HR such as those of the FANC pathway. FancD2 knockout animals (Navarro et al. 2006) display a reduced number of HSCs, and FancC/FancG double mutant animals develop haematological disease in the form of myelodysplasia (Pulliam-Leath et al. 2010). More evidence for the association of DNA damage with loss of HSC function with ageing comes from work using an artificial model of physiological stress. This model recapitulates the levels of physiological stress that HSCs would experience over the lifetime of an organism to measure the effects of this on HSC function. Repeated injections of animals with polyinosinic:polycytidylic acid to mimic a viral infection revealed a rapid transition of HSCs from a quiescent state to actively cycling (Walter et al. 2015). This was associated with elevated levels of DNA damage not observed in the initial quiescent HSC population and highlights that quiescence is also potentially a defence mechanism used by HSCs to combat excessive levels of DNA damage.

DNA repair and neural stem cell ageing

Throughout adulthood, neural stem cells reside in certain areas of the mammalian brain where they function to sustain a low level of neurogenesis through proceeding slowly through the cell cycle. They are present at a very low frequency and are surrounded by large numbers of post-mitotic neurons. The number of neural stem cells in the brain decreases with ageing and is associated with a decline in neurogenesis (Maslov et al. 2004); however, in vitro culture of these cells reveals no functional impairment (Ahlenius et al. 2009). This indicates that decreased rates of neurogenesis in vivo are likely to be due to environmental changes within the neural stem cell niche. Furthermore, increased numbers of senescent neural stem cells can be observed during ageing (Molofsky et al. 2006) and are believed to contribute to the cognitive decline observed during ageing.

DNA repair in neural stem cells is crucial in order to maintain their stem cell properties such as self-renewal. For example, when neural stem cells are exposed to IR, this treatment induces senescence and differentiation via a DNA damage signalling pathway involving STAT3, BMP2 and ATM (Schneider et al. 2013). Furthermore, studies of neural cells located in the subventricular zone of the brain have shown that following exposure to radiation, neural stem cells and neural progenitor cells respond very differently (Barazzuol et al. 2017). In the case of progenitor cells, radiation induces high levels of apoptosis within this cell population. In contrast, neural stem cells were shown to be activated from their quiescent state and induced to cycle in response to radiation. These responses were both shown to be dependent on ATM, and it has been suggested that this minimises radiation-induced DNA damage to promote the derivation of neuroblasts from neural stem cells to compensate for loss of neural progenitors. Neural stem cells are also particularly sensitive to DNA damage by ROS and rely on careful regulation of their oxidative state. For example, mice deficient in Rb1cc1 required for autophagy and oxidative regulation display loss of neural stem cells in the brain and a reduced differentiation capacity, which could be rescued by increasing levels of antioxidants (Wang et al. 2013).

DNA repair in neuronal stem cells is essential and key features of many diseases with defective DNA repair proteins are progressive neurodegeneration and premature ageing (Kulkarni and Wilson 2008). Neurogenesis in the adult brain is sensitive to DNA damage, and it has been shown in mice that a single dose of radiation significantly decreased numbers of immature neurons and proliferating progenitor cells (Mizumatsu et al. 2003), which in vivo could contribute to cognitive decline similar to that observed during ageing. The ATM protein in particular is crucial for the maintenance of neural stem cell function within the adult brain, and is an absolute requirement for cell survival and determining cell fate decisions (Allen et al. 2001). The development of in vitro experimental models of neural stem cell ageing using DNA damage-inducing agents has revealed that high levels of DNA damage may underlie a lot of the phenotypes observed in the ageing brain. For example, treatment of neural stem cells with hydroxyurea in culture induces senescence, which is accompanied by decreased levels of expression of XRCC2, XRCC3 and Ku70, key proteins required for conventional DNA repair (Dong et al. 2014).

DNA repair and intestinal stem cell ageing

The gut epithelium is turned over at a high rate and as such requires intestinal stem cells to support this process. There are two main populations of intestinal stem cells in mammals: (1) highly proliferative cells located in the base of the crypt and (2) quiescent cells which can be found just above the base of the crypt (Potten et al. 2001, Takeda et al. 2011). Regardless of organism age, it appears that the highly proliferative cells located in the base of the crypt are much more sensitive to DNA damage when compared to the quiescent population located further above (Barker 2014). Upon the induction of DNA damage, quiescent intestinal stem cells are induced to cycle via a metabolic switch mediated by musashi RNA binding protein 1 (Yousefi et al. 2016). WNT signalling in particular appears to be important in the response of intestinal stem cells to DNA damage induced by radiation in both quiescent and proliferating cell populations. Hyper activation of WNT in proliferating intestinal stem cells sensitises this population to radiation-induced DNA damage via LGR5 (Tao et al. 2015). As well as this, increasing WNT signalling in quiescent intestinal stem cells appears to enhance their regenerative capacity following radiation damage through the upregulation DNA ligase IV and activity of the NHEJ DNA repair pathway (Jun et al. 2016). In vivo, WNT derived from marcophages is thought to be a crucial mediator of this regeneration process following radiation injury (Saha et al. 2016).

Studies using IR have suggested that with ageing, the intestinal stem cells become more sensitive to DNA damage as measured by an increase in apoptosis, but appear to increase in numbers following such an insult in older animals. Aside from this, little is known about how intestinal stem cell numbers and function change with ageing in mammals, but information gained from work in *Drosophila* could be an important starting point to direct future research in higher organisms. In this system, intestinal stem cells can be identified through expression of notch ligand delta and transcription factor escargot (Micchelli and Perrimon 2006). Using these markers, it has been shown that cell numbers increase with age (Choi et al. 2008), and this is reminiscent of the increase observed during infection of the gut with bacteria (Chatterjee and Ip 2009). It has been proposed that a major mechanism leading to intestinal stem cell functional decline in *Drosophila* could be mitochondrial dysfunction, as enhancing mitochondrial activity through an overexpression system was proven to delay age-related changes to the intestine structure and also extend organism lifespan (Rera et al. 2011).

With regard to direct measurement of DNA repair, one study measured this in a stem cell subpopulation in the small intestine of mice known as crypt base columnar stem cells and found that these highly proliferative cells preferentially repair their DNA using HR in response to radiation-induced DNA damage (Hua et al. 2012). It was concluded that DNA repair via this pathway was crucial for repopulation of the crypt and that survival of this stem cell population could be used directly as a predictor of crypt regeneration. Work using mouse models has also shown that intestinal stem cells may respond differently to DNA damage according to their external environment. This is well illustrated by a study in which mice were treated with the chemotherapeutic drug etoposide either after feeding or fasting. It was shown that fasting animals were able to maintain their intestinal stem cell populations, through more efficient DNA repair and decreased rates of apoptosis, when compared to animals which had been fed (Tinkum et al. 2015). This has important implications in clinical practice for sparing intestinal stem cells and overall intestinal architecture in patients being treated with chemotherapy.

DNA repair and muscle stem cell ageing

Skeletal muscle fibres are regenerated by a population of muscle stem cells known as satellite cells in response to stressors such as muscle injury. It is observed that satellite cell number decreases with ageing (Gibson and Schultz 1983), along with ability to proliferate in vitro (Bernet et al. 2014) and the ability to engraft and regenerate lost muscle mass upon transplantation in vivo (Cosgrove et al. 2014). These changes are accompanied by altered differentiation potential, in which satellite cells are more likely to differentiate to cells of the fibrogenic lineage and less likely to differentiate into cells of the myogenic lineage (Brack et al. 2007). This altered potential can give rise to increased muscle tissue fibrosis and, as such, a loss of muscle function and strength as observed during ageing. This process is thought to be in part due to overactivation of the WNT signalling pathway in satellite cells caused by changes in the niche environment (Brack et al. 2007). Indeed, a number of studies have transplanted satellite cells from old animals into young animals and shown that cell extrinsic factors and signalling changes are the likely cause of the ageing satellite cell phenotype in terms of decreased regenerative potential (Carlson and

Faulkner 1989). In contrast, cell-intrinsic factors such as increased cellular senescence appear to underlie self-renewal defects in ageing satellite cells (Sousa-Victor et al. 2014), as these are still apparent even after transplantation into a young stem cell niche environment.

There is little evidence supporting a role for DNA repair deficiency in the ageing of satellite cells when directly comparing markers of DNA damage between young and old individuals. In ageing mice, for instance, satellite cells display similar levels of the DNA DSB marker gamma-H2AX following muscle injury when compared to cells obtained from young animals, suggesting older cells do not display excessive accumulation of DNA DSBs (Cousin et al. 2013). Furthermore, through the use of DNA DSB repair–deficient animals (Cousin et al. 2013), it was shown that muscle regeneration following injury was not influenced by the DNA repair ability of satellite cells, again highlighting that a decline in DNA repair with ageing is unlikely to contribute directly to satellite cell functional decline. In contrast to satellite cells, terminally differentiated muscle cells have been shown to be highly sensitive to DNA damage in the form of ROS (Narciso et al. 2007). This was characterised by defective BER and accumulation of DNA single-strand breaks, which over time in vivo could contribute to the muscle degeneration occurring during ageing.

DNA repair and skin stem cell ageing

The skin contains several different types of stem cells, each with its own unique niche. Two of the most important types are hair follicle stem cells which regulate hair growth and melanocyte stem cells which produce cells responsible for skin pigmentation. Hair follicle stem cells do not continuously regenerate, but go through cycles of proliferation, degeneration and rest in order to sustain hair growth. During the ageing process, hair follicle stem cell numbers do not change (Giangreco et al. 2008), but they increase periods of rest culminating in a functional loss of continual hair growth mediated by changes in signalling by Nfatc1 (Keyes et al. 2013). In contrast, melanocyte stem cell numbers decrease significantly with ageing (Ortonne 1990), due to alterations in their self-renewal properties and increasing levels of cell differentiation. Exposure of melanocyte stem cells obtained from young individuals to IR also induces a differentiation phenotype similar to that observed in aged cells (Inomata et al. 2009), suggesting that genomic instability caused by DNA damage may play a role in reduced hair growth and hair greying observed during ageing.

Studies of murine hair follicle stem cells have indicated that in response to DSBs these cells upregulate the anti-apoptotic protein BCL2 and initiate DNA repair by NHEJ, mainly mediated by DNA-PK (Sotiropoulou et al. 2010). Using mouse models, it has been shown that during ageing, hair follicle stem cells display increased markers of DNA DSBs under basal conditions, and also in response to radiation, as measured by an increased number of 53BP1 foci (Schuler and Rube 2013). This could be due to the fact that the skin in particular is constantly exposed to DNA-damaging agents within the environment due to its external location. For example, frequent sun exposure can induce UV damage to the skin. This can result in apoptosis of melanocytes (Arck et al. 2006) and in more severe cases a malignant transformation in the form of skin cancer (Ahmed et al. 1999). Indeed, evidence from mouse models has shown that even low doses of IR causes irreparable DNA damage to melanocyte stem cells (Inomata et al. 2009), resulting in activation of the melanocyte stem cell differentiation programme and hair greying.

DISEASES OF STEM CELL AGEING

As discussed, there are many physiological features of stem cell ageing which occur as a natural part of organism ageing in the absence of disease. However, there is a fine balance between normal ageing-associated changes to stem cell function, and changes which occur as part of the so-called diseases of stem cell ageing. Adult stem cells in a number of locations have been implicated in the development of age-related diseases; from neural stem cells located in the brain to HSCs located in the BM. In fact, much attention is being paid to reverse age-associated changes in stem cell function in order to prolong human life and promote healthy ageing. There are three main key areas of research with regard to understanding and treating diseases of ageing: (1) understanding the basic mechanisms underlying ageing and diseases, (2) developing new drugs to reverse age-associated changes and (3) identifying ways to repair tissues affected by ageing using surgical techniques until the ageing process itself can be stopped or reversed. If all of these aspects of ageing can be understood and treatments implemented, then all of the diseases of ageing could potentially be eradicated in the near future. However, in the meantime, much research is being conducted to fully characterise the process of stem cell ageing in particular.

It is a widely accepted concept that one possible explanation for disease-associated loss of stem cell functions with ageing is due to the accumulation of DNA damage over the lifetime of these long-lived cell populations (Behrens et al. 2014). Indeed, stem cells taken from aged organisms display increased markers of DNA damage and significant telomere shortening (Vaziri et al. 1994, Rossi et al. 2007). Telomeres are of particular interest, as these function to protect the genome and prevent unwanted repair and recombination events which could be detrimental to the cell. Like other cell types, human stem cells express the enzyme telomerase required for the maintenance of telomere length, but still experience telomere shortening with age which is associated with an increased likelihood of cellular senescence and potentially apoptosis (Henriques and Ferreira 2012). There is significant evidence for the involvement of telomere shortening in age-associated diseases including diabetes, cardiovascular diseases and increased risk of cancer development (Rizvi et al. 2014).

Aside from dysfunction of telomerase, stem cells can accumulate DNA damage and genomic instability from external sources. One of the most prominent being exposure to damaging agents such as oxidative stress. Oxidative stress refers to damage caused by free radicals derived from oxygen and one of the main sources of this is the production of ATP by oxidative phosphorylation within the mitochondria (Lenaz 1998). Although this is an important physiological process, overactivation and thus dysfunction of this pathway has been proposed to play an important role in the ageing process. For example, senescent cells rapidly increase in numbers during ageing and contribute to the production of large amounts of ROS which cause damage to their DNA and proteins (Chen et al. 1995). Many adult stem cells are quiescent and rely on glycolysis for their energy needs so are somewhat protected from oxidative damage; however, secretion of factors by neighbouring senescent cells could induce oxidative damage which could negatively affect stem cell functions. The ATM protein, involved in repair of DNA DSBs, is also crucial to preventing oxidative damage in stem cells. Studies in ATM knockout mice have confirmed that loss of ATM in HSCs results in increased levels of ROS and a loss of self-renewal capacity, which is reversed following administration of antioxidants to counteract such oxidative damage (Ito et al. 2004).

Cancer and stem cell ageing

Phenotypic features of ageing stem cells have been associated with cancer development and the concept of so-called cancer stem cells is becoming accepted, for example, in haematological malignancies such as leukaemia (Wang and Dick 2005). Leukemic stem cells are specifically characterised by their ability to regenerate human acute myeloid leukaemia cell populations in immunocompromised mouse transplantation models (Lapidot et al. 1994). The prevalence of cancer stem cells in the BM of leukaemia patients is variable but provides a useful prognostic indicator. For example, it has been observed that patients who present with high numbers of leukemic stem cells are more likely to relapse following treatment and have increased mortality (van Rhenen et al. 2005), highlighting the importance of these cells to disease pathogenesis.

The term 'cancer stem cells' originates from the fact that these cells share two of the most important characteristics of normal adult stem cells, these being the ability to self-renew and the ability to differentiate into cells of multiple lineages (Verga Falzacappa et al. 2012). Similarly to normal stem cells, cancer stem cells can exist in a state of dormancy and quiescence or in a state of activity and proliferation. This property may explain the persistence of cancer following treatment, as dormant cancer stem cells could become activated and re-initiate disease (Sosa et al. 2014). With regard to DNA damage responses, cancer stem cells display a more robust DNA DSB response when compared to more differentiated malignant cells, which is believed to contribute to their increased resistance to antineoplastic agents currently used for the treatment of cancer. For example, CD133-expressing cells have been proposed as a population of cancer stem cells contributing to disease initiation and maintenance in many types of cancer. CD133+ cell lines generated from lung cancer patients display enhanced repair of IR-induced DNA DSBs and upregulation of genes involved in DSB resolution (Desai et al. 2014). These properties conferred an overall radioresistance phenotype which could explain why cancer stem cells persist even after DNA damaging treatments such as chemotherapy. Similar results were found in the case of cancer stem cells in glioblastoma, where CD133+ cells showed preferential activation of the DNA damage checkpoint response and more efficient DNA repair when compared to CD133– cells (Bao et al. 2006). As well as this, cancer stem cells taken from mammary tumours display a rapid DNA repair response characterised by high levels of NHEJ and increased expression of 53BP1 (Chang et al. 2015). Overall, an efficient DNA repair response appears to be advantageous for the survival and persistence of cancer stem cells in many types of disease.

Another similarity between normal stem cells and some cancer stem cells is the ability to maintain low levels of intracellular ROS, which is thought to be crucial to preserving stem cell function. It is known that ROS have an important role in the induction of cell death following treatment with IR (Powell and McMillan 1990), and thus it has been suggested that minimising levels of ROS contributes to the radioresistance exhibited by cancer stem cells. This indeed can be observed in lung cancer stem cells which display metabolic properties typical of normal stem cells, including low levels of mitochondrial DNA, decreased oxygen and glucose consumption and low levels of intracellular ROS (Ye et al. 2011). Similar work has also been conducted using breast and liver cancer stem cells isolated from human patients and has yielded similar results, with cells exhibiting a higher survival rate and decreased levels of DNA damage following IR (Diehn et al. 2009, Kim et al. 2012).

In contrast to normal stem cells, cancer stem cells display extreme heterogeneity in terms of phenotype, functions and plasticity. This heterogeneity is believed to be caused by the abnormal way in which cancer stem cells are generated, through the acquisition of mutations which result in their perturbed functioning. Moreover, heterogeneity within cancerous tumours containing cancer stem cells exists in many forms, including differences in cellular growth rates, response to therapy, genetic abnormalities and cell surface marker expression (Kreso and Dick 2014). Despite this heterogeneity, cancer stem cells often express the same markers as normal stem cells present within the same anatomical location. For example, leukaemic stem cells isolated from patient BM have been shown to express CD34, CD38, CD71 and HLA-DR, similar to normal HSCs (Warner et al. 2004, Dick 2005). This makes it difficult to target a cancer treatment based on cell surface marker expression alone, as healthy cells would also be affected in this case. However, there are some differences between cell surface marker expression when comparing normal stem cells to leukemic stem cells, including a loss of CD117 expression upon leukaemic transformation (Blair and Sutherland 2000), suggesting further profiling may reveal leukaemic stem cell specific markers which could be targeted for treatment of the disease.

Cancer development and progression require oncogenic transformation of cells through the induction of increasing numbers of mutations which take time to develop. Given the long-lived nature of stem cells, it could be suggested that they are the ideal cell type in which to harbour such mutations which eventually lead to disease. Stem cells have many defences against potentially mutagenic DNA damaging agents; however, if these defences fail in only a small number of cells, this can be sufficient to induce tumorigenesis. This becomes problematic in the treatment of such cancers, as chemotherapeutic agents are unlikely to remove all of these rare cell populations which may be responsible for the initiation of the disease. In fact, transplantation models using CD133+ cells have shown that as few as 100 cancer stem cells isolated from human brain tumours are sufficient in order to recapitulate the disease when transplanted into the brains of mice (Singh et al. 2004). Current cancer treatments usually consist of a combination of chemotherapy, radiotherapy and surgery to attempt to (1) remove cancer cells directly by surgical techniques or (2) effect a treatment indirectly by inducing high levels of DNA damage resulting in cell death. However, there is mounting evidence that DNA damage itself can induce cell transformation, for example, patients being treated with chemotherapy for unrelated cancers can develop therapy-related blood disorders and cancers which were not present before the treatment (Yin et al. 2015). This leads to the question of whether inducing more DNA damage during patient treatment would be beneficial and supports the development of more targeted approaches.

The importance of the DNA damage response in cancer is evident from patients suffering with inherited disorders of DNA repair. For example, patients with xeroderma pigmentosum display defects in DNA repair following exposure to UV, which is associated with an increased risk of skin cancer (Lehmann et al. 2011). A large proportion of these patients have been shown to have mutations in genes necessary for NER (Cleaver 1969), illustrating that changes in the components of the DNA damage response can directly contribute to relative cancer risk. Another example is patients with Lynch syndrome, a type of hereditary colorectal cancer known to be caused by mutations in proteins required for DNA MMR (Papadopoulos et al. 1994), which subsequently results in microsatellite repeat instability characteristic of these cancer cells. Evidently, the DNA damage response has an important role in cancer susceptibility, but in the context of stem cells, this has not been well studied in all cancer types.

Diseases of haematopoietic stem cell ageing

As previously discussed, one system in which the effects of ageing in the context of stem cells and DNA damage have been well defined is the haematopoietic system. This is also the case in terms of disease phenotypes which have their origin in HSCs such as myelodysplastic syndromes and leukaemias. The effects of ageing in this system are thought to promote conditions within the BM niche that favour genomic instability within the HSC population, resulting in the persistence of stem cells with damaged DNA that contribute to disease development. The emergence of clonal haematopoiesis is thought to be one of the reasons why the incidence of blood disorders increases with ageing, as this process could increase the likelihood of HSCs acquiring additional mutations important for the initiation of disease. Similarly, the concept of clonal haematopoiesis of indeterminate potential (CHIP) is becoming important in the diagnosis and understanding of MDS, a common blood disorder which in many cases can progress to acute myeloid leukaemia (Steensma et al. 2015). CHIP is a term used to describe individuals who have a mutation in their blood or BM cells which is associated with haematological malignancy, but do not display any features of disease. Identifying these individuals appears to be important, as many people with CHIP go on to develop a haematological malignancy, supporting the ideas that acquisition of mutations and clonal haematopoiesis are two important factors involved in the development of MDS.

MDS is characterised by excess BM cellularity and blood cytopenias, and patients present with symptoms caused by a degeneration of haematopoietic tissue. These include recurrent infections, chronic fatigue and unexplainable bruises and bleeding. MDS is quickly being considered as a disease of ageing HSCs showing signs of genomic instability and is most frequent in individuals aged 60 or over (Ma 2012). Furthermore, a large number of MDS patients progress to acute myeloid leukaemia via a presently unknown mechanism. It has been suggested that genomic instability and changes in the DNA damage response could be important mechanisms responsible for disease progression and MDS initiation. For example, MDS can also arise in patients being treated with chemotherapy or radiotherapy for other types of cancer (Takahashi et al. 2017), indicating that DNA damage itself can increase the likelihood of MDS development. There is a growing body of evidence that leukaemic stem cells can be found in cases of AML (Bonnet and Dick 1997), but a distinct lack of evidence for this in MDS, as illustrated by a lack of engraftment and no signs of MDS following transplantation of MDS cells in mouse models (Nilsson et al. 2000, 2002). Hence, there is still work to be done to identify the exact cell of origin in MDS, but study of HSCs is directing current research.

Work comparing the gene expression profiles of HSCs in a sub-group of MDS patients has shown that indeed they resemble more closely the profile of HSCs taken from healthy individuals and are dissimilar to progenitor cells (Nilsson et al. 2007). This provides support for the hypothesis that HSCs may indeed be the origin of MDS (Figure 14.4). Furthermore, it has been observed that in a sub-group of patients displaying 5q-syndrome, 92%–100% of HSCs had this chromosomal abnormality indicating that this disease-associated genetic change could be passed down from HSCs to progenitor cells (Nilsson et al. 2007).

Figure 14.4 **Model for how DNA damage in haematopoietic stem cells may contribute to the development of the symptoms of myelodysplastic syndrome.** With ageing, increased levels of DNA damage accumulate within the haematopoietic stem cell population, contributing to the generation of mutated cells. This DNA damage persists when haematopoietic stem cells differentiate into more mature progenitor cells. Many of these mutated progenitor cells die through the process of apoptosis, but some persist and contribute to the production of mature myeloid cells including red blood cells, platelets and white blood cells. Overall, this results in a decrease in the production of normal functional blood cells which is associated with the development of the symptoms of myelodysplastic syndrome. HSC, haematopoietic stem cell; MDS, myelodysplastic syndrome.

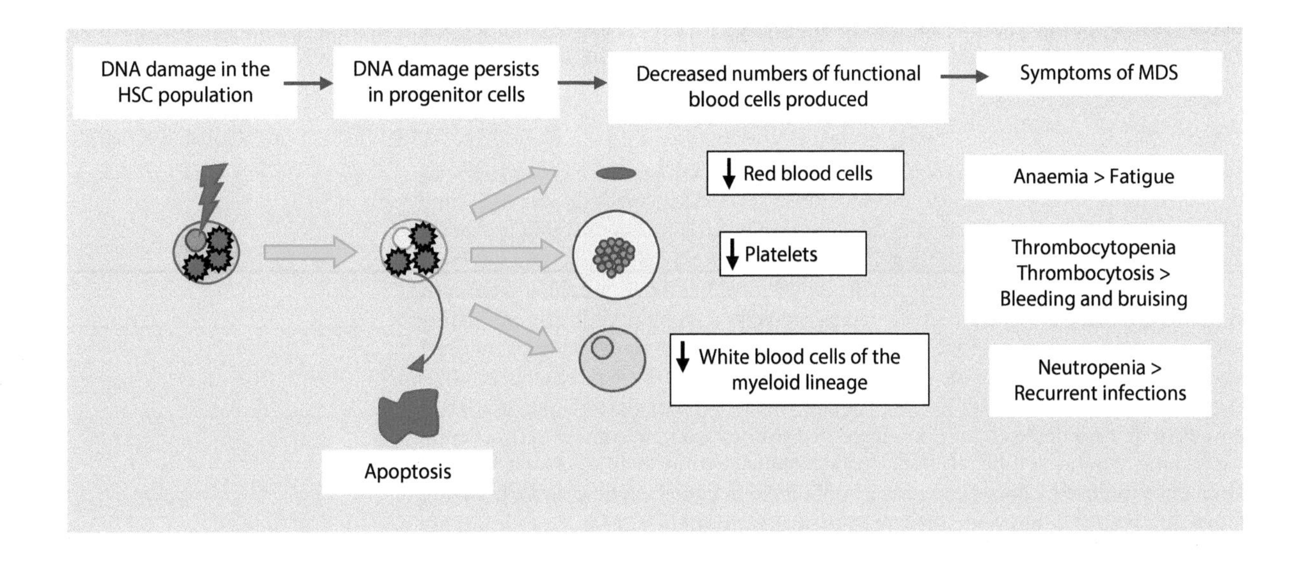

Genomic instability is without a doubt one of the hallmarks of the MDS phenotype and evidence of the accumulation of mutations and excessive amounts of DNA damage have both been implicated in this. Increased numbers of micronuclei can be observed in lymphocytes isolated from MDS patients when compared to age-matched healthy controls, and their numbers can be correlated with disease severity in these patients (Kuramoto et al. 2002). Furthermore, a multitude of mutations in specific genes and chromosome abnormalities are common in MDS, many of which are associated with genes involved in RNA splicing machinery (Papaemmanuil et al. 2013). This increase in mutational burden in MDS is accompanied by a parallel increase in levels of DNA damage in the form of oxidised nucleotides even in low-risk patients. This is the case for both cells of the BM (Novotna et al. 2009) as well as the peripheral blood (Jankowska et al. 2008). Work using purified populations of BM cells has furthered these findings by identifying that this DNA damage could only be observed in the stem cells containing the CD34+ cell population of MDS patients and not in the CD34− population (Peddie et al. 1997). This study provides more evidence that DNA damage in the HSC compartment could be important for disease initiation. As previously mentioned, a proportion of MDS patients go through the process of disease progression to AML. The prognosis for AML patients is variable and largely depends on the cytogenetic classification. Defects in DNA repair could represent a significant contributor to AML development by providing a scenario to allow the occurrence of chromosomal translocations typical of the disease. This idea is supported by the fact that many inherited disorders of DNA repair such as Bloom syndrome and Fanconi anaemia are associated with an increased risk of AML development (Auerbach 1992, Poppe et al. 2001).

Genomic instability can arise in MDS and AML from defects in any one of the DNA damage response pathways. First, it has been observed that myeloid leukaemia cells display elevated levels of DNA DSB repair by NHEJ when compared to CD34+ cells isolated from healthy individuals (Gaymes et al. 2002). Over-activation of repair by NHEJ was shown to result in incorrect repair and induced large deletions within the genomes of the leukaemic cells, providing evidence for the role of altered DNA repair pathways in leukaemia development. In addition to this, a commonly occurring mutation in AML patients known as FLT3/ITD has been shown to result in changes in the expression of proteins involved in the NHEJ pathway (Fan et al. 2010). This is characterised by a decrease in expression of Ku and an increase in expression of DNA ligase IIIα, characteristic of promoting DNA repair by alternative NHEJ, which is known to be highly error-prone. Mutations in the Nras and Bcl2 genes are also common in human AML patients, and data from mouse models have shown that, similar to the FLT3/ITD mutation, these mutations also cause changes in DNA repair by NHEJ, which in this case are accompanied by elevated levels of ROS (Rassool et al. 2007). Aside from NHEJ, repair of DNA DSBs by HR is also implicated in AML development. This association between HR and AML has been identified in FA patients who present with BM failure and an increased pre-disposition to AML (Michl et al. 2016). FA patients have inherited mutations in one of proteins involved in the FA pathway, a DNA repair pathway that coordinates several DNA repair processes, including HR, to repair DNA interstrand cross-links. Chromosome abnormalities and the presence of polymorphisms within genes essential for HR are also important determinants for AML risk, including RAD51 and XRCC3. By studying the frequency of two polymorphisms, namely RAD51 (G135C) and XRCC3 (T241M), in AML patients and healthy controls, it was found that risk of AML was significantly increased when these were present in both de novo and therapy-related AML (Seedhouse et al. 2004). These results suggest that changes in DNA repair by HR and related pathways is a likely contributor to the pathogenesis of AML.

Diseases of neural stem cell ageing

Some of the most prevalent age-related diseases are associated with degeneration of the nervous system, and three prime examples are Parkinson's disease (PD), Alzheimer's disease (AD) and stroke. In each of these diseases, a specific population of neural cells within the brain is affected and has the potential to be regenerated using stem cells. Studying stem cells within the adult brain is problematic, as these cells constitute a very small proportion of brain cells overall. It is believed that neural stem cells may only play a small role in age-related diseases of the brain, mainly through their loss of self-renewal with age. For example, an early symptom of AD is a loss of smell which is caused by loss of neurons in the olfactory region of the brain (Devanand et al. 2000). It has been suggested that this loss of neurons could be due to a loss of proliferation and migration of neural stem cells to this region, thus contributing to the disease process. In addition to this, neural stem cells have been shown to proliferate following a stroke (Arvidsson et al. 2002), which could explain some of the brain functional improvement observed in patients following such a brain injury. Despite this response, however, it has been estimated that only 0.2% of the cells lost are replaced, illustrating that the potential of neural stem cells to regenerate is apparent, but, perhaps due to age-associated changes in their self-renewal capacity, is not realised. Last, to date there is no evidence that changes in neural stem cell function are associated with PD, nor that neural stem cells proliferate in response to neural cell loss associated with the development of the disease.

Although there is a distinct lack of evidence for DNA repair in neural stem cells being the root cause of age-related neurodegenerative diseases, many studies have identified that inefficient DNA repair in the adult brain could make a significant contribution to the progression of the disease. In fact, it has been suggested that DNA repair defects could represent a unifying mechanism in many neurodegenerative diseases, highlighting the potential for therapeutic intervention. For example, mutations in XRCC1, required for the repair of DNA single-strand breaks, have been associated with multiple neurological symptoms, including cerebellar ataxia. Cells from individuals with XRCC1 mutations display reduced rates of SSB repair, elevated levels of ADP-ribosylation and overactivation of PARP1, which overall resulted in the persistence of unrepaired DNA SSBs (Hoch et al. 2017). This persistence of DNA SSBs is believed to have a role in the pathology of cerebellar ataxia in patients with XRCC1 mutations and identifies inhibition of PARP1 as a potentially good therapeutic target in such patients. This could be promising, as clinical trials using PARP inhibitors are already taking place for other diseases such as breast and ovarian cancer (Fong et al. 2009). Defects in DNA repair have also been identified in patients suffering with Huntington's disease, an inherited and progressive neurodegenerative disorder characterised by uncontrolled movements and loss of cognition due to loss of neural cells in the brain. DNA repair pathways have been identified as the most important modifiers affecting the age of disease onset in Huntington's patients, suggesting manipulation of these processes could be of benefit in delaying the disease process. The huntingtin protein affected by the disease is known to localize to sites of DNA damage in an ATM-dependent manner, where it acts as a scaffold for proteins involved in the DNA damage response pathway following oxidative stress (Maiuri et al. 2017). This role for huntingtin in DNA repair in response to oxidative stress is important, as elevated levels of ROS observed during ageing could potentially accelerate the onset of disease in Huntington's patients. These studies show that DNA repair is an important determinant of neurological diseases, and although it has not been formally shown, age-related defects in DNA repair in neural stem cells could contribute to the loss of neurons with ageing via a decrease in regeneration capacity following apoptosis of mature neurons.

TREATMENTS FOR DISEASES OF STEM CELL AGEING

As mentioned, changes in the function of stem cells associated with ageing can directly contribute to the development of some human disease phenotypes such as those related to the haematopoietic system. In contrast, there is little evidence for the contribution of stem cell functional decline in certain other age-related diseases, such as those of the nervous system including PD, AD and stroke. Despite these differences, most age-associated diseases have the potential to be cured by stem cells, which could be used to replace cells which have been lost or damaged during the course of the disease. Furthermore, the development of treatments targeted towards DNA repair pathways to try and induce 'synthetic lethality' within diseased cells is currently being investigated. The principle of this concept has been well studied in the context of cancer (Kaelin 2005). Many cancer cells already have a DNA repair defect and it is considered to be potentially beneficial to inhibit other DNA repair pathways to induce death of the cancer cells whilst leaving normal cells unaffected. This allows a targeted treatment approach which spares normal, healthy cells, unlike many chemotherapeutic agents used in the clinic today.

Diseases of the haematopoietic system

Blood disorders of stem cells within the myeloid compartment such as MDS and AML have been treated with stem cells for a number of years through a procedure known as a bone marrow transplant (BMT). Although other treatments are currently being used, a BMT is considered to be the only curative treatment for these conditions. One potential problem with using BMT as a treatment for MDS in particular is that studies have identified that the success rates of this procedure decreases considerably with age (Gupta et al. 2010). With most MDS patients being over 60 years of age, there clearly need to be some improvements to procedures to increase this success rate.

At present, there are three types of BM stem cell transplant: (1) autologous, (2) allogenic and (3) umbilical cord blood. Autologous transplant involves extracting the patient's own BM stem cells before chemotherapy and IR treatment, and then transplanting these cells to repopulate the BM. This is not ideal, as although the cells will not be rejected, it is possible that the cells causing the blood cancer will return following transplant. Second, patients can have an allogenic transplant in which cells are obtained from a matched donor. This proves more successful than autologous transplantation, as it can be certain that the transplanted cells are healthy. However finding a matched donor can sometimes be problematic. Last, cells can be obtained from umbilical cord blood following the birthing process and used for transplantation. In this case, the number of stem cells obtained is very low and the recovery of normal blood cell numbers can be a long process, but these cells require less donor matching, which is an advantage.

If we take BMT as an example, it could be considered that this branch of medicine could benefit significantly from advances in stem cell research such as reprogramming somatic cells to iPSCs. iPSCs are the most recently generated type of stem cells (Takahasi and Yamanaka, 2006) and are adult cells which have been genetically reprogrammed to express genes required for the cells to return to an embryonic-like, undifferentiated state. Mouse iPSCs were first generated in 2006 by Yamanaka and colleagues (Takahashi and Yamanaka 2006) through the overexpression of just four genes (Oct3/4, Sox2, Klf4 and c-Myc) in adult cells, and the first human iPSCs followed shortly after in 2007 (Takahashi et al. 2007, Yu et al. 2007). iPSCs are most similar to embryonic stem cells in that they are pluripotent and can give rise to all of the cells of the body, including HSCs that could be used for BMT. Use of iPSC would eliminate the requirement for donors of adult and cord blood stem cells, as there would be no issues with immune rejection and large cell numbers could be obtained. In fact, it is hoped that cells could be taken from a non-diseased tissue of a blood cancer patient, for example, the skin, reprogrammed into iPSCs and then expanded and differentiated in vitro into BM stem cells to be used for transplantation. This sort of procedure would revolutionise transplantation medicine as a blood cancer treatment and illustrates the importance of stem cell research.

So far, the myeloid compartment of the haematopoietic system has been discussed, as with age, the differentiation of HSCs is skewed towards this cell lineage, resulting in a decreased production of lymphoid cells (Kim et al. 2003, Rossi et al. 2005). However, it is also important to consider the lymphoid compartment, as DNA repair processes in particular have an important role in B and T cell differentiation. There is little direct evidence suggesting that age-related changes in HSCs contribute to the development of lymphoid diseases, and, in fact, the majority of paediatric leukaemia cases tend to be lymphoid in origin, whereas leukaemia in older individuals is more likely to be of myeloid origin. This suggests that these two types of disease have different mechanisms of development and depend highly upon the developmental potential of HSCs at different stages of life. This is illustrated by a study looking at ectopic expression of the BCR-ABL oncogene, known to be a driver of chronic myeloid leukaemia in humans (Signer et al. 2007). Expression of this gene in young mice resulted in the development of myeloproliferative disease accompanied by B cell leukaemia, whereas in older animals, no lymphoid involvement was observed, demonstrating the altered potential of HSCs with age.

DNA repair in cells of the lymphoid lineage is particularly important for the process of V(D)J recombination to generate B and T cells with antigen-specific receptors (Jung et al. 2006). This is mediated by the generation and repair of DSBs using the NHEJ machinery, including the Artemis nuclease and DNA-PK (Ma et al. 2005) (further discussed in Chapter 7). If this process is perturbed by mutations in Artemis, this results in severe combined immunodeficiency in humans, possessing a NHEJ defect (Moshous et al. 2001). This presents clinically with an increase in opportunistic infections and failure to thrive within the first few months of life. Furthermore, mutations in other DNA repair genes such as DNA ligase IV can result in a similar phenotype. Patients with DNA ligase IV mutations have a more variable presentation but include characteristics of radiosensitive leukaemia, immunodeficiency, growth retardation and microcephaly (Riballo et al. 1999, O'Driscoll et al. 2001, van der Burg et al. 2006). There are distinct differences in the DNA repair kinetics, depending upon which mutation is present within the disease. For example, by measuring gamma-H2AX recruitment, it was shown that DNA ligase IV-deficient cells display slow DSB repair kinetics, whereas cells deficient in Artemis show a normal initial repair response but fail to repair 15%–20% of DNA DSBs at later time points (Riballo et al. 2004, Wang et al. 2005, van der Burg et al. 2006, Darroudi et al. 2007). This indicates that the variable patient phenotypes in terms of immunodeficiency could be explained by the different effects genetic mutations have on DNA repair kinetics. Similarly to diseases of the myeloid compartment, diseases of the lymphoid compartment can also be treated using BMT and have the potential to be cured by the use of iPSCs.

Diseases of the central nervous system

In contrast to blood disorders, the use of stem cells for the treatment of age-related diseases of the central nervous system is not well established. As there is little evidence that aberrant neural stem cell function is associated with the pathology of AD, PD and stroke, research in these areas is more directed at the potential for transplantation of more mature neural cells to replace neurons lost in the disease process. It is hoped that these mature neurons could be produced from iPSCs generated from the patient's own cells and contribute to restoring brain function. In theory, replacement of lost neurons with functional neurons should alleviate symptoms of PD, which indeed can be seen when human fetel ventral mesencephalic tissue containing developing nigral dopaminergic cells is transplanted into some patients (Kordower et al. 1995, Piccini et al. 1999). Despite these promising results with fetel tissue, there is still a long way to go in terms of developing a stem cell-based treatment for PD. Cells with some of the properties of dopaminergic neuroblasts have been generated in vitro from a variety of stem cell sources and appear to have some functional effects in animal models of the disease (Dezawa et al. 2004, Takagi et al. 2005). However, not all of the functional properties of these cells have been established, for example, the restoration of dopamine release in vivo, and this work would need to be completed before human trials with neural stem cells could commence. The outlook for the use of stem cells in the treatment of AD is not as promising. Patients suffering from AD have much more widespread brain damage involving many cell types, and it is possible that the signalling mechanisms required for proper neural stem cell differentiation and maintenance of mature neurons would not be intact, preventing the induction of neuronal differentiation and functions. In the case of stroke patients, there are currently no available treatments to promote recovery, so if treatment using stem cells could induce even minor improvements, this would be advantageous. Animal studies have yielded positive results, with transplantation of neural stem cells and progenitors improving migration of new neurons towards the damaged part of the brain and improving motor functions in rats and mice following a stroke (Kelly et al. 2004, Ikeda et al. 2005, Hayashi et al. 2006). Human trials are a little behind these advances, but have shown promising results with transplantation of neurons generated from a teratocarcinoma cell line (Kondziolka et al. 2000, Meltzer et al. 2001), demonstrating the potential benefits of stem cell transplantation for human stroke patients.

SUMMARY

In summary, this chapter has discussed the importance of DNA repair mechanisms used by stem cells to maintain genomic stability. DNA repair is of great importance in all cell types, but is arguably most important in stem cells. This is because they provide a lifelong resource for replacing lost or damaged cells and must remain functional for the entire lifespan of an organism; therefore, their genomic integrity must be preserved. Evidence has been presented which indicates a loss of function of many stem cell compartments with ageing, which recapitulates phenotypes observed in inherited disorders of DNA repair. The interplay between stem cell function and DNA repair may underlie the age-associated decline in stem cell function, but also may have a role to play in the development of age-related diseases. The hypothesis that ineffective DNA repair, combined with increased levels of DNA damage, makes a significant contribution to the ageing process is still an active area of research.

Furthermore, the interplay between DNA damage and DNA repair in the context of stem cells is not well characterised. Studies of the haematopoietic system and HSCs are leading this field and will provide a platform for similar studies to be carried out in other systems such as the central nervous system. One key aspect is to understand exactly how and why DNA repair mechanisms in stem cells change during ageing and if these changes can potentially be reversed to combat the ageing process. The use of iPSCs will be an important part of this development, as they can be used as an in vitro model to study such mechanisms and in the future may become useful as a stem cell replacement therapy in ageing and diseased individuals. In the context of diseases including cancer, increased knowledge of the DNA repair mechanisms used by stem cells will provide clinical benefits. For example, the use of DNA damaging treatments may be reconsidered and the use of more targeted treatments against specific components of the DNA damage response pathway may become a more common practice. This would hopefully reduce off-target effects and promote the development of more effective cancer treatments which reduce the risk of relapse. Overall, study of the DNA repair mechanisms used by stem cells will ultimately increase our knowledge and potential to manipulate these processes for the treatment of human diseases.

REFERENCES

Adams, B. R. et al. 2010. ATM-independent, high-fidelity nonhomologous end joining predominates in human embryonic stem cells. *Aging (Albany NY)* 2(9): 582–596.

Adams, P. D. et al. 2015. Aging-induced stem cell mutations as drivers for disease and cancer. *Cell Stem Cell* 16(6): 601–612.

Ahlenius, H. et al. 2009. Neural stem and progenitor cells retain their potential for proliferation and differentiation into functional neurons despite lower number in aged brain. *J Neurosci* 29(14): 4408–4419.

Ahmed, N. U. et al. 1999. High levels of 8-hydroxy-2′-deoxyguanosine appear in normal human epidermis after a single dose of ultraviolet radiation. *Br J Dermatol* 140(2): 226–231.

Allen, D. M. et al. 2001. Ataxia telangiectasia mutated is essential during adult neurogenesis. *Genes Dev* 15(5): 554–566.

Arck, P. C. et al. 2006. Towards a 'free radical theory of graying': Melanocyte apoptosis in the aging human hair follicle is an indicator of oxidative stress induced tissue damage. *FASEB J* 20(9): 1567–1569.

Arvidsson, A. et al. 2002. Neuronal replacement from endogenous precursors in the adult brain after stroke. *Nat Med* 8(9): 963–970.

Auerbach, A. D. 1992. Fanconi anemia and leukemia: Tracking the genes. *Leukemia* 6(Suppl 1): 1–4.

Aul, C. et al. 2001. Epidemiological features of myelodysplastic syndromes: Results from regional cancer surveys and hospital-based statistics. *Int J Hematol* 73(4): 405–410.

Bao, S. et al. 2006. Glioma stem cells promote radioresistance by preferential activation of the DNA damage response. *Nature* 4447120: 756–760.

Barazzuol, L. et al. 2017. A coordinated DNA damage response promotes adult quiescent neural stem cell activation. *PLoS Biol* 15(5): e2001264.

Barker, N. 2014. Adult intestinal stem cells: Critical drivers of epithelial homeostasis and regeneration. *Nat Rev Mol Cell Biol* 15(1): 19–33.

Beckman, K. B. and B. N. Ames 1998. The free radical theory of aging matures. *Physiol Rev* 78(2): 547–581.

Beerman, I. et al. 2014. Quiescent hematopoietic stem cells accumulate DNA damage during aging that is repaired upon entry into cell cycle. *Cell Stem Cell* 15(1): 37–50.

Behrens, A. et al. 2014. Impact of genomic damage and ageing on stem cell function. *Nat Cell Biol* 16(3): 201–207.

Bernet, J. D. et al. 2014. p38 MAPK signaling underlies a cell-autonomous loss of stem cell self-renewal in skeletal muscle of aged mice. *Nat Med* 20(3): 265–271.

Bernstein, C. et al. 2013. DNA damage, DNA repair and cancer. In C. Chen, editor, *New Research Directions in DNA Repair*, Chap. 16. InTech. doi: 10.5772/53919.

Blair, A. and H. J. Sutherland 2000. Primitive acute myeloid leukemia cells with long-term proliferative ability *in vitro* and *in vivo* lack surface expression of c-kit (CD117). *Exp Hematol* 28(6): 660–671.

Blaisdell, J. O. et al. 2001. Base excision repair processing of radiation-induced clustered DNA lesions. *Radiat Prot Dosimetry* 97(1): 25–31.

Bogomazova, A. N. et al. 2011. Error-prone nonhomologous end joining repair operates in human pluripotent stem cells during late G2. *Aging (Albany NY)* 3(6): 584–596.

Bonnet, D. and J. E. Dick 1997. Human acute myeloid leukemia is organized as a hierarchy that originates from a primitive hematopoietic cell. *Nat Med* 3(7): 730–737.

Brack, A. S. et al. 2007. Increased Wnt signaling during aging alters muscle stem cell fate and increases fibrosis. *Science* 317(5839): 807–810.

Branzei, D. and M. Foiani 2008. Regulation of DNA repair throughout the cell cycle. *Nat Rev Mol Cell Biol* 9(4): 297–308.

Caldecott, K. W. 2008. Single-strand break repair and genetic disease. *Nat Rev Genet* 9(8): 619–631.

Carlson, B. M. and J. A. Faulkner 1989. Muscle transplantation between young and old rats: Age of host determines recovery. *Am J Physiol* 256(6 Pt 1): C1262–1266.

Casorelli, I. et al. 2007. Methylation damage response in hematopoietic progenitor cells. *DNA Repair (Amst)* 6(8): 1170–1178.

Chance, B. et al. 1979. Hydroperoxide metabolism in mammalian organs. *Physiol Rev* 59(3): 527–605.

Chang, C. H. et al. 2015. Mammary stem cells and tumor-initiating cells are more resistant to apoptosis and exhibit increased DNA repair activity in response to DNA damage. *Stem Cell Reports* 5(3): 378–391.

Chatterjee, M. and Y. T. Ip 2009. Pathogenic stimulation of intestinal stem cell response in Drosophila. *J Cell Physiol* 220(3): 664–671.

Chen, Q. et al. 1995. Oxidative DNA damage and senescence of human diploid fibroblast cells. *Proc Natl Acad Sci USA* 92(10): 4337–4341.

Cheung, T. H. and T. A. Rando 2013. Molecular regulation of stem cell quiescence. *Nat Rev Mol Cell Biol* 14(6): 329–340.

Choi, N. H. et al. 2008. Age-related changes in *Drosophila* midgut are associated with PVF2, a PDGF/VEGF-like growth factor. *Aging Cell* 7(3): 318–334.

Cleaver, J. E. 1969. Xeroderma pigmentosum: A human disease in which an initial stage of DNA repair is defective. *Proc Natl Acad Sci USA* 63(2): 428–435.

Coppe, J. P. et al. 2010. The senescence-associated secretory phenotype: The dark side of tumor suppression. *Annu Rev Pathol* 5: 99–118.

Cosgrove, B. D. et al. 2014. Rejuvenation of the muscle stem cell population restores strength to injured aged muscles. *Nat Med* 20(3): 255–264.

Cousin, W. et al. 2013. Regenerative capacity of old muscle stem cells declines without significant accumulation of DNA damage. *PLOS ONE* 8(5): e63528.

Darroudi, F. et al. 2007. Role of Artemis in DSB repair and guarding chromosomal stability following exposure to ionizing radiation at different stages of cell cycle. *Mutat Res* 615(1–2): 111–124.

Davis, A. J. and D. J. Chen 2013. DNA double strand break repair via non-homologous end-joining. *Transl Cancer Res* 2(3): 130–143.

Desai, A. et al. 2014. CD133+ cells contribute to radioresistance via altered regulation of DNA repair genes in human lung cancer cells. *Radiother Oncol* 110(3): 538–545.

Devanand, D. P. et al. 2000. Olfactory deficits in patients with mild cognitive impairment predict Alzheimer's disease at follow-up. *Am J Psychiatry* 157(9): 1399–1405.

de Waard, H. et al. 2008. Cell-type-specific consequences of nucleotide excision repair deficiencies: Embryonic stem cells versus fibroblasts. *DNA Repair (Amst)* 7(10): 1659–1669.

Dezawa, M. et al. 2004. Specific induction of neuronal cells from bone marrow stromal cells and application for autologous transplantation. *J Clin Invest* 113(12): 1701–1710.

Dick, J. E. 2005. Acute myeloid leukemia stem cells. *Ann N Y Acad Sci* 1044: 1–5.

Diehn, M. et al. 2009. Association of reactive oxygen species levels and radioresistance in cancer stem cells. *Nature* 4587239: 780–783.

Dong, C. M. et al. 2014. A stress-induced cellular aging model with post-natal neural stem cells. *Cell Death Dis* 5: e1116.

Drummond-Barbosa, D. 2008. Stem cells, their niches and the systemic environment: An aging network. *Genetics* 180(4): 1787–1797.

Fan, J. et al. 2010. Cells expressing FLT3/ITD mutations exhibit elevated repair errors generated through alternative NHEJ pathways: Implications for genomic instability and therapy. *Blood* 116(24): 5298–5305.

Flach, J. et al. 2014. Replication stress is a potent driver of functional decline in ageing haematopoietic stem cells. *Nature* 512(7513): 198–202.

Fong, P. C. et al. 2009. Inhibition of poly(ADP-ribose) polymerase in tumors from BRCA mutation carriers. *N Engl J Med* 361(2): 123–134.

Fousteri, M. and L. H. Mullenders 2008. Transcription-coupled nucleotide excision repair in mammalian cells: Molecular mechanisms and biological effects. *Cell Res* 18(1): 73–84.

Francis, R. and C. Richardson 2007. Multipotent hematopoietic cells susceptible to alternative double-strand break repair pathways that promote genome rearrangements. *Genes Dev* 21(9): 1064–1074.

Gaymes, T. J. et al. 2002. Myeloid leukemias have increased activity of the nonhomologous end-joining pathway and concomitant DNA misrepair that is dependent on the Ku70/86 heterodimer. *Cancer Res* 62(10): 2791–2797.

Geiger, H. et al. 2013. The ageing haematopoietic stem cell compartment. *Nat Rev Immunol* 13(5): 376–389.

Giangreco, A. et al. 2008. Epidermal stem cells are retained *in vivo* throughout skin aging. *Aging Cell* 7(2): 250–259.

Gibson, M. C. and E. Schultz 1983. Age-related differences in absolute numbers of skeletal muscle satellite cells. *Muscle Nerve* 6(8): 574–580.

Gonzalez, F. et al. 2013. Homologous recombination DNA repair genes play a critical role in reprogramming to a pluripotent state. *Cell Rep* 3(3): 651–660.

Gupta, V. et al. 2010. Impact of age on outcomes after bone marrow transplantation for acquired aplastic anemia using HLA-matched sibling donors. *Haematologica* 95(12): 2119–2125.

Hayashi, J. et al. 2006. Primate embryonic stem cell-derived neuronal progenitors transplanted into ischemic brain. *J Cereb Blood Flow Metab* 26(7): 906–914.

Hayflick, L. and P. S. Moorhead 1961. The serial cultivation of human diploid cell strains. *Exp Cell Res* 25: 585–621.

Henriques, C. M. and M. G. Ferreira 2012. Consequences of telomere shortening during lifespan. *Curr Opin Cell Biol* 24(6): 804–808.

Hildrestrand, G. A. et al. 2007. The capacity to remove 8-oxoG is enhanced in newborn neural stem/progenitor cells and decreases in juvenile mice and upon cell differentiation. *DNA Repair (Amst)* 6(6): 723–732.

Hoch, N. C. et al. 2017. XRCC1 mutation is associated with PARP1 hyperactivation and cerebellar ataxia. *Nature* 541(7635): 87–91.

Hsu, P. H. et al. 2007. Nucleotide excision repair phenotype of human acute myeloid leukemia cell lines at various stages of differentiation. *Mutat Res* 614(1–2): 3–15.

Hua, G. et al. 2012. Crypt base columnar stem cells in small intestines of mice are radioresistant. *Gastroenterology* 143(5): 1266–1276.

Hung, W. W. et al. 2011. Recent trends in chronic disease, impairment and disability among older adults in the United States. *BMC Geriatr* 11: 47.

Ikeda, R. et al. 2005. Transplantation of neural cells derived from retinoic acid-treated cynomolgus monkey embryonic stem cells successfully improved motor function of hemiplegic mice with experimental brain injury. *Neurobiol Dis* 20(1): 38–48.

Inomata, K. et al. 2009. Genotoxic stress abrogates renewal of melanocyte stem cells by triggering their differentiation. *Cell* 137(6): 1088–1099.

Ito, K. et al. 2004. Regulation of oxidative stress by ATM is required for self-renewal of haematopoietic stem cells. *Nature* 431(7011): 997–1002.

Jankowska, A. M. et al. 2008. Base excision repair dysfunction in a subgroup of patients with myelodysplastic syndrome. *Leukemia* 22(3): 551–558.

Jun, S. et al. 2016. LIG4 mediates Wnt signalling-induced radioresistance. *Nat Commun* 7: 10994.

Jung, D. et al. 2006. Mechanism and control of V(D)J recombination at the immunoglobulin heavy chain locus. *Annu Rev Immunol* 24: 541–570.

Kaelin, W. G., Jr. 2005. The concept of synthetic lethality in the context of anticancer therapy. *Nat Rev Cancer* 5(9): 689–698.

Kelly, S. et al. 2004. Transplanted human fetal neural stem cells survive, migrate, and differentiate in ischemic rat cerebral cortex. *Proc Natl Acad Sci USA* 101(32): 11839–11844.

Keyes, B. E. et al. 2013. Nfatc1 orchestrates aging in hair follicle stem cells. *Proc Natl Acad Sci USA* 110(51): E4950–4959.

Kim, H. M. et al. 2012. Increased CD13 expression reduces reactive oxygen species, promoting survival of liver cancer stem cells via an epithelial-mesenchymal transition-like phenomenon. *Ann Surg Oncol* 19(Suppl 3): S539–548.

Kim, M. et al. 2003. Major age-related changes of mouse hematopoietic stem/progenitor cells. *Ann N Y Acad Sci* 996: 195–208.

Kim, Y. J. and D. M. Wilson 3rd, 2012. Overview of base excision repair biochemistry. *Curr Mol Pharmacol* 5(1): 3–13.

Knoch, J. et al. 2012. Rare hereditary diseases with defects in DNA-repair. *Eur J Dermatol* 22(4): 443–455.

Kondziolka, D. et al. 2000. Transplantation of cultured human neuronal cells for patients with stroke. *Neurology* 55(4): 565–569.

Kordower, J. H. et al. 1995. Neuropathological evidence of graft survival and striatal reinnervation after the transplantation of fetal mesencephalic tissue in a patient with Parkinson's disease. *N Engl J Med* 332(17): 1118–1124.

Kreso, A. and J. E. Dick 2014. Evolution of the cancer stem cell model. *Cell Stem Cell* 14(3): 275–291.

Kulju, K. S. and J. M. Lehman 1995. Increased p53 protein associated with aging in human diploid fibroblasts. *Exp Cell Res* 217(2): 336–345.

Kulkarni, A. and D. M. Wilson 3rd, 2008. The involvement of DNA-damage and -repair defects in neurological dysfunction. *Am J Hum Genet* 82(3): 539–566.

Kunkel, T. A. and D. A. Erie 2005. DNA mismatch repair. *Annu Rev Biochem* 74: 681–710.

Kuramoto, K. et al. 2002. Chromosomal instability and radiosensitivity in myelodysplastic syndrome cells. *Leukemia* 16(11): 2253–2258.

Lapidot, T. et al. 1994. A cell initiating human acute myeloid leukaemia after transplantation into SCID mice. *Nature* 367: 645–648.

Lehmann, A. R. et al. 2011. Xeroderma pigmentosum. *Orphanet J Rare Dis* 6: 70.

Lenaz, G. 1998. Role of mitochondria in oxidative stress and ageing. *Biochim Biophys Acta* 1366(1–2): 53–67.

Li, L. and H. Clevers 2010. Coexistence of quiescent and active adult stem cells in mammals. *Science* 327(5965): 542–545.

Lin, B. et al. 2014. Human pluripotent stem cells have a novel mismatch repair-dependent damage response. *J Biol Chem* 289(35): 24314–24324.

Lopez-Otin, C. et al. 2013. The hallmarks of aging. *Cell* 153(6): 1194–1217.

Ma, X. 2012. Epidemiology of myelodysplastic syndromes. *Am J Med* 125(7 Suppl): S2–5.

Ma, Y. et al. 2005. The DNA-dependent protein kinase catalytic subunit phosphorylation sites in human Artemis. *J Biol Chem* 280(40): 33839–33846.

Mahaney, B. L. et al. 2009. Repair of ionizing radiation-induced DNA double-strand breaks by non-homologous end-joining. *Biochem J* 417(3): 639–650.

Maiuri, T. et al. 2017. Huntingtin is a scaffolding protein in the ATM oxidative DNA damage response complex. *Hum Mol Genet* 26(2): 395–406.

Mandal, P. K. et al. 2011. DNA damage response in adult stem cells: Pathways and consequences. *Nat Rev Mol Cell Biol* 12(3): 198–202.

Maslov, A. Y. et al. 2004. Neural stem cell detection, characterization, and age-related changes in the subventricular zone of mice. *J Neurosci* 24(7): 1726–1733.

Maynard, S. et al. 2008. Human embryonic stem cells have enhanced repair of multiple forms of DNA damage. *Stem Cells* 26(9): 2266–2274.

Meltzer, C. C. et al. 2001. Serial [18F] fluorodeoxyglucose positron emission tomography after human neuronal implantation for stroke. *Neurosurgery* 49(3): 586–591, discussion 591–582.

Micchelli, C. A. and N. Perrimon 2006. Evidence that stem cells reside in the adult Drosophila midgut epithelium. *Nature* 439(7075): 475–479.

Michl, J. et al. 2016. Interplay between Fanconi anemia and homologous recombination pathways in genome integrity. *EMBO J* 35(9): 909–923.

Mizumatsu, S. et al. 2003. Extreme sensitivity of adult neurogenesis to low doses of X-irradiation. *Cancer Res* 63(14): 4021–4027.

Moehrle, B. M. et al. 2015. Stem cell-specific mechanisms ensure genomic fidelity within HSCs and upon aging of HSCs. *Cell Rep* 13(11): 2412–2424.

Mohrin, M. et al. 2010. Hematopoietic stem cell quiescence promotes error-prone DNA repair and mutagenesis. *Cell Stem Cell* 7(2): 174–185.

Molofsky, A. V. et al. 2006. Increasing p16INK4a expression decreases forebrain progenitors and neurogenesis during ageing. *Nature* 443(7110): 448–452.

Morrison, S. J. and D. T. Scadden 2014. The bone marrow niche for haematopoietic stem cells. *Nature* 505(7483): 327–334.

Moshous, D. et al. 2001. Artemis, a novel DNA double-strand break repair/V(D)J recombination protein, is mutated in human severe combined immune deficiency. *Cell* 105(2): 177–186.

Narciso, L. et al. 2007. Terminally differentiated muscle cells are defective in base excision DNA repair and hypersensitive to oxygen injury. *Proc Natl Acad Sci USA* 104(43): 17010–17015.

Narita, M. et al. 2003. Rb-mediated heterochromatin formation and silencing of E2F target genes during cellular senescence. *Cell* 113(6): 703–716.

Navarro, S. et al. 2006. Hematopoietic dysfunction in a mouse model for Fanconi anemia group D1. *Mol Ther* 14(4): 525–535.

Ng, Y. Y. et al. 2009. Isolation of human and mouse hematopoietic stem cells. *Methods Mol Biol* 506: 13–21.

Niccoli, T. and L. Partridge 2012. Ageing as a risk factor for disease. *Curr Biol* 22(17): R741–752.

Nijnik, A. et al. 2007. DNA repair is limiting for haematopoietic stem cells during ageing. *Nature* 447(7145): 686–690.

Nilsson, L. et al. 2000. Isolation and characterization of hematopoietic progenitor/stem cells in 5q-deleted myelodysplastic syndromes: Evidence for involvement at the hematopoietic stem cell level. *Blood* 96(6): 2012–2021.

Nilsson, L. et al. 2002. Involvement and functional impairment of the CD34(+)CD38(-)Thy-1(+) hematopoietic stem cell pool in myelodysplastic syndromes with trisomy 8. *Blood* 100(1): 259–267.

Nilsson, L. et al. 2007. The molecular signature of MDS stem cells supports a stem-cell origin of 5q myelodysplastic syndromes. *Blood* 110(8): 3005–3014.

Nouspikel, T. 2007. DNA repair in differentiated cells: Some new answers to old questions. *Neuroscience* 145(4): 1213–1221.

Nouspikel, T. and P. C. Hanawalt 2000. Terminally differentiated human neurons repair transcribed genes but display attenuated global DNA repair and modulation of repair gene expression. *Mol Cell Biol* 20(5): 1562–1570.

Novotna, B. et al. 2009. Oxidative DNA damage in bone marrow cells of patients with low-risk myelodysplastic syndrome. *Leuk Res* 33(2): 340–343.

O'Driscoll, M. et al. 2001. DNA ligase IV mutations identified in patients exhibiting developmental delay and immunodeficiency. *Mol Cell* 8(6): 1175–1185.

Ortonne, J. P. 1990. Pigmentary changes of the ageing skin. *Br J Dermatol* 122(Suppl 35): 21–28.

Pang, W. W. et al. 2011. Human bone marrow hematopoietic stem cells are increased in frequency and myeloid-biased with age. *Proc Natl Acad Sci USA* 108(50): 20012–20017.

Papadopoulos, N. et al. 1994. Mutation of a mutL homolog in hereditary colon cancer. *Science* 263(5153): 1625–1629.

Papaemmanuil, E. et al. 2013. Clinical and biological implications of driver mutations in myelodysplastic syndromes. *Blood* 122(22): 3616–3627; quiz 3699.

Peddie, C. M. et al. 1997. Oxidative DNA damage in CD34+ myelodysplastic cells is associated with intracellular redox changes and elevated plasma tumour necrosis factor-alpha concentration. *Br J Haematol* 99(3): 625–631.

Petruseva, I. O. et al. 2014. Molecular mechanism of global genome nucleotide excision repair. *Acta Naturae* 6(1): 23–34.

Piccini, P. et al. 1999. Dopamine release from nigral transplants visualized *in vivo* in a Parkinson's patient. *Nat Neurosci* 2(12): 1137–1140.

Poppe, B. et al. 2001. Chromosomal aberrations in Bloom syndrome patients with myeloid malignancies. *Cancer Genet Cytogenet* 128(1): 39–42.

Potten, C. S. et al. 2001. Ageing of murine small intestinal stem cells. *Novartis Found Symp* 235: 66–79; discussion 79–84, 101–104.

Powell, S. and T. J. McMillan 1990. DNA damage and repair following treatment with ionizing radiation. *Radiother Oncol* 19(2): 95–108.

Pulliam-Leath, A. C. et al. 2010. Genetic disruption of both Fancc and Fancg in mice recapitulates the hematopoietic manifestations of Fanconi anemia. *Blood* 116(16): 2915–2920.

Rassool, F. V. et al. 2007. Reactive oxygen species, DNA damage, and error-prone repair: A model for genomic instability with progression in myeloid leukemia? *Cancer Res* 67(18): 8762–8771.

Renkawitz, J. et al. 2014. Mechanisms and principles of homology search during recombination. *Nat Rev Mol Cell Biol* 15(6): 369–383.

Rera, M. et al. 2011. Modulation of longevity and tissue homeostasis by the *Drosophila* PGC-1 homolog. *Cell Metab* 14(5): 623–634.

Riballo, E. et al. 1999. Identification of a defect in DNA ligase IV in a radiosensitive leukaemia patient. *Curr Biol* 9(13): 699–702.

Riballo, E. et al. 2004. A pathway of double-strand break rejoining dependent upon ATM, Artemis, and proteins locating to gamma-H2AX foci. *Mol Cell* 16(5): 715–724.

Risitano, A. M. et al. 2007. Function and malfunction of hematopoietic stem cells in primary bone marrow failure syndromes. *Curr Stem Cell Res Ther* 2(1): 39–52.

Rizvi, S. et al. 2014. Telomere length variations in aging and age-related diseases. *Curr Aging Sci* 7(3): 161–167.

Rossi, D. J. et al. 2005. Cell intrinsic alterations underlie hematopoietic stem cell aging. *Proc Natl Acad Sci USA* 102(26): 9194–9199.

Rossi, D. J. et al. 2007. Deficiencies in DNA damage repair limit the function of haematopoietic stem cells with age. *Nature* 447(7145): 725–729.

Rube, C. E. et al. 2011. Accumulation of DNA damage in hematopoietic stem and progenitor cells during human aging. *PLOS ONE* 6(3): e17487.

Saha, S. et al. 2016. Macrophage-derived extracellular vesicle-packaged WNTs rescue intestinal stem cells and enhance survival after radiation injury. *Nat Commun* 7: 13096.

Saretzki, G. et al. 2008. Downregulation of multiple stress defense mechanisms during differentiation of human embryonic stem cells. *Stem Cells* 26(2): 455–464.

Savatier, P. et al. 2002. Analysis of the cell cycle in mouse embryonic stem cells. *Methods Mol Biol* 185: 27–33.

Schneider, L. et al. 2013. DNA damage in mammalian neural stem cells leads to astrocytic differentiation mediated by BMP2 signaling through JAK-STAT. *Stem Cell Reports* 1(2): 123–138.

Schuler, N. and C. E. Rube 2013. Accumulation of DNA damage-induced chromatin alterations in tissue-specific stem cells: The driving force of aging? *PLOS ONE* 8(5): e63932.

Seedhouse, C. et al. 2004. Polymorphisms in genes involved in homologous recombination repair interact to increase the risk of developing acute myeloid leukemia. *Clin Cancer Res* 10(8): 2675–2680.

Shrivastav, M. et al. 2008. Regulation of DNA double-strand break repair pathway choice. *Cell Res* 18(1): 134–147.

Signer, R. A. et al. 2007. Age-related defects in B lymphopoiesis underlie the myeloid dominance of adult leukemia. *Blood* 110(6): 1831–1839.

Simsek, T. et al. 2010. The distinct metabolic profile of hematopoietic stem cells reflects their location in a hypoxic niche. *Cell Stem Cell* 7(3): 380–390.

Singh, S. K. et al. 2004. Identification of human brain tumour initiating cells. *Nature* 432(7015): 396–401.

Sosa, M. S. et al. 2014. Mechanisms of disseminated cancer cell dormancy: An awakening field. *Nat Rev Cancer* 14(9): 611–622.

Sotiropoulou, P. A. et al. 2010. Bcl-2 and accelerated DNA repair mediates resistance of hair follicle bulge stem cells to DNA-damage-induced cell death. *Nat Cell Biol* 12(6): 572–582.

Sousa-Victor, P. et al. 2014. Geriatric muscle stem cells switch reversible quiescence into senescence. *Nature* 506(7488): 316–321.

Spencer, J. A. et al. 2014. Direct measurement of local oxygen concentration in the bone marrow of live animals. *Nature* 508(7495): 269–273.

Steensma, D. P. et al. 2015. Clonal hematopoiesis of indeterminate potential and its distinction from myelodysplastic syndromes. *Blood* 126(1): 9–16.

Stojic, L. et al. 2004. Mismatch repair and DNA damage signalling. *DNA Repair (Amst)* 3(8–9): 1091–1101.

Sudo, K. et al. 2000. Age-associated characteristics of murine hematopoietic stem cells. *J Exp Med* 192(9): 1273–1280.

Takagi, Y. et al. 2005. Dopaminergic neurons generated from monkey embryonic stem cells function in a Parkinson primate model. *J Clin Invest* 115(1): 102–109.

Takahashi, K. et al. 2007. Induction of pluripotent stem cells from adult human fibroblasts by defined factors. *Cell* 131(5): 861–872.

Takahashi, K. et al. 2017. Preleukaemic clonal haemopoiesis and risk of therapy-related myeloid neoplasms: A case-control study. *Lancet Oncol* 18(1): 100–111.

Takahashi, K. and S. Yamanaka 2006. Induction of pluripotent stem cells from mouse embryonic and adult fibroblast cultures by defined factors. *Cell* 126(4): 663–676.

Takeda, N. et al. 2011. Interconversion between intestinal stem cell populations in distinct niches. *Science* 334(6061): 1420–1424.
Tao, S. et al. 2015. Wnt activity and basal niche position sensitize intestinal stem and progenitor cells to DNA damage. *EMBO J* 34(5): 624–640.
Tichy, E. D. et al. 2011. Mismatch and base excision repair proficiency in murine embryonic stem cells. *DNA Repair (Amst)* 10(4): 445–451.
Tinkum, K. L. et al. 2015. Fasting protects mice from lethal DNA damage by promoting small intestinal epithelial stem cell survival. *Proc Natl Acad Sci USA* 112(51): E7148–7154.
van der Burg, M. et al. 2006. A new type of radiosensitive T-B-NK+ severe combined immunodeficiency caused by a LIG4 mutation. *J Clin Invest* 116(1): 137–145.
van Rhenen, A. et al. 2005. High stem cell frequency in acute myeloid leukemia at diagnosis predicts high minimal residual disease and poor survival. *Clin Cancer Res* 11(18): 6520–6527.
Vaupel, J. W. 2010. Biodemography of human ageing. *Nature* 464(7288): 536–542.
Vaziri, H. et al. 1994. Evidence for a mitotic clock in human hematopoietic stem cells: Loss of telomeric DNA with age. *Proc Natl Acad Sci USA* 91(21): 9857–9860.
Verga Falzacappa, M. V. et al. 2012. Regulation of self-renewal in normal and cancer stem cells. *FEBS J* 279(19): 3559–3572.
Walter, D. et al. 2015. Exit from dormancy provokes DNA-damage-induced attrition in haematopoietic stem cells. *Nature* 520(7548): 549–552.
Wang, C. et al. 2013. FIP200 is required for maintenance and differentiation of postnatal neural stem cells. *Nat Neurosci* 16(5): 532–542.
Wang, J. et al. 2005. Artemis deficiency confers a DNA double-strand break repair defect and Artemis phosphorylation status is altered by DNA damage and cell cycle progression. *DNA Repair (Amst)* 4(5): 556–570.
Wang, J. C. and J. E. Dick 2005. Cancer stem cells: Lessons from leukemia. *Trends Cell Biol* 15(9): 494–501.
Warner, J. K. et al. 2004. Concepts of human leukemic development. *Oncogene* 23(43): 7164–7177.
Weeden, C. E. et al. 2017. Lung basal stem cells rapidly repair DNA damage using the error-prone nonhomologous end-joining pathway. *PLoS Biol* 15(1): e2000731.
Wilmoth, J. R. et al. 2000. Increase of maximum life-span in Sweden, 1861–1999. *Science* 289(5488): 2366–2368.
Yang, Y. G. et al. 2004. Ablation of PARP-1 does not interfere with the repair of DNA double-strand breaks, but compromises the reactivation of stalled replication forks. *Oncogene* 23(21): 3872–3882.
Ye, X. Q. et al. 2011. Mitochondrial and energy metabolism-related properties as novel indicators of lung cancer stem cells. *Int J Cancer* 129(4): 820–831.
Yin, C. C. et al. 2015. Clinical significance of newly emerged isolated del(20q) in patients following cytotoxic therapies. *Mod Pathol* 28(8): 1014–1022.
Yousefi, M. et al. 2016. Msi RNA-binding proteins control reserve intestinal stem cell quiescence. *J Cell Biol* 215(3): 401–413.
Yu, J. et al. 2007. Induced pluripotent stem cell lines derived from human somatic cells. *Science* 318(5858): 1917–1920.

Targeting Replication Stress in Sporadic Tumours

Marwan Kwok and Tatjana Stankovic

15

INTRODUCTION

Targeting replication stress by inhibitors of the ATR pathway represents a new approach to the management of cancer chemoresistance. Preclinical results obtained with ATR and Chk1 inhibitors have shown significant promise. The understanding of the mechanistic basis of ATR pathway inhibition is far from complete, and the clinical development of ATR pathway inhibitors is still in the initial stages. Consequently, identification of predictive biomarkers of responses underpinned by full mechanistic understanding remains a research priority. The rational selection of patients for clinical studies based on these biomarkers could translate into better therapeutic outcomes. It is likely that identification of these biomarkers could unravel other components of the replication stress responses that can also be targeted. In conclusion, the development of ATR pathway inhibitors exemplifies the emerging role of precision medicine in cancer treatment where vulnerabilities in cancer cells are systematically identified and targeted.

Replication stress (RS) is a state of interrupted DNA replication that frequently occurs in cancer cells (also discussed in Chapters 2, 6 & 12). During RS, the genome is exposed to replication defects where, despite the presence of regions of unreplicated DNA, cells continue to cycle (Zeman and Cimprich 2014). RS is characterised by a spectrum of DNA lesions that are sequentially generated in under-replicated genomic loci during proliferation. These 'markers' of replication stress include formation of ultrafine anaphase bridges and exposure of common fragile sites during mitosis, as well as the formation of structures in the G1 phase of the cell cycle, such as micronucleation and 53BP1 bodies, that shield DNA against excessive degradation. RS is counteracted by a number of responsive proteins and pathways including ATR, DNA helicases (BLM, WRN), homologous recombination repair (HR) proteins (RPA, Rad 51, BRCA1) and the Fanconi's anaemia complex that, depending upon the nature and degree of induced DNA lesions, associates with 53BP1 bodies.

If not properly counteracted, RS leads to DNA double-strand breaks (DSBs) and activation of the ATM/p53 pathway. The ATM protein kinase regulates the response to DNA DSBs by triggering signalling which synchronises DNA repair, cell-cycle arrest and apoptosis (Shiloh and Ziv 2013). In the presence of low levels of DSBs, ATM activates p53, leading to G1/S cell cycle arrest, and this facilitates HR through direct or indirect activation of a number of DNA repair proteins. Severe or persistent DNA damage leads to activation of ATM/p53-induced apoptosis. Consequently, loss of ATM/p53 function is beneficial to tumour progression, and RS represents an important cellular force driving ATM/p53 functional loss. Once selected for, ATM/p53 functional loss can further enhance the consequences of RS, exacerbate genetic alterations and facilitate tumour progression. Thus, one mechanism by which RS can drive tumourigenesis involves the following steps: presence of unreplicated DNA and accumulation of DNA damage, activation and functional loss of ATM/p53, followed by genomic instability. This sequence of events operates under oncogene activation in solid tumours but is also likely to represent a general principle of RS-driven tumourigenesis (Bartkova et al. 2006, Lucas et al. 2011, Barlow et al. 2013, Burrell et al. 2013). Consistent with this concept, *ATM* and *TP53* genomic alterations occur late in CLL tumourigenesis and are associated with genomic complexity, a footprint of genomic instability (Zeman and Cimprich 2014).

Although the full extent of the mechanism underlying replication stress following oncogene activation in cancer cells remains unclear, it has been suggested that the accelerated use of RPA protein during uncontrolled tumour proliferation could be a contributing factor (Toledo et al. 2013). Other potential mechanisms, such as the deregulation of replication initiation, have also been postulated (Hills and Diffley 2014, Macheret and Halazonetis 2015). Irrespective of the underlying mechanism, high constitutive levels of cellular replication stress render cancer cells particularly sensitive to suppression of replication stress responses.

The ATR pathway is a principal regulator of responses triggered by replication stress. Until recently, it has been assumed that ATR is physiologically indispensable for cellular survival and not suitable for therapeutic targeting (Brown and Baltimore 2003). Consistent with this notion, *ATR* mutations are uncommon in human cancers. Among the few cancer types where *ATR* mutations have been reported, the acquisition of these mutations has been associated with tumour progression and poor prognosis (Zighelboim et al. 2009). Indeed, in accordance with ATR's role in fundamental cellular processes, such as replication, abrogation of the ATR gene is embryonically lethal, while its deletion in adult mice results in rapid aging and stem cell loss (de Klein et al. 2000, Ruzankina et al. 2007). Furthermore, in patients with Seckel syndrome, in whom ATR signalling is defective due to hypomorphic germ line mutations of both ATR alleles, developmental impairment is a typical manifestation (O'Driscoll et al. 2003, Murga et al. 2009).

Consequently, whereas complete or near-complete abolition of ATR activity results in toxicity to healthy tissues, partial suppression of ATR activity is tolerated in healthy cells but not in tumour cells. In a study by Schoppy and colleagues (2012), reduction of ATR in mice to 10% of normal levels, by deleting one ATR allele and replacing the other with a Seckel mutation, resulted in minimal adverse effect in healthy haematopoietic and intestinal tissues. In contrast, suppression of ATR to this level restricted the growth of fibrosarcomas driven by *H-Ras* and *p53* loss, as well as acute myeloid leukaemia driven by *N-Ras* and the MLL-ENL translocation (Gilad et al. 2010). Such a reduction of ATR level similarly prevented the development of Myc-driven lymphomas and pancreatic tumours with high levels of replication stress (Murga et al. 2011). Given this differential sensitivity of healthy and tumour cells to ATR inhibition, it follows that the ATR pathway represents a potential therapeutic target.

THE ATR PATHWAY AND THE CONCEPT OF SYNTHETIC LETHALITY

The substantial functional redundancy within cellular DDR (DNA damage response) pathways has been recognised for some time. This is a situation where more than one pathway is involved in supporting a particular cellular function. Physiologically, this provides protection against disruption of efficient DDR, particularly in cancer cells where DDR genes are mutated with high frequency. However, when one pathway is disrupted, cells become heavily dependent on collateral pathways. Thus, in a scenario where two independent pathways regulate an essential DDR process, the absence of one pathway is compatible with cell survival, whereas the absence of both results in cell death. This gives rise to an emerging therapeutic concept known as synthetic lethality, in which collaborating pathways are abolished in cancer cells with an already existing DDR defect to induce cytotoxicity (Shaheen et al. 2011, McLornan et al. 2014).

From the aforementioned studies, it is apparent that tumours with increased levels of replication stress, such as those driven by Myc or Ras oncogenes, are particularly sensitive to ATR inhibition. Therefore, understanding other circumstances under which tumour cells become dependent on ATR is essential in order to identify the full extent of predictive biomarkers of sensitivity to ATR inhibition. There is evidence that such an 'addiction' to ATR may occur in tumour cells with *TP53* or *ATM* functional loss, genetic events that are frequent across various tumour types.

The loss of the G1/S cell cycle checkpoint due to *TP53* functional impairment imposes a dependence on the G2/M cell cycle checkpoint, controlled primarily through ATR via its downstream kinase Chk1. Consequently, ATR inhibition has synthetically lethal interaction with p53 deficiency. Nghiem and colleagues (2001) provided one of the earliest examples demonstrating hypersensitivity of G1/S checkpoint-deficient cells to ATR loss. In this study, investigators observed that the osteosarcoma cell line U2OS, in the absence of ATR function, displayed premature chromatin condensation in response to hydroxyurea or ultraviolet radiation–induced replication stress. Premature chromatin condensation is indicative of mitotic catastrophe, a form of cell death occurring during mitosis as a consequence of the accumulation of intolerable levels of DNA damage (Castedo et al. 2004). Moreover, repression of p53 function by overexpression of MDM2, a negative regulator of p53, markedly potentiated the lethal effect of ATR inhibition in these cells (Nghiem et al. 2001). These findings were recapitulated in a later study showing that a p53-deficient colorectal cell line, rendered ATR deficient by replacing both wild-type ATR alleles with Seckel alleles, exhibited marked sensitivity to hydroxyurea or cisplatin-induced replication stress. Notably, restoration of p53 function reduced its sensitivity to these agents (Sangster-Guity et al. 2011).

These in vitro studies were complemented by in vivo experiments, in which *ATR* knockout murine models were generated on the background of *TP53* knockout animals. Mice with concomitant loss of both ATR and p53 displayed markedly reduced survival compared to those harbouring only loss of ATR. In addition, animals carrying a combined loss of ATR and p53, but not those with loss of ATR or p53 alone, exhibited high levels of DNA damage (Ruzankina et al. 2009). In a further study using xenotransplantation models of both p53 wild-type and p53 mutant triple-negative breast cancer, inhibition of Chk1, a downstream ATR target, potentiated chemotherapy-induced apoptosis in p53 mutant xenografts but not in the p53 wild-type counterparts. Combining Chk1 inhibition with the topoisomerase inhibitor irinotecan resulted in suppression of tumour growth and prolongation of survival in p53 mutant, but not wild-type xenografts (Ma et al. 2012).

ATR inhibition leads to stalled replication forks, induction of replication stress and DNA damage. In the absence of ATR, cells become dependent on ATM for DNA repair. As a result, ATR deficiency is synthetically lethal with ATM functional loss. Consistent with this assumption, Reaper and colleagues (2011) observed that ATR inhibition was invariably cytotoxic to ATM-defective tumour cell lines. Moreover, ATR inhibition was found to be synergistic with replication stress-inducing genotoxic agents such as cisplatin and carboplatin in ATM-defective cells compared to those without such defects. Similar observations were made in cells subjected to pharmacological inhibition of ATM compared to those without ATM inhibition (Reaper et al. 2011).

A number of additional DNA repair defects confer enhanced sensitivity to ATR pathway inhibition. These defects include deficiency in the DNA repair proteins such as XRCC1, ERCC1, POLD1 or PRIM1 (Sultana et al. 2013, Mohni et al. 2014, Hocke et al. 2016). XRCC1 is a protein involved in base excision repair and single-strand break repair pathways, whereas ERCC1 mediates the repair of several types of DNA lesions including bulky adducts, DSBs and interstrand cross-links, and facilitates the separation of sister chromatids at fragile sites. POLD1 and PRIM1 are involved in DNA replication synthesis. Recently, an RNAi screen identified AT-rich interaction domain 1A (ARID1A) deficiency as an additional marker of sensitivity to ATR inhibition. Mutations in ARID1A are common in cancers and lead to disrupted topoisomerase localization and cell cycle progression, thus imposing on cancer cells a dependence on the ATR signalling pathway (Williamson et al. 2016).

Finally, a study by Flynn et al. (2015) suggested that the hypersensitivity to ATR inhibition in cancer cells is reliant on a mechanism of telomere maintenance known as alternative lengthening of telomeres (ALT), where telomeres are elongated through recombination. The absence of ATR leads to abrogation of ALT, compromising telomere stability in ALT-dependent cancer cells, and resulting in DNA damage, telomere loss as well as selective lethality of these cells.

PRE-CLINICAL STUDIES WITH SMALL MOLECULE INHIBITORS OF ATR AND CHK1

Given the atypical nature of PIKKs, the design and development of selective ATR kinase inhibitors has not been straightforward, and early inhibitors of ATR were not specific. However, more recently, two drug companies have manufactured potent and highly specific ATP-competitive ATR kinase inhibitors. They are, respectively, VE-821 and its analogue VX-970 produced by Vertex Pharmaceuticals, as well as AZ20 and AZD6738 developed by AstraZeneca (Charrier et al. 2011, Foote et al. 2013).

Pre-clinical data with these inhibitors are very encouraging. For example, in vitro, exposure of *TP53* mutant multiple myeloma cell lines to 1 μM VE-821 for 3 days resulted in a reduction in cell viability by >50% (Cottini et al. 2015). VE-821 monotherapy was also shown to be cytotoxic to glioma and lymphoma cell lines, to rectal carcinoma cell lines under hypoxic conditions, and to breast cancer and non-small cell lung cancer lines with an ERCC1 defect (Pires et al. 2012, Mohni et al. 2014, Middleton et al. 2015, Muralidharan et al. 2016). Colon cancer HCT166 cells, particularly those rendered deficient in ARID1A, responded to VE-821, VX-970 and AZ20, as reflected by a marked reduction (by >90%) in viability and clonogenic survival upon treatment with ATR inhibitor at doses of ≤1 μM (Williamson et al. 2016). In acute myeloid leukaemia (AML), AZ20 doses of 350 nM to 1.4 μM were sufficient to reduce viability of AML cell lines by 50% (EC_{50}), whereas in primary AML samples this was achieved with AZ20 doses of 800 nM to 27 μM (Ma et al. 2017). With AZD6738, single agent efficacy was seen in cell lines of HER2-positive breast cancer and ATM-deficient gastric cancer (Kim et al. 2017, Min et al. 2017). Finally, in chronic lymphocytic leukaemia (CLL), our group demonstrated that AZD6738 treatment for 4 days led to lethality in *TP53* or *ATM*-defective CLL cell lines as well as proliferating primary CLL samples, with an average EC_{50} of 1.4 μM in cell lines and 8.5 μM in primary CLL cells (Kwok et al. 2016).

In vivo, VX-970 suppressed tumour growth and prevented tumour establishment in mice xenografted with ARID1A-deficient HCT116 or ovarian cancer TOV-21G cells, while AZ20 showed efficacy in reducing tumour infiltration and improving survival of murine models transplanted with MLL-rearranged AML cells (Morgado-Palacin et al. 2016, Williamson et al. 2016). Moreover, we showed that AZD6738 monotherapy, administered orally for 2 weeks, led to a marked reduction in tumour load in PDX of biallelic *ATM* or *TP53* inactivated CLL, accompanied by a reduction in the proportion of CLL cells with these defects (Kwok et al. 2016). AZD6738 also reduced tumour growth and induced apoptosis in a murine xenograft model of ATM-deficient gastric cancer (Min et al. 2017).

There have been a number of pre-clinical studies suggesting therapeutic utility of inhibition of the ATR target Chk1. In vitro, single agent Chk1 inhibition with ≤1 μM AZD7762 led to >90% reduction in the viability of radioresistant breast cancer cell lines, as well as viability reductions of varying magnitudes in metastatic melanoma cell lines (Magnussen et al. 2015, Zhang et al. 2016). With MK-8776 (Sch900776) as a single agent, cytotoxic effects were observed in neuroblastoma lines (median EC_{50} 900 nM) and in myeloid leukaemia cell lines as well as BRCA-mutant ovarian cancer (Yuan et al. 2014, Kim et al. 2016). Cell lines that were sensitive to MK-8776 monotherapy were found to display aberrant Cdk2 activation in S phase upon Chk1 inhibition, leading to DSBs, possibly because these cells depended highly on constitutive suppression of Cdc25 phosphatases by Chk1 (Sakurikar et al. 2016). Moreover, LY2603618 exerted single-agent cytotoxicity on AML cell lines (EC_{50} 0.1–1.6 μM) and patient samples ($EC_{50} < 9$ μM), as well as in osteosarcoma cell lines (Duan et al. 2014, Zhao et al. 2016). Furthermore, single-agent PF-00477736 activity was evident in T cell acute lymphoblastic leukaemia (T-ALL) cell lines (EC_{50} 20–109 nM) and primary cells ($EC_{50} < 200$ nM), which overexpress Chk1 (Sarmento et al. 2015). PF-00477736 also exerted single-agent activity in cell lines of mantle cell lymphoma (EC_{50} 0.68 nM), germinal centre B-cell–like (GCB), diffuse large B-cell lymphoma (DLBCL; EC_{50} 10.2 nM) and activated B-cell (ABC) DLBCL (EC_{50} 87.3 nM) (Chila et al. 2015). Finally, V158411 inhibited cell proliferation and induced caspase activation in breast and ovarian cancer cell lines (Bryant et al. 2014).

In vivo, AZD7762 treatment for 3 days resulted in a delay in tumour growth in MCF-7/C6 radioresistant breast cancer xenografts (Zhang et al. 2016). In addition, MK-8776 (Sch900776) reduced tumour volume in MDA-MB-231 breast cancer xenografts (Zhou et al. 2017). Furthermore, in a xenotransplantation model of T-ALL, 30 days of treatment with PF-00477736 yielded a significant reduction in tumour growth compared to vehicle treated controls (Sarmento et al. 2015).

A summary of the major preclinical studies involving ATR inhibitors and Chk1 inhibitors is presented in Tables 15.1 and 15.2.

CHEMOTHERAPEUTIC COMBINATIONS INVOLVING ATR AND CHK1 INHIBITORS

ATR and Chk1 inhibitors can be readily combined with conventional chemotherapeutic agents. Whereas the effectiveness of ATR/Chk1 inhibitor monotherapy is often dependent on the acquisition of specific tumour phenotypes, such as p53 deficiency or Myc overexpression, ATR and Chk1 inhibitors synergistically enhance the effect of chemotherapy across a broad range of cancer phenotypes, DDR defective or otherwise. This can be attributed to the potentiation of chemotherapy-induced RS by ATR/Chk1 inhibitors. The ability of ATR pathway inhibitors to re-sensitize chemoresistant tumour cells to conventional chemotherapeutic agents is of particular clinical significance.

Table 15.1 **Preclinical studies involving ATR inhibitors.**

Malignancy	Study	Inhibitor	Experimental model	Biomarker of sensitivity	Treatments potentiated by Chk1 inhibitor
AML	Ma et al. 2017	AZ20	Cell lines, primary cells	–	Cytarabine
	Chauduri et al. 2014	VE-821	Cell lines	–	Wee1 inhibitor
Breast	Yazinski et al. 2017	AZ20, VE-821	Cell lines, primary cells	BRCA1 deficiency	PARP1 inhibitor
	Kim et al. 2017	AZD6738	Cell lines	–	Cisplatin
Cervical	Teng et al. 2015	ETP-46464	Cell lines	–	Cisplatin, radiotherapy
CLL	Kwok et al. 2016	AZD6738	Cell lines, primary cells, PDX	ATM, p53 deficiency	Chlorambucil, fludarabine, cyclophosphamide, bendamustine, ibrutinib
Colorectal	Hocke et al. 2016	VX-970, NU-6027	Cell lines	POLD1 deficiency, PRIM1 deficiency	–
Endometrial	Teng et al. 2015	ETP-46464	Cell lines	–	Cisplatin, radiotherapy
Esophageal	Leszczynska et al. 2016	VX-970	Cell lines, cell line xenografts	–	Cisplatin, carboplatin, radiotherapy
Gastric	Min et al. 2017	AZD6738	Cell lines, cell line xenografts	ATM deficiency	–
Lung	Vendetti et al. 2015	AZD6738	Cell lines, cell line xenografts	–	Cisplatin, gemcitabine, radiotherapy
	Hall et al. 2014	VX-970	Cell lines, primary cells, PDX	–	Cisplatin
Lymphoma	Muralidharan et al. 2016	AZ20, VE-821	Cell lines, cell line xenografts	–	BET inhibitor
	Menezes et al. 2015	WO2010/073034	Cell lines, cell line xenografts	ATM deficiency	–
Myeloma	Cottini et al. 2015	VE-821	Cell lines, primary cells	Myc expression, p53 deficiency	Piperlongumine
Ovarian	Kim et al. 2016	AZD6738	Cell lines, PDX	–	PARP1 inhibitor
	Huntoon et al. 2013	VE-821	Cell lines	–	Cisplatin, gemcitabine, PARP inhibitor
Pancreatic	Prevo et al. 2012	VE-821	Cell lines, primary cells	–	Gemcitabine, radiotherapy
	Fokas et al. 2012	VX-970	Cell lines, cell line xenografts	–	Gemcitabine, radiotherapy
Various	Toledo et al. 2011	ETP-46464	Cell lines	p53 deficiency, cyclin E overexpression	Hydroxyurea, Chk1 inhibitor
	Sultana et al. 2013	NU-6027	Cell lines	XRCC1 deficiency	Cisplatin
	Repear et al. 2011	VE-821	Cell lines	ATM, p53 deficiency	Cisplatin
	Pires et al. 2012		Cell lines	–	Radiotherapy
	Mohni et al. 2015		Cell lines	TLS polymerase, 53BP1	Cisplatin
	Krajewska et al. 2015		Cell lines	HRR deficient, Rad51 deficiency	–
	Middelton et al. 2015		Cell lines	HR/BER defects	–
	Josse et al. 2014	VE-821, VX-970	Cell lines, cell line xenografts	–	Topoisomerase inhibitor
	Williamson et al. 2016		Cell lines, cell line xenografts	ARID1A deficiency	–
	Mohni et al. 2014	VE-821	Cell lines	ERCC1 deficiency	–

Notes: AML, acute myeloid leukemia; CLL, chronic lymphocytic leukemia; PDX, patient-derived xenografts; TLS, translesion synthesis.

Table 15.2 Pre-clinical studies involving Chk1 inhibitors.

Malignancy	Study	Inhibitor	Experimental model	Biomarker of sensitivity	Treatments potentiated by Chk1 inhibitor
ALL	Sarmento et al. (2015)	PF477736	Cell lines, cell line xenografts, primary cells	–	–
AML	Zhao et al. (2016)	LY2603618	Cell lines, primary cells	–	Bcl-2 inhibitor
	Chaudhuri et al. (2014)	MK-8776	Cell lines	–	Wee1 inhibitor
	Yuan et al. (2014)	SCH900766, AZD7762	Cell lines	–	–
Bladder	Wang et al. (2015)	Gö6976	Cell lines	–	Gemcitabine
Breast	Zhang et al. (2016)	AZD7762	Cell lines, cell line xenografts	–	–
	Zhou et al. (2017)	MK-8776	Cell lines, cell line xenografts	–	Radiotherapy
	Ma et al. (2012)	UCN-01, AZD7762	PDX	p53 deficiency	Topoisomerase inhibitor
	Tang et al. (2012)		Cell lines, cell line xenografts	–	PARP inhibitor, radiotherapy
	Bryant et al. (2014)	V158411, PF477736, AZD7762	Cell lines	–	Gemcitabine, cisplatin
Colon	Martino-Echarri et al. (2014)	MK-8776, AZD7762	Cell lines	–	5-fluorouracil
	Origanti et al. (2013)	UCN-01	Murine models, orthotropic models	p53 deficiency, p21 deficiency	Topoisomerase inhibitor
Head and neck	Gadhikar et al. (2013)	AZD7762	Cell lines	p53 deficiency	Cisplatin
	Barker et al. (2016)	CCT244747	Cell lines, cell line xenografts, primary cells	–	Paclitaxel
Lung	Bartucci et al. (2012)	SB218078, AZD7762	Primary cells, PDX	–	Gemcitabine, cisplatin, paclitaxel
Lymphoma	Chila et al. (2015)	PF477736	Cell lines, cell line xenografts	–	Wee1 inhibitor
	Zemanova et al. (2016)	SCH900776	Cell lines, primary cells, murine models	p53 deficiency	Fludarabine, cytarabine, gemcitabine
	Murga et al. (2011)	UCN-01	Murine models	Myc expression	–
Melanoma	Magnussen et al. (2015)	AZD7762	Cell lines, cell line xenografts	–	Wee1 inhibitor
Myeloma	Pei et al. (2014)	CEP3891	Cell lines, primary cells	–	MEK1/2 inhibitor
	Dai et al. (2011)	UCN-01	Cell lines, primary cells, PDX	–	SRC inhibitor
Nasopharyngeal	Mak et al. (2015)	AZD7762	Cell lines, cell line xenografts	–	Wee1 inhibitor
Neuroblastoma	Russell et al. (2013)	MK-8776	Cell lines, cell line xenografts	–	Wee1 inhibitor
	Cole et al. (2011)	SB218078, TCS2312	Cell lines, primary cells	Myc expression	–
Oral SCC	Sankunny et al. (2014)	PF477736	Cell lines	ATM deficiency	Radiotherapy
Osteosarcoma	Duan et al. (2014)	LY2603618	Cell lines	–	Cisplatin
Ovarian	Kim et al. (2016)	MK-8776	Cell lines, PDX	–	PARP inhibitor
	Bryant et al. (2014)	V158411, PF477736, AZD7762	Cell lines	–	Gemcitabine, cisplatin
Pancreatic	Morgan et al. (2010)	AZD7762	Cell lines, cell line xenografts, PDX	–	Radiotherapy
	Vance et al. (2011)		Cell lines	p53 deficiency	PARP inhibitor
	Engelke et al. (2013)	MK-8776	Cell lines, cell line xenografts	HRR deficiency	Gemcitabine, radiotherapy

Table 15.2 *(Continued)* **Pre-clinical studies involving Chk1 inhibitors.**

Malignancy	Study	Inhibitor	Experimental model	Biomarker of sensitivity	Treatments potentiated by Chk1 inhibitor
Various	McNeely et al. (2010)	AZD7762	Cell lines	BRCA2 deficiency, XRCC3 deficiency, DNA-PK deficiency	Gemcitabine
	Krajewska et al. (2015)		Cell lines	HRR deficient, Rad51 deficiency	–
	Sakurikar et al. (2016)	MK-8776	Cell lines	CDK2 activation in S phase	–
	Dietlein et al. (2015)	PF477736	Cell lines, cell line xenografts, primary cells	KRAS mutation	MK2 inhibitor

Abbreviations: ALL, acute lymphoblastic leukaemia; AML, acute myeloid leukaemia; SCC, squamous cell carcinoma; PDX, patient-derived xenografts.

Among the numerous chemotherapeutic agents that synergize with ATR/Chk1 inhibitors, cisplatin and gemcitabine are the most frequently studied. Cisplatin causes DNA breaks and cross-links, whereas gemcitabine induces RS by decreasing the deoxyribonucleotide triphosphate (dNTP) required for DNA replication. ATR inhibitors were shown to sensitize a panel of lung cancer cell lines, cell line xenografts and PDX to cisplatin or gemcitabine, as well as ovarian cancer cell lines to cisplatin (Hall et al. 2014, Vendetti et al. 2015). The sensitization to cisplatin by ATR inhibitors is especially profound in tumour cells with deficiency in ATM, p53 or XRCC1, or with loss of 53BP1 or polymerase ζ (involved in translesion synthesis) (Reaper et al. 2011, Sangster-Guity et al. 2011, Sultana et al. 2013, Mohni et al. 2015). Other DNA damaging agents that synergize with ATR inhibitors include cytarabine, as observed in AML cell lines and primary cells, as well as chlorambucil, bendamustine, fludarabine and cyclophosphamide, as observed in CLL cell lines and primary CLL cells. Furthermore, ATR inhibitors sensitise tumour cells to topoisomerase inhibitors through disruption of DNA replication initiation and fork elongation processes (Josse et al. 2014).

It is important to note that ATR pathway inhibitors display synergy across a range of novel agents. Two studies, for example, have highlighted the synergistic interaction between poly(ADP-ribose) polymerase (PARP) inhibitors and ATR or Chk1 inhibitors in ovarian and breast cancer, respectively, using cell lines complemented by primary tumour samples or PDX (patient-derived xenografts) (Kim et al. 2016, Yazinski et al. 2017). PARP inhibitor treatment led to increased accumulation of cells in G2 phase, thus intensifying their dependence on the ATR pathway for checkpoint control and the maintenance of genome stability. ATR and/or Chk1 inhibitors have also shown synergy with inhibitors of other DNA repair molecules, such as Wee1, which contributes to the regulation of the G2/M checkpoint, and MK2, which is critical for prolonged checkpoint maintenance (Russell et al. 2013, Chaudhuri et al. 2014, Chila et al. 2015, Dietlein et al. 2015, Magnussen et al. 2015, Mak et al. 2015). Finally, therapeutic combinations with ATR pathway inhibitors are not restricted to agents targeting DDR. For example, ATR inhibitors are capable of enhancing the effect of a BET bromodomain inhibitor in Myc-driven lymphoma and a B cell receptor signalling inhibitor in CLL (Kwok et al. 2016, Muralidharan et al. 2016).

CLINICAL TRIALS OF SMALL MOLECULE INHIBITORS TARGETING THE ATR PATHWAY

Both AZD6738 and VX-970 have now entered phase I/II clinical testing. However, most of these studies (11 clinical trials on ATR inhibitors are ongoing for a range of malignancies) have been recently initiated, and no results have yet been reported. These studies are addressing the use of the ATR inhibitor either alone or in combination with a range of conventional and novel therapeutic agents, including cisplatin, carboplatin, gemcitabine, etoposide, irinotecan, PARP inhibitors olaparib and veliparib, and the programmed death-ligand 1 (PD-L1) immune checkpoint inhibitor durvalumab.

To explore the use of selective Chk1 inhibitors in oncology practice, several clinical studies have been carried out using AZD7762, LY2603618, LY2606368 and MK-8776 (Sch900776). These were phase I or II trials with the aim to assess safety and tolerability, to establish the maximum tolerable dose and, within certain studies, to determine response rate and durability. Unfortunately, the clinical development of AZD7762 was hampered by drug-related cardiac toxicity, which was dose limiting in some cases (Seto et al. 2013, Sausville et al. 2014). In contrast, LY2603618, LY2606368 and MK-8776 (Sch900776) had an acceptable toxicity profile, with which serious adverse events were rare (Karp et al. 2012, Daud et al. 2015, Doi et al. 2015, Hong et al. 2016). Moreover, early evidence of clinical activity was seen with these agents. In particular, 8 (out of 24) refractory AML patients achieved complete remission, and 15 (out of 30) individuals with advanced solid malignancy had partial response or stable disease following treatment with MK-8776 (Sch900776), either alone or in combination with chemotherapy (Karp et al. 2012, Daud et al. 2015). Similarly, a partial response was seen in two patients with advanced cancer following LY2606368 monotherapy (Hong et al. 2016).

Despite these promising responses, two phase II trials investigating LY2603618 in advanced lung and pancreatic cancer reported an absence of effect of LY2603618, used in combination with premetrexed or gemcitabine, respectively, when compared to premetrexed or gemcitabine alone (Scagliotti et al. 2016, Laquente et al. 2017). However, since the patients enrolled on these trials were unselected, it is possible that a substantial proportion of tumour cells were lacking RS that would yield synthetic lethality with ATR/Chk1 inhibition. Indeed, genomic analysis of an exceptional responder, who was cured of her invasive ureteric cancer with AZD7762 in combination with irinotecan, revealed mutation within *RAD50*, which attenuated ATM signalling and enforced dependence on the ATR pathway (Al-Ahmadie et al. 2014). This underscores the importance of focusing the use of ATR pathway inhibitors on patients with the appropriate genotype who are most likely to benefit from such treatment.

Unfortunately, the clinical development of AZD7762, LY2603618 and MK-8776 has been suspended. Currently, the Chk1 inhibitors under active clinical investigation include LY2606368 and CCT245737.

SUMMARY

While ATR and Chk1 act in the same pathway, ATR is upstream of Chk1 and therefore controls additional downstream processes. Targeting ATR may therefore produce more wide-ranging effects than targeting Chk1, and this is reflected in the ability of ATR inhibitors to potentiate a broader range of chemotherapeutic agents. It remains to be determined whether ATR inhibitors are clinically more effective than Chk1 inhibitors. This might depend on the type of malignancy or, indeed, on the phenotypes within a tumour type. For example, tumours with ATM deficiency might benefit more from ATR inhibition than from Chk1 inhibition, given the functional crosstalk between the ATM and ATR pathways.

A fundamental assumption underpinning models of synthetic lethality is that there are only two major pathways regulating a process. Therefore, if one pathway is defective, cellular demise is assured when the other pathway is blocked. However, this notion is an over-simplification of the range of collateral pathways regulating a single cellular process, many of which are not fully elucidated. Consequently, when one collateral pathway is therapeutically inhibited, tumour cells may upregulate an alternative pathway to mitigate the effects of the initial block, thereby resulting in therapeutic resistance.

Alongside ATM, DNA-dependent protein kinase (DNA-PK) plays a role in DSB repair through control of NHEJ. Recently, there have been reports pointing to a possible functional redundancy between ATR and DNA-PK in regulating downstream Chk1 activity (Lin et al. 2014). Buisson et al. (2015) demonstrated the existence of a DNA-PK-Chk1 backup pathway that can mediate resistance to ATR inhibitors. In this model, ATR inhibition is cytotoxic to a proportion of tumour cells with the highest levels of replication stress, but those with moderate levels are protected through the backup pathway. Simultaneous targeting of ATR and Chk1 or DNA-PK could potentially overcome this problem. Indeed, a potentiating interaction between ATR and Chk1 inhibitors has been reported, suggesting that the combined use of these inhibitors could be more efficacious than either agent alone (Sanjiv et al. 2016). However, it is important to bear in mind adverse effects and the potential toxicities of any combined use of ATR and Chk1 inhibitors.

Finally, both ATR and Chk1 inhibitors are kinase inhibitors. As observed from other small molecule kinase inhibitors that are currently in clinical use, there are a number of mechanisms through which a cancer cell can become resistant to such a strategy. One of the notorious problems is development of point mutations in the target kinase. This represents a serious clinical problem and usually requires targeting other components of the same pathway.

The long-term effects and toxicity of ATR and Chk1 inhibition are unknown, and this is often a source of contention. Targeting a process of such importance to genome integrity can be potentially dangerous if it also affects healthy cells. Moreover, sub-lethal targeting of tumour cells with ATR pathway inhibitors can promote tumourigenesis and clonal evolution, as these cells accumulate replication stress and DNA damage, but escape death (Gilad et al. 2010). It is therefore important that patients be selected for ATR pathway inhibitor treatment according to their tumour characteristics: whether they possess features that make them differentially sensitive to these inhibitors. By selecting patients whose tumours are hypersensitive to these inhibitors, tumour killing can be maximized while minimizing treatment duration, thereby reducing both toxicity to healthy tissues and the risk of tumour evolution. While a number of biomarkers of sensitivity to ATR pathway inhibition have already been discussed, it is likely that many others have yet to be discovered. Therefore, a search for identify of predictive biomarkers of sensitivity to ATR and Chk1 inhibitors should be an absolute priority.

The question remains whether ATR/Chk1 inhibitors should be administered as a single agent or in combination with chemotherapy or other targeted therapies. If so, what is the ideal therapeutic regimen to be used? An argument in favour of combined use with chemotherapy is that the addition of the latter would increase the potency of ATR pathway inhibitors and prevent resistance to single agents. This, however, needs to be balanced by a possible increase in toxicity. Whatever the choice, it is obvious that the ideal chemotherapeutic agent to be used in combination in different tumour types will need to be assessed through mechanistic and clinical studies. The application in the context of other targeted treatments may be of particular relevance to a number of malignancies where a substantial proportion of tumour cells is quiescent. These cells are unlikely to be susceptible to ATR pathway inhibition and may give rise to residual tumour cells from which therapeutic resistance can arise. Therefore, a combination of ATR pathway inhibitors with agents targeting pathways that are independent of replication could allow the simultaneous eradication of both proliferating and quiescent tumour populations. With this in mind, it is important to consider the scheduling and sequencing of ATR pathway inhibitors within therapeutic regimens.

REFERENCES

Al-Ahmadie, H., Iyer, G., Hohl, M., Asthana, S., Inagaki, A., Schultz, N., Hanrahan, A. J. et al. 2014. Synthetic lethality in ATM-deficient RAD50-mutant tumors underlies outlier response to cancer therapy. *Cancer Discov* 4: 1014–21.

Barker, H. E., Patel, R., McLaughlin, M., Schick, U., Zaidi, S., Nutting, C. M., Newbold, K. L., Bhide, S., Harrington, K. J. et al. 2016. CHK1 inhibition radiosensitizes head and neck cancers to paclitaxel-based chemoradiotherapy. *Mol Cancer Ther* 15: 2042–54.

Barlow, J. H., Faryabi, R. B., Callén, E., Wong, N., Malhowski, A., Chen, H. T., Gutierrez-Cruz, G. et al. 2013. Identification of early replicating fragile sites that contribute to genome instability. *Cell* 152(3): 620–32.

Bartkova, J., Rezaei, N., Liontos, M., Karakaidos, P., Kletsas, D., Issaeva, N., Vassiliou, L. V. et al. 2006. Oncogene-induced senescence is part of the tumorigenesis barrier imposed by DNA damage checkpoints. *Nature* 444(7119): 633–7.

Bartucci, M., Svensson, S., Romania, P., Datillo, R., Patrizii, M., Signore, M., Navarra, S. et al. 2012. Therapeutic targeting of Chk1 in NSCLC stem cells during chemotherapy. *Cell Death Differ* 19: 768–78.

Brown, E. J., Baltimore, D. 2003. Essential and dispensable roles of ATR in cell cycle arrest and genome maintenance. *Genes Dev* 17: 615–28.

Bryant, C., Rawlinson, R., Massey, A. J. 2014. Chk1 inhibition as a novel therapeutic strategy for treating triple-negative breast and ovarian cancers. *BMC Cancer* 14: 570.

Buisson, R., Boisvert, J. L., Benes, C. H., Zou, L. 2015. Distinct but concerted roles of ATR, DNA-PK, and Chk1 in countering replication stress during S phase. *Mol Cell* 59: 1011–24.

Burrell, R. A., McClelland, S. E., Endesfelder, D., Groth, P., Weller, M. C., Shaikh, N., Domingo, E. et al. 2013. Replication stress links structural and numerical cancer chromosomal instability. *Nature* 494(7438): 492–6.

Castedo, M., Perfettini, J. L., Roumier, T., Andreau, K., Medema, R., Kroemer, G. 2004. Cell death by mitotic catastrophe: A molecular definition. *Oncogene* 23: 2825–37.

Charrier, J. D., Durrant, S. J., Golec, J. M., Kay, D. P., Knegtel, R. M., MacCormick, S., Mortimore, M. et al. 2011. Discovery of potent and selective inhibitors of ataxia telangiectasia mutated and Rad3 related (ATR) protein kinase as potential anticancer agents. *J Med Chem* 54: 2320–30.

Chaudhuri, L., Vincelette, N. D., Koh, B. D., Naylor, R. M., Flatten, K. S., Peterson, K. L., McNally, A. et al. 2014. CHK1 and WEE1 inhibition combine synergistically to enhance therapeutic efficacy in acute myeloid leukemia *ex vivo*. *Haematologica* 99: 688–96.

Chila, R., Basana, A., Lupi, M., Guffanti, F., Gaudio, E., Rinaldi, A., Cascione, L. et al. 2015. Combined inhibition of Chk1 and Wee1 as a new therapeutic strategy for mantle cell lymphoma. *Oncotarget* 6: 3394–408.

Cole, K. A., Huggins, J., Laquaglia, M., Hulderman, C. E., Russell, M. R., Bosse, K., Diskin, S. J. et al. 2011. RNAi screen of the protein kinome identifies checkpoint kinase 1 (CHK1) as a therapeutic target in neuroblastoma. *Proc Natl Acad Sci USA* 108: 3336–41.

Cottini, F., Hideshima, T., Suzuki, R., Tai, Y. T., Bianchini, G., Richardson, P. G., Anderson, K. C., Tonon, G. 2015. Synthetic lethal approaches exploiting DNA damage in aggressive myeloma. *Cancer Discov* 5: 972–87.

Dai, Y., Chen, S., Shah, R., Pey, X. Y., Wang, L., Almenara, J. A., Kramer, L. B., Dent, P., Grant, S. 2011. Disruption of Src function potentiates Chk1-inhibitor-induced apoptosis in human multiple myeloma cells *in vitro* and *in vivo*. *Blood* 117: 1947–57.

Daud, A. I., Ashworth, M. T., Strosberg, J., Goldman, J. W., Mendelson, D., Springett, G., Venook, A. P. et al. 2015. Phase I dose-escalation trial of checkpoint kinase 1 inhibitor MK-8776 as monotherapy and in combination with gemcitabine in patients with advanced solid tumors. *J Clin Oncol* 33: 1060–6.

de Klein, A., Muijtjens, M., van Os, R., Verhoeven, Y., Smit, B., Carr, A. M., Lehmann, A. R., Hoeijmakers, J. H. 2000. Targeted disruption of the cell-cycle checkpoint gene ATR leads to early embryonic lethality in mice. *Curr Biol* 10: 479–82.

Dietlein, F., Kalb, B., Jokic, M., Noll, E. M., Strong, A., Tharun, L., Ozretic, L. et al. 2015. A synergistic interaction between Chk1- and MK2 inhibitors in KRAS-mutant cancer. *Cell* 162: 146–59.

Doi, T., Yoshino, T., Shitara, K., Matsubara, N., Fuse, N., Naito, Y., Uenaka, K., Nakamura, T., Hynes, S. M., Lin, A. B. 2015. Phase I study of LY2603618, a CHK1 inhibitor, in combination with gemcitabine in Japanese patients with solid tumors. *Anticancer Drugs* 26: 1043–53.

Duan, L., Perez, R. E., Hansen, M., Gitelis, S., Maki, C. G. 2014. Increasing cisplatin sensitivity by schedule-dependent inhibition of AKT and Chk1. *Cancer Biol Ther* 15: 1600–12.

Engelke, C. G., Parsels, L. A., Qian, Y., Zhang, Q., Karnak, D., Robertson, J. R., Tanska, D. M. et al. 2013. Sensitization of pancreatic cancer to chemoradiation by the Chk1 inhibitor MK8776. *Clin Cancer Res* 19: 4412–21.

Flynn, R. L., Cox, K. E., Jeitany, M., Wakimoto, H., Bryll, A. R., Ganem, N. J., Bersani, F. et al. 2015. Alternative lengthening of telomeres renders cancer cells hypersensitive to ATR inhibitors. *Science* 347: 273–7.

Foote, K. M., Blades, K., Cronin, A., Fillery, S., Guichard, S. S., Hassall, L., Hickson, I. et al. 2013. Discovery of 4-{4-[(3R)-3-Methylmorpholin-4-yl]-6-[1-(methylsulfonyl)cyclopropyl]pyrimidin-2-y l}-1H-indole (AZ20): A potent and selective inhibitor of ATR protein kinase with monotherapy *in vivo* antitumor activity. *J Med Chem* 56: 2125–38.

Gadhikar, M. A., Sciuto, M. R., Alves, M. V., Pickering, C. R., Osman, A. A., Neskey, D. M., Zhao, M., Fitzgerald, A. L., Myers, J. N., Frederick, M. J. 2013. Chk1/2 inhibition overcomes the cisplatin resistance of head and neck cancer cells secondary to the loss of functional p53. *Mol Cancer Ther* 12: 1860–73.

Gilad, O., Nabet, B. Y., Ragland, R. L., Schoppy, D. W., Smith, K. D., Durham, A. C., Brown, E. J. 2010. Combining ATR suppression with oncogenic Ras synergistically increases genomic instability, causing synthetic lethality or tumorigenesis in a dosage-dependent manner. *Cancer Res* 70: 9693–702.

Hall, A. B., Newsome, D., Wang, Y., Boucher, D. M., Eustace, B., Gu, Y., Hare, B. et al. 2014. Potentiation of tumor responses to DNA damaging therapy by the selective ATR inhibitor VX-970. *Oncotarget* 5: 5674–85.

Hills, S. A., Diffley, J. F. 2014. DNA replication and oncogene-induced replicative stress. *Curr Biol* 24: R435–44.

Hocke, S., Guo, Y., Job, A., Orth, M., Ziesch, A., Lauber, K., De Toni, E. N. et al. 2016. A synthetic lethal screen identifies ATR-inhibition as a novel therapeutic approach for POLD1-deficient cancers. *Oncotarget* 7: 7080–95.

Hong, D., Infante, J., Janku, F., Jones, S., Nguyen, L. M., Burris, H., Naing, A. et al. 2016. Phase I study of LY2606368, a checkpoint kinase 1 inhibitor, in patients with advanced cancer. *J Clin Oncol* 34: 1764–71.

Huntoon, C. J., Flatten, K. S., Wahner Hendrickson, A. E., Huehls, A. M., Sutor, T. L., Kaufmann, S. H., Karnitz, L. M. 2013. ATR inhibition broadly sensitizes ovarian cancer cells to chemotherapy independent of BRCA status. *Cancer Res* 73: 3683–91.

Josse, R., Martin, S. E., Guha, R., Ormanoglu, P., Pfister, T. D., Reaper, P. M., Barnes, C. S. et al. 2014. ATR inhibitors VE-821 and VX-970 sensitize cancer cells to topoisomerase I inhibitors by disabling DNA replication initiation and fork elongation responses. *Cancer Res* 74: 6968–79.

Karp, J. E., Thomas, B. M., Greer, J. M., Sorge, C., Gore, S. D., Pratz, K. W., Smith, B. D. et al. 2012. Phase I and pharmacologic trial of cytosine arabinoside with the selective checkpoint 1 inhibitor Sch 900776 in refractory acute leukemias. *Clin Cancer Res* 18: 6723–31.

Kim, H., George, E., Ragland, R. L., Rafail, S., Zhang, R., Krepler, C., Morgan, M., Herlyn, M., Brown, E. J., Simpkins, F. 2016. Targeting the ATR/CHK1 axis with PARP inhibition results in tumor regression in BRCA mutant ovarian cancer models. *Clin Cancer Res* 23(12): 3097–108.

Kim, H. J., Min, A., Im, S. A., Jang, H., Lee, K. H., Lau, A., Lee, M. et al. 2017. Anti-tumor activity of the ATR inhibitor AZD6738 in HER2 positive breast cancer cells. *Int J Cancer* 140: 109–19.

Krajewska, M., Fehrmann, R. S., Schoonen, P. M., Labib, S., de Vries, E. G., Franke, L., van Vugt, M. A. 2015. ATR inhibition preferentially targets homologous recombination-deficient tumor cells. *Oncogene* 34: 3474–81.

Kwok, M., Davies, N., Agathanggelou, A., Smith, E., Oldreive, C., Petermann, E., Stewart, G. et al. 2016. ATR inhibition induces synthetic lethality and overcomes chemoresistance in TP53- or ATM-defective chronic lymphocytic leukemia cells. *Blood* 127: 582–95.

Laquente, B., Lopez-Martin, J., Richards, D., Illerhaus, G., Chang, D. Z., Kim, G., Stella, P. et al. 2017. A phase II study to evaluate LY2603618 in combination with gemcitabine in pancreatic cancer patients. *BMC Cancer* 17: 137.

Leszczynska, K. B., Dobrynin, G., Leslie, R. E., Ient, J., Boumelha, A. J., Senra, J. M., Hawkins, M. A., Maughan, T., Mukherjee, S., Hammond, E. M. 2016. Preclinical testing of an Atr inhibitor demonstrates improved response to standard therapies for esophageal cancer. *Radiother Oncol* 121: 232–8.

Lukas, C., Savic, V., Bekker-Jensen, S., Doil, C., Neumann, B., Pedersen, R. S., Grofte, M. et al. 2011. 53BP1 nuclear bodies form around DNA lesions generated by mitotic transmission of chromosomes under replication stress. *Nat Cell Biol* 13:243–53.

Lin, Y. F., Shih, H. Y., Shang, Z., Matsunaga, S., Chen, B. P. 2014. DNA-PKcs is required to maintain stability of Chk1 and Claspin for optimal replication stress response. *Nucleic Acids Res* 42: 4463–73.

Ma, C. X., Cai, S., Li, S., Ryan, C. E., Guo, Z., Schaiff, W. T., Lin, L. et al. 2012. Targeting Chk1 in p53-deficient triple-negative breast cancer is therapeutically beneficial in human-in-mouse tumor models. *J Clin Invest* 122: 1541–52.

Ma, J., Li, X., Su, Y., Zhao, J., Luedtke, D. A., Epshteyn, V., Edwards, H. et al. 2017. Mechanisms responsible for the synergistic antileukemic interactions between ATR inhibition and cytarabine in acute myeloid leukemia cells. *Sci Rep* 7: 41950.

Macheret, M., Halazonetis, T. D. 2015. DNA replication stress as a hallmark of cancer. *Annu Rev Pathol* 10: 425–48.

Magnussen, G. I., Emilsen, E., Giller Fleten, K., Engesaeter, B., Nahse-Kumpf, V., Fjaer, R., Slipicevic, A., Florenes, V. A. 2015. Combined inhibition of the cell cycle related proteins Wee1 and Chk1/2 induces synergistic anti-cancer effect in melanoma. *BMC Cancer* 15: 462.

Mak, J. P., Man, W. Y., Chow, J. P., Ma, H. T., Poon, R. Y. 2015. Pharmacological inactivation of CHK1 and WEE1 induces mitotic catastrophe in nasopharyngeal carcinoma cells. *Oncotarget* 6: 21074–84.

Martino-Echarri, E., Henderson, B. R., Brocardo, M. G. 2014. Targeting the DNA replication checkpoint by pharmacologic inhibition of Chk1 kinase: A strategy to sensitize APC mutant colon cancer cells to 5-fluorouracil chemotherapy. *Oncotarget* 5: 9889–900.

McLornan, D. P., List, A., Mufti, G. J. 2014. Applying synthetic lethality for the selective targeting of cancer. *N Engl J Med* 371: 1725–35.

McNeely, S., Conti, C., Sheikh, T., Patel, H., Zabludoff, S., Pommier, Y., Schwartz, G., Tse, A. 2010. Chk1 inhibition after replicative stress activates a double strand break response mediated by ATM and DNA-dependent protein kinase. *Cell Cycle* 9: 995–1004.

Menezes, D. L., Holt, J., Tang, Y., Feng, J., Barsanti, P., Pan, Y., Ghoddusi, M. et al. 2015. A synthetic lethal screen reveals enhanced sensitivity to ATR inhibitor treatment in mantle cell lymphoma with ATM loss-of-function. *Mol Cancer Res* 13: 120–9.

Middleton, F. K., Patterson, M. J., Elstob, C. J., Fordham, S., Herriott, A., Wade, M. A., McCormick, A. et al. 2015. Common cancer-associated imbalances in the DNA damage response confer sensitivity to single agent ATR inhibition. *Oncotarget* 6: 32396–409.

Min, A., Im, S. A., Jang, H., Kim, S., Lee, M., Kim, D. K., Yang, Y. et al. 2017. AZD6738, a novel oral inhibitor of ATR, induces synthetic lethality with ATM-deficiency in gastric cancer cells. *Mol Cancer Ther* 16(4): 566–77.

Mohni, K. N., Kavanaugh, G. M., Cortez, D. 2014. ATR pathway inhibition is synthetically lethal in cancer cells with ERCC1 deficiency. *Cancer Res* 74: 2835–45.

Mohni, K. N., Thompson, P. S., Luzwick, J. W., Glick, G. G., Pendleton, C. S., Lehmann, B. D., Pietenpol, J. A., Cortez, D. 2015. A synthetic lethal screen identifies DNA repair pathways that sensitize cancer cells to combined ATR inhibition and cisplatin treatments. *PLOS ONE* 10: e0125482.

Morgado-Palacin, I., Day, A., Murga, M., Lafarga, V., Anton, M. E., Tubbs, A., Chen, H. T. et al. 2016. Targeting the kinase activities of ATR and ATM exhibits antitumoral activity in mouse models of MLL-rearranged AML. *Sci Signal* 9: ra91.

Morgan, M. A., Parsels, L. A., Zhao, L., Parsels, J. D., Davis, M. A., Hassan, M. C., Arumugarajah, S. 2010. Mechanism of radiosensitization by the Chk1/2 inhibitor AZD7762 involves abrogation of the G2 checkpoint and inhibition of homologous recombinational DNA repair. *Cancer Res* 70(12):4972–81.

Muralidharan, S. V., Bhadury, J., Nilsson, L. M., Green, L. C., McLure, K. G., Nilsson, J. A. 2016. BET bromodomain inhibitors synergize with ATR inhibitors to induce DNA damage, apoptosis, senescence-associated secretory pathway and ER stress in Myc-induced lymphoma cells. *Oncogene* 35: 4689–97.

Murga, M., Bunting, S., Montana, M. F., Soria, R., Mulero, F., Canamero, M., Lee, Y., McKinnon, P. J., Nussenzweig, A., Fernandez-Capetillo, O. 2009. A mouse model of ATR-Seckel shows embryonic replicative stress and accelerated aging. *Nat Genet* 41: 891–8.

Murga, M., Campaner, S., Lopez-Contreras, A. J., Toledo, L. I., Soria, R., Montana, M. F., D'Artista, L. et al. 2011. Exploiting oncogene-induced replicative stress for the selective killing of Myc-driven tumors. *Nat Struct Mol Biol* 18: 1331–5.

Nghiem, P., Park, P. K., Kim, Y., Vaziri, C., Schreiber, S. L. 2001. ATR inhibition selectively sensitizes G1 checkpoint-deficient cells to lethal premature chromatin condensation. *Proc Natl Acad Sci USA* 98: 9092–7.

O'Driscoll, M., Ruiz-Perez, V. L., Woods, C. G., Jeggo, P. A., Goodship, J. A. 2003. A splicing mutation affecting expression of ataxia-telangiectasia and Rad3-related protein (ATR) results in Seckel syndrome. *Nat Genet* 33: 497–501.

Origanti, S., Cai, S. R., Munir, A. Z., White, L. S., Piwnica-Worms, H. 2013. Synthetic lethality of Chk1 inhibition combined with p53 and/or p21 loss during a DNA damage response in normal and tumor cells. *Oncogene* 32: 577–88.

Pei, X. Y., Dai, Y., Felthousen, J., Chen, S., Takabatake, Y., Zhou, L., Youssefian, L. E. et al. 2014. Circumvention of Mcl-1-dependent drug resistance by simultaneous Chk1 and MEK1/2 inhibition in human multiple myeloma cells. *PLOS ONE* 9: e89064.

Pires, I. M., Olcina, M. M., Anbalagan, S., Pollard, J. R., Reaper, P. M., Charlton, P. A., McKenna, W. G., Hammond, E. M. 2012. Targeting radiation-resistant hypoxic tumour cells through ATR inhibition. *Br J Cancer* 107: 291–9.

Prevo, R., Fokas, E., Reaper, P. M., Charlton, P. A., Pollard, J. R., McKenna, W. G., Muschel, R. J., Brunner, T. B. 2012. The novel ATR inhibitor VE-821 increases sensitivity of pancreatic cancer cells to radiation and chemotherapy. *Cancer Biol Ther* 13: 1072–81.

Reaper, P. M., Griffiths, M. R., Long, J. M., Charrier, J. D., Maccormick, S., Charlton, P. A., Golec, J. M., Pollard, J. R. 2011. Selective killing of ATM- or p53-deficient cancer cells through inhibition of ATR. *Nat Chem Biol* 7: 428–30.

Russell, M. R., Levin, K., Rader, J., Belcastro, L., Li, Y., Martinez, D., Pawel, B., Shumway, S. D., Maris, J. M., Cole, K. A. 2013. Combination therapy targeting the Chk1 and Wee1 kinases shows therapeutic efficacy in neuroblastoma. *Cancer Res* 73: 776–84.

Ruzankina, Y., Pinzon-Guzman, C., Asare, A., Ong, T., Pontano, L., Cotsarelis, G., Zediak, V. P., Velez, M., Bhandoola, A., Brown, E. J. 2007. Deletion of the developmentally essential gene ATR in adult mice leads to age-related phenotypes and stem cell loss. *Cell Stem Cell* 1: 113–26.

Ruzankina, Y., Schoppy, D. W., Asare, A., Clark, C. E., Vonderheide, R. H., Brown, E. J. 2009. Tissue regenerative delays and synthetic lethality in adult mice after combined deletion of Atr and Trp53. *Nat Genet* 41: 1144–9.

Sakurikar, N., Thompson, R., Montano, R., Eastman, A. 2016. A subset of cancer cell lines is acutely sensitive to the Chk1 inhibitor MK-8776 as monotherapy due to CDK2 activation in S phase. *Oncotarget* 7: 1380–94.

Sangster-Guity, N., Conrad, B. H., Papadopoulos, N., Bunz, F. 2011. ATR mediates cisplatin resistance in a p53 genotype-specific manner. *Oncogene* 30: 2526–33.

Sanjiv, K., Hagenkort, A., Calderon-Montano, J. M., Koolmeister, T., Reaper, P. M., Mortusewicz, O., Jacques, S. A. et al. 2016. Cancer-specific synthetic lethality between ATR and CHK1 kinase activities. *Cell Rep* 17: 3407–16.

Sankunny, M., Parikh, R. A., Lewis, D. W., Gooding, W. E., Saunders, W. S., Gollin, S. M. 2014. Targeted inhibition of ATR or CHEK1 reverses radioresistance in oral squamous cell carcinoma cells with distal chromosome arm 11q loss. *Genes Chromosomes Cancer* 53: 129–43.

Sarmento, L. M., Povoa, V., Nascimento, R., Real, G., Antunes, I., Martins, L. R., Moita, C. et al. 2015. CHK1 overexpression in T-cell acute lymphoblastic leukemia is essential for proliferation and survival by preventing excessive replication stress. *Oncogene* 34: 2978–90.

Sausville, E., Lorusso, P., Carducci, M., Carter, J., Quinn, M. F., Malburg, L., Azad, N. et al. 2014. Phase I dose-escalation study of AZD7762, a checkpoint kinase inhibitor, in combination with gemcitabine in US patients with advanced solid tumors. *Cancer Chemother Pharmacol* 73: 539–49.

Scagliotti, G., Kang, J. H., Smith, D., Rosenberg, R., Park, K., Kim, S. W., Su, W. C. et al. 2016. Phase II evaluation of LY2603618, a first-generation CHK1 inhibitor, in combination with pemetrexed in patients with advanced or metastatic non-small cell lung cancer. *Invest New Drugs* 34: 625–35.

Schoppy, D. W., Ragland, R. L., Gilad, O., Shastri, N., Peters, A. A., Murga, M., Fernandez-Capetillo, O., Diehl, J. A., Brown, E. J. 2012. Oncogenic stress sensitizes murine cancers to hypomorphic suppression of ATR. *J Clin Invest* 122: 241–52.

Seto, T., Esaki, T., Hirai, F., Arita, S., Nosaki, K., Makiyama, A., Kometani, T. et al. 2013. Phase I, dose-escalation study of AZD7762 alone and in combination with gemcitabine in Japanese patients with advanced solid tumours. *Cancer Chemother Pharmacol* 72: 619–27.

Shaheen, M., Allen, C., Nickoloff, J. A., Hromas, R. 2011. Synthetic lethality: Exploiting the addiction of cancer to DNA repair. *Blood* 117: 6074–82.

Shiloh, Y., Ziv, Y. 2013. The ATM protein kinase: Regulating the cellular response to genotoxic stress, and more. *Nat Rev Mol Cell Biol* 14(4):197–210.

Sultana, R., Abdel-Fatah, T., Perry, C., Moseley, P., Albarakti, N., Mohan, V., Seedhouse, C., Chan, S., Madhusudan, S. 2013. Ataxia telangiectasia mutated and Rad3 related (ATR) protein kinase inhibition is synthetically lethal in XRCC1 deficient ovarian cancer cells. *PLOS ONE* 8: e57098.

Tang, Y., Hamed, H. A., Poklepovic, A., Dai, Y., Grant, S., Dent, P. 2012. Poly(ADP-ribose) polymerase 1 modulates the lethality of CHK1 inhibitors in mammary tumors. *Mol Pharmacol* 82: 322–32.

Teng, P. N., Bateman, N. W., Darcy, K. M., Hamilton, C. A., Maxwell, G. L., Bakkenist, C. J., Conrads, T. P. 2015. Pharmacologic inhibition of ATR and ATM offers clinically important distinctions to enhancing platinum or radiation response in ovarian, endometrial, and cervical cancer cells. *Gynecol Oncol* 136: 554–61.

Toledo, L. I., Altmeyer, M., Rask, M. B., Lukas, C., Larsen, D. H., Povlsen, L. K., Bekker-Jensen, S., Mailand, N., Bartek, J., Lukas, J. 2013. ATR prohibits replication catastrophe by preventing global exhaustion of RPA. *Cell* 155: 1088–103.

Vance, S., Liu, E., Zhao, L., Parsels, J. D., Parsels, L. A., Brown, J. L., Maybaum, J., Lawrence, T. S., Morgan, M. A. 2011. Selective radiosensitization of p53 mutant pancreatic cancer cells by combined inhibition of Chk1 and PARP1. *Cell Cycle* 10: 4321–9.

Vendetti, F. P., Lau, A., Schamus, S., Conrads, T. P., O'Connor, M. J., Bakkenist, C. J. 2015. The orally active and bioavailable ATR kinase inhibitor AZD6738 potentiates the anti-tumor effects of cisplatin to resolve ATM-deficient non-small cell lung cancer *in vivo*. *Oncotarget* 6: 44289–305.

Wang, W. T., Catto, J. W., Meuth, M. 2015. Differential response of normal and malignant urothelial cells to CHK1 and ATM inhibitors. *Oncogene* 34: 2887–96.

Williamson, C. T., Miller, R., Pemberton, H. N., Jones, S. E., Campbell, J., Konde, A., Badham, N. et al. 2016. ATR inhibitors as a synthetic lethal therapy for tumours deficient in ARID1A. *Nat Commun* 7: 13837.

Yazinski, S. A., Comaills, V., Buisson, R., Genois, M. M., Nguyen, H. D., Ho, C. K., Todorova Kwan, T. et al. 2017. ATR inhibition disrupts rewired homologous recombination and fork protection pathways in PARP inhibitor-resistant BRCA-deficient cancer cells. *Genes Dev* 31: 318–32.

Yuan, L. L., Green, A., David, L., Dozier, C., Recher, C., Didier, C., Tamburini, J., Manenti, S. 2014. Targeting CHK1 inhibits cell proliferation in FLT3-ITD positive acute myeloid leukemia. *Leuk Res* 38: 1342–9.

Zeman, M. K., Cimprich, K. A. 2014. Causes and consequences of replication stress. *Nat Cell Biol* 16(1):2–9.

Zemanova, J., Hylse, O., Collakova, J., Vesely, P., Oltova, A., Borsky, M., Zaprazna, K. et al. 2016. Chk1 inhibition significantly potentiates activity of nucleoside analogs in TP53-mutated B-lymphoid cells. *Oncotarget* 7: 62091–106.

Zhang, Y., Lai, J., Du, Z., Gao, J., Yang, S., Gorityala, S., Xiong, X. et al. 2016. Targeting radioresistant breast cancer cells by single agent CHK1 inhibitor via enhancing replication stress. *Oncotarget* 7: 34688–702.

Zhao, J., Niu, X., Li, X., Edwards, H., Wang, G., Wang, Y., Taub, J. W., Lin, H., Ge, Y. 2016. Inhibition of CHK1 enhances cell death induced by the Bcl-2-selective inhibitor ABT-199 in acute myeloid leukemia cells. *Oncotarget* 7: 34785–99.

Zhou, Z. R., Yang, Z. Z., Wang, S. J., Zhang, L., Luo, J. R., Feng, Y., Yu, X. L., Chen, X. X., Guo, X. M. 2017. The Chk1 inhibitor MK-8776 increases the radiosensitivity of human triple-negative breast cancer by inhibiting autophagy. *Acta Pharmacol Sin* 38(4): 513–23.

Zighelboim, I., Schmidt, A. P., Gao, F., Thaker, P. H., Powell, M. A., Rader, J. S., Gibb, R. K., Mutch, D. G., Goodfellow, P. J. 2009. ATR mutation in endometrioid endometrial cancer is associated with poor clinical outcomes. *J Clin Oncol* 27: 3091–6.

A Few of the Many Outstanding Questions

16

John J. Reynolds and Roger J. A. Grand

Over the past few decades, our knowledge of the mechanisms and functions of the many DNA replication and repair pathways that exist in cells has grown exponentially. Not only are we still regularly identifying new DNA repair factors, and even occasionally discovering a new DNA repair pathway, but we are also gaining deeper understanding of the molecular mechanisms underpinning existing pathways. This progress has been driven by building on the ever-expanding collective knowledge and experience of the field, by the discovery and study of inherited human diseases associated with defective DNA replication and repair and by significant advances in new technologies. From this, we have gained a much greater appreciation of the roles that the DNA damage response pathways play in the maintenance of human health and the prevention of disease.

Furthermore, we are now also coming to terms with the fact that nothing happens in isolation within the cell, or even within an organism, and that many pathways/factors are linked in ways that we had not previously recognised. Much of the research performed thus far has been limited to in vitro experiments looking at naked DNA, using depletion/knockout experiments on common cancer cell lines, employing easy to obtain and/or grow patient-derived cell lines or studying simple model organisms. Whilst the necessities of research have enforced this, and these approaches have rewarded us with much of the knowledge within this book, it is clear that as we look to the future we will need more powerful and all-encompassing experiments and model systems. Indeed, we are now at the point where it is feasible to perform experiments on nucleosomes in vitro, and attempts to purify and study the whole replisome in vitro have also been made. Furthermore, organoid cell cultures that more closely model disease-affected tissues are becoming more common, and new unbiased 'omic' approaches are illuminating new complexities within cellular/disease processes.

The increased rate of discovery of human diseases caused by mutations in DNA repair genes has been helped by the continual improvement and widespread use of whole genome/exome sequencing. However, although it is now relatively easier to provide a genetic diagnosis, it is much more difficult to understand the molecular defects that underlie these diseases, and harder still to know how to treat them or if treatment is even possible. As we gain further understanding of the disease pathologies, new questions and aims are constantly arising.

Despite advances, we are still left with many unanswered questions. Why is there such phenotypic variability associated with some diseases? Whilst some variability may be explained by the nature of the disease-causing mutations or perhaps some known genetic modifiers, it is clear that this cannot explain everything. Also, why are some tissue types/organ systems sensitive to the mutation or loss of specific DNA repair genes and not others?

How can we reliably treat genetic diseases? If we understood what the underlying defect was, with further advances in gene targeting techniques, could we correct tissue-specific defects? Also, are there any strategies to treat/prevent the progression of neurodegeneration in patients? Indeed, is neurodegeneration always attributable to loss of the DDR functions of the proteins with which they have been associated? How can we better use our knowledge of DNA replication and repair pathways to design more targeted therapies for cancer? In particular, a lot of effort is being devoted to discovering novel synthetic lethality or viability treatment strategies, but is this the best approach from a clinical viewpoint? Significantly, impairment of

DDR gene function contributes to cancer progression. Are these problems inherently different to those encountered by patients with inherited cancer predispositions?

Although we have made significant progress in the treatments for some diseases, this can sometimes uncover other complexities. For example, Fanconi anaemia was once considered to be a childhood disease with a very poor prognosis, but with recent advances in the treatment of bone marrow failure, there is now an increasing number of patients surviving into adulthood. Whilst this is a happy situation, adult Fanconi anaemia patients face an extra set of complications, such as an elevated risk of developing cancers. This also raises another question: How do we treat cancer in patients who are hypersensitive to DNA-damaging chemotherapies or radiotherapy? Some novel immunotherapies are showing promise, but there is still a long way to go.

These are only a few of the many questions we need to answer, and there is no doubt that as we continue to understand more about the pathways of genome maintenance, we will assuredly encounter more questions/problems/complexities. However, with the combined efforts of researchers within the DNA replication and DNA repair fields, and the clinicians who diagnose and treat patients, we will continue to advance our understanding, improve our treatments and extend the life expectancies of patients.

Index